EIGHTH EDITION

8

FRANK H. NETTER, MD

NETTER ATLAS
of HUMAN
ANATOMY

A Systems Approach

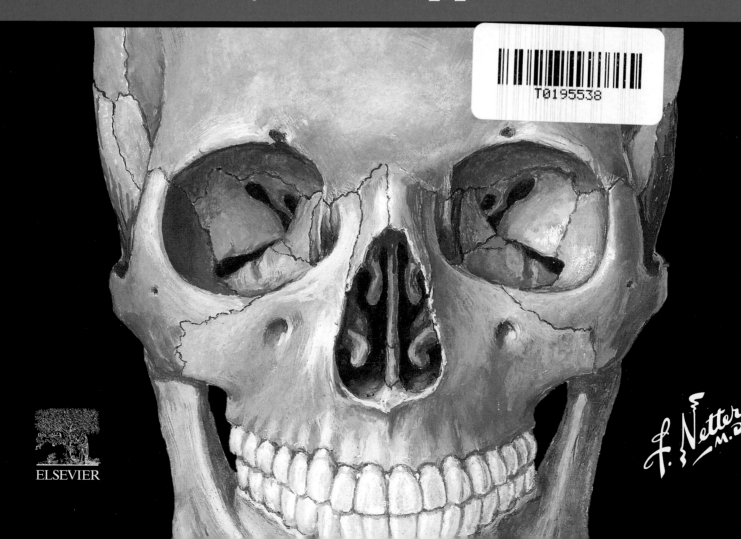

T0195538

ELSEVIER

F. Netter
M.D.

ELSEVIER
1600 John F. Kennedy Blvd.
Ste. 1600
Philadelphia, PA 19103-2899

**ATLAS OF HUMAN ANATOMY : A SYSTEMS APPROACH,
EIGHTH EDITION**

Standard Edition: 978-0-323-76028-7

Copyright © 2023 by Elsevier Inc.
Previous editions copyrighted 2019, 2014, 2011, 2006, 2003, 1997, 1989

All rights reserved. No part of this publication may be reproduced or transmitted in any form or by any means, electronic or mechanical, including photocopying, recording, or any information storage and retrieval system, without permission in writing from the publisher. Details on how to seek permission, further information about the Publisher's permissions policies and our arrangements with organizations such as the Copyright Clearance Center and the Copyright Licensing Agency can be found at our website: www.elsevier.com/permissions.

This book and the individual contributions contained in it are protected under copyright by the Publisher (other than as may be noted herein).

Permission to use Netter Art figures may be sought through the website *NetterImages.com* or by emailing Elsevier's Licensing Department at H.Licensing@elsevier.com.

Notices

Knowledge and best practice in this field are constantly changing. As new research and experience broaden our understanding, changes in research methods, professional practices, or medical treatment may become necessary.

Practitioners and researchers must always rely on their own experience and knowledge in evaluating and using any information, methods, compounds, or experiments described herein. In using such information or methods they should be mindful of their own safety and the safety of others, including parties for whom they have a professional responsibility.

With respect to any drug or pharmaceutical products identified, readers are advised to check the most current information provided (i) on procedures featured or (ii) by the manufacturer of each product to be administered, to verify the recommended dose or formula, the method and duration of administration, and contraindications. It is the responsibility of practitioners, relying on their own experience and knowledge of their patients, to make diagnoses, to determine dosages and the best treatment for each individual patient, and to take all appropriate safety precautions.

To the fullest extent of the law, neither the Publisher nor the authors, contributors, or editors, assume any liability for any injury and/or damage to persons or property as a matter of products liability, negligence or otherwise, or from any use or operation of any methods, products, instructions, or ideas contained in the material herein.

International Standard Book Number: 978-0-323-76028-7

Publisher: Elyse O'Grady
Senior Content Strategist: Marybeth Thiel
Publishing Services Manager: Catherine Jackson
Senior Project Manager/Specialist: Carrie Stetz
Book Design: Renee Duenow

Printed in China

9 8 7 6 5 4 3 2 1

 Working together
to grow libraries in
developing countries

www.elsevier.com • www.bookaid.org

CONSULTING EDITORS

Chief Contributing Illustrator and Art Load Editor

Carlos A.G. Machado, MD

Terminology Content Lead Editors

Paul E. Neumann, MD
Professor, Department of Medical Neuroscience
Faculty of Medicine
Dalhousie University
Halifax, Nova Scotia
Canada

R. Shane Tubbs, MS, PA-C, PhD
Professor of Neurosurgery, Neurology, Surgery, and Structural
    and Cellular Biology
Director of Surgical Anatomy, Tulane University School of Medicine
Program Director of Anatomical Research, Clinical
    Neuroscience Research Center, Center for Clinical
    Neurosciences
Department of Neurosurgery, Tulane University School of
    Medicine, New Orleans, Louisiana
Department of Neurology, Tulane University School of
    Medicine, New Orleans, Louisiana
Department of Structural and Cellular Biology, Tulane University
    School of Medicine, New Orleans, Louisiana
Professor, Department of Neurosurgery, and Ochsner
    Neuroscience Institute, Ochsner Health System, New
    Orleans, Louisiana
Professor of Anatomy, Department of Anatomical Sciences,
    St. George's University, Grenada
Honorary Professor, University of Queensland, Brisbane, Australia
Faculty, National Skull Base Center of California, Thousand
    Oakes, California

Electronic Content Lead Editors

Brion Benninger, MD, MBChB, MSc
Professor of Medical Innovation, Technology, & Research
Director, Ultrasound
Professor of Clinical Anatomy
Executive Director, Medical Anatomy Center
Department of Medical Anatomical Sciences
Faculty, COMP and COMP-Northwest
Faculty College of Dentistry, Western University of Health
    Sciences, Lebanon, Oregon and Pomona, California
Faculty, Sports Medicine, Orthopaedic & General Surgery
    Residencies
Samaritan Health Services, Corvallis, Oregon
Faculty, Surgery, Orthopedics & Rehabilitation, and Oral
    Maxillofacial Surgery
Oregon Health & Science University, Portland, Oregon
Visiting Professor of Medical Innovation and Clinical Anatomy,
    School of Basic Medicine, Peking Union Medical College,
    Beijing, China
Professor of Medical Innovation and Clinical Anatomy Post
    Graduate Diploma Surgical Anatomy, Otago University,
    Dunedin, New Zealand

Todd M. Hoagland, PhD
Clinical Professor of Biomedical Sciences and Occupational
    Therapy
Marquette University College of Health Sciences
Milwaukee, Wisconsin

Educational Content Lead Editors

Jennifer K. Brueckner-Collins, PhD
Distinguished Teaching Professor
Vice Chair for Educational Programs
Department of Anatomical Sciences and Neurobiology
University of Louisville School of Medicine
Louisville, Kentucky

Martha Johnson Gdowski, PhD
Associate Professor and Associate Chair of Medical Education,
    Department of Neuroscience
University of Rochester School of Medicine and Dentistry
Rochester, NY

Virginia T. Lyons, PhD
Associate Professor of Medical Education
Associate Dean for Preclinical Education
Geisel School of Medicine at Dartmouth
Hanover, New Hampshire

Peter J. Ward, PhD
Professor
Department of Biomedical Sciences
West Virginia School of Osteopathic Medicine
Lewisburg, West Virginia

Emeritus Editor

John T. Hansen, PhD
Professor Emeritus of Neuroscience and former Schmitt Chair
    of Neurobiology and Anatomy and Associate Dean for
    Admissions University of Rochester Medical Center
Rochester, New York

EDITORS OF PREVIOUS EDITIONS

First Edition
Sharon Colacino, PhD

Second Edition
Arthur F. Dalley II, PhD

Third Edition
Carlos A.G. Machado, MD

John T. Hansen, PhD

Fourth Edition
Carlos A.G. Machado, MD
John T. Hansen, PhD
Jennifer K. Brueckner, PhD
Stephen W. Carmichael, PhD, DSc
Thomas R. Gest, PhD
Noelle A. Granger, PhD
Anil H. Waljii, MD, PhD

Fifth Edition
Carlos A.G. Machado, MD
John T. Hansen, PhD
Brion Benninger, MD, MS
Jennifer K. Brueckner, PhD
Stephen W. Carmichael, PhD, DSc
Noelle A. Granger, PhD
R. Shane Tubbs, MS, PA-C, PhD

Sixth Edition
Carlos A.G. Machado, MD
John T. Hansen, PhD
Brion Benninger, MD, MS
Jennifer Brueckner-Collins, PhD
Todd M. Hoagland, PhD
R. Shane Tubbs, MS, PA-C, PhD

Seventh Edition
Carlos A.G. Machado, MD
John T. Hansen, PhD
Brion Benninger, MD, MS
Jennifer Brueckner-Collins, PhD
Todd M. Hoagland, PhD
R. Shane Tubbs, MS, PA-C, PhD

OTHER CONTRIBUTING ILLUSTRATORS

Rob Duckwall, MA (DragonFly Media Group)
Kristen Wienandt Marzejon, MS, MFA
Tiffany S. DaVanzo, MA, CMI
James A. Perkins, MS, MFA

INTERNATIONAL ADVISORY BOARD

Nihal Apaydin, MD, PhD
Professor of Anatomy
Faculty of Medicine, Department of
    Anatomy
Ankara University;
Chief, Department of Multidisciplinary
    Neuroscience
Institute of Health Sciences
Ankara, Turkey

Hassan Amiralli, MD, MS, FUICC
Professor and Chair
Department of Anatomy
American University of Antigua College
    of Medicine
Antigua, West Indies;
Former Professor of Surgery
Muhimbili University of Health Sciences
Daressalaam, Tanzania

Belinda R. Beck, BHMS(Ed), MS, PhD,
Professor of Anatomy and Exercise
    Science
Director, Bone Densitometry Research
    Laboratory
Griffith University, Gold Coast Campus
Queensland, Australia

Jonathan Campbell, MD, FAAOS
Assistant Professor of Orthopaedic
    Surgery
Division of Sports Medicine
Medical College of Wisconsin
Milwaukee, Wisconsin

Francisco J. Caycedo, MD, FAAOS
St. Vincent's Hospital
Birmingham, Alabama

Thazhumpal Chacko Mathew, MSc, PhD,
    FRCPath
Professor
Faculty of Allied Health Sciences
Health Sciences Centre
Kuwait University
Kuwait City, Kuwait

Eduardo Cotecchia Ribeiro, MS, PhD
Associate Professor of Descriptive and
    Topographic Anatomy
School of Medicine
Federal University of São Paulo
São Paulo, Brazil

William E. Cullinan, PhD
Professor and Dean
College for Health Sciences
Marquette University
Milwaukee, Wisconsin

Elisabeth Eppler, MD
University Lecturer
Institute of Anatomy
University of Berne
Berne, Switzerland

Christopher Kelly, MD, MS
North Carolina Heart and Vascular
Raleigh, North Carolina

Michelle D. Lazarus, PhD
Director, Centre for Human Anatomy
    Education
Monash Centre for Scholarship in Health
    Education (MCSHE) Curriculum
    Integration Network Lead
Monash Education Academy Fellow
Monash University
Clayton, Victoria, Australia

Robert G. Louis, MD, FAANS
Empower360 Endowed Chair for
    Skull Base and Minimally Invasive
    Neurosurgery
Chair, Division of Neurosurgery
Pickup Family Neurosciences Institute
Hoag Memorial Hospital
Newport Beach, California

Chao Ma, MD
Department of Human Anatomy,
    Histology & Embryology
Peking Union Medical College
Beijing, China

Diego Pineda Martínez, MD, PhD
Chief, Department of Innovation in
    Human Biological Material
Professor of Anatomy
Faculty of Medicine of the National
    Autonomous University of Mexico
President, Mexican Society of Anatomy
Mexico City, Mexico

William J. Swartz, PhD
Emeritus Professor of Cell Biology and
    Anatomy
Louisiana State University Health
    Sciences Center
New Orleans, Louisiana

Kimberly S. Topp, PT, PhD, FAAA
Professor and Chair Emeritus
Department of Physical Therapy and
    Rehabilitation Science
School of Medicine
University of California San Francisco
San Francisco, California

Ivan Varga, PhD
Professor of Anatomy, Histology, and
    Embryology
Faculty of Medicine
Comenius University
Bratislava, Slovak Republic

Robert J. Ward, MD, CCD, DABR
Chief Executive Officer
Sullivan's Island Imaging, LLC
Sullivan's Island, South Carolina;
Professor of Radiology
Saint Georges University
Grenada, West Indies

Alexandra L. Webb, BSc, MChiro, PhD
Associate Professor
Deputy Director, Medical School
College of Health and Medicine
Australian National University
Canberra, ACT, Australia

PREFACE

The illustrations comprising the *Netter Atlas of Human Anatomy* were painted by physician-artists, Frank H. Netter, MD, and Carlos Machado, MD. Dr. Netter was a surgeon and Dr. Machado is a cardiologist. Their clinical insights and perspectives have informed their approaches to these works of art. The collective expertise of the anatomists, educators, and clinicians guiding the selection, arrangement, labeling, and creation of the illustrations ensures the accuracy, relevancy, and educational power of this outstanding collection.

You have a copy of the **Systems Approach 8th edition** with English-language terminology. This is a new organization, available for the first time. Traditionally, the Netter Atlas has only been offered as a regionally organized Atlas. This arrangement is still available (with English or Latin terminology options), but this new systems organization reflects the needs of a growing number of programs that approach anatomy within a body systems context. In all cases, the same beautiful and instructive Art Plates and Table information are included.

New to this Edition

New Art
More than 20 new illustrations have been added and over 30 art modifications have been made throughout this edition. Highlights include new views of the temporal and infratemporal fossa, pelvic fascia, nasal cavity and paranasal sinuses, plus multiple new perspectives of the heart, a cross-section of the foot, enhanced surface anatomy plates, and overviews of many body systems. In these pages you will find the most robust illustrated coverage to date for modern clinical anatomy courses.

Terminology and Label Updates
This 8th edition incorporates terms of the *Terminologia Anatomica* (2nd edition), as published by the Federative International Programme on Anatomical Terminology in 2019 (https://fipat.library.dal.ca/ta2) and adopted by the International Federation of Associations of Anatomy in 2020. A fully searchable database of the updated *Terminologia Anatomica* can be accessed at https://ta2viewer.openanatomy.org. Common clinical eponyms and former terminologies are selectively included, parenthetically, for clarity. In addition, a strong effort has been made to reduce label text on the page while maximizing label information through the use of abbreviations (muscle/s = m./mm., artery/ies = a./aa.; vein/s = v./vv.; and nerve/s = n./nn.) and focusing on the labels most relevant to the subject of each Plate.

Nerve Tables
The muscle tables and clinical tables of previous editions have been so positively received that new tables have been added to cover four major nerve groups: cranial nerves and the nerves of the cervical, brachial, and lumbosacral plexuses.

The Future of the Netter Anatomy Atlas
As the Netter Atlas continues to evolve to meet the needs of students, educators, and clinicians, we welcome suggestions! Please use the following form to provide your feedback:
https://tinyurl.com/NetterAtlas8

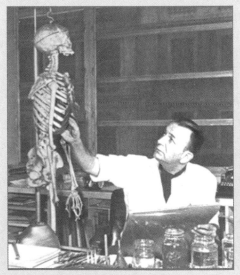

To my dear wife, Vera

PREFACE TO THE FIRST EDITION

I have often said that my career as a medical artist for almost 50 years has been a sort of "command performance" in the sense that it has grown in response to the desires and requests of the medical profession. Over these many years, I have produced almost 4,000 illustrations, mostly for *The CIBA* (now *Netter*) *Collection of Medical Illustrations* but also for *Clinical Symposia*. These pictures have been concerned with the varied subdivisions of medical knowledge such as gross anatomy, histology, embryology, physiology, pathology, diagnostic modalities, surgical and therapeutic techniques, and clinical manifestations of a multitude of diseases. As the years went by, however, there were more and more requests from physicians and students for me to produce an atlas purely of gross anatomy. Thus, this atlas has come about, not through any inspiration on my part but rather, like most of my previous works, as a fulfillment of the desires of the medical profession.

It involved going back over all the illustrations I had made over so many years, selecting those pertinent to gross anatomy, classifying them and organizing them by system and region, adapting them to page size and space, and arranging them in logical sequence. Anatomy of course does not change, but our understanding of anatomy and its clinical significance does change, as do anatomical terminology and nomenclature. This therefore required much updating of many of the older pictures and even revision of a number of them in order to make them more pertinent to today's ever-expanding scope of medical and surgical practice. In addition, I found that there were gaps in the portrayal of medical knowledge as pictorialized in the illustrations I had previously done, and this necessitated my making a number of new pictures that are included in this volume.

In creating an atlas such as this, it is important to achieve a happy medium between complexity and simplification. If the pictures are too complex, they may be difficult and confusing to read; if oversimplified, they may not be adequately definitive or may even be misleading. I have therefore striven for a middle course of realism without the clutter of confusing minutiae. I hope that the students and members of the medical and allied professions will find the illustrations readily understandable, yet instructive and useful.

At one point, the publisher and I thought it might be nice to include a foreword by a truly outstanding and renowned anatomist, but there are so many in that category that we could not make a choice. We did think of men like Vesalius, Leonardo da Vinci, William Hunter, and Henry Gray, who of course are unfortunately unavailable, but I do wonder what their comments might have been about this atlas.

Frank H. Netter, MD
(1906–1991)

FRANK H. NETTER, MD

Frank H. Netter was born in New York City in 1906. He studied art at the Art Students League and the National Academy of Design before entering medical school at New York University, where he received his Doctor of Medicine degree in 1931. During his student years, Dr. Netter's notebook sketches attracted the attention of the medical faculty and other physicians, allowing him to augment his income by illustrating articles and textbooks. He continued illustrating as a sideline after establishing a surgical practice in 1933, but he ultimately opted to give up his practice in favor of a full-time commitment to art. After service in the United States Army during World War II, Dr. Netter began his long collaboration with the CIBA Pharmaceutical Company (now Novartis Pharmaceuticals). This 45-year partnership resulted in the production of the extraordinary collection of medical art so familiar to physicians and other medical professionals worldwide.

Icon Learning Systems acquired the Netter Collection in July 2000 and continued to update Dr. Netter's original paintings and to add newly commissioned paintings by artists trained in the style of Dr. Netter. In 2005, Elsevier Inc. purchased the Netter Collection and all publications from Icon Learning Systems. There are now over 50 publications featuring the art of Dr. Netter available through Elsevier Inc.

Dr. Netter's works are among the finest examples of the use of illustration in the teaching of medical concepts. The 13-book *Netter Collection of Medical Illustrations*, which includes the greater part of the more than 20,000 paintings created by Dr. Netter, became and remains one of the most famous medical works ever published. *The Netter Atlas of Human Anatomy*, first published in 1989, presents the anatomic paintings from the Netter Collection. Now translated into 16 languages, it is the anatomy atlas of choice among medical and health professions students the world over.

The Netter illustrations are appreciated not only for their aesthetic qualities, but, more importantly, for their intellectual content. As Dr. Netter wrote in 1949 "clarification of a subject is the aim and goal of illustration. No matter how beautifully painted, how delicately and subtly rendered a subject may be, it is of little value as a *medical illustration* if it does not serve to make clear some medical point." Dr. Netter's planning, conception, point of view, and approach are what inform his paintings and what make them so intellectually valuable.

Frank H. Netter, MD, physician and artist, died in 1991.

ABOUT THE EDITORS

Carlos A.G. Machado, MD was chosen by Novartis to be Dr. Netter's successor. He continues to be the main artist who contributes to the Netter collection of medical illustrations.

Self-taught in medical illustration, cardiologist Carlos Machado has contributed meticulous updates to some of Dr. Netter's original plates and has created many paintings of his own in the style of Netter as an extension of the Netter collection. Dr. Machado's photorealistic expertise and his keen insight into the physician/patient relationship inform his vivid and unforgettable visual style. His dedication to researching each topic and subject he paints places him among the premier medical illustrators at work today.

Learn more about his background and see more of his art at: https://netterimages.com/artist-carlos-a-g-machado.html

Paul E. Neumann, MD was clinically trained in anatomical pathology and neuropathology. Most of his research publications have been in mouse neurogenetics and molecular human genetics. In the past several years, he has concentrated on the anatomical sciences, and has frequently written about anatomical terminology and anatomical ontology in the journal *Clinical Anatomy*. As an officer of the Federative International Programme for Anatomical Terminology (FIPAT), he participated in the production of Terminologia Anatomica (2nd edition), Terminologia Embryologica (2nd edition), and Terminologia Neuroanatomica. In addition to serving as the lead Latin editor of the 8th edition of Netter's Atlas, he was a contributor to the 33rd edition of *Dorland's Illustrated Medical Dictionary*.

R. Shane Tubbs, MS, PA-C, PhD is a native of Birmingham, Alabama and a clinical anatomist. His research interests are centered around clinical/surgical problems that are identified and solved with anatomical studies. This investigative paradigm in anatomy as resulted in over 1,700 peer reviewed publications. Dr. Tubbs' laboratory has made novel discoveries in human anatomy including a new nerve to the skin of the lower eyelid, a new space of the face, a new venous sinus over the spinal cord, new connections between the parts of the sciatic nerve, new ligaments of the neck, a previously undescribed cutaneous branch of the inferior gluteal nerve, and an etiology for postoperative C5 nerve palsies. Moreover, many anatomical feasibility studies from Dr. Tubbs' laboratory have gone on to be used by surgeons from around the world and have thus resulted in new surgical/clinical procedures such as treating hydrocephalus by shunting cerebrospinal fluid into various bones, restoration of upper limb function in paralyzed patients with neurotization procedures using the contralateral spinal accessory nerve, and harvesting of clavicle for anterior cervical discectomy and fusion procedures in patients with cervical instability or degenerative spine disease.

Dr. Tubbs sits on the editorial board of over 15 anatomical journals and has reviewed for over 150 scientific journals.

He has been a visiting professor to major institutions in the United States and worldwide. Dr. Tubbs has authored over 40 books and over 75 book chapters. His published books by Elsevier include *Gray's Anatomy Review*, *Gray's Clinical Photographic Dissector of the Human Body*, *Netter's Introduction to Clinical Procedures*, and *Nerves and Nerve Injuries* volumes I and II. He is an editor for the 41st and 42nd editions of the over 150-year-old *Gray's Anatomy*, the 5th through 8th editions of *Netter's Atlas of Anatomy*, and is the editor-in-chief of the journal *Clinical Anatomy*. He is the Chair of the Federative International Programme on Anatomical Terminologies (FIPAT).

Jennifer K. Brueckner-Collins, PhD is a proud Kentucky native. She pursued her undergraduate and graduate training at the University of Kentucky. During her second year of graduate school there, she realized that her professional calling was not basic science research in skeletal muscle biology, but was instead helping medical students master the anatomical sciences. She discovered this during a required teaching assistantship in medical histology, where working with students at the 10-headed microscope changed her career path.

The next semester of graduate school, she assisted in teaching dissection-based gross anatomy, although she had taken anatomy when the lab component was prosection based. After teaching in the first lab, she knew that she needed to learn anatomy more thoroughly through dissection on her own, so she dissected one to two labs ahead of the students that semester; that was when she really learned anatomy and was inspired to teach this discipline as a profession. All of this occurred in the early 1990s, when pursuing a teaching career was frowned upon by many; it was thought that you only pursued this track if you were unsuccessful in research. She taught anatomy part-time during the rest of her graduate training, on her own time, to gain requisite experience to ultimately secure a faculty position.

Dr. Brueckner-Collins spent 10 years at the University of Kentucky as a full-time faculty member teaching dissection-based gross anatomy to medical, dental, and allied health students. Then, after meeting the love of her life, she moved to the University of Louisville and has taught medical and dental students there for more than a decade. Over 20 years of teaching full time at two medical schools in the state, her teaching efforts have been recognized through receipt of the highest teaching honor at each medical school in the state, the Provost's Teaching Award at University of Kentucky, and the Distinguished Teaching Professorship at University of Louisville.

Martha Johnson Gdowski, PhD earned her BS in Biology cum laude from Gannon University in 1990, followed by a PhD in Anatomy from the Pennsylvania State University College of Medicine in 1995. She completed postdoctoral fellowships at the Cleveland Clinic and Northwestern University School of Medicine prior to accepting a faculty position in the Department of Neuroscience at the University of Rochester School of Medicine and Dentistry in 2001. Previous research interests include the development of an adult model of hydrocephalus, sensorimotor integration in the basal ganglia, and sensorimotor integration in normal and pathological aging.

Her passion throughout her career has been in her service as an educator. Her teaching has encompassed a variety of learning formats, including didactic lecture, laboratory, journal club, and problem-based learning. She has taught for four academic institutions in different capacities (The Pennsylvania State University School of Medicine, Northwestern University School of Medicine, Ithaca College, and The University of Rochester School of Medicine and Dentistry). She has taught in the following curricula: Undergraduate and Graduate Neuroscience, Graduate Neuroanatomy, Graduate Human Anatomy and Physiology for Physical Therapists, Undergraduate Medical Human Anatomy and Histology, and Undergraduate and Graduate Human Anatomy. These experiences have provided an opportunity to instruct students that vary in age, life experience, race, ethnicity, and economic background, revealing how diversity in student populations enriches learning environments in ways that benefit everyone. She has been honored to be the recipient of numerous awards for her teaching and mentoring of students during their undergraduate medical education. Martha enjoys gardening, hiking, and swimming with her husband, Greg Gdowski, PhD, and their dogs, Sophie and Ivy.

Virginia T. Lyons, PhD is an Associate Professor of Medical Education and the Associate Dean for Preclinical Education at the Geisel School of Medicine at Dartmouth. She received her BS in Biology from Rochester Institute of Technology and her PhD in Cell Biology and Anatomy from the University of North Carolina at Chapel Hill. Dr. Lyons has devoted her career to education in the anatomical sciences, teaching gross anatomy, histology, embryology, and neuroanatomy to medical students and other health professions students. She has led courses and curricula in human gross anatomy and embryology for more than 20 years and is a strong advocate for incorporating engaged pedagogies into preclinical medical education. Dr. Lyons has been recognized with numerous awards for teaching and mentoring students and was elected to the Dartmouth chapter of the Alpha Omega Alpha Honor Medical Society. She is the author of *Netter's Essential Systems-Based Anatomy* and co-author of the Human Anatomy Learning Modules website accessed by students worldwide. Dr. Lyons also serves as the Discipline Editor for Anatomy on the Aquifer Sciences Curriculum Editorial Board, working to integrate anatomical concepts into virtual patient cases that are used in multiple settings including clerkships and residency training.

Peter J. Ward, PhD grew up in Casper, Wyoming, graduating from Kelly Walsh High School and then attending Carnegie Mellon University in Pittsburgh, Pennsylvania. He began graduate school at Purdue University, where he first encountered gross anatomy, histology, embryology, and neuroanatomy. Having found a course of study that engrossed him, he helped teach those courses in the veterinary and medical programs at Purdue. Dr. Ward completed a PhD program in anatomy education and, in 2005, he joined the faculty at the West Virginia School of Osteopathic Medicine (WVSOM) in Lewisburg, West Virginia. There he has taught gross anatomy, embryology, neuroscience, histology, and the history of medicine. Dr. Ward has received numerous teaching awards, including the WVSOM Golden Key Award, the Basmajian Award from the American Association of Anatomists, and has been a two-time finalist in the West Virginia Merit Foundation's Professor of the Year selection. Dr. Ward has also been director of the WVSOM plastination facility, coordinator of the anatomy graduate teaching assistants, chair of the curriculum committee, chair of the faculty council, creator and director of a clinical anatomy elective course, and host of many anatomy-centered events between WVSOM and two Japanese Colleges of Osteopathy. Dr. Ward has also served as council member and association secretary for the American Association of Clinical Anatomists. In conjunction with Bone Clones, Inc., Dr. Ward has produced tactile models that mimic the feel of anatomical structures when intact and when ruptured during the physical examination. He created the YouTube channel Clinical Anatomy Explained! and continues to pursue interesting ways to present the anatomical sciences to the public. Dr. Ward was the Senior Associate Editor for the three volumes of *The Netter Collection: The Digestive System*, 2nd Edition, a contributor to *Gray's Anatomy,* 42nd Edition, and is author of *Netter's Integrated Musculoskeletal System: Clinical Anatomy Explained.*

Brion Benninger, MD, MBChB, MSc currently teaches surgical, imaging, and dynamic anatomy to medical students and residents in several countries (United States, New Zealand, China, Japan, Korea, The Caribbean, Mexico). He develops, invents, and assesses ultrasound probes, medical equipment, simulations, and software while identifying dynamic anatomy. He enjoys mixing educational techniques integrating macro imaging and surgical anatomy. Dr. Benninger developed the teaching theory of anatomy deconstruction/reconstruction and was the first to combine ultrasound with Google Glass during physical examination, coining the term "triple feedback examination." An early user of ultrasound, he continues to develop eFAST teaching and training techniques, has developed and shares a patent on a novel ultrasound finger probe, and is currently developing a new revolutionary ultrasound probe for breast screening. He is a reviewer for several ultrasound, clinical anatomy, surgical, and radiology journals and edits and writes medical textbooks. His research interests integrate clinical anatomy with conventional and emerging technologies to improve training techniques in situ and simulation. Dr. Benninger pioneered and coined the term "dynamic anatomy," developed a technique to deliver novel contrast medium to humans, and was the first to reveal vessels and nerves not previously seen using CT and MRI imaging. He has mentored more than 200 students on over 350 research projects presented at national and international conferences and has received numerous awards for projects related to emergency procedures, ultrasound, sports medicine, clinical anatomy, medical simulation, reverse translational research, medical education, and technology. He is proud to have received medical teaching awards from several countries and institutions, including being the first recipient in more than 25 years to receive the Commendation Medal Award from the Commission of Osteopathic Accreditation for innovative clinical anatomy teaching that he designed and facilitated in Lebanon, Oregon. Dr. Benninger has received sports medicine accolades from Sir Roger Bannister regarding his medical invention on shoulder proprioception. He is also Executive Director of the Medical Anatomy Center and collaborates with colleagues globally from surgical and nonsurgical specialties. He is also an invited course speaker for surgical anatomy in New Zealand. Dr. Benninger collects medical history books, loves mountains and sports, and is an anonymous restaurant critic. British mentors directly responsible for his training include Prof. Peter Bell (surgery), Prof. Sir Alec Jeffreys (genetic fingerprinting), Profs. David deBono and Tony Gershlick (cardiology), Prof. Roger Greenhalgh (vascular surgery), Profs. Chris Colton, John Webb, and Angus Wallace (orthopaedics), Prof. Harold Ellis CBE (surgery and clinical anatomy), and Prof. Susan Standring (Guys Hospital/Kings College).

Todd M. Hoagland, PhD is Clinical Professor of Biomedical Sciences and Occupational Therapy at Marquette University in the College of Health Sciences. Previously he was Professor of Anatomy at the Medical College of Wisconsin (MCW). Prior to MCW, Dr. Hoagland was at Boston University School of Medicine (BUSM) and he still holds an adjunct faculty position at Boston University Goldman School of Dental Medicine. Dr. Hoagland is a passionate teacher and is dedicated to helping students achieve their goals. He believes in being a strong steward of the anatomical sciences, which involves teaching it to students while contemporaneously developing resources to improve the transfer of knowledge and preparing the next generation to be even better teachers. While at BUSM, Dr. Hoagland was a leader for the Carnegie Initiative on the Doctorate in Neuroscience and helped develop the Vesalius Program (teacher training) for graduate students. The program ensures that graduate students learn about effective teaching, receive authentic experiences in the classroom, and understand how to share what they learn via scholarship.

Dr. Hoagland's dedication to health professions education has been richly rewarded by numerous teaching awards from the University of Notre Dame, BUSM, and MCW. Dr. Hoagland received the Award for Outstanding Ethical Leadership in 2009, was inducted into the Alpha Omega Alpha Honor Medical Society in 2010, received the American Association of Anatomists Basmajian Award in 2012, and was inducted into the Society of Teaching Scholars in 2012 and was their director from 2016–2020.

Dr. Hoagland's scholarly activity centers on (1) evaluating content and instructional/learning methodology in Clinical Human Anatomy and Neuroanatomy courses, especially as relevant to clinical practice, (2) translating basic anatomical science research findings into clinically meaningful information, and (3) evaluating professionalism in students to enhance their self-awareness and improve patient care outcomes. Dr. Hoagland is also consulting editor for *Netter's Atlas of Human Anatomy,* co-author for the digital anatomy textbook *AnatomyOne,* and lead author for *Clinical Human Anatomy Dissection Guide.*

ACKNOWLEDGMENTS

Carlos A. G. Machado, MD

With the completion of this 8th edition, I celebrate 27 years contributing to the Netter brand of educational products, 25 years of which have been dedicated to the update—seven editions—of this highly prestigious, from birth, *Atlas of Human Anatomy*. For these 25 years I have had the privilege and honor of working with some of the most knowledgeable anatomists, educators, and consulting editors—my treasured friends—from whom I have learned considerably.

For the last 16 years it has also been a great privilege to be part of the Elsevier team and be under the skillful coordination and orientation of Marybeth Thiel, Elsevier's Senior Content Development Specialist, and Elyse O'Grady, Executive Content Strategist. I thank both for their friendship, support, sensibility, and very dedicated work.

Once more I thank my wife Adriana and my daughter Beatriz for all their love and encouragement, and for patiently steering me back on track when I get lost in philosophical divagations about turning scientific research into artistic inspiration—and vice-versa!

It is impossible to put in words how thankful I am to my much-loved parents, Carlos and Neide, for their importance in my education and in the formation of my moral and ethical values.

I am eternally grateful to the body donors for their inestimable contribution to the correct understanding of human anatomy; to the students, teachers, health professionals, colleagues, educational institutions, and friends who have, anonymously or not, directly or indirectly, been an enormous source of motivation and invaluable scientific references, constructive comments, and relevant suggestions.

My last thanks, but far from being the least, go to my teachers Eugênio Cavalcante, Mário Fortes, and Paulo Carneiro, for their inspiring teachings on the practical application of the knowledge of anatomy.

Paul E. Neumann, MD

It has been a privilege to work on the English and Latin editions of *Netter's Atlas of Human Anatomy*. I thank the staff at Elsevier (especially Elyse O'Grady, Marybeth Thiel and Carrie Stetz), Dr. Carlos Machado, and the other editors for their efforts to produce a new, improved edition. I am also grateful to my wife, Sandra Powell, and my daughter, Eve, for their support of my academic work.

R. Shane Tubbs, MS, PA-C, PhD

I thank Elyse O'Grady and Marybeth Thiel for their dedication and hard work on this edition. As always, I thank my wife, Susan, and son, Isaiah, for their patience with me on such projects. Additionally, I thank Drs. George and Frank Salter who inspired and encouraged me along my path to anatomy.

Jennifer K. Brueckner-Collins, PhD

Reba McEntire once said "To succeed in life, you need three things: a wishbone, a backbone and a funny bone."

My work with the *Netter Atlas* and the people associated with it over the past 15 years has played an instrumental role in helping me develop and sustain these three metaphorical bones in my professional and personal life.

I am forever grateful to John Hansen, who believed in my ability to serve as an editor starting with the 4th edition.

I extend my sincere thanks to Marybeth Thiel and Elyse O'Grady for not only being the finest of colleagues, but part of my professional family as well. Thanks to you both for your professionalism, support, patience, and collegiality.

To Carlos Machado, you continue to amaze me and inspire me with your special gift of bringing anatomy to life through your art.

For this edition, I also count in my blessings, the ability to work closely with the talented team of educational leaders, including Martha Gdowski, Virginia Lyons, and Peter Ward. It is humbling to work with such brilliant and dedicated teachers as we collectively assembled the systems-based *Netter Atlas* concept.

Finally, I dedicate my work on this edition with unconditional and infinite love to Kurt, Lincoln, my Dad in Heaven, as well as my dog boys, Bingo and Biscuit.

Martha Johnson Gdowski, PhD

I am grateful for the honor to work with the team of editors that Elsevier has selected for the preparation of this 8th edition; they are exceptional in their knowledge, passion as educators, and collegiality. I especially would like to thank Elyse O'Grady and Marybeth Thiel, who have been outstanding in their expertise, patience, and guidance. I am grateful to John T. Hansen, PhD, for his guidance, mentorship, and friendship as a colleague at the University of Rochester and for giving me the opportunity to participate in this work. He continues to be an outstanding role model who has shaped my career as an anatomical sciences educator. Special thanks to Carlos Machado for his gift for making challenging anatomical dissections and difficult concepts accessible to students of anatomy through his artistry, research of the details, and thoughtful discussions. I am indebted to the selfless individuals who have gifted their bodies for anatomical study, the students of anatomy, and my colleagues at the University of Rochester, all of whom motivate me to work to be the best educator I can be. I am most grateful for my loving husband and best friend, Greg, who is my greatest source of support and inspiration.

Virginia T. Lyons, PhD

It has been a joy to work with members of the editorial team on the iconic *Atlas of Human Anatomy* by Frank Netter. I would like to thank Elyse O'Grady and Marybeth Thiel for their expert guidance and ability to nourish the creative process while also keeping us focused (otherwise we would have reveled in debating anatomy minutiae for hours!). I am amazed by the talent of Carlos Machado, who is able to transform our ideas into beautiful, detailed illustrations that simplify concepts for students. I appreciate the patience and support of my husband, Patrick, and my children, Sean and Nora, who keep me sane when things get busy. Finally, I am grateful for the opportunity to teach and learn from the outstanding medical students at the Geisel School of Medicine at Dartmouth. I am fulfilled by their energy, curiosity, and love of learning.

Peter J. Ward, PhD

It is a thrill and honor to contribute to the 8th edition of *Netter's Atlas of Human Anatomy*. It still amazes me that I am helping to showcase the incomparable illustrations of Frank Netter and Carlos Machado. I hope that this atlas continues to bring these works of medical art to a new generation of students as they begin investigating the awesome enigma of the human body. Thanks to all the amazing contributors and to the hardworking team at Elsevier, especially Marybeth Thiel and Elyse O'Grady, for keeping all of us moving forward. Thank you especially to Todd Hoagland for recommending me to the team. I have immense gratitude to James Walker and Kevin Hannon, who introduced me to the world of anatomy. They both seamlessly combined high expectations for their students along with enthusiastic teaching that made the topic fascinating and rewarding. Great thanks to my parents, Robert and Lucinda Ward, for their lifelong support of my education and for the many formative museum trips to stare at dinosaur bones. Sarah, Archer, and Dashiell, you are all the reason I work hard and try to make the world a slightly better place. Your love and enthusiasm mean everything to me.

Brion Benninger, MD, MBChB, MSc

I thank all the healthcare institutions worldwide and the allopathic and osteopathic associations who have provided me the privilege to wake up each day and focus on how to improve our knowledge of teaching and healing the anatomy of the mind, body, and soul while nurturing humanism. I am grateful and fortunate to have my lovely wife, Alison, and thoughtful son, Jack, support my efforts during late nights and long weekends. Their laughs and experiences complete my life. I thank Elsevier, especially Marybeth Thiel, Elyse O'Grady, and Madelene Hyde for expecting the highest standards and providing guidance, enabling my fellow coeditors to work in a fluid diverse environment. Many thanks to Carlos Machado and Frank Netter: the world is proud. I thank clinicians who trained me, especially my early gifted surgeon/anatomist/teacher mentors, Drs. Gerald Tressidor and Harold Ellis CBE (Cambridge & Guy's Hospital); Dr. S. Standring and Dr. M. England, who embody professionalism; Drs. P. Crone, E. Szeto, and J. Heatherington, for supporting innovative medical education; my past, current, and future students and patients; and clinical colleagues from all corners of the world who keep medicine and anatomy dynamic, fresh, and wanting. Special thanks to Drs. J.L. Horn, S. Echols, J. Anderson, and J. Underwood, friends, mentors and fellow visionaries who also see "outside the box," challenging the status quo. Heartfelt tribute to my late mentors, friends, and sister, Jim McDaniel, Bill Bryan, and Gail Hendricks, who represent what is good in teaching, caring, and healing. They made this world a wee bit better. Lastly, I thank my mother for her love of education and equality and my father for his inquisitive and creative mind.

Todd M. Hoagland, PhD

It is a privilege to teach clinical human anatomy, and I am eternally grateful to all the body donors and their families for enabling healthcare professionals to train in the dissection laboratory. It is my honor to work with occupational therapy and health professions students and colleagues at Marquette University. I am grateful to John Hansen and the professionals of the Elsevier team for the opportunity to be a steward of the incomparable *Netter Atlas*. Marybeth Thiel and Elyse O'Grady were especially helpful and a pleasure to work with. It was an honor to collaborate with the brilliant Carlos Machado and all the consulting editors. I thank Dave Bolender, Brian Bear, and Rebecca Lufler for being outstanding colleagues, and I thank all the graduate students I've worked with for helping me grow as a person; it is such a pleasure to see them flourish. I am deeply appreciative of Stan Hillman and Jack O'Malley for inspiring me with masterful teaching and rigorous expectations. I am indebted to Gary Kolesari and Richard Hoyt Jr for helping me become a competent clinical anatomist, and to Rob Bouchie for the intangibles and his camaraderie. I am most grateful to my brother, Bill, for his unwavering optimism and for always being there. I thank my mother, Liz, for her dedication and love, and for instilling a strong work ethic. I am humbled by my three awesome children, Ella, Caleb, and Gregory, for helping me redefine love, wonder, and joy. Olya, ty moye solntse!

CONTENTS

7th Edition to 8th Edition Plate Number Conversion Chart Available Online at https://tinyurl.com/Netter7to8conversion

Nerves of Upper Limb • Plates S–106 to S–112

Nerves of Lower Limb • Plates S–113 to S–119

Structures with High Clinical Significance • Tables 2.1–2.3

Cranial Nerves • Tables 2.4– 2.7

Branches of Cervical Plexus • Table 2.8

Nerves of Brachial Plexus • Tables 2.9–2.11

Nerves of Lumbosacral Plexus • Tables 2.12–2.14

Electronic Bonus Plates • S–BP 3 to S–BP 18

SECTION 3   SKELETAL SYSTEM • Plates S–120 to S–182

Upper Limb • Plates S–227 to S–264

Lower Limb • Plates S–265 to S–301

Structures with High Clinical Significance • Tables 4.1–4.3

Electronic Bonus Plates • Plates S–BP 34 to S–BP 43

SECTION 5 — CARDIOVASCULAR SYSTEM • Plates S–302 to S–342

Overview • Plates S–302 to S–306

Pericardium • Plates S–307 to S–308

Heart • Plates S–309 to S–321

SECTION 6   LYMPH VESSELS AND LYMPHOID ORGANS • Plates S–343 to S–357

SECTION 7   RESPIRATORY SYSTEM • Plates S–358 to S–393

Vasculature and Innervation of Mediastinum • Plates S–389 to S–393

Structures with High Clinical Significance • Table 7.1

Electronic Bonus Plates • Plates S–BP 59 to S–BP 64

SECTION 8   DIGESTIVE SYSTEM • Plates S–394 to S–457

Overview • Plates S–394 to S–395

Mouth • Plates S–396 to S–406

Pharynx • Plates S–407 to S–409

Viscera (Esophagus, Stomach, Intestines, Liver, Pancreas) • Plates S–410 to S–434

Visceral Vasculature • Plates S–435 to S–446

Visceral Innervation • Plates S–447 to S–457

Structures with High Clinical Significance · Table 8.1

Electronic Bonus Plates · Plates S–BP 65 to S–BP 82

SECTION 9   URINARY SYSTEM · Plates S–458 to S–471

Overview · Plate S–458

Kidneys and Ureter · Plates S–459 to S–463

Urinary Bladder and Urethra · Plates S–464 to S–466

Vasculature and Innervation • Plates S–467 to S–471

Structures with High Clinical Significance • Table 9.1

Electronic Bonus Plates • Plates S–BP 83 to S–BP 88

SECTION 10   REPRODUCTIVE SYSTEM • Plates S–472 to S–521

Overview • Plate S–472

Mammary Glands • Plates S–473 to S–474

Bony Pelvis • Plates S–475 to S–478

Pelvic Diaphragm and Pelvic Cavity • Plates S–479 to S–484

Female Internal Genitalia • Plates S–485 to S–490

Structures with High Clinical Significance • Table 11.1

Electronic Bonus Plates · Plates S–BP 98 to S–BP 112

APPENDIX

Appendix A Muscles   Tables A.1 to A.18

References

Index

Available Online (see inside front cover)

Bonus Plates
Study Guides
Self-Assessments

ELECTRONIC BONUS PLATES

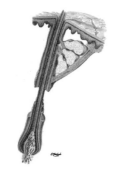

S–BP 1 Pilosebaceous Apparatus

S–BP 2 Major Body Cavities

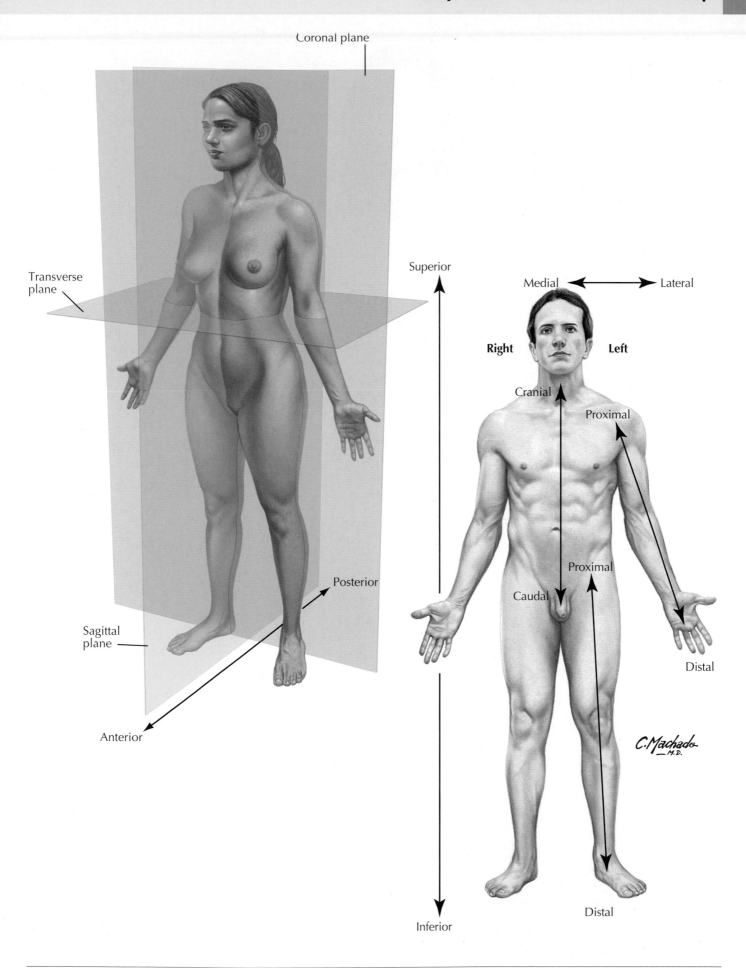

Coronal plane

Transverse plane

Sagittal plane

Superior

Medial — Lateral

Right   Left

Cranial

Proximal

Posterior

Proximal

Caudal

Anterior

Distal

Inferior

Distal

C. Machado
—M.D.

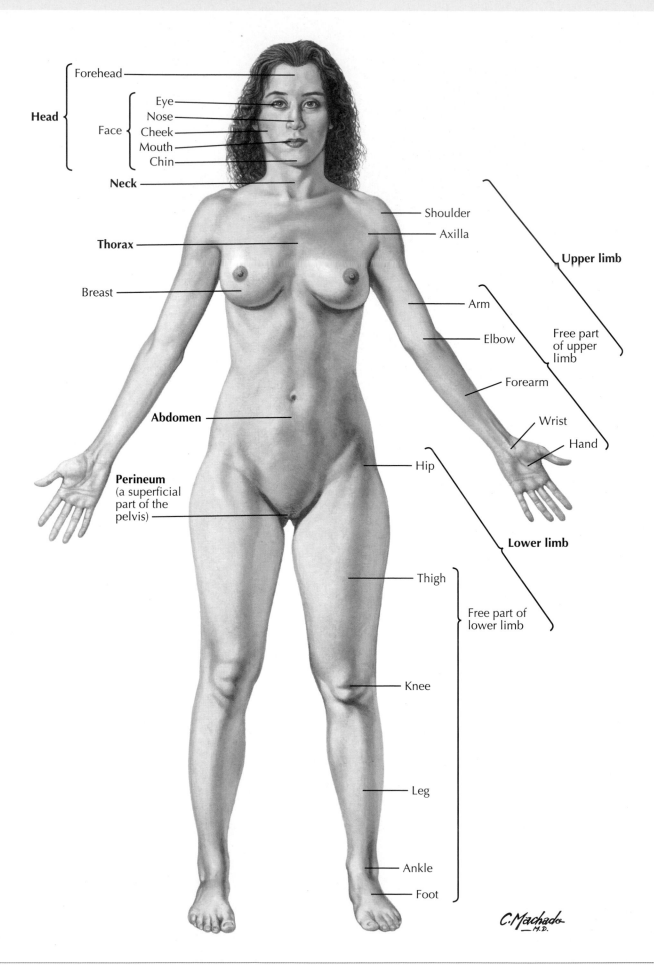

Head
- Forehead
- Face
  - Eye
  - Nose
  - Cheek
  - Mouth
  - Chin

Neck

Thorax

Breast

Abdomen

Perineum
(a superficial part of the pelvis)

Shoulder

Axilla

Upper limb

Arm

Elbow

Free part of upper limb

Forearm

Wrist

Hand

Hip

Lower limb

Thigh

Free part of lower limb

Knee

Leg

Ankle

Foot

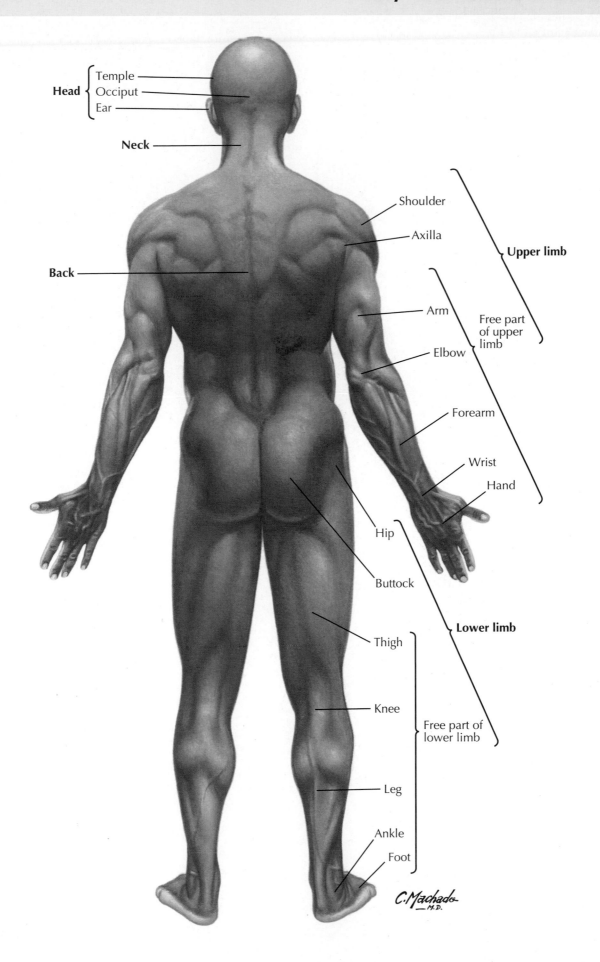

Head
  Temple
  Occiput
  Ear

Neck

Shoulder

Axilla

Back

Upper limb

Arm

Free part of upper limb

Elbow

Forearm

Wrist

Hand

Hip

Buttock

Lower limb

Thigh

Free part of lower limb

Knee

Leg

Ankle

Foot

C. Machado
_M.D.

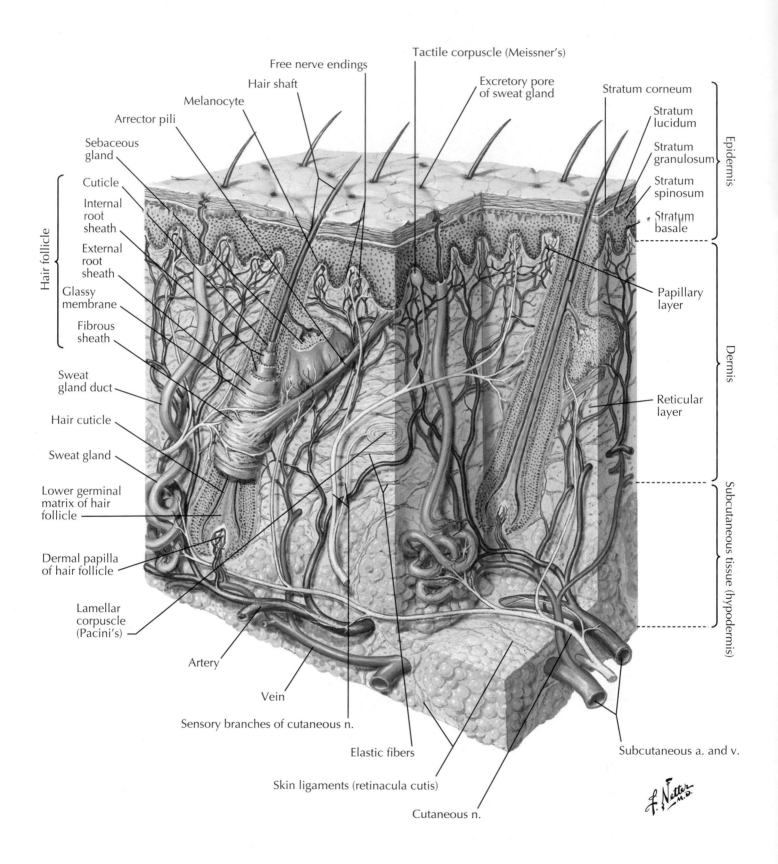

Free nerve endings

Tactile corpuscle (Meissner's)

Hair shaft

Excretory pore of sweat gland

Stratum corneum

Melanocyte

Stratum lucidum

Arrector pili

Stratum granulosum

Sebaceous gland

Stratum spinosum

Cuticle

Stratum basale

Internal root sheath

Epidermis

Hair follicle

External root sheath

Papillary layer

Glassy membrane

Dermis

Fibrous sheath

Reticular layer

Sweat gland duct

Hair cuticle

Sweat gland

Subcutaneous tissue (hypodermis)

Lower germinal matrix of hair follicle

Dermal papilla of hair follicle

Lamellar corpuscle (Pacini's)

Artery

Vein

Sensory branches of cutaneous n.

Elastic fibers

Subcutaneous a. and v.

Skin ligaments (retinacula cutis)

Cutaneous n.

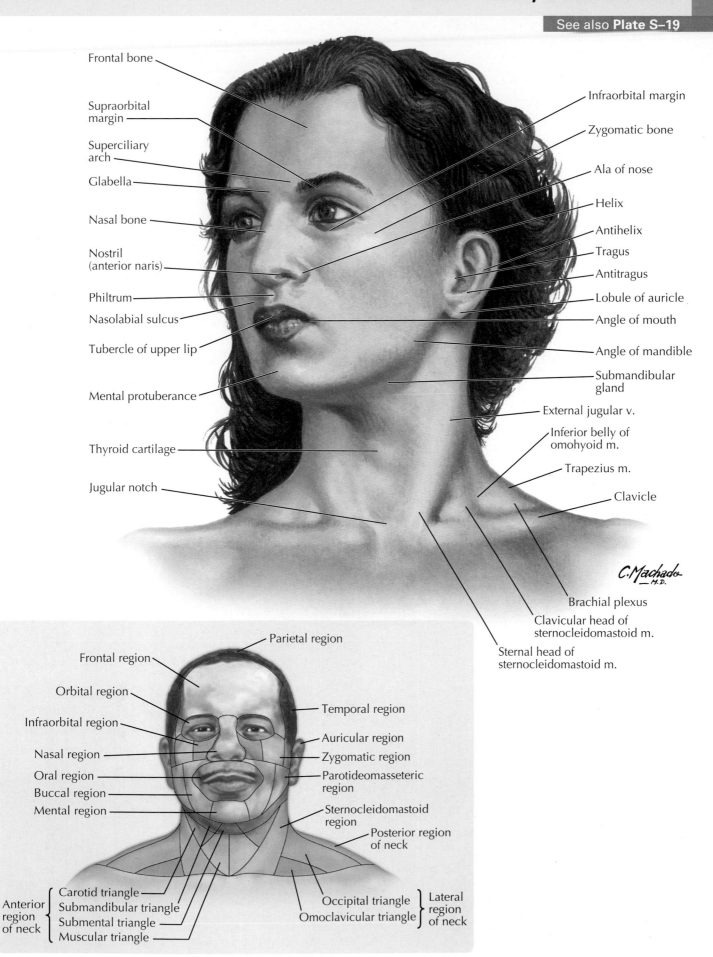

Frontal bone

Supraorbital margin

Superciliary arch

Glabella

Nasal bone

Nostril (anterior naris)

Philtrum

Nasolabial sulcus

Tubercle of upper lip

Mental protuberance

Thyroid cartilage

Jugular notch

Infraorbital margin

Zygomatic bone

Ala of nose

Helix

Antihelix

Tragus

Antitragus

Lobule of auricle

Angle of mouth

Angle of mandible

Submandibular gland

External jugular v.

Inferior belly of omohyoid m.

Trapezius m.

Clavicle

Brachial plexus

Clavicular head of sternocleidomastoid m.

Sternal head of sternocleidomastoid m.

C. Machado —M.D.

Parietal region

Frontal region

Orbital region

Infraorbital region

Nasal region

Oral region

Buccal region

Mental region

Temporal region

Auricular region

Zygomatic region

Parotideomasseteric region

Sternocleidomastoid region

Posterior region of neck

Anterior region of neck
{ Carotid triangle
Submandibular triangle
Submental triangle
Muscular triangle }

Occipital triangle
Omoclavicular triangle
} Lateral region of neck

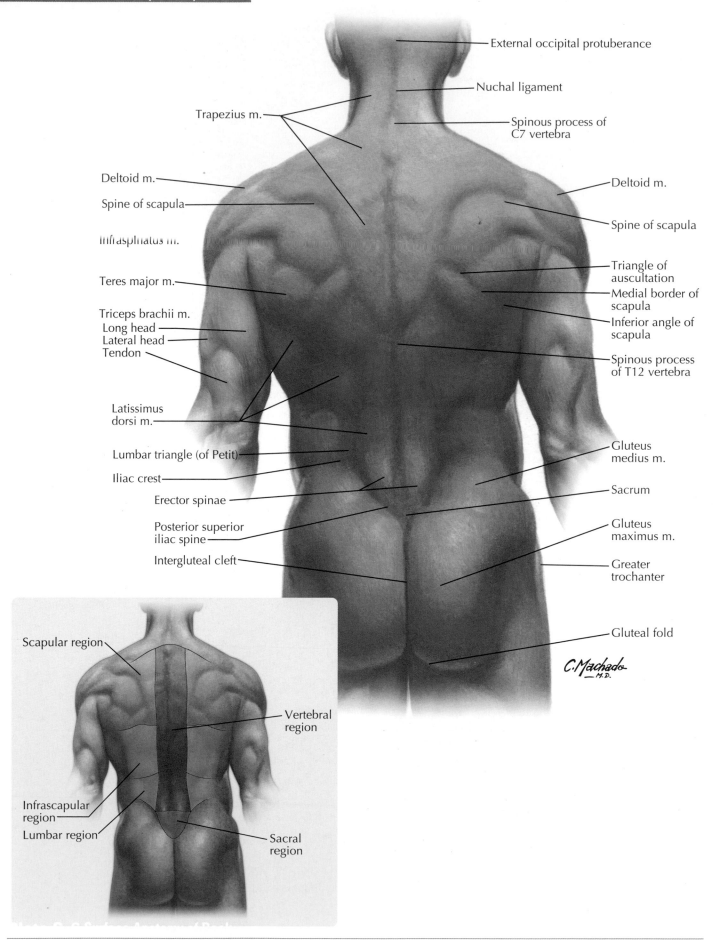

External occipital protuberance

Nuchal ligament

Trapezius m.

Spinous process of C7 vertebra

Deltoid m.

Deltoid m.

Spine of scapula

Spine of scapula

Infraspinatus m.

Triangle of auscultation

Teres major m.

Medial border of scapula

Triceps brachii m.
Long head
Lateral head
Tendon

Inferior angle of scapula

Spinous process of T12 vertebra

Latissimus dorsi m.

Lumbar triangle (of Petit)

Gluteus medius m.

Iliac crest

Sacrum

Erector spinae

Posterior superior iliac spine

Gluteus maximus m.

Intergluteal cleft

Greater trochanter

Gluteal fold

C. Machado
M.D.

Scapular region

Vertebral region

Infrascapular region

Lumbar region

Sacral region

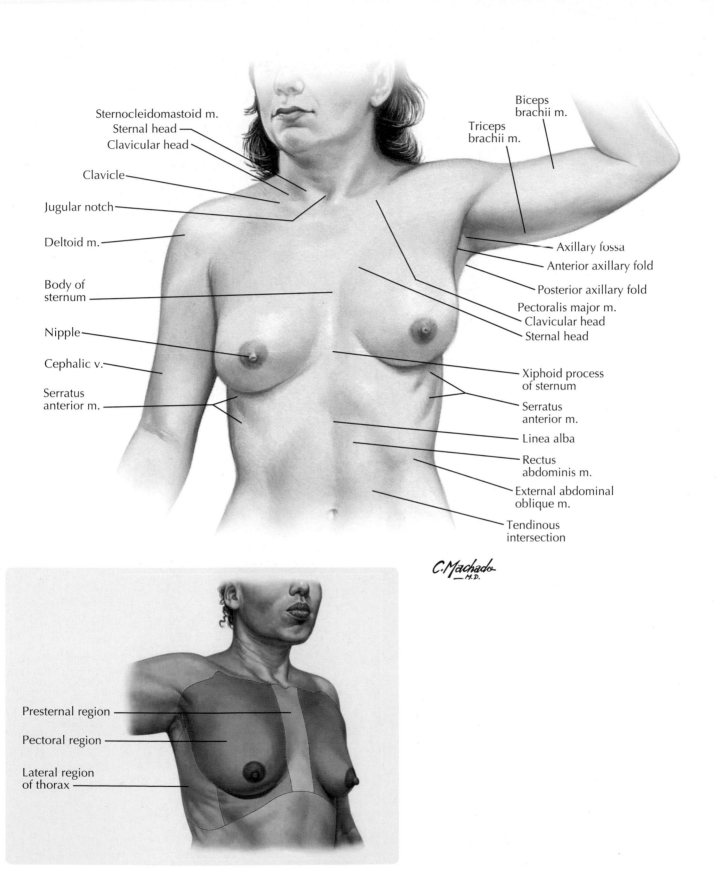

Sternocleidomastoid m.
Sternal head
Clavicular head

Clavicle

Jugular notch

Deltoid m.

Body of
sternum

Nipple

Cephalic v.

Serratus
anterior m.

Biceps
brachii m.

Triceps
brachii m.

Axillary fossa

Anterior axillary fold

Posterior axillary fold

Pectoralis major m.
Clavicular head
Sternal head

Xiphoid process
of sternum

Serratus
anterior m.

Linea alba

Rectus
abdominis m.

External abdominal
oblique m.

Tendinous
intersection

C. Machado
M.D.

Presternal region

Pectoral region

Lateral region
of thorax

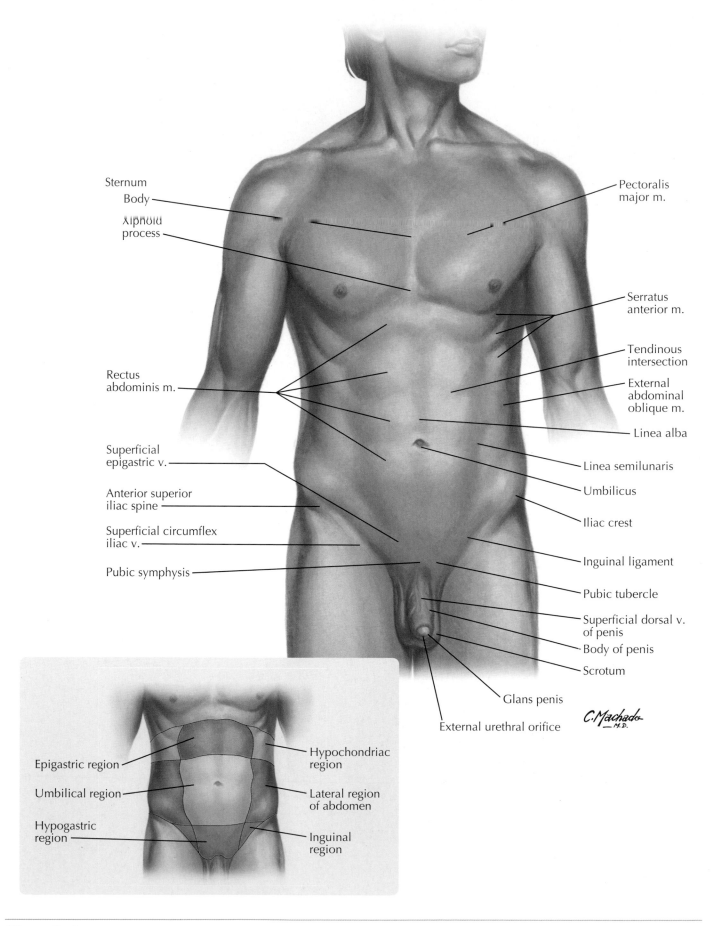

Sternum

Body

Xiphoid process

Rectus abdominis m.

Superficial epigastric v.

Anterior superior iliac spine

Superficial circumflex iliac v.

Pubic symphysis

Pectoralis major m.

Serratus anterior m.

Tendinous intersection

External abdominal oblique m.

Linea alba

Linea semilunaris

Umbilicus

Iliac crest

Inguinal ligament

Pubic tubercle

Superficial dorsal v. of penis

Body of penis

Scrotum

Glans penis

External urethral orifice

C.Machado
— M.D.

Epigastric region

Umbilical region

Hypogastric region

Hypochondriac region

Lateral region of abdomen

Inguinal region

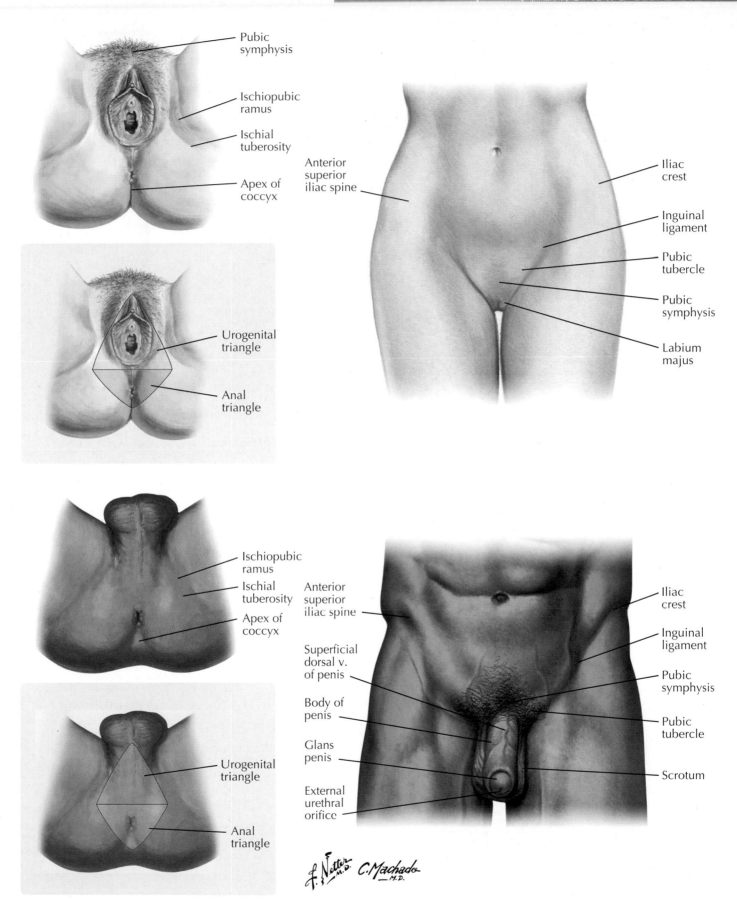

Pubic symphysis

Ischiopubic ramus

Ischial tuberosity

Apex of coccyx

Urogenital triangle

Anal triangle

Anterior superior iliac spine

Iliac crest

Inguinal ligament

Pubic tubercle

Pubic symphysis

Labium majus

Ischiopubic ramus

Ischial tuberosity

Apex of coccyx

Urogenital triangle

Anal triangle

Anterior superior iliac spine

Superficial dorsal v. of penis

Body of penis

Glans penis

External urethral orifice

Iliac crest

Inguinal ligament

Pubic symphysis

Pubic tubercle

Scrotum

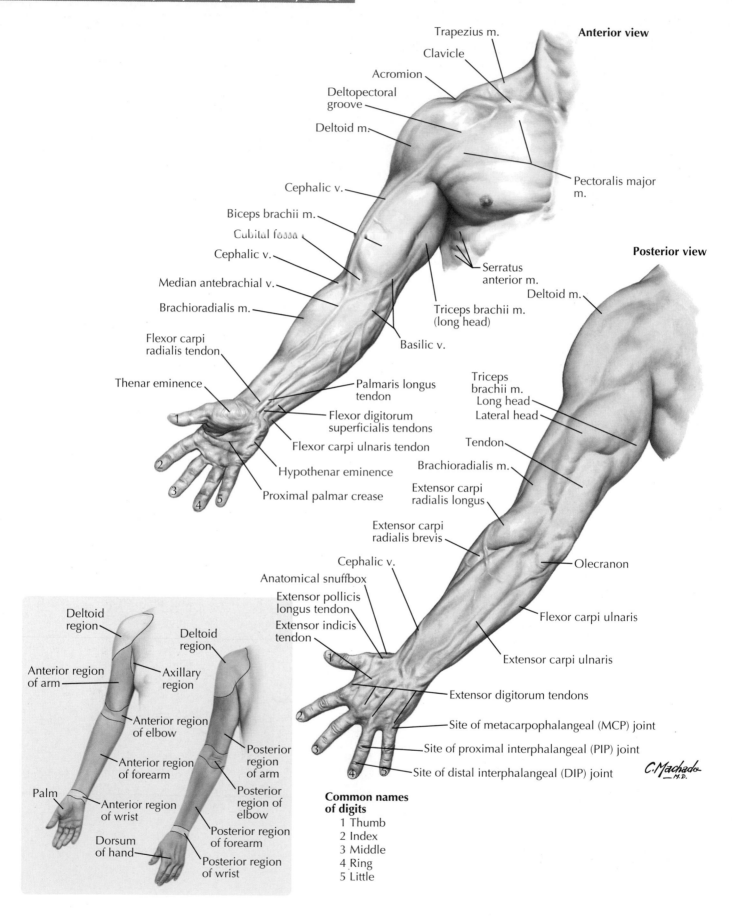

Anterior view

Trapezius m.

Clavicle

Acromion

Deltopectoral groove

Deltoid m.

Cephalic v.

Biceps brachii m.

Cubital fossa

Cephalic v.

Median antebrachial v.

Brachioradialis m.

Flexor carpi radialis tendon

Thenar eminence

Pectoralis major m.

Serratus anterior m.

Triceps brachii m. (long head)

Basilic v.

Palmaris longus tendon

Flexor digitorum superficialis tendons

Flexor carpi ulnaris tendon

Hypothenar eminence

Proximal palmar crease

1
2
3
4 5

Posterior view

Deltoid m.

Triceps brachii m.
Long head
Lateral head

Tendon

Brachioradialis m.

Extensor carpi radialis longus

Extensor carpi radialis brevis

Cephalic v.

Anatomical snuffbox

Extensor pollicis longus tendon

Extensor indicis tendon

Olecranon

Flexor carpi ulnaris

Extensor carpi ulnaris

Extensor digitorum tendons

Site of metacarpophalangeal (MCP) joint

Site of proximal interphalangeal (PIP) joint

Site of distal interphalangeal (DIP) joint

1
2
3
4 5

C. Machado
M.D.

Common names of digits
1 Thumb
2 Index
3 Middle
4 Ring
5 Little

Deltoid region

Deltoid region

Anterior region of arm

Axillary region

Anterior region of elbow

Posterior region of arm

Anterior region of forearm

Posterior region of elbow

Palm

Anterior region of wrist

Posterior region of forearm

Dorsum of hand

Posterior region of wrist

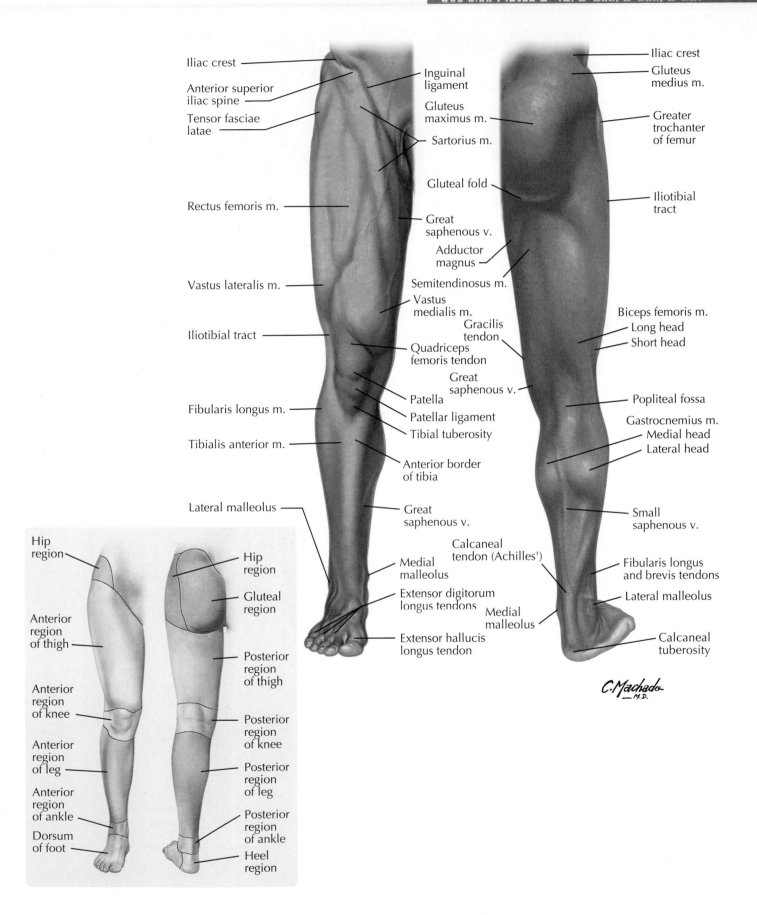

Iliac crest

Anterior superior
iliac spine

Tensor fasciae
latae

Rectus femoris m.

Vastus lateralis m.

Iliotibial tract

Fibularis longus m.

Tibialis anterior m.

Lateral malleolus

Inguinal
ligament

Gluteus
maximus m.

Sartorius m.

Gluteal fold

Great
saphenous v.

Adductor
magnus

Semitendinosus m.

Vastus
medialis m.

Gracilis
tendon

Quadriceps
femoris tendon

Great
saphenous v.

Patella

Patellar ligament

Tibial tuberosity

Anterior border
of tibia

Great
saphenous v.

Calcaneal
tendon (Achilles')

Medial
malleolus

Extensor digitorum
longus tendons

Extensor hallucis
longus tendon

Medial
malleolus

Iliac crest

Gluteus
medius m.

Greater
trochanter
of femur

Iliotibial
tract

Biceps femoris m.
Long head
Short head

Popliteal fossa

Gastrocnemius m.
Medial head
Lateral head

Small
saphenous v.

Fibularis longus
and brevis tendons

Lateral malleolus

Calcaneal
tuberosity

C. Machado
M.D.

Hip
region

Anterior
region
of thigh

Anterior
region
of knee

Anterior
region
of leg

Anterior
region
of ankle

Dorsum
of foot

Hip
region

Gluteal
region

Posterior
region
of thigh

Posterior
region
of knee

Posterior
region
of leg

Posterior
region
of ankle

Heel
region

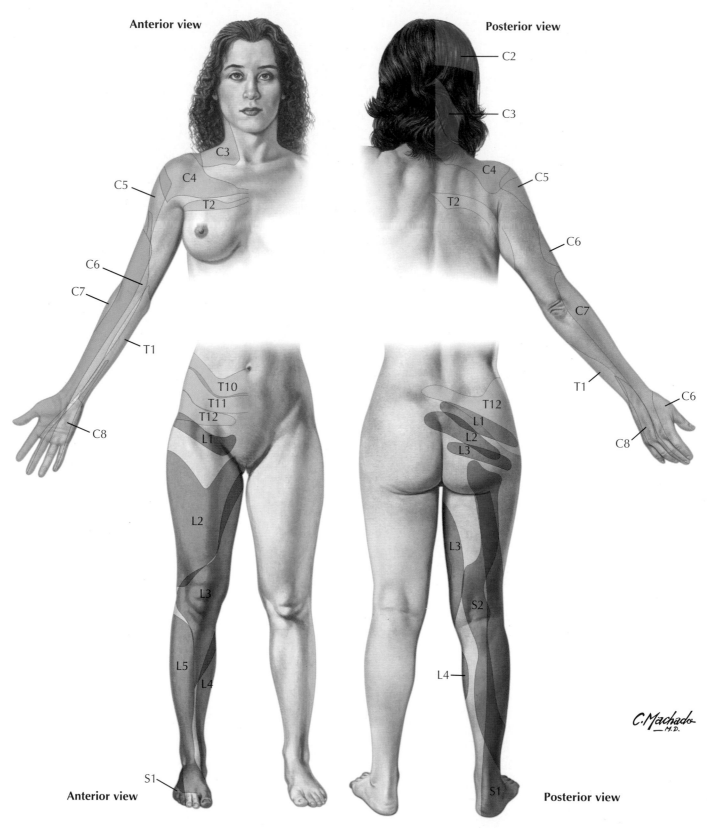

Anterior view

Posterior view

C2

C3

C3

C4

C4

C5

C5

C5

T2

T2

C6

C6

C6

C7

C7

C7

T1

T1

T1

T10

T11

T12

C8

C8

T12

L1

C6

L1

L2

C8

L3

L2

L3

L3

L2

S2

L3

L4

L5

L4

L4

S1

S1

C.Machado
M.D.

Anterior view

Posterior view

Schematic based on Lee MW, McPhee RW, Stringer MD. An evidence-based approach to human dermatomes. Clin Anat. 2008;21(5):363–373. doi: 10.1002/ca.20636. PMID: 18470936. Please note that these areas are not absolute and vary from person to person. S3, S4, S5, and Co supply the perineum but are not shown for reasons of clarity. Of note, the dermatomes are larger than illustrated as the figure is based on best evidence; gaps represent areas in which the data are inconclusive.

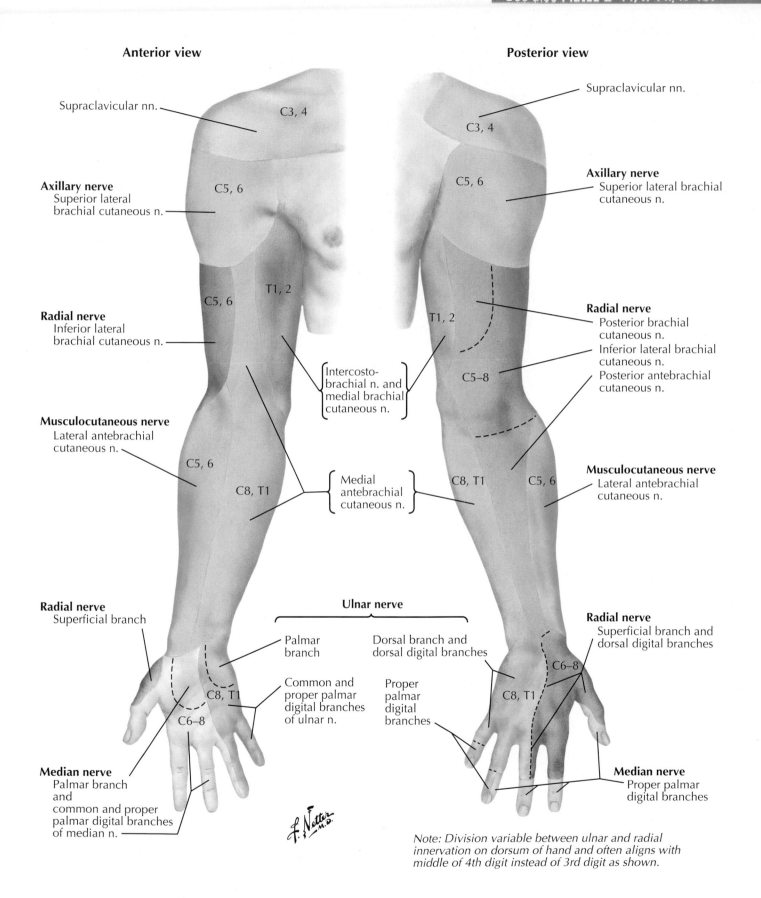

Anterior view

Posterior view

Supraclavicular nn.

Supraclavicular nn.

C3, 4

C3, 4

Axillary nerve
Superior lateral
brachial cutaneous n.

Axillary nerve
Superior lateral brachial
cutaneous n.

C5, 6

C5, 6

C5, 6

T1, 2

Radial nerve
Inferior lateral
brachial cutaneous n.

T1, 2

Radial nerve
Posterior brachial
cutaneous n.
Inferior lateral brachial
cutaneous n.
Posterior antebrachial
cutaneous n.

C5–8

Intercosto-
brachial n. and
medial brachial
cutaneous n.

Musculocutaneous nerve
Lateral antebrachial
cutaneous n.

C5, 6

Medial
antebrachial
cutaneous n.

C8, T1

C5, 6

Musculocutaneous nerve
Lateral antebrachial
cutaneous n.

C8, T1

Radial nerve
Superficial branch

Ulnar nerve

Radial nerve
Superficial branch and
dorsal digital branches

Palmar
branch

Dorsal branch and
dorsal digital branches

C6–8

C8, T1

Common and
proper palmar
digital branches
of ulnar n.

Proper
palmar
digital
branches

C8, T1

C6–8

Median nerve
Palmar branch
and
common and proper
palmar digital branches
of median n.

Median nerve
Proper palmar
digital branches

f. Netter M.D.

*Note: Division variable between ulnar and radial
innervation on dorsum of hand and often aligns with
middle of 4th digit instead of 3rd digit as shown.*

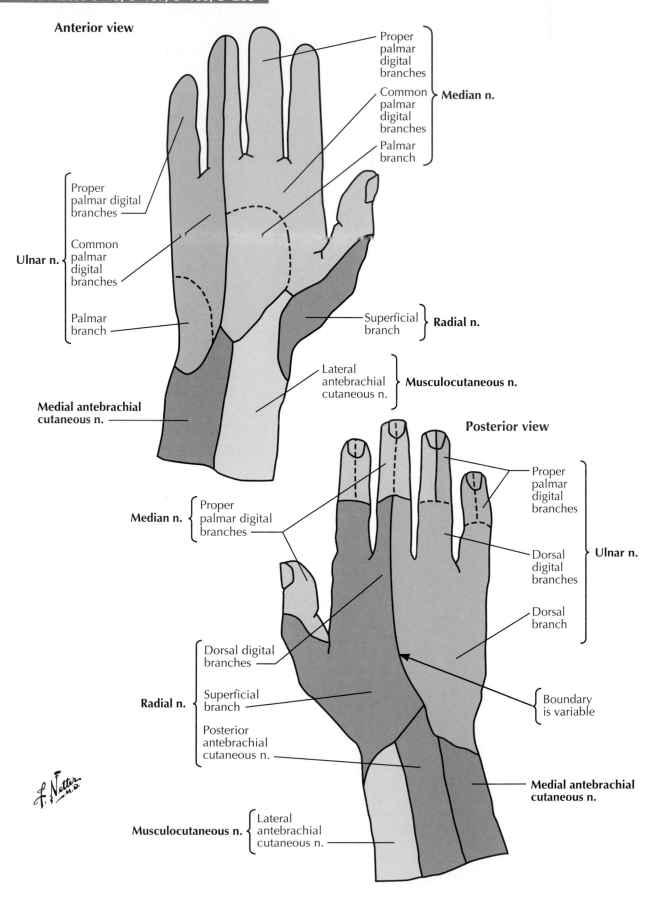

Anterior view

Proper
palmar
digital
branches

Common
palmar
digital
branches

Palmar
branch

} **Median n.**

Proper
palmar digital
branches

Common
palmar
digital
branches

Palmar
branch

Ulnar n. {

Superficial
branch

} **Radial n.**

Lateral
antebrachial
cutaneous n.

} **Musculocutaneous n.**

**Medial antebrachial
cutaneous n.**

Posterior view

Median n. {
Proper
palmar digital
branches

Proper
palmar
digital
branches

Dorsal
digital
branches

Dorsal
branch

} **Ulnar n.**

Boundary
is variable

Dorsal digital
branches

Radial n.
Superficial
branch

Posterior
antebrachial
cutaneous n.

**Medial antebrachial
cutaneous n.**

Musculocutaneous n. {
Lateral
antebrachial
cutaneous n.

NERVOUS SYSTEM AND SENSE ORGANS

2

ELECTRONIC BONUS PLATES

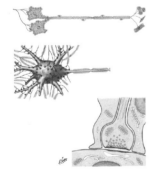

S–BP 3 Neurons and Synapses

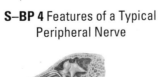

S–BP 4 Features of a Typical Peripheral Nerve

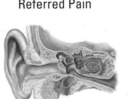

S–BP 5 Sites of Visceral Referred Pain

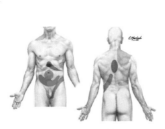

S–BP 6 Afferent Innervation of Oral Cavity and Pharynx

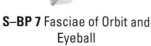

S–BP 7 Fasciae of Orbit and Eyeball

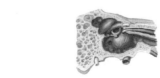

S–BP 8 Tympanic Cavity: Medial and Lateral Views

S–BP 9 Anatomy of the Pediatric Ear

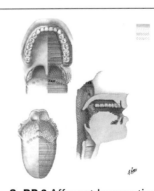
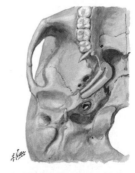

S–BP 10 Auditory Tube (Eustachian)

ELECTRONIC BONUS PLATES—*cont'd*

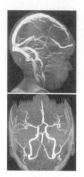

S–BP 11 Cranial Imaging (MRV and MRA)

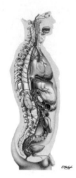

S–BP 12 Axial and Coronal MRIs of Brain

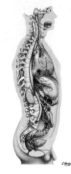

S–BP 13 Sympathetic Nervous System: General Topography

S–BP 14 Parasympathetic Nervous System: General Topography

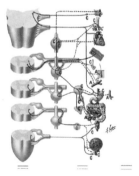

S–BP 15 Cholinergic and Adrenergic Synapses: Schema

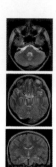
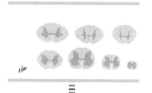

S–BP 16 Spinal Cord Cross Sections: Fiber Tracts

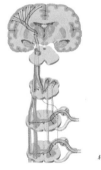

S–BP 17 Somatosensory System: Trunk and Limbs

S–BP 18 Pyramidal System

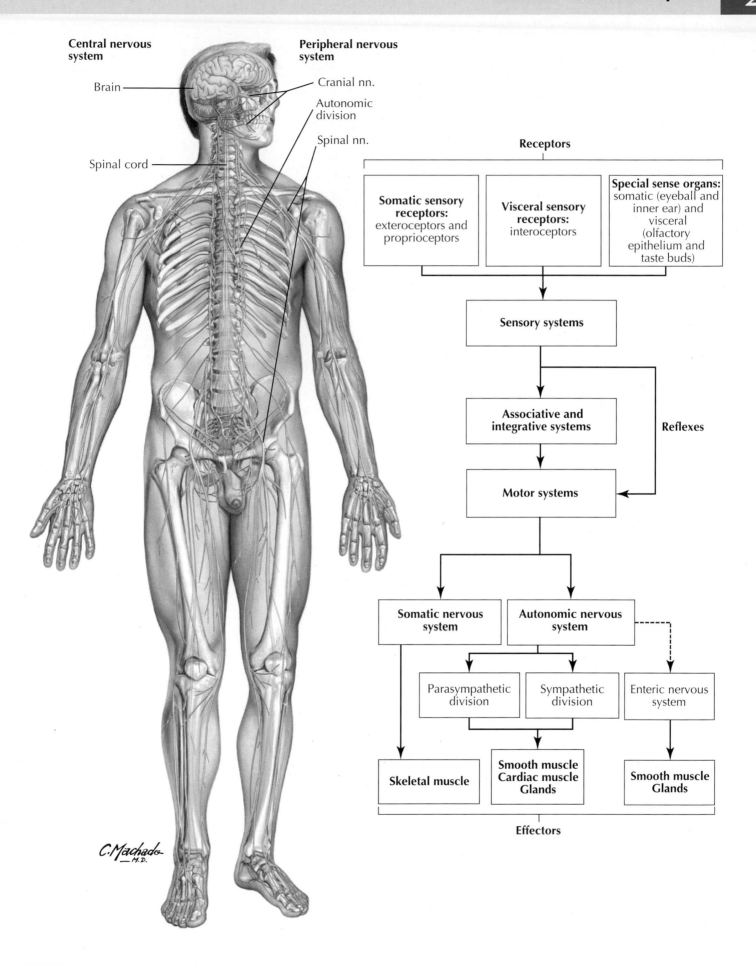

Central nervous system

Brain

Spinal cord

Peripheral nervous system

Cranial nn.

Autonomic division

Spinal nn.

Receptors

| Somatic sensory receptors: exteroceptors and proprioceptors | Visceral sensory receptors: interoceptors | Special sense organs: somatic (eyeball and inner ear) and visceral (olfactory epithelium and taste buds) |

Sensory systems

Associative and integrative systems

Reflexes

Motor systems

| Somatic nervous system | Autonomic nervous system |

| Parasympathetic division | Sympathetic division | Enteric nervous system |

| Skeletal muscle | Smooth muscle Cardiac muscle Glands | Smooth muscle Glands |

Effectors

C. Machado
—M.D.

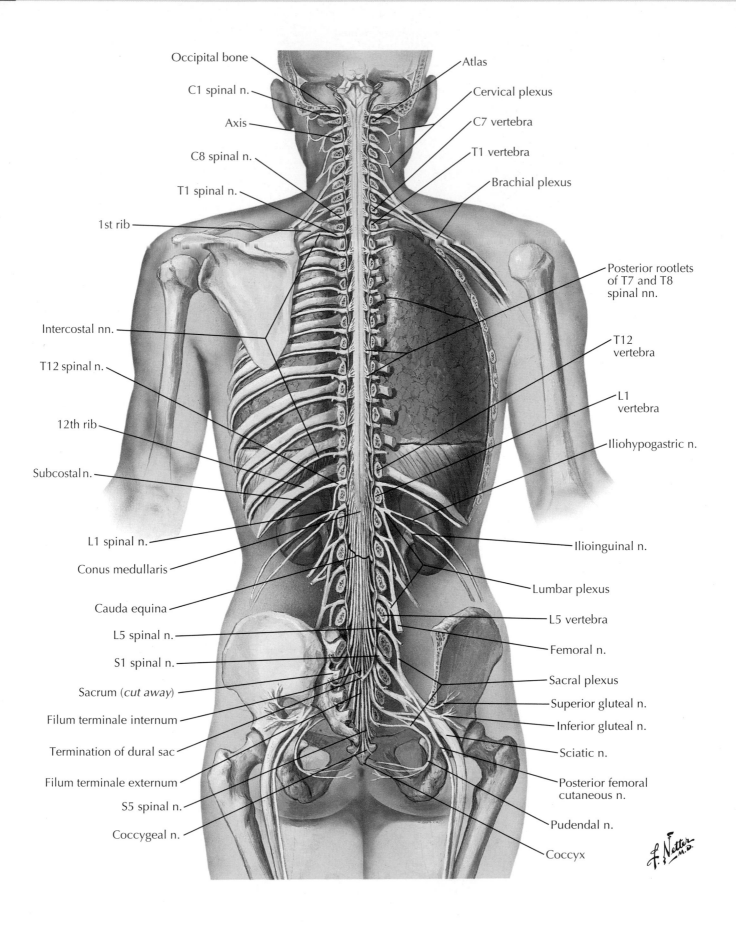

Occipital bone

C1 spinal n.

Axis

C8 spinal n.

T1 spinal n.

1st rib

Intercostal nn.

T12 spinal n.

12th rib

Subcostal n.

L1 spinal n.

Conus medullaris

Cauda equina

L5 spinal n.

S1 spinal n.

Sacrum (*cut away*)

Filum terminale internum

Termination of dural sac

Filum terminale externum

S5 spinal n.

Coccygeal n.

Atlas

Cervical plexus

C7 vertebra

T1 vertebra

Brachial plexus

Posterior rootlets of T7 and T8 spinal nn.

T12 vertebra

L1 vertebra

Iliohypogastric n.

Ilioinguinal n.

Lumbar plexus

L5 vertebra

Femoral n.

Sacral plexus

Superior gluteal n.

Inferior gluteal n.

Sciatic n.

Posterior femoral cutaneous n.

Pudendal n.

Coccyx

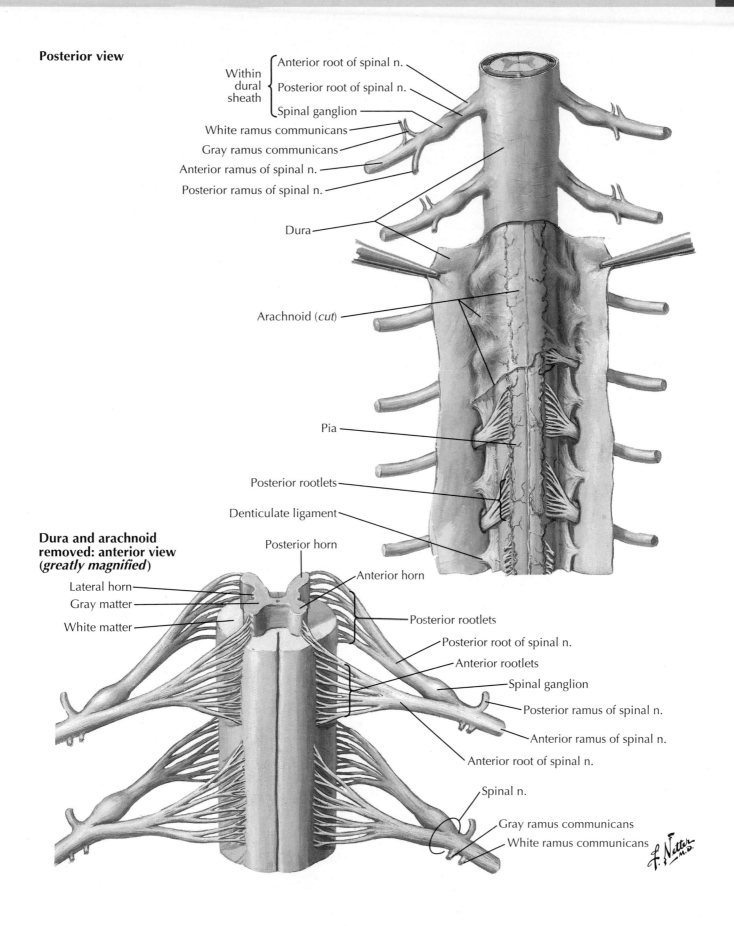

Posterior view

Within dural sheath {
Anterior root of spinal n.
Posterior root of spinal n.
Spinal ganglion

White ramus communicans
Gray ramus communicans
Anterior ramus of spinal n.
Posterior ramus of spinal n.

Dura

Arachnoid (*cut*)

Pia

Posterior rootlets

Denticulate ligament

**Dura and arachnoid
removed: anterior view
(*greatly magnified*)**

Posterior horn

Lateral horn
Gray matter
White matter

Anterior horn

Posterior rootlets

Posterior root of spinal n.

Anterior rootlets

Spinal ganglion

Posterior ramus of spinal n.

Anterior ramus of spinal n.

Anterior root of spinal n.

Spinal n.

Gray ramus communicans
White ramus communicans

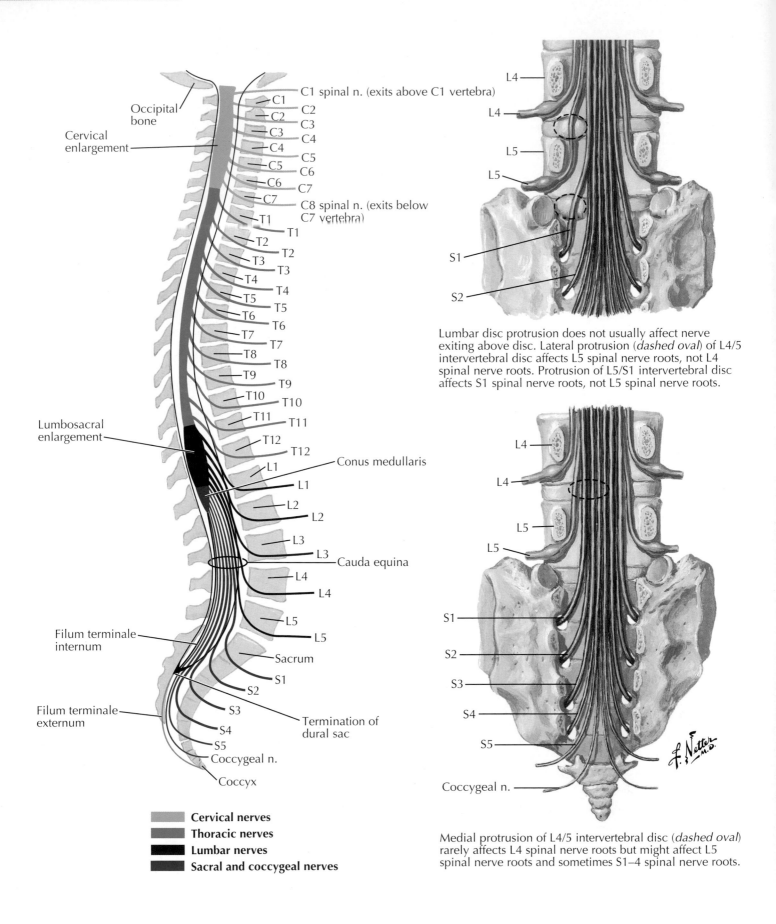

Occipital bone

Cervical enlargement

C1
C2
C3
C4
C5
C6
C7
T1
T2
T3
T4
T5
T6
T7
T8
T9
T10
T11
T12

Lumbosacral enlargement

Filum terminale internum

Filum terminale externum

C1 spinal n. (exits above C1 vertebra)
C2
C3
C4
C5
C6
C7
C8 spinal n. (exits below C7 vertebra)
T1
T2
T3
T4
T5
T6
T7
T8
T9
T10
T11
T12

Conus medullaris

L1
L2
L3
Cauda equina
L4
L5
Sacrum
S1
S2
S3
S4
S5
Coccygeal n.
Coccyx

L1
L2
L3
L4
L5

Termination of dural sac

Cervical nerves
Thoracic nerves
Lumbar nerves
Sacral and coccygeal nerves

L4
L4
L5
L5
S1
S2

Lumbar disc protrusion does not usually affect nerve exiting above disc. Lateral protrusion (*dashed oval*) of L4/5 intervertebral disc affects L5 spinal nerve roots, not L4 spinal nerve roots. Protrusion of L5/S1 intervertebral disc affects S1 spinal nerve roots, not L5 spinal nerve roots.

L4
L4
L5
L5
S1
S2
S3
S4
S5
Coccygeal n.

Medial protrusion of L4/5 intervertebral disc (*dashed oval*) rarely affects L4 spinal nerve roots but might affect L5 spinal nerve roots and sometimes S1–4 spinal nerve roots.

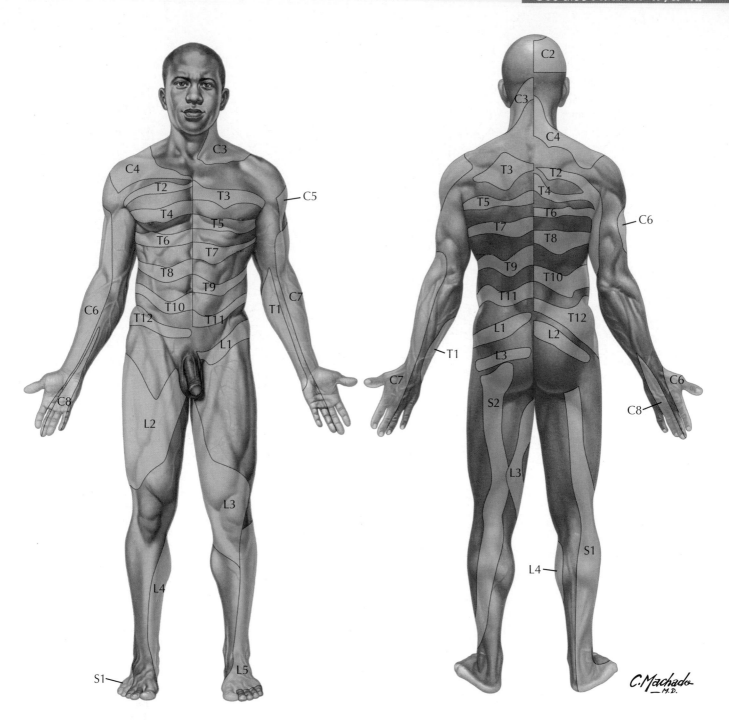

Levels of principal dermatomes

C4	Level of clavicles	**T10**	Level of umbilicus
C5, C6, C7	Lateral surfaces of upper limbs	**L1**	Inguinal region and proximal anterior thigh
C8, T1	Medial surfaces of upper limbs	**L1, L2, L3, L4**	Anteromedial lower limb and gluteal region
C6	Lateral digits	**L4, L5, S1**	Foot
C6, C7, C8	Hand	**L4**	Medial leg
C8	Medial digits	**L5, S1**	Posterolateral lower limb and dorsum of foot
T4	Level of nipples	**S1**	Lateral foot

Schematic based on Lee MW, McPhee RW, Stringer MD. An evidence-based approach to human dermatomes. Clin Anat. 2008; 21(5):363–373. doi: 10.1002/ca.20636. PMID: 18470936. Please note that these areas are not absolute and vary from person to person. S3, S4, S5, and Co supply the perineum but are not shown for reasons of clarity. Of note, the dermatomes are larger than illustrated as the figure is based on best evidence; gaps represent areas in which the data are inconclusive.

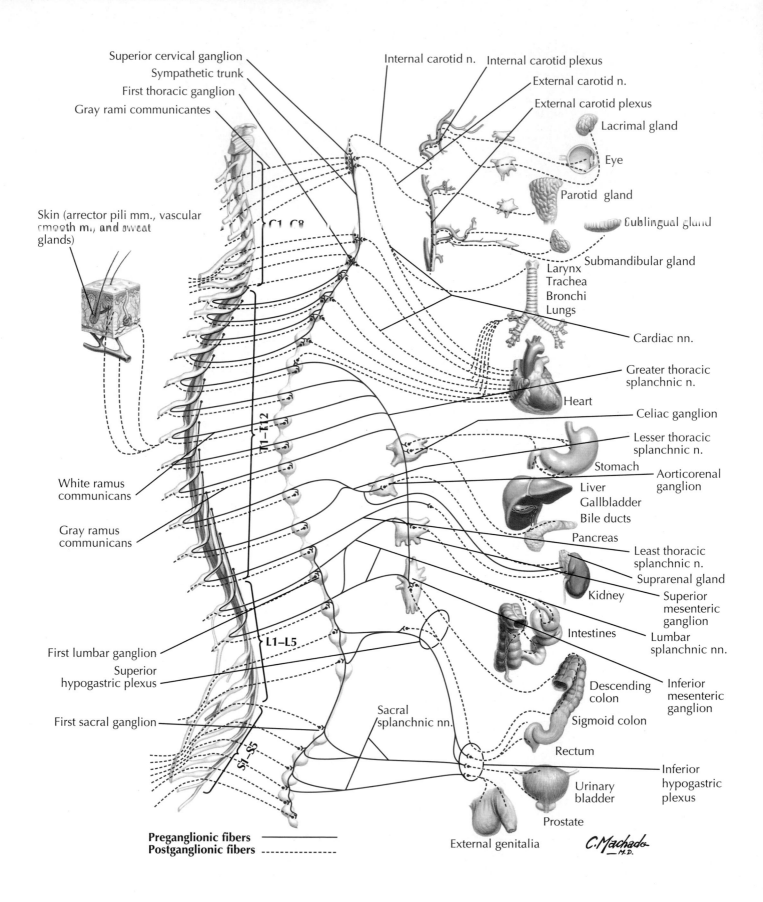

Superior cervical ganglion
Sympathetic trunk
First thoracic ganglion
Gray rami communicantes

Internal carotid n.
Internal carotid plexus
External carotid n.
External carotid plexus
Lacrimal gland
Eye
Parotid gland
Sublingual gland
Submandibular gland

Skin (arrector pili mm., vascular smooth m., and sweat glands)

C1-C8

Larynx
Trachea
Bronchi
Lungs

Cardiac nn.

Greater thoracic splanchnic n.

Heart

T1-T12

Celiac ganglion

Lesser thoracic splanchnic n.

Stomach
Aorticorenal ganglion

White ramus communicans

Liver
Gallbladder
Bile ducts

Pancreas

Gray ramus communicans

Least thoracic splanchnic n.

Suprarenal gland

Kidney

Superior mesenteric ganglion

Intestines

Lumbar splanchnic nn.

L1-L5

First lumbar ganglion

Descending colon

Inferior mesenteric ganglion

Superior hypogastric plexus

Sigmoid colon

First sacral ganglion

Sacral splanchnic nn.

Rectum

S1-S5

Inferior hypogastric plexus

Urinary bladder

Prostate

Preganglionic fibers ————
Postganglionic fibers ------------

External genitalia

C.Machado
M.D.

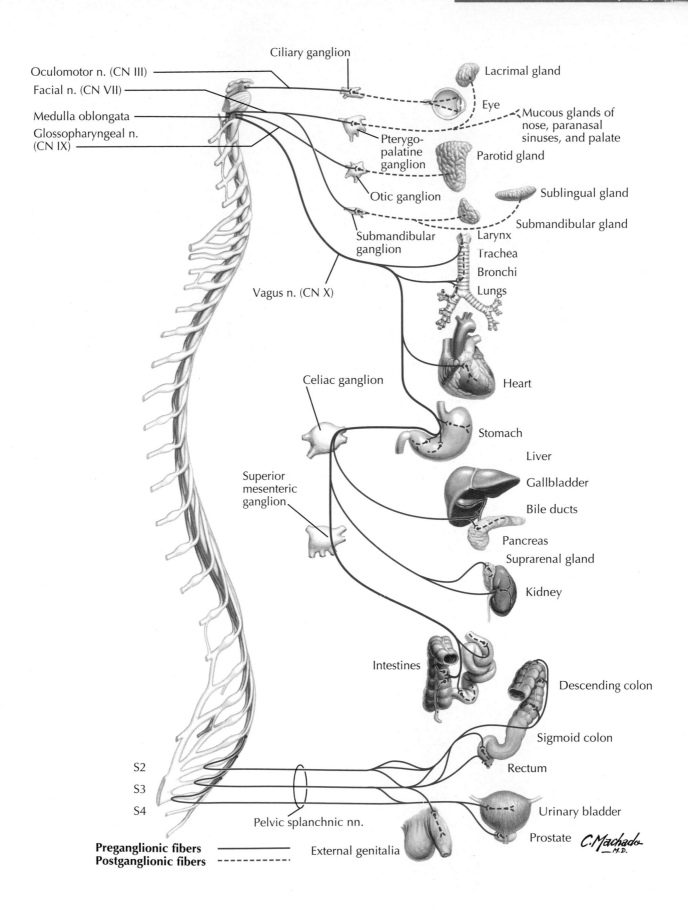

Ciliary ganglion

Oculomotor n. (CN III)

Facial n. (CN VII)

Medulla oblongata

Glossopharyngeal n. (CN IX)

Pterygo-palatine ganglion

Otic ganglion

Submandibular ganglion

Vagus n. (CN X)

Lacrimal gland

Eye

Mucous glands of nose, paranasal sinuses, and palate

Parotid gland

Sublingual gland

Submandibular gland

Larynx

Trachea

Bronchi

Lungs

Heart

Celiac ganglion

Stomach

Liver

Gallbladder

Bile ducts

Pancreas

Suprarenal gland

Kidney

Superior mesenteric ganglion

Intestines

Descending colon

Sigmoid colon

Rectum

S2

S3

S4

Urinary bladder

Prostate

Pelvic splanchnic nn.

Preganglionic fibers ————

Postganglionic fibers - - - - -

External genitalia

C. Machado
— M.D.

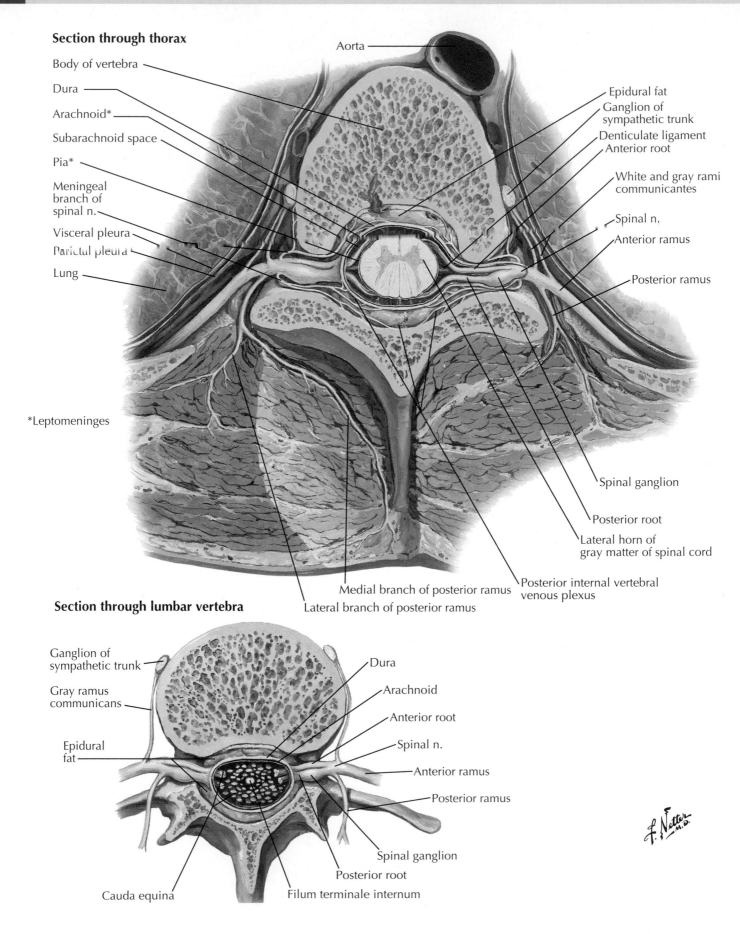

Section through thorax

Body of vertebra

Dura

Arachnoid*

Subarachnoid space

Pia*

Meningeal branch of spinal n.

Visceral pleura

Parietal pleura

Lung

*Leptomeninges

Aorta

Epidural fat

Ganglion of sympathetic trunk

Denticulate ligament

Anterior root

White and gray rami communicantes

Spinal n.

Anterior ramus

Posterior ramus

Spinal ganglion

Posterior root

Lateral horn of gray matter of spinal cord

Medial branch of posterior ramus

Lateral branch of posterior ramus

Posterior internal vertebral venous plexus

Section through lumbar vertebra

Ganglion of sympathetic trunk

Gray ramus communicans

Epidural fat

Dura

Arachnoid

Anterior root

Spinal n.

Anterior ramus

Posterior ramus

Spinal ganglion

Posterior root

Filum terminale internum

Cauda equina

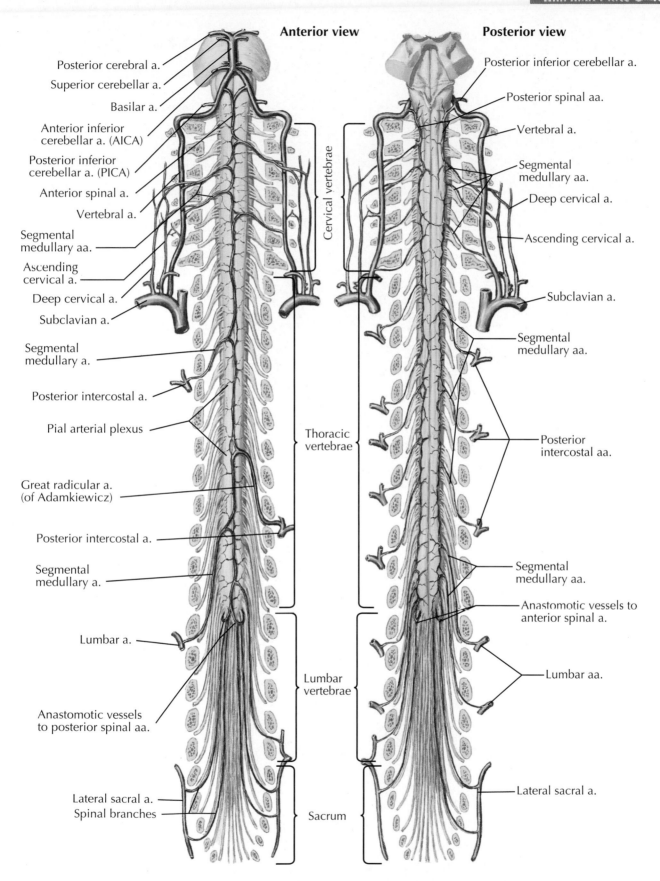

Anterior view

Posterior view

Posterior cerebral a.

Superior cerebellar a.

Basilar a.

Anterior inferior cerebellar a. (AICA)

Posterior inferior cerebellar a. (PICA)

Anterior spinal a.

Vertebral a.

Segmental medullary aa.

Ascending cervical a.

Deep cervical a.

Subclavian a.

Segmental medullary a.

Posterior intercostal a.

Pial arterial plexus

Great radicular a. (of Adamkiewicz)

Posterior intercostal a.

Segmental medullary a.

Lumbar a.

Anastomotic vessels to posterior spinal aa.

Lateral sacral a.
Spinal branches

Cervical vertebrae

Thoracic vertebrae

Lumbar vertebrae

Sacrum

Posterior inferior cerebellar a.

Posterior spinal aa.

Vertebral a.

Segmental medullary aa.

Deep cervical a.

Ascending cervical a.

Subclavian a.

Segmental medullary aa.

Posterior intercostal aa.

Segmental medullary aa.

Anastomotic vessels to anterior spinal a.

Lumbar aa.

Lateral sacral a.

Note: All spinal nerve roots have associated radicular or segmental medullary arteries. Most roots have radicular arteries (see Plate S–24). Both types of arteries run along roots, but radicular arteries end before reaching anterior or posterior spinal arteries; larger segmental medullary arteries continue on to supply a segment of these arteries.

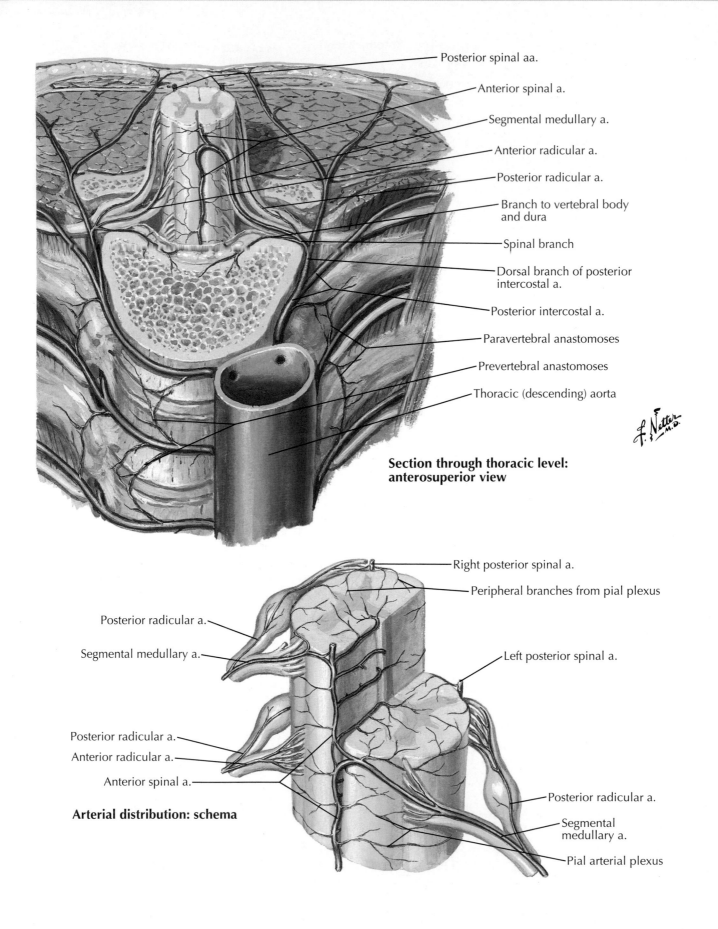

Posterior spinal aa.

Anterior spinal a.

Segmental medullary a.

Anterior radicular a.

Posterior radicular a.

Branch to vertebral body and dura

Spinal branch

Dorsal branch of posterior intercostal a.

Posterior intercostal a.

Paravertebral anastomoses

Prevertebral anastomoses

Thoracic (descending) aorta

Section through thoracic level: anterosuperior view

Right posterior spinal a.

Peripheral branches from pial plexus

Posterior radicular a.

Segmental medullary a.

Left posterior spinal a.

Posterior radicular a.

Anterior radicular a.

Anterior spinal a.

Posterior radicular a.

Segmental medullary a.

Arterial distribution: schema

Pial arterial plexus

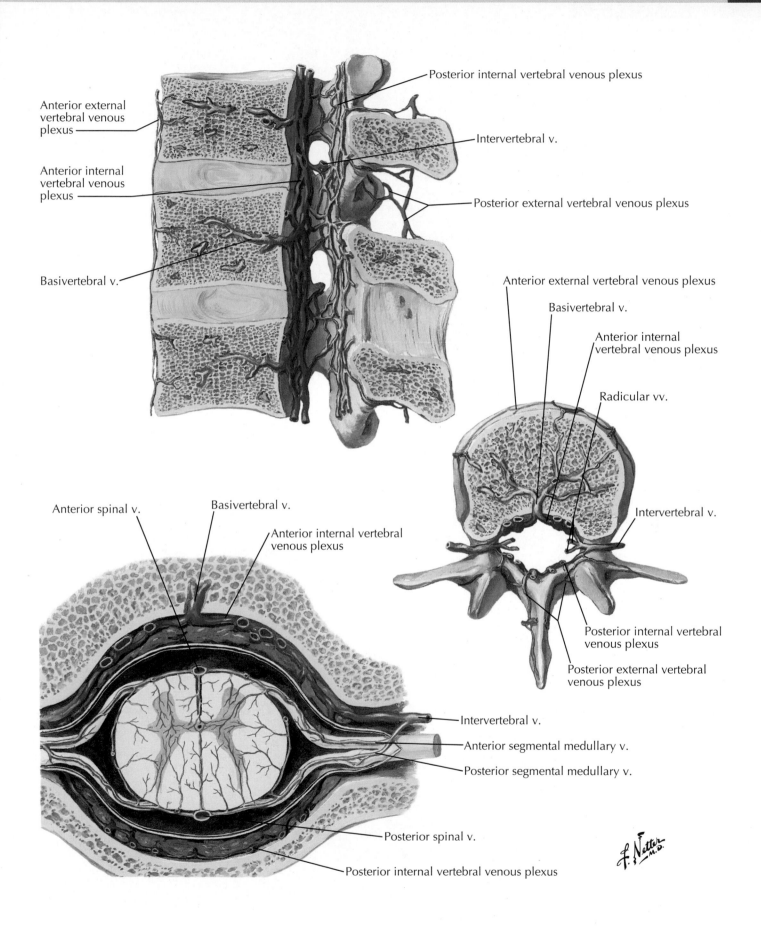

Anterior external vertebral venous plexus

Anterior internal vertebral venous plexus

Basivertebral v.

Posterior internal vertebral venous plexus

Intervertebral v.

Posterior external vertebral venous plexus

Anterior external vertebral venous plexus

Basivertebral v.

Anterior internal vertebral venous plexus

Radicular vv.

Intervertebral v.

Posterior internal vertebral venous plexus

Posterior external vertebral venous plexus

Anterior spinal v.

Basivertebral v.

Anterior internal vertebral venous plexus

Intervertebral v.

Anterior segmental medullary v.

Posterior segmental medullary v.

Posterior spinal v.

Posterior internal vertebral venous plexus

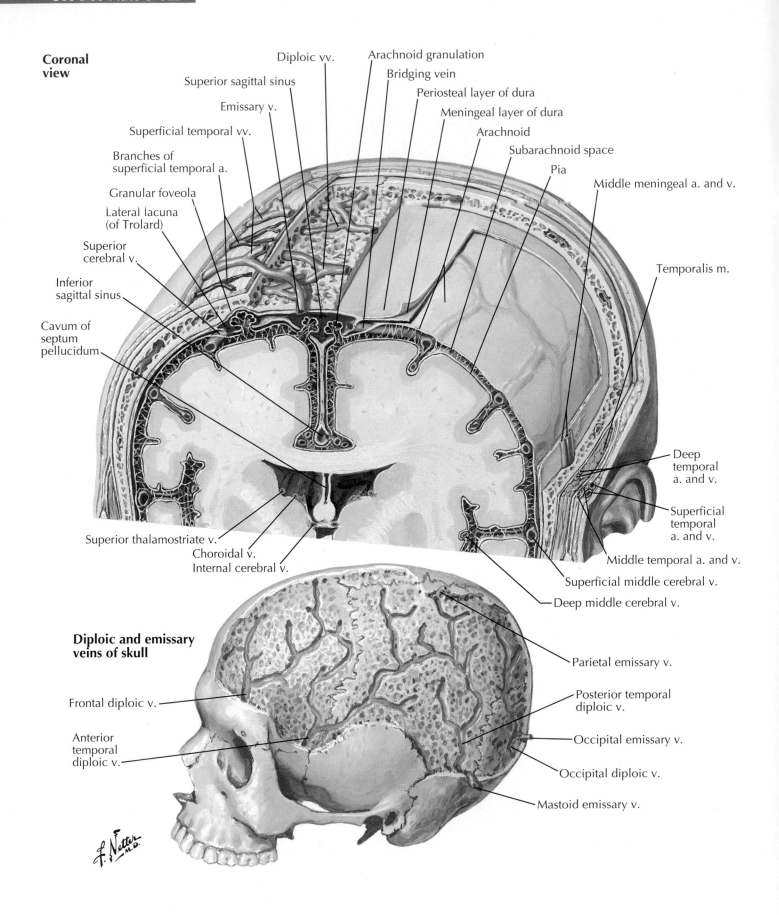

Coronal view

Diploic vv.

Arachnoid granulation

Superior sagittal sinus

Bridging vein

Periosteal layer of dura

Emissary v.

Meningeal layer of dura

Superficial temporal vv.

Arachnoid

Branches of superficial temporal a.

Subarachnoid space

Pia

Granular foveola

Middle meningeal a. and v.

Lateral lacuna (of Trolard)

Superior cerebral v.

Temporalis m.

Inferior sagittal sinus

Cavum of septum pellucidum

Deep temporal a. and v.

Superior thalamostriate v.

Superficial temporal a. and v.

Choroidal v.

Middle temporal a. and v.

Internal cerebral v.

Superficial middle cerebral v.

Deep middle cerebral v.

Diploic and emissary veins of skull

Parietal emissary v.

Frontal diploic v.

Posterior temporal diploic v.

Anterior temporal diploic v.

Occipital emissary v.

Occipital diploic v.

Mastoid emissary v.

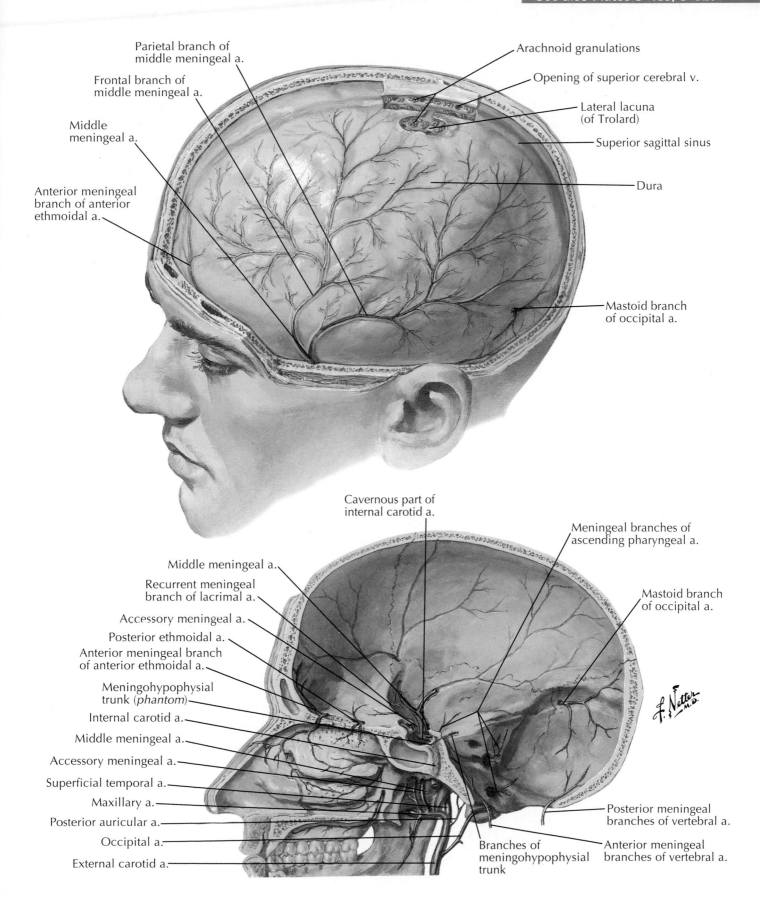

Parietal branch of
middle meningeal a.

Frontal branch of
middle meningeal a.

Middle
meningeal a.

Anterior meningeal
branch of anterior
ethmoidal a.

Arachnoid granulations

Opening of superior cerebral v.

Lateral lacuna
(of Trolard)

Superior sagittal sinus

Dura

Mastoid branch
of occipital a.

Cavernous part of
internal carotid a.

Middle meningeal a.

Recurrent meningeal
branch of lacrimal a.

Accessory meningeal a.

Posterior ethmoidal a.

Anterior meningeal branch
of anterior ethmoidal a.

Meningohypophysial
trunk (phantom)

Internal carotid a.

Middle meningeal a.

Accessory meningeal a.

Superficial temporal a.

Maxillary a.

Posterior auricular a.

Occipital a.

External carotid a.

Meningeal branches of
ascending pharyngeal a.

Mastoid branch
of occipital a.

Posterior meningeal
branches of vertebral a.

Anterior meningeal
branches of vertebral a.

Branches of
meningohypophysial
trunk

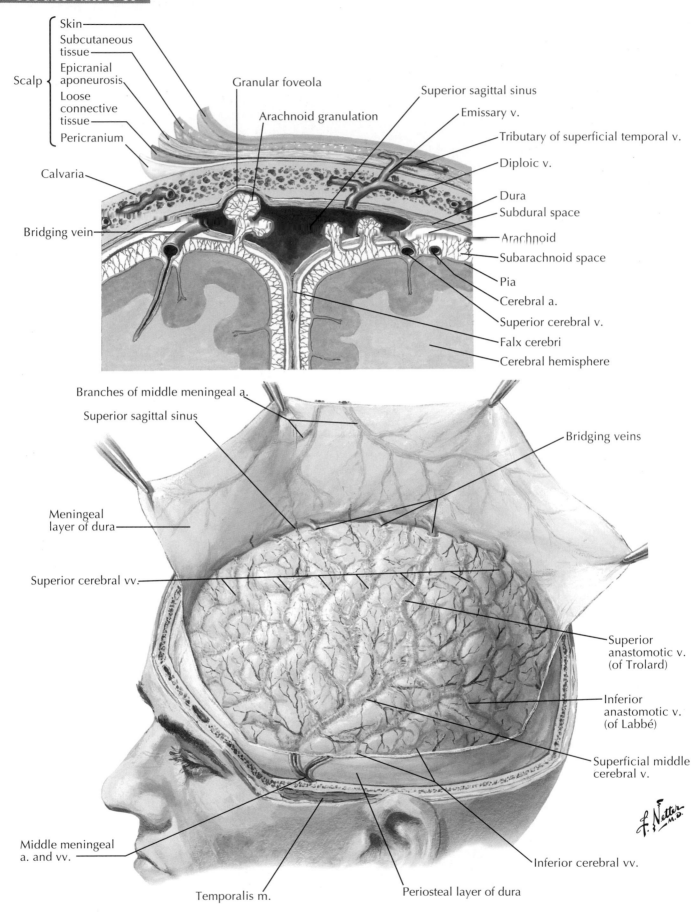

Scalp
- Skin
- Subcutaneous tissue
- Epicranial aponeurosis
- Loose connective tissue
- Pericranium

Granular foveola

Arachnoid granulation

Calvaria

Bridging vein

Superior sagittal sinus

Emissary v.

Tributary of superficial temporal v.

Diploic v.

Dura

Subdural space

Arachnoid

Subarachnoid space

Pia

Cerebral a.

Superior cerebral v.

Falx cerebri

Cerebral hemisphere

Branches of middle meningeal a.

Superior sagittal sinus

Meningeal layer of dura

Superior cerebral vv.

Bridging veins

Superior anastomotic v. (of Trolard)

Inferior anastomotic v. (of Labbé)

Superficial middle cerebral v.

Inferior cerebral vv.

Middle meningeal a. and vv.

Temporalis m.

Periosteal layer of dura

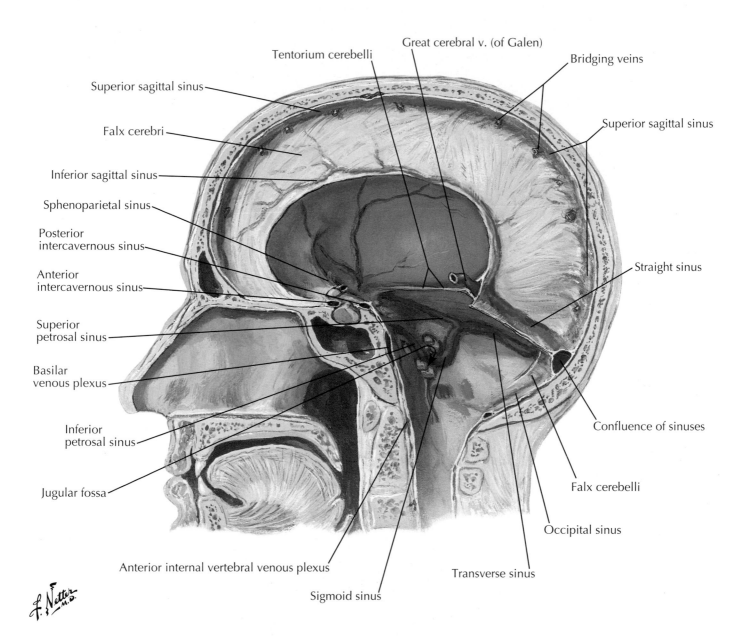

Tentorium cerebelli

Great cerebral v. (of Galen)

Bridging veins

Superior sagittal sinus

Superior sagittal sinus

Falx cerebri

Inferior sagittal sinus

Sphenoparietal sinus

Straight sinus

Posterior intercavernous sinus

Anterior intercavernous sinus

Superior petrosal sinus

Basilar venous plexus

Confluence of sinuses

Inferior petrosal sinus

Falx cerebelli

Jugular fossa

Occipital sinus

Anterior internal vertebral venous plexus

Transverse sinus

Sigmoid sinus

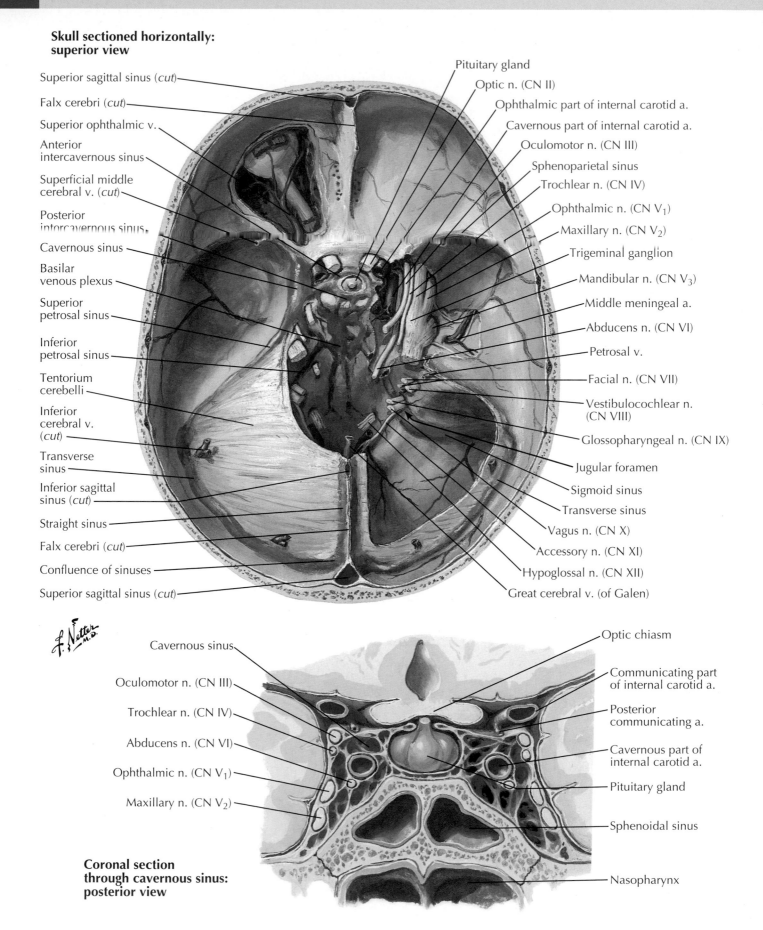

Skull sectioned horizontally: superior view

Superior sagittal sinus (*cut*)

Falx cerebri (*cut*)

Superior ophthalmic v.

Anterior intercavernous sinus

Superficial middle cerebral v. (*cut*)

Posterior intercavernous sinus

Cavernous sinus

Basilar venous plexus

Superior petrosal sinus

Inferior petrosal sinus

Tentorium cerebelli

Inferior cerebral v. (*cut*)

Transverse sinus

Inferior sagittal sinus (*cut*)

Straight sinus

Falx cerebri (*cut*)

Confluence of sinuses

Superior sagittal sinus (*cut*)

Pituitary gland

Optic n. (CN II)

Ophthalmic part of internal carotid a.

Cavernous part of internal carotid a.

Oculomotor n. (CN III)

Sphenoparietal sinus

Trochlear n. (CN IV)

Ophthalmic n. (CN V_1)

Maxillary n. (CN V_2)

Trigeminal ganglion

Mandibular n. (CN V_3)

Middle meningeal a.

Abducens n. (CN VI)

Petrosal v.

Facial n. (CN VII)

Vestibulocochlear n. (CN VIII)

Glossopharyngeal n. (CN IX)

Jugular foramen

Sigmoid sinus

Transverse sinus

Vagus n. (CN X)

Accessory n. (CN XI)

Hypoglossal n. (CN XII)

Great cerebral v. (of Galen)

Cavernous sinus

Oculomotor n. (CN III)

Trochlear n. (CN IV)

Abducens n. (CN VI)

Ophthalmic n. (CN V_1)

Maxillary n. (CN V_2)

Optic chiasm

Communicating part of internal carotid a.

Posterior communicating a.

Cavernous part of internal carotid a.

Pituitary gland

Sphenoidal sinus

Nasopharynx

Coronal section through cavernous sinus: posterior view

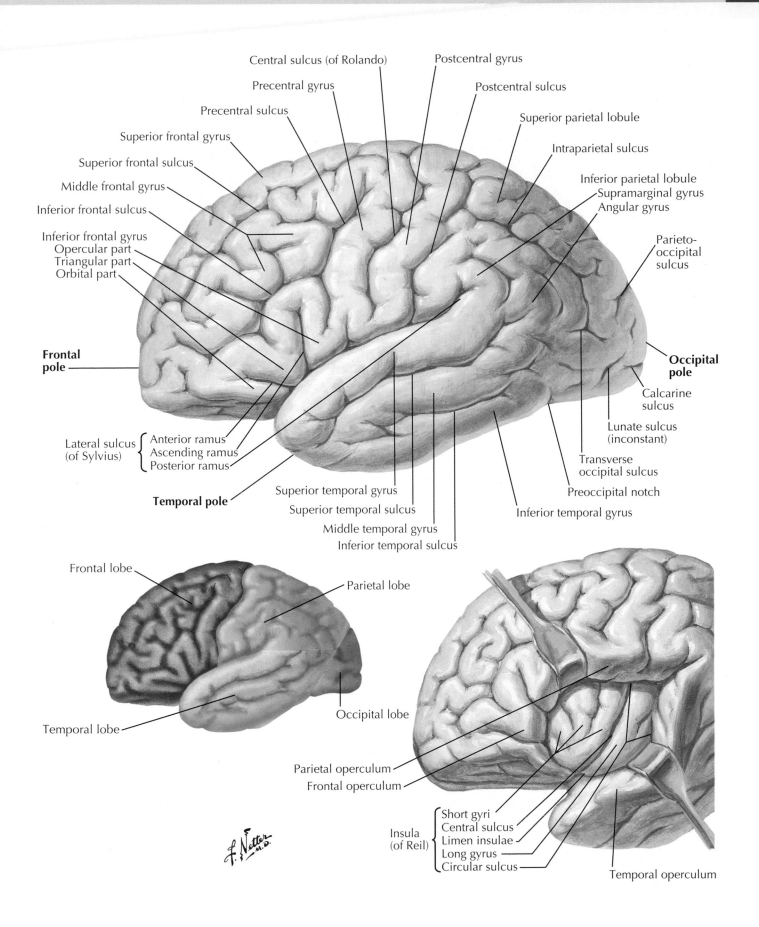

Central sulcus (of Rolando)

Postcentral gyrus

Precentral gyrus

Postcentral sulcus

Precentral sulcus

Superior parietal lobule

Superior frontal gyrus

Intraparietal sulcus

Superior frontal sulcus

Inferior parietal lobule

Middle frontal gyrus

Supramarginal gyrus

Inferior frontal sulcus

Angular gyrus

Inferior frontal gyrus

Parieto-occipital sulcus

Opercular part

Triangular part

Orbital part

Frontal pole

Occipital pole

Calcarine sulcus

Lunate sulcus (inconstant)

Lateral sulcus (of Sylvius) { Anterior ramus / Ascending ramus / Posterior ramus

Transverse occipital sulcus

Preoccipital notch

Temporal pole

Superior temporal gyrus

Superior temporal sulcus

Inferior temporal gyrus

Middle temporal gyrus

Inferior temporal sulcus

Frontal lobe

Parietal lobe

Temporal lobe

Occipital lobe

Parietal operculum

Frontal operculum

Insula (of Reil) { Short gyri / Central sulcus / Limen insulae / Long gyrus / Circular sulcus

Temporal operculum

f. Netter M.D.

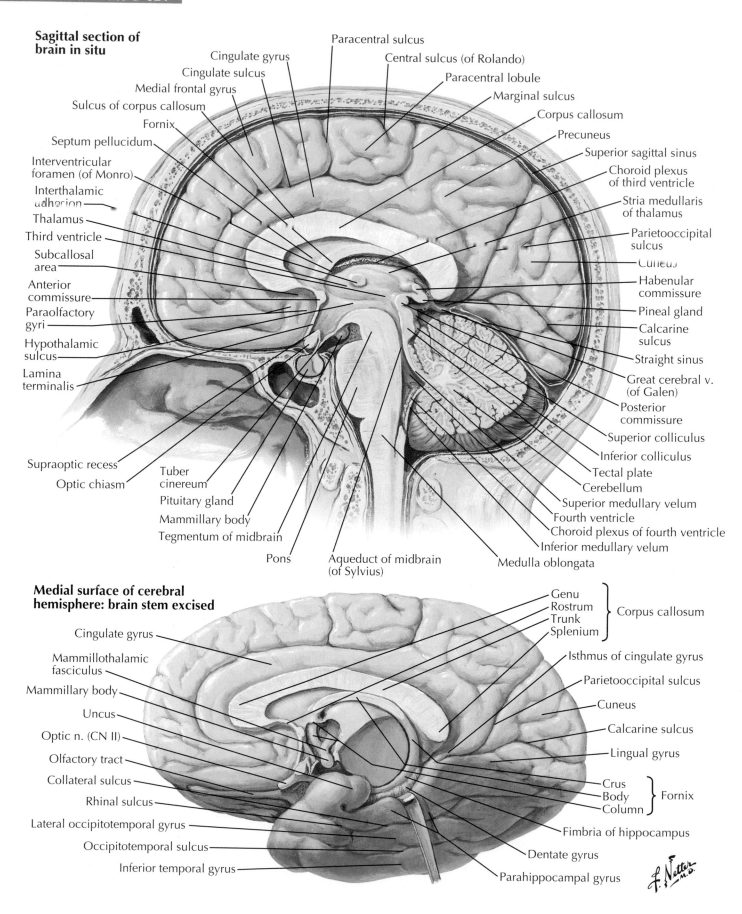

Sagittal section of brain in situ

Paracentral sulcus
Central sulcus (of Rolando)
Cingulate gyrus
Paracentral lobule
Cingulate sulcus
Marginal sulcus
Medial frontal gyrus
Corpus callosum
Sulcus of corpus callosum
Precuneus
Fornix
Superior sagittal sinus
Septum pellucidum
Choroid plexus of third ventricle
Interventricular foramen (of Monro)
Stria medullaris of thalamus
Interthalamic adhesion
Parietooccipital sulcus
Thalamus
Cuneus
Third ventricle
Habenular commissure
Subcallosal area
Pineal gland
Anterior commissure
Calcarine sulcus
Paraolfactory gyri
Straight sinus
Hypothalamic sulcus
Great cerebral v. (of Galen)
Lamina terminalis
Posterior commissure
Superior colliculus
Inferior colliculus
Supraoptic recess
Tectal plate
Tuber cinereum
Cerebellum
Optic chiasm
Superior medullary velum
Pituitary gland
Fourth ventricle
Mammillary body
Choroid plexus of fourth ventricle
Tegmentum of midbrain
Inferior medullary velum
Pons
Aqueduct of midbrain (of Sylvius)
Medulla oblongata

Medial surface of cerebral hemisphere: brain stem excised

Genu
Rostrum
Corpus callosum
Trunk
Splenium
Cingulate gyrus
Isthmus of cingulate gyrus
Mammillothalamic fasciculus
Parietooccipital sulcus
Mammillary body
Cuneus
Uncus
Calcarine sulcus
Optic n. (CN II)
Lingual gyrus
Olfactory tract
Collateral sulcus
Crus
Rhinal sulcus
Body
Fornix
Column
Lateral occipitotemporal gyrus
Fimbria of hippocampus
Occipitotemporal sulcus
Dentate gyrus
Inferior temporal gyrus
Parahippocampal gyrus

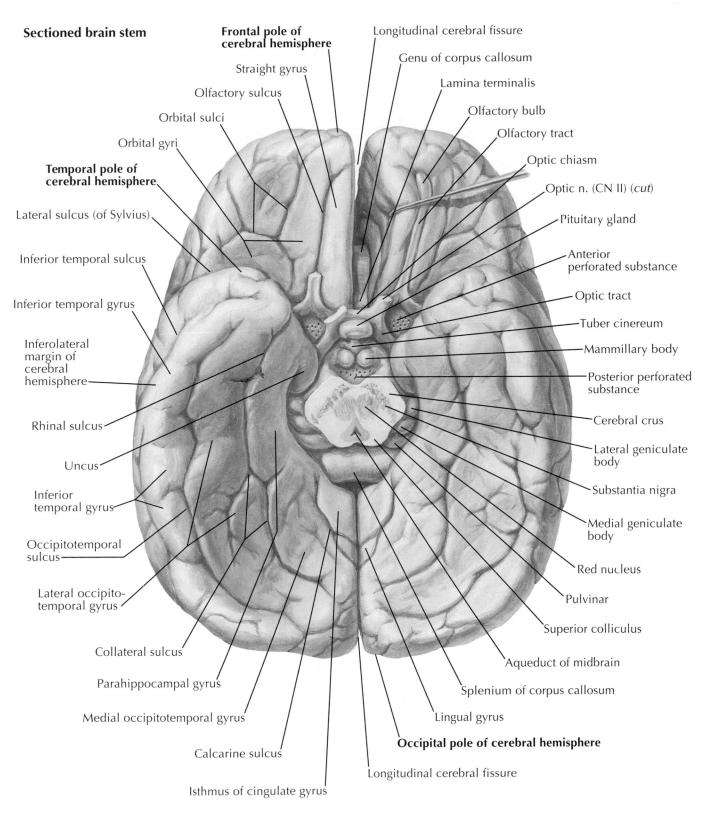

Sectioned brain stem

Frontal pole of cerebral hemisphere

Straight gyrus

Olfactory sulcus

Orbital sulci

Orbital gyri

Temporal pole of cerebral hemisphere

Lateral sulcus (of Sylvius)

Inferior temporal sulcus

Inferior temporal gyrus

Inferolateral margin of cerebral hemisphere

Rhinal sulcus

Uncus

Inferior temporal gyrus

Occipitotemporal sulcus

Lateral occipito-temporal gyrus

Collateral sulcus

Parahippocampal gyrus

Medial occipitotemporal gyrus

Calcarine sulcus

Isthmus of cingulate gyrus

Longitudinal cerebral fissure

Genu of corpus callosum

Lamina terminalis

Olfactory bulb

Olfactory tract

Optic chiasm

Optic n. (CN II) (cut)

Pituitary gland

Anterior perforated substance

Optic tract

Tuber cinereum

Mammillary body

Posterior perforated substance

Cerebral crus

Lateral geniculate body

Substantia nigra

Medial geniculate body

Red nucleus

Pulvinar

Superior colliculus

Aqueduct of midbrain

Splenium of corpus callosum

Lingual gyrus

Occipital pole of cerebral hemisphere

Longitudinal cerebral fissure

Left lateral phantom view

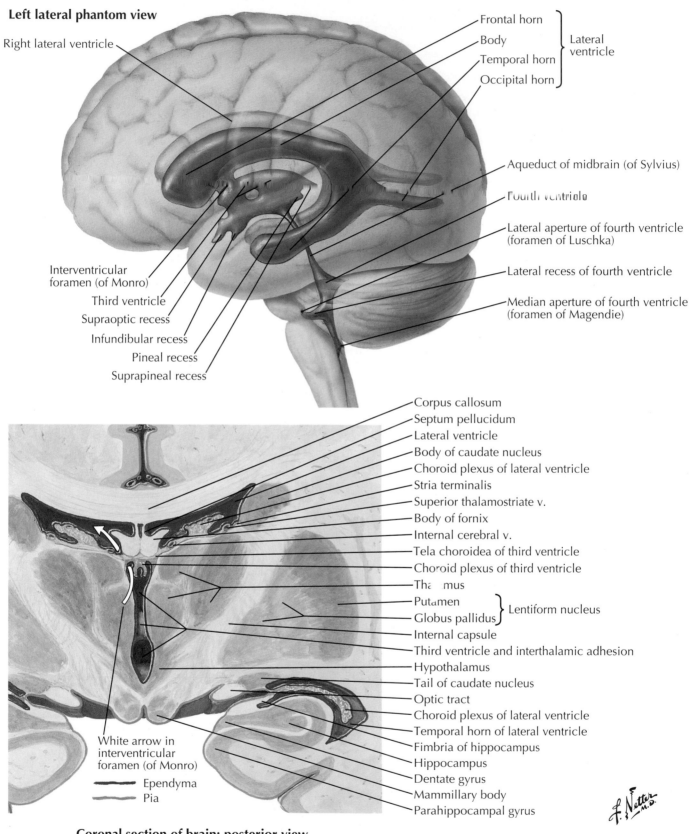

Right lateral ventricle

Frontal horn
Body
Temporal horn — Lateral ventricle
Occipital horn

Aqueduct of midbrain (of Sylvius)

Fourth ventricle

Lateral aperture of fourth ventricle
(foramen of Luschka)

Lateral recess of fourth ventricle

Median aperture of fourth ventricle
(foramen of Magendie)

Interventricular
foramen (of Monro)

Third ventricle

Supraoptic recess

Infundibular recess

Pineal recess

Suprapineal recess

Corpus callosum
Septum pellucidum
Lateral ventricle
Body of caudate nucleus
Choroid plexus of lateral ventricle
Stria terminalis
Superior thalamostriate v.
Body of fornix
Internal cerebral v.
Tela choroidea of third ventricle
Choroid plexus of third ventricle
Thalamus
Putamen ⎱ Lentiform nucleus
Globus pallidus ⎰
Internal capsule
Third ventricle and interthalamic adhesion
Hypothalamus
Tail of caudate nucleus
Optic tract
Choroid plexus of lateral ventricle
Temporal horn of lateral ventricle
Fimbria of hippocampus
Hippocampus
Dentate gyrus
Mammillary body
Parahippocampal gyrus

White arrow in
interventricular
foramen (of Monro)

Ependyma
Pia

Coronal section of brain: posterior view

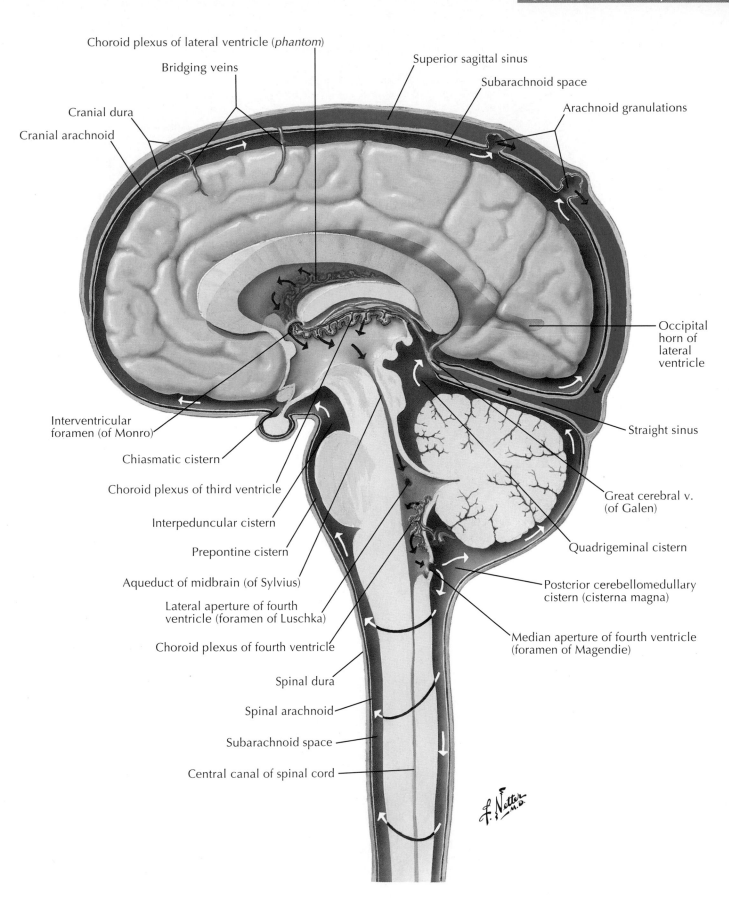

Choroid plexus of lateral ventricle (*phantom*)

Bridging veins

Cranial dura

Cranial arachnoid

Superior sagittal sinus

Subarachnoid space

Arachnoid granulations

Occipital horn of lateral ventricle

Straight sinus

Interventricular foramen (of Monro)

Chiasmatic cistern

Choroid plexus of third ventricle

Interpeduncular cistern

Prepontine cistern

Aqueduct of midbrain (of Sylvius)

Lateral aperture of fourth ventricle (foramen of Luschka)

Choroid plexus of fourth ventricle

Spinal dura

Spinal arachnoid

Subarachnoid space

Central canal of spinal cord

Great cerebral v. (of Galen)

Quadrigeminal cistern

Posterior cerebellomedullary cistern (cisterna magna)

Median aperture of fourth ventricle (foramen of Magendie)

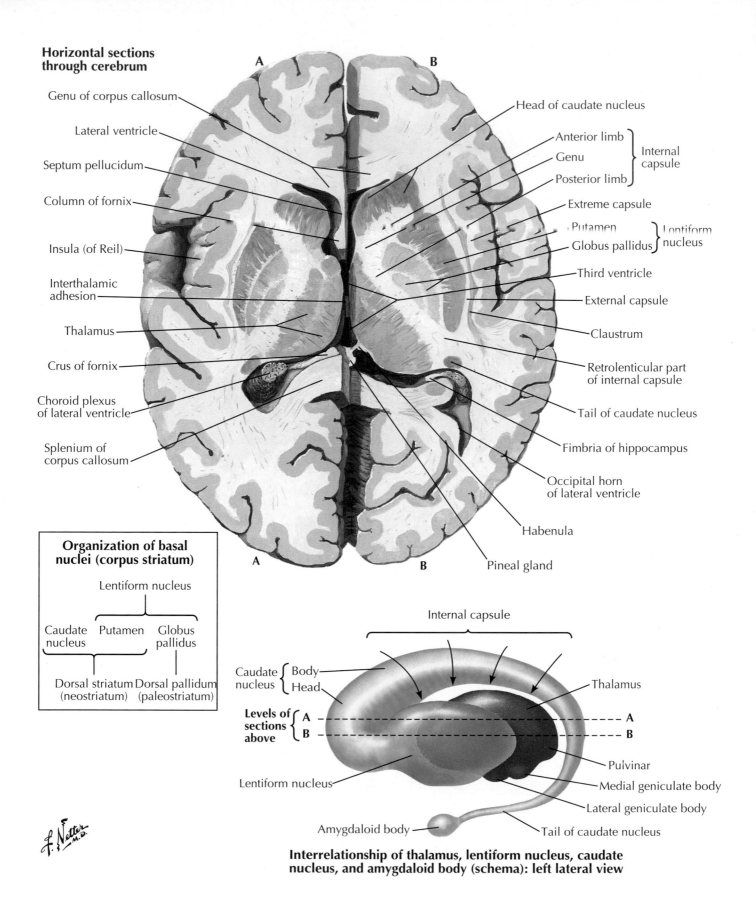

Horizontal sections through cerebrum

Genu of corpus callosum

Lateral ventricle

Septum pellucidum

Column of fornix

Insula (of Reil)

Interthalamic adhesion

Thalamus

Crus of fornix

Choroid plexus of lateral ventricle

Splenium of corpus callosum

Head of caudate nucleus

Anterior limb
Genu
Posterior limb
} Internal capsule

Extreme capsule

Putamen
Globus pallidus
} Lentiform nucleus

Third ventricle

External capsule

Claustrum

Retrolenticular part of internal capsule

Tail of caudate nucleus

Fimbria of hippocampus

Occipital horn of lateral ventricle

Habenula

Pineal gland

Organization of basal nuclei (corpus striatum)

Lentiform nucleus

Caudate nucleus | Putamen | Globus pallidus

Dorsal striatum (neostriatum) | Dorsal pallidum (paleostriatum)

Internal capsule

Caudate { Body
nucleus { Head

Thalamus

Levels of sections above { A
{ B

A
B

Pulvinar

Medial geniculate body

Lateral geniculate body

Lentiform nucleus

Amygdaloid body

Tail of caudate nucleus

Interrelationship of thalamus, lentiform nucleus, caudate nucleus, and amygdaloid body (schema): left lateral view

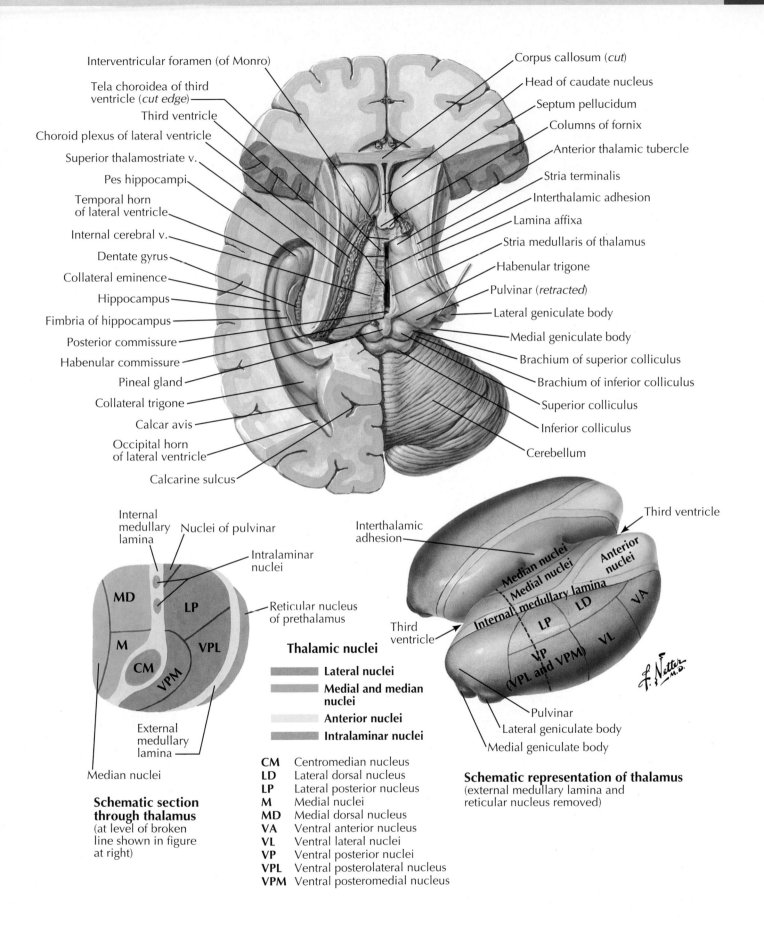

Interventricular foramen (of Monro)
Tela choroidea of third ventricle (*cut edge*)
Third ventricle
Choroid plexus of lateral ventricle
Superior thalamostriate v.
Pes hippocampi
Temporal horn of lateral ventricle
Internal cerebral v.
Dentate gyrus
Collateral eminence
Hippocampus
Fimbria of hippocampus
Posterior commissure
Habenular commissure
Pineal gland
Collateral trigone
Calcar avis
Occipital horn of lateral ventricle
Calcarine sulcus

Corpus callosum (*cut*)
Head of caudate nucleus
Septum pellucidum
Columns of fornix
Anterior thalamic tubercle
Stria terminalis
Interthalamic adhesion
Lamina affixa
Stria medullaris of thalamus
Habenular trigone
Pulvinar (*retracted*)
Lateral geniculate body
Medial geniculate body
Brachium of superior colliculus
Brachium of inferior colliculus
Superior colliculus
Inferior colliculus
Cerebellum

Internal medullary lamina
Nuclei of pulvinar
Intralaminar nuclei
Reticular nucleus of prethalamus

MD
LP
M
VPL
CM
VPM

External medullary lamina
Median nuclei

Schematic section through thalamus
(at level of broken line shown in figure at right)

Interthalamic adhesion
Third ventricle

Median nuclei
Medial nuclei
Anterior nuclei
Internal medullary lamina
LP
LD
VA
VP (VPL and VPM)
VL
Pulvinar
Lateral geniculate body
Medial geniculate body

Third ventricle

Schematic representation of thalamus
(external medullary lamina and reticular nucleus removed)

Thalamic nuclei

Lateral nuclei
Medial and median nuclei
Anterior nuclei
Intralaminar nuclei

CM	Centromedian nucleus
LD	Lateral dorsal nucleus
LP	Lateral posterior nucleus
M	Medial nuclei
MD	Medial dorsal nucleus
VA	Ventral anterior nucleus
VL	Ventral lateral nuclei
VP	Ventral posterior nuclei
VPL	Ventral posterolateral nucleus
VPM	Ventral posteromedial nucleus

Superior view

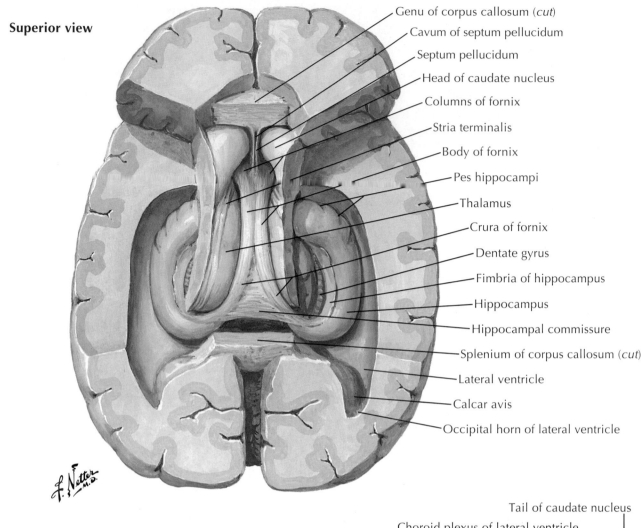

Genu of corpus callosum (*cut*)

Cavum of septum pellucidum

Septum pellucidum

Head of caudate nucleus

Columns of fornix

Stria terminalis

Body of fornix

Pes hippocampi

Thalamus

Crura of fornix

Dentate gyrus

Fimbria of hippocampus

Hippocampus

Hippocampal commissure

Splenium of corpus callosum (*cut*)

Lateral ventricle

Calcar avis

Occipital horn of lateral ventricle

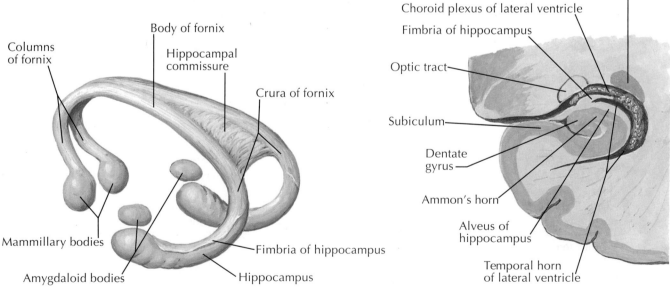

Columns
of fornix

Body of fornix

Hippocampal
commissure

Crura of fornix

Mammillary bodies

Amygdaloid bodies

Fimbria of hippocampus

Hippocampus

Fornix: schema

Tail of caudate nucleus

Choroid plexus of lateral ventricle

Fimbria of hippocampus

Optic tract

Subiculum

Dentate
gyrus

Ammon's horn

Alveus of
hippocampus

Temporal horn
of lateral ventricle

Coronal section: posterior view

T1-weighted MRI, sagittal view

Body of corpus callosum

Genu of corpus callosum

Rostrum of corpus callosum

Mammillary body

Optic chiasm

Infundibular stalk

Crista galli

Pituitary gland

Sphenoidal sinus

Clivus

Pharyngeal tonsil

Hard palate

Soft palate

Tongue

Splenium of corpus callosum

Third ventricle

Aqueduct of midbrain

Tectum

Midbrain

Cerebellum

Fourth ventricle

Pons

Medulla oblongata

Anterior arch of atlas

Body of axis

Cervical part of spinal cord

T2-weighted MRI, axial views without contrast

Eyeball

Ethmoidal cells

Sphenoidal sinus

Temporal lobe

Trigeminal cave

Internal carotid a.

Basilar a.

Trigeminal n. (CN V)

Pons

Middle cerebellar peduncle

Fourth ventricle

Cerebellum

Gray matter (cerebral cortex)

White matter of telencephalon

Head of caudate nucleus

Anterior limb of internal capsule

External capsule

Putamen

Genu of internal capsule

Posterior limb of internal capsule

Thalamus

Longitudinal cerebral fissure

Genu of corpus callosum

Frontal horn of lateral ventricle

Interventricular foramen (of Monro)

Third ventricle

Occipital horn of lateral ventricle

Splenium of corpus callosum

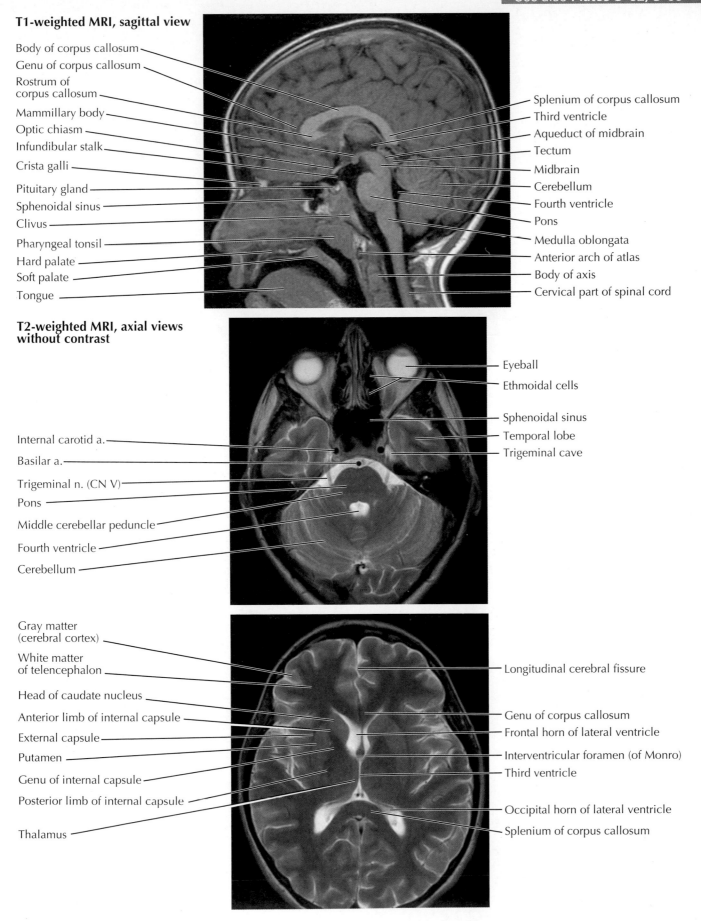

Brain

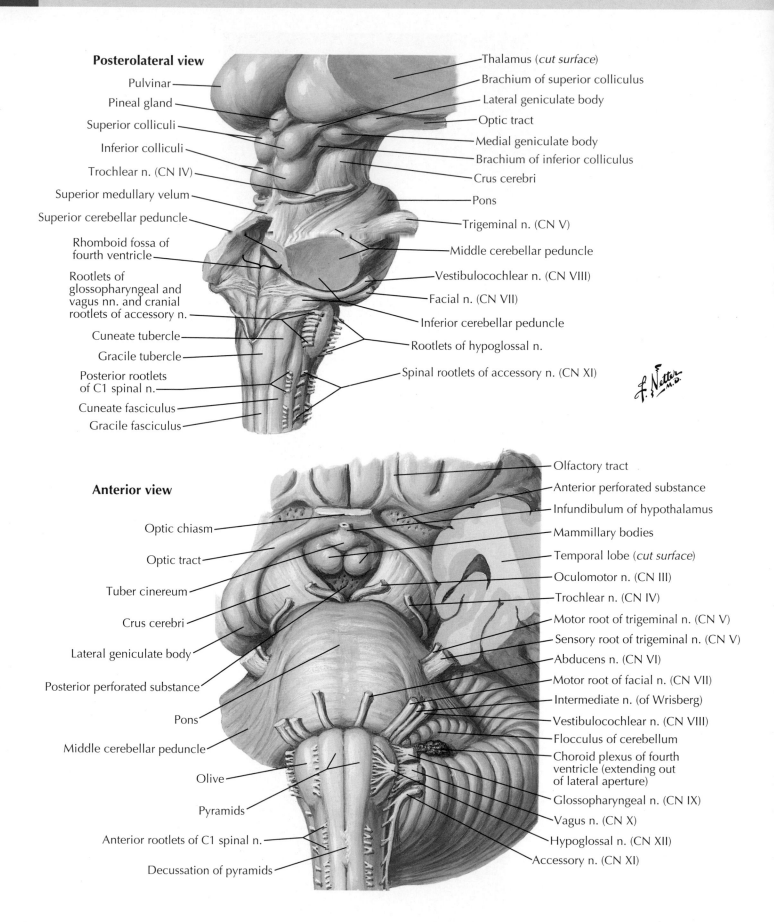

Posterolateral view

Pulvinar

Pineal gland

Superior colliculi

Inferior colliculi

Trochlear n. (CN IV)

Superior medullary velum

Superior cerebellar peduncle

Rhomboid fossa of fourth ventricle

Rootlets of glossopharyngeal and vagus nn. and cranial rootlets of accessory n.

Cuneate tubercle

Gracile tubercle

Posterior rootlets of C1 spinal n.

Cuneate fasciculus

Gracile fasciculus

Thalamus (*cut surface*)

Brachium of superior colliculus

Lateral geniculate body

Optic tract

Medial geniculate body

Brachium of inferior colliculus

Crus cerebri

Pons

Trigeminal n. (CN V)

Middle cerebellar peduncle

Vestibulocochlear n. (CN VIII)

Facial n. (CN VII)

Inferior cerebellar peduncle

Rootlets of hypoglossal n.

Spinal rootlets of accessory n. (CN XI)

Anterior view

Optic chiasm

Optic tract

Tuber cinereum

Crus cerebri

Lateral geniculate body

Posterior perforated substance

Pons

Middle cerebellar peduncle

Olive

Pyramids

Anterior rootlets of C1 spinal n.

Decussation of pyramids

Olfactory tract

Anterior perforated substance

Infundibulum of hypothalamus

Mammillary bodies

Temporal lobe (*cut surface*)

Oculomotor n. (CN III)

Trochlear n. (CN IV)

Motor root of trigeminal n. (CN V)

Sensory root of trigeminal n. (CN V)

Abducens n. (CN VI)

Motor root of facial n. (CN VII)

Intermediate n. (of Wrisberg)

Vestibulocochlear n. (CN VIII)

Flocculus of cerebellum

Choroid plexus of fourth ventricle (extending out of lateral aperture)

Glossopharyngeal n. (CN IX)

Vagus n. (CN X)

Hypoglossal n. (CN XII)

Accessory n. (CN XI)

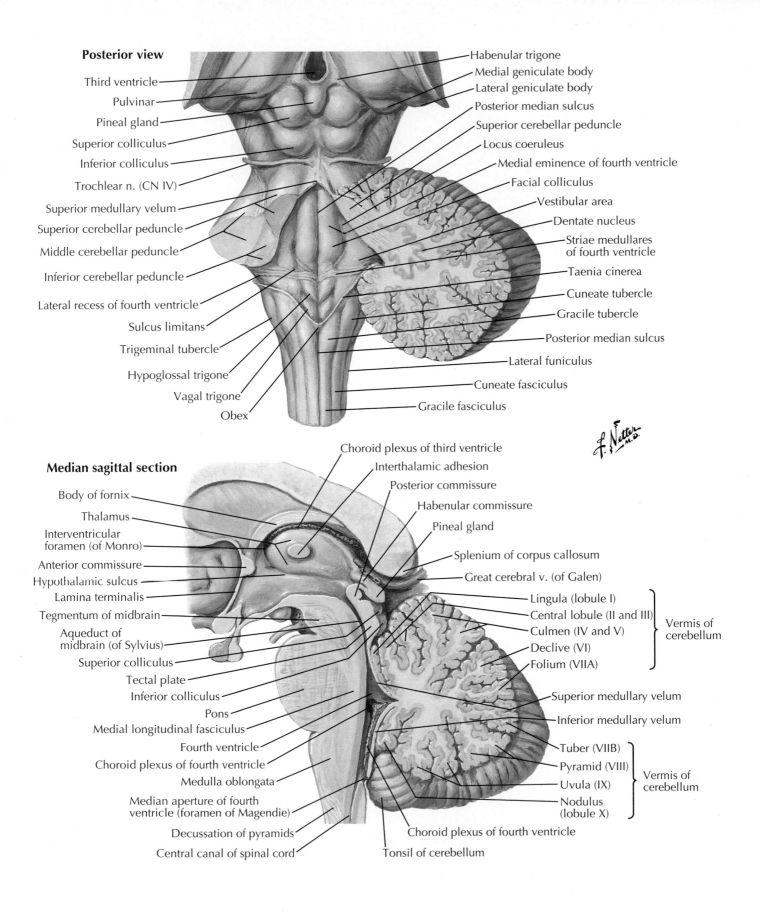

Posterior view

Third ventricle
Pulvinar
Pineal gland
Superior colliculus
Inferior colliculus
Trochlear n. (CN IV)
Superior medullary velum
Superior cerebellar peduncle
Middle cerebellar peduncle
Inferior cerebellar peduncle
Lateral recess of fourth ventricle
Sulcus limitans
Trigeminal tubercle
Hypoglossal trigone
Vagal trigone
Obex

Habenular trigone
Medial geniculate body
Lateral geniculate body
Posterior median sulcus
Superior cerebellar peduncle
Locus coeruleus
Medial eminence of fourth ventricle
Facial colliculus
Vestibular area
Dentate nucleus
Striae medullares of fourth ventricle
Taenia cinerea
Cuneate tubercle
Gracile tubercle
Posterior median sulcus
Lateral funiculus
Cuneate fasciculus
Gracile fasciculus

Median sagittal section

Body of fornix
Thalamus
Interventricular foramen (of Monro)
Anterior commissure
Hypothalamic sulcus
Lamina terminalis
Tegmentum of midbrain
Aqueduct of midbrain (of Sylvius)
Superior colliculus
Tectal plate
Inferior colliculus
Pons
Medial longitudinal fasciculus
Fourth ventricle
Choroid plexus of fourth ventricle
Medulla oblongata
Median aperture of fourth ventricle (foramen of Magendie)
Decussation of pyramids
Central canal of spinal cord

Choroid plexus of third ventricle
Interthalamic adhesion
Posterior commissure
Habenular commissure
Pineal gland
Splenium of corpus callosum
Great cerebral v. (of Galen)
Lingula (lobule I)
Central lobule (II and III)
Culmen (IV and V)
Declive (VI)
Folium (VIIA)
Vermis of cerebellum
Superior medullary velum
Inferior medullary velum
Tuber (VIIB)
Pyramid (VIII)
Uvula (IX)
Nodulus (lobule X)
Vermis of cerebellum
Choroid plexus of fourth ventricle
Tonsil of cerebellum

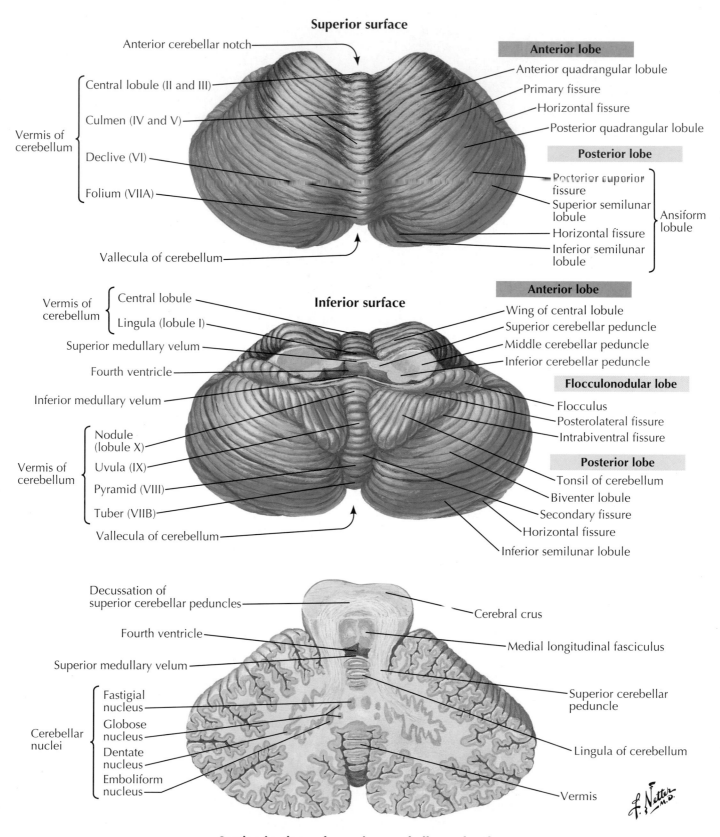

Superior surface

Anterior cerebellar notch

Anterior lobe

Central lobule (II and III)

Anterior quadrangular lobule

Primary fissure

Culmen (IV and V)

Horizontal fissure

Posterior quadrangular lobule

Vermis of cerebellum

Declive (VI)

Posterior lobe

Posterior superior fissure

Folium (VIIA)

Superior semilunar lobule

Horizontal fissure

Inferior semilunar lobule

Ansiform lobule

Vallecula of cerebellum

Inferior surface

Vermis of cerebellum

Central lobule

Lingula (lobule I)

Anterior lobe

Wing of central lobule

Superior cerebellar peduncle

Superior medullary velum

Middle cerebellar peduncle

Fourth ventricle

Inferior cerebellar peduncle

Inferior medullary velum

Flocculonodular lobe

Flocculus

Posterolateral fissure

Intrabiventral fissure

Nodule (lobule X)

Vermis of cerebellum

Uvula (IX)

Posterior lobe

Tonsil of cerebellum

Pyramid (VIII)

Biventer lobule

Secondary fissure

Tuber (VIIB)

Horizontal fissure

Vallecula of cerebellum

Inferior semilunar lobule

Decussation of superior cerebellar peduncles

Cerebral crus

Fourth ventricle

Medial longitudinal fasciculus

Superior medullary velum

Fastigial nucleus

Superior cerebellar peduncle

Globose nucleus

Cerebellar nuclei

Dentate nucleus

Lingula of cerebellum

Emboliform nucleus

Vermis

Section in plane of superior cerebellar peduncle

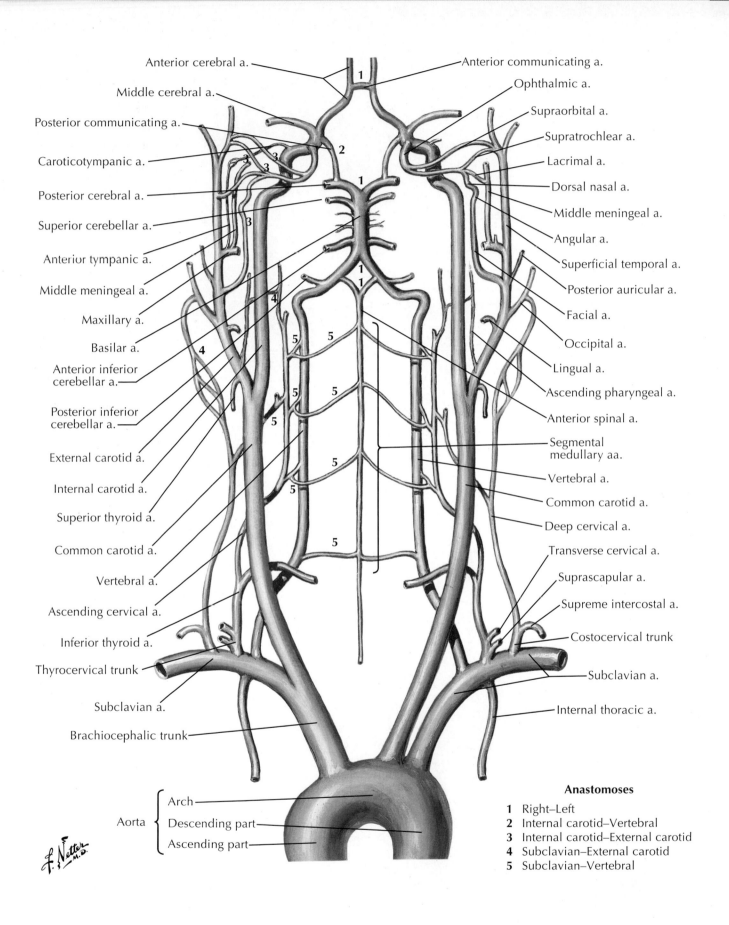

Anterior cerebral a.

Middle cerebral a.

Posterior communicating a.

Caroticotympanic a.

Posterior cerebral a.

Superior cerebellar a.

Anterior tympanic a.

Middle meningeal a.

Maxillary a.

Basilar a.

Anterior inferior cerebellar a.

Posterior inferior cerebellar a.

External carotid a.

Internal carotid a.

Superior thyroid a.

Common carotid a.

Vertebral a.

Ascending cervical a.

Inferior thyroid a.

Thyrocervical trunk

Subclavian a.

Brachiocephalic trunk

Anterior communicating a.

Ophthalmic a.

Supraorbital a.

Supratrochlear a.

Lacrimal a.

Dorsal nasal a.

Middle meningeal a.

Angular a.

Superficial temporal a.

Posterior auricular a.

Facial a.

Occipital a.

Lingual a.

Ascending pharyngeal a.

Anterior spinal a.

Segmental medullary aa.

Vertebral a.

Common carotid a.

Deep cervical a.

Transverse cervical a.

Suprascapular a.

Supreme intercostal a.

Costocervical trunk

Subclavian a.

Internal thoracic a.

Aorta
Arch
Descending part
Ascending part

Anastomoses

1 Right–Left
2 Internal carotid–Vertebral
3 Internal carotid–External carotid
4 Subclavian–External carotid
5 Subclavian–Vertebral

F. Netter M.D.

Cerebral Vasculature   **Plate S–43**

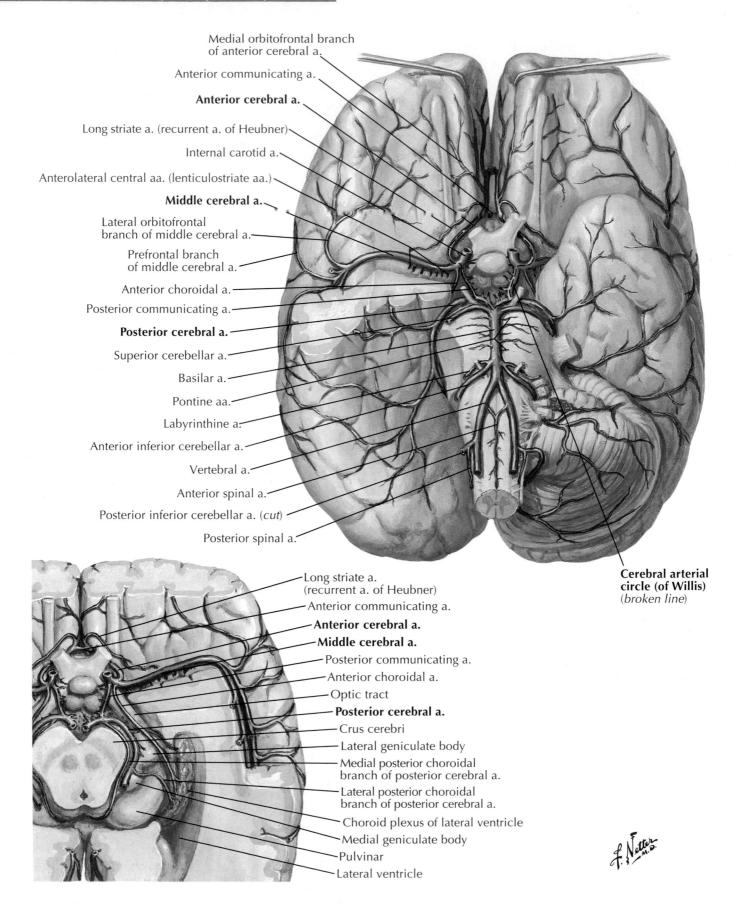

Medial orbitofrontal branch
of anterior cerebral a.

Anterior communicating a.

Anterior cerebral a.

Long striate a. (recurrent a. of Heubner)

Internal carotid a.

Anterolateral central aa. (lenticulostriate aa.)

Middle cerebral a.

Lateral orbitofrontal
branch of middle cerebral a.

Prefrontal branch
of middle cerebral a.

Anterior choroidal a.

Posterior communicating a.

Posterior cerebral a.

Superior cerebellar a.

Basilar a.

Pontine aa.

Labyrinthine a.

Anterior inferior cerebellar a.

Vertebral a.

Anterior spinal a.

Posterior inferior cerebellar a. (*cut*)

Posterior spinal a.

**Cerebral arterial
circle (of Willis)**
(*broken line*)

Long striate a.
(recurrent a. of Heubner)

Anterior communicating a.

Anterior cerebral a.

Middle cerebral a.

Posterior communicating a.

Anterior choroidal a.

Optic tract

Posterior cerebral a.

Crus cerebri

Lateral geniculate body

Medial posterior choroidal
branch of posterior cerebral a.

Lateral posterior choroidal
branch of posterior cerebral a.

Choroid plexus of lateral ventricle

Medial geniculate body

Pulvinar

Lateral ventricle

Vessels dissected out: inferior view

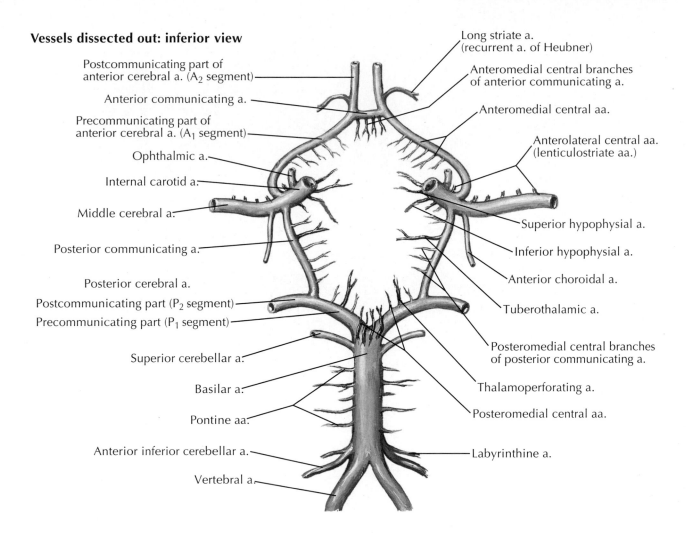

Postcommunicating part of anterior cerebral a. (A₂ segment)

Anterior communicating a.

Precommunicating part of anterior cerebral a. (A₁ segment)

Ophthalmic a.

Internal carotid a.

Middle cerebral a.

Posterior communicating a.

Posterior cerebral a.

Postcommunicating part (P₂ segment)

Precommunicating part (P₁ segment)

Superior cerebellar a.

Basilar a.

Pontine aa.

Anterior inferior cerebellar a.

Vertebral a.

Long striate a. (recurrent a. of Heubner)

Anteromedial central branches of anterior communicating a.

Anteromedial central aa.

Anterolateral central aa. (lenticulostriate aa.)

Superior hypophysial a.

Inferior hypophysial a.

Anterior choroidal a.

Tuberothalamic a.

Posteromedial central branches of posterior communicating a.

Thalamoperforating a.

Posteromedial central aa.

Labyrinthine a.

Vessels in situ: inferior view

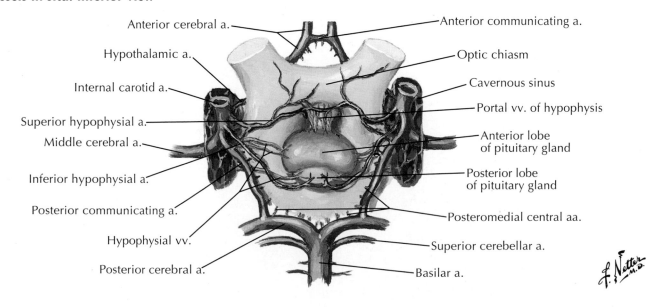

Anterior cerebral a.

Hypothalamic a.

Internal carotid a.

Superior hypophysial a.

Middle cerebral a.

Inferior hypophysial a.

Posterior communicating a.

Hypophysial vv.

Posterior cerebral a.

Anterior communicating a.

Optic chiasm

Cavernous sinus

Portal vv. of hypophysis

Anterior lobe of pituitary gland

Posterior lobe of pituitary gland

Posteromedial central aa.

Superior cerebellar a.

Basilar a.

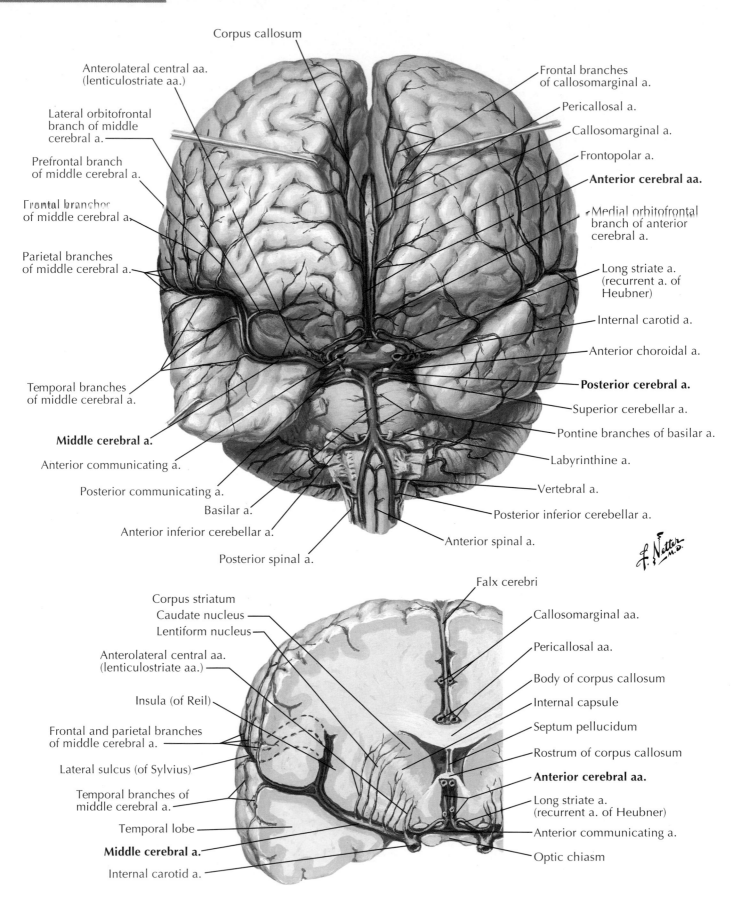

Corpus callosum

Anterolateral central aa. (lenticulostriate aa.)

Lateral orbitofrontal branch of middle cerebral a.

Prefrontal branch of middle cerebral a.

Frontal branches of middle cerebral a.

Parietal branches of middle cerebral a.

Temporal branches of middle cerebral a.

Middle cerebral a.

Anterior communicating a.

Posterior communicating a.

Basilar a.

Anterior inferior cerebellar a.

Posterior spinal a.

Frontal branches of callosomarginal a.

Pericallosal a.

Callosomarginal a.

Frontopolar a.

Anterior cerebral aa.

Medial orbitofrontal branch of anterior cerebral a.

Long striate a. (recurrent a. of Heubner)

Internal carotid a.

Anterior choroidal a.

Posterior cerebral a.

Superior cerebellar a.

Pontine branches of basilar a.

Labyrinthine a.

Vertebral a.

Posterior inferior cerebellar a.

Anterior spinal a.

Falx cerebri

Corpus striatum

Caudate nucleus

Lentiform nucleus

Anterolateral central aa. (lenticulostriate aa.)

Insula (of Reil)

Frontal and parietal branches of middle cerebral a.

Lateral sulcus (of Sylvius)

Temporal branches of middle cerebral a.

Temporal lobe

Middle cerebral a.

Internal carotid a.

Callosomarginal aa.

Pericallosal aa.

Body of corpus callosum

Internal capsule

Septum pellucidum

Rostrum of corpus callosum

Anterior cerebral aa.

Long striate a. (recurrent a. of Heubner)

Anterior communicating a.

Optic chiasm

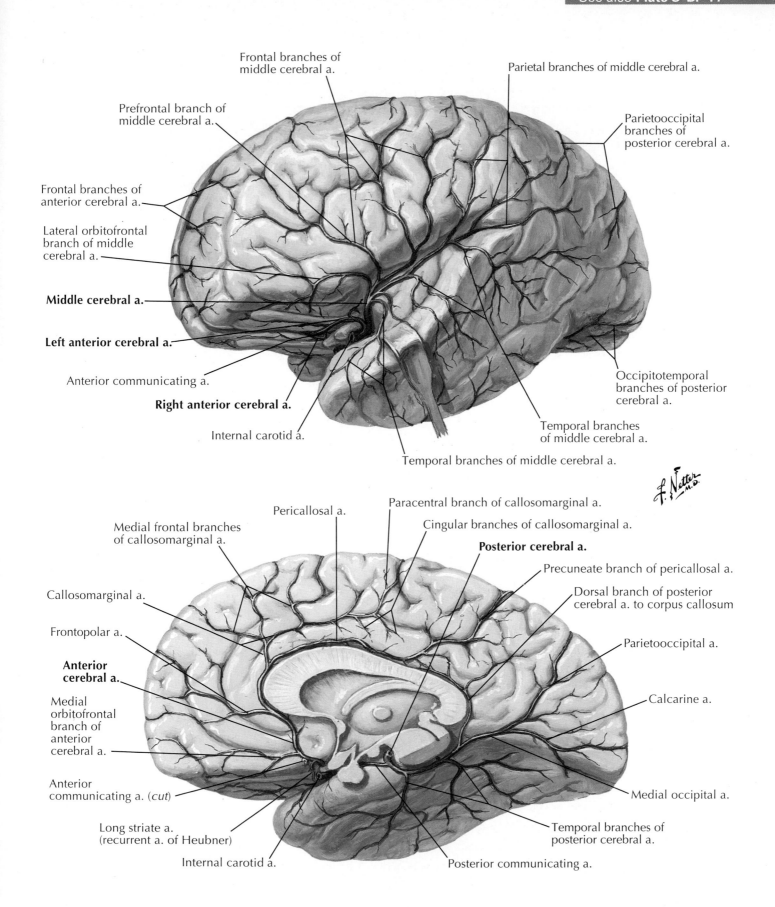

Frontal branches of middle cerebral a.

Parietal branches of middle cerebral a.

Prefrontal branch of middle cerebral a.

Parietooccipital branches of posterior cerebral a.

Frontal branches of anterior cerebral a.

Lateral orbitofrontal branch of middle cerebral a.

Middle cerebral a.

Left anterior cerebral a.

Anterior communicating a.

Right anterior cerebral a.

Internal carotid a.

Occipitotemporal branches of posterior cerebral a.

Temporal branches of middle cerebral a.

Temporal branches of middle cerebral a.

Pericallosal a.

Paracentral branch of callosomarginal a.

Cingular branches of callosomarginal a.

Medial frontal branches of callosomarginal a.

Posterior cerebral a.

Precuneate branch of pericallosal a.

Callosomarginal a.

Dorsal branch of posterior cerebral a. to corpus callosum

Frontopolar a.

Anterior cerebral a.

Parietooccipital a.

Medial orbitofrontal branch of anterior cerebral a.

Calcarine a.

Anterior communicating a. (*cut*)

Medial occipital a.

Long striate a. (recurrent a. of Heubner)

Temporal branches of posterior cerebral a.

Internal carotid a.

Posterior communicating a.

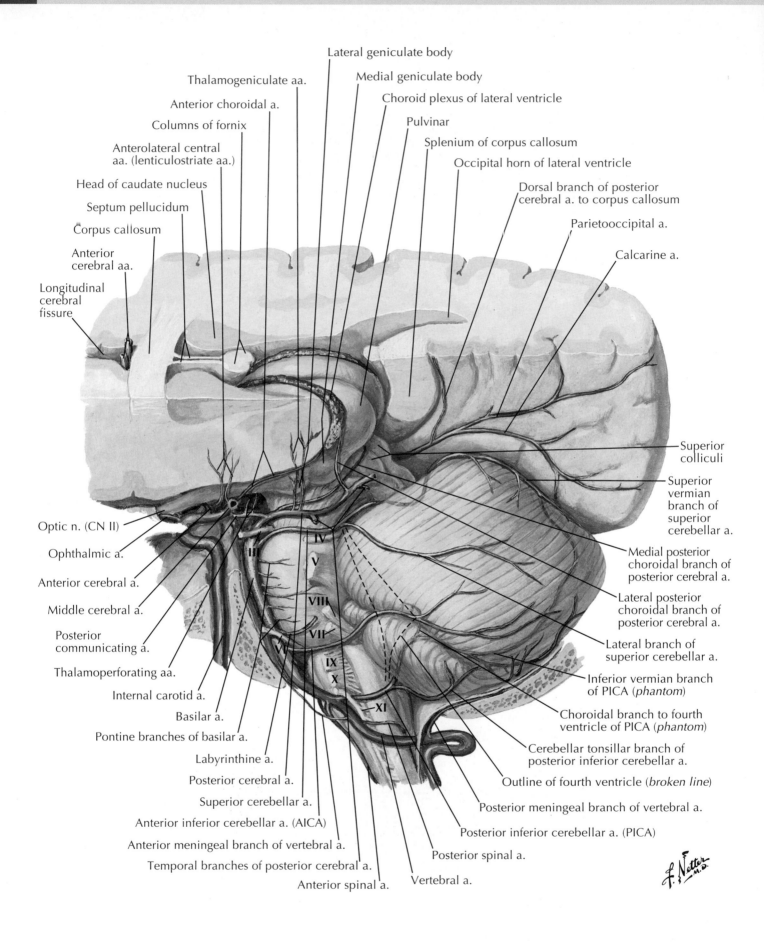

Lateral geniculate body

Medial geniculate body

Thalamogeniculate aa.

Choroid plexus of lateral ventricle

Anterior choroidal a.

Pulvinar

Columns of fornix

Splenium of corpus callosum

Anterolateral central aa. (lenticulostriate aa.)

Occipital horn of lateral ventricle

Head of caudate nucleus

Dorsal branch of posterior cerebral a. to corpus callosum

Septum pellucidum

Parietooccipital a.

Corpus callosum

Calcarine a.

Anterior cerebral aa.

Longitudinal cerebral fissure

Superior colliculi

Superior vermian branch of superior cerebellar a.

Optic n. (CN II)

Ophthalmic a.

Medial posterior choroidal branch of posterior cerebral a.

Anterior cerebral a.

Lateral posterior choroidal branch of posterior cerebral a.

Middle cerebral a.

Posterior communicating a.

Lateral branch of superior cerebellar a.

Thalamoperforating aa.

Inferior vermian branch of PICA (phantom)

Internal carotid a.

Choroidal branch to fourth ventricle of PICA (phantom)

Basilar a.

Pontine branches of basilar a.

Cerebellar tonsillar branch of posterior inferior cerebellar a.

Labyrinthine a.

Outline of fourth ventricle (broken line)

Posterior cerebral a.

Posterior meningeal branch of vertebral a.

Superior cerebellar a.

Anterior inferior cerebellar a. (AICA)

Posterior inferior cerebellar a. (PICA)

Anterior meningeal branch of vertebral a.

Posterior spinal a.

Temporal branches of posterior cerebral a.

Anterior spinal a.

Vertebral a.

III   IV   V   VIII   VII   VI   IX   X   XI

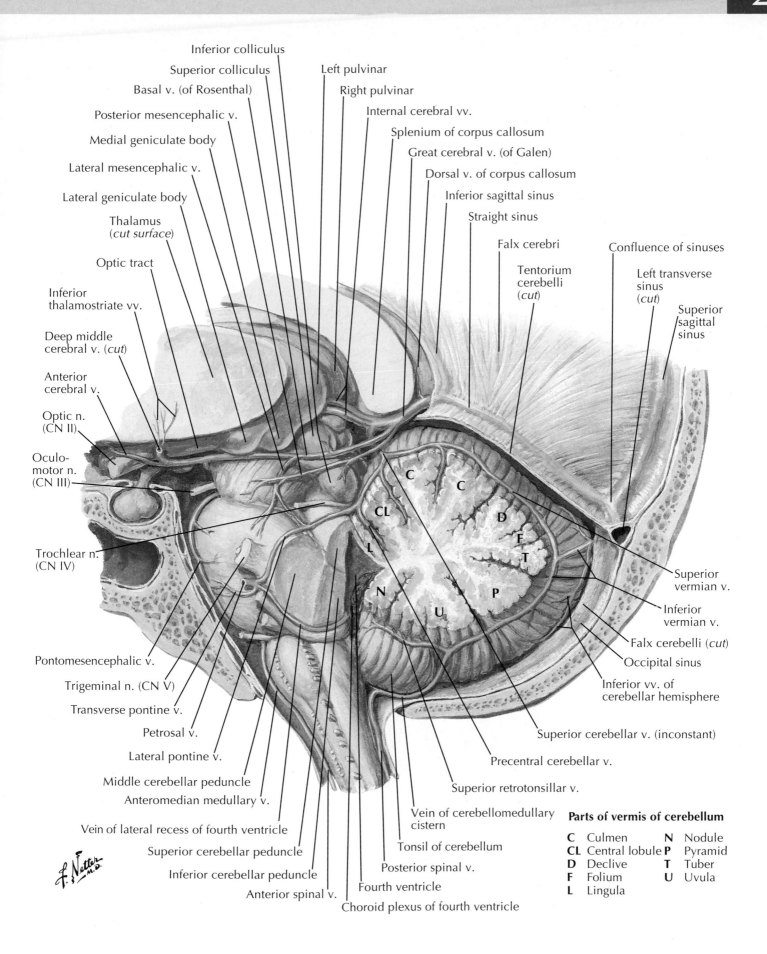

Inferior colliculus
Superior colliculus
Basal v. (of Rosenthal)
Posterior mesencephalic v.
Medial geniculate body
Lateral mesencephalic v.
Lateral geniculate body
Thalamus (*cut surface*)
Optic tract
Inferior thalamostriate vv.
Deep middle cerebral v. (*cut*)
Anterior cerebral v.
Optic n. (CN II)
Oculo-motor n. (CN III)
Trochlear n. (CN IV)
Pontomesencephalic v.
Trigeminal n. (CN V)
Transverse pontine v.
Petrosal v.
Lateral pontine v.
Middle cerebellar peduncle
Anteromedian medullary v.
Vein of lateral recess of fourth ventricle
Superior cerebellar peduncle
Inferior cerebellar peduncle
Anterior spinal v.

Left pulvinar
Right pulvinar
Internal cerebral vv.
Splenium of corpus callosum
Great cerebral v. (of Galen)
Dorsal v. of corpus callosum
Inferior sagittal sinus
Straight sinus
Falx cerebri
Tentorium cerebelli (*cut*)
Confluence of sinuses
Left transverse sinus (*cut*)
Superior sagittal sinus
Superior vermian v.
Inferior vermian v.
Falx cerebelli (*cut*)
Occipital sinus
Inferior vv. of cerebellar hemisphere
Superior cerebellar v. (inconstant)
Precentral cerebellar v.
Superior retrotonsillar v.
Vein of cerebellomedullary cistern
Tonsil of cerebellum
Posterior spinal v.
Fourth ventricle
Choroid plexus of fourth ventricle

Parts of vermis of cerebellum

C	Culmen	**N**	Nodule
CL	Central lobule	**P**	Pyramid
D	Declive	**T**	Tuber
F	Folium	**U**	Uvula
L	Lingula		

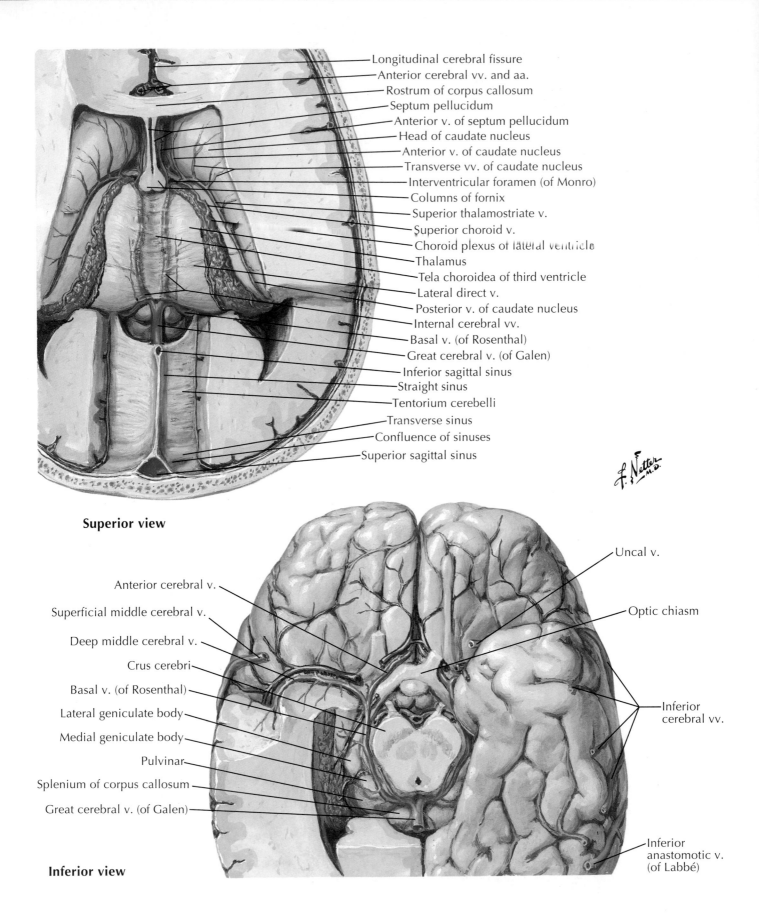

Longitudinal cerebral fissure
Anterior cerebral vv. and aa.
Rostrum of corpus callosum
Septum pellucidum
Anterior v. of septum pellucidum
Head of caudate nucleus
Anterior v. of caudate nucleus
Transverse vv. of caudate nucleus
Interventricular foramen (of Monro)
Columns of fornix
Superior thalamostriate v.
Superior choroid v.
Choroid plexus of lateral ventricle
Thalamus
Tela choroidea of third ventricle
Lateral direct v.
Posterior v. of caudate nucleus
Internal cerebral vv.
Basal v. (of Rosenthal)
Great cerebral v. (of Galen)
Inferior sagittal sinus
Straight sinus
Tentorium cerebelli
Transverse sinus
Confluence of sinuses
Superior sagittal sinus

Superior view

Uncal v.

Anterior cerebral v.

Superficial middle cerebral v.

Optic chiasm

Deep middle cerebral v.

Crus cerebri

Basal v. (of Rosenthal)

Lateral geniculate body

Medial geniculate body

Inferior
cerebral vv.

Pulvinar

Splenium of corpus callosum

Great cerebral v. (of Galen)

Inferior
anastomotic v.
(of Labbé)

Inferior view

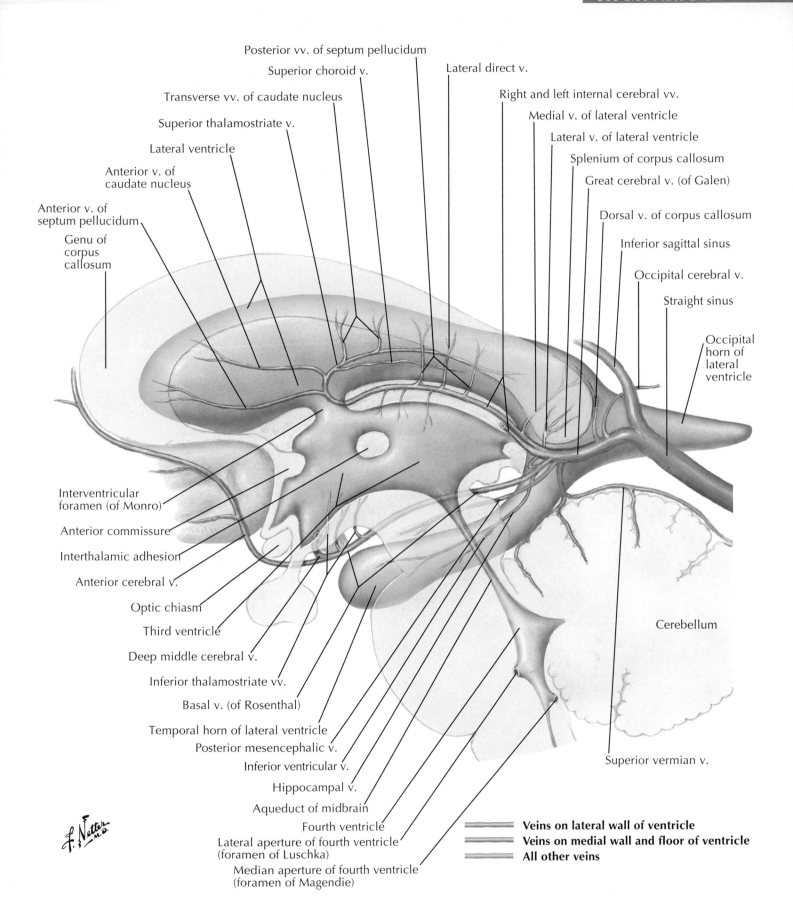

Posterior vv. of septum pellucidum

Superior choroid v.

Transverse vv. of caudate nucleus

Superior thalamostriate v.

Lateral ventricle

Anterior v. of caudate nucleus

Anterior v. of septum pellucidum

Genu of corpus callosum

Lateral direct v.

Right and left internal cerebral vv.

Medial v. of lateral ventricle

Lateral v. of lateral ventricle

Splenium of corpus callosum

Great cerebral v. (of Galen)

Dorsal v. of corpus callosum

Inferior sagittal sinus

Occipital cerebral v.

Straight sinus

Occipital horn of lateral ventricle

Interventricular foramen (of Monro)

Anterior commissure

Interthalamic adhesion

Anterior cerebral v.

Optic chiasm

Third ventricle

Deep middle cerebral v.

Inferior thalamostriate vv.

Basal v. (of Rosenthal)

Temporal horn of lateral ventricle

Posterior mesencephalic v.

Inferior ventricular v.

Hippocampal v.

Aqueduct of midbrain

Fourth ventricle

Lateral aperture of fourth ventricle (foramen of Luschka)

Median aperture of fourth ventricle (foramen of Magendie)

Cerebellum

Superior vermian v.

Veins on lateral wall of ventricle
Veins on medial wall and floor of ventricle
All other veins

Magnetic resonance angiography (MRA) at level of cerebral arterial circle (3D time-of-flight image without contrast)

Postcommunicating part of anterior cerebral a. (A2 segment)

Middle cerebral a. (M2 segments)

Anterior communicating a.

Internal carotid a.

Superior cerebellar a.

Basilar a.

Precommunicating part of anterior cerebral a. (A1 segment)

Middle cerebral a. (M1 segment)

Posterior communicating a.

Posterior cerebral a.

Anterior inferior cerebellar a.

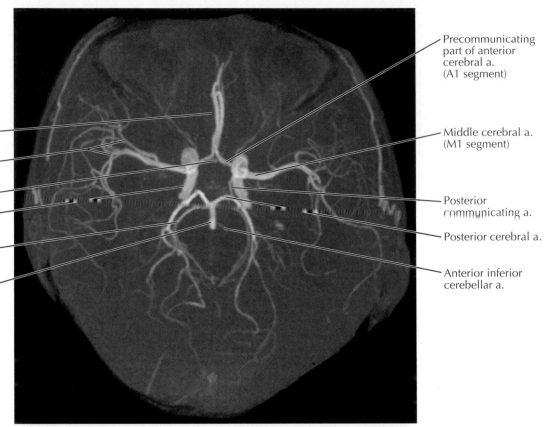

Magnetic resonance venography (MRV) (2D time-of-flight image without contrast)

Superior cerebral v.

Superior sagittal sinus

Internal cerebral v.

Great cerebral v. (of Galen)

Straight sinus

Confluence of sinuses

Transverse sinus

Sigmoid sinus

Internal jugular v.

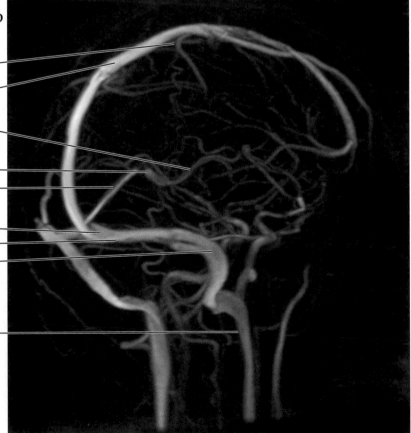

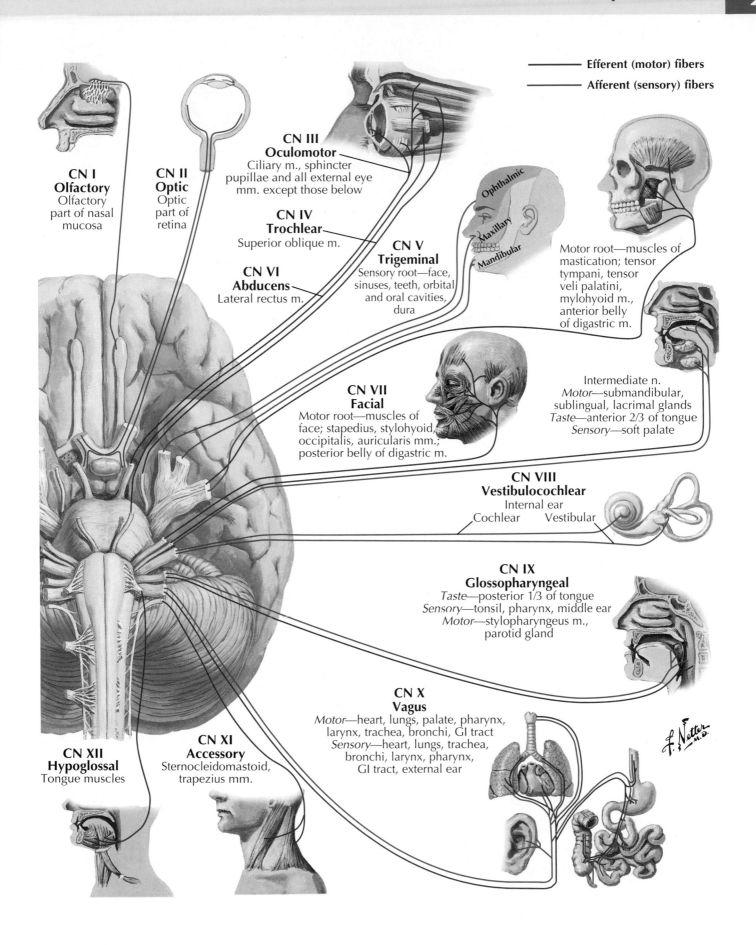

Efferent (motor) fibers
Afferent (sensory) fibers

CN III
Oculomotor
Ciliary m., sphincter pupillae and all external eye mm. except those below

CN IV
Trochlear
Superior oblique m.

CN I
Olfactory
Olfactory part of nasal mucosa

CN II
Optic
Optic part of retina

CN V
Trigeminal
Sensory root—face, sinuses, teeth, orbital and oral cavities, dura

Ophthalmic
Maxillary
Mandibular

CN VI
Abducens
Lateral rectus m.

Motor root—muscles of mastication; tensor tympani, tensor veli palatini, mylohyoid m., anterior belly of digastric m.

Intermediate n.
Motor—submandibular, sublingual, lacrimal glands
Taste—anterior 2/3 of tongue
Sensory—soft palate

CN VII
Facial
Motor root—muscles of face; stapedius, stylohyoid, occipitalis, auricularis mm.; posterior belly of digastric m.

CN VIII
Vestibulocochlear
Internal ear
Cochlear   Vestibular

CN IX
Glossopharyngeal
Taste—posterior 1/3 of tongue
Sensory—tonsil, pharynx, middle ear
Motor—stylopharyngeus m., parotid gland

CN X
Vagus
Motor—heart, lungs, palate, pharynx, larynx, trachea, bronchi, GI tract
Sensory—heart, lungs, trachea, bronchi, larynx, pharynx, GI tract, external ear

CN XII
Hypoglossal
Tongue muscles

CN XI
Accessory
Sternocleidomastoid, trapezius mm.

F. Netter
M.D.

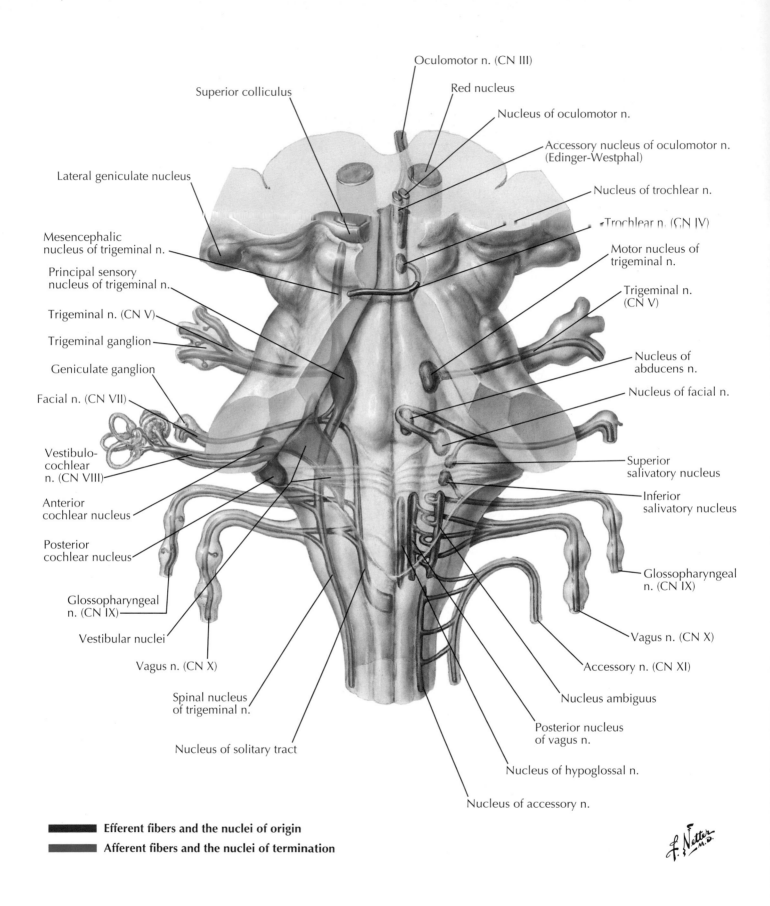

Oculomotor n. (CN III)

Superior colliculus

Red nucleus

Nucleus of oculomotor n.

Accessory nucleus of oculomotor n. (Edinger-Westphal)

Lateral geniculate nucleus

Nucleus of trochlear n.

Trochlear n. (CN IV)

Mesencephalic nucleus of trigeminal n.

Motor nucleus of trigeminal n.

Principal sensory nucleus of trigeminal n.

Trigeminal n. (CN V)

Trigeminal n. (CN V)

Trigeminal ganglion

Nucleus of abducens n.

Geniculate ganglion

Nucleus of facial n.

Facial n. (CN VII)

Vestibulo-cochlear n. (CN VIII)

Superior salivatory nucleus

Inferior salivatory nucleus

Anterior cochlear nucleus

Posterior cochlear nucleus

Glossopharyngeal n. (CN IX)

Glossopharyngeal n. (CN IX)

Vagus n. (CN X)

Vestibular nuclei

Accessory n. (CN XI)

Vagus n. (CN X)

Nucleus ambiguus

Spinal nucleus of trigeminal n.

Posterior nucleus of vagus n.

Nucleus of solitary tract

Nucleus of hypoglossal n.

Nucleus of accessory n.

▬▬▬▬ Efferent fibers and the nuclei of origin

▬▬▬▬ Afferent fibers and the nuclei of termination

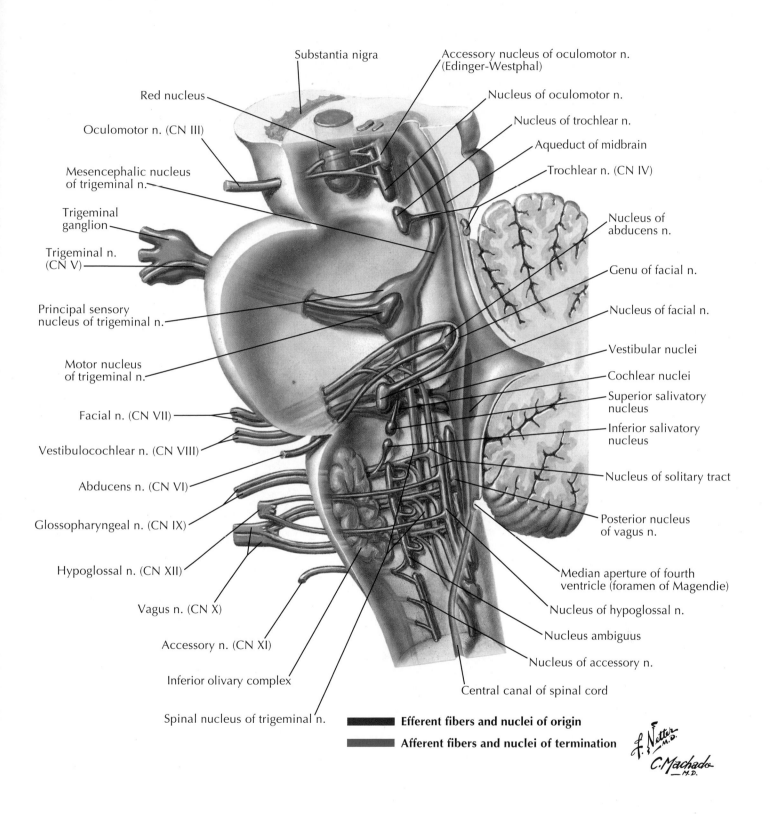

Substantia nigra

Accessory nucleus of oculomotor n. (Edinger-Westphal)

Red nucleus

Nucleus of oculomotor n.

Oculomotor n. (CN III)

Nucleus of trochlear n.

Aqueduct of midbrain

Mesencephalic nucleus of trigeminal n.

Trochlear n. (CN IV)

Trigeminal ganglion

Nucleus of abducens n.

Trigeminal n. (CN V)

Genu of facial n.

Principal sensory nucleus of trigeminal n.

Nucleus of facial n.

Vestibular nuclei

Motor nucleus of trigeminal n.

Cochlear nuclei

Superior salivatory nucleus

Facial n. (CN VII)

Inferior salivatory nucleus

Vestibulocochlear n. (CN VIII)

Nucleus of solitary tract

Abducens n. (CN VI)

Glossopharyngeal n. (CN IX)

Posterior nucleus of vagus n.

Hypoglossal n. (CN XII)

Median aperture of fourth ventricle (foramen of Magendie)

Vagus n. (CN X)

Nucleus of hypoglossal n.

Accessory n. (CN XI)

Nucleus ambiguus

Inferior olivary complex

Nucleus of accessory n.

Central canal of spinal cord

Spinal nucleus of trigeminal n.

Efferent fibers and nuclei of origin

Afferent fibers and nuclei of termination

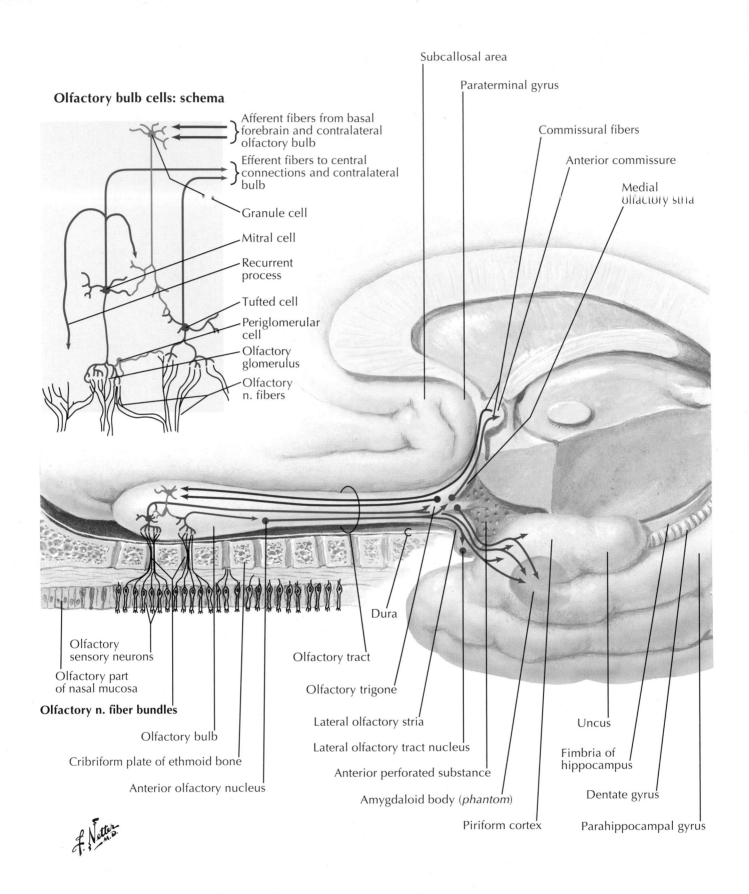

Olfactory bulb cells: schema

Afferent fibers from basal forebrain and contralateral olfactory bulb

Efferent fibers to central connections and contralateral bulb

Granule cell

Mitral cell

Recurrent process

Tufted cell

Periglomerular cell

Olfactory glomerulus

Olfactory n. fibers

Subcallosal area

Paraterminal gyrus

Commissural fibers

Anterior commissure

Medial olfactory stria

Olfactory sensory neurons

Olfactory part of nasal mucosa

Olfactory n. fiber bundles

Olfactory bulb

Cribriform plate of ethmoid bone

Anterior olfactory nucleus

Dura

Olfactory tract

Olfactory trigone

Lateral olfactory stria

Lateral olfactory tract nucleus

Anterior perforated substance

Amygdaloid body (*phantom*)

Piriform cortex

Uncus

Fimbria of hippocampus

Dentate gyrus

Parahippocampal gyrus

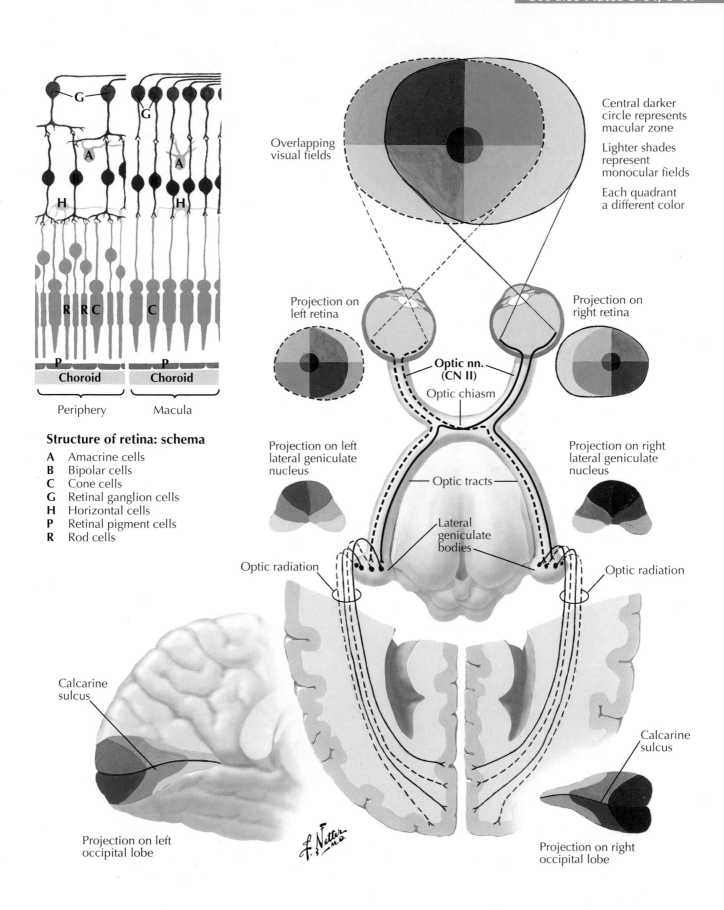

Overlapping
visual fields

Central darker
circle represents
macular zone

Lighter shades
represent
monocular fields

Each quadrant
a different color

Projection on
left retina

Projection on
right retina

**Optic nn.
(CN II)**

Optic chiasm

Projection on left
lateral geniculate
nucleus

Projection on right
lateral geniculate
nucleus

Optic tracts

Lateral
geniculate
bodies

Optic radiation

Optic radiation

Calcarine
sulcus

Calcarine
sulcus

Projection on left
occipital lobe

Projection on right
occipital lobe

G

G

A

A

B

H

H

R R C

C

P

P

Choroid

Choroid

Periphery

Macula

Structure of retina: schema

A Amacrine cells
B Bipolar cells
C Cone cells
G Retinal ganglion cells
H Horizontal cells
P Retinal pigment cells
R Rod cells

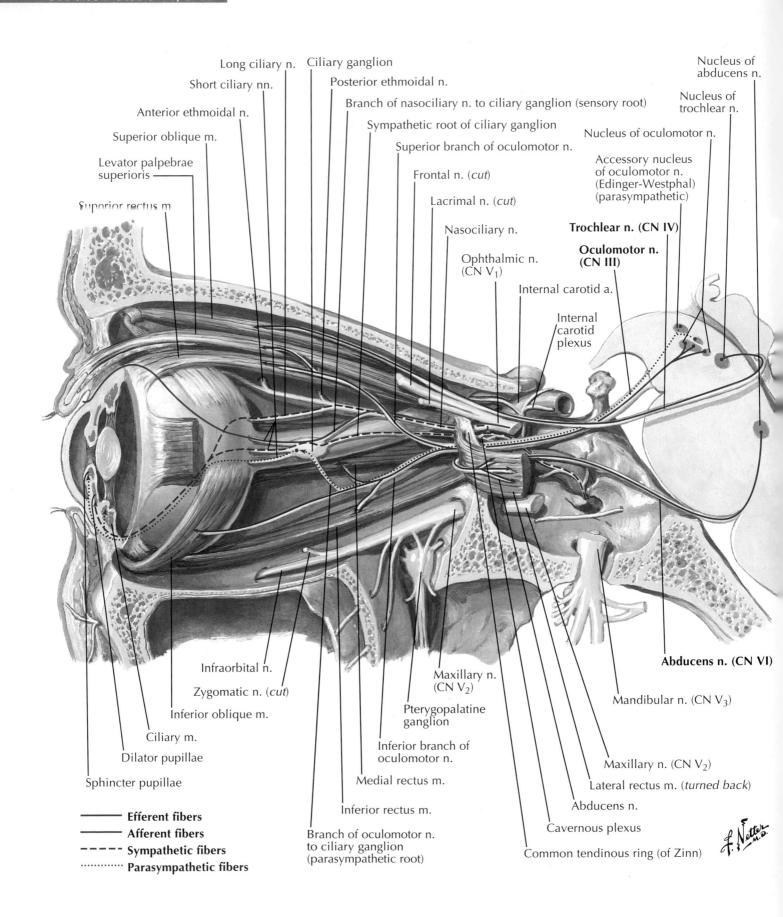

Long ciliary n.

Short ciliary nn.

Anterior ethmoidal n.

Superior oblique m.

Levator palpebrae superioris

Superior rectus m

Ciliary ganglion

Posterior ethmoidal n.

Branch of nasociliary n. to ciliary ganglion (sensory root)

Sympathetic root of ciliary ganglion

Superior branch of oculomotor n.

Frontal n. (*cut*)

Lacrimal n. (*cut*)

Nasociliary n.

Ophthalmic n. (CN V$_1$)

Internal carotid a.

Internal carotid plexus

Nucleus of abducens n.

Nucleus of trochlear n.

Nucleus of oculomotor n.

Accessory nucleus of oculomotor n. (Edinger-Westphal) (parasympathetic)

Trochlear n. (CN IV)

Oculomotor n. (CN III)

Infraorbital n.

Zygomatic n. (*cut*)

Inferior oblique m.

Ciliary m.

Dilator pupillae

Sphincter pupillae

Maxillary n. (CN V$_2$)

Pterygopalatine ganglion

Inferior branch of oculomotor n.

Medial rectus m.

Inferior rectus m.

Branch of oculomotor n. to ciliary ganglion (parasympathetic root)

Abducens n. (CN VI)

Mandibular n. (CN V$_3$)

Maxillary n. (CN V$_2$)

Lateral rectus m. (*turned back*)

Abducens n.

Cavernous plexus

Common tendinous ring (of Zinn)

——	**Efferent fibers**
——	**Afferent fibers**
----	**Sympathetic fibers**
····	**Parasympathetic fibers**

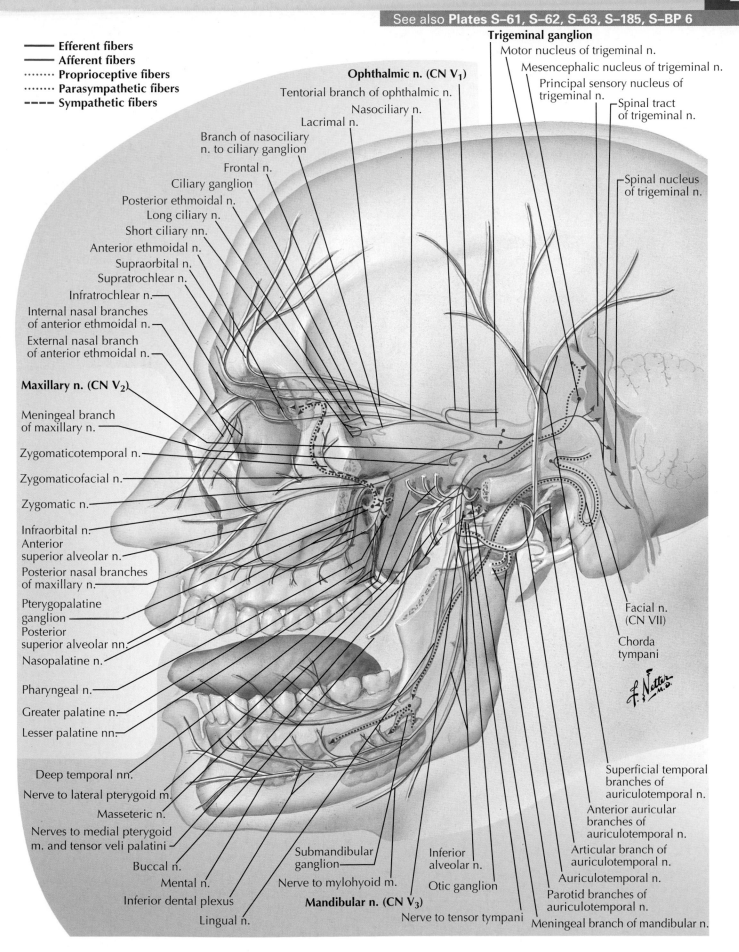

Efferent fibers
Afferent fibers
Proprioceptive fibers
Parasympathetic fibers
Sympathetic fibers

Trigeminal ganglion
Motor nucleus of trigeminal n.
Mesencephalic nucleus of trigeminal n.
Principal sensory nucleus of trigeminal n.
Spinal tract of trigeminal n.
Spinal nucleus of trigeminal n.

Ophthalmic n. (CN V₁)
Tentorial branch of ophthalmic n.
Nasociliary n.
Lacrimal n.
Branch of nasociliary n. to ciliary ganglion
Frontal n.
Ciliary ganglion
Posterior ethmoidal n.
Long ciliary n.
Short ciliary nn.
Anterior ethmoidal n.
Supraorbital n.
Supratrochlear n.
Infratrochlear n.
Internal nasal branches of anterior ethmoidal n.
External nasal branch of anterior ethmoidal n.

Maxillary n. (CN V₂)
Meningeal branch of maxillary n.
Zygomaticotemporal n.
Zygomaticofacial n.
Zygomatic n.
Infraorbital n.
Anterior superior alveolar n.
Posterior nasal branches of maxillary n.
Pterygopalatine ganglion
Posterior superior alveolar nn.
Nasopalatine n.
Pharyngeal n.
Greater palatine n.
Lesser palatine nn.

Deep temporal nn.
Nerve to lateral pterygoid m.
Masseteric n.
Nerves to medial pterygoid m. and tensor veli palatini
Buccal n.
Mental n.
Inferior dental plexus
Lingual n.

Submandibular ganglion
Nerve to mylohyoid m.
Mandibular n. (CN V₃)

Inferior alveolar n.
Otic ganglion
Nerve to tensor tympani

Facial n. (CN VII)
Chorda tympani

Superficial temporal branches of auriculotemporal n.
Anterior auricular branches of auriculotemporal n.
Articular branch of auriculotemporal n.
Auriculotemporal n.
Parotid branches of auriculotemporal n.
Meningeal branch of mandibular n.

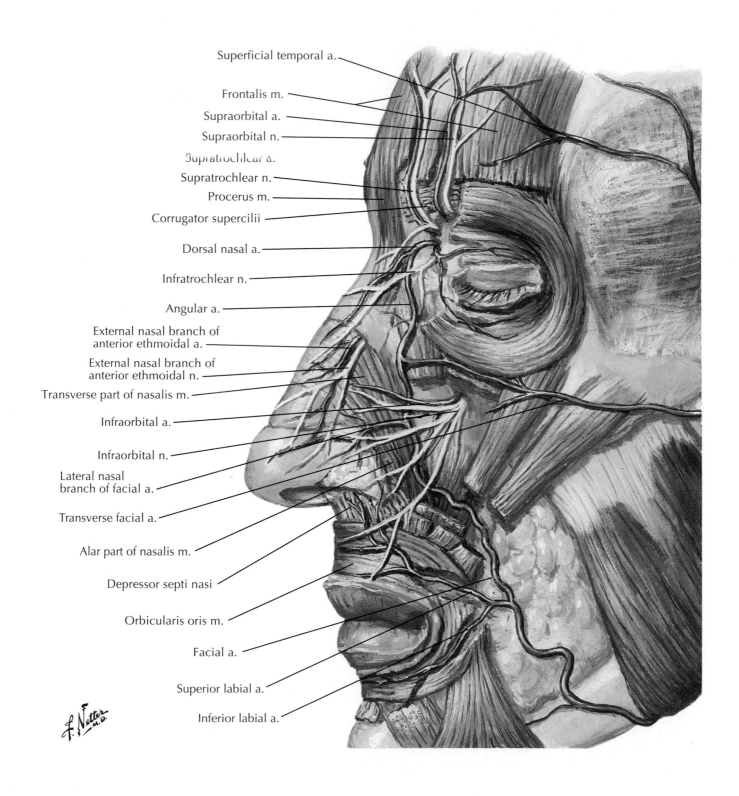

Superficial temporal a.

Frontalis m.

Supraorbital a.

Supraorbital n.

Supratrochlear a.

Supratrochlear n.

Procerus m.

Corrugator supercilii

Dorsal nasal a.

Infratrochlear n.

Angular a.

External nasal branch of
anterior ethmoidal a.

External nasal branch of
anterior ethmoidal n.

Transverse part of nasalis m.

Infraorbital a.

Infraorbital n.

Lateral nasal
branch of facial a.

Transverse facial a.

Alar part of nasalis m.

Depressor septi nasi

Orbicularis oris m.

Facial a.

Superior labial a.

Inferior labial a.

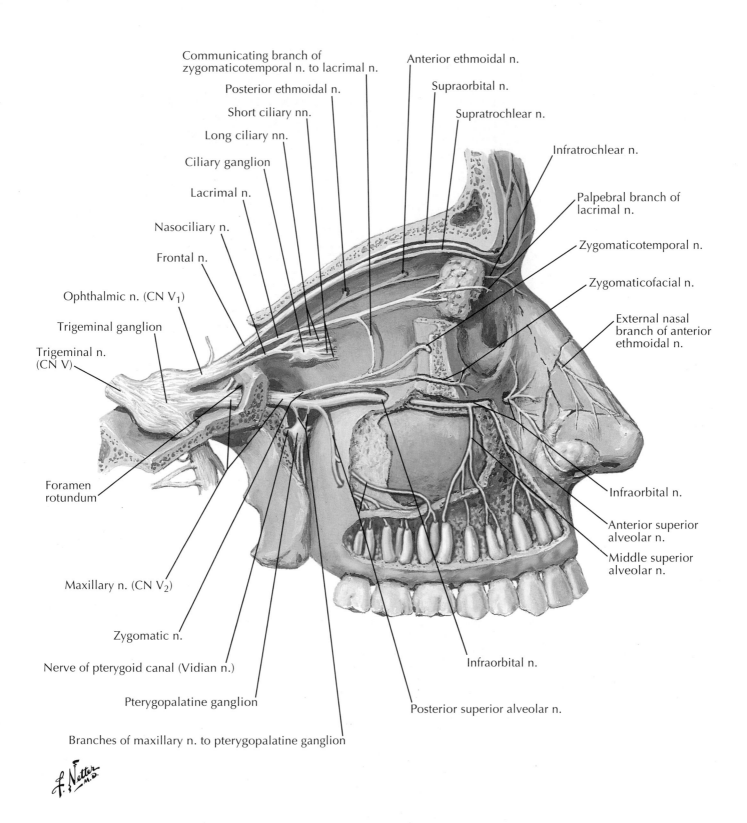

Communicating branch of zygomaticotemporal n. to lacrimal n.

Posterior ethmoidal n.

Short ciliary nn.

Long ciliary nn.

Ciliary ganglion

Lacrimal n.

Nasociliary n.

Frontal n.

Ophthalmic n. (CN V₁)

Trigeminal ganglion

Trigeminal n. (CN V)

Foramen rotundum

Maxillary n. (CN V₂)

Zygomatic n.

Nerve of pterygoid canal (Vidian n.)

Pterygopalatine ganglion

Branches of maxillary n. to pterygopalatine ganglion

Anterior ethmoidal n.

Supraorbital n.

Supratrochlear n.

Infratrochlear n.

Palpebral branch of lacrimal n.

Zygomaticotemporal n.

Zygomaticofacial n.

External nasal branch of anterior ethmoidal n.

Infraorbital n.

Anterior superior alveolar n.

Middle superior alveolar n.

Infraorbital n.

Posterior superior alveolar n.

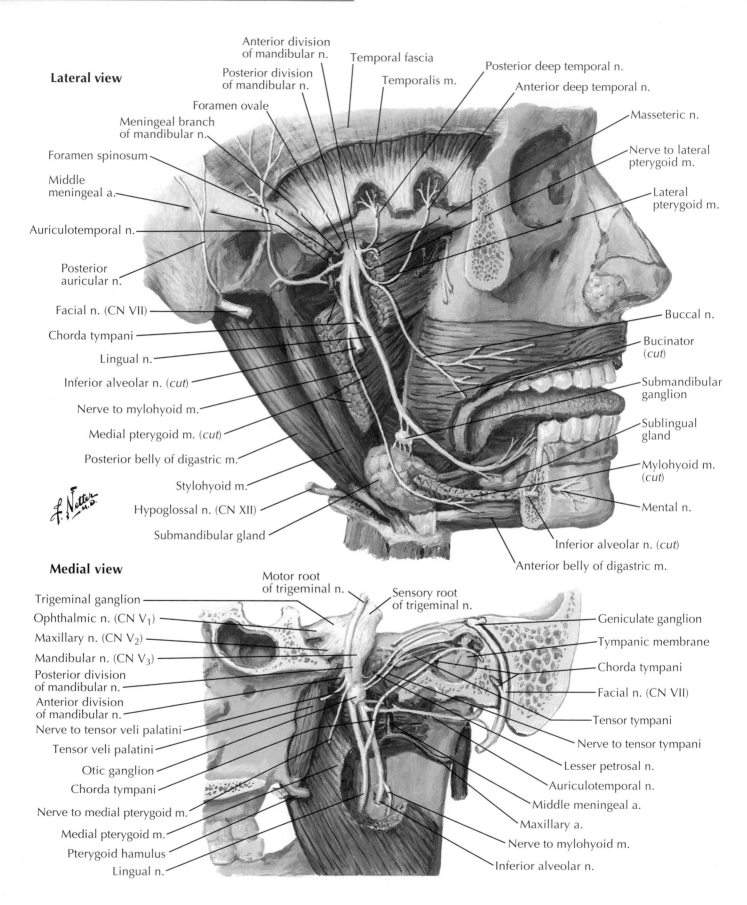

Lateral view

Anterior division of mandibular n.

Posterior division of mandibular n.

Temporal fascia

Temporalis m.

Posterior deep temporal n.

Anterior deep temporal n.

Masseteric n.

Nerve to lateral pterygoid m.

Foramen ovale

Meningeal branch of mandibular n.

Foramen spinosum

Middle meningeal a.

Lateral pterygoid m.

Auriculotemporal n.

Posterior auricular n.

Buccal n.

Facial n. (CN VII)

Buccinator (cut)

Chorda tympani

Lingual n.

Submandibular ganglion

Inferior alveolar n. (cut)

Sublingual gland

Nerve to mylohyoid m.

Medial pterygoid m. (cut)

Mylohyoid m. (cut)

Posterior belly of digastric m.

Stylohyoid m.

Mental n.

Hypoglossal n. (CN XII)

Submandibular gland

Inferior alveolar n. (cut)

Anterior belly of digastric m.

Medial view

Motor root of trigeminal n.

Sensory root of trigeminal n.

Trigeminal ganglion

Geniculate ganglion

Ophthalmic n. (CN V₁)

Tympanic membrane

Maxillary n. (CN V₂)

Chorda tympani

Mandibular n. (CN V₃)

Facial n. (CN VII)

Posterior division of mandibular n.

Tensor tympani

Anterior division of mandibular n.

Nerve to tensor tympani

Nerve to tensor veli palatini

Lesser petrosal n.

Tensor veli palatini

Auriculotemporal n.

Otic ganglion

Middle meningeal a.

Chorda tympani

Maxillary a.

Nerve to medial pterygoid m.

Nerve to mylohyoid m.

Medial pterygoid m.

Inferior alveolar n.

Pterygoid hamulus

Lingual n.

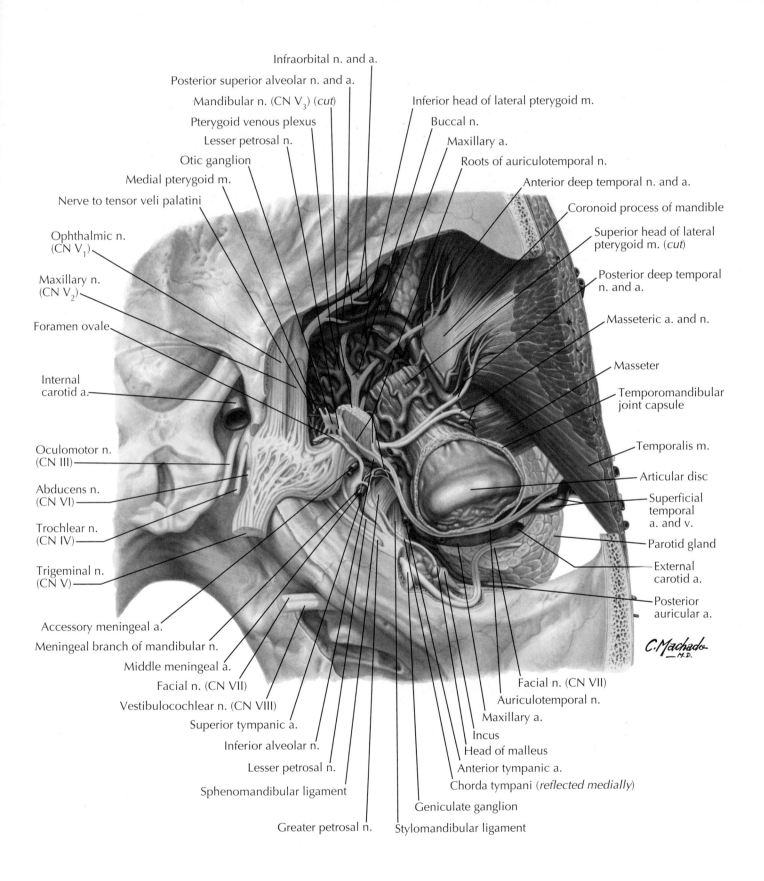

Infraorbital n. and a.

Posterior superior alveolar n. and a.

Mandibular n. (CN V₃) (*cut*)

Pterygoid venous plexus

Lesser petrosal n.

Otic ganglion

Medial pterygoid m.

Nerve to tensor veli palatini

Ophthalmic n. (CN V₁)

Maxillary n. (CN V₂)

Foramen ovale

Internal carotid a.

Oculomotor n. (CN III)

Abducens n. (CN VI)

Trochlear n. (CN IV)

Trigeminal n. (CN V)

Accessory meningeal a.

Meningeal branch of mandibular n.

Middle meningeal a.

Facial n. (CN VII)

Vestibulocochlear n. (CN VIII)

Superior tympanic a.

Inferior alveolar n.

Lesser petrosal n.

Sphenomandibular ligament

Greater petrosal n.

Inferior head of lateral pterygoid m.

Buccal n.

Maxillary a.

Roots of auriculotemporal n.

Anterior deep temporal n. and a.

Coronoid process of mandible

Superior head of lateral pterygoid m. (*cut*)

Posterior deep temporal n. and a.

Masseteric a. and n.

Masseter

Temporomandibular joint capsule

Temporalis m.

Articular disc

Superficial temporal a. and v.

Parotid gland

External carotid a.

Posterior auricular a.

Facial n. (CN VII)

Auriculotemporal n.

Maxillary a.

Incus

Head of malleus

Anterior tympanic a.

Chorda tympani (*reflected medially*)

Geniculate ganglion

Stylomandibular ligament

C. Machado M.D.

Cranial and Cervical Nerves

Plate S–63

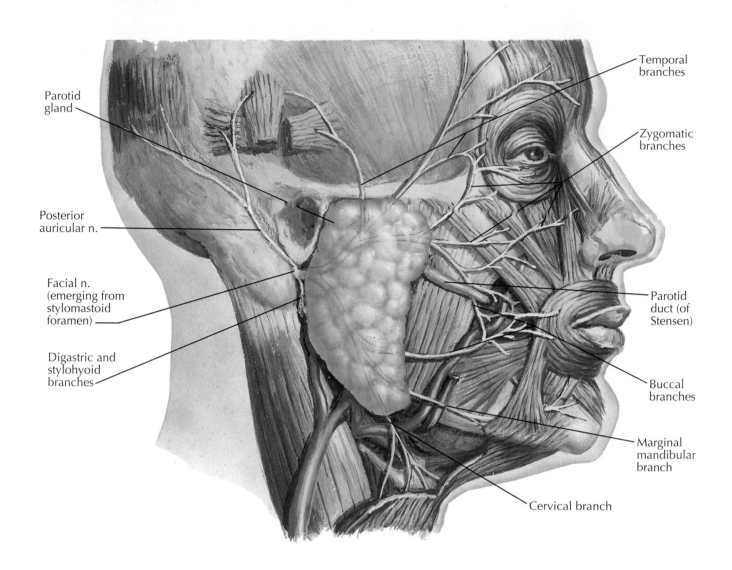

Parotid gland

Temporal branches

Zygomatic branches

Posterior auricular n.

Facial n. (emerging from stylomastoid foramen)

Parotid duct (of Stensen)

Digastric and stylohyoid branches

Buccal branches

Marginal mandibular branch

Cervical branch

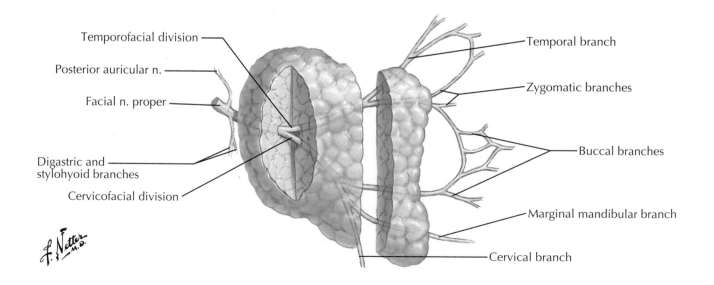

Temporofacial division

Posterior auricular n.

Facial n. proper

Digastric and stylohyoid branches

Cervicofacial division

Temporal branch

Zygomatic branches

Buccal branches

Marginal mandibular branch

Cervical branch

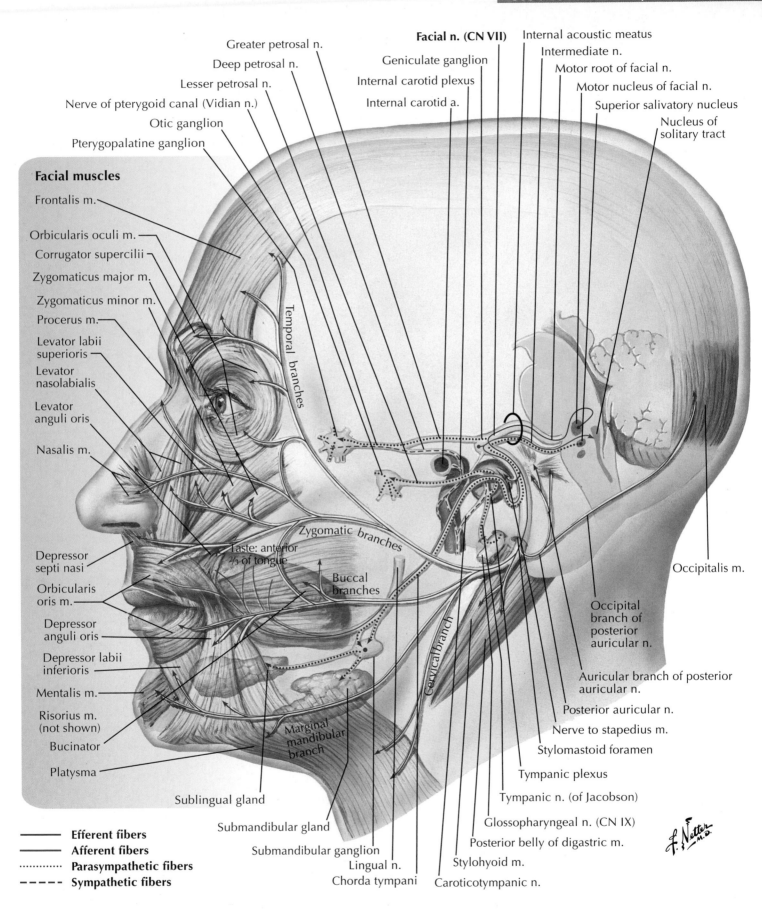

Greater petrosal n.

Deep petrosal n.

Lesser petrosal n.

Nerve of pterygoid canal (Vidian n.)

Otic ganglion

Pterygopalatine ganglion

Facial n. (CN VII)

Geniculate ganglion

Internal carotid plexus

Internal carotid a.

Internal acoustic meatus

Intermediate n.

Motor root of facial n.

Motor nucleus of facial n.

Superior salivatory nucleus

Nucleus of solitary tract

Facial muscles

Frontalis m.

Orbicularis oculi m.

Corrugator supercilii

Zygomaticus major m.

Zygomaticus minor m.

Procerus m.

Levator labii superioris

Levator nasolabialis

Levator anguli oris

Nasalis m.

Depressor septi nasi

Orbicularis oris m.

Depressor anguli oris

Depressor labii inferioris

Mentalis m.

Risorius m. (not shown)

Bucinator

Platysma

Temporal branches

Zygomatic branches

Taste: anterior ⅔ of tongue

Buccal branches

Marginal mandibular branch

Cervical branch

Sublingual gland

Submandibular gland

Submandibular ganglion

Lingual n.

Chorda tympani

Occipitalis m.

Occipital branch of posterior auricular n.

Auricular branch of posterior auricular n.

Posterior auricular n.

Nerve to stapedius m.

Stylomastoid foramen

Tympanic plexus

Tympanic n. (of Jacobson)

Glossopharyngeal n. (CN IX)

Posterior belly of digastric m.

Stylohyoid m.

Caroticotympanic n.

——— Efferent fibers

——— Afferent fibers

········· Parasympathetic fibers

- - - - Sympathetic fibers

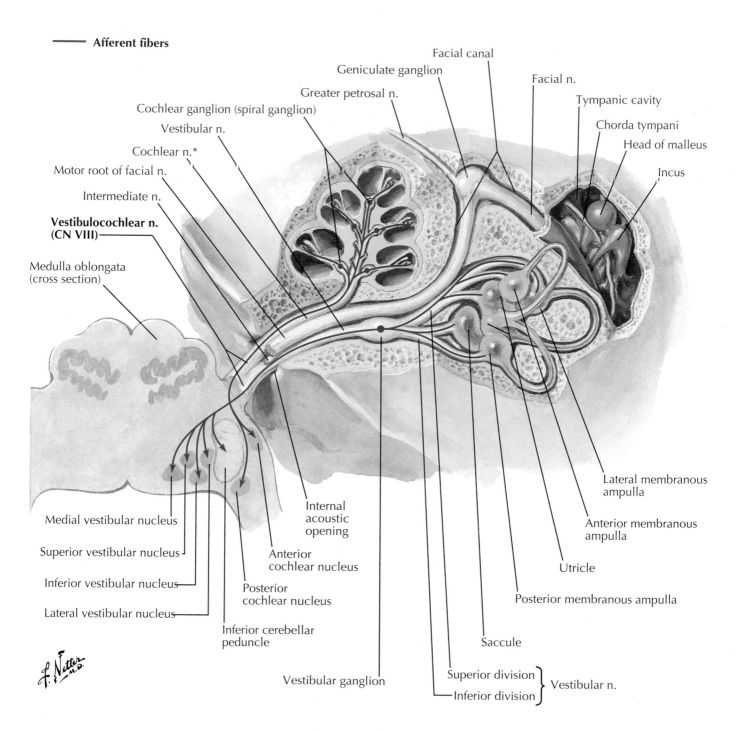

Afferent fibers

Facial canal

Geniculate ganglion

Greater petrosal n.

Facial n.

Cochlear ganglion (spiral ganglion)

Tympanic cavity

Vestibular n.

Chorda tympani

Cochlear n.*

Head of malleus

Motor root of facial n.

Incus

Intermediate n.

Vestibulocochlear n. (CN VIII)

Medulla oblongata (cross section)

Lateral membranous ampulla

Anterior membranous ampulla

Internal acoustic opening

Medial vestibular nucleus

Superior vestibular nucleus

Anterior cochlear nucleus

Utricle

Inferior vestibular nucleus

Posterior cochlear nucleus

Posterior membranous ampulla

Lateral vestibular nucleus

Inferior cerebellar peduncle

Saccule

Superior division

Inferior division

} Vestibular n.

Vestibular ganglion

*Note: The cochlear nerve also contains efferent fibers to the sensory epithelium.
These fibers are derived from the vestibular nerve while in the internal acoustic meatus.*

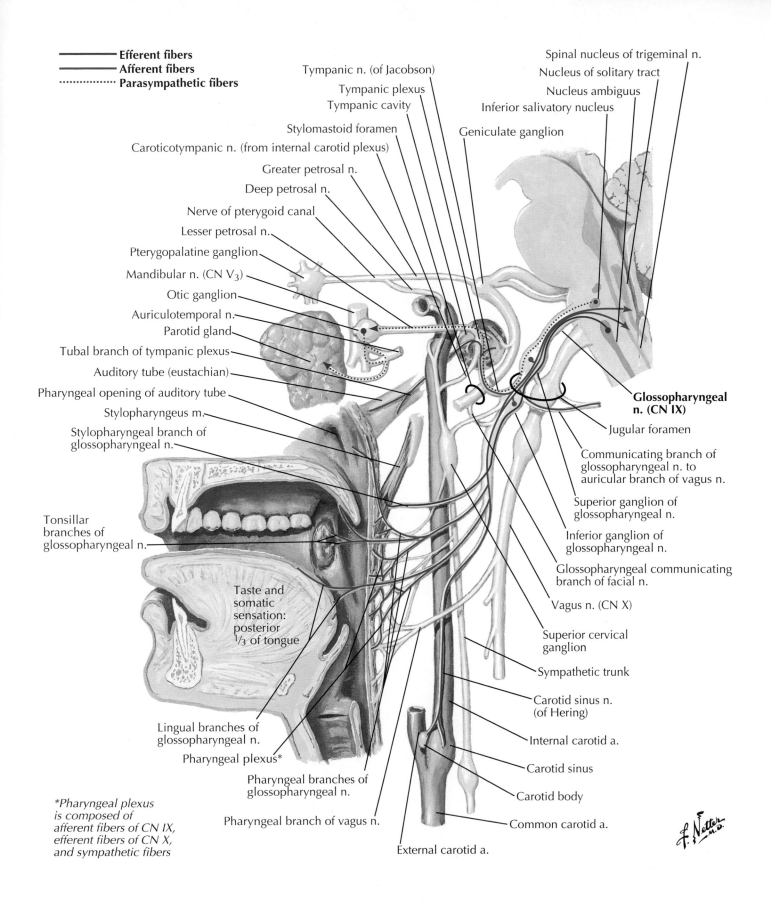

Efferent fibers
Afferent fibers
Parasympathetic fibers

Tympanic n. (of Jacobson)
Tympanic plexus
Tympanic cavity
Stylomastoid foramen
Caroticotympanic n. (from internal carotid plexus)
Greater petrosal n.
Deep petrosal n.
Nerve of pterygoid canal
Lesser petrosal n.
Pterygopalatine ganglion
Mandibular n. (CN V₃)
Otic ganglion
Auriculotemporal n.
Parotid gland
Tubal branch of tympanic plexus
Auditory tube (eustachian)
Pharyngeal opening of auditory tube
Stylopharyngeus m.
Stylopharyngeal branch of glossopharyngeal n.

Spinal nucleus of trigeminal n.
Nucleus of solitary tract
Nucleus ambiguus
Inferior salivatory nucleus
Geniculate ganglion

Tonsillar branches of glossopharyngeal n.

Taste and somatic sensation: posterior ⅓ of tongue

Lingual branches of glossopharyngeal n.

Pharyngeal plexus*

Pharyngeal branches of glossopharyngeal n.

Pharyngeal branch of vagus n.

External carotid a.

Pharyngeal plexus is composed of afferent fibers of CN IX, efferent fibers of CN X, and sympathetic fibers

Glossopharyngeal n. (CN IX)
Jugular foramen
Communicating branch of glossopharyngeal n. to auricular branch of vagus n.
Superior ganglion of glossopharyngeal n.
Inferior ganglion of glossopharyngeal n.
Glossopharyngeal communicating branch of facial n.
Vagus n. (CN X)
Superior cervical ganglion
Sympathetic trunk
Carotid sinus n. (of Hering)
Internal carotid a.
Carotid sinus
Carotid body
Common carotid a.

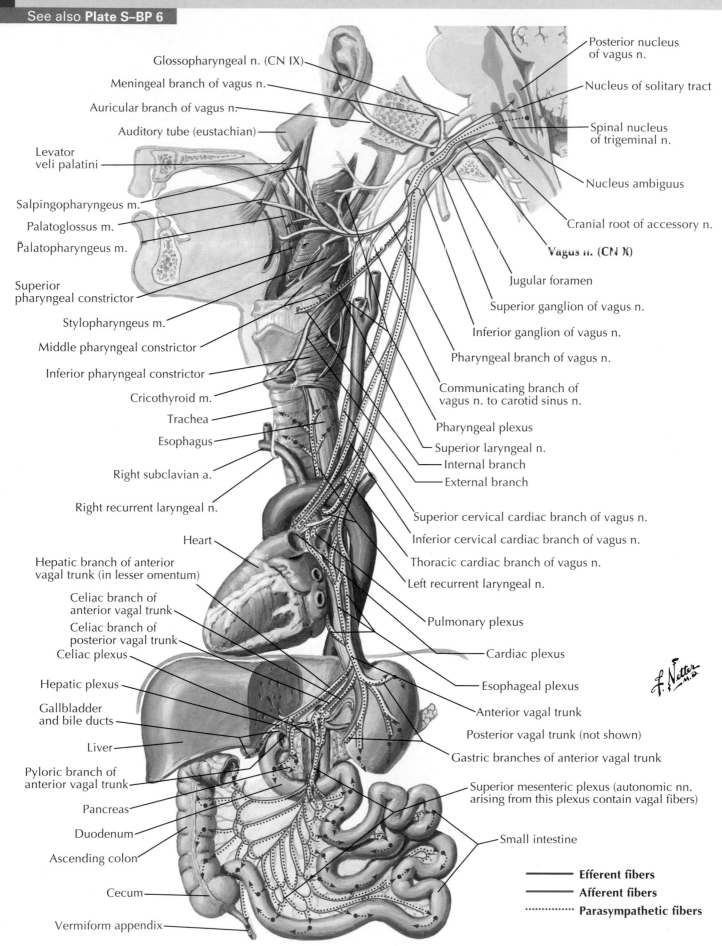

Glossopharyngeal n. (CN IX)

Meningeal branch of vagus n.

Auricular branch of vagus n.

Auditory tube (eustachian)

Levator veli palatini

Salpingopharyngeus m.

Palatoglossus m.

Palatopharyngeus m.

Superior pharyngeal constrictor

Stylopharyngeus m.

Middle pharyngeal constrictor

Inferior pharyngeal constrictor

Cricothyroid m.

Trachea

Esophagus

Right subclavian a.

Right recurrent laryngeal n.

Heart

Hepatic branch of anterior vagal trunk (in lesser omentum)

Celiac branch of anterior vagal trunk

Celiac branch of posterior vagal trunk

Celiac plexus

Hepatic plexus

Gallbladder and bile ducts

Liver

Pyloric branch of anterior vagal trunk

Pancreas

Duodenum

Ascending colon

Cecum

Vermiform appendix

Posterior nucleus of vagus n.

Nucleus of solitary tract

Spinal nucleus of trigeminal n.

Nucleus ambiguus

Cranial root of accessory n.

Vagus n. (CN X)

Jugular foramen

Superior ganglion of vagus n.

Inferior ganglion of vagus n.

Pharyngeal branch of vagus n.

Communicating branch of vagus n. to carotid sinus n.

Pharyngeal plexus

Superior laryngeal n.

Internal branch

External branch

Superior cervical cardiac branch of vagus n.

Inferior cervical cardiac branch of vagus n.

Thoracic cardiac branch of vagus n.

Left recurrent laryngeal n.

Pulmonary plexus

Cardiac plexus

Esophageal plexus

Anterior vagal trunk

Posterior vagal trunk (not shown)

Gastric branches of anterior vagal trunk

Superior mesenteric plexus (autonomic nn. arising from this plexus contain vagal fibers)

Small intestine

— **Efferent fibers**

— **Afferent fibers**

····· **Parasympathetic fibers**

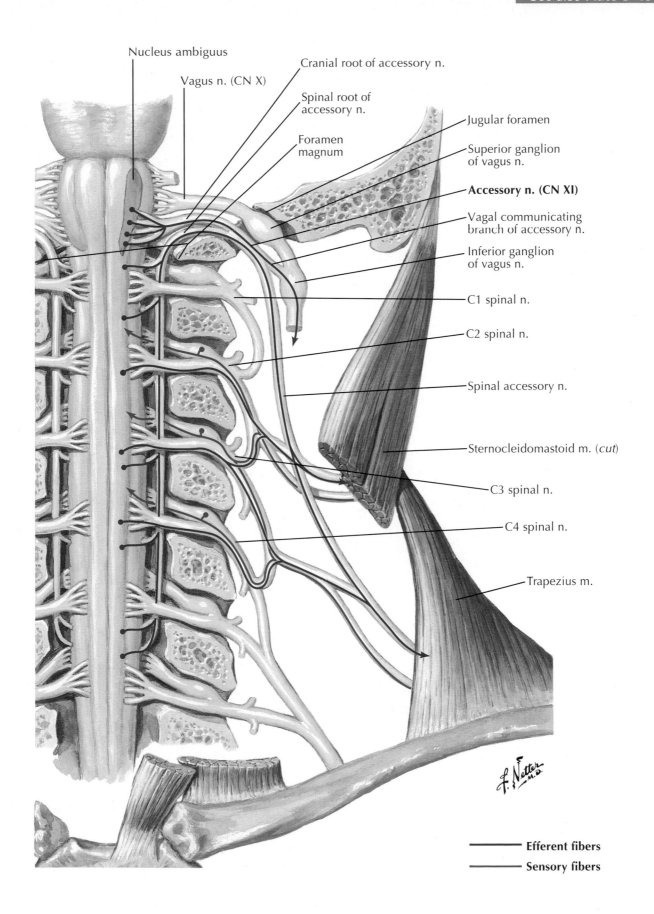

Nucleus ambiguus

Vagus n. (CN X)

Cranial root of accessory n.

Spinal root of accessory n.

Foramen magnum

Jugular foramen

Superior ganglion of vagus n.

Accessory n. (CN XI)

Vagal communicating branch of accessory n.

Inferior ganglion of vagus n.

C1 spinal n.

C2 spinal n.

Spinal accessory n.

Sternocleidomastoid m. (*cut*)

C3 spinal n.

C4 spinal n.

Trapezius m.

—— **Efferent fibers**

—— **Sensory fibers**

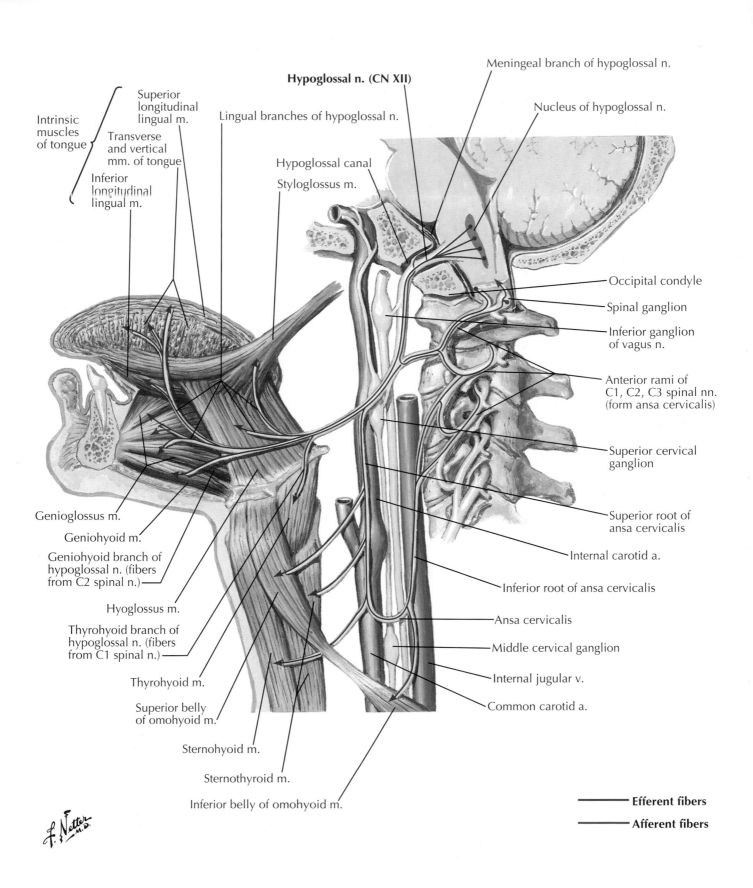

Intrinsic muscles of tongue
- Superior longitudinal lingual m.
- Transverse and vertical mm. of tongue
- Inferior longitudinal lingual m.

Lingual branches of hypoglossal n.

Hypoglossal n. (CN XII)

Hypoglossal canal

Styloglossus m.

Meningeal branch of hypoglossal n.

Nucleus of hypoglossal n.

Occipital condyle

Spinal ganglion

Inferior ganglion of vagus n.

Anterior rami of C1, C2, C3 spinal nn. (form ansa cervicalis)

Superior cervical ganglion

Superior root of ansa cervicalis

Internal carotid a.

Inferior root of ansa cervicalis

Ansa cervicalis

Middle cervical ganglion

Internal jugular v.

Common carotid a.

Genioglossus m.

Geniohyoid m.

Geniohyoid branch of hypoglossal n. (fibers from C2 spinal n.)

Hyoglossus m.

Thyrohyoid branch of hypoglossal n. (fibers from C1 spinal n.)

Thyrohyoid m.

Superior belly of omohyoid m.

Sternohyoid m.

Sternothyroid m.

Inferior belly of omohyoid m.

Efferent fibers

Afferent fibers

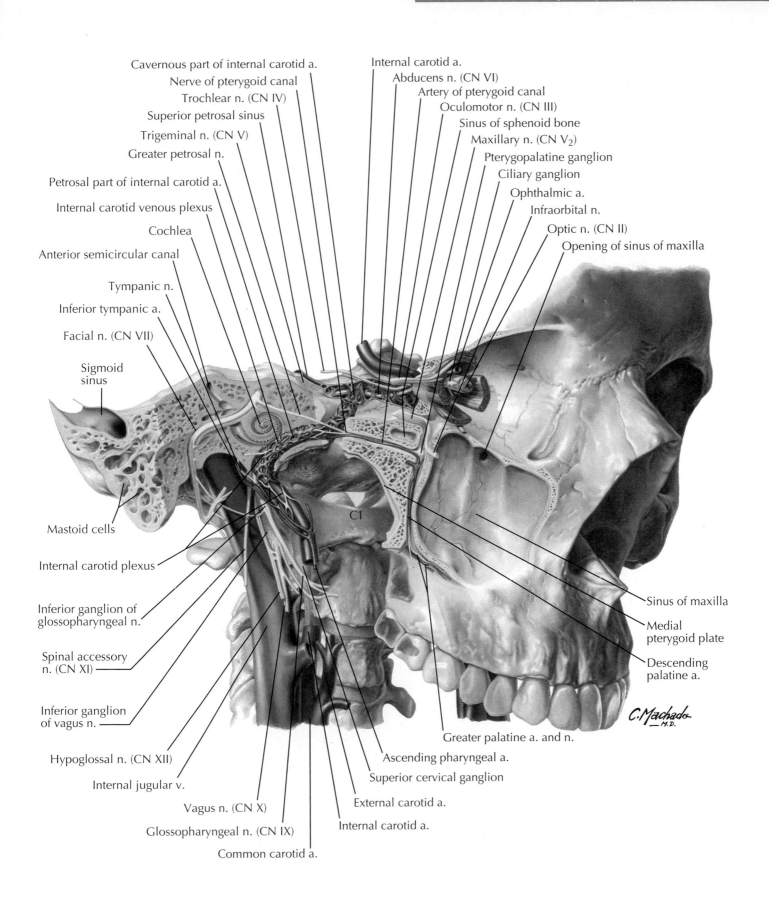

Cavernous part of internal carotid a.
Nerve of pterygoid canal
Trochlear n. (CN IV)
Superior petrosal sinus
Trigeminal n. (CN V)
Greater petrosal n.
Petrosal part of internal carotid a.
Internal carotid venous plexus
Cochlea
Anterior semicircular canal
Tympanic n.
Inferior tympanic a.
Facial n. (CN VII)
Sigmoid sinus
Mastoid cells
Internal carotid plexus
Inferior ganglion of glossopharyngeal n.
Spinal accessory n. (CN XI)
Inferior ganglion of vagus n.
Hypoglossal n. (CN XII)
Internal jugular v.
Vagus n. (CN X)
Glossopharyngeal n. (CN IX)
Common carotid a.

Internal carotid a.
Abducens n. (CN VI)
Artery of pterygoid canal
Oculomotor n. (CN III)
Sinus of sphenoid bone
Maxillary n. (CN V$_2$)
Pterygopalatine ganglion
Ciliary ganglion
Ophthalmic a.
Infraorbital n.
Optic n. (CN II)
Opening of sinus of maxilla

C1

Sinus of maxilla
Medial pterygoid plate
Descending palatine a.

Greater palatine a. and n.
Ascending pharyngeal a.
Superior cervical ganglion
External carotid a.
Internal carotid a.

C. Machado
M.D.

Cranial and Cervical Nerves

Plate S–71

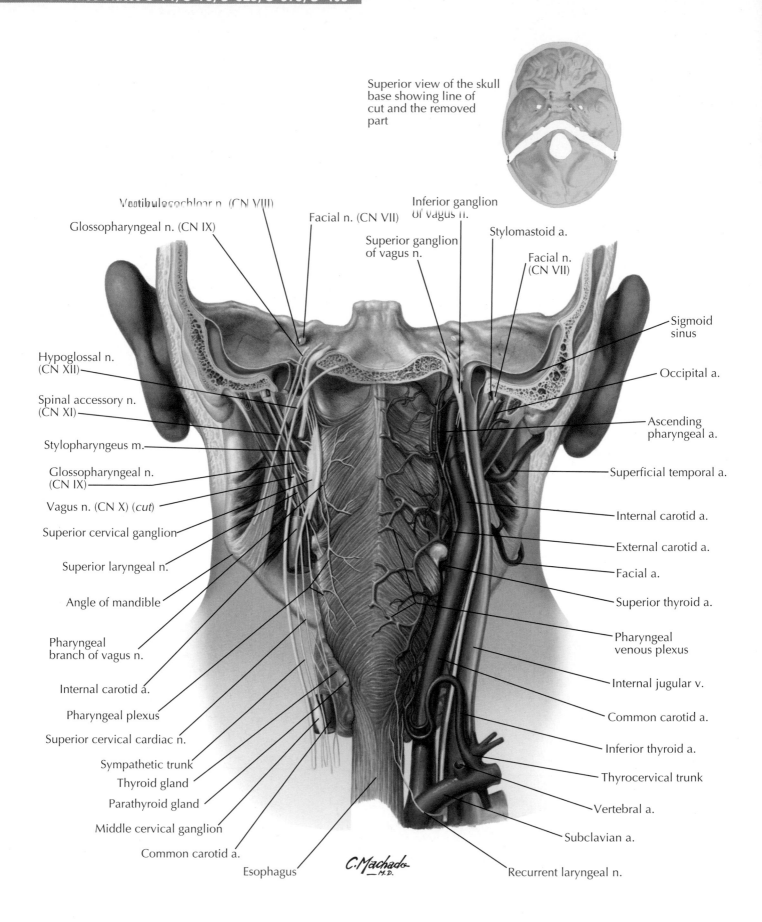

Superior view of the skull base showing line of cut and the removed part

Vestibulocochlear n. (CN VIII)

Glossopharyngeal n. (CN IX)

Facial n. (CN VII)

Inferior ganglion of vagus n.

Superior ganglion of vagus n.

Stylomastoid a.

Facial n. (CN VII)

Sigmoid sinus

Hypoglossal n. (CN XII)

Spinal accessory n. (CN XI)

Occipital a.

Ascending pharyngeal a.

Stylopharyngeus m.

Glossopharyngeal n. (CN IX)

Superficial temporal a.

Vagus n. (CN X) (cut)

Internal carotid a.

Superior cervical ganglion

External carotid a.

Superior laryngeal n.

Facial a.

Angle of mandible

Superior thyroid a.

Pharyngeal branch of vagus n.

Pharyngeal venous plexus

Internal carotid a.

Internal jugular v.

Pharyngeal plexus

Common carotid a.

Superior cervical cardiac n.

Inferior thyroid a.

Sympathetic trunk

Thyroid gland

Thyrocervical trunk

Parathyroid gland

Vertebral a.

Middle cervical ganglion

Subclavian a.

Common carotid a.

Esophagus

Recurrent laryngeal n.

C. Machado
M.D.

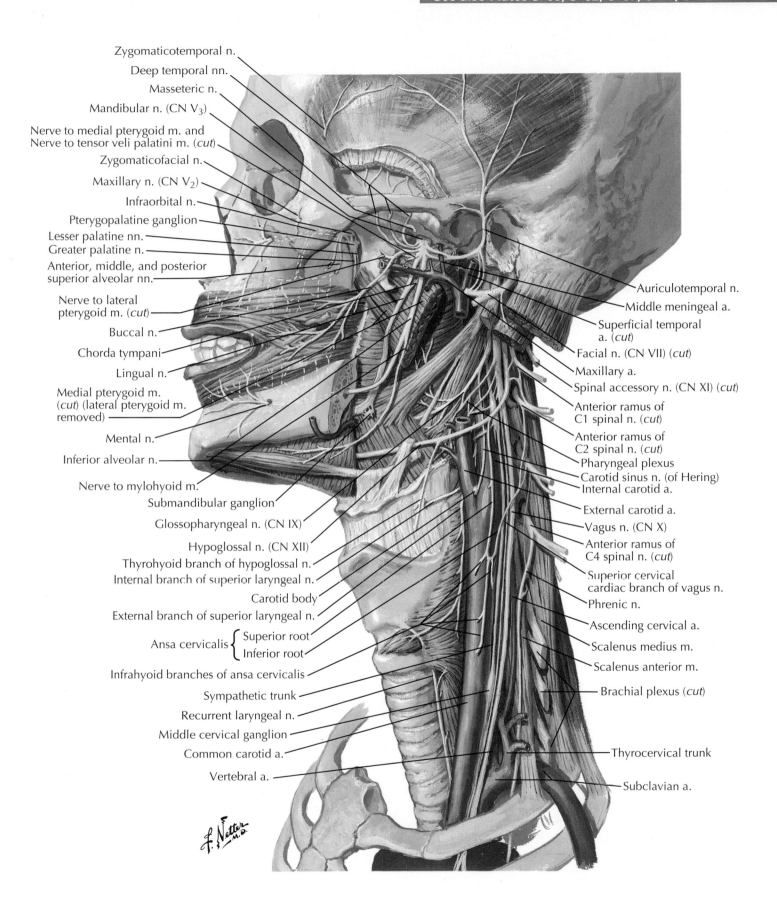

Zygomaticotemporal n.

Deep temporal nn.

Masseteric n.

Mandibular n. (CN V₃)

Nerve to medial pterygoid m. and
Nerve to tensor veli palatini m. (cut)

Zygomaticofacial n.

Maxillary n. (CN V₂)

Infraorbital n.

Pterygopalatine ganglion

Lesser palatine nn.

Greater palatine n.

Anterior, middle, and posterior
superior alveolar nn.

Nerve to lateral
pterygoid m. (cut)

Buccal n.

Chorda tympani

Lingual n.

Medial pterygoid m.
(cut) (lateral pterygoid m.
removed)

Mental n.

Inferior alveolar n.

Nerve to mylohyoid m.

Submandibular ganglion

Glossopharyngeal n. (CN IX)

Hypoglossal n. (CN XII)

Thyrohyoid branch of hypoglossal n.

Internal branch of superior laryngeal n.

Carotid body

External branch of superior laryngeal n.

Ansa cervicalis { Superior root
                  Inferior root

Infrahyoid branches of ansa cervicalis

Sympathetic trunk

Recurrent laryngeal n.

Middle cervical ganglion

Common carotid a.

Vertebral a.

Auriculotemporal n.

Middle meningeal a.

Superficial temporal
a. (cut)

Facial n. (CN VII) (cut)

Maxillary a.

Spinal accessory n. (CN XI) (cut)

Anterior ramus of
C1 spinal n. (cut)

Anterior ramus of
C2 spinal n. (cut)

Pharyngeal plexus

Carotid sinus n. (of Hering)

Internal carotid a.

External carotid a.

Vagus n. (CN X)

Anterior ramus of
C4 spinal n. (cut)

Superior cervical
cardiac branch of vagus n.

Phrenic n.

Ascending cervical a.

Scalenus medius m.

Scalenus anterior m.

Brachial plexus (cut)

Thyrocervical trunk

Subclavian a.

f. Netter
M.D.

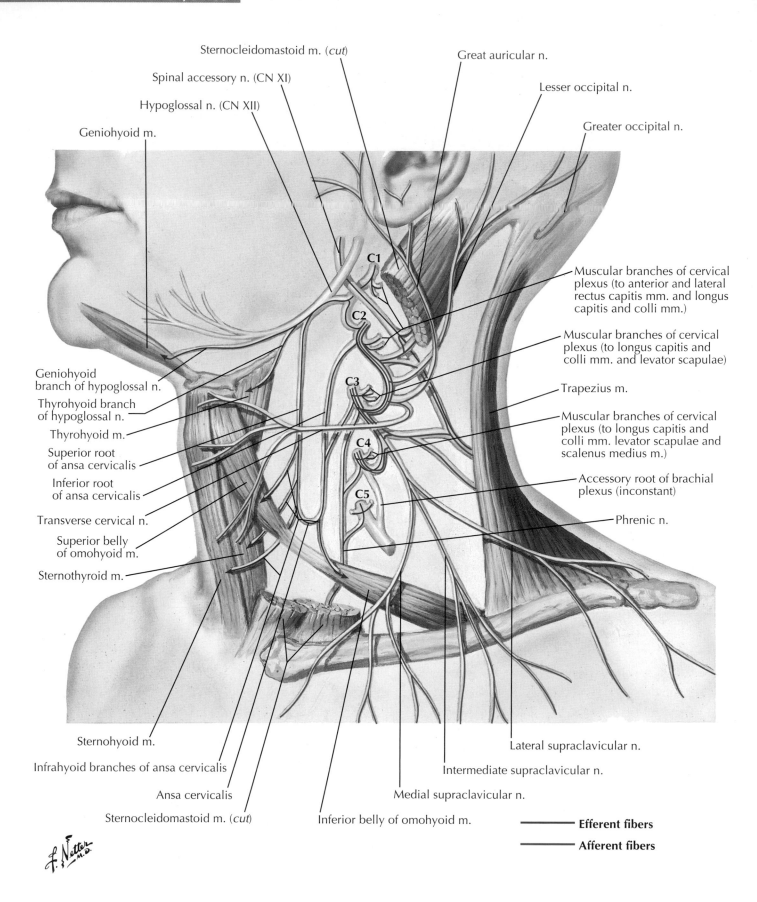

Sternocleidomastoid m. (*cut*)

Spinal accessory n. (CN XI)

Hypoglossal n. (CN XII)

Geniohyoid m.

Great auricular n.

Lesser occipital n.

Greater occipital n.

C1

C2

C3

C4

C5

Muscular branches of cervical plexus (to anterior and lateral rectus capitis mm. and longus capitis and colli mm.)

Muscular branches of cervical plexus (to longus capitis and colli mm. and levator scapulae)

Trapezius m.

Muscular branches of cervical plexus (to longus capitis and colli mm. levator scapulae and scalenus medius m.)

Accessory root of brachial plexus (inconstant)

Phrenic n.

Geniohyoid branch of hypoglossal n.

Thyrohyoid branch of hypoglossal n.

Thyrohyoid m.

Superior root of ansa cervicalis

Inferior root of ansa cervicalis

Transverse cervical n.

Superior belly of omohyoid m.

Sternothyroid m.

Sternohyoid m.

Infrahyoid branches of ansa cervicalis

Ansa cervicalis

Sternocleidomastoid m. (*cut*)

Inferior belly of omohyoid m.

Medial supraclavicular n.

Intermediate supraclavicular n.

Lateral supraclavicular n.

Efferent fibers

Afferent fibers

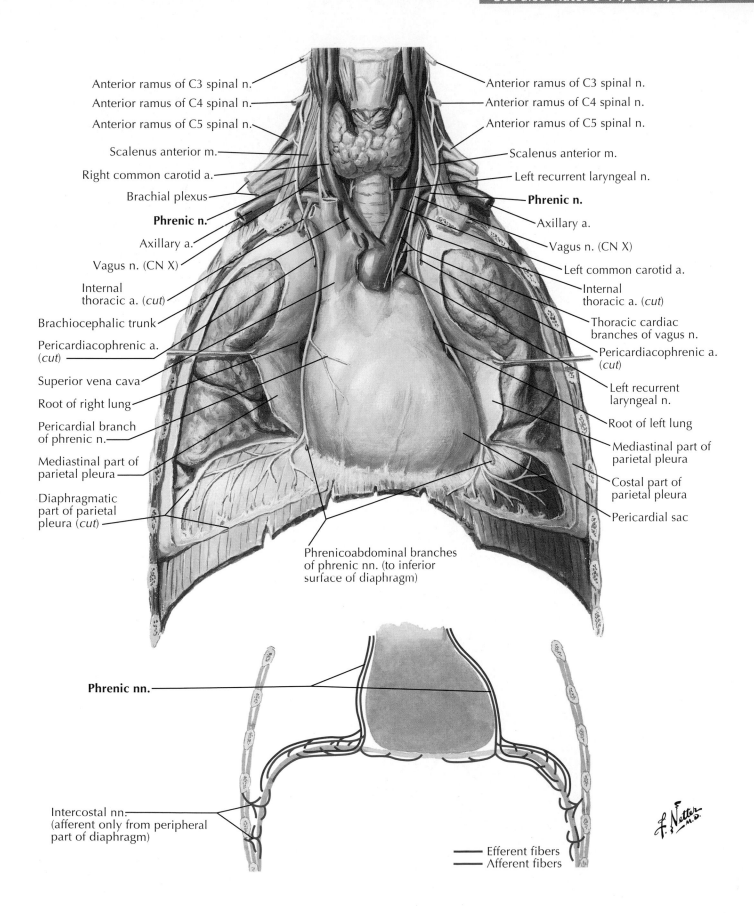

Anterior ramus of C3 spinal n.

Anterior ramus of C4 spinal n.

Anterior ramus of C5 spinal n.

Scalenus anterior m.

Right common carotid a.

Brachial plexus

Phrenic n.

Axillary a.

Vagus n. (CN X)

Internal
thoracic a. (*cut*)

Brachiocephalic trunk

Pericardiacophrenic a.
(*cut*)

Superior vena cava

Root of right lung

Pericardial branch
of phrenic n.

Mediastinal part of
parietal pleura

Diaphragmatic
part of parietal
pleura (*cut*)

Anterior ramus of C3 spinal n.

Anterior ramus of C4 spinal n.

Anterior ramus of C5 spinal n.

Scalenus anterior m.

Left recurrent laryngeal n.

Phrenic n.

Axillary a.

Vagus n. (CN X)

Left common carotid a.

Internal
thoracic a. (*cut*)

Thoracic cardiac
branches of vagus n.

Pericardiacophrenic a.
(*cut*)

Left recurrent
laryngeal n.

Root of left lung

Mediastinal part of
parietal pleura

Costal part of
parietal pleura

Pericardial sac

Phrenicoabdominal branches
of phrenic nn. (to inferior
surface of diaphragm)

Phrenic nn.

Intercostal nn.
(afferent only from peripheral
part of diaphragm)

Efferent fibers
Afferent fibers

f. Netter
M.D.

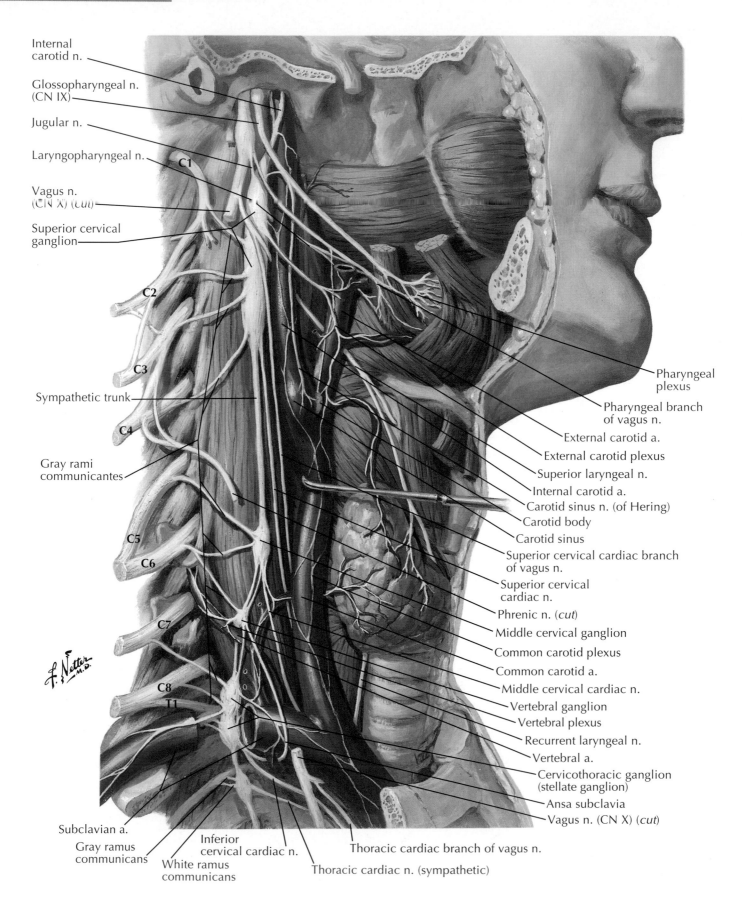

Internal carotid n.

Glossopharyngeal n. (CN IX)

Jugular n.

Laryngopharyngeal n.

C1

Vagus n. (CN X) (cut)

Superior cervical ganglion

C2

C3

Sympathetic trunk

C4

Gray rami communicantes

C5

C6

C7

C8

T1

Subclavian a.

Gray ramus communicans

White ramus communicans

Inferior cervical cardiac n.

Pharyngeal plexus

Pharyngeal branch of vagus n.

External carotid a.

External carotid plexus

Superior laryngeal n.

Internal carotid a.

Carotid sinus n. (of Hering)

Carotid body

Carotid sinus

Superior cervical cardiac branch of vagus n.

Superior cervical cardiac n.

Phrenic n. (cut)

Middle cervical ganglion

Common carotid plexus

Common carotid a.

Middle cervical cardiac n.

Vertebral ganglion

Vertebral plexus

Recurrent laryngeal n.

Vertebral a.

Cervicothoracic ganglion (stellate ganglion)

Ansa subclavia

Vagus n. (CN X) (cut)

Thoracic cardiac branch of vagus n.

Thoracic cardiac n. (sympathetic)

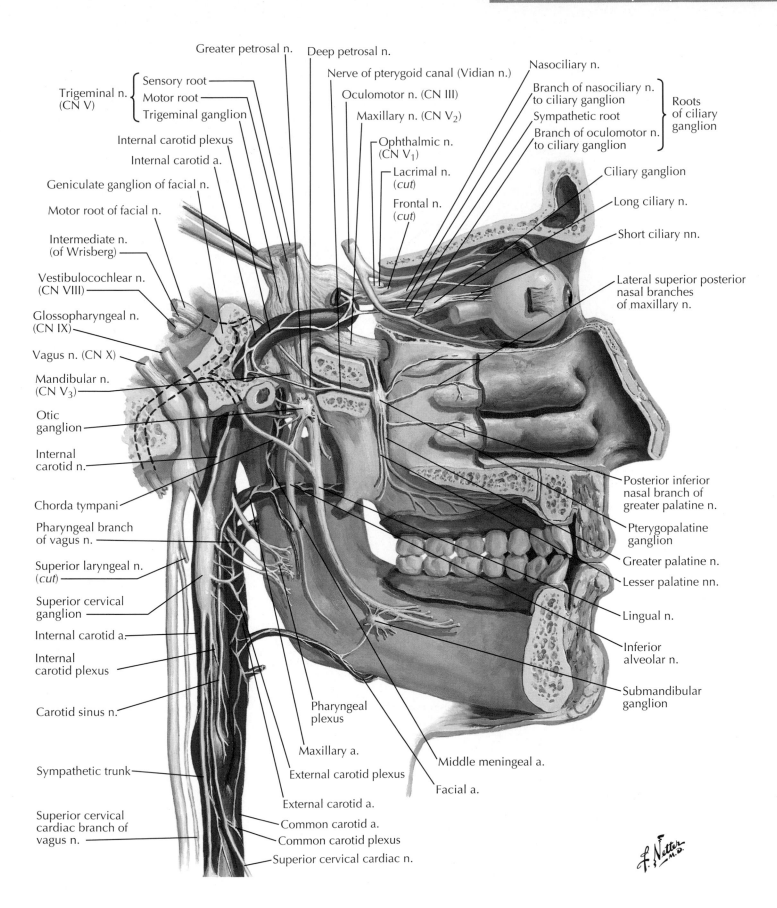

Greater petrosal n.

Deep petrosal n.

Nerve of pterygoid canal (Vidian n.)

Nasociliary n.

Trigeminal n. (CN V)
- Sensory root
- Motor root
- Trigeminal ganglion

Oculomotor n. (CN III)

Maxillary n. (CN V₂)

Branch of nasociliary n. to ciliary ganglion
Sympathetic root
Branch of oculomotor n. to ciliary ganglion

Roots of ciliary ganglion

Internal carotid plexus

Ophthalmic n. (CN V₁)

Internal carotid a.

Lacrimal n. (*cut*)

Ciliary ganglion

Geniculate ganglion of facial n.

Frontal n. (*cut*)

Long ciliary n.

Motor root of facial n.

Short ciliary nn.

Intermediate n. (of Wrisberg)

Vestibulocochlear n. (CN VIII)

Lateral superior posterior nasal branches of maxillary n.

Glossopharyngeal n. (CN IX)

Vagus n. (CN X)

Mandibular n. (CN V₃)

Otic ganglion

Internal carotid n.

Posterior inferior nasal branch of greater palatine n.

Chorda tympani

Pterygopalatine ganglion

Pharyngeal branch of vagus n.

Greater palatine n.

Superior laryngeal n. (*cut*)

Lesser palatine nn.

Superior cervical ganglion

Lingual n.

Internal carotid a.

Inferior alveolar n.

Internal carotid plexus

Submandibular ganglion

Carotid sinus n.

Pharyngeal plexus

Sympathetic trunk

Maxillary a.

Middle meningeal a.

External carotid plexus

Superior cervical cardiac branch of vagus n.

Facial a.

External carotid a.

Common carotid a.

Common carotid plexus

Superior cervical cardiac n.

f. Netter

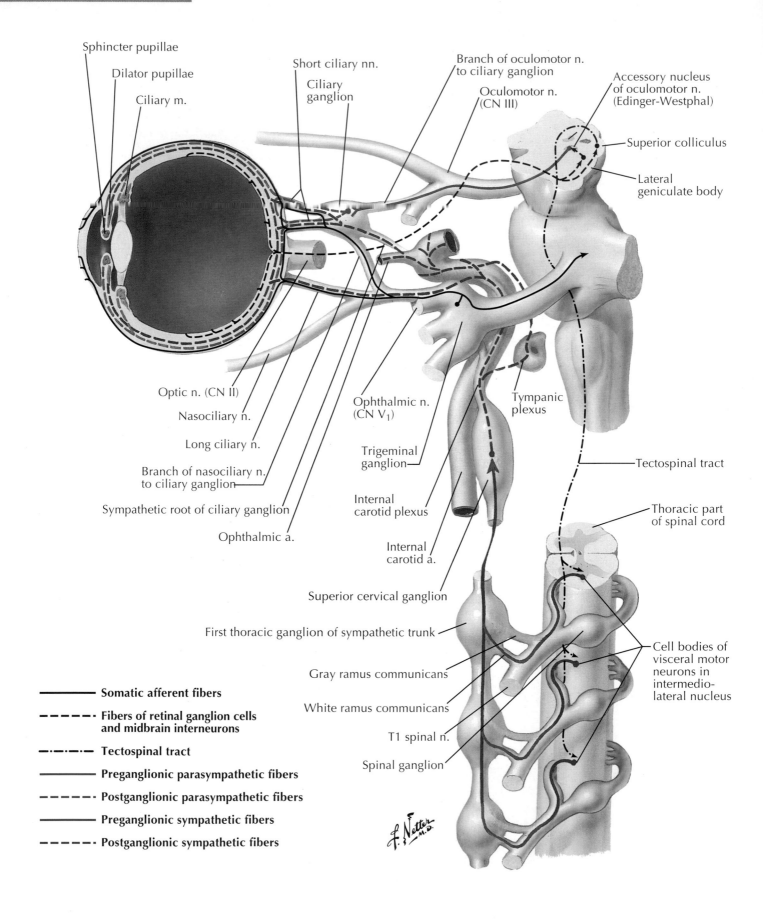

Sphincter pupillae

Dilator pupillae

Ciliary m.

Short ciliary nn.

Ciliary ganglion

Branch of oculomotor n. to ciliary ganglion

Oculomotor n. (CN III)

Accessory nucleus of oculomotor n. (Edinger-Westphal)

Superior colliculus

Lateral geniculate body

Optic n. (CN II)

Nasociliary n.

Long ciliary n.

Branch of nasociliary n. to ciliary ganglion

Sympathetic root of ciliary ganglion

Ophthalmic a.

Ophthalmic n. (CN V$_1$)

Trigeminal ganglion

Internal carotid plexus

Internal carotid a.

Tympanic plexus

Tectospinal tract

Thoracic part of spinal cord

Superior cervical ganglion

First thoracic ganglion of sympathetic trunk

Gray ramus communicans

White ramus communicans

T1 spinal n.

Spinal ganglion

Cell bodies of visceral motor neurons in intermedio-lateral nucleus

——————— Somatic afferent fibers

‐ ‐ ‐ ‐ ‐ Fibers of retinal ganglion cells and midbrain interneurons

‐·‐·‐·‐ Tectospinal tract

——————— Preganglionic parasympathetic fibers

‐ ‐ ‐ ‐ ‐ Postganglionic parasympathetic fibers

——————— Preganglionic sympathetic fibers

‐ ‐ ‐ ‐ ‐ Postganglionic sympathetic fibers

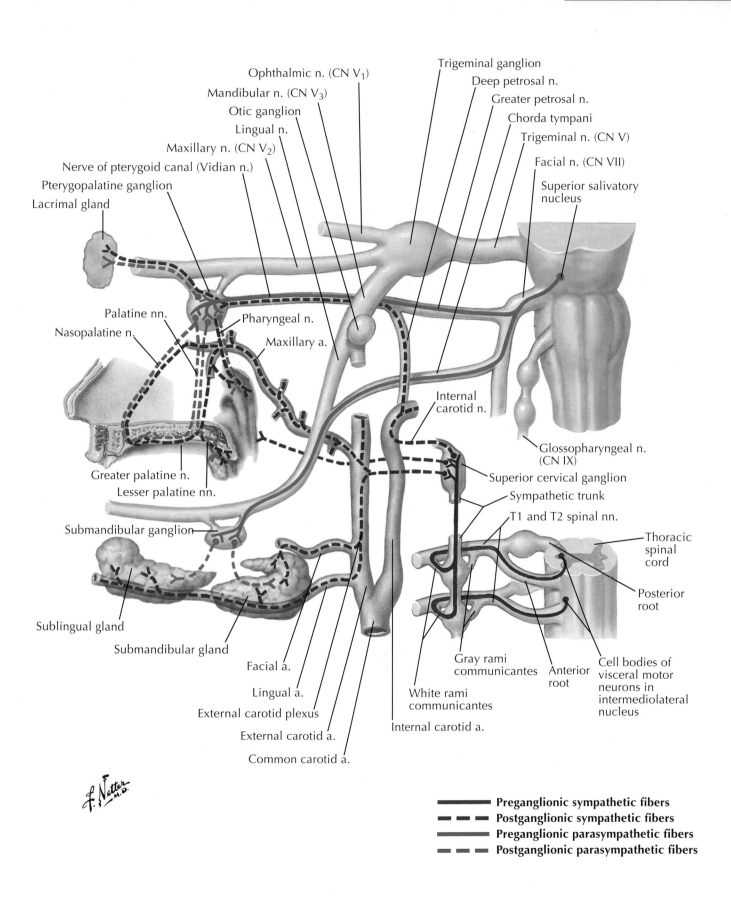

Ophthalmic n. (CN V_1)
Mandibular n. (CN V_3)
Otic ganglion
Lingual n.
Maxillary n. (CN V_2)
Nerve of pterygoid canal (Vidian n.)
Pterygopalatine ganglion
Lacrimal gland

Trigeminal ganglion
Deep petrosal n.
Greater petrosal n.
Chorda tympani
Trigeminal n. (CN V)
Facial n. (CN VII)
Superior salivatory nucleus

Palatine nn.
Nasopalatine n.
Pharyngeal n.
Maxillary a.
Internal carotid n.
Greater palatine n.
Lesser palatine nn.
Glossopharyngeal n. (CN IX)
Superior cervical ganglion
Sympathetic trunk
T1 and T2 spinal nn.
Thoracic spinal cord
Posterior root
Submandibular ganglion
Sublingual gland
Submandibular gland
Facial a.
Lingual a.
External carotid plexus
External carotid a.
Common carotid a.
White rami communicantes
Internal carotid a.
Gray rami communicantes
Anterior root
Cell bodies of visceral motor neurons in intermediolateral nucleus

Preganglionic sympathetic fibers
Postganglionic sympathetic fibers
Preganglionic parasympathetic fibers
Postganglionic parasympathetic fibers

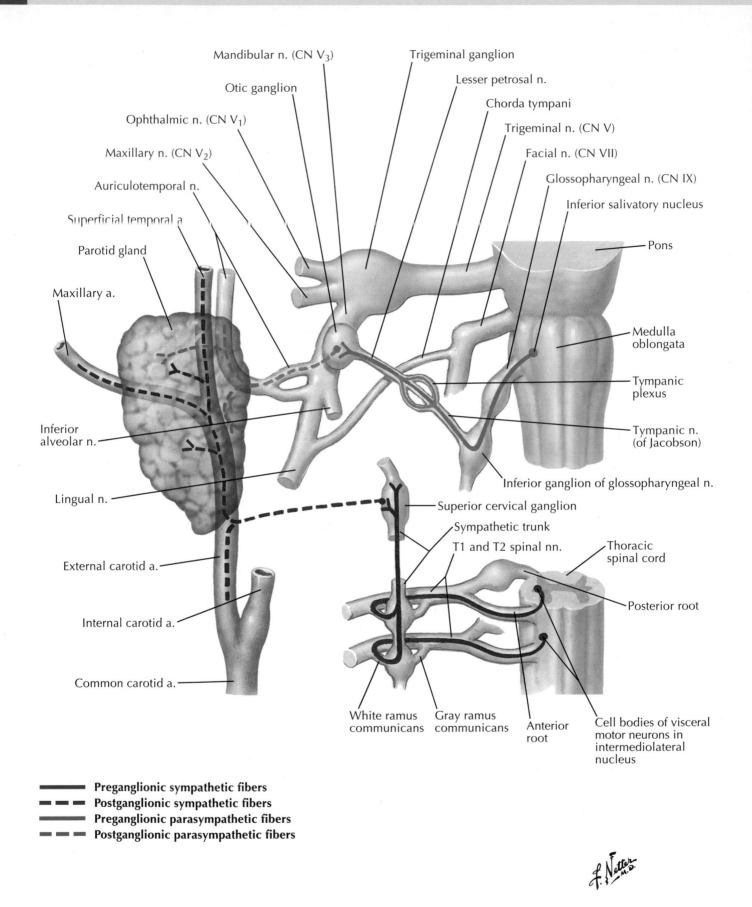

Mandibular n. (CN V₃)

Otic ganglion

Ophthalmic n. (CN V₁)

Maxillary n. (CN V₂)

Auriculotemporal n.

Superficial temporal a.

Parotid gland

Maxillary a.

Inferior alveolar n.

Lingual n.

External carotid a.

Internal carotid a.

Common carotid a.

Trigeminal ganglion

Lesser petrosal n.

Chorda tympani

Trigeminal n. (CN V)

Facial n. (CN VII)

Glossopharyngeal n. (CN IX)

Inferior salivatory nucleus

Pons

Medulla oblongata

Tympanic plexus

Tympanic n. (of Jacobson)

Inferior ganglion of glossopharyngeal n.

Superior cervical ganglion

Sympathetic trunk

T1 and T2 spinal nn.

Thoracic spinal cord

Posterior root

White ramus communicans

Gray ramus communicans

Anterior root

Cell bodies of visceral motor neurons in intermediolateral nucleus

━━━━━━ **Preganglionic sympathetic fibers**
┅ ┅ ┅ **Postganglionic sympathetic fibers**
══════ **Preganglionic parasympathetic fibers**
╌ ╌ ╌ **Postganglionic parasympathetic fibers**

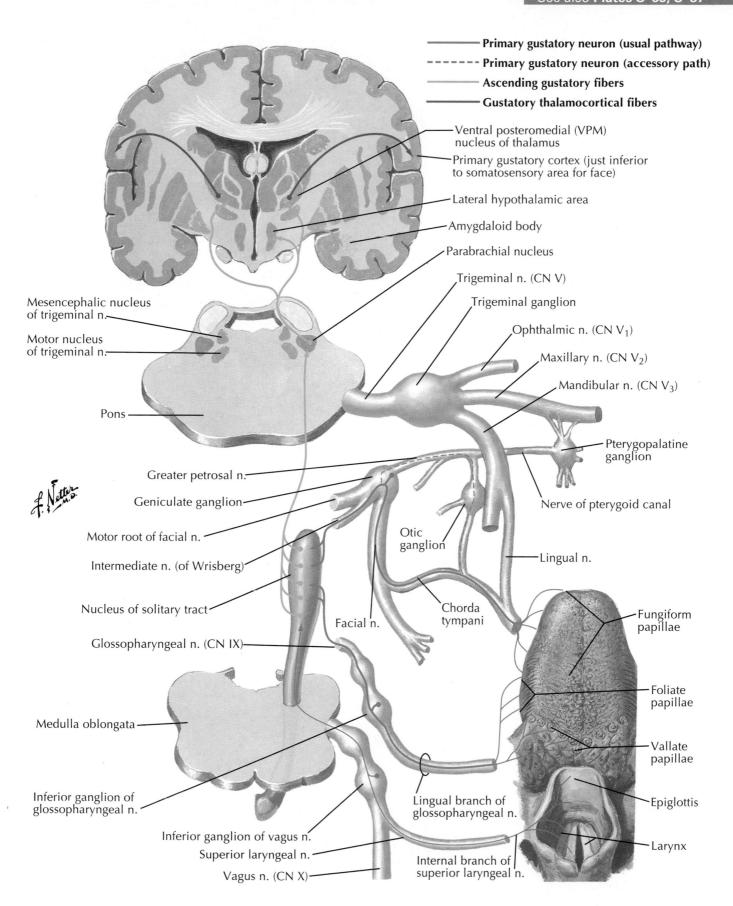

Primary gustatory neuron (usual pathway)
Primary gustatory neuron (accessory path)
Ascending gustatory fibers
Gustatory thalamocortical fibers

Ventral posteromedial (VPM) nucleus of thalamus
Primary gustatory cortex (just inferior to somatosensory area for face)
Lateral hypothalamic area
Amygdaloid body
Parabrachial nucleus
Trigeminal n. (CN V)
Trigeminal ganglion
Ophthalmic n. (CN V₁)
Maxillary n. (CN V₂)
Mandibular n. (CN V₃)
Pterygopalatine ganglion
Nerve of pterygoid canal

Mesencephalic nucleus of trigeminal n.
Motor nucleus of trigeminal n.
Pons

Greater petrosal n.
Geniculate ganglion
Motor root of facial n.
Intermediate n. (of Wrisberg)
Nucleus of solitary tract
Glossopharyngeal n. (CN IX)

Otic ganglion
Lingual n.
Facial n.
Chorda tympani
Fungiform papillae
Foliate papillae

Medulla oblongata

Vallate papillae

Inferior ganglion of glossopharyngeal n.
Lingual branch of glossopharyngeal n.
Epiglottis

Inferior ganglion of vagus n.
Superior laryngeal n.
Vagus n. (CN X)
Internal branch of superior laryngeal n.
Larynx

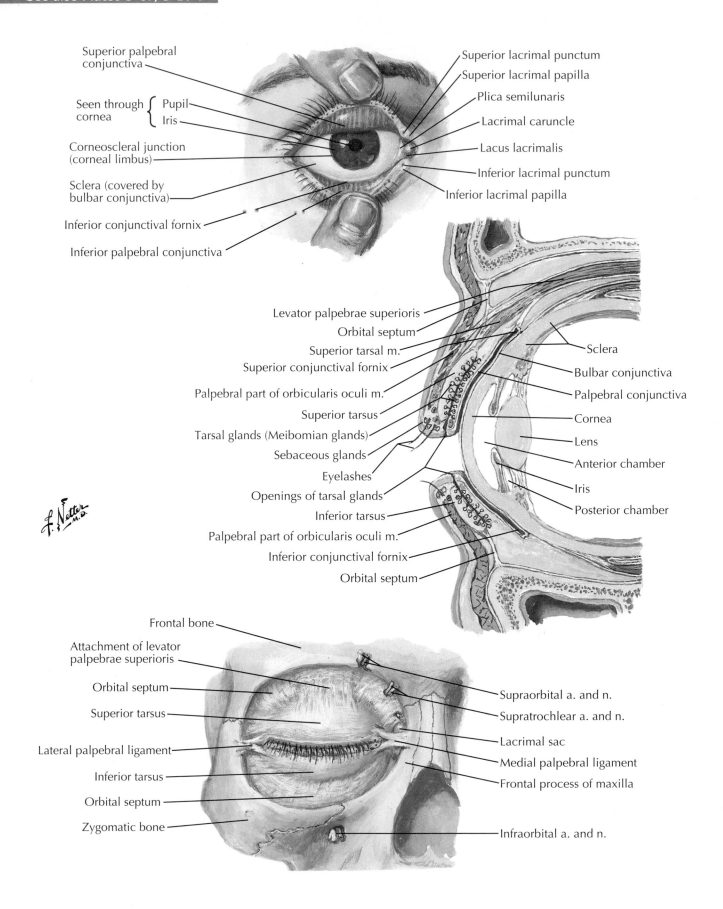

Superior palpebral conjunctiva

Seen through cornea { Pupil / Iris

Corneoscleral junction (corneal limbus)

Sclera (covered by bulbar conjunctiva)

Inferior conjunctival fornix

Inferior palpebral conjunctiva

Superior lacrimal punctum

Superior lacrimal papilla

Plica semilunaris

Lacrimal caruncle

Lacus lacrimalis

Inferior lacrimal punctum

Inferior lacrimal papilla

Levator palpebrae superioris

Orbital septum

Superior tarsal m.

Superior conjunctival fornix

Palpebral part of orbicularis oculi m.

Superior tarsus

Tarsal glands (Meibomian glands)

Sebaceous glands

Eyelashes

Openings of tarsal glands

Inferior tarsus

Palpebral part of orbicularis oculi m.

Inferior conjunctival fornix

Orbital septum

Sclera

Bulbar conjunctiva

Palpebral conjunctiva

Cornea

Lens

Anterior chamber

Iris

Posterior chamber

Frontal bone

Attachment of levator palpebrae superioris

Orbital septum

Superior tarsus

Lateral palpebral ligament

Inferior tarsus

Orbital septum

Zygomatic bone

Supraorbital a. and n.

Supratrochlear a. and n.

Lacrimal sac

Medial palpebral ligament

Frontal process of maxilla

Infraorbital a. and n.

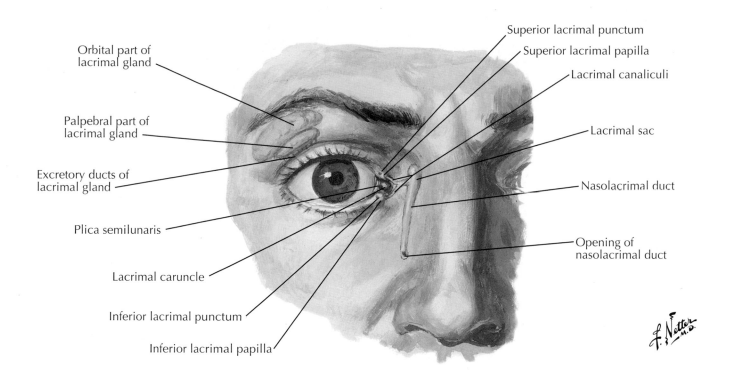

Orbital part of lacrimal gland

Palpebral part of lacrimal gland

Excretory ducts of lacrimal gland

Plica semilunaris

Lacrimal caruncle

Inferior lacrimal punctum

Inferior lacrimal papilla

Superior lacrimal punctum

Superior lacrimal papilla

Lacrimal canaliculi

Lacrimal sac

Nasolacrimal duct

Opening of nasolacrimal duct

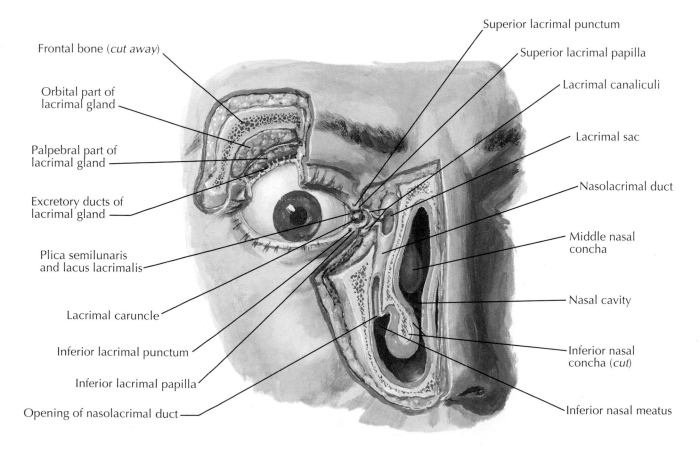

Frontal bone (*cut away*)

Orbital part of lacrimal gland

Palpebral part of lacrimal gland

Excretory ducts of lacrimal gland

Plica semilunaris and lacus lacrimalis

Lacrimal caruncle

Inferior lacrimal punctum

Inferior lacrimal papilla

Opening of nasolacrimal duct

Superior lacrimal punctum

Superior lacrimal papilla

Lacrimal canaliculi

Lacrimal sac

Nasolacrimal duct

Middle nasal concha

Nasal cavity

Inferior nasal concha (*cut*)

Inferior nasal meatus

Eye

Right lateral view

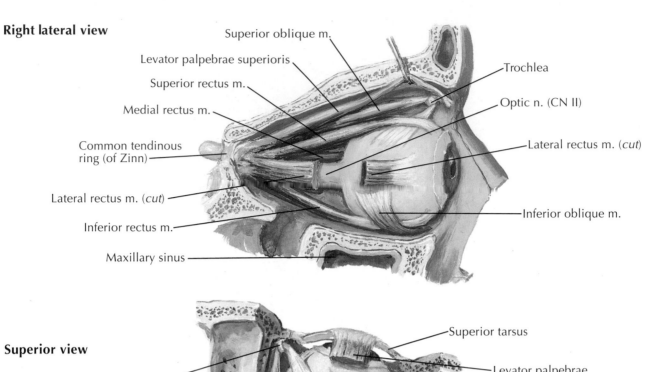

Superior oblique m.

Levator palpebrae superioris

Superior rectus m.

Medial rectus m.

Common tendinous ring (of Zinn)

Lateral rectus m. (*cut*)

Inferior rectus m.

Maxillary sinus

Trochlea

Optic n. (CN II)

Lateral rectus m. (*cut*)

Inferior oblique m.

Superior view

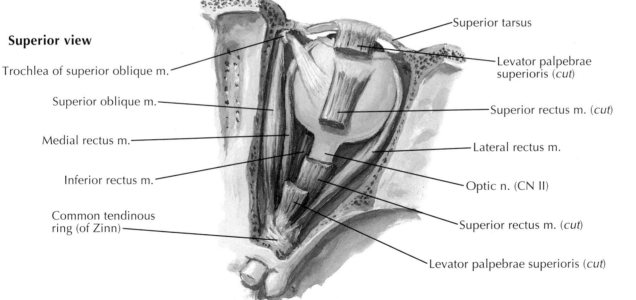

Trochlea of superior oblique m.

Superior oblique m.

Medial rectus m.

Inferior rectus m.

Common tendinous ring (of Zinn)

Superior tarsus

Levator palpebrae superioris (*cut*)

Superior rectus m. (*cut*)

Lateral rectus m.

Optic n. (CN II)

Superior rectus m. (*cut*)

Levator palpebrae superioris (*cut*)

Innervation of extrinsic eye muscles: anterior view

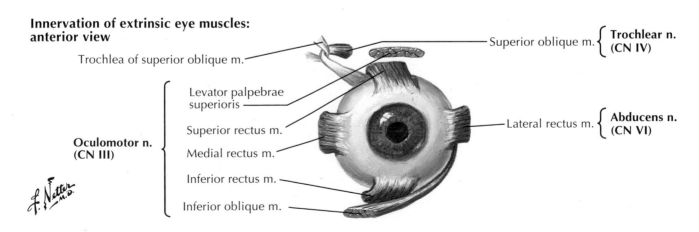

Trochlea of superior oblique m.

Levator palpebrae superioris

Superior rectus m.

Medial rectus m.

Inferior rectus m.

Inferior oblique m.

Oculomotor n. (CN III)

Superior oblique m. { **Trochlear n. (CN IV)**

Lateral rectus m. { **Abducens n. (CN VI)**

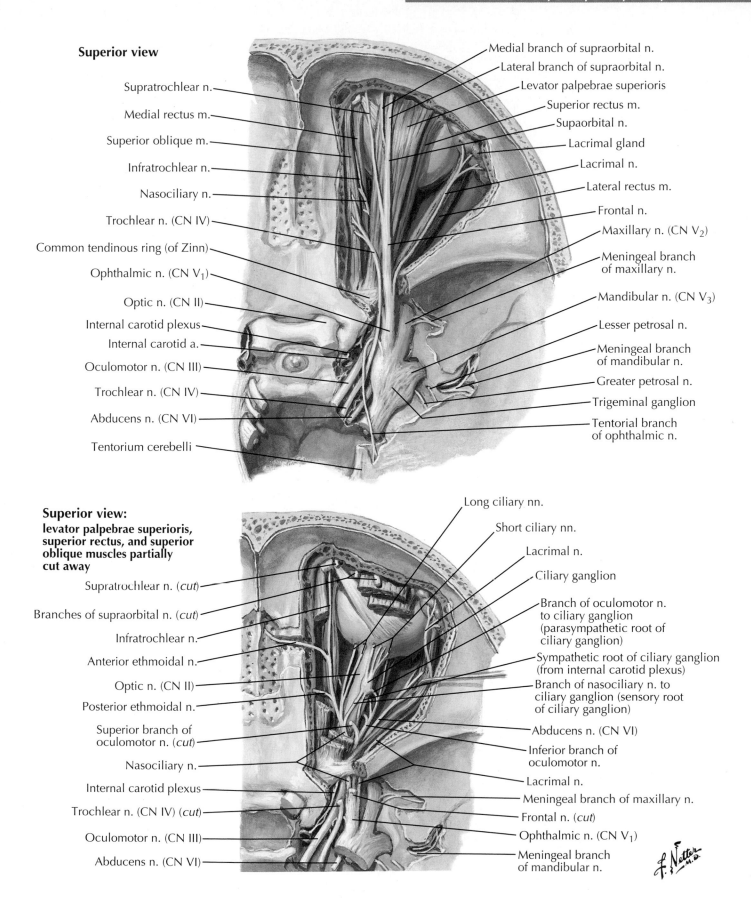

Superior view

Supratrochlear n.

Medial rectus m.

Superior oblique m.

Infratrochlear n.

Nasociliary n.

Trochlear n. (CN IV)

Common tendinous ring (of Zinn)

Ophthalmic n. (CN V₁)

Optic n. (CN II)

Internal carotid plexus

Internal carotid a.

Oculomotor n. (CN III)

Trochlear n. (CN IV)

Abducens n. (CN VI)

Tentorium cerebelli

Medial branch of supraorbital n.

Lateral branch of supraorbital n.

Levator palpebrae superioris

Superior rectus m.

Supaorbital n.

Lacrimal gland

Lacrimal n.

Lateral rectus m.

Frontal n.

Maxillary n. (CN V₂)

Meningeal branch of maxillary n.

Mandibular n. (CN V₃)

Lesser petrosal n.

Meningeal branch of mandibular n.

Greater petrosal n.

Trigeminal ganglion

Tentorial branch of ophthalmic n.

Superior view:
levator palpebrae superioris, superior rectus, and superior oblique muscles partially cut away

Supratrochlear n. (cut)

Branches of supraorbital n. (cut)

Infratrochlear n.

Anterior ethmoidal n.

Optic n. (CN II)

Posterior ethmoidal n.

Superior branch of oculomotor n. (cut)

Nasociliary n.

Internal carotid plexus

Trochlear n. (CN IV) (cut)

Oculomotor n. (CN III)

Abducens n. (CN VI)

Long ciliary nn.

Short ciliary nn.

Lacrimal n.

Ciliary ganglion

Branch of oculomotor n. to ciliary ganglion (parasympathetic root of ciliary ganglion)

Sympathetic root of ciliary ganglion (from internal carotid plexus)

Branch of nasociliary n. to ciliary ganglion (sensory root of ciliary ganglion)

Abducens n. (CN VI)

Inferior branch of oculomotor n.

Lacrimal n.

Meningeal branch of maxillary n.

Frontal n. (cut)

Ophthalmic n. (CN V₁)

Meningeal branch of mandibular n.

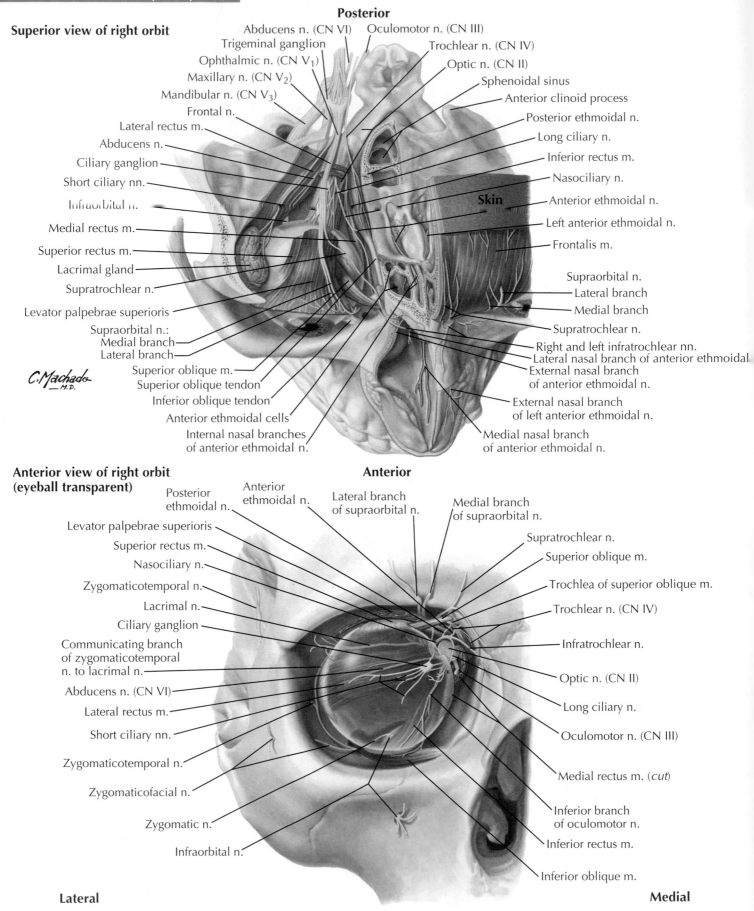

Superior view of right orbit

Posterior

Abducens n. (CN VI)
Trigeminal ganglion
Ophthalmic n. (CN V₁)
Maxillary n. (CN V₂)
Mandibular n. (CN V₃)
Frontal n.
Lateral rectus m.
Abducens n.
Ciliary ganglion
Short ciliary nn.
Infraorbital n.
Medial rectus m.
Superior rectus m.
Lacrimal gland
Supratrochlear n.
Levator palpebrae superioris
Supraorbital n.:
 Medial branch
 Lateral branch

C. Machado M.D.

Superior oblique m.
Superior oblique tendon
Inferior oblique tendon
Anterior ethmoidal cells
Internal nasal branches
of anterior ethmoidal n.

Oculomotor n. (CN III)
Trochlear n. (CN IV)
Optic n. (CN II)
Sphenoidal sinus
Anterior clinoid process
Posterior ethmoidal n.
Long ciliary n.
Inferior rectus m.
Nasociliary n.

Skin

Anterior ethmoidal n.
Left anterior ethmoidal n.
Frontalis m.

Supraorbital n.
Lateral branch
Medial branch
Supratrochlear n.
Right and left infratrochlear nn.
Lateral nasal branch of anterior ethmoidal
External nasal branch
of anterior ethmoidal n.
External nasal branch
of left anterior ethmoidal n.
Medial nasal branch
of anterior ethmoidal n.

Anterior view of right orbit
(eyeball transparent)

Anterior

Posterior
ethmoidal n.
Anterior
ethmoidal n.
Lateral branch
of supraorbital n.
Medial branch
of supraorbital n.

Levator palpebrae superioris
Superior rectus m.
Nasociliary n.
Zygomaticotemporal n.
Lacrimal n.
Ciliary ganglion
Communicating branch
of zygomaticotemporal
n. to lacrimal n.
Abducens n. (CN VI)
Lateral rectus m.
Short ciliary nn.
Zygomaticotemporal n.
Zygomaticofacial n.
Zygomatic n.
Infraorbital n.

Supratrochlear n.
Superior oblique m.
Trochlea of superior oblique m.
Trochlear n. (CN IV)
Infratrochlear n.
Optic n. (CN II)
Long ciliary n.
Oculomotor n. (CN III)
Medial rectus m. (*cut*)
Inferior branch
of oculomotor n.
Inferior rectus m.
Inferior oblique m.

Lateral

Medial

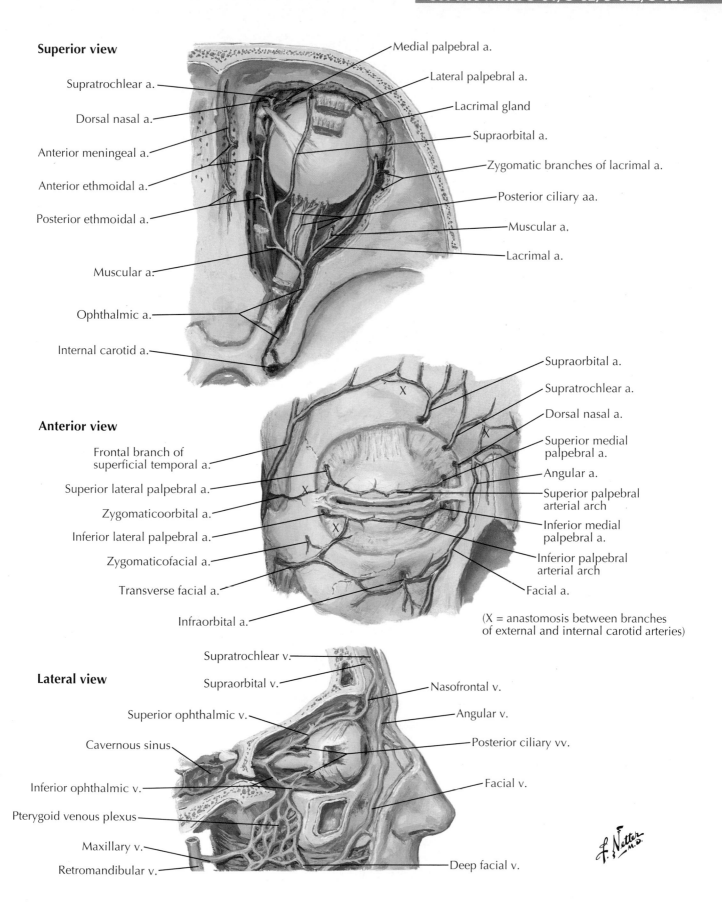

Superior view

Medial palpebral a.

Supratrochlear a.

Lateral palpebral a.

Dorsal nasal a.

Lacrimal gland

Anterior meningeal a.

Supraorbital a.

Anterior ethmoidal a.

Zygomatic branches of lacrimal a.

Posterior ethmoidal a.

Posterior ciliary aa.

Muscular a.

Lacrimal a.

Muscular a.

Ophthalmic a.

Internal carotid a.

Anterior view

Supraorbital a.

Supratrochlear a.

Dorsal nasal a.

Frontal branch of superficial temporal a.

Superior medial palpebral a.

Superior lateral palpebral a.

Angular a.

Zygomaticoorbital a.

Superior palpebral arterial arch

Inferior lateral palpebral a.

Inferior medial palpebral a.

Zygomaticofacial a.

Inferior palpebral arterial arch

Transverse facial a.

Facial a.

Infraorbital a.

(X = anastomosis between branches of external and internal carotid arteries)

Lateral view

Supratrochlear v.

Supraorbital v.

Nasofrontal v.

Superior ophthalmic v.

Angular v.

Cavernous sinus

Posterior ciliary vv.

Inferior ophthalmic v.

Facial v.

Pterygoid venous plexus

Maxillary v.

Retromandibular v.

Deep facial v.

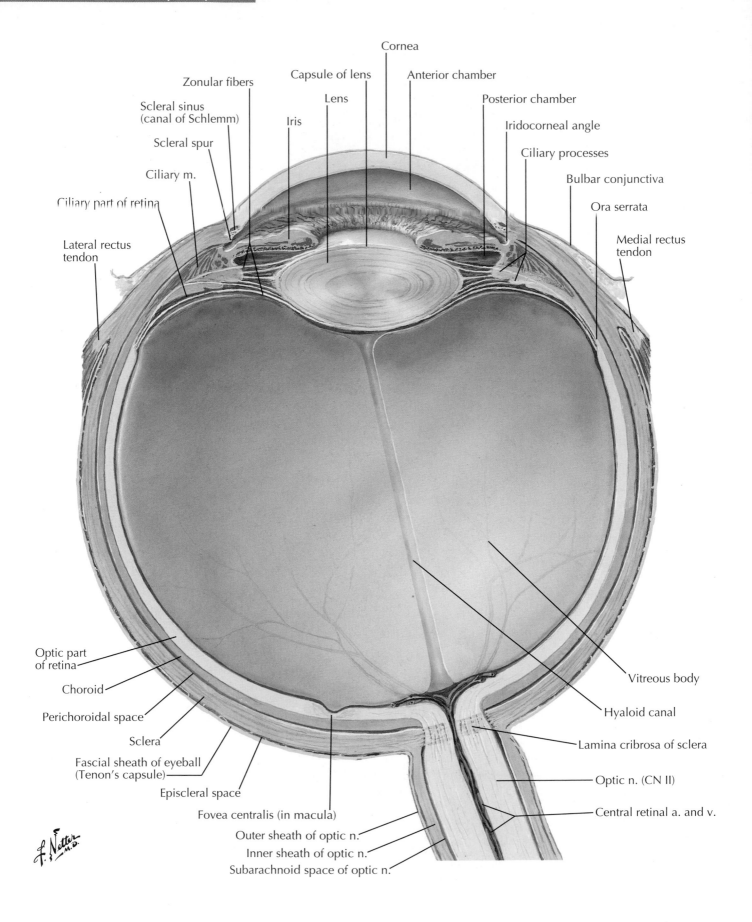

Cornea

Capsule of lens

Anterior chamber

Zonular fibers

Posterior chamber

Lens

Scleral sinus
(canal of Schlemm)

Iridocorneal angle

Scleral spur

Iris

Ciliary processes

Ciliary m.

Bulbar conjunctiva

Ciliary part of retina

Ora serrata

Lateral rectus
tendon

Medial rectus
tendon

Optic part
of retina

Vitreous body

Choroid

Hyaloid canal

Perichoroidal space

Sclera

Lamina cribrosa of sclera

Fascial sheath of eyeball
(Tenon's capsule)

Optic n. (CN II)

Episcleral space

Fovea centralis (in macula)

Central retinal a. and v.

Outer sheath of optic n.

Inner sheath of optic n.

Subarachnoid space of optic n.

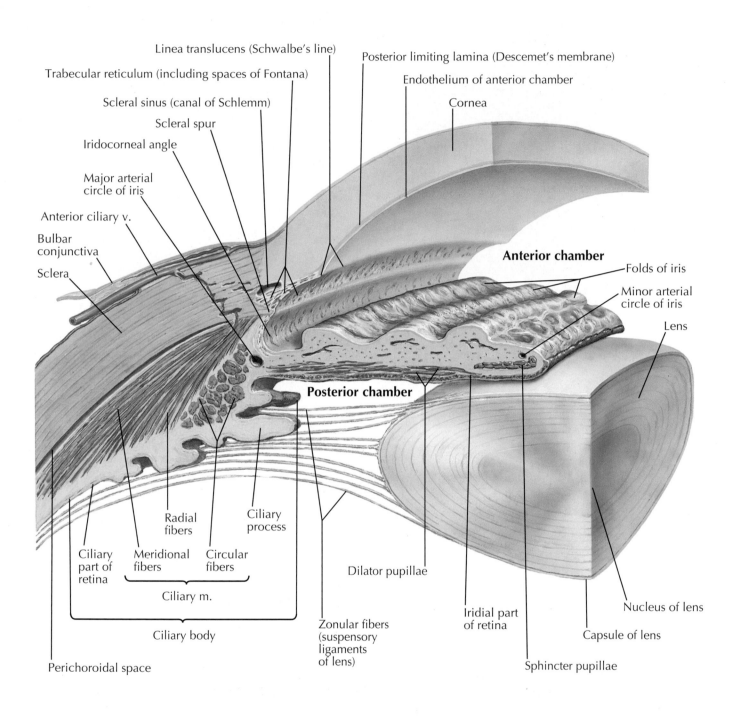

Linea translucens (Schwalbe's line)

Trabecular reticulum (including spaces of Fontana)

Scleral sinus (canal of Schlemm)

Scleral spur

Iridocorneal angle

Major arterial circle of iris

Anterior ciliary v.

Bulbar conjunctiva

Sclera

Posterior limiting lamina (Descemet's membrane)

Endothelium of anterior chamber

Cornea

Anterior chamber

Folds of iris

Minor arterial circle of iris

Lens

Posterior chamber

Radial fibers

Ciliary process

Ciliary part of retina

Meridional fibers

Circular fibers

Dilator pupillae

Iridial part of retina

Nucleus of lens

Capsule of lens

Ciliary m.

Ciliary body

Perichoroidal space

Zonular fibers (suspensory ligaments of lens)

Sphincter pupillae

Note: For clarity, only single plane of zonular fibers shown; actually, fibers surround entire circumference of lens.

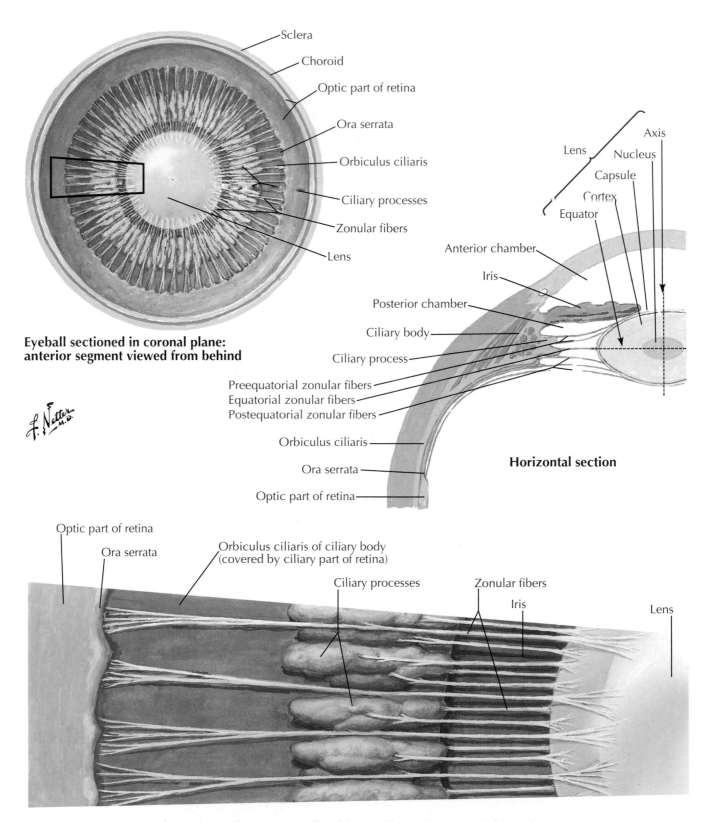

Sclera

Choroid

Optic part of retina

Ora serrata

Orbiculus ciliaris

Ciliary processes

Zonular fibers

Lens

**Eyeball sectioned in coronal plane:
anterior segment viewed from behind**

Lens

Axis

Nucleus

Capsule

Cortex

Equator

Anterior chamber

Iris

Posterior chamber

Ciliary body

Ciliary process

Preequatorial zonular fibers

Equatorial zonular fibers

Postequatorial zonular fibers

Orbiculus ciliaris

Ora serrata

Optic part of retina

Horizontal section

Optic part of retina

Ora serrata

Orbiculus ciliaris of ciliary body
(covered by ciliary part of retina)

Ciliary processes

Zonular fibers

Iris

Lens

Enlargement of segment outlined in top illustration (semischematic)

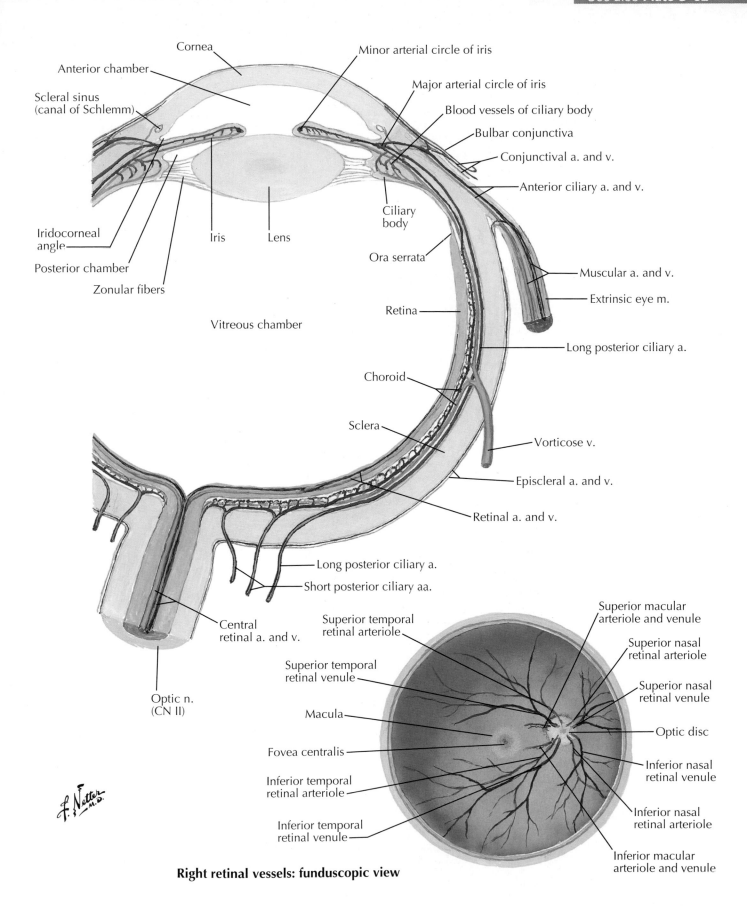

Cornea

Anterior chamber

Scleral sinus
(canal of Schlemm)

Minor arterial circle of iris

Major arterial circle of iris

Blood vessels of ciliary body

Bulbar conjunctiva

Conjunctival a. and v.

Anterior ciliary a. and v.

Iridocorneal angle

Posterior chamber

Zonular fibers

Iris

Lens

Ciliary body

Ora serrata

Muscular a. and v.

Extrinsic eye m.

Retina

Long posterior ciliary a.

Vitreous chamber

Choroid

Sclera

Vorticose v.

Episcleral a. and v.

Retinal a. and v.

Long posterior ciliary a.

Short posterior ciliary aa.

Central retinal a. and v.

Optic n. (CN II)

Superior temporal retinal arteriole

Superior temporal retinal venule

Macula

Fovea centralis

Inferior temporal retinal arteriole

Inferior temporal retinal venule

Superior macular arteriole and venule

Superior nasal retinal arteriole

Superior nasal retinal venule

Optic disc

Inferior nasal retinal venule

Inferior nasal retinal arteriole

Inferior macular arteriole and venule

Right retinal vessels: funduscopic view

Vascular arrangements within the vascular layer of eyeball

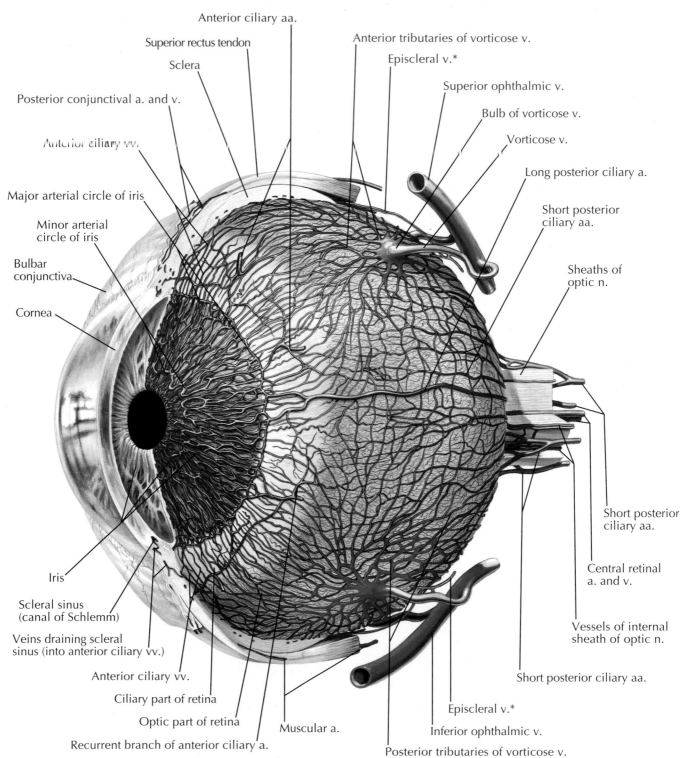

Anterior ciliary aa.

Superior rectus tendon

Sclera

Posterior conjunctival a. and v.

Anterior ciliary vv.

Major arterial circle of iris

Minor arterial circle of iris

Bulbar conjunctiva

Cornea

Iris

Scleral sinus (canal of Schlemm)

Veins draining scleral sinus (into anterior ciliary vv.)

Anterior ciliary vv.

Ciliary part of retina

Optic part of retina

Recurrent branch of anterior ciliary a.

Muscular a.

Anterior tributaries of vorticose v.

Episcleral v.*

Superior ophthalmic v.

Bulb of vorticose v.

Vorticose v.

Long posterior ciliary a.

Short posterior ciliary aa.

Sheaths of optic n.

Short posterior ciliary aa.

Central retinal a. and v.

Vessels of internal sheath of optic n.

Short posterior ciliary aa.

Episcleral v.*

Inferior ophthalmic v.

Posterior tributaries of vorticose v.

The episcleral veins are shown here anastomosing with the vorticose veins, which they do; however, they also drain into the anterior ciliary veins.

C. Machado
—M.D.

See also Plates S-94, S-96, S-BP 9, S-BP 10

Frontal section
Facial n. (CN VII) (cut)
Base of stapes
Vestibular window (oval window)
Vestibule
Limbs of stapes
Semicircular ducts
Prominence of lateral semicircular canal
Tegmen tympani
Incus
Facial n. (CN VII) (cut)
Arcuate eminence
Head of malleus
Vestibular n.
Cochlear n.
Epitympanic recess
Membranous ampullae
Internal acoustic meatus
Vestibulocochlear n. (CN VIII)
Auricle
Utricle
Saccule
External acoustic meatus
Tympanic membrane
Parotid gland
Tympanic cavity
Promontory of tympanic cavity
Internal jugular v.
Auditory tube (eustachian)
Cochlear window (round window)
Cochlear duct
Nasopharynx
Helicotrema
Scala vestibuli
Scala tympani
Cochlea
F. Netter M.D.

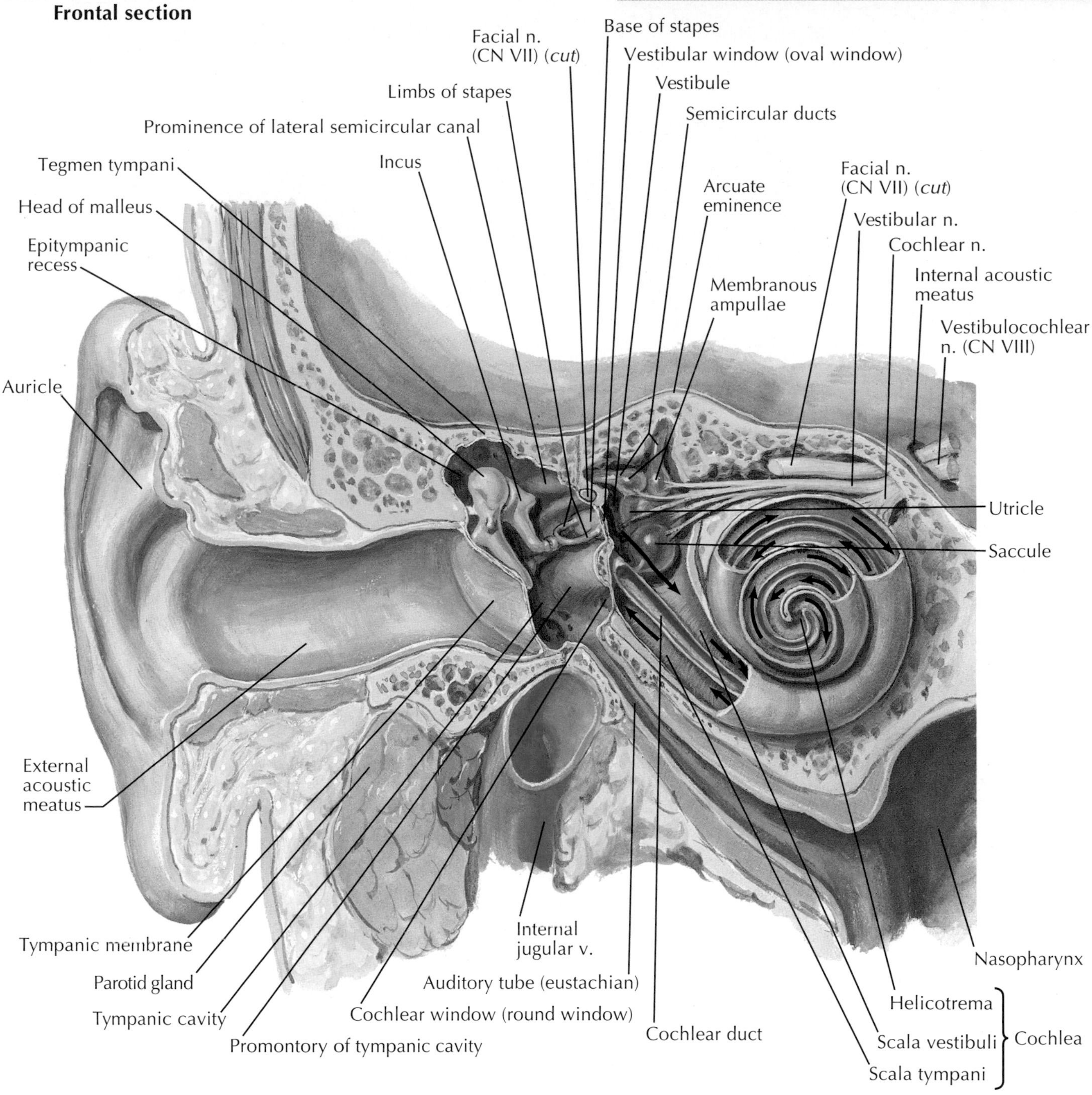

The auditory tube is shorter and more horizontal in children than in adults

Adult
Auditory tube

Child
Auditory tube

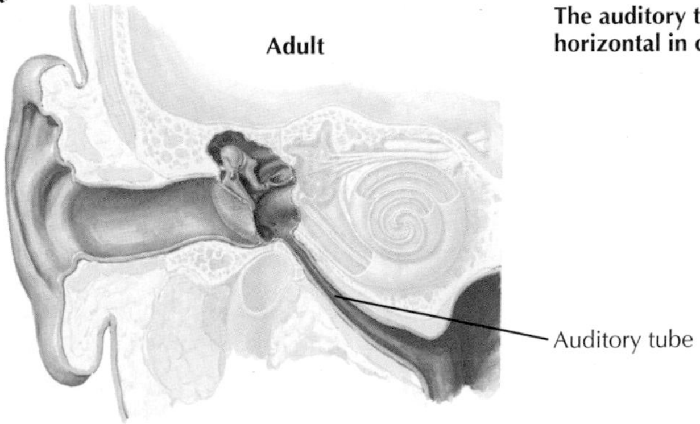

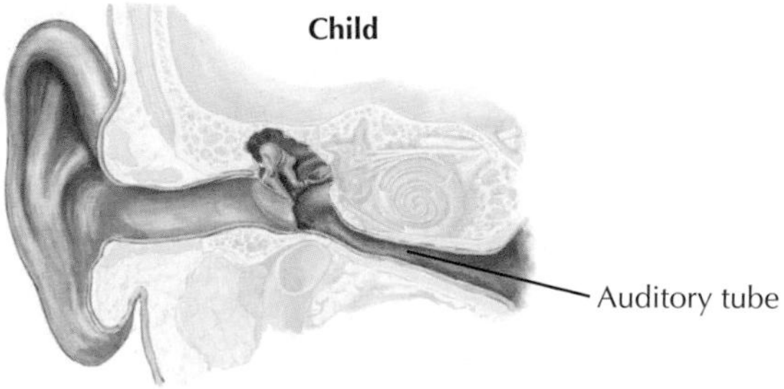

Right auricle

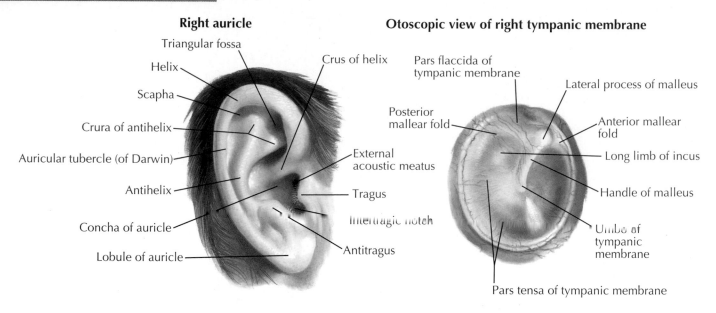

- Triangular fossa
- Helix
- Scapha
- Crura of antihelix
- Auricular tubercle (of Darwin)
- Antihelix
- Concha of auricle
- Lobule of auricle
- Crus of helix
- External acoustic meatus
- Tragus
- Intertragic notch
- Antitragus

Otoscopic view of right tympanic membrane

- Pars flaccida of tympanic membrane
- Posterior mallear fold
- Lateral process of malleus
- Anterior mallear fold
- Long limb of incus
- Handle of malleus
- Umbo of tympanic membrane
- Pars tensa of tympanic membrane

Coronal oblique section of external acoustic meatus and middle ear

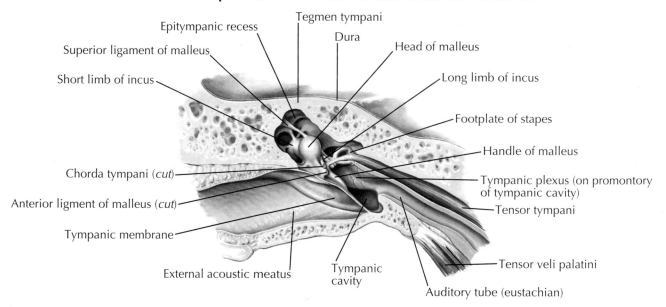

- Epitympanic recess
- Superior ligament of malleus
- Short limb of incus
- Tegmen tympani
- Dura
- Head of malleus
- Long limb of incus
- Footplate of stapes
- Handle of malleus
- Tympanic plexus (on promontory of tympanic cavity)
- Tensor tympani
- Chorda tympani (*cut*)
- Anterior ligment of malleus (*cut*)
- Tympanic membrane
- External acoustic meatus
- Tympanic cavity
- Auditory tube (eustachian)
- Tensor veli palatini

Right tympanic cavity after removal of tympanic membrane (lateral view)

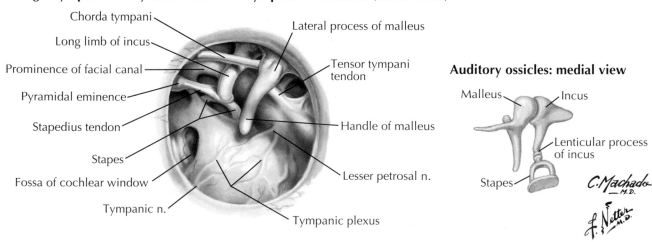

- Chorda tympani
- Long limb of incus
- Prominence of facial canal
- Pyramidal eminence
- Stapedius tendon
- Stapes
- Fossa of cochlear window
- Tympanic n.
- Lateral process of malleus
- Tensor tympani tendon
- Handle of malleus
- Lesser petrosal n.
- Tympanic plexus

Auditory ossicles: medial view

- Malleus
- Incus
- Lenticular process of incus
- Stapes

C. Machado, M.D.

F. Netter, M.D.

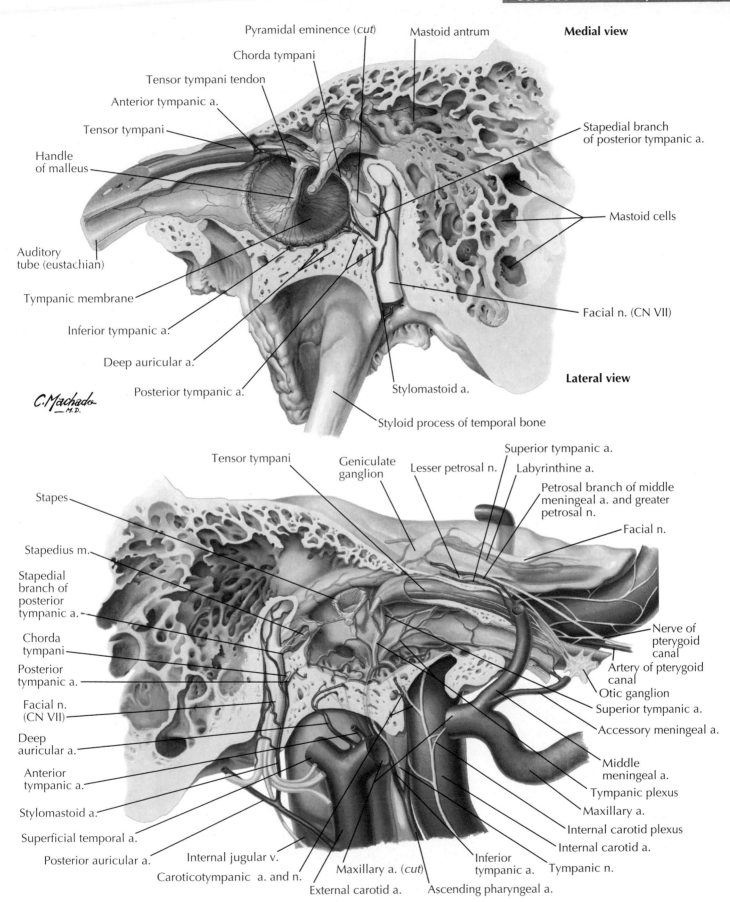

Pyramidal eminence (*cut*)

Chorda tympani

Tensor tympani tendon

Anterior tympanic a.

Tensor tympani

Handle of malleus

Auditory tube (eustachian)

Tympanic membrane

Inferior tympanic a.

Deep auricular a.

Posterior tympanic a.

C. Machado—M.D.

Mastoid antrum

Medial view

Stapedial branch of posterior tympanic a.

Mastoid cells

Facial n. (CN VII)

Lateral view

Stylomastoid a.

Styloid process of temporal bone

Tensor tympani

Geniculate ganglion

Lesser petrosal n.

Superior tympanic a.

Labyrinthine a.

Petrosal branch of middle meningeal a. and greater petrosal n.

Facial n.

Stapes

Stapedius m.

Stapedial branch of posterior tympanic a.

Chorda tympani

Posterior tympanic a.

Facial n. (CN VII)

Deep auricular a.

Anterior tympanic a.

Stylomastoid a.

Superficial temporal a.

Posterior auricular a.

Internal jugular v.

Caroticotympanic a. and n.

Maxillary a. (*cut*)

External carotid a.

Ascending pharyngeal a.

Inferior tympanic a.

Tympanic n.

Nerve of pterygoid canal

Artery of pterygoid canal

Otic ganglion

Superior tympanic a.

Accessory meningeal a.

Middle meningeal a.

Tympanic plexus

Maxillary a.

Internal carotid plexus

Internal carotid a.

Ear

Plate S–95

Right bony labyrinth (otic capsule), anterolateral view: surrounding cancellous bone removed

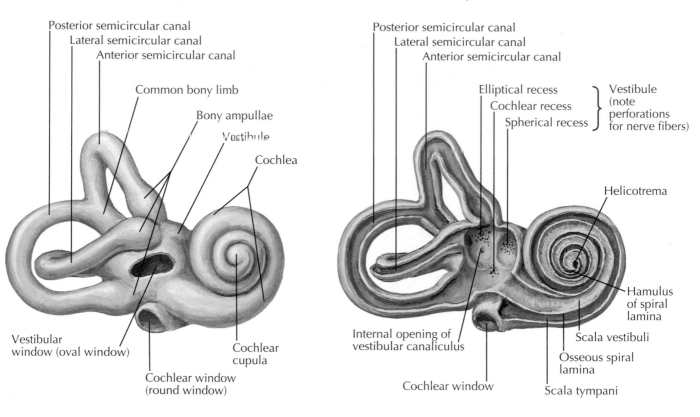

Posterior semicircular canal
Lateral semicircular canal
Anterior semicircular canal

Common bony limb

Bony ampullae

Vestibule

Cochlea

Vestibular window (oval window)

Cochlear cupula

Cochlear window (round window)

Dissected right bony labyrinth (otic capsule): membranous labyrinth removed

Posterior semicircular canal
Lateral semicircular canal
Anterior semicircular canal

Elliptical recess
Cochlear recess
Spherical recess

Vestibule (note perforations for nerve fibers)

Helicotrema

Hamulus of spiral lamina

Scala vestibuli

Osseous spiral lamina

Scala tympani

Internal opening of vestibular canaliculus

Cochlear window

Right membranous labyrinth with nerves: medial view

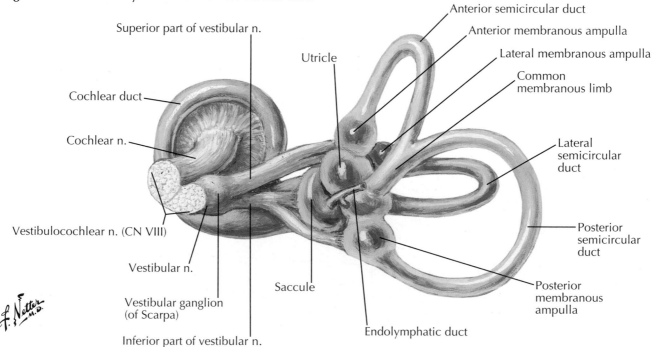

Superior part of vestibular n.

Anterior semicircular duct

Anterior membranous ampulla

Lateral membranous ampulla

Common membranous limb

Utricle

Cochlear duct

Cochlear n.

Lateral semicircular duct

Vestibulocochlear n. (CN VIII)

Posterior semicircular duct

Vestibular n.

Vestibular ganglion (of Scarpa)

Saccule

Posterior membranous ampulla

Endolymphatic duct

Inferior part of vestibular n.

Bony and membranous labyrinths: schema

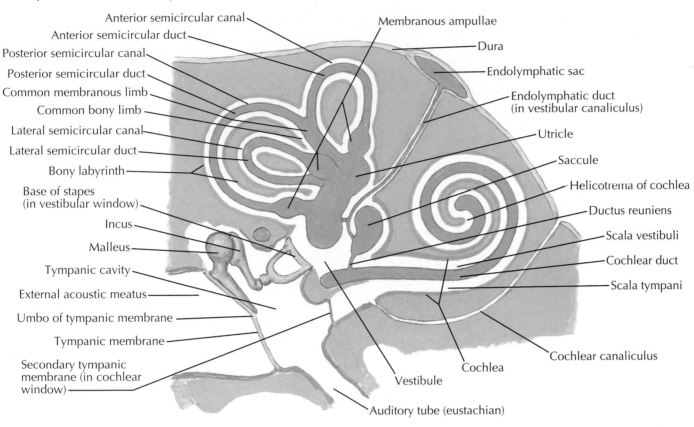

Anterior semicircular canal
Anterior semicircular duct
Posterior semicircular canal
Posterior semicircular duct
Common membranous limb
Common bony limb
Lateral semicircular canal
Lateral semicircular duct
Bony labyrinth
Base of stapes (in vestibular window)
Incus
Malleus
Tympanic cavity
External acoustic meatus
Umbo of tympanic membrane
Tympanic membrane
Secondary tympanic membrane (in cochlear window)

Membranous ampullae
Dura
Endolymphatic sac
Endolymphatic duct (in vestibular canaliculus)
Utricle
Saccule
Helicotrema of cochlea
Ductus reuniens
Scala vestibuli
Cochlear duct
Scala tympani
Cochlear canaliculus
Cochlea
Vestibule
Auditory tube (eustachian)

Section through turn of cochlea

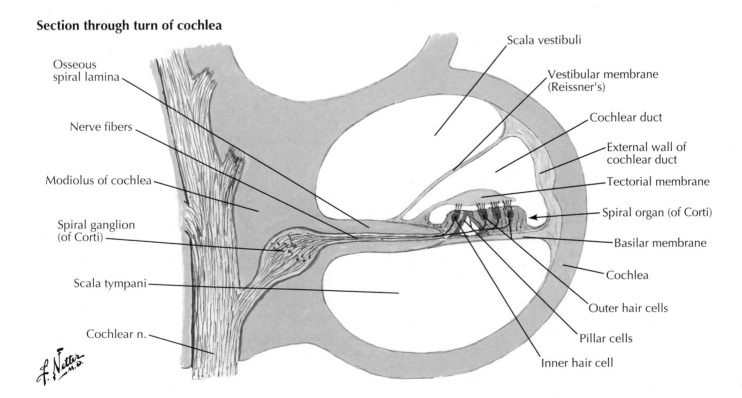

Osseous spiral lamina
Nerve fibers
Modiolus of cochlea
Spiral ganglion (of Corti)
Scala tympani
Cochlear n.

Scala vestibuli
Vestibular membrane (Reissner's)
Cochlear duct
External wall of cochlear duct
Tectorial membrane
Spiral organ (of Corti)
Basilar membrane
Cochlea
Outer hair cells
Pillar cells
Inner hair cell

Superior projection of right bony labyrinth on floor of cranium

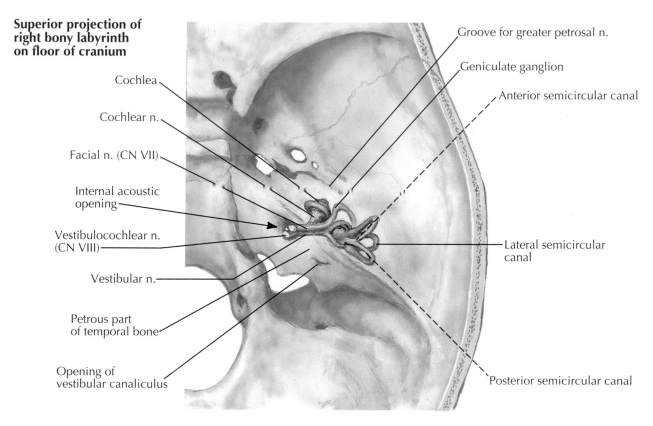

Cochlea

Cochlear n.

Facial n. (CN VII)

Internal acoustic opening

Vestibulocochlear n. (CN VIII)

Vestibular n.

Petrous part of temporal bone

Opening of vestibular canaliculus

Groove for greater petrosal n.

Geniculate ganglion

Anterior semicircular canal

Lateral semicircular canal

Posterior semicircular canal

Lateral projection of right membranous labyrinth

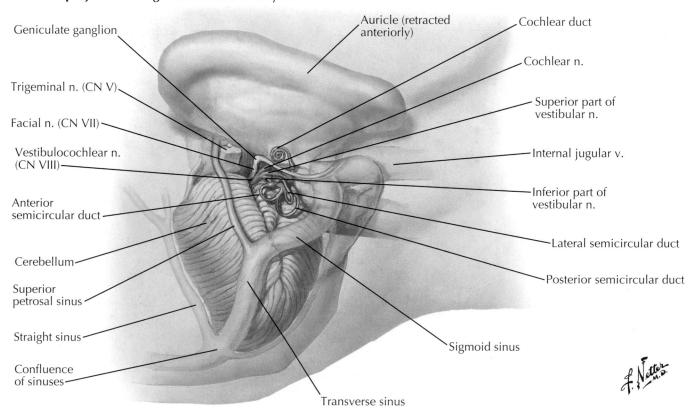

Geniculate ganglion

Trigeminal n. (CN V)

Facial n. (CN VII)

Vestibulocochlear n. (CN VIII)

Anterior semicircular duct

Cerebellum

Superior petrosal sinus

Straight sinus

Confluence of sinuses

Auricle (retracted anteriorly)

Cochlear duct

Cochlear n.

Superior part of vestibular n.

Internal jugular v.

Inferior part of vestibular n.

Lateral semicircular duct

Posterior semicircular duct

Sigmoid sinus

Transverse sinus

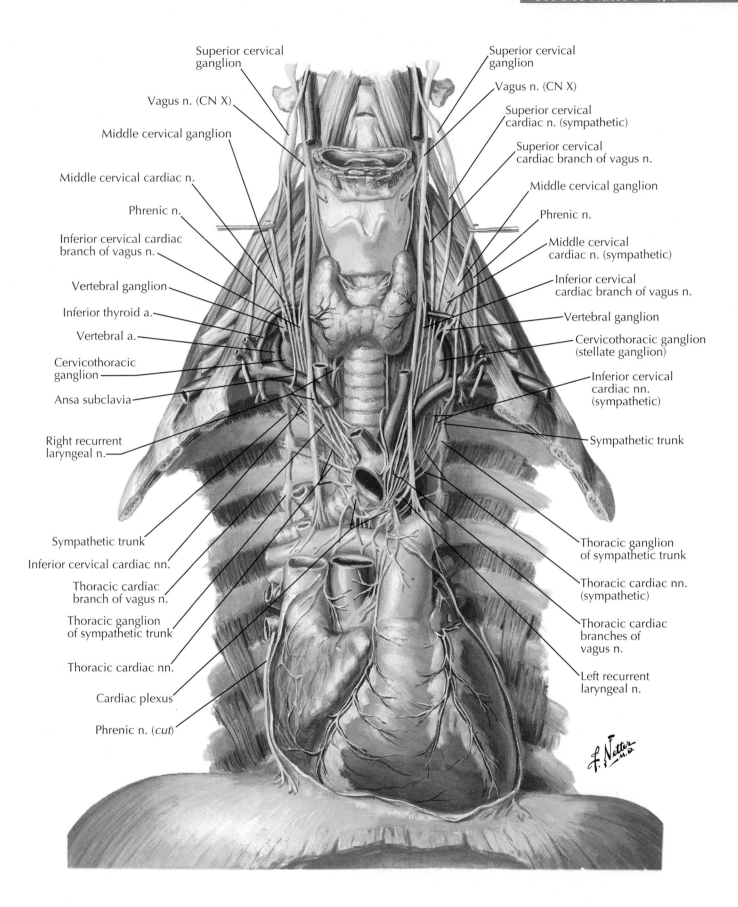

Superior cervical ganglion

Vagus n. (CN X)

Middle cervical ganglion

Middle cervical cardiac n.

Phrenic n.

Inferior cervical cardiac branch of vagus n.

Vertebral ganglion

Inferior thyroid a.

Vertebral a.

Cervicothoracic ganglion

Ansa subclavia

Right recurrent laryngeal n.

Sympathetic trunk

Inferior cervical cardiac nn.

Thoracic cardiac branch of vagus n.

Thoracic ganglion of sympathetic trunk

Thoracic cardiac nn.

Cardiac plexus

Phrenic n. (*cut*)

Superior cervical ganglion

Vagus n. (CN X)

Superior cervical cardiac n. (sympathetic)

Superior cervical cardiac branch of vagus n.

Middle cervical ganglion

Phrenic n.

Middle cervical cardiac n. (sympathetic)

Inferior cervical cardiac branch of vagus n.

Vertebral ganglion

Cervicothoracic ganglion (stellate ganglion)

Inferior cervical cardiac nn. (sympathetic)

Sympathetic trunk

Thoracic ganglion of sympathetic trunk

Thoracic cardiac nn. (sympathetic)

Thoracic cardiac branches of vagus n.

Left recurrent laryngeal n.

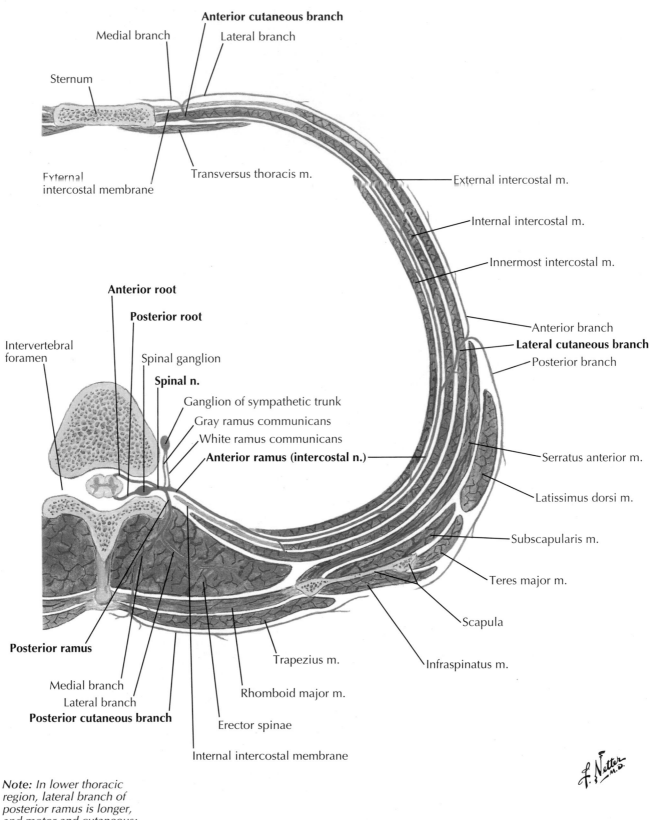

Anterior cutaneous branch

Medial branch

Lateral branch

Sternum

External
intercostal membrane

Transversus thoracis m.

External intercostal m.

Internal intercostal m.

Innermost intercostal m.

Anterior root

Posterior root

Intervertebral
foramen

Spinal ganglion

Spinal n.

Ganglion of sympathetic trunk

Gray ramus communicans

White ramus communicans

Anterior ramus (intercostal n.)

Anterior branch

Lateral cutaneous branch

Posterior branch

Serratus anterior m.

Latissimus dorsi m.

Subscapularis m.

Teres major m.

Scapula

Posterior ramus

Medial branch

Lateral branch

Posterior cutaneous branch

Erector spinae

Internal intercostal membrane

Trapezius m.

Rhomboid major m.

Infraspinatus m.

Note: *In lower thoracic region, lateral branch of posterior ramus is longer, and motor and cutaneous; medial branch is shorter and motor only.*

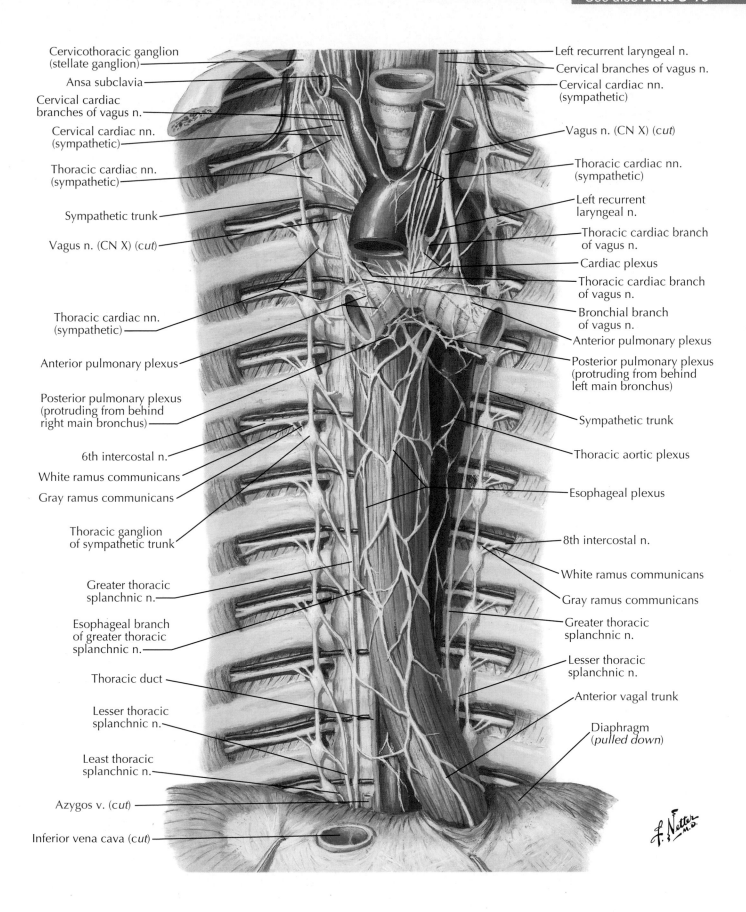

Cervicothoracic ganglion (stellate ganglion)

Ansa subclavia

Cervical cardiac branches of vagus n.

Cervical cardiac nn. (sympathetic)

Thoracic cardiac nn. (sympathetic)

Sympathetic trunk

Vagus n. (CN X) (cut)

Thoracic cardiac nn. (sympathetic)

Anterior pulmonary plexus

Posterior pulmonary plexus (protruding from behind right main bronchus)

6th intercostal n.

White ramus communicans

Gray ramus communicans

Thoracic ganglion of sympathetic trunk

Greater thoracic splanchnic n.

Esophageal branch of greater thoracic splanchnic n.

Thoracic duct

Lesser thoracic splanchnic n.

Least thoracic splanchnic n.

Azygos v. (cut)

Inferior vena cava (cut)

Left recurrent laryngeal n.

Cervical branches of vagus n.

Cervical cardiac nn. (sympathetic)

Vagus n. (CN X) (cut)

Thoracic cardiac nn. (sympathetic)

Left recurrent laryngeal n.

Thoracic cardiac branch of vagus n.

Cardiac plexus

Thoracic cardiac branch of vagus n.

Bronchial branch of vagus n.

Anterior pulmonary plexus

Posterior pulmonary plexus (protruding from behind left main bronchus)

Sympathetic trunk

Thoracic aortic plexus

Esophageal plexus

8th intercostal n.

White ramus communicans

Gray ramus communicans

Greater thoracic splanchnic n.

Lesser thoracic splanchnic n.

Anterior vagal trunk

Diaphragm (pulled down)

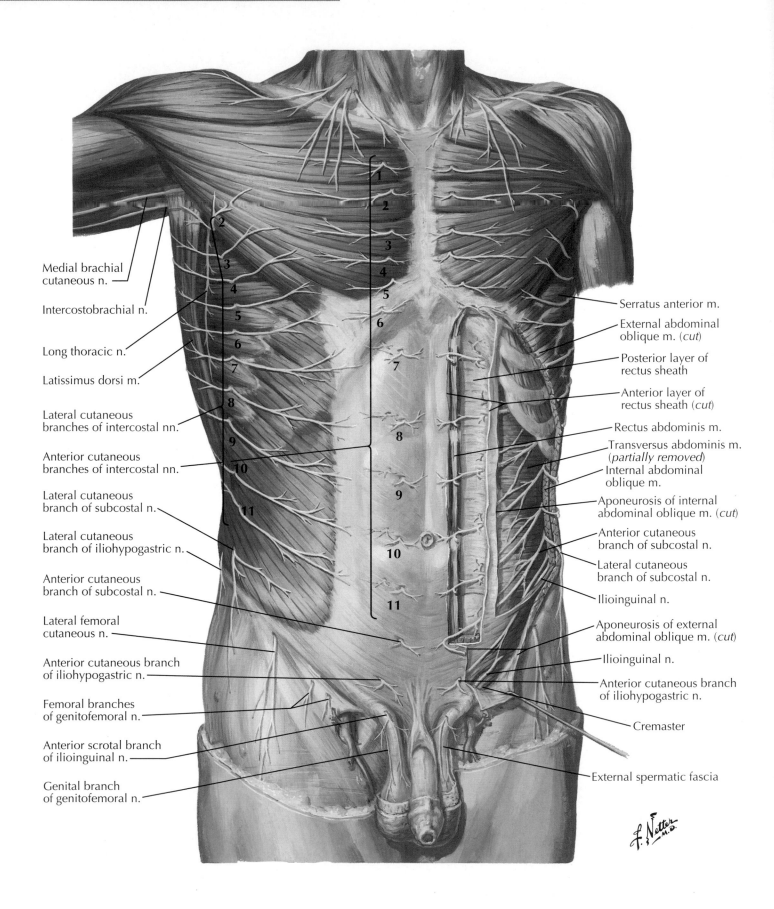

Medial brachial
cutaneous n.

Intercostobrachial n.

Long thoracic n.

Latissimus dorsi m.

Lateral cutaneous
branches of intercostal nn.

Anterior cutaneous
branches of intercostal nn.

Lateral cutaneous
branch of subcostal n.

Lateral cutaneous
branch of iliohypogastric n.

Anterior cutaneous
branch of subcostal n.

Lateral femoral
cutaneous n.

Anterior cutaneous branch
of iliohypogastric n.

Femoral branches
of genitofemoral n.

Anterior scrotal branch
of ilioinguinal n.

Genital branch
of genitofemoral n.

Serratus anterior m.

External abdominal
oblique m. (cut)

Posterior layer of
rectus sheath

Anterior layer of
rectus sheath (cut)

Rectus abdominis m.

Transversus abdominis m.
(partially removed)

Internal abdominal
oblique m.

Aponeurosis of internal
abdominal oblique m. (cut)

Anterior cutaneous
branch of subcostal n.

Lateral cutaneous
branch of subcostal n.

Ilioinguinal n.

Aponeurosis of external
abdominal oblique m. (cut)

Ilioinguinal n.

Anterior cutaneous branch
of iliohypogastric n.

Cremaster

External spermatic fascia

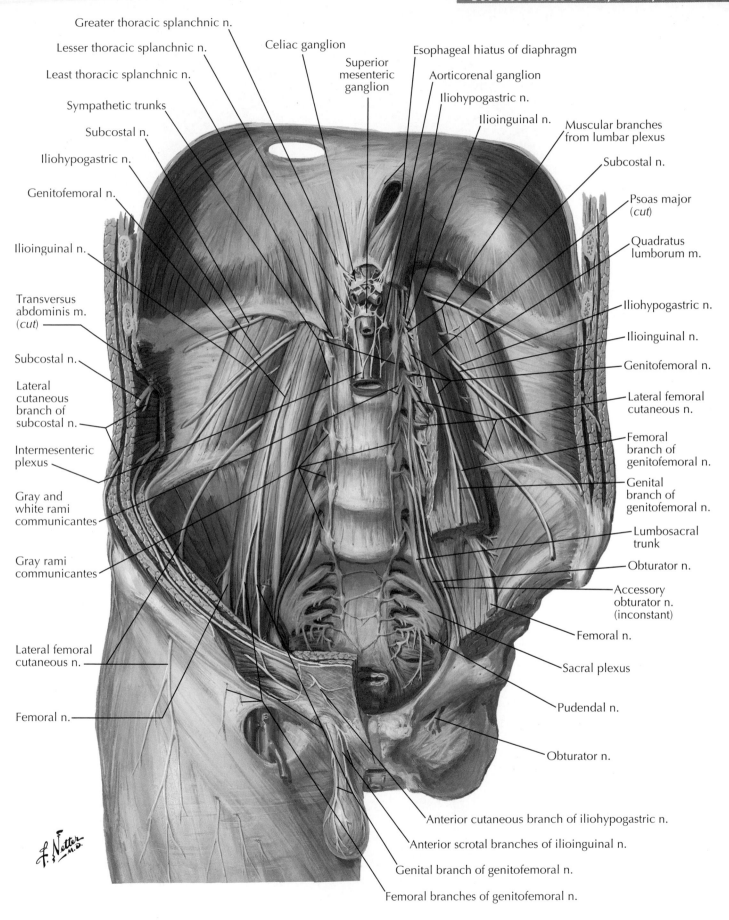

Greater thoracic splanchnic n.

Lesser thoracic splanchnic n.

Least thoracic splanchnic n.

Sympathetic trunks

Subcostal n.

Iliohypogastric n.

Genitofemoral n.

Ilioinguinal n.

Transversus abdominis m. (cut)

Subcostal n.

Lateral cutaneous branch of subcostal n.

Intermesenteric plexus

Gray and white rami communicantes

Gray rami communicantes

Lateral femoral cutaneous n.

Femoral n.

Celiac ganglion

Superior mesenteric ganglion

Esophageal hiatus of diaphragm

Aorticorenal ganglion

Iliohypogastric n.

Ilioinguinal n.

Muscular branches from lumbar plexus

Subcostal n.

Psoas major (cut)

Quadratus lumborum m.

Iliohypogastric n.

Ilioinguinal n.

Genitofemoral n.

Lateral femoral cutaneous n.

Femoral branch of genitofemoral n.

Genital branch of genitofemoral n.

Lumbosacral trunk

Obturator n.

Accessory obturator n. (inconstant)

Femoral n.

Sacral plexus

Pudendal n.

Obturator n.

Anterior cutaneous branch of iliohypogastric n.

Anterior scrotal branches of ilioinguinal n.

Genital branch of genitofemoral n.

Femoral branches of genitofemoral n.

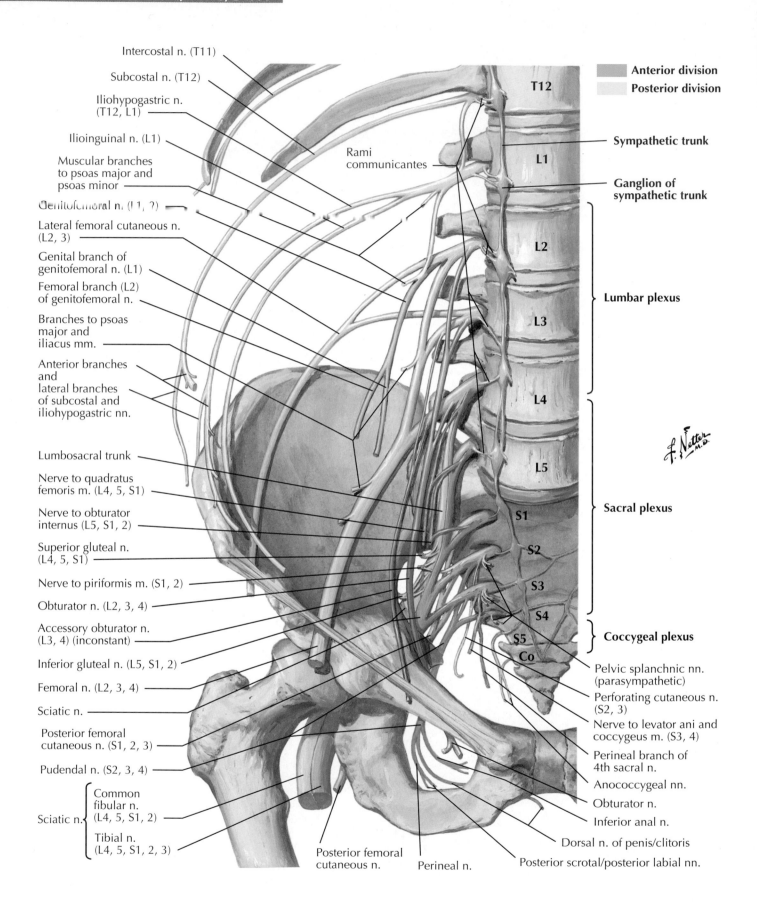

Intercostal n. (T11)

Subcostal n. (T12)

Iliohypogastric n. (T12, L1)

Ilioinguinal n. (L1)

Muscular branches to psoas major and psoas minor

Genitofemoral n. (L1, 2)

Lateral femoral cutaneous n. (L2, 3)

Genital branch of genitofemoral n. (L1)

Femoral branch (L2) of genitofemoral n.

Branches to psoas major and iliacus mm.

Anterior branches and lateral branches of subcostal and iliohypogastric nn.

Lumbosacral trunk

Nerve to quadratus femoris m. (L4, 5, S1)

Nerve to obturator internus (L5, S1, 2)

Superior gluteal n. (L4, 5, S1)

Nerve to piriformis m. (S1, 2)

Obturator n. (L2, 3, 4)

Accessory obturator n. (L3, 4) (inconstant)

Inferior gluteal n. (L5, S1, 2)

Femoral n. (L2, 3, 4)

Sciatic n.

Posterior femoral cutaneous n. (S1, 2, 3)

Pudendal n. (S2, 3, 4)

Sciatic n. { Common fibular n. (L4, 5, S1, 2)

Tibial n. (L4, 5, S1, 2, 3)

Rami communicantes

T12

L1

L2

L3

L4

L5

S1

S2

S3

S4

S5

Co

Anterior division
Posterior division

Sympathetic trunk

Ganglion of sympathetic trunk

Lumbar plexus

Sacral plexus

Coccygeal plexus

Pelvic splanchnic nn. (parasympathetic)

Perforating cutaneous n. (S2, 3)

Nerve to levator ani and coccygeus m. (S3, 4)

Perineal branch of 4th sacral n.

Anococcygeal nn.

Obturator n.

Inferior anal n.

Dorsal n. of penis/clitoris

Posterior scrotal/posterior labial nn.

Posterior femoral cutaneous n.

Perineal n.

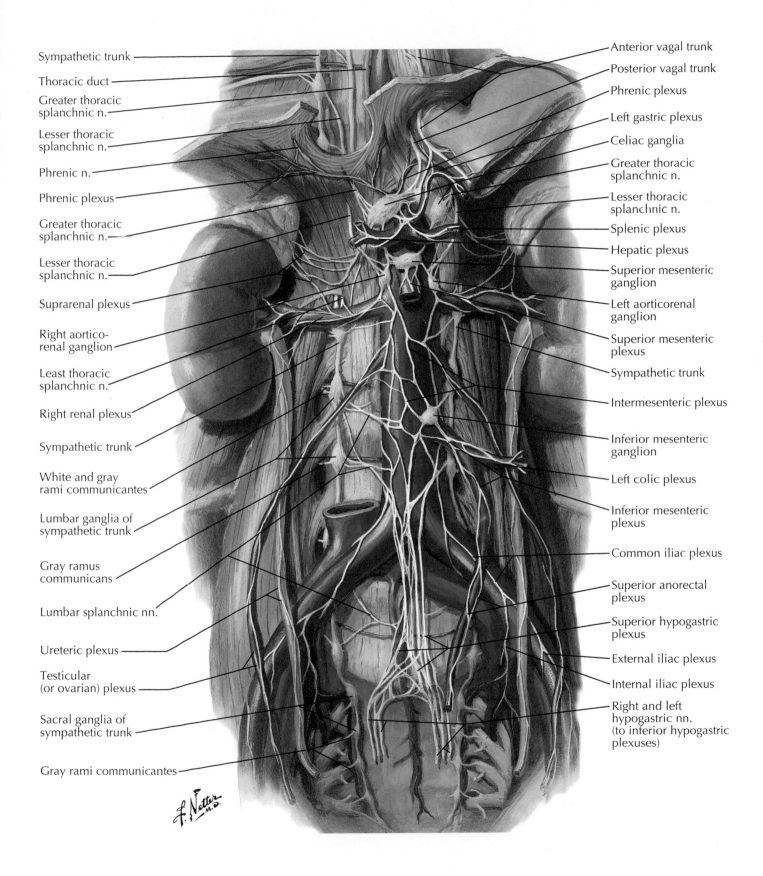

Sympathetic trunk

Thoracic duct

Greater thoracic
splanchnic n.

Lesser thoracic
splanchnic n.

Phrenic n.

Phrenic plexus

Greater thoracic
splanchnic n.

Lesser thoracic
splanchnic n.

Suprarenal plexus

Right aortico-
renal ganglion

Least thoracic
splanchnic n.

Right renal plexus

Sympathetic trunk

White and gray
rami communicantes

Lumbar ganglia of
sympathetic trunk

Gray ramus
communicans

Lumbar splanchnic nn.

Ureteric plexus

Testicular
(or ovarian) plexus

Sacral ganglia of
sympathetic trunk

Gray rami communicantes

Anterior vagal trunk

Posterior vagal trunk

Phrenic plexus

Left gastric plexus

Celiac ganglia

Greater thoracic
splanchnic n.

Lesser thoracic
splanchnic n.

Splenic plexus

Hepatic plexus

Superior mesenteric
ganglion

Left aorticorenal
ganglion

Superior mesenteric
plexus

Sympathetic trunk

Intermesenteric plexus

Inferior mesenteric
ganglion

Left colic plexus

Inferior mesenteric
plexus

Common iliac plexus

Superior anorectal
plexus

Superior hypogastric
plexus

External iliac plexus

Internal iliac plexus

Right and left
hypogastric nn.
(to inferior hypogastric
plexuses)

F. Netter, M.D.

Note: Prefixed plexus has large C4 contribution but lacks T1. Postfixed plexus lacks C5 but has T2 contribution.

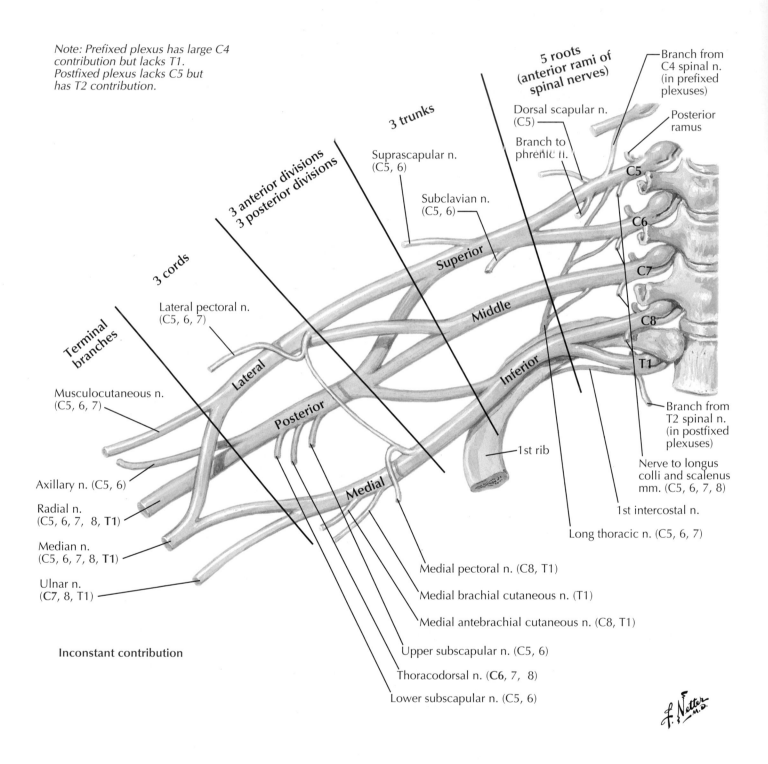

5 roots (anterior rami of spinal nerves)

Branch from C4 spinal n. (in prefixed plexuses)

Posterior ramus

Dorsal scapular n. (C5)

Branch to phrenic n.

3 trunks

Suprascapular n. (C5, 6)

Subclavian n. (C5, 6)

C5

C6

C7

C8

T1

3 anterior divisions
3 posterior divisions

Superior

Middle

Inferior

3 cords

Lateral pectoral n. (C5, 6, 7)

Lateral

Posterior

Terminal branches

Musculocutaneous n. (C5, 6, 7)

Medial

1st rib

Branch from T2 spinal n. (in postfixed plexuses)

Nerve to longus colli and scalenus mm. (C5, 6, 7, 8)

1st intercostal n.

Long thoracic n. (C5, 6, 7)

Axillary n. (C5, 6)

Radial n. (C5, 6, 7, **T1**)

Median n. (C5, 6, 7, 8, **T1**)

Ulnar n. (**C7**, 8, T1)

Medial pectoral n. (C8, T1)

Medial brachial cutaneous n. (T1)

Medial antebrachial cutaneous n. (C8, T1)

Upper subscapular n. (C5, 6)

Thoracodorsal n. (**C6**, 7, 8)

Lower subscapular n. (C5, 6)

Inconstant contribution

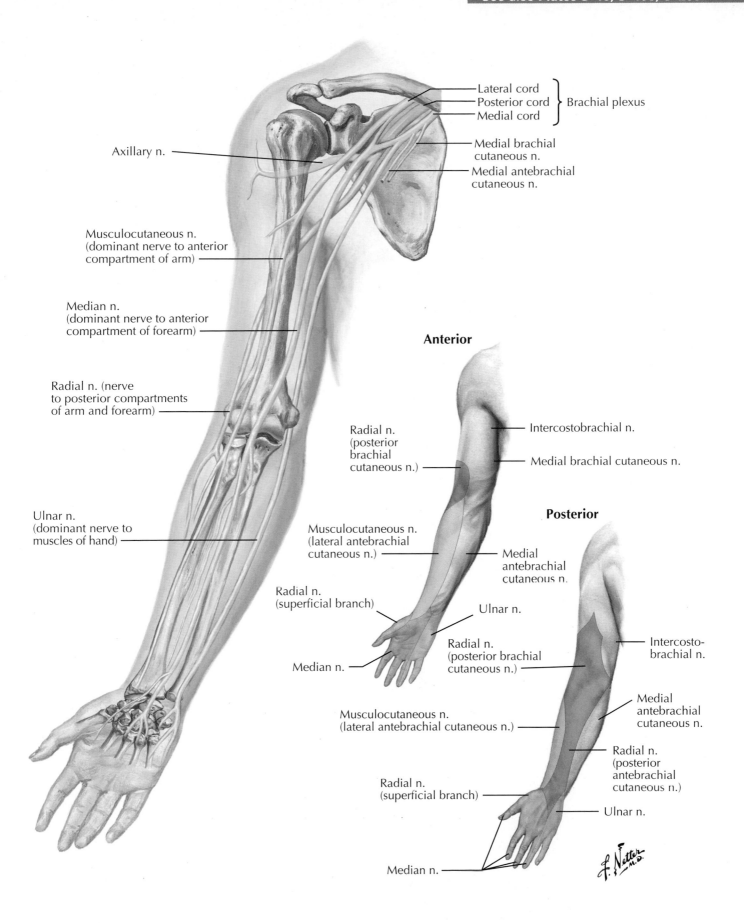

Lateral cord
Posterior cord } Brachial plexus
Medial cord

Axillary n.

Medial brachial cutaneous n.

Medial antebrachial cutaneous n.

Musculocutaneous n. (dominant nerve to anterior compartment of arm)

Median n. (dominant nerve to anterior compartment of forearm)

Anterior

Radial n. (nerve to posterior compartments of arm and forearm)

Radial n. (posterior brachial cutaneous n.)

Intercostobrachial n.

Medial brachial cutaneous n.

Posterior

Ulnar n. (dominant nerve to muscles of hand)

Musculocutaneous n. (lateral antebrachial cutaneous n.)

Medial antebrachial cutaneous n.

Radial n. (superficial branch)

Ulnar n.

Radial n. (posterior brachial cutaneous n.)

Intercosto-brachial n.

Median n.

Musculocutaneous n. (lateral antebrachial cutaneous n.)

Medial antebrachial cutaneous n.

Radial n. (posterior antebrachial cutaneous n.)

Radial n. (superficial branch)

Ulnar n.

Median n.

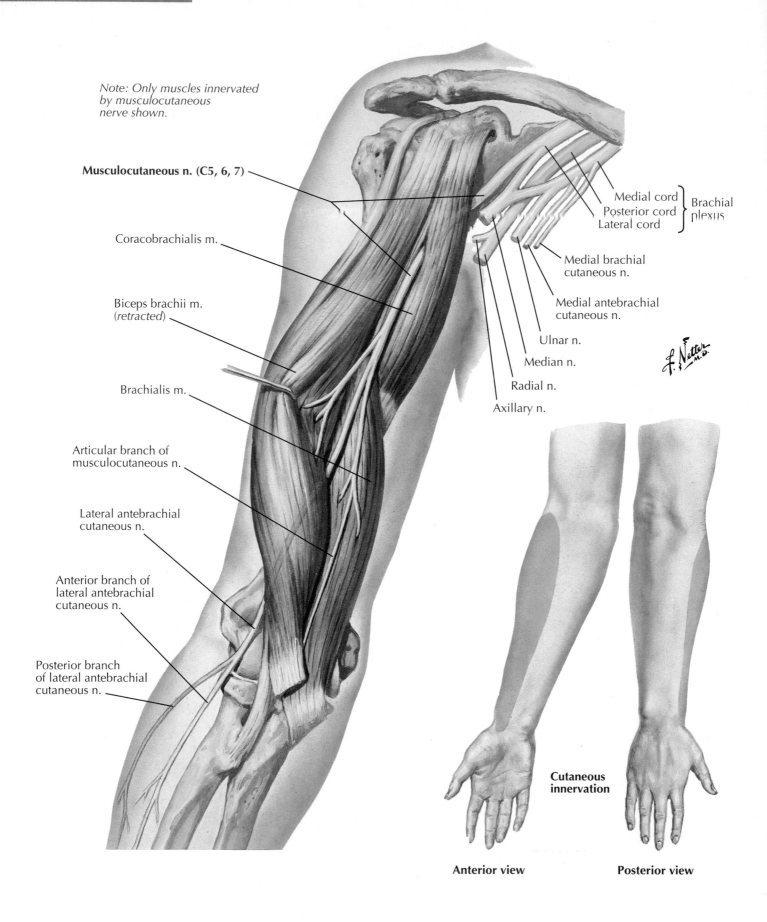

Note: Only muscles innervated by musculocutaneous nerve shown.

Musculocutaneous n. (C5, 6, 7)

Coracobrachialis m.

Biceps brachii m. (*retracted*)

Brachialis m.

Articular branch of musculocutaneous n.

Lateral antebrachial cutaneous n.

Anterior branch of lateral antebrachial cutaneous n.

Posterior branch of lateral antebrachial cutaneous n.

Medial cord
Posterior cord
Lateral cord
} Brachial plexus

Medial brachial cutaneous n.

Medial antebrachial cutaneous n.

Ulnar n.

Median n.

Radial n.

Axillary n.

Cutaneous innervation

Anterior view

Posterior view

Note: Only muscles innervated by median nerve shown.

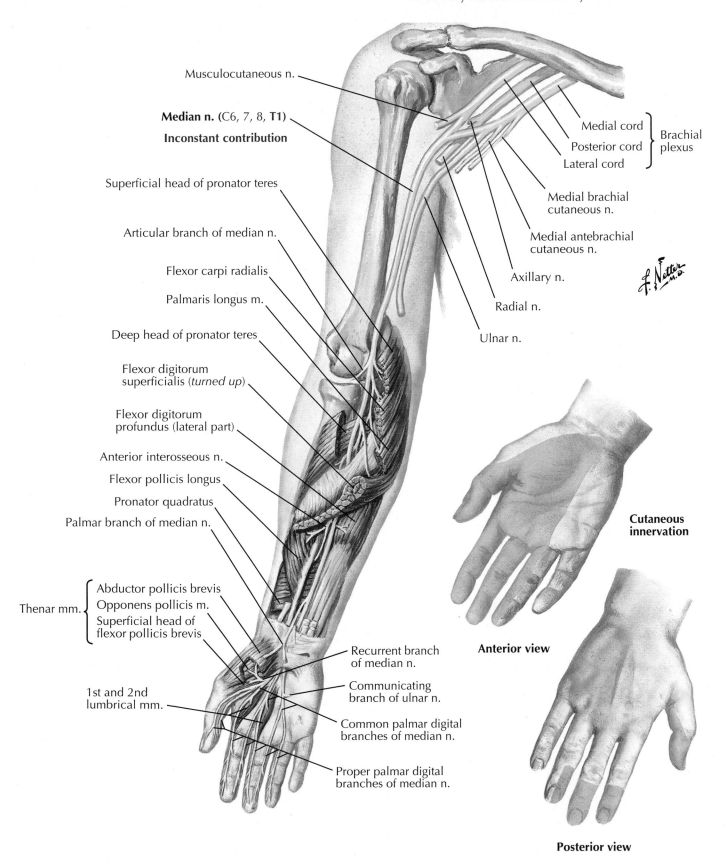

Musculocutaneous n.

Median n. (C6, 7, 8, **T1**)
Inconstant contribution

Superficial head of pronator teres

Articular branch of median n.

Flexor carpi radialis

Palmaris longus m.

Deep head of pronator teres

Flexor digitorum
superficialis (*turned up*)

Flexor digitorum
profundus (lateral part)

Anterior interosseous n.

Flexor pollicis longus

Pronator quadratus

Palmar branch of median n.

Thenar mm. {
Abductor pollicis brevis
Opponens pollicis m.
Superficial head of
flexor pollicis brevis
}

1st and 2nd
lumbrical mm.

Medial cord } Brachial
Posterior cord } plexus
Lateral cord

Medial brachial
cutaneous n.

Medial antebrachial
cutaneous n.

f. Netter
M.D.

Axillary n.

Radial n.

Ulnar n.

Recurrent branch
of median n.

Communicating
branch of ulnar n.

Common palmar digital
branches of median n.

Proper palmar digital
branches of median n.

**Cutaneous
innervation**

Anterior view

Posterior view

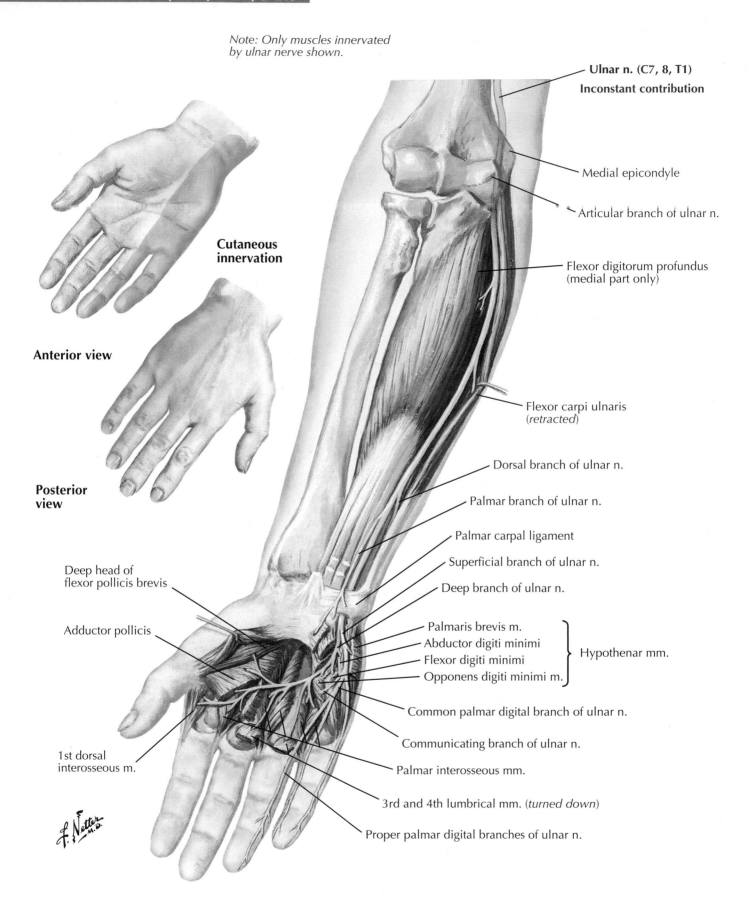

Note: Only muscles innervated by ulnar nerve shown.

Cutaneous innervation

Anterior view

Posterior view

Ulnar n. (C7, 8, T1)
Inconstant contribution

Medial epicondyle

Articular branch of ulnar n.

Flexor digitorum profundus (medial part only)

Flexor carpi ulnaris (*retracted*)

Dorsal branch of ulnar n.

Palmar branch of ulnar n.

Palmar carpal ligament

Superficial branch of ulnar n.

Deep branch of ulnar n.

Palmaris brevis m.
Abductor digiti minimi
Flexor digiti minimi
Opponens digiti minimi m.

Hypothenar mm.

Common palmar digital branch of ulnar n.

Communicating branch of ulnar n.

Palmar interosseous mm.

3rd and 4th lumbrical mm. (*turned down*)

Proper palmar digital branches of ulnar n.

Deep head of flexor pollicis brevis

Adductor pollicis

1st dorsal interosseous m.

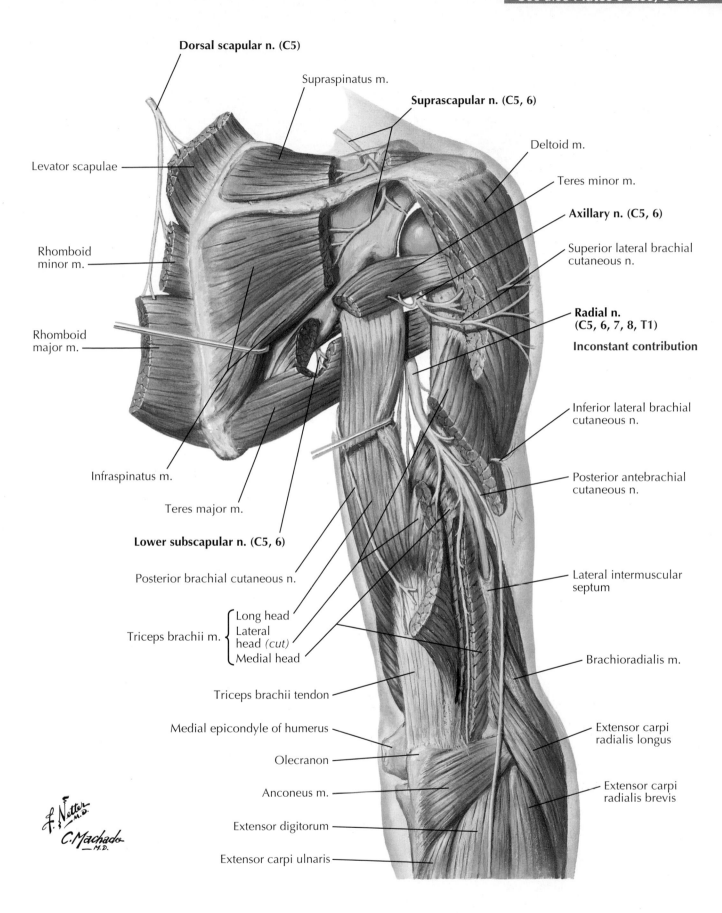

Dorsal scapular n. (C5)

Supraspinatus m.

Suprascapular n. (C5, 6)

Deltoid m.

Levator scapulae

Teres minor m.

Axillary n. (C5, 6)

Superior lateral brachial cutaneous n.

Rhomboid minor m.

Radial n. (C5, 6, 7, 8, T1)

Inconstant contribution

Rhomboid major m.

Inferior lateral brachial cutaneous n.

Posterior antebrachial cutaneous n.

Infraspinatus m.

Teres major m.

Lower subscapular n. (C5, 6)

Posterior brachial cutaneous n.

Lateral intermuscular septum

Long head
Lateral head (cut)
Medial head

Triceps brachii m.

Brachioradialis m.

Triceps brachii tendon

Medial epicondyle of humerus

Extensor carpi radialis longus

Olecranon

Anconeus m.

Extensor carpi radialis brevis

Extensor digitorum

Extensor carpi ulnaris

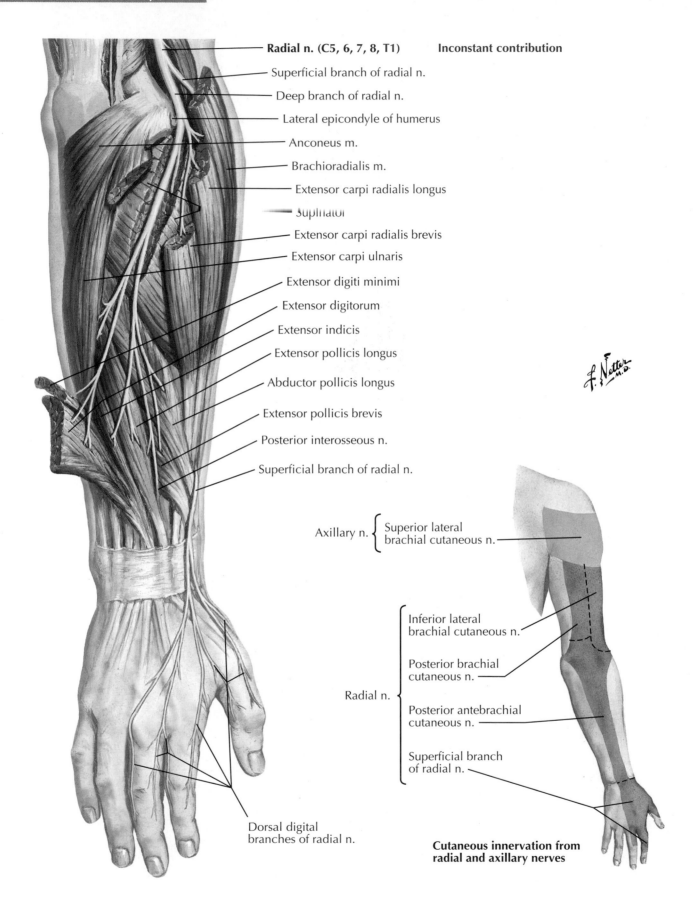

Radial n. (C5, 6, 7, 8, T1)   Inconstant contribution

Superficial branch of radial n.

Deep branch of radial n.

Lateral epicondyle of humerus

Anconeus m.

Brachioradialis m.

Extensor carpi radialis longus

Supinator

Extensor carpi radialis brevis

Extensor carpi ulnaris

Extensor digiti minimi

Extensor digitorum

Extensor indicis

Extensor pollicis longus

Abductor pollicis longus

Extensor pollicis brevis

Posterior interosseous n.

Superficial branch of radial n.

Dorsal digital branches of radial n.

Axillary n. { Superior lateral brachial cutaneous n.

Inferior lateral brachial cutaneous n.

Posterior brachial cutaneous n.

Radial n. {

Posterior antebrachial cutaneous n.

Superficial branch of radial n.

Cutaneous innervation from radial and axillary nerves

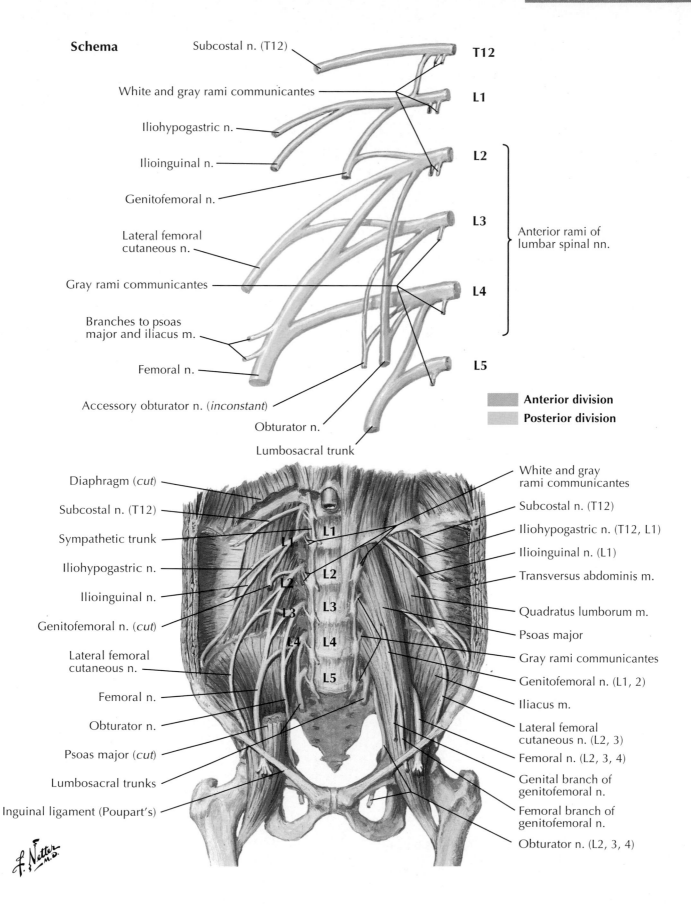

Schema

Subcostal n. (T12)

T12

White and gray rami communicantes

L1

Iliohypogastric n.

Ilioinguinal n.

L2

Genitofemoral n.

L3

Anterior rami of lumbar spinal nn.

Lateral femoral cutaneous n.

Gray rami communicantes

L4

Branches to psoas major and iliacus m.

Femoral n.

L5

Accessory obturator n. (*inconstant*)

Obturator n.

Lumbosacral trunk

Anterior division

Posterior division

Diaphragm (*cut*)

White and gray rami communicantes

Subcostal n. (T12)

Subcostal n. (T12)

Sympathetic trunk

Iliohypogastric n. (T12, L1)

Iliohypogastric n.

Ilioinguinal n. (L1)

Ilioinguinal n.

Transversus abdominis m.

Genitofemoral n. (*cut*)

Quadratus lumborum m.

Lateral femoral cutaneous n.

Psoas major

Femoral n.

Gray rami communicantes

Obturator n.

Genitofemoral n. (L1, 2)

Psoas major (*cut*)

Iliacus m.

Lumbosacral trunks

Lateral femoral cutaneous n. (L2, 3)

Inguinal ligament (Poupart's)

Femoral n. (L2, 3, 4)

Genital branch of genitofemoral n.

Femoral branch of genitofemoral n.

Obturator n. (L2, 3, 4)

Nerves of Lower Limb

Plate S–113

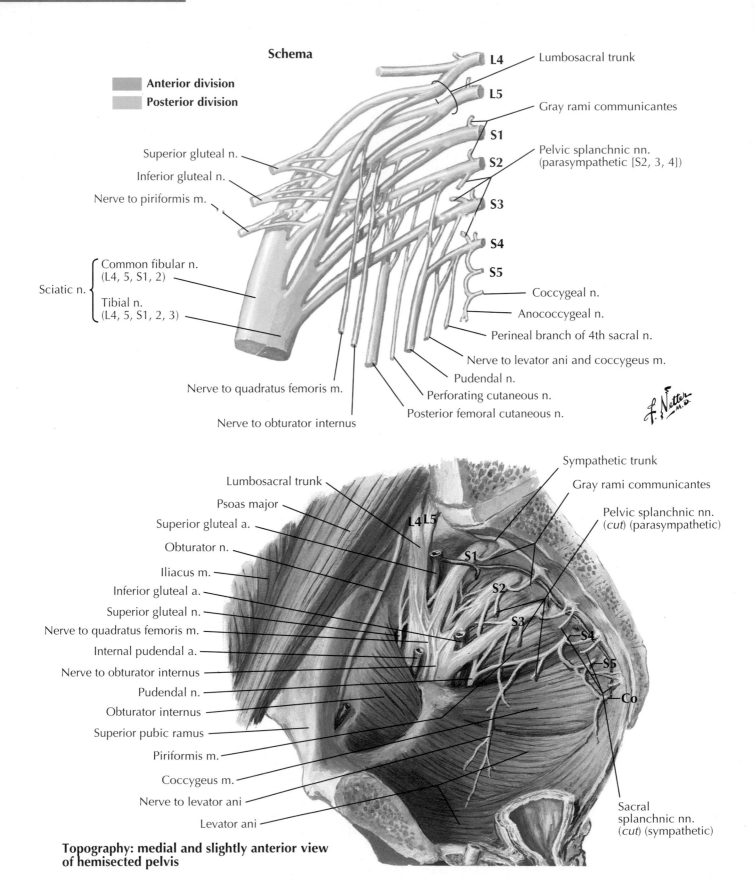

Schema

Anterior division
Posterior division

L4 — Lumbosacral trunk

L5

Gray rami communicantes

S1

Superior gluteal n.

S2 — Pelvic splanchnic nn. (parasympathetic [S2, 3, 4])

Inferior gluteal n.

Nerve to piriformis m.

S3

S4

S5

Common fibular n. (L4, 5, S1, 2)

Sciatic n.

Tibial n. (L4, 5, S1, 2, 3)

Coccygeal n.

Anococcygeal n.

Perineal branch of 4th sacral n.

Nerve to levator ani and coccygeus m.

Nerve to quadratus femoris m.

Pudendal n.

Perforating cutaneous n.

Posterior femoral cutaneous n.

Nerve to obturator internus

Lumbosacral trunk

Psoas major

Superior gluteal a.

Obturator n.

Iliacus m.

Inferior gluteal a.

Superior gluteal n.

Nerve to quadratus femoris m.

Internal pudendal a.

Nerve to obturator internus

Pudendal n.

Obturator internus

Superior pubic ramus

Piriformis m.

Coccygeus m.

Nerve to levator ani

Levator ani

Sympathetic trunk

Gray rami communicantes

Pelvic splanchnic nn. (cut) (parasympathetic)

L4 L5

S1

S2

S3

S4

S5

Co

Sacral splanchnic nn. (cut) (sympathetic)

Topography: medial and slightly anterior view of hemisected pelvis

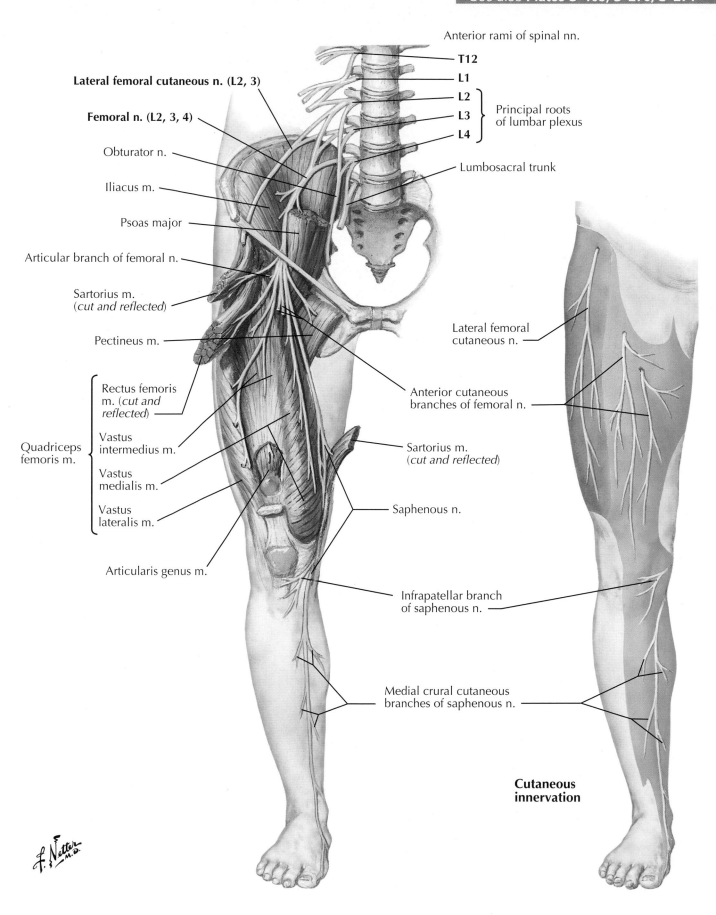

Anterior rami of spinal nn.

T12

Lateral femoral cutaneous n. (L2, 3)

L1

L2

Femoral n. (L2, 3, 4)

L3

Principal roots of lumbar plexus

L4

Obturator n.

Iliacus m.

Lumbosacral trunk

Psoas major

Articular branch of femoral n.

Sartorius m. (*cut and reflected*)

Pectineus m.

Lateral femoral cutaneous n.

Rectus femoris m. (*cut and reflected*)

Anterior cutaneous branches of femoral n.

Quadriceps femoris m.

Vastus intermedius m.

Sartorius m. (*cut and reflected*)

Vastus medialis m.

Vastus lateralis m.

Saphenous n.

Articularis genus m.

Infrapatellar branch of saphenous n.

Medial crural cutaneous branches of saphenous n.

Cutaneous innervation

Nerves of Lower Limb

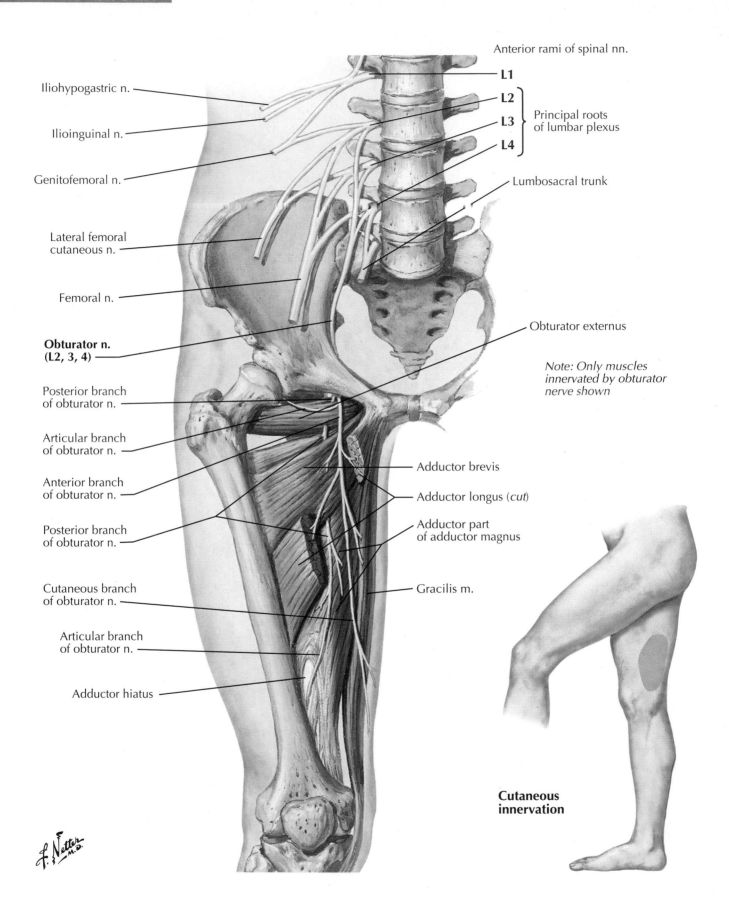

Anterior rami of spinal nn.

L1

L2

L3 } Principal roots
of lumbar plexus

L4

Iliohypogastric n.

Ilioinguinal n.

Genitofemoral n.

Lumbosacral trunk

Lateral femoral
cutaneous n.

Femoral n.

Obturator externus

**Obturator n.
(L2, 3, 4)**

*Note: Only muscles
innervated by obturator
nerve shown*

Posterior branch
of obturator n.

Articular branch
of obturator n.

Adductor brevis

Anterior branch
of obturator n.

Adductor longus (*cut*)

Adductor part
of adductor magnus

Posterior branch
of obturator n.

Cutaneous branch
of obturator n.

Gracilis m.

Articular branch
of obturator n.

Adductor hiatus

**Cutaneous
innervation**

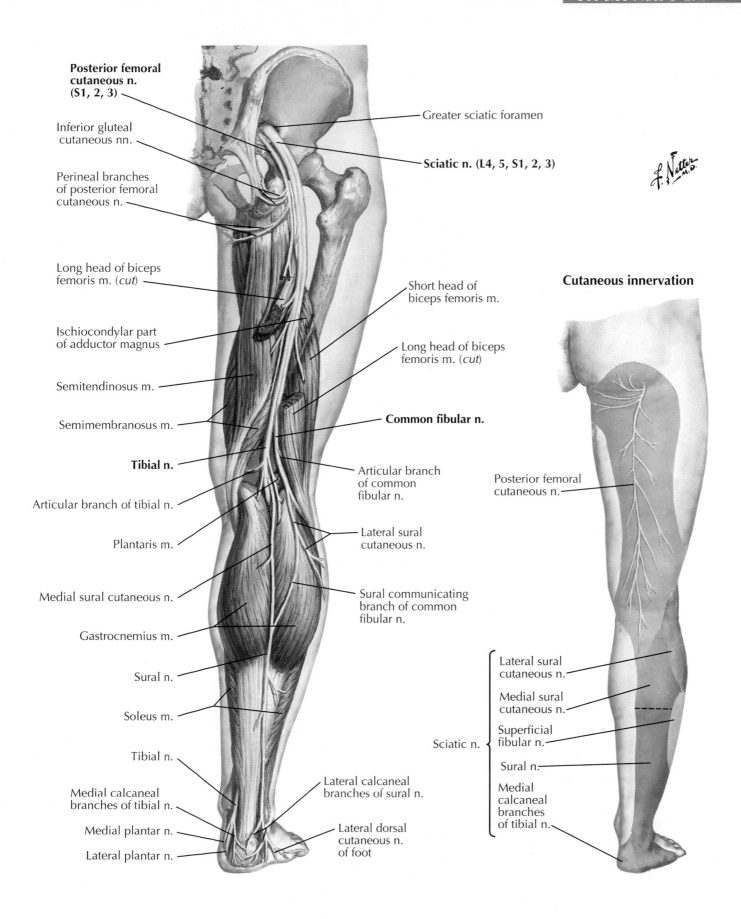

Posterior femoral cutaneous n. (S1, 2, 3)

Inferior gluteal cutaneous nn.

Perineal branches of posterior femoral cutaneous n.

Long head of biceps femoris m. (*cut*)

Ischiocondylar part of adductor magnus

Semitendinosus m.

Semimembranosus m.

Tibial n.

Articular branch of tibial n.

Plantaris m.

Medial sural cutaneous n.

Gastrocnemius m.

Sural n.

Soleus m.

Tibial n.

Medial calcaneal branches of tibial n.

Medial plantar n.

Lateral plantar n.

Greater sciatic foramen

Sciatic n. (L4, 5, S1, 2, 3)

Short head of biceps femoris m.

Long head of biceps femoris m. (*cut*)

Common fibular n.

Articular branch of common fibular n.

Lateral sural cutaneous n.

Sural communicating branch of common fibular n.

Lateral calcaneal branches of sural n.

Lateral dorsal cutaneous n. of foot

Cutaneous innervation

Posterior femoral cutaneous n.

Lateral sural cutaneous n.

Medial sural cutaneous n.

Superficial fibular n.

Sural n.

Medial calcaneal branches of tibial n.

Sciatic n.

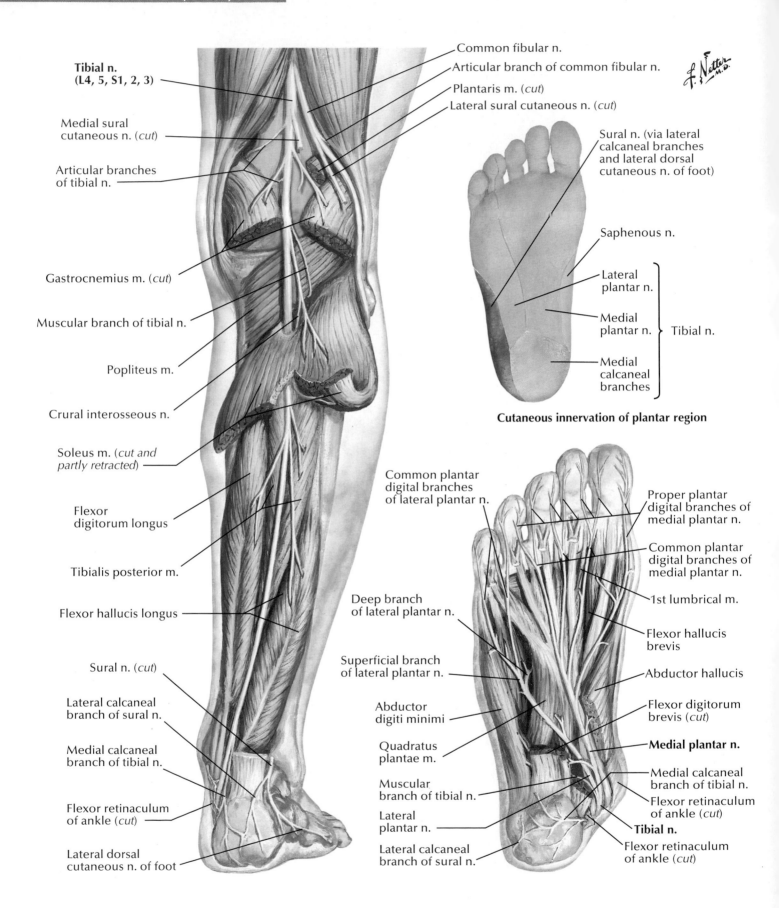

Tibial n.
(L4, 5, S1, 2, 3)

Medial sural
cutaneous n. (cut)

Articular branches
of tibial n.

Gastrocnemius m. (cut)

Muscular branch of tibial n.

Popliteus m.

Crural interosseous n.

Soleus m. (cut and
partly retracted)

Flexor
digitorum longus

Tibialis posterior m.

Flexor hallucis longus

Sural n. (cut)

Lateral calcaneal
branch of sural n.

Medial calcaneal
branch of tibial n.

Flexor retinaculum
of ankle (cut)

Lateral dorsal
cutaneous n. of foot

Common fibular n.

Articular branch of common fibular n.

Plantaris m. (cut)

Lateral sural cutaneous n. (cut)

Sural n. (via lateral
calcaneal branches
and lateral dorsal
cutaneous n. of foot)

Saphenous n.

Lateral
plantar n.

Medial
plantar n.   } Tibial n.

Medial
calcaneal
branches

Cutaneous innervation of plantar region

Common plantar
digital branches
of lateral plantar n.

Deep branch
of lateral plantar n.

Superficial branch
of lateral plantar n.

Abductor
digiti minimi

Quadratus
plantae m.

Muscular
branch of tibial n.

Lateral
plantar n.

Lateral calcaneal
branch of sural n.

Proper plantar
digital branches of
medial plantar n.

Common plantar
digital branches of
medial plantar n.

1st lumbrical m.

Flexor hallucis
brevis

Abductor hallucis

Flexor digitorum
brevis (cut)

Medial plantar n.

Medial calcaneal
branch of tibial n.

Flexor retinaculum
of ankle (cut)

Tibial n.

Flexor retinaculum
of ankle (cut)

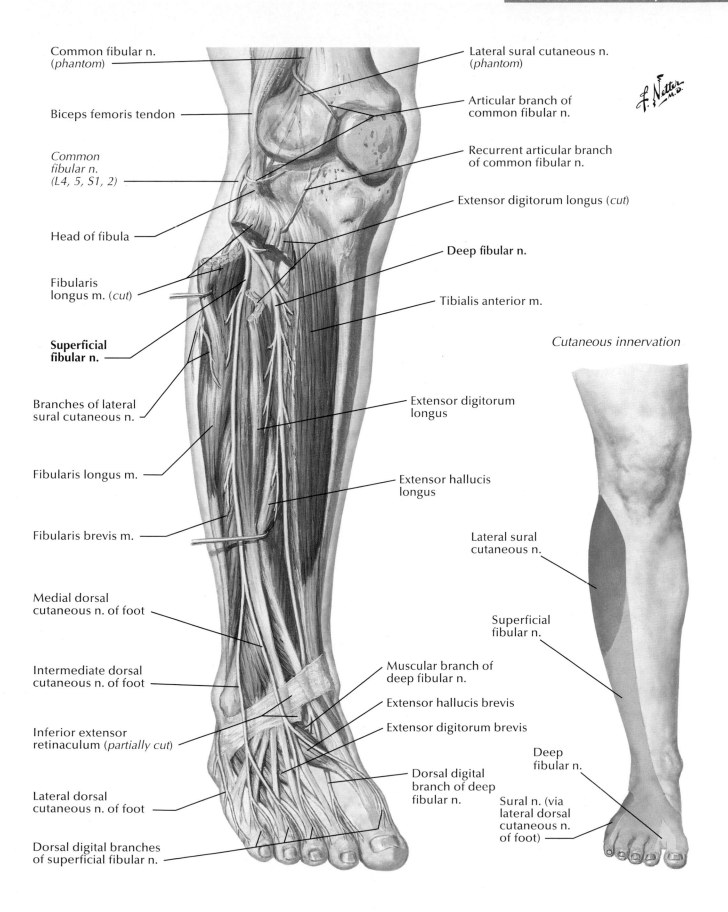

Common fibular n. (phantom)

Biceps femoris tendon

Common fibular n. (L4, 5, S1, 2)

Head of fibula

Fibularis longus m. (cut)

Superficial fibular n.

Branches of lateral sural cutaneous n.

Fibularis longus m.

Fibularis brevis m.

Medial dorsal cutaneous n. of foot

Intermediate dorsal cutaneous n. of foot

Inferior extensor retinaculum (partially cut)

Lateral dorsal cutaneous n. of foot

Dorsal digital branches of superficial fibular n.

Lateral sural cutaneous n. (phantom)

Articular branch of common fibular n.

Recurrent articular branch of common fibular n.

Extensor digitorum longus (cut)

Deep fibular n.

Tibialis anterior m.

Extensor digitorum longus

Extensor hallucis longus

Muscular branch of deep fibular n.

Extensor hallucis brevis

Extensor digitorum brevis

Dorsal digital branch of deep fibular n.

Cutaneous innervation

Lateral sural cutaneous n.

Superficial fibular n.

Deep fibular n.

Sural n. (via lateral dorsal cutaneous n. of foot)

Structures with High* Clinical Significance

ANATOMIC STRUCTURES	CLINICAL IMPORTANCE	PLATE NUMBERS
Head and Neck		
Spinal accessory nerve (CN XI)	Lymph node biopsy in posterior cervical triangle can cause iatrogenic injury of CN XI	S–194
Cervical plexus	Cervical plexus blocks are performed for neck procedures	S–194, S–195
Trigeminal nerve (CN V)	Branches of CN V are anesthetized for procedures on face or anterior scalp; compression of the nerve may result in painful condition known as trigeminal neuralgia	S–59, S–60
Olfactory nerve (CN I)	One of most commonly injured cranial nerves; can be avulsed at cribriform plate following falls, resulting in anosmia	S–56, S–364
Facial nerve (CN VII)	Idiopathic unilateral facial nerve palsy (Bell's palsy) can result in inability to fully close eye and result in desiccated cornea ipsilaterally	S–64
Recurrent laryngeal nerve	May be compressed or damaged by procedures in neck (e.g., thyroidectomy), aortic arch aneurysm, or lung cancer, producing hoarseness of voice; identified by (posterior) suspensory ligament of thyroid and/or inferior thyroid artery and/or tracheoesophageal groove	S–526, S–527
Oculomotor (CN III), trochlear (CN IV), and abducens (CN VI) nerves	Cavernous sinus thrombosis can result in dysfunction of extraocular muscles caused by compression of one, two, or all three nerves; abducens nerve is most commonly affected	S–30
Superior colliculus and aqueduct of midbrain	Tumor of midbrain can result in compression of aqueduct of midbrain, with resultant hydrocephalus	S–35
Optic nerve (CN II)	A pituitary gland mass may cause compression at the optic chiasm and resulting bitemporal hemianopsia; optic radiations (Meyer's loop) can be affected by temporal lobe tumors; early sign of ophthalmic artery aneurysm is visual loss due to compression of overlying optic nerve	S–45, S–57
Fovea centralis	Location of highest density of retinal cones, making this part of the macula lutea site of greatest visual acuity and color vision	S–88
Utricle and saccule	Location of calcium carbonate crystals known as otoconia; accumulation of otoconia in semicircular canals is most common cause of vertigo and is known as benign paroxysmal positional vertigo (BPPV)	S–96, S–97
Lens	Degeneration and opacification, known as cataract, may lead to progressive vision loss	S–88
Orbital septum	Infections anterior to this structure are known as preseptal/periorbital cellulitis and are milder than infections extending posterior to it, known as orbital cellulitis	S–82
Anterior segment of eyeball	Ciliary body produces aqueous humor, which flows through iris into anterior chamber and drains through scleral sinus; anterior displacement of lens may obstruct flow of aqueous humor and cause elevated intraocular pressure, a painful, vision-threatening condition known as acute angle closure glaucoma	S–89, S–91
Tympanic membrane	Visualized with otoscope; bulging indicates middle ear infection with effusion; rupture and otorrhea may occur in severe infection; tympanostomy tubes (T-tubes) may be placed in children with recurrent effusive infections	S–94
External acoustic meatus	May become infected or inflamed in children, a condition known as otitis externa; can be diagnosed with auricle (pinna) pull technique	S–93

Table 2.1 **Structures with High Clinical Significance**

ANATOMIC STRUCTURES	CLINICAL IMPORTANCE	PLATE NUMBERS
Back		
Conus medullaris	Is inferior limit of spinal cord; can lie as inferior as L4 vertebra in neonates and as superior as T12 in adults (average is L1–L2); necessary to determine its location in procedures such as lumbar puncture; in adults, start at L2 or lower	S–16
Cauda equina	Lumbar and sacral nerve roots may be anesthetized with anesthesia injected into subarachnoid space (spinal block)	S–16, S–18
Spinal meninges	Access to epidural and subarachnoid spaces is necessary for clinical procedures such as epidural anesthesia and lumbar puncture; meningitis is a life-threatening infection	S–22, S–138
Thorax		
Long thoracic nerve	May be damaged during chest tube placement or mastectomy, resulting in winged scapula (denervation of serratus anterior muscle)	S–203, S–474
Intercostal nerve	Site of local anesthetic nerve block for procedures such as thoracostomy or to alleviate pain caused by herpes zoster (shingles)	S–205, S–206
Spinal (posterior root) ganglion	Can house dormant varicella zoster virus, which, when activated, can result in herpes zoster (shingles)	S–206
Phrenic nerve	Surgical injury to phrenic nerve may cause ipsilateral paralysis of diaphragm; diaphragmatic irritation may manifest as shoulder pain because of referral to C3–C5 spinal levels	S–75, S–208, S–309
Recurrent laryngeal nerve	The left nerve takes a circuitous path around the aorta and may rarely become compressed by a large thoracic aortic aneurysm or enlarged left atrium, producing hoarse voice (Ortner's syndrome); more often, these branches are affected by malignancies, such as those of the thyroid gland	S–391, S–410
Thoracic cardiac nerves (sympathetic)	Pain of myocardial ischemia referred to upper thoracic dermatomes; may be perceived as somatic pain in thorax and medial upper limb	S–320
Abdomen		
Ilioinguinal and genitofemoral nerves	Mediate cremasteric reflex; femoral branch of genitofemoral nerve provides cutaneous innervation to skin over femoral triangle	S–103
Intercostal, subcostal, and iliohypogastric nerves	Convey well-localized pain sensations from abdominal wall and parietal peritoneum; pain in dermatomal distribution indicates problem with spinal nerves (e.g., herpes zoster infection)	S–102
Celiac ganglion	Some patients with medically intractable pain from chronic pancreatitis or advanced pancreatic malignancy undergo celiac ganglion block; located typically at upper or lower L1 vertebral level	S–448, S–457
Thoracic and lumbar splanchnic nerves (sympathetic)	Convey pain sensations from abdominal viscera that are often referred to other sites; quadrant in which pain is located and site of radiation provide clues to source of pain	S–450, S–453
Iliohypogastric nerve	Nephrectomy through quadratus lumborum muscle can damage iliohypogastric nerve, with resultant anesthesia superior to the pubic symphysis	S–460
Pelvis		
Pudendal nerve	Pudendal block is performed to anesthetize the perineum for childbirth or minor surgical procedures in the perineum	S–516
Inferior anal (rectal) nerve	Anesthetized in ischioanal fossa for surgical excision of external hemorrhoids	S–520

Structures with High Clinical Significance

Table 2.2

ANATOMIC STRUCTURES	CLINICAL IMPORTANCE	PLATE NUMBERS
Pelvis—Continued		
Prostatic plexus and cavernous nerves	Disruption of these nerves during procedures (e.g., prostate surgery) can produce inability to achieve erection	S–517
Upper Limb		
Long thoracic nerve	Injury may produce "winged scapula" caused by denervation of serratus anterior muscle; can be injured with repetitive overhead motion	S–235, S–237
Axillary nerve	Position of nerve close to surgical neck of humerus makes it vulnerable to injury with fractures or dislocations of humerus; poorly fitting crutches can also compress the axillary nerve	S–240
Median nerve	Compressed in carpal tunnel syndrome, producing pain and paresthesia in lateral three and one-half digits; major risk factors include obesity, pregnancy, diabetes, and hypothyroidism	S–109, S–251
Recurrent branch of median nerve	May be injured in superficial lacerations of palm over thenar eminence	S–250
Ulnar nerve	Vulnerable to compression or injury where it passes posterior to medial epicondyle of humerus, and at wrist in ulnar tunnel (Guyon's canal)	S–110, S–227
Radial nerve	Vulnerable to compression or injury where it lies against humerus in radial groove (e.g., with humerus fracture); common symptom is wrist drop due to weakness of wrist extensors; poorly fitting crutches can also compress the radial nerve	S–111, S–112
Lower Limb		
Sural nerve	Nerve is commonly biopsied for peripheral neuropathies and commonly used as donor graft in neurotization procedures	S–117, S–267
Common fibular nerve	Injury to this nerve from blunt trauma or compression by leg cast weakens dorsiflexion and results in foot drop	S–115, S–117, S–119
Obturator nerve	Nerve is blocked or transected for adductor muscle spasticity in cerebral palsy; may be injured during pelvic fractures or pelvic surgical procedures such as lymphadenectomy	S–116
Femoral nerve	Can be compressed from femoral artery hematoma and can be anesthetized for procedures of the lower limb just below inguinal ligament	S–113, S–115, S–270
Saphenous nerve	Can be anesthetized in the adductor canal to provide pain relief after knee replacement surgery	S–115, S–266, S–270
Lateral femoral cutaneous nerve	Compression at the inguinal ligament leads to meralgia paresthetica, a pain and paresthesia syndrome of the anterolateral thigh; risk factors include obesity, pregnancy, and tight-fitting waistbands	S–115, S–270

*Selections are based largely on clinical data and commonly discussed clinical correlations in macroscopic ("gross") anatomy courses.

Table 2.3   **Structures with High Clinical Significance**

Cranial nerves are traditionally described as tree-like structures that emerge from the brain and branch peripherally. This matches the direction actions potentials travel in efferent fibers in the nerve. It must be remembered that action potentials travel in the opposite direction in afferent fibers within these nerves.

NERVE	ORIGIN	COURSE	BRANCHES	MOTOR	SENSORY
Olfactory nerve (CN I)	Olfactory bulb	Neurons of olfactory mucosa send approximately 20 axon bundles through cribriform plate to synapse on olfactory bulb neurons			SVA (smell): olfactory epithelium
Optic nerve (CN II)	Optic chiasm	Axons of retinal ganglion neurons exit orbit through optic canal to enter cranial cavity			SSA (vision): optic part of retina
Oculomotor nerve (CN III)	Interpeduncular fossa of midbrain	Exits midbrain into posterior cranial fossa and then middle cranial fossa; traverses cavernous sinus to enter orbit via superior orbital fissure	Superior and inferior branches	GSE: medial, superior, and inferior rectus, and inferior oblique muscles, and levator palpebrae superioris GVE: ciliary ganglion	
Trochlear nerve (CN IV)	Posterior surface of midbrain	Exits dorsal midbrain, coursing lateral to cerebral peduncle to anterior surface of brain stem; follows medial edge of tentorium cerebelli to enter middle cranial fossa; traverses cavernous sinus to enter orbit via superior orbital fissure		GSE: superior oblique muscle	
Trigeminal nerve (CN V)	Motor and sensory roots (arising from the pons)	Exits anterolateral pons into posterior cranial fossa; large sensory and small motor roots enter middle cranial fossa; sensory root forms trigeminal ganglion, which will give rise to ophthalmic, maxillary, and mandibular nerves; motor root passes deep to ganglion and contributes only to the mandibular nerve	Ophthalmic, maxillary, and mandibular nerves	SVE: see branches	GSA: see branches
Ophthalmic nerve (CN V₁)	Trigeminal nerve	Exits anterior border of trigeminal ganglion and traverses cavernous sinus to leave cranium through superior orbital fissure into orbit	Lacrimal, frontal, and nasociliary nerves, meningeal branch		GSA: forehead, upper eyelid, conjunctiva
Maxillary nerve (CN V₂)	Trigeminal nerve	Exits trigeminal ganglion, passes through foramen rotundum into pterygopalatine fossa, and enters orbit via inferior orbital fissure	Nasopalatine, pharyngeal, palatine, zygomatic, posterior superior alveolar and infraorbital nerves, superior posterior nasal and meningeal branches		GSA: midface, nasal cavity, paranasal sinuses, palate, maxillary teeth
Mandibular nerve (CN V₃)	Trigeminal nerve	Exits trigeminal ganglion inferiorly, leaves the cranium through foramen ovale, and enters infratemporal fossa.	Deep temporal, buccal, auriculotemporal, lingual, and inferior alveolar nerves, meningeal branch	SVE: muscles of mastication, mylohyoid and digastric (anterior belly) muscles, tensor tympani, tensor veli palatini	GSA: mandibular teeth, anterior tongue, floor of oral cavity, temporomandibular joint

Continued

NERVE	ORIGIN	COURSE	BRANCHES	MOTOR	SENSORY
Abducens nerve (CN VI)	Medullopontine sulcus (medial to CN VII)	Exits between pons and medulla near midline, pierces dura on clivus and grooves on petrous part of temporal bone to access middle cranial fossa; traverses cavernous sinus to enter orbit through superior orbital fissure		GSE: lateral rectus muscle	
Facial nerve (CN VII)	Motor root and intermediate nerve (sensory root)	Exits laterally between pons and medulla oblongata as a larger motor root and smaller intermediate nerve (carrying GVA, GVE, and GSA fibers); both roots traverse internal acoustic meatus to enter facial canal in which a sharp turn (genu) occurs just before the geniculate (facial) ganglion; facial nerve gives off several branches within facial canal, before it exits through stylomastoid foramen and forms terminal branches within parotid gland	Greater petrosal and posterior auricular nerves, chorda tympani, and temporal, zygomatic, buccal, marginal mandibular, and cervical branches	SVE: muscles of facial expression (including epicranial muscles and platysma), digastric (posterior belly), stylohyoid, and stapedius muscles GVE: pterygopalatine and submandibular ganglia	SVA (taste): anterior tongue GSA: part of external ear
Vestibulocochlear nerve (CN VIII)	Medullopontine sulcus (lateral to CN VII)	Exits brain stem between pons and medulla oblongata, near inferior cerebellar peduncle, and divides into two branches as it traverses posterior cranial fossa	Vestibular and cochlear nerves		SSA: internal ear (see branches)
Cochlear nerve	Vestibulocochlear nerve	Contains processes of cochlear ganglion cells; passes through internal acoustic meatus into internal ear			SSA (hearing): spiral organ of cochlear duct
Vestibular nerve	Vestibulocochlear nerve	Contains processes of vestibular ganglion cells; passes through internal acoustic meatus into internal ear	Superior and inferior branches		SSA (equilibrium and motion): maculae and cristae ampullares of vestibular labyrinth
Glossopharyngeal nerve (CN IX)	Retroolivary groove of medulla oblongata	Emerges from upper medulla between olive and inferior cerebral peduncle and exits cranium by traversing jugular foramen with vagus and accessory nerves; superior and inferior ganglia for afferent components of CN IX are located just inferior to jugular foramen; passes inferiorly, innervating and following stylopharyngeus muscle, ultimately sending a branch into posterior oral cavity that passes deep to hyoglossus muscle and a branch that enters pharynx by passing between superior and middle pharyngeal constrictors	Tympanic nerve	SVE: stylopharyngeus muscle GVE: otic ganglion	SVA (taste): posterior tongue GVA: carotid body and sinus GSA: posterior tongue, oropharynx, middle ear

Table 2.5   **Cranial Nerves**

NERVE	ORIGIN	COURSE	BRANCHES	MOTOR	SENSORY
Vagus nerve (CN X)	Retro-olivary groove of medulla oblongata (between CN IX and cranial roots of CN XI)	Exits from retro-olivary groove of medulla oblongata to traverse jugular foramen with accessory and glossopharyngeal nerves; initially runs between internal carotid artery and internal jugular vein; courses of right and left vagus nerves differ, with right vagus nerve coursing inferiorly between subclavian artery and vein, at which point right recurrent laryngeal nerve ascends and vagus nerve descends along trachea and root of right lung, forming pulmonary and esophageal plexuses; left vagus nerve descends between left subclavian and common carotid arteries posteriorly to left brachiocephalic vein, anteriorly to aortic arch, at which point left recurrent laryngeal nerve ascends posteriorly to aorta and rest of vagus nerve descends posteriorly to root of lung, forming pulmonary and esophageal plexuses; vagus nerve fibers continue into abdomen via esophageal plexuses and vagal trunks	Pharyngeal branch, superior and recurrent laryngeal nerves	GVE: thoracic and abdominal visceral ganglia SVE: see branches	GVA: aortic arch and bodies SVA (taste): epiglottis, oropharynx GSA: external ear (also see branches)
Pharyngeal branch of vagus nerve	Vagus nerve	Passes between carotid arteries superficial to middle pharyngeal constrictor		SVE: pharyngeal constrictors, palatoglossus, palatopharyngeus, and salpingopharyngeus muscles, and levator veli palatini	
Superior laryngeal nerve	Vagus nerve	Exits inferior ganglion, passing inferiorly and medially deep to internal carotid artery, dividing into external (motor) branch and internal (sensory) branch that pierces thyrohyoid membrane	Internal and external branches	SVE: cricothyroid muscle	GSA: superior larynx

Continued

NERVE	ORIGIN	COURSE	BRANCHES	MOTOR	SENSORY
Recurrent laryngeal nerve	Vagus nerve	On right side, arises anterior to subclavian artery and spirals to run superiorly and posteriorly to common carotid and inferior thyroid arteries, lateral to trachea; on left side, arises inferior to aortic arch, passing posterior to arch lateral to ligamentum arteriosum to ascend superiorly lateral to trachea; both right and left nerves enter larynx at junction of esophagus and inferior pharyngeal constrictor		SVE: intrinsic muscles of larynx (except cricothyroid), striated muscle of esophagus	GSA: inferior larynx
Accessory nerve (CN XI)	Cranial and spinal roots	Spinal roots from C1–C5 segments of spinal cord ascend and enter cranium through foramen magnum, where they course with cranial root arising from retro-olivary groove of medulla oblongata; spinal accessory nerve exits from cranium via jugular foramen; it passes inferiorly and posteriorly into upper third of sternocleidomastoid muscle then inferiorly, crossing occipital triangle to enter trapezius muscle; vagal communicating branch joins vagus nerve at jugular foramen to innervate larynx, palate, and pharyngeal muscles	Spinal accessory nerve, vagal communicating branch	GSE: trapezius and sternocleidomastoid muscles	
Hypoglossal nerve (CN XII)	Hypoglossal rootlets and hypoglossal communicating branch of C1 spinal nerve	Rootlets from anterolateral sulcus of medulla oblongata exit cranium through hypoglossal canal, then pass inferiorly and anteriorly between vagus and spinal accessory nerves to inferior border of posterior belly of digastric muscle eventually entering oral cavity by passing between mylohyoid and hyoglossus muscles	Lingual branches and branches conveying fibers from C1 spinal nerve (thyrohyoid and geniohyoid branches, and superior root of ansa cervicalis)	GSE: intrinsic muscles of tongue, and genioglossus, hyoglossus, and styloglossus muscles C1: thyrohyoid, geniohyoid, and omohyoid (superior belly) muscles	

Table 2.7   **Cranial Nerves**

The roots of the cervical plexus are the anterior rami of C1–C4 spinal nerves.

NERVE	ORIGIN	COURSE	BRANCHES	MOTOR	CUTANEOUS SENSORY
Hypoglossal communicating branch	Anterior ramus of C1 spinal nerve	Emerges and briefly adheres to hypoglossal nerve; superior root of ansa cervicalis exits just posterior to greater horn of hyoid bone and descends along carotid sheath where it joins inferior root at C4–C5 level	Superior root of ansa cervicalis, thyrohyoid and geniohyoid branches of hypoglossal nerve	Omohyoid (superior belly), thyrohyoid, and geniohyoid muscles	
Inferior root of ansa cervicalis	Anterior rami of C2–C3 spinal nerves	Descends along anterolateral carotid sheath, joining superior root at C4–C5 level	Infrahyoid branches	Omohyoid (inferior belly), sternohyoid, and sternothyroid muscles	
Muscular branches of cervical plexus	Anterior rami of C1–C4 spinal nerves	Three loops form along C1–C4 vertebrae that course laterally between levator scapulae and scalenus medius muscle, deep to sternocleidomastoid muscle		Rectus anterior and lateralis capitis, longus capitis and longus colli, scalenus anterior, medius and posterior muscles; levator scapulae	
Phrenic nerve	Anterior rami of C3–C5 spinal nerves	Descends on scalenus anterior muscle deep to inferior belly of omohyoid muscle and transverse cervical and suprascapular vessels; enters thorax between subclavian vein and artery, passing anterior to root of lung and along lateral border of pericardium to pierce diaphragm		Diaphragm	
Lesser occipital nerve	Anterior ramus of C2 spinal nerve	Formed in posterior cervical triangle deep to sternocleidomastoid muscle, ascending along its posterior border; perforates deep fascia at mastoid process, ascending posterior to ear			Temporal, auricular, and mastoid regions
Great auricular nerve	Anterior rami of C2–C3 spinal nerves	Formed in posterior cervical triangle deep to sternocleidomastoid muscle, ascending obliquely between that muscle and platysma	Anterior and posterior branches		Parotid, auricular, and mastoid regions
Transverse cervical nerve	Anterior rami of C2–C3 spinal nerves	Formed in posterior cervical triangle deep to sternocleidomastoid muscle, runs superficial to that muscle, passing deep to external jugular vein	Superior and inferior branches		Anterior and lateral regions of neck
Supraclavicular nerve	Anterior rami of C3–C4 spinal nerves	Formed in posterior cervical triangle deep to mid-third of sternocleidomastoid muscle, passes lateral to external jugular vein, and descends just inferior to clavicle	Medial, intermediate, and lateral supraclavicular nerves		Clavicular and infraclavicular regions

The roots of the brachial plexus are typically the anterior rami of C5–T1 spinal nerves. Variations in the spinal nerve contributions to the plexus, and the nerves that arise from this plexus, are common due to prefixed ("high") and postfixed ("low") plexuses.

NERVE	ORIGIN	COURSE	BRANCHES	MOTOR	CUTANEOUS
Dorsal scapular nerve	Anterior ramus of C5 spinal nerve	Pierces scalenus medius muscle to run posteriorly and inferiorly on levator scapulae along vertebral border of scapula		Rhomboid major and minor muscles, levator scapulae	
Long thoracic nerve	Anterior rami of C5–C7 spinal nerves	C5–C6 join within scalenus medius muscle, and at 1st rib are joined by C7; runs inferiorly and posterior to brachial plexus and axillary vessels; follows midaxillary line on surface of serratus anterior muscle		Serratus anterior muscle	
Suprascapular nerve	Superior trunk (C5–C6)	Traverses posterior cervical triangle, coursing posterior to inferior belly of omohyoid muscle and border of trapezius muscle to pass through scapular notch deep to superior transverse scapular ligament; continues laterally and then through spinoglenoid notch into infraspinous fossa		Supraspinatus and infraspinatus muscles	
Subclavian nerve	Superior trunk (C5–C6)	Runs inferiorly at distal aspect of anterior rami		Subclavius muscle	
Lateral pectoral nerve	Lateral cord (C5–C7)	Emerges lateral and superficial to axillary artery and vein, coursing just medial to pectoralis minor muscle		Pectoralis major and minor muscles	
Musculocutaneous nerve	Lateral cord (C5–C7)	Emerges at inferior border of pectoralis minor muscle, pierces coracobrachialis muscle to run between brachialis and biceps brachii muscles; just proximal to elbow, pierces deep fascia to continue as lateral antebrachial cutaneous nerve	Muscular branches, lateral antebrachial cutaneous nerve	Anterior compartment of arm	See lateral antebrachial cutaneous nerve
Lateral antebrachial cutaneous nerve	Musculocutaneous nerve	Runs posterior to cephalic vein and travels along lateral surface of forearm	Divides at elbow joint into anterior and posterior branches		Lateral forearm
Subscapular nerves	Posterior cord (C5–C6)	Upper and lower subscapular nerves emerge to traverse anterior surface of subscapularis muscle; lower subscapular nerve ends in teres major muscle		Teres major and subscapularis muscles	

Table 2.9

Nerves of Brachial Plexus

NERVE	ORIGIN	COURSE	BRANCHES	MOTOR	CUTANEOUS
Thoracodorsal nerve	Posterior cord (C6–C8)	Emerges between upper and lower subscapular nerves, courses with thoracodorsal artery along posterior axillary wall, diving deep to latissimus dorsi muscle		Latissimus dorsi muscle	
Radial nerve	Posterior cord (C5–T1)	Runs anterior to latissimus dorsi muscle to inferior border of teres major muscle, where it accompanies deep brachial artery along radial groove of humerus to course between medial and lateral heads of triceps brachii muscle	Posterior and inferior lateral brachial cutaneous nerves, posterior antebrachial cutaneous nerve, muscular, deep, and superficial branches and posterior interosseous nerve	Triceps brachii, anconeus, and brachioradialis muscles, extensor carpi radialis longus and brevis, supinator (also see posterior interosseous nerve)	Lateral part of dorsum of hand (also see cutaneous branches)
Posterior brachial cutaneous nerve	Radial nerve	Emerges from radial nerve in medial axilla			Posterior part of medial arm
Inferior lateral brachial cutaneous nerve	Radial nerve	Perforates lateral head of triceps brachii muscle below deltoid tuberosity, coursing anteriorly with cephalic vein			Distal part of lateral arm
Posterior antebrachial cutaneous nerve	Radial nerve	Emerges from plane between lateral and medial heads of triceps brachii muscle to become cutaneous			Posterior part of lateral forearm
Posterior interosseous nerve	Deep branch of radial nerve	Continuation of deep radial nerve courses under cover of supinator distally along posterior surface of interosseous membrane of forearm		Posterior compartment of forearm (some exceptions)	
Axillary nerve	Posterior cord (C5–C6)	Passes anterior to subscapularis muscle to exit axilla with posterior circumflex humeral artery through quadrangular space	Muscular branches, superior lateral brachial cutaneous nerve	Deltoid and teres minor muscles	See superior lateral brachial cutaneous nerve
Superior lateral brachial cutaneous nerve	Axillary nerve	Pierces deep fascia at posteroinferior edge of deltoid muscle to become cutaneous			Proximal part of lateral arm
Medial pectoral nerve	Medial cord (C8–T1)	Emerges and runs between axillary artery and vein to pierce pectoralis minor muscle en route to pectoralis major muscle		Pectoralis minor and major muscles	
Medial brachial cutaneous nerve	Medial cord (T1)	Emerges and traverses axilla anterior to latissimus dorsi muscle, running posteromedial with axillary vein, piercing deep fascia to descend with basilic vein	Anterior and posterior branches		Anterior part of medial arm
Medial antebrachial cutaneous nerve	Medial cord (C8–T1)	Emerges medial to axillary artery, traverses axilla to pierce deep fascia supplying anterior arm, and continues on ulnar side of forearm with basilic vein	Anterior and posterior branches		Anterior arm, medial part of forearm

Continued

Nerves of Brachial Plexus

Table 2.10

NERVE	ORIGIN	COURSE	BRANCHES	MOTOR	CUTANEOUS
Ulnar nerve	Medial cord (C7–T1)	Emerges medial to axillary artery, continuing medial to brachial artery along medial head of triceps brachii muscle in groove for ulnar nerve between olecranon and medial epicondyle; enters forearm between heads of flexor carpi ulnaris; runs distally between flexor carpi ulnaris and flexor digitorum profundus, giving off dorsal branch before entering hand	Muscular, dorsal, palmar, superficial, and deep branches	Flexor carpi ulnaris, flexor digitorum profundus (medial half), adductor pollicis, hypothenar muscles, dorsal and palmar interosseous muscles, lumbrical muscles (medial two)	Medial part of palm and dorsum of hand, 5th finger and part of 4th
Median nerve	Medial and lateral cords (C6–T1)	Emerges and runs distally with brachial artery to enter forearm between heads of pronator teres; courses distally on deep surface of flexor digitorum superficialis to become superficial at flexor retinaculum of wrist; traverses carpal tunnel deep to flexor retinaculum of wrist	Anterior interosseous nerve, muscular, palmar, recurrent, and common palmar digital branches	Anterior compartment of forearm (some exceptions), lumbricals (lateral two), and thenar muscles (also see anterior interosseous nerve)	Lateral part of palm, thumb, 2nd and 3rd fingers and part of 4th finger
Anterior interosseous nerve	Median nerve	At elbow runs distally with anterior interosseous artery along anterior surface of interosseous membrane of forearm		Flexor pollicis longus, pronator quadratus, flexor digitorum profundus (lateral half)	

Table 2.11   **Nerves of Brachial Plexus**

The lumbosacral plexus includes the lumbar, sacral, and coccygeal plexuses, whose roots are the anterior rami of spinal nerves (lumbar plexus: typically L2–L4 with a small contribution from L1; sacral plexus: L4–S4; coccygeal plexus: S4–Co). Variations in the spinal nerve contributions to the plexuses, and the nerves that arise from these plexuses, are common and can be due to a prefixed ("high") or postfixed ("low") plexus.

NERVE	ORIGIN	COURSE	BRANCHES	MOTOR	CUTANEOUS
Lateral femoral cutaneous nerve	Posterior divisions of lumbar plexus (L2–L3)	Courses posterior to or through inguinal ligament medial to anterior superior iliac spine, then travels superficial in lateral thigh			Lateral thigh
Femoral nerve	Posterior divisions of lumbar plexus (L2–L4)	Passes posterior to inguinal ligament to lie on iliacus muscle in femoral triangle	Muscular and anterior cutaneous branches, saphenous nerve	Anterior compartment of thigh, iliacus and pectineus muscles	Anterior thigh
Saphenous nerve	Femoral nerve	Leaves femoral nerve in adductor canal, pierces fascia lata to travel superficially with great saphenous vein	Infrapatellar and medial crural cutaneous branches		Medial knee, leg, ankle, and foot
Genitofemoral nerve	Anterior divisions of lumbar plexus (L1–L2)	Runs on anterior surface of psoas major; genital branch traverses inguinal canal; femoral branch courses into femoral triangle by passing deep to inguinal ligament	Genital and femoral branches	Cremaster	Lateral part of femoral triangle, anterior scrotum/vulva
Obturator nerve	Anterior divisions of lumbar plexus (L2–L4)	At L5 vertebra passes deep to common iliac vessels, enters medial thigh via obturator foramen	Anterior and posterior branches	Medial compartment of thigh	Medial thigh
Superior gluteal nerve	Posterior divisions of sacral plexus (L4–S1)	Exits pelvis via greater sciatic foramen superior to piriformis muscle		Gluteus medius and minimus muscles, tensor fasciae latae	
Inferior gluteal nerve	Posterior divisions of sacral plexus (L5–S2)	Exits pelvis via greater sciatic foramen inferior to piriformis muscle		Gluteus maximus muscle	
Nerve to piriformis muscle	Posterior divisions of sacral plexus (S1–S2)			Piriformis muscle	
Perforating cutaneous nerve	Posterior divisions of sacral plexus (S2–S3)	Exits pelvis and pierces sacrotuberous ligament			Inferomedial gluteal region
Nerve to obturator internus	Anterior divisions of sacral plexus (L5–S2)	Exits pelvis via greater sciatic foramen, inferior to piriformis muscle; re-enters pelvis via lesser sciatic foramen to innervate obturator internus		Obturator internus and superior gemellus muscle	
Nerve to quadratus femoris muscle	Anterior divisions of sacral plexus (L4–S1)	Exits pelvis via greater sciatic foramen inferior to piriformis muscle		Quadratus femoris and inferior gemellus muscles	
Nerve to levator ani	Anterior divisions of sacral plexus (S3–S4)	From its source, travels on superior surface of levator ani		Iliococcygeus and pubococcygeus muscles	
Nerve to coccygeus muscle	Anterior divisions of sacral plexus (S3–S4)	From its source, travels on superior surface of coccygeus muscle		Coccygeus (ischiococcygeus) muscle	

Continued

Nerves of Lumbosacral Plexus

NERVE	ORIGIN	COURSE	BRANCHES	MOTOR	CUTANEOUS
Pudendal nerve	Anterior divisions of sacral plexus (S2–S4)	Exits pelvis via greater sciatic foramen inferior to piriformis muscle; enters perineum via lesser sciatic foramen, coursing through ischioanal fossa and pudendal (Alcock's) canal	Inferior anal and perineal nerves, dorsal nerve of clitoris or penis	Perineal muscles, external anal sphincter, pubococcygeus muscle	Posterior scrotum/vulva, clitoris or penis
Posterior femoral cutaneous nerve	Anterior and posterior divisions of sacral plexus (S1–S3)	Exits pelvis via greater sciatic foramen inferior to piriformis muscle; runs inferiorly to gluteus maximus muscle and continues distally to popliteal fossa	Inferior gluteal cutaneous nerves, perineal branches		Inferior gluteal region, posterior thigh, popliteal region
Sciatic nerve	Anterior and posterior divisions of sacral plexus (L4–S3)	Exits pelvis via greater sciatic foramen inferior to piriformis muscle, passing superficial to lateral rotators and deep to gluteus medius muscle to access thigh between greater trochanter of femur and ischial tuberosity	Muscular branches, tibial and common fibular nerves	Posterior compartment of thigh (also see branches)	See branches
Tibial nerve	Sciatic nerve	Courses through popliteal fossa and sural region deep to soleus muscle, traversing tarsal tunnel posterior to medial malleolus	Muscular and medial calcaneal branches, crural interosseous nerve, medial sural cutaneous nerve, medial and lateral plantar nerves	Posterior compartment of leg (also see branches)	Heel (also see branches)
Common fibular nerve	Sciatic nerve	Courses medial to biceps femoris muscle and lateral to lateral head of gastrocnemius muscle and neck of fibula	Superficial and deep fibular nerves, lateral sural cutaneous nerve, sural communicating branch	See branches	See branches
Deep fibular nerve	Common fibular nerve	Courses deep to fibularis longus muscle and extensor digitorum longus on surface of interosseous membrane of leg deep to extensor retinaculu	Muscular and dorsal digital branches	Anterior compartment of leg	Dorsal aspect of adjacent parts of toes 1 and 2
Superficial fibular nerve	Common fibular nerve	Courses between fibularis longus and brevis muscles in lateral compartment of leg, piercing crural fascia distally	Muscular branches, medial and intermediate dorsal cutaneous nerves of foot	Lateral compartment of leg	Inferior part of anterior leg, dorsum of foot and toes
Lateral sural cutaneous nerve	Common fibular nerve	Branches from common fibular nerve just proximal to plantaris muscle			Lateral leg
Medial sural cutaneous nerve	Tibial nerve	Branches from tibial nerve just proximal to plantaris muscle, runs between both heads of gastrocnemius muscle, pierces crural fascia	Sural nerve		Proximal posterolateral leg
Sural nerve	Union of medial sural cutaneous nerve and sural communicating branch of common fibular nerve	Descends sural region along lateral calcaneal tendon and into foot between lateral malleolus and calcaneus	Lateral dorsal cutaneous nerve of foot		Distal posterolateral leg, lateral foot

Table 2.13

NERVE	ORIGIN	COURSE	BRANCHES	MOTOR	CUTANEOUS
Lateral plantar nerve	Tibial nerve	Passes deep to abductor hallucis brevis and travels between flexor digitorum brevis and quadratus plantae muscle	Superficial and deep branches	Abductor digiti minimi, adductor hallucis, flexor digiti minimi, quadratus plantae muscle, dorsal and plantar interossei muscles, lumbrical muscles (lateral three)	Lateral sole, plantar aspect of 5th toe and part of 4th toe
Medial plantar nerve	Tibial nerve	Passes deep to abductor hallucis brevis and runs anteriorly along medial border of flexor digitorum brevis	Muscular and common plantar digital branches	Abductor hallucis, flexor digitorum brevis, flexor hallucis brevis, lumbrical muscles (most medial one)	Medial sole, plantar aspect of toes 1–3 and part of 4th toe

SKELETAL SYSTEM 3

ELECTRONIC BONUS PLATES

S–BP 19 3D Skull Reconstruction CTs

S–BP 20 Degenerative Changes in Cervical Vertebrae

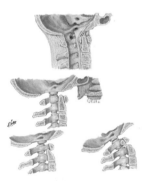

S–BP 21 Atlantooccipital Junction

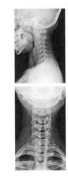

S–BP 22 Cervical Spine: Radiographs

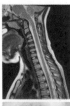

S–BP 23 Cervical Spine: MRI and Radiograph

S–BP 24 Lumbar Vertebrae: Radiographs

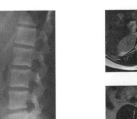

S–BP 25 Lumbar Spine: MRIs

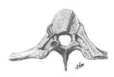

S–BP 26 Ligaments of Vertebral Column

ELECTRONIC BONUS PLATES—*cont'd*

S–BP 27 Vertebral Veins: Detail Showing Venous Communications

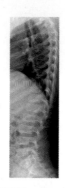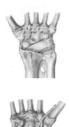

S–BP 28 Thoracolumbar Spine: Lateral Radiograph

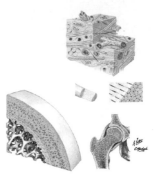

S–BP 29 Ligaments of Wrist: Posterior and Anterior Views

S–BP 30 Joints: Connective Tissues and Articular Cartilage

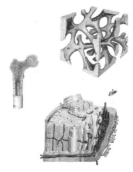

S–BP 31 Architecture of Bone

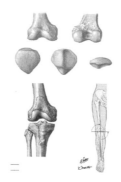

S–BP 32 Osteology of Knee

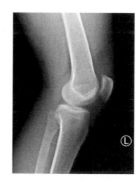

S–BP 33 Knee Radiograph: Lateral View

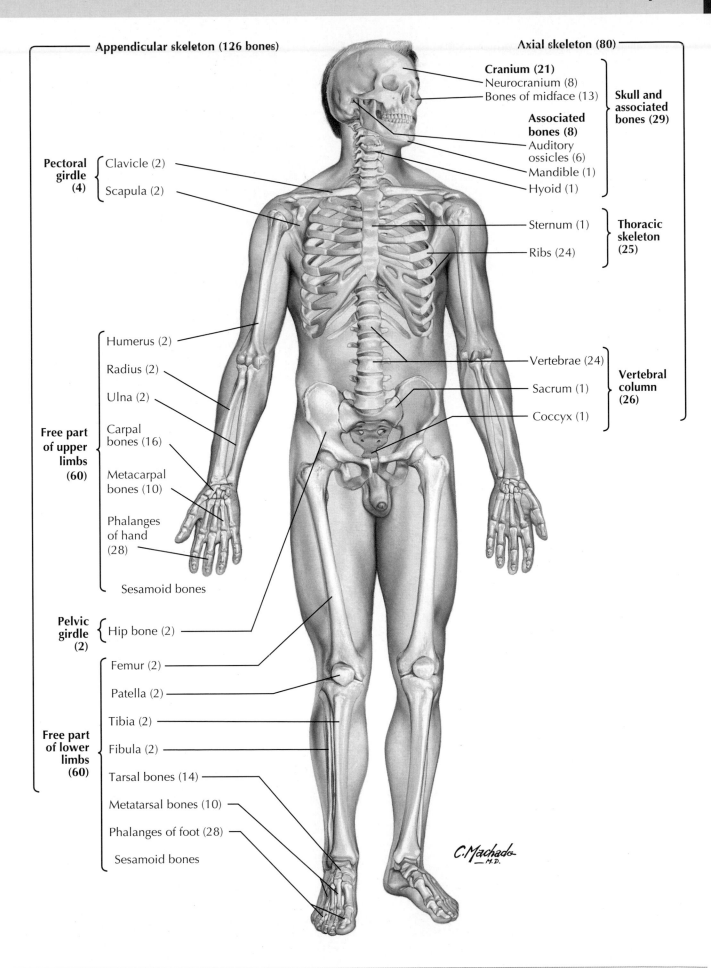

Appendicular skeleton (126 bones)

Axial skeleton (80)

Cranium (21)
Neurocranium (8)
Bones of midface (13)

Associated bones (8)
Auditory ossicles (6)
Mandible (1)
Hyoid (1)

Skull and associated bones (29)

Pectoral girdle (4)
Clavicle (2)
Scapula (2)

Sternum (1)
Ribs (24)

Thoracic skeleton (25)

Free part of upper limbs (60)
Humerus (2)
Radius (2)
Ulna (2)
Carpal bones (16)
Metacarpal bones (10)
Phalanges of hand (28)
Sesamoid bones

Vertebrae (24)
Sacrum (1)
Coccyx (1)

Vertebral column (26)

Pelvic girdle (2)
Hip bone (2)

Free part of lower limbs (60)
Femur (2)
Patella (2)
Tibia (2)
Fibula (2)
Tarsal bones (14)
Metatarsal bones (10)
Phalanges of foot (28)
Sesamoid bones

C. Machado
—M.D.

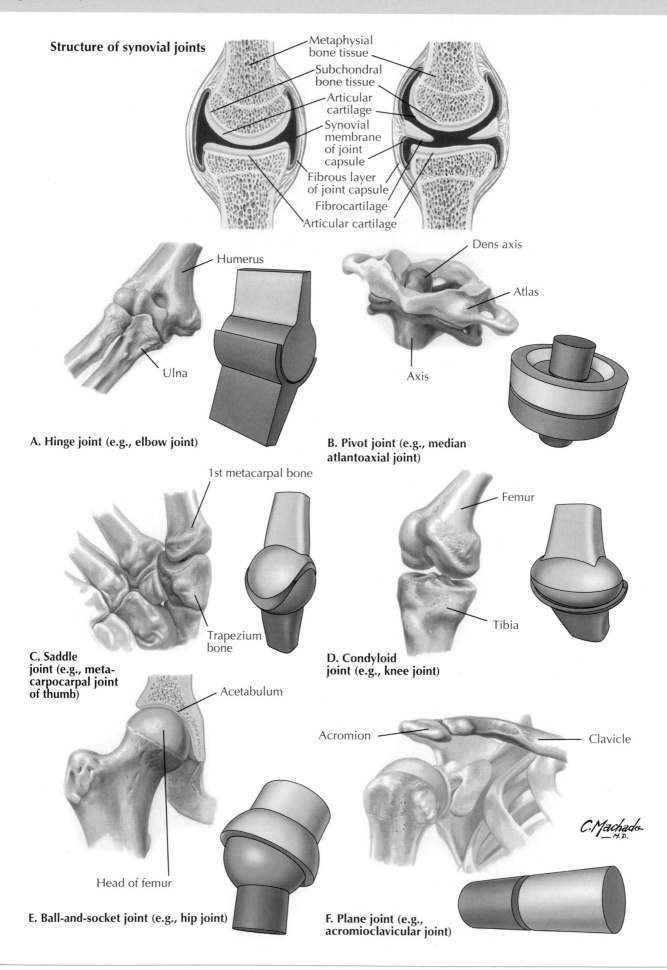

Structure of synovial joints

Metaphysial bone tissue
Subchondral bone tissue
Articular cartilage
Synovial membrane of joint capsule
Fibrous layer of joint capsule
Fibrocartilage
Articular cartilage

Humerus

Ulna

A. Hinge joint (e.g., elbow joint)

Dens axis

Atlas

Axis

B. Pivot joint (e.g., median atlantoaxial joint)

1st metacarpal bone

Trapezium bone

C. Saddle joint (e.g., meta-carpocarpal joint of thumb)

Femur

Tibia

D. Condyloid joint (e.g., knee joint)

Acetabulum

Head of femur

E. Ball-and-socket joint (e.g., hip joint)

Acromion

Clavicle

F. Plane joint (e.g., acromioclavicular joint)

Anterolateral view

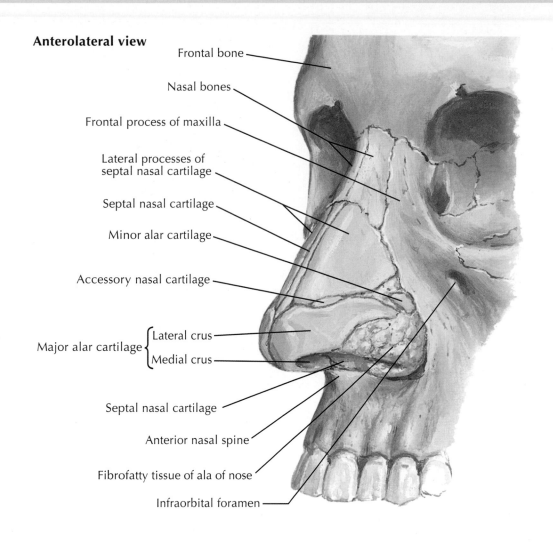

Frontal bone

Nasal bones

Frontal process of maxilla

Lateral processes of
septal nasal cartilage

Septal nasal cartilage

Minor alar cartilage

Accessory nasal cartilage

Major alar cartilage { Lateral crus

Medial crus }

Septal nasal cartilage

Anterior nasal spine

Fibrofatty tissue of ala of nose

Infraorbital foramen

Inferior view

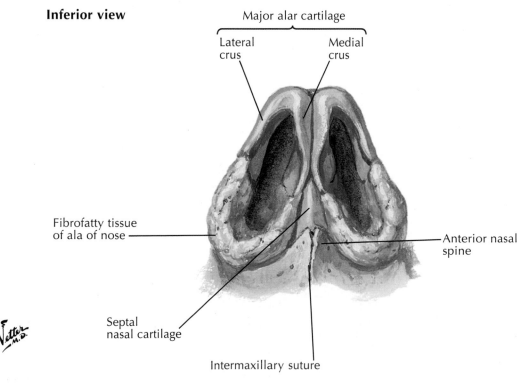

Major alar cartilage

Lateral
crus

Medial
crus

Fibrofatty tissue
of ala of nose

Anterior nasal
spine

Septal
nasal cartilage

Intermaxillary suture

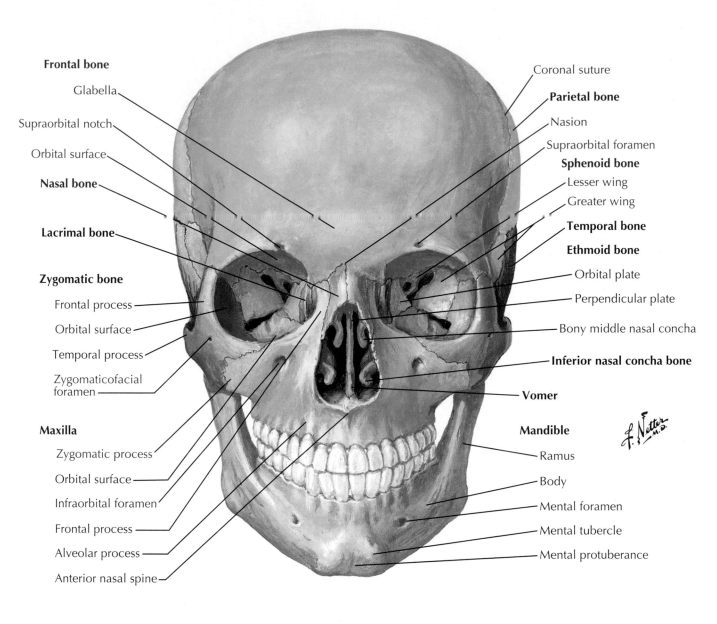

Frontal bone

Glabella

Supraorbital notch

Orbital surface

Nasal bone

Lacrimal bone

Zygomatic bone

Frontal process

Orbital surface

Temporal process

Zygomaticofacial foramen

Maxilla

Zygomatic process

Orbital surface

Infraorbital foramen

Frontal process

Alveolar process

Anterior nasal spine

Coronal suture

Parietal bone

Nasion

Supraorbital foramen

Sphenoid bone

Lesser wing

Greater wing

Temporal bone

Ethmoid bone

Orbital plate

Perpendicular plate

Bony middle nasal concha

Inferior nasal concha bone

Vomer

Mandible

Ramus

Body

Mental foramen

Mental tubercle

Mental protuberance

Right orbit: frontal view

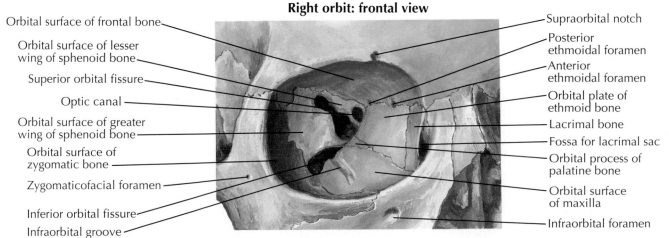

Orbital surface of frontal bone

Orbital surface of lesser wing of sphenoid bone

Superior orbital fissure

Optic canal

Orbital surface of greater wing of sphenoid bone

Orbital surface of zygomatic bone

Zygomaticofacial foramen

Inferior orbital fissure

Infraorbital groove

Supraorbital notch

Posterior ethmoidal foramen

Anterior ethmoidal foramen

Orbital plate of ethmoid bone

Lacrimal bone

Fossa for lacrimal sac

Orbital process of palatine bone

Orbital surface of maxilla

Infraorbital foramen

Posterior anterior view

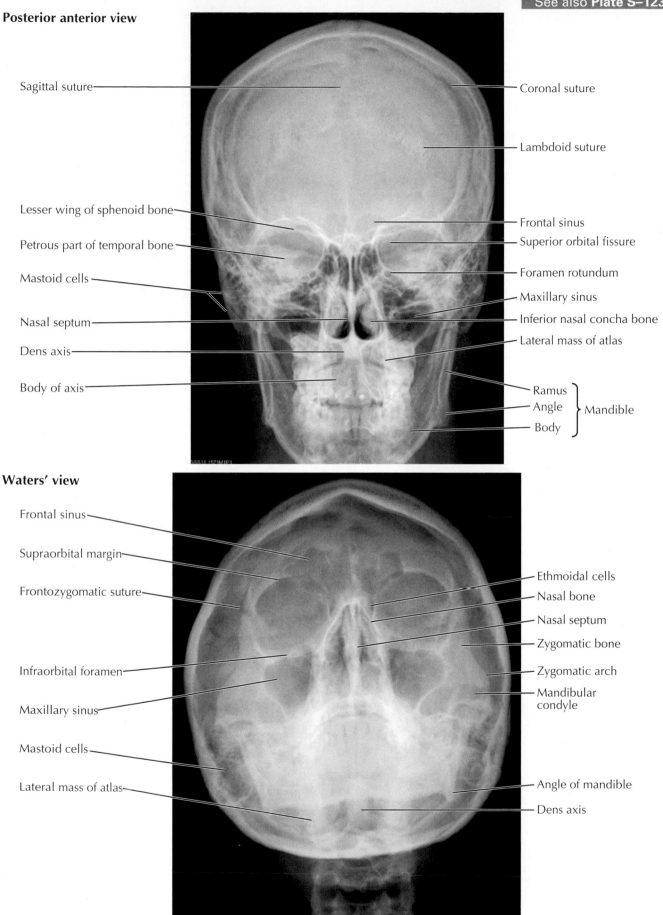

Sagittal suture

Lesser wing of sphenoid bone

Petrous part of temporal bone

Mastoid cells

Nasal septum

Dens axis

Body of axis

Coronal suture

Lambdoid suture

Frontal sinus

Superior orbital fissure

Foramen rotundum

Maxillary sinus

Inferior nasal concha bone

Lateral mass of atlas

Ramus

Angle

Body

} Mandible

Waters' view

Frontal sinus

Supraorbital margin

Frontozygomatic suture

Infraorbital foramen

Maxillary sinus

Mastoid cells

Lateral mass of atlas

Ethmoidal cells

Nasal bone

Nasal septum

Zygomatic bone

Zygomatic arch

Mandibular condyle

Angle of mandible

Dens axis

Cranium, Mandible, and Temporomandibular Joint

Plate S–124

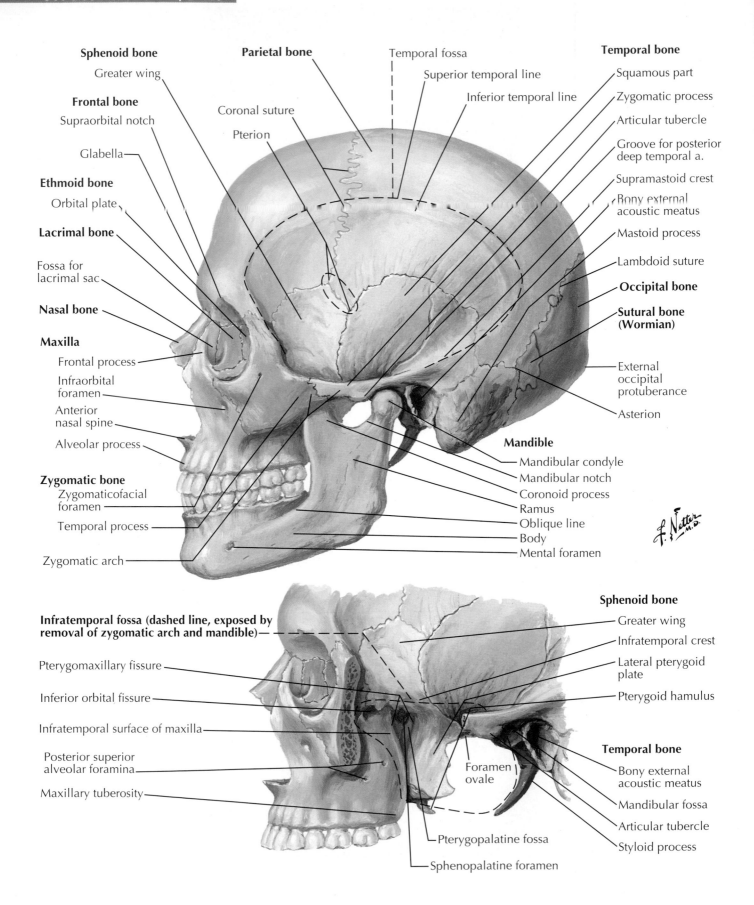

Sphenoid bone
Greater wing

Frontal bone
Supraorbital notch

Glabella

Ethmoid bone
Orbital plate

Lacrimal bone

Fossa for lacrimal sac

Nasal bone

Maxilla
Frontal process
Infraorbital foramen
Anterior nasal spine
Alveolar process

Zygomatic bone
Zygomaticofacial foramen
Temporal process

Zygomatic arch

Parietal bone

Coronal suture
Pterion

Temporal fossa
Superior temporal line
Inferior temporal line

Temporal bone
Squamous part
Zygomatic process
Articular tubercle
Groove for posterior deep temporal a.
Supramastoid crest
Bony external acoustic meatus
Mastoid process
Lambdoid suture

Occipital bone

Sutural bone (Wormian)

External occipital protuberance

Asterion

Mandible
Mandibular condyle
Mandibular notch
Coronoid process
Ramus
Oblique line
Body
Mental foramen

Infratemporal fossa (dashed line, exposed by removal of zygomatic arch and mandible)

Pterygomaxillary fissure

Inferior orbital fissure

Infratemporal surface of maxilla

Posterior superior alveolar foramina

Maxillary tuberosity

Foramen ovale

Pterygopalatine fossa

Sphenopalatine foramen

Sphenoid bone
Greater wing
Infratemporal crest
Lateral pterygoid plate
Pterygoid hamulus

Temporal bone
Bony external acoustic meatus
Mandibular fossa
Articular tubercle
Styloid process

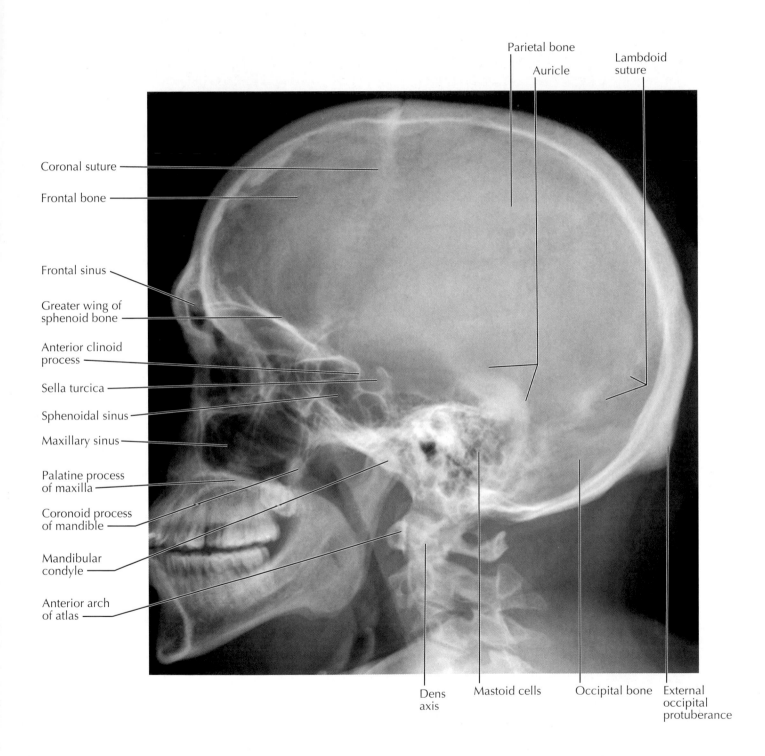

Parietal bone

Auricle

Lambdoid
suture

Coronal suture

Frontal bone

Frontal sinus

Greater wing of
sphenoid bone

Anterior clinoid
process

Sella turcica

Sphenoidal sinus

Maxillary sinus

Palatine process
of maxilla

Coronoid process
of mandible

Mandibular
condyle

Anterior arch
of atlas

Dens
axis

Mastoid cells

Occipital bone

External
occipital
protuberance

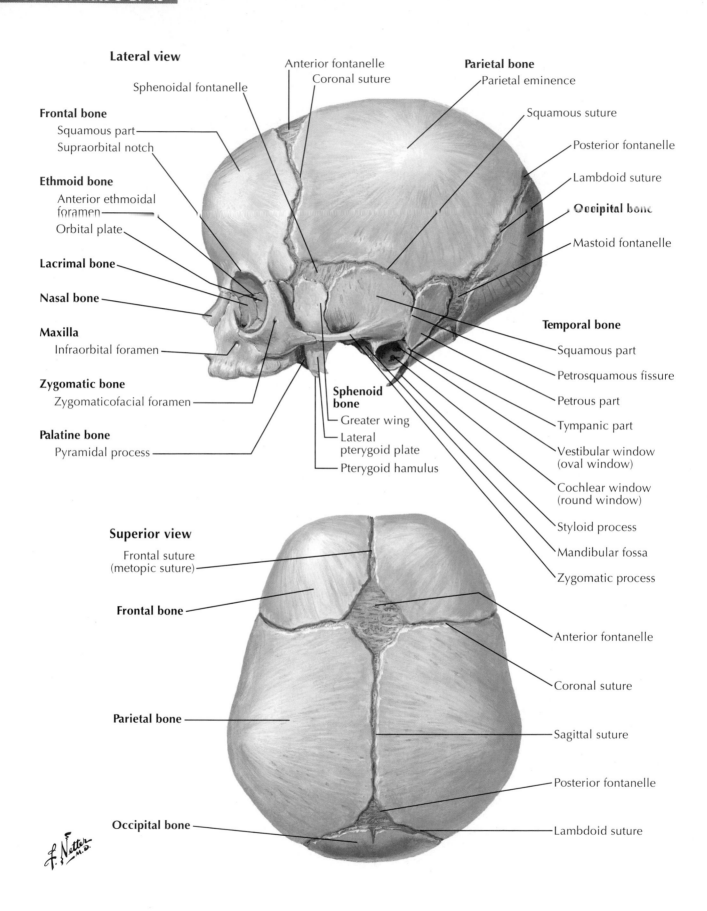

Lateral view

Anterior fontanelle
Coronal suture

Parietal bone
Parietal eminence

Sphenoidal fontanelle

Squamous suture

Frontal bone
Squamous part
Supraorbital notch

Posterior fontanelle

Lambdoid suture

Ethmoid bone
Anterior ethmoidal foramen
Orbital plate

Occipital bone

Mastoid fontanelle

Lacrimal bone

Nasal bone

Temporal bone

Maxilla
Infraorbital foramen

Squamous part

Petrosquamous fissure

Zygomatic bone
Zygomaticofacial foramen

Sphenoid bone
Greater wing
Lateral pterygoid plate
Pterygoid hamulus

Petrous part

Tympanic part

Palatine bone
Pyramidal process

Vestibular window (oval window)

Cochlear window (round window)

Styloid process

Mandibular fossa

Zygomatic process

Superior view
Frontal suture (metopic suture)

Frontal bone

Anterior fontanelle

Parietal bone

Coronal suture

Sagittal suture

Occipital bone

Posterior fontanelle

Lambdoid suture

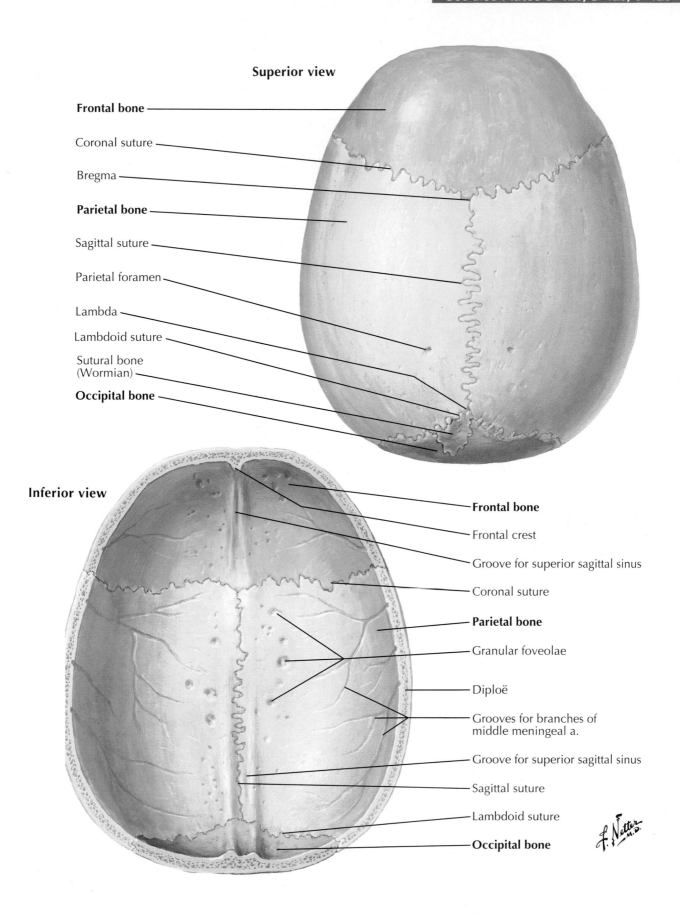

Superior view

Frontal bone

Coronal suture

Bregma

Parietal bone

Sagittal suture

Parietal foramen

Lambda

Lambdoid suture

Sutural bone
(Wormian)

Occipital bone

Inferior view

Frontal bone

Frontal crest

Groove for superior sagittal sinus

Coronal suture

Parietal bone

Granular foveolae

Diploë

Grooves for branches of
middle meningeal a.

Groove for superior sagittal sinus

Sagittal suture

Lambdoid suture

Occipital bone

Cranium, Mandible, and Temporomandibular Joint

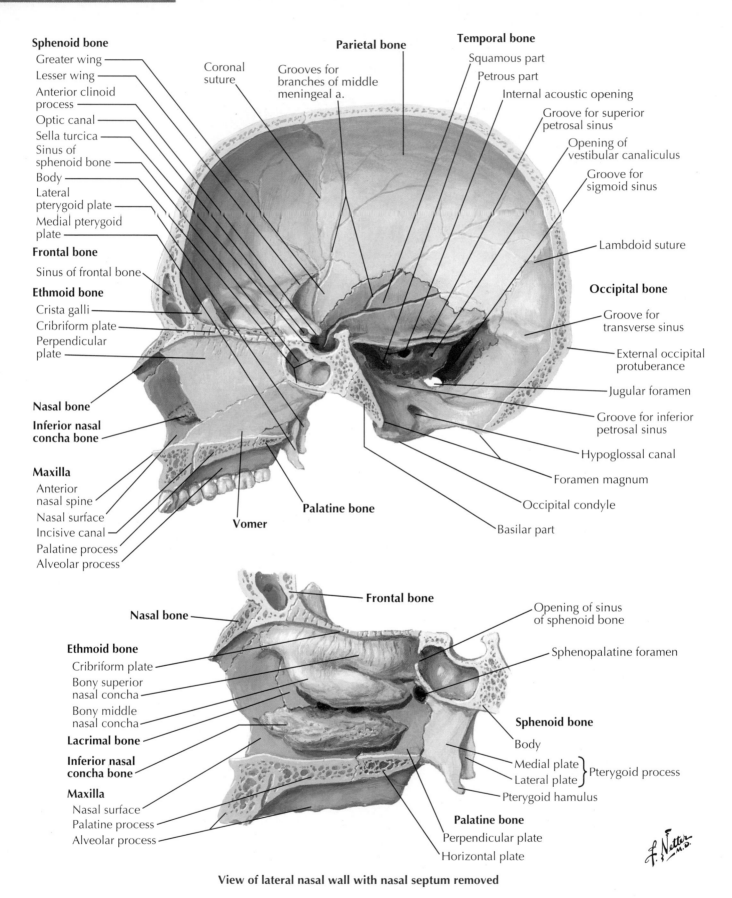

Sphenoid bone
Greater wing
Lesser wing
Anterior clinoid process
Optic canal
Sella turcica
Sinus of sphenoid bone
Body
Lateral pterygoid plate
Medial pterygoid plate

Frontal bone
Sinus of frontal bone

Ethmoid bone
Crista galli
Cribriform plate
Perpendicular plate

Nasal bone

Inferior nasal concha bone

Maxilla
Anterior nasal spine
Nasal surface
Incisive canal
Palatine process
Alveolar process

Coronal suture

Parietal bone
Grooves for branches of middle meningeal a.

Temporal bone
Squamous part
Petrous part
Internal acoustic opening
Groove for superior petrosal sinus
Opening of vestibular canaliculus
Groove for sigmoid sinus

Lambdoid suture

Occipital bone
Groove for transverse sinus
External occipital protuberance
Jugular foramen
Groove for inferior petrosal sinus
Hypoglossal canal
Foramen magnum
Occipital condyle
Basilar part

Palatine bone

Vomer

Nasal bone

Ethmoid bone
Cribriform plate
Bony superior nasal concha
Bony middle nasal concha

Lacrimal bone

Inferior nasal concha bone

Maxilla
Nasal surface
Palatine process
Alveolar process

Frontal bone

Opening of sinus of sphenoid bone

Sphenopalatine foramen

Sphenoid bone
Body
Medial plate
Lateral plate } Pterygoid process
Pterygoid hamulus

Palatine bone
Perpendicular plate
Horizontal plate

View of lateral nasal wall with nasal septum removed

Posterior view

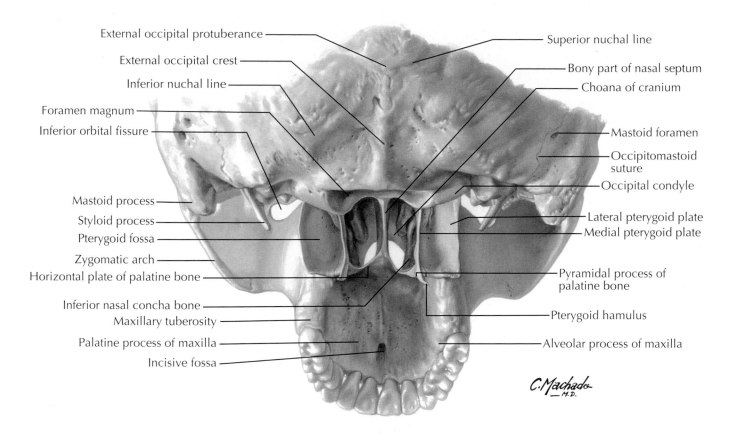

External occipital protuberance

External occipital crest

Inferior nuchal line

Foramen magnum

Inferior orbital fissure

Mastoid process

Styloid process

Pterygoid fossa

Zygomatic arch

Horizontal plate of palatine bone

Inferior nasal concha bone

Maxillary tuberosity

Palatine process of maxilla

Incisive fossa

Superior nuchal line

Bony part of nasal septum

Choana of cranium

Mastoid foramen

Occipitomastoid suture

Occipital condyle

Lateral pterygoid plate

Medial pterygoid plate

Pyramidal process of palatine bone

Pterygoid hamulus

Alveolar process of maxilla

C. Machado
—M.D.

Lateral view

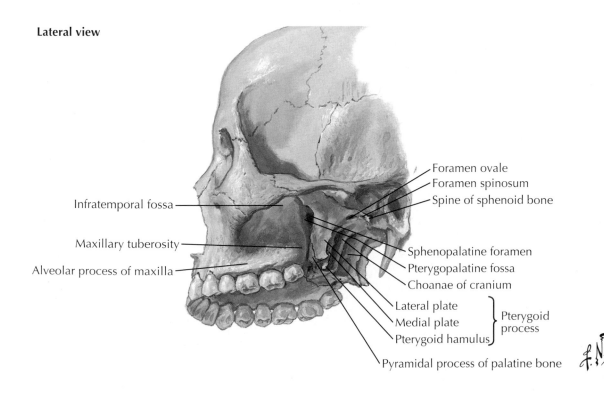

Infratemporal fossa

Maxillary tuberosity

Alveolar process of maxilla

Foramen ovale

Foramen spinosum

Spine of sphenoid bone

Sphenopalatine foramen

Pterygopalatine fossa

Choanae of cranium

Lateral plate

Medial plate   } Pterygoid process

Pterygoid hamulus

Pyramidal process of palatine bone

f. Netter
—M.D.

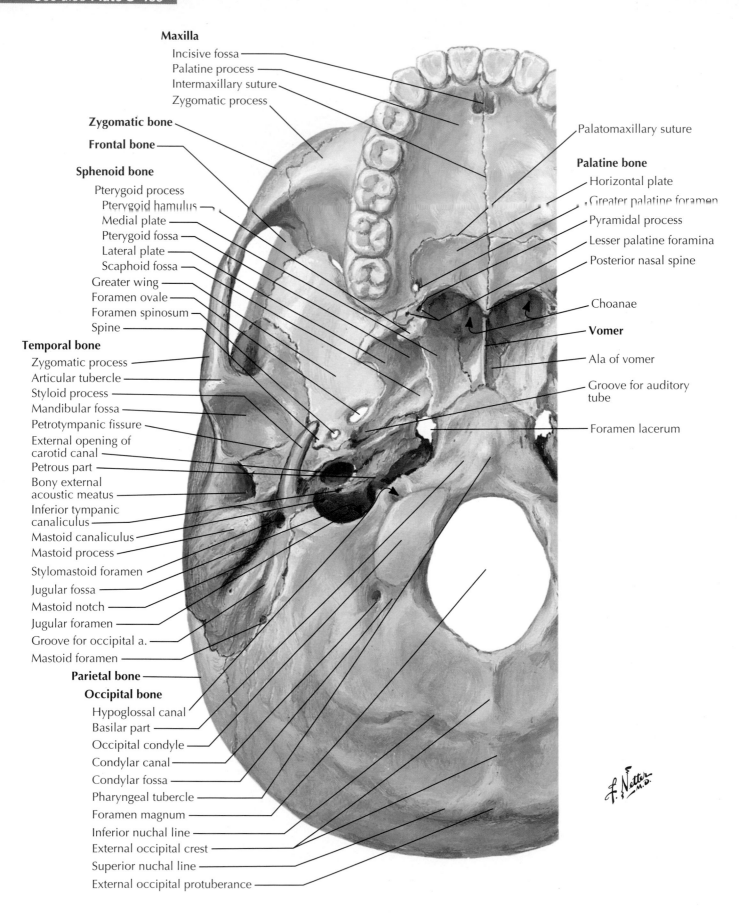

Maxilla
Incisive fossa
Palatine process
Intermaxillary suture
Zygomatic process

Zygomatic bone

Frontal bone

Sphenoid bone
Pterygoid process
Pterygoid hamulus
Medial plate
Pterygoid fossa
Lateral plate
Scaphoid fossa
Greater wing
Foramen ovale
Foramen spinosum
Spine

Temporal bone
Zygomatic process
Articular tubercle
Styloid process
Mandibular fossa
Petrotympanic fissure
External opening of
carotid canal
Petrous part
Bony external
acoustic meatus
Inferior tympanic
canaliculus
Mastoid canaliculus
Mastoid process
Stylomastoid foramen
Jugular fossa
Mastoid notch
Jugular foramen
Groove for occipital a.
Mastoid foramen

Parietal bone

Occipital bone
Hypoglossal canal
Basilar part
Occipital condyle
Condylar canal
Condylar fossa
Pharyngeal tubercle
Foramen magnum
Inferior nuchal line
External occipital crest
Superior nuchal line
External occipital protuberance

Palatomaxillary suture

Palatine bone
Horizontal plate
Greater palatine foramen
Pyramidal process
Lesser palatine foramina
Posterior nasal spine

Choanae

Vomer

Ala of vomer

Groove for auditory
tube

Foramen lacerum

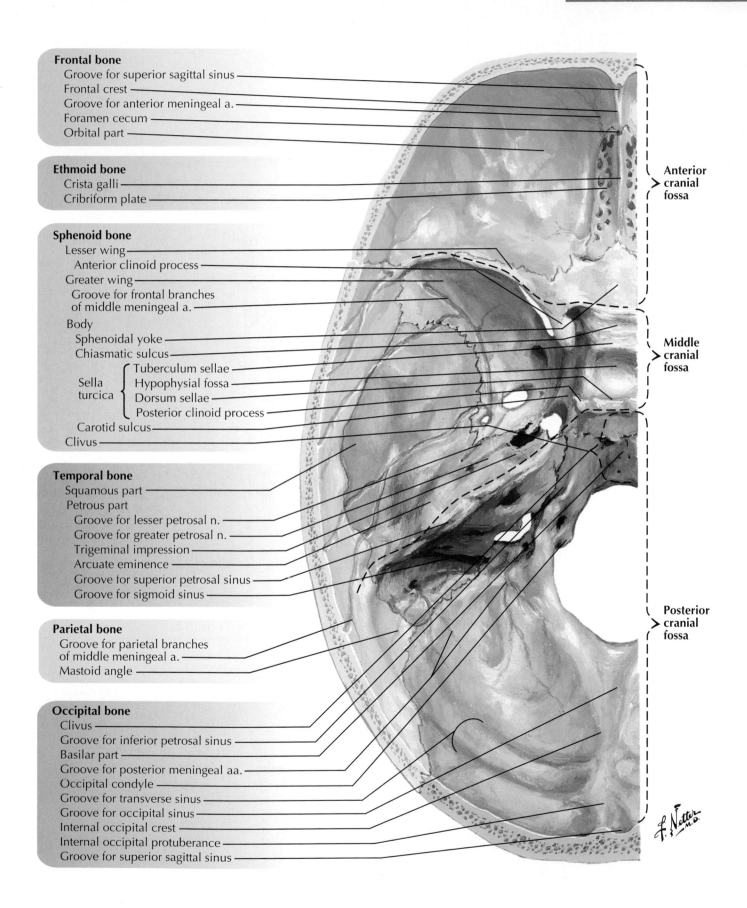

Frontal bone
Groove for superior sagittal sinus
Frontal crest
Groove for anterior meningeal a.
Foramen cecum
Orbital part

Ethmoid bone
Crista galli
Cribriform plate

Sphenoid bone
Lesser wing
Anterior clinoid process
Greater wing
Groove for frontal branches
of middle meningeal a.
Body
Sphenoidal yoke
Chiasmatic sulcus
Sella turcica:
Tuberculum sellae
Hypophysial fossa
Dorsum sellae
Posterior clinoid process
Carotid sulcus
Clivus

Temporal bone
Squamous part
Petrous part
Groove for lesser petrosal n.
Groove for greater petrosal n.
Trigeminal impression
Arcuate eminence
Groove for superior petrosal sinus
Groove for sigmoid sinus

Parietal bone
Groove for parietal branches
of middle meningeal a.
Mastoid angle

Occipital bone
Clivus
Groove for inferior petrosal sinus
Basilar part
Groove for posterior meningeal aa.
Occipital condyle
Groove for transverse sinus
Groove for occipital sinus
Internal occipital crest
Internal occipital protuberance
Groove for superior sagittal sinus

Anterior cranial fossa

Middle cranial fossa

Posterior cranial fossa

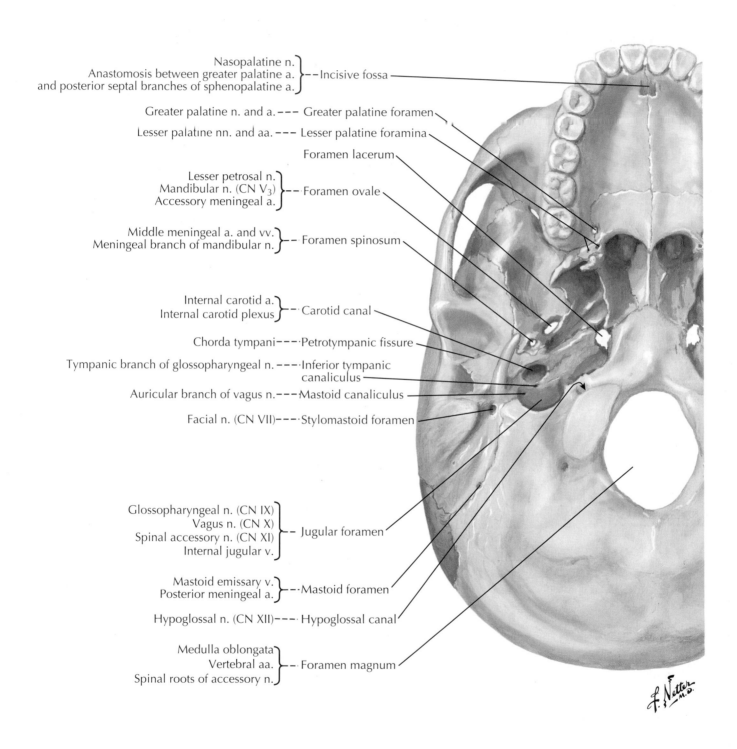

Nasopalatine n.
Anastomosis between greater palatine a.
and posterior septal branches of sphenopalatine a. } --- Incisive fossa

Greater palatine n. and a. --- Greater palatine foramen

Lesser palatine nn. and aa. --- Lesser palatine foramina

Foramen lacerum

Lesser petrosal n.
Mandibular n. (CN V₃) } --- Foramen ovale
Accessory meningeal a.

Middle meningeal a. and vv.
Meningeal branch of mandibular n. } --- Foramen spinosum

Internal carotid a.
Internal carotid plexus } --- Carotid canal

Chorda tympani --- Petrotympanic fissure

Tympanic branch of glossopharyngeal n. --- Inferior tympanic canaliculus

Auricular branch of vagus n. --- Mastoid canaliculus

Facial n. (CN VII) --- Stylomastoid foramen

Glossopharyngeal n. (CN IX)
Vagus n. (CN X)
Spinal accessory n. (CN XI) } --- Jugular foramen
Internal jugular v.

Mastoid emissary v.
Posterior meningeal a. } --- Mastoid foramen

Hypoglossal n. (CN XII) --- Hypoglossal canal

Medulla oblongata
Vertebral aa. } --- Foramen magnum
Spinal roots of accessory n.

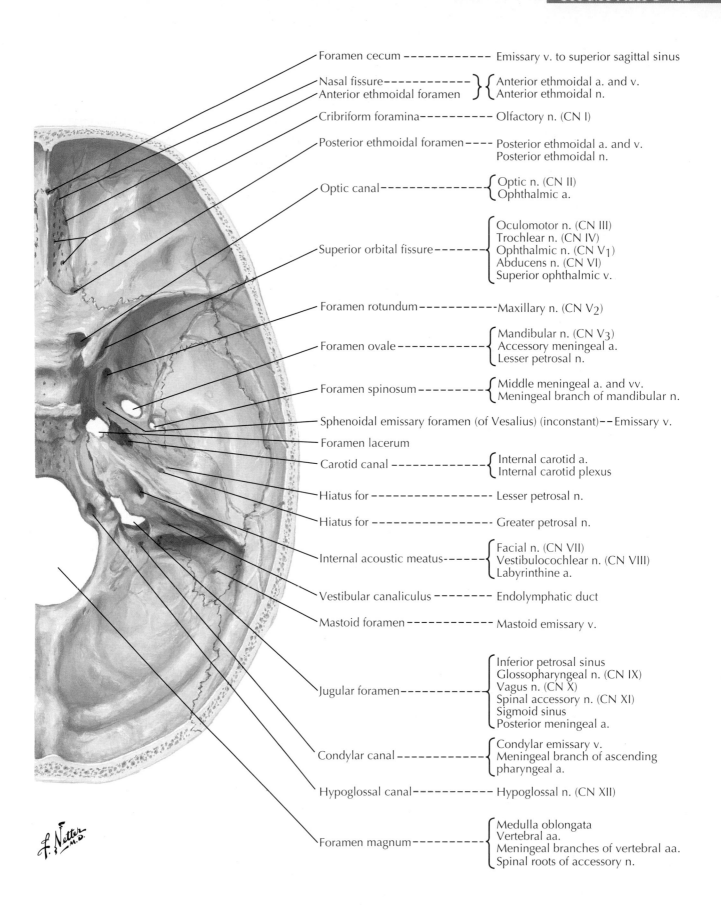

Foramen cecum --------- Emissary v. to superior sagittal sinus

Nasal fissure ---------- } { Anterior ethmoidal a. and v.
Anterior ethmoidal foramen } { Anterior ethmoidal n.

Cribriform foramina -------- Olfactory n. (CN I)

Posterior ethmoidal foramen ---- Posterior ethmoidal a. and v.
Posterior ethmoidal n.

Optic canal ----------- { Optic n. (CN II)
{ Ophthalmic a.

Superior orbital fissure ------ { Oculomotor n. (CN III)
{ Trochlear n. (CN IV)
{ Ophthalmic n. (CN V$_1$)
{ Abducens n. (CN VI)
{ Superior ophthalmic v.

Foramen rotundum ---------- Maxillary n. (CN V$_2$)

Foramen ovale ----------- { Mandibular n. (CN V$_3$)
{ Accessory meningeal a.
{ Lesser petrosal n.

Foramen spinosum --------- { Middle meningeal a. and vv.
{ Meningeal branch of mandibular n.

Sphenoidal emissary foramen (of Vesalius) (inconstant) -- Emissary v.

Foramen lacerum

Carotid canal ----------- { Internal carotid a.
{ Internal carotid plexus

Hiatus for --------------- Lesser petrosal n.

Hiatus for --------------- Greater petrosal n.

Internal acoustic meatus ------ { Facial n. (CN VII)
{ Vestibulocochlear n. (CN VIII)
{ Labyrinthine a.

Vestibular canaliculus ------- Endolymphatic duct

Mastoid foramen --------- Mastoid emissary v.

Jugular foramen --------- { Inferior petrosal sinus
{ Glossopharyngeal n. (CN IX)
{ Vagus n. (CN X)
{ Spinal accessory n. (CN XI)
{ Sigmoid sinus
{ Posterior meningeal a.

Condylar canal ----------- { Condylar emissary v.
{ Meningeal branch of ascending
{ pharyngeal a.

Hypoglossal canal --------- Hypoglossal n. (CN XII)

Foramen magnum --------- { Medulla oblongata
{ Vertebral aa.
{ Meningeal branches of vertebral aa.
{ Spinal roots of accessory n.

F. Netter
M.D.

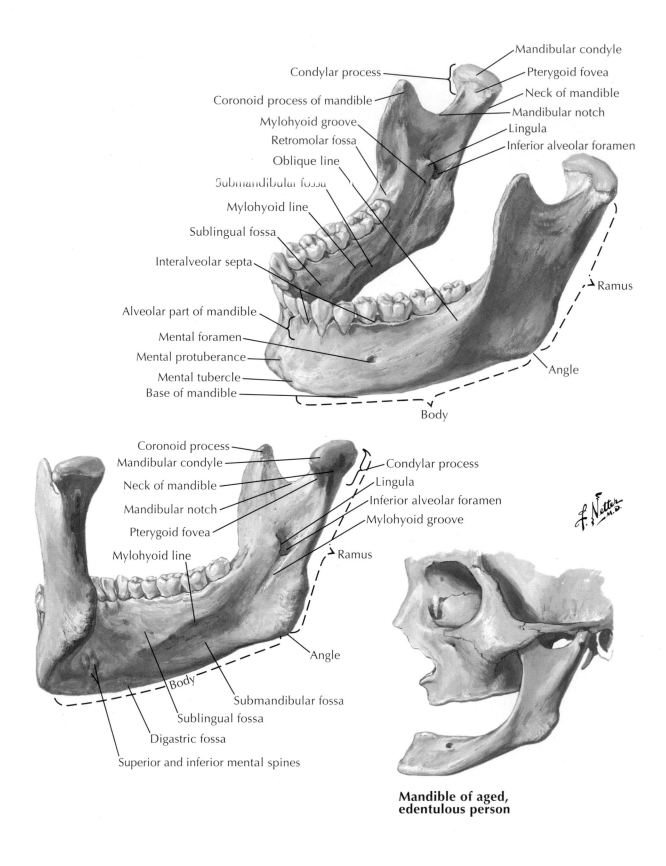

Condylar process

Coronoid process of mandible

Mylohyoid groove

Retromolar fossa

Oblique line

Submandibular fossa

Mylohyoid line

Sublingual fossa

Interalveolar septa

Alveolar part of mandible

Mental foramen

Mental protuberance

Mental tubercle

Base of mandible

Mandibular condyle

Pterygoid fovea

Neck of mandible

Mandibular notch

Lingula

Inferior alveolar foramen

Ramus

Angle

Body

Coronoid process

Mandibular condyle

Neck of mandible

Mandibular notch

Pterygoid fovea

Mylohyoid line

Condylar process

Lingula

Inferior alveolar foramen

Mylohyoid groove

Ramus

Angle

Body

Submandibular fossa

Sublingual fossa

Digastric fossa

Superior and inferior mental spines

Mandible of aged, edentulous person

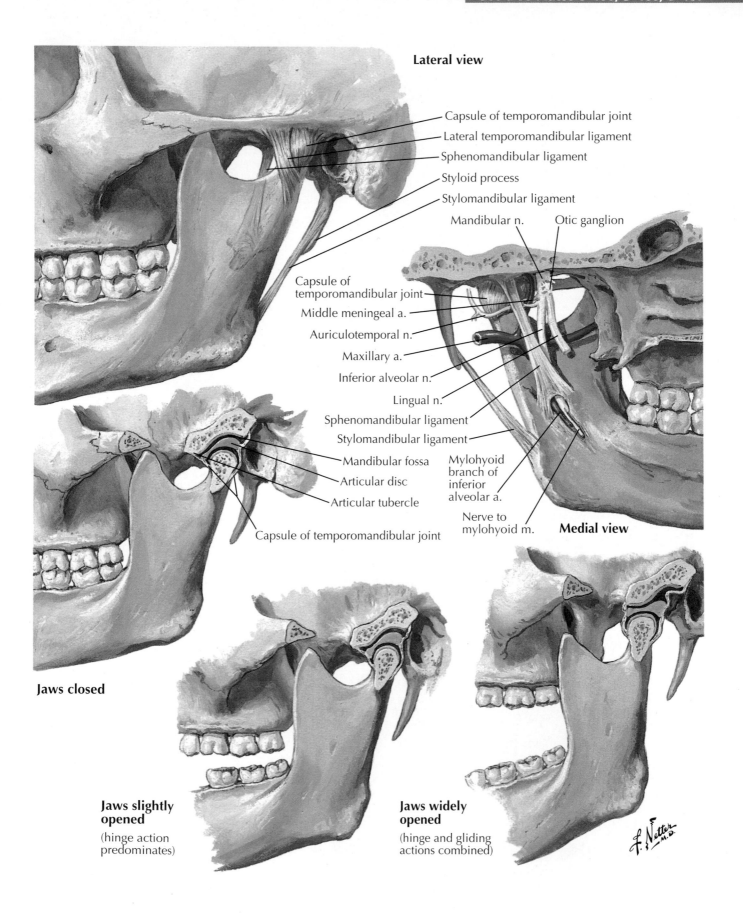

Lateral view

Capsule of temporomandibular joint
Lateral temporomandibular ligament
Sphenomandibular ligament
Styloid process
Stylomandibular ligament
Mandibular n.　Otic ganglion

Capsule of temporomandibular joint
Middle meningeal a.
Auriculotemporal n.
Maxillary a.
Inferior alveolar n.
Lingual n.
Sphenomandibular ligament
Stylomandibular ligament

Mandibular fossa
Articular disc
Articular tubercle

Capsule of temporomandibular joint

Mylohyoid branch of inferior alveolar a.

Nerve to mylohyoid m.　**Medial view**

Jaws closed

Jaws slightly opened

(hinge action predominates)

Jaws widely opened

(hinge and gliding actions combined)

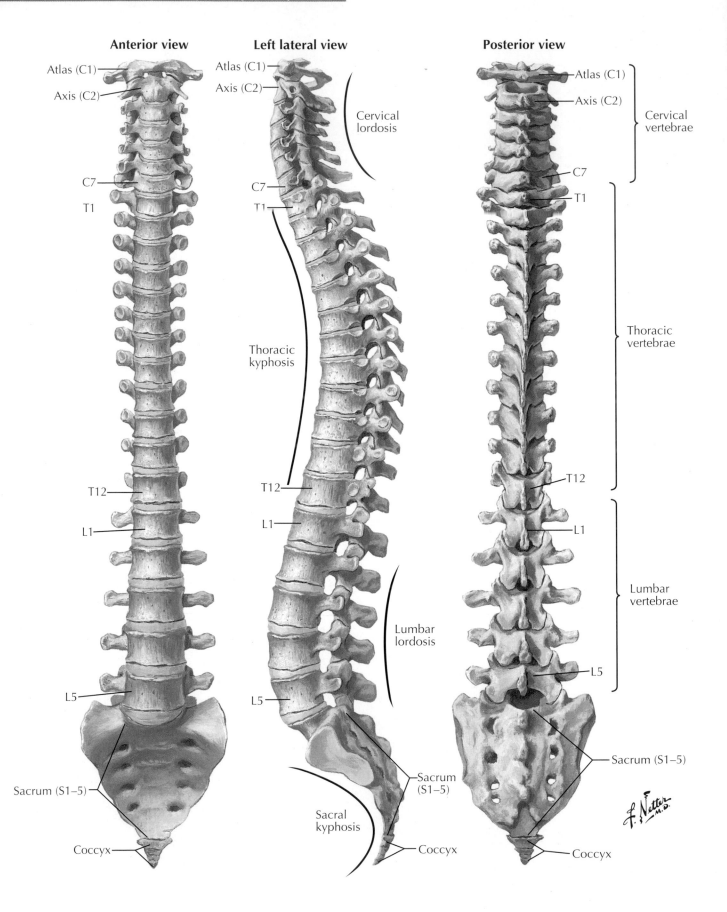

Anterior view

Atlas (C1)
Axis (C2)

C7
T1

T12

L1

L5

Sacrum (S1–5)

Coccyx

Left lateral view

Atlas (C1)
Axis (C2)

Cervical
lordosis

C7
T1

Thoracic
kyphosis

T12

L1

Lumbar
lordosis

L5

Sacrum
(S1–5)

Sacral
kyphosis

Coccyx

Posterior view

Atlas (C1)
Axis (C2)

Cervical
vertebrae

C7
T1

Thoracic
vertebrae

T12

L1

Lumbar
vertebrae

L5

Sacrum (S1–5)

Coccyx

Anteroposterior radiograph of thoracolumbar spine

T9 vertebra

12th rib

Transverse process of L1 vertebra

Superior articular process of L2 vertebra

Inferior articular process of L2 vertebra

S2 segment of sacrum

Lamina of L1 vertebra

Spinous process of L2 vertebra

Pedicle of L4 vertebra

Ilium

Sacroiliac joint

T2-weighted sagittal MRI of lumbar spine

T12 vertebral body

L2–L3 intervertebral disc

Cauda equina

S1 segment of sacrum

Rectum

Urinary bladder

Conus medullaris

Spinous process of L1 vertebra

Skin

Subcutaneous fat

Supraspinous ligament

Ligamentum flavum

Epidural fat

Dura

Cerebrospinal fluid

Termination of dural sac

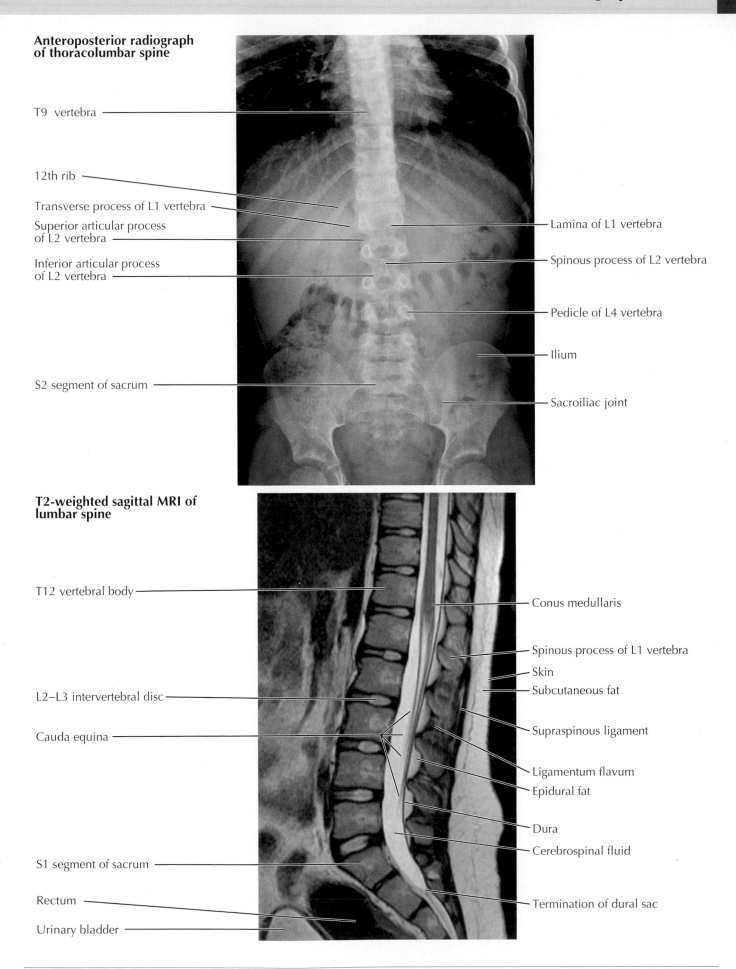

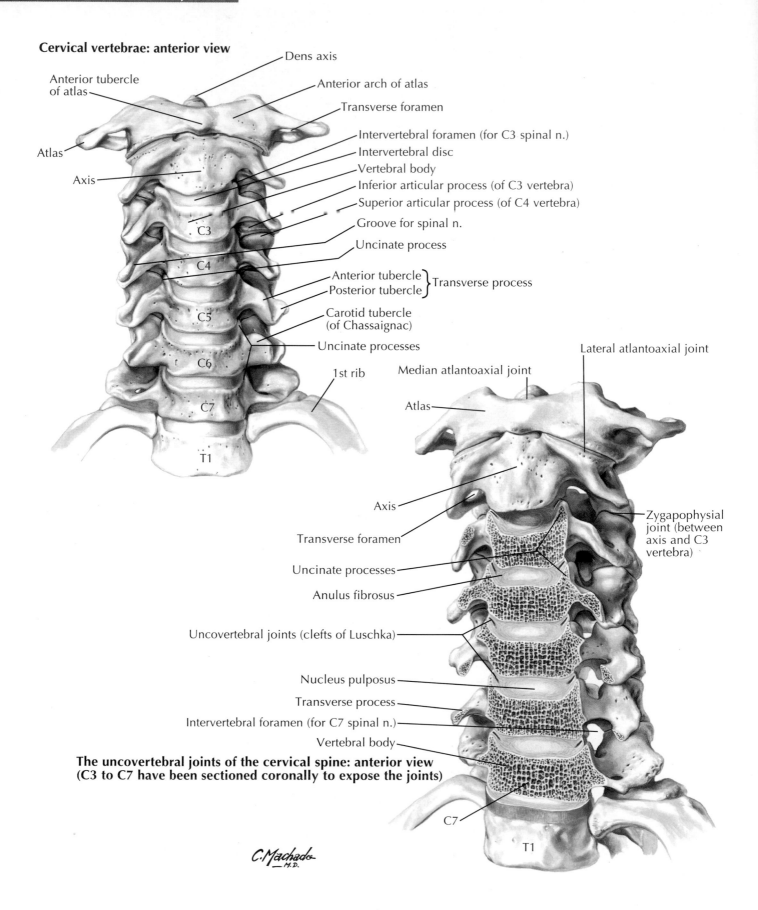

Cervical vertebrae: anterior view

Dens axis

Anterior tubercle of atlas

Anterior arch of atlas

Transverse foramen

Intervertebral foramen (for C3 spinal n.)

Intervertebral disc

Vertebral body

Inferior articular process (of C3 vertebra)

Superior articular process (of C4 vertebra)

Groove for spinal n.

Uncinate process

Anterior tubercle } Transverse process
Posterior tubercle }

Carotid tubercle (of Chassaignac)

Uncinate processes

1st rib

Atlas

Axis

C3

C4

C5

C6

C7

T1

Median atlantoaxial joint

Lateral atlantoaxial joint

Atlas

Axis

Transverse foramen

Uncinate processes

Anulus fibrosus

Uncovertebral joints (clefts of Luschka)

Nucleus pulposus

Transverse process

Intervertebral foramen (for C7 spinal n.)

Vertebral body

Zygapophysial joint (between axis and C3 vertebra)

C7

T1

The uncovertebral joints of the cervical spine: anterior view (C3 to C7 have been sectioned coronally to expose the joints)

C.Machado
M.D.

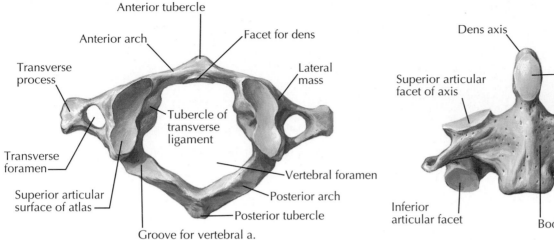

Anterior tubercle

Anterior arch

Facet for dens

Transverse process

Lateral mass

Facet for dens

Tubercle of transverse ligament

Transverse foramen

Vertebral foramen

Superior articular surface of atlas

Posterior arch

Posterior tubercle

Groove for vertebral a.

Atlas: superior view

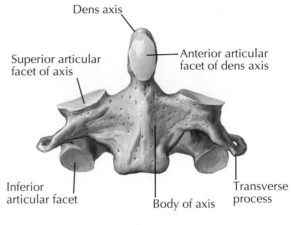

Dens axis

Superior articular facet of axis

Anterior articular facet of dens axis

Inferior articular facet

Body of axis

Transverse process

Axis: anterior view

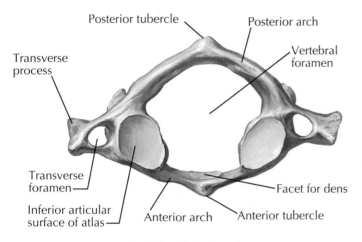

Posterior tubercle

Posterior arch

Transverse process

Vertebral foramen

Transverse foramen

Facet for dens

Inferior articular surface of atlas

Anterior arch

Anterior tubercle

Atlas: inferior view

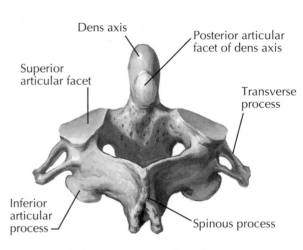

Dens axis

Posterior articular facet of dens axis

Superior articular facet

Transverse process

Inferior articular process

Spinous process

Axis: posterosuperior view

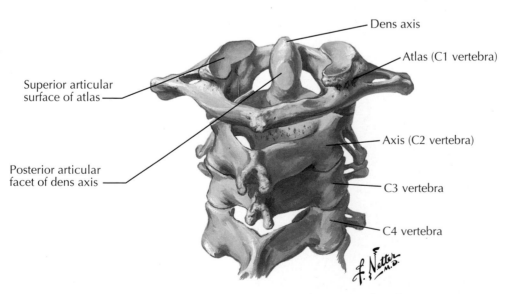

Dens axis

Atlas (C1 vertebra)

Superior articular surface of atlas

Axis (C2 vertebra)

Posterior articular facet of dens axis

C3 vertebra

C4 vertebra

Upper cervical vertebrae: posterosuperior view

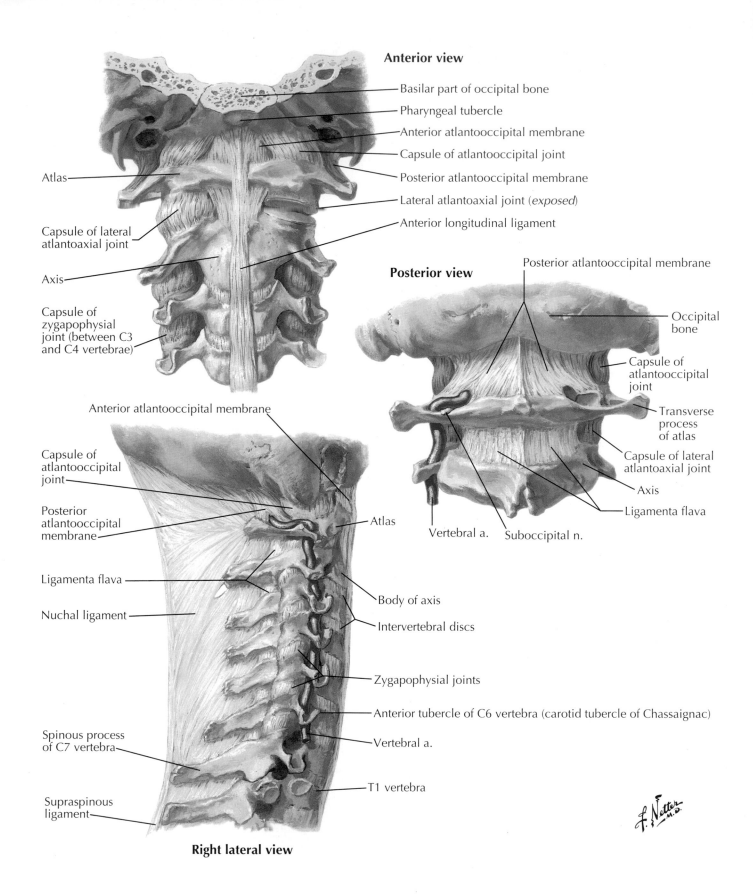

Anterior view

Basilar part of occipital bone

Pharyngeal tubercle

Anterior atlantooccipital membrane

Capsule of atlantooccipital joint

Posterior atlantooccipital membrane

Lateral atlantoaxial joint (*exposed*)

Anterior longitudinal ligament

Atlas

Capsule of lateral atlantoaxial joint

Axis

Capsule of zygapophysial joint (between C3 and C4 vertebrae)

Posterior view

Posterior atlantooccipital membrane

Occipital bone

Capsule of atlantooccipital joint

Transverse process of atlas

Capsule of lateral atlantoaxial joint

Axis

Ligamenta flava

Vertebral a.

Suboccipital n.

Anterior atlantooccipital membrane

Capsule of atlantooccipital joint

Posterior atlantooccipital membrane

Ligamenta flava

Nuchal ligament

Atlas

Body of axis

Intervertebral discs

Zygapophysial joints

Anterior tubercle of C6 vertebra (carotid tubercle of Chassaignac)

Vertebral a.

Spinous process of C7 vertebra

T1 vertebra

Supraspinous ligament

Right lateral view

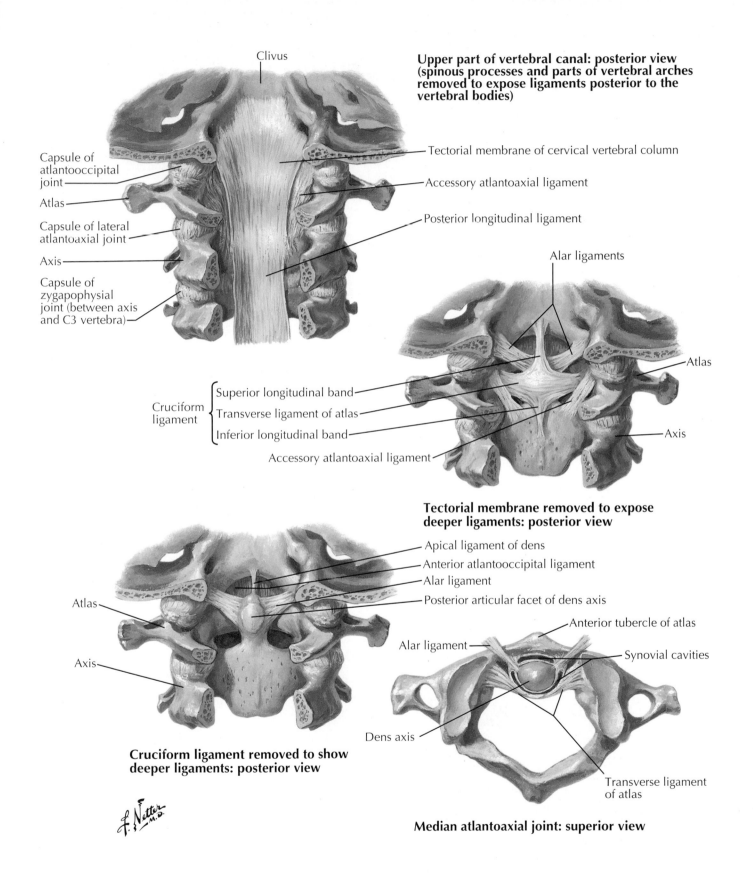

Clivus

Upper part of vertebral canal: posterior view (spinous processes and parts of vertebral arches removed to expose ligaments posterior to the vertebral bodies)

Tectorial membrane of cervical vertebral column

Accessory atlantoaxial ligament

Posterior longitudinal ligament

Capsule of atlantooccipital joint

Atlas

Capsule of lateral atlantoaxial joint

Axis

Capsule of zygapophysial joint (between axis and C3 vertebra)

Alar ligaments

Cruciform ligament
{
Superior longitudinal band
Transverse ligament of atlas
Inferior longitudinal band
}

Accessory atlantoaxial ligament

Atlas

Axis

Tectorial membrane removed to expose deeper ligaments: posterior view

Atlas

Axis

Apical ligament of dens
Anterior atlantooccipital ligament
Alar ligament
Posterior articular facet of dens axis

Anterior tubercle of atlas

Alar ligament

Synovial cavities

Dens axis

Transverse ligament of atlas

Cruciform ligament removed to show deeper ligaments: posterior view

Median atlantoaxial joint: superior view

Inferior aspect of C3 vertebra and superior aspect of C4 vertebra showing the sites of the articular surfaces of the uncovertebral joints

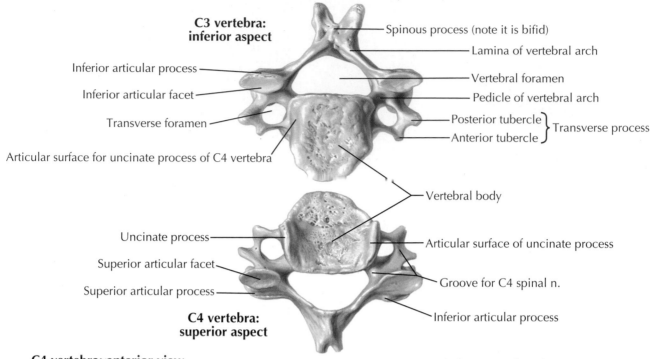

C3 vertebra: inferior aspect

- Inferior articular process
- Inferior articular facet
- Transverse foramen
- Articular surface for uncinate process of C4 vertebra

- Spinous process (note it is bifid)
- Lamina of vertebral arch
- Vertebral foramen
- Pedicle of vertebral arch
- Posterior tubercle } Transverse process
- Anterior tubercle }

- Vertebral body
- Uncinate process
- Superior articular facet
- Superior articular process

- Articular surface of uncinate process
- Groove for C4 spinal n.
- Inferior articular process

C4 vertebra: superior aspect

C4 vertebra: anterior view

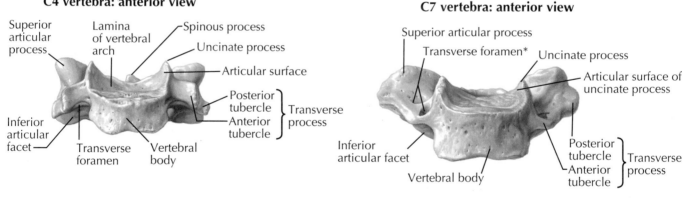

- Superior articular process
- Lamina of vertebral arch
- Spinous process
- Uncinate process
- Articular surface
- Posterior tubercle } Transverse process
- Anterior tubercle }
- Inferior articular facet
- Transverse foramen
- Vertebral body

C7 vertebra: anterior view

- Superior articular process
- Transverse foramen*
- Uncinate process
- Articular surface of uncinate process
- Inferior articular facet
- Posterior tubercle } Transverse process
- Anterior tubercle }
- Vertebral body

C7 vertebra (vertebra prominens): superior view

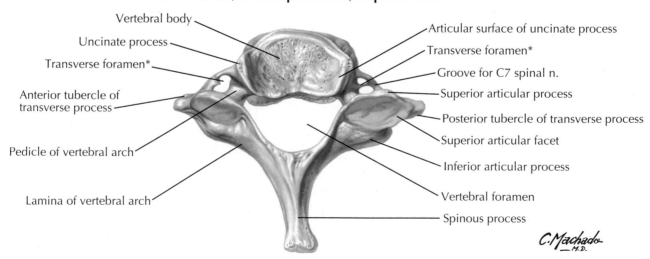

- Vertebral body
- Uncinate process
- Transverse foramen*
- Anterior tubercle of transverse process
- Pedicle of vertebral arch
- Lamina of vertebral arch

- Articular surface of uncinate process
- Transverse foramen*
- Groove for C7 spinal n.
- Superior articular process
- Posterior tubercle of transverse process
- Superior articular facet
- Inferior articular process
- Vertebral foramen
- Spinous process

C.Machado M.D.

The transverse foramina of C7 transmit vertebral veins, but usually not the vertebral artery, and are asymmetrical in these drawings. Note the right transverse foramen is septated.

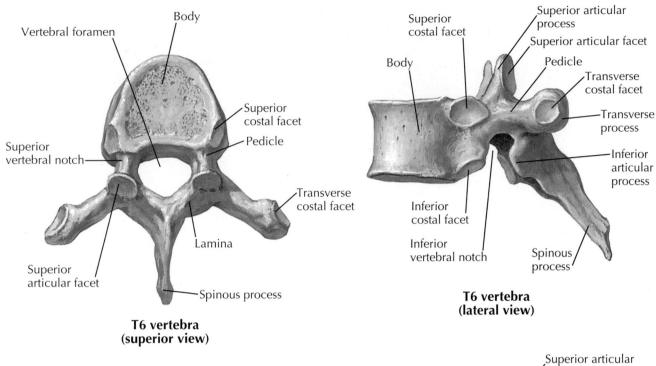

Body

Vertebral foramen

Superior costal facet

Pedicle

Superior vertebral notch

Transverse costal facet

Superior articular facet

Lamina

Spinous process

**T6 vertebra
(superior view)**

Superior costal facet

Body

Superior articular process

Superior articular facet

Pedicle

Transverse costal facet

Transverse process

Inferior articular process

Inferior costal facet

Inferior vertebral notch

Spinous process

**T6 vertebra
(lateral view)**

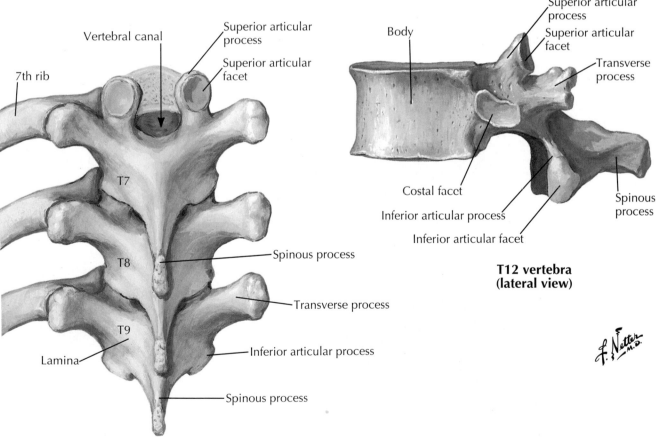

Vertebral canal

7th rib

Superior articular process

Superior articular facet

T7

T8

T9

Lamina

Spinous process

Transverse process

Inferior articular process

Spinous process

**T7, T8, and T9 vertebrae
(posterior view)**

Body

Superior articular process

Superior articular facet

Transverse process

Costal facet

Inferior articular process

Inferior articular facet

Spinous process

**T12 vertebra
(lateral view)**

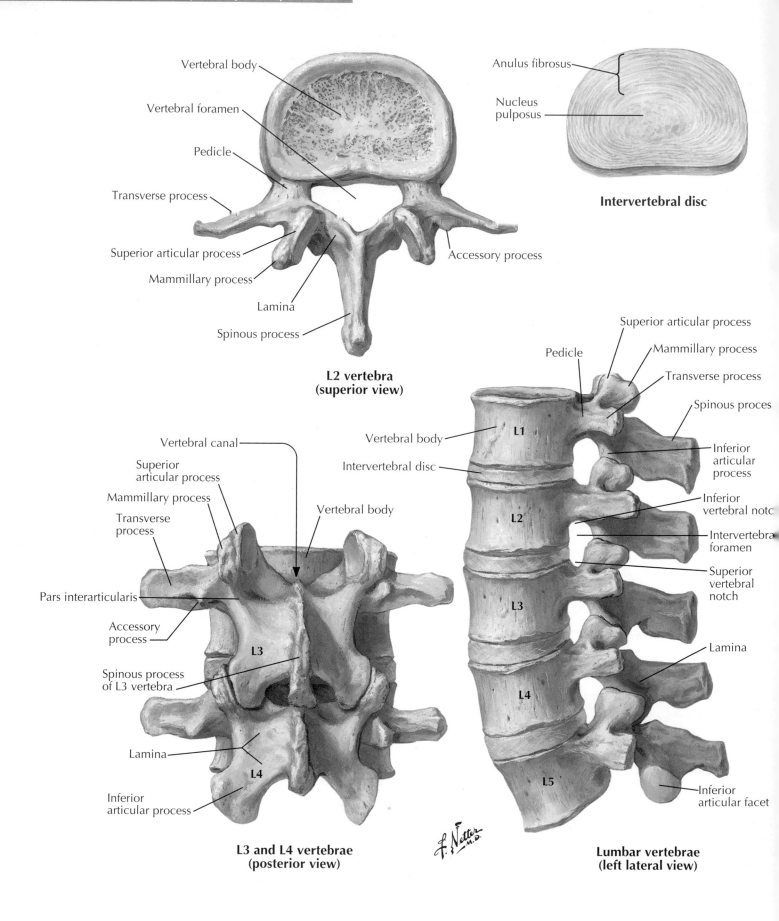

Vertebral body

Vertebral foramen

Pedicle

Transverse process

Superior articular process

Mammillary process

Lamina

Spinous process

**L2 vertebra
(superior view)**

Anulus fibrosus

Nucleus pulposus

Intervertebral disc

Vertebral canal

Superior articular process

Mammillary process

Transverse process

Pars interarticularis

Accessory process

Spinous process of L3 vertebra

Lamina

Inferior articular process

Accessory process

Vertebral body

L3

L4

**L3 and L4 vertebrae
(posterior view)**

Pedicle

Superior articular process

Mammillary process

Transverse process

Spinous proces

Vertebral body

Intervertebral disc

L1

L2

L3

L4

L5

Inferior articular process

Inferior vertebral notc

Intervertebra foramen

Superior vertebral notch

Lamina

Inferior articular facet

**Lumbar vertebrae
(left lateral view)**

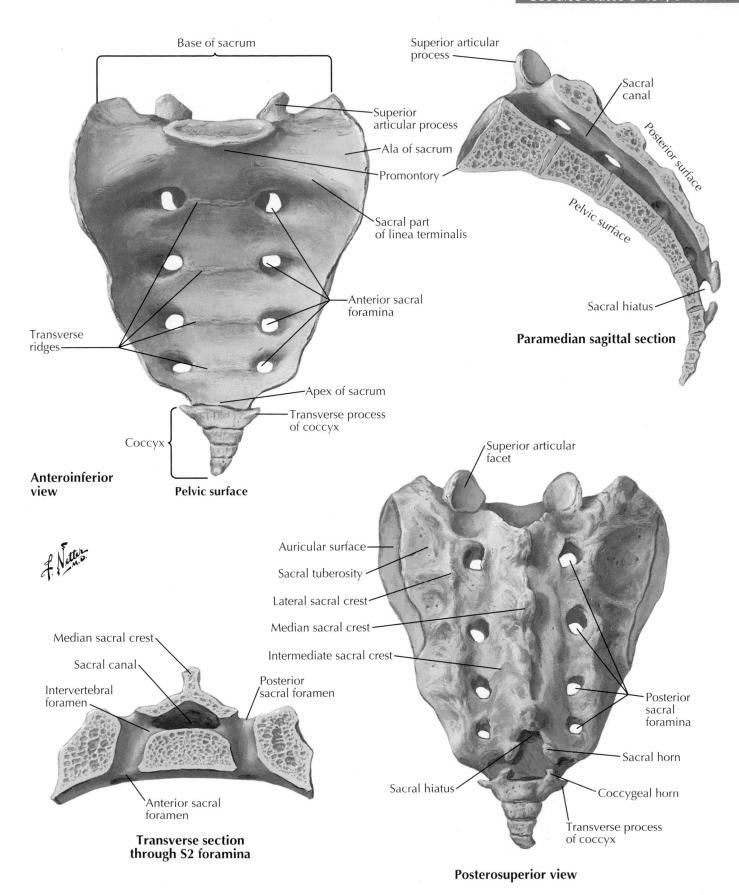

Base of sacrum

Superior
articular
process

Ala of sacrum

Promontory

Sacral part
of linea terminalis

Anterior sacral
foramina

Transverse
ridges

Apex of sacrum

Transverse process
of coccyx

Coccyx

**Anteroinferior
view**

Pelvic surface

Superior articular
process

Sacral
canal

Posterior surface

Pelvic surface

Sacral hiatus

Paramedian sagittal section

Superior articular
facet

Auricular surface

Sacral tuberosity

Lateral sacral crest

Median sacral crest

Intermediate sacral crest

Posterior
sacral
foramina

Sacral horn

Coccygeal horn

Sacral hiatus

Transverse process
of coccyx

Posterosuperior view

Median sacral crest

Sacral canal

Intervertebral
foramen

Posterior
sacral foramen

Anterior sacral
foramen

**Transverse section
through S2 foramina**

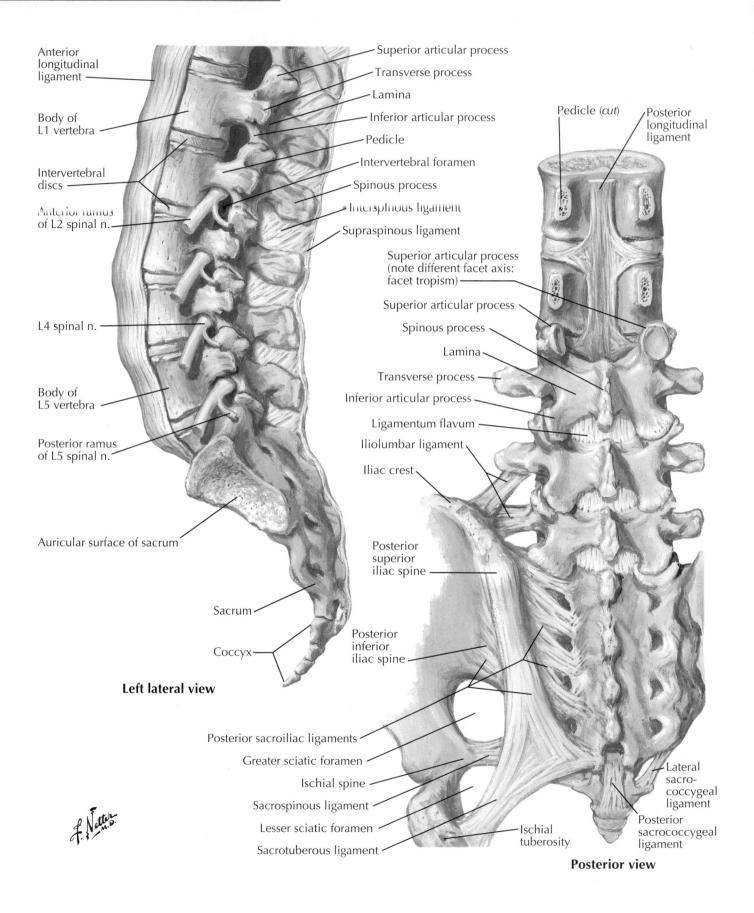

Anterior longitudinal ligament

Body of L1 vertebra

Intervertebral discs

Anterior ramus of L2 spinal n.

L4 spinal n.

Body of L5 vertebra

Posterior ramus of L5 spinal n.

Auricular surface of sacrum

Sacrum

Coccyx

Left lateral view

Superior articular process

Transverse process

Lamina

Inferior articular process

Pedicle

Intervertebral foramen

Spinous process

Interspinous ligament

Supraspinous ligament

Pedicle (*cut*)

Posterior longitudinal ligament

Superior articular process (note different facet axis: facet tropism)

Superior articular process

Spinous process

Lamina

Transverse process

Inferior articular process

Ligamentum flavum

Iliolumbar ligament

Iliac crest

Posterior superior iliac spine

Posterior inferior iliac spine

Posterior sacroiliac ligaments

Greater sciatic foramen

Ischial spine

Sacrospinous ligament

Lesser sciatic foramen

Sacrotuberous ligament

Ischial tuberosity

Lateral sacrococcygeal ligament

Posterior sacrococcygeal ligament

Posterior view

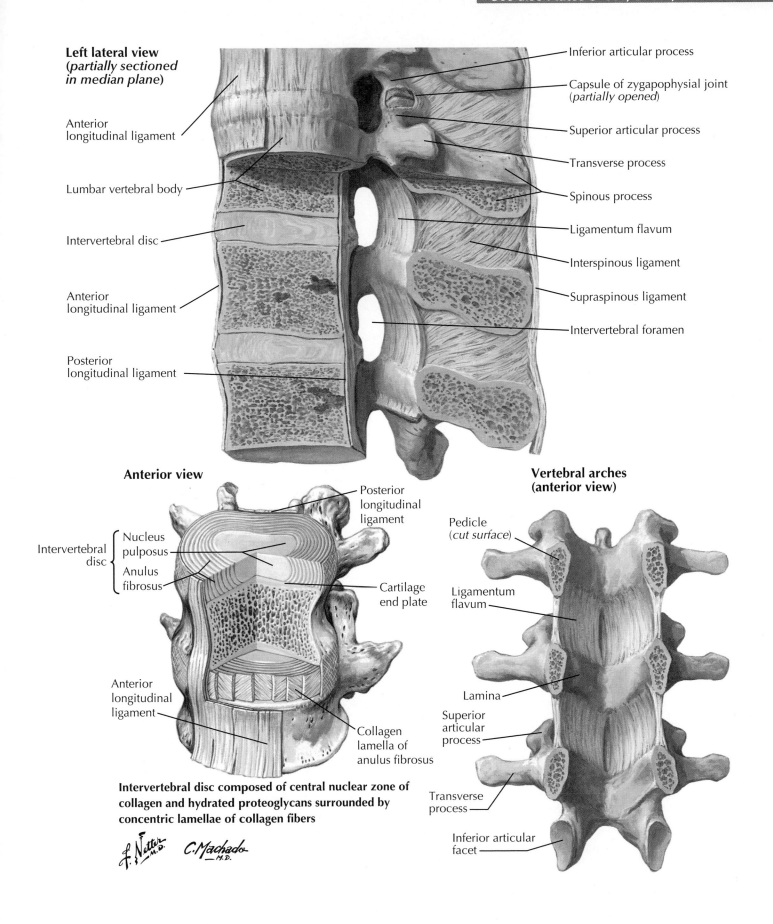

Left lateral view (*partially sectioned in median plane*)

Anterior longitudinal ligament

Lumbar vertebral body

Intervertebral disc

Anterior longitudinal ligament

Posterior longitudinal ligament

Inferior articular process

Capsule of zygapophysial joint (*partially opened*)

Superior articular process

Transverse process

Spinous process

Ligamentum flavum

Interspinous ligament

Supraspinous ligament

Intervertebral foramen

Anterior view

Posterior longitudinal ligament

Intervertebral disc {
Nucleus pulposus
Anulus fibrosus

Cartilage end plate

Anterior longitudinal ligament

Collagen lamella of anulus fibrosus

Intervertebral disc composed of central nuclear zone of collagen and hydrated proteoglycans surrounded by concentric lamellae of collagen fibers

Vertebral arches (anterior view)

Pedicle (*cut surface*)

Ligamentum flavum

Lamina

Superior articular process

Transverse process

Inferior articular facet

F. Netter, M.D.   C. Machado, M.D.

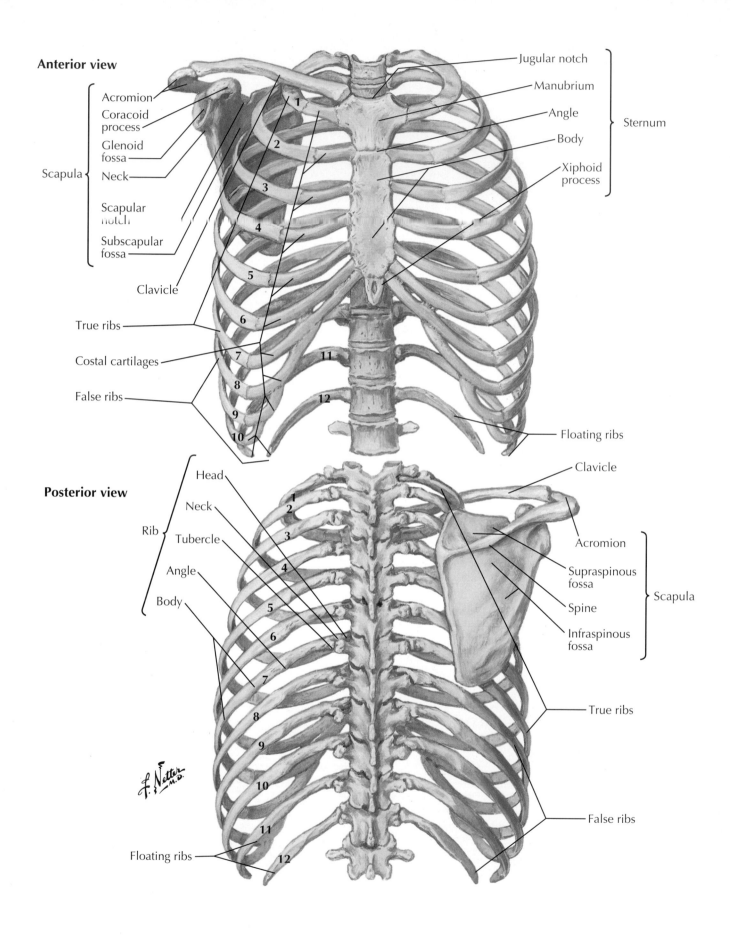

Anterior view

Acromion

Coracoid process

Glenoid fossa

Neck

Scapular notch

Subscapular fossa

Scapula

Clavicle

True ribs

Costal cartilages

False ribs

1
2
3
4
5
6
7
8
9
10
11
12

Jugular notch

Manubrium

Angle

Body

Xiphoid process

Sternum

Floating ribs

Posterior view

Head

Neck

Tubercle

Angle

Body

Rib

1
2
3
4
5
6
7
8
9
10
11
12

Clavicle

Acromion

Supraspinous fossa

Spine

Infraspinous fossa

Scapula

True ribs

False ribs

Floating ribs

Thoracic Skeleton

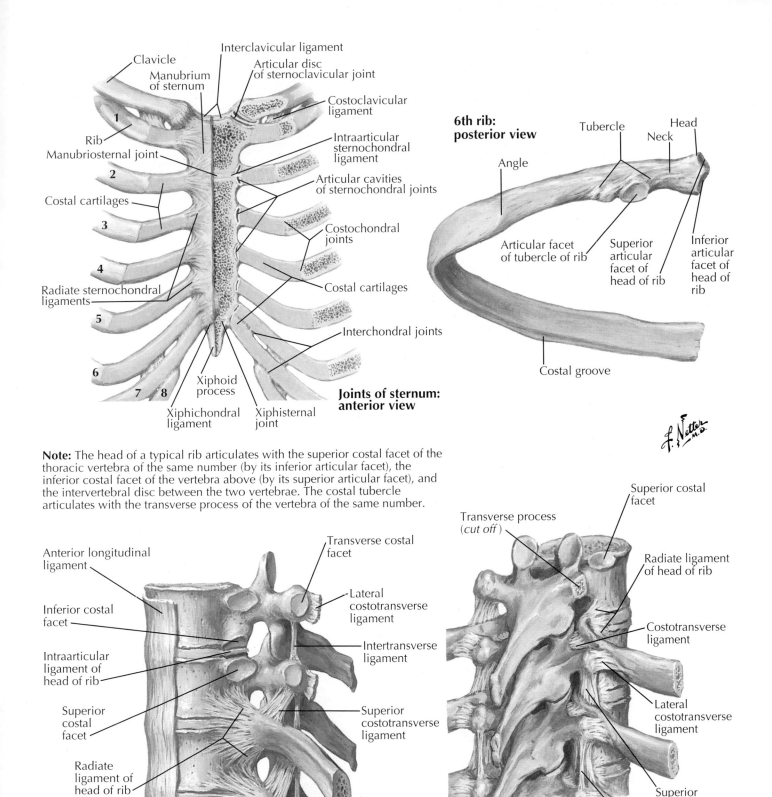

Clavicle
Interclavicular ligament
Manubrium of sternum
Articular disc of sternoclavicular joint
Costoclavicular ligament
1
Rib
Manubriosternal joint
Intraarticular sternochondral ligament
2
Articular cavities of sternochondral joints
Costal cartilages
Costochondral joints
3
4
Costal cartilages
Radiate sternochondral ligaments
5
Interchondral joints
6
7   8
Xiphoid process
Xiphichondral ligament
Xiphisternal joint
Joints of sternum: anterior view

6th rib: posterior view
Angle
Tubercle
Neck
Head
Articular facet of tubercle of rib
Superior articular facet of head of rib
Inferior articular facet of head of rib
Costal groove

Note: The head of a typical rib articulates with the superior costal facet of the thoracic vertebra of the same number (by its inferior articular facet), the inferior costal facet of the vertebra above (by its superior articular facet), and the intervertebral disc between the two vertebrae. The costal tubercle articulates with the transverse process of the vertebra of the same number.

Anterior longitudinal ligament
Transverse costal facet
Lateral costotransverse ligament
Inferior costal facet
Intraarticular ligament of head of rib
Intertransverse ligament
Superior costal facet
Superior costotransverse ligament
Radiate ligament of head of rib

Left anterolateral view

Superior costal facet
Transverse process (cut off)
Radiate ligament of head of rib
Costotransverse ligament
Lateral costotransverse ligament
Superior costotransverse ligament
Intertransverse ligament

Right posterolateral view

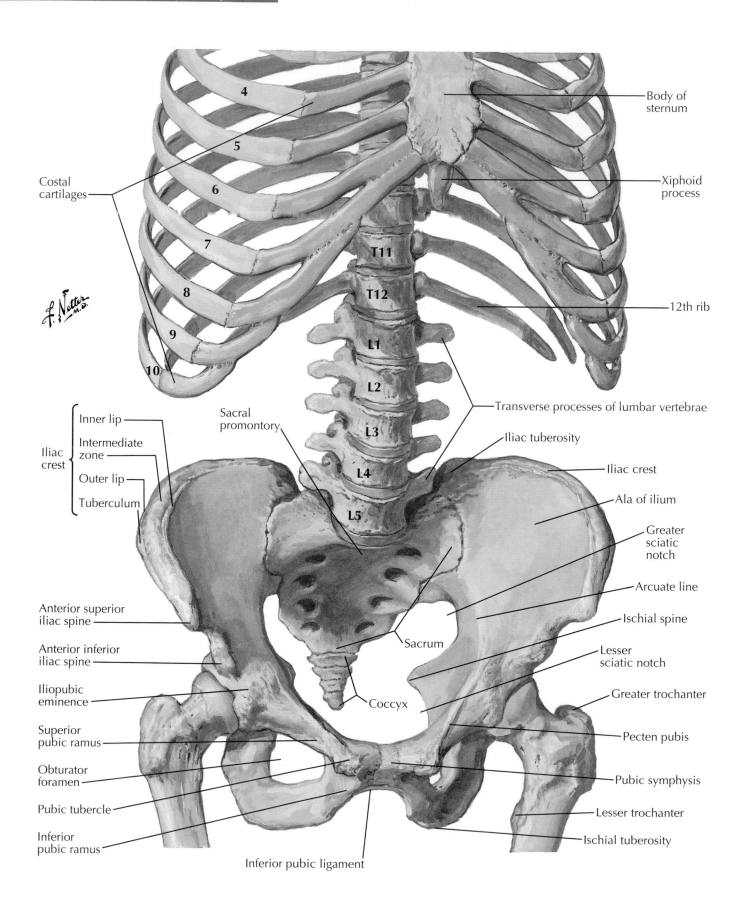

Costal cartilages

Inner lip
Intermediate zone
Outer lip
Tuberculum

Iliac crest

Anterior superior iliac spine

Anterior inferior iliac spine

Iliopubic eminence

Superior pubic ramus

Obturator foramen

Pubic tubercle

Inferior pubic ramus

Inferior pubic ligament

Sacral promontory

Sacrum

Coccyx

Body of sternum

Xiphoid process

12th rib

Transverse processes of lumbar vertebrae

Iliac tuberosity

Iliac crest

Ala of ilium

Greater sciatic notch

Arcuate line

Ischial spine

Lesser sciatic notch

Greater trochanter

Pecten pubis

Pubic symphysis

Lesser trochanter

Ischial tuberosity

4
5
6
7
8
9
10

T11
T12
L1
L2
L3
L4
L5

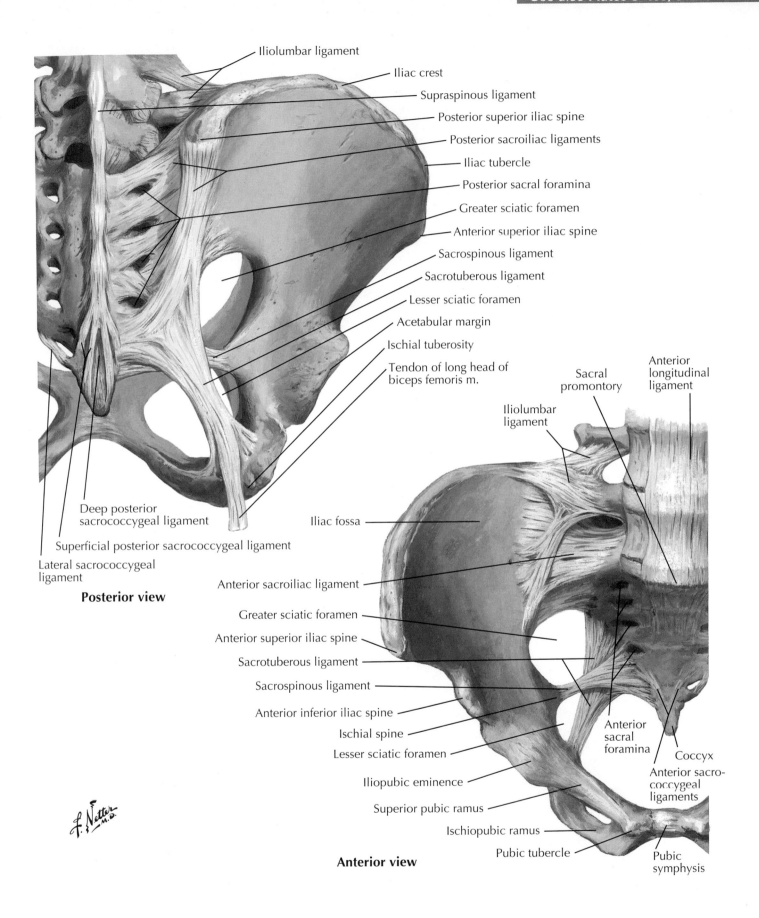

Iliolumbar ligament

Iliac crest

Supraspinous ligament

Posterior superior iliac spine

Posterior sacroiliac ligaments

Iliac tubercle

Posterior sacral foramina

Greater sciatic foramen

Anterior superior iliac spine

Sacrospinous ligament

Sacrotuberous ligament

Lesser sciatic foramen

Acetabular margin

Ischial tuberosity

Tendon of long head of biceps femoris m.

Deep posterior sacrococcygeal ligament

Superficial posterior sacrococcygeal ligament

Lateral sacrococcygeal ligament

Posterior view

Sacral promontory

Anterior longitudinal ligament

Iliolumbar ligament

Iliac fossa

Anterior sacroiliac ligament

Greater sciatic foramen

Anterior superior iliac spine

Sacrotuberous ligament

Sacrospinous ligament

Anterior inferior iliac spine

Ischial spine

Lesser sciatic foramen

Iliopubic eminence

Superior pubic ramus

Ischiopubic ramus

Pubic tubercle

Anterior sacral foramina

Coccyx

Anterior sacrococcygeal ligaments

Pubic symphysis

Anterior view

See also **Plates** S–151, S–152, S–154

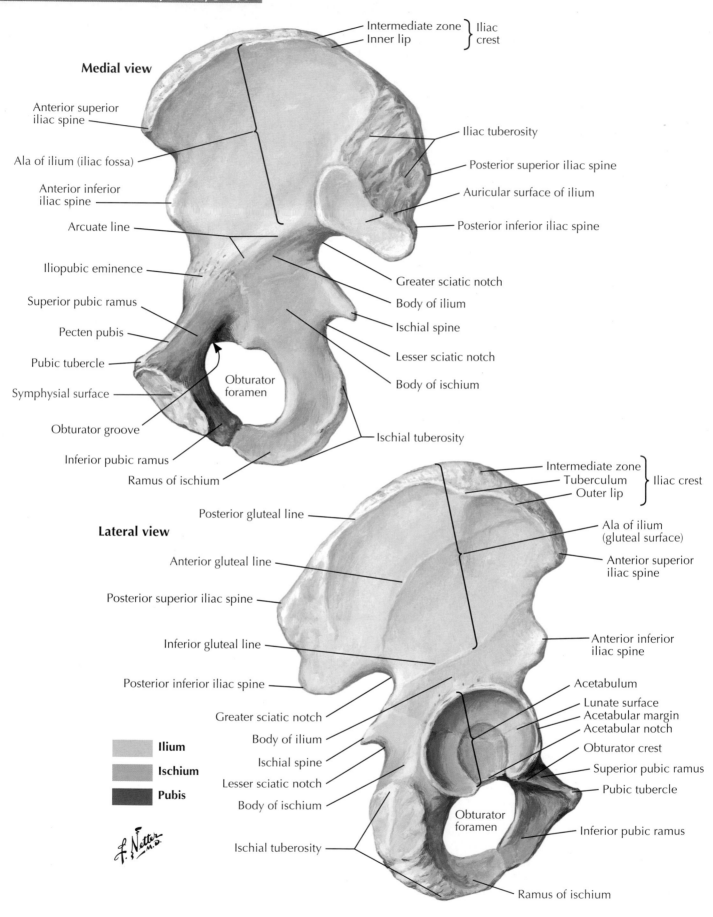

Medial view

Intermediate zone ⎱ Iliac
Inner lip ⎰ crest

Anterior superior iliac spine

Ala of ilium (iliac fossa)

Anterior inferior iliac spine

Arcuate line

Iliopubic eminence

Superior pubic ramus

Pecten pubis

Pubic tubercle

Symphysial surface

Obturator groove

Inferior pubic ramus

Ramus of ischium

Obturator foramen

Iliac tuberosity

Posterior superior iliac spine

Auricular surface of ilium

Posterior inferior iliac spine

Greater sciatic notch

Body of ilium

Ischial spine

Lesser sciatic notch

Body of ischium

Ischial tuberosity

Lateral view

Posterior gluteal line

Anterior gluteal line

Posterior superior iliac spine

Inferior gluteal line

Posterior inferior iliac spine

Greater sciatic notch

Body of ilium

Ischial spine

Lesser sciatic notch

Body of ischium

Ischial tuberosity

Intermediate zone ⎱
Tuberculum ⎬ Iliac crest
Outer lip ⎰

Ala of ilium (gluteal surface)

Anterior superior iliac spine

Anterior inferior iliac spine

Acetabulum

Lunate surface

Acetabular margin

Acetabular notch

Obturator crest

Superior pubic ramus

Pubic tubercle

Inferior pubic ramus

Obturator foramen

Ramus of ischium

Ilium

Ischium

Pubis

See also **Plates S–154, S–217, S–219, S–221, S–273, S–275**

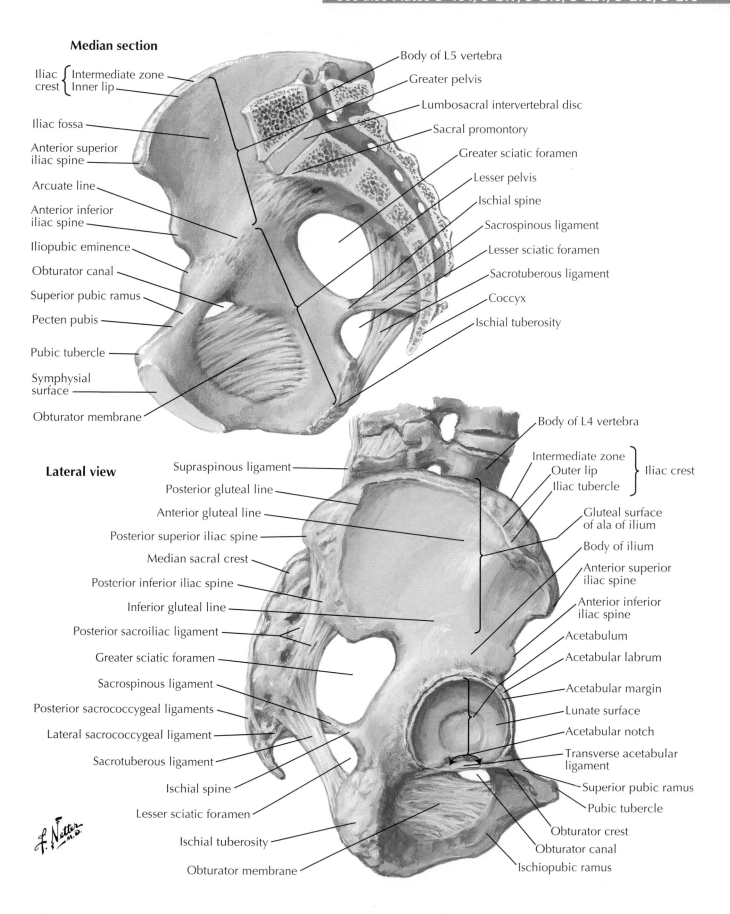

Median section

Iliac crest { Intermediate zone / Inner lip

Iliac fossa

Anterior superior iliac spine

Arcuate line

Anterior inferior iliac spine

Iliopubic eminence

Obturator canal

Superior pubic ramus

Pecten pubis

Pubic tubercle

Symphysial surface

Obturator membrane

Body of L5 vertebra

Greater pelvis

Lumbosacral intervertebral disc

Sacral promontory

Greater sciatic foramen

Lesser pelvis

Ischial spine

Sacrospinous ligament

Lesser sciatic foramen

Sacrotuberous ligament

Coccyx

Ischial tuberosity

Lateral view

Supraspinous ligament

Posterior gluteal line

Anterior gluteal line

Posterior superior iliac spine

Median sacral crest

Posterior inferior iliac spine

Inferior gluteal line

Posterior sacroiliac ligament

Greater sciatic foramen

Sacrospinous ligament

Posterior sacrococcygeal ligaments

Lateral sacrococcygeal ligament

Sacrotuberous ligament

Ischial spine

Lesser sciatic foramen

Ischial tuberosity

Obturator membrane

Body of L4 vertebra

Intermediate zone / Outer lip / Iliac tubercle } Iliac crest

Gluteal surface of ala of ilium

Body of ilium

Anterior superior iliac spine

Anterior inferior iliac spine

Acetabulum

Acetabular labrum

Acetabular margin

Lunate surface

Acetabular notch

Transverse acetabular ligament

Superior pubic ramus

Pubic tubercle

Obturator crest

Obturator canal

Ischiopubic ramus

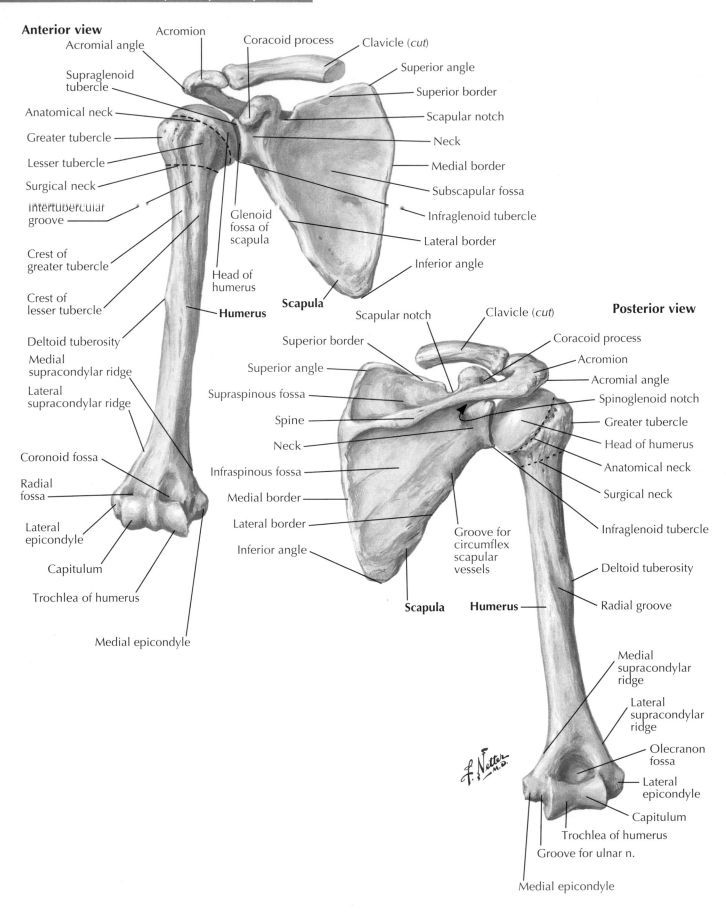

Anterior view

Acromion
Acromial angle
Coracoid process
Clavicle (*cut*)
Supraglenoid tubercle
Superior angle
Superior border
Anatomical neck
Scapular notch
Greater tubercle
Neck
Lesser tubercle
Medial border
Surgical neck
Subscapular fossa
Intertubercular groove
Glenoid fossa of scapula
Infraglenoid tubercle
Crest of greater tubercle
Lateral border
Crest of lesser tubercle
Head of humerus
Inferior angle
Deltoid tuberosity
Scapula
Humerus
Medial supracondylar ridge
Lateral supracondylar ridge
Coronoid fossa
Radial fossa
Lateral epicondyle
Capitulum
Trochlea of humerus
Medial epicondyle

Posterior view

Scapular notch
Clavicle (*cut*)
Coracoid process
Superior border
Acromion
Superior angle
Acromial angle
Supraspinous fossa
Spinoglenoid notch
Spine
Greater tubercle
Neck
Head of humerus
Infraspinous fossa
Anatomical neck
Medial border
Surgical neck
Lateral border
Infraglenoid tubercle
Groove for circumflex scapular vessels
Deltoid tuberosity
Scapula
Humerus
Radial groove
Inferior angle
Medial supracondylar ridge
Lateral supracondylar ridge
Olecranon fossa
Lateral epicondyle
Capitulum
Trochlea of humerus
Groove for ulnar n.
Medial epicondyle

Right clavicle

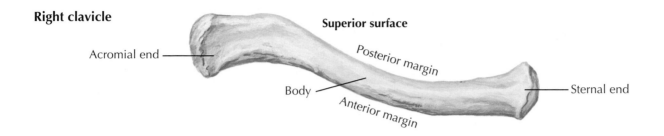

Superior surface

Acromial end

Posterior margin

Body

Anterior margin

Sternal end

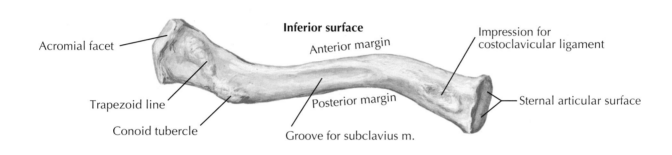

Inferior surface

Acromial facet

Anterior margin

Impression for costoclavicular ligament

Trapezoid line

Posterior margin

Sternal articular surface

Conoid tubercle

Groove for subclavius m.

Sternoclavicular joint

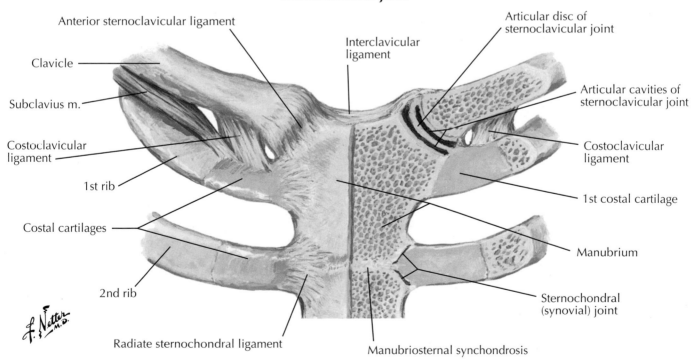

Anterior sternoclavicular ligament

Interclavicular ligament

Articular disc of sternoclavicular joint

Clavicle

Subclavius m.

Articular cavities of sternoclavicular joint

Costoclavicular ligament

Costoclavicular ligament

1st rib

1st costal cartilage

Costal cartilages

Manubrium

2nd rib

Sternochondral (synovial) joint

Radiate sternochondral ligament

Manubriosternal synchondrosis

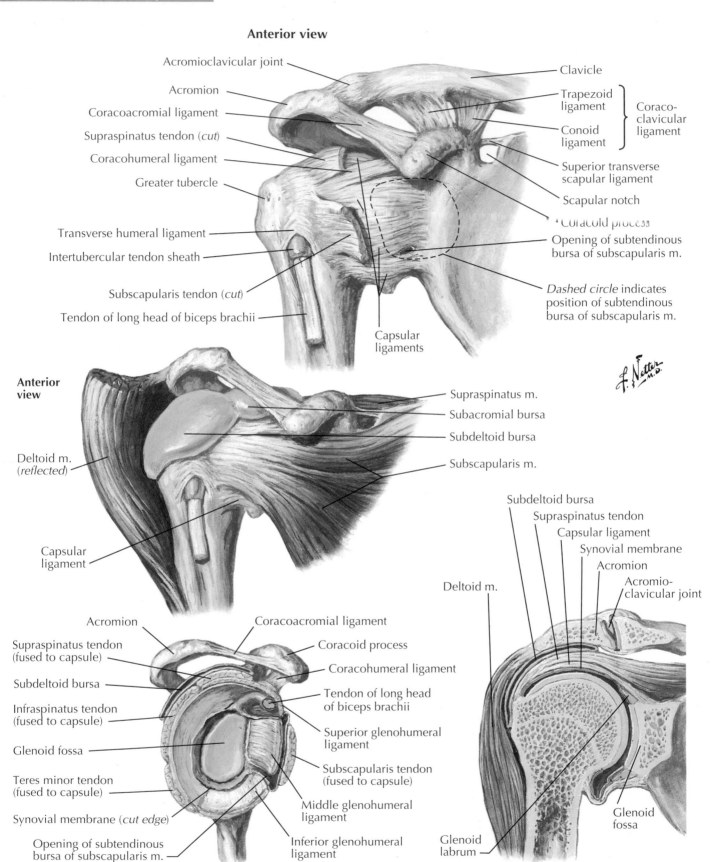

Anterior view

Acromioclavicular joint

Acromion

Coracoacromial ligament

Supraspinatus tendon (*cut*)

Coracohumeral ligament

Greater tubercle

Transverse humeral ligament

Intertubercular tendon sheath

Subscapularis tendon (*cut*)

Tendon of long head of biceps brachii

Clavicle

Trapezoid ligament ⎱ Coraco-
Conoid ligament ⎰ clavicular ligament

Superior transverse scapular ligament

Scapular notch

Coracoid process

Opening of subtendinous bursa of subscapularis m.

Dashed circle indicates position of subtendinous bursa of subscapularis m.

Capsular ligaments

Anterior view

Deltoid m. (*reflected*)

Capsular ligament

Supraspinatus m.

Subacromial bursa

Subdeltoid bursa

Subscapularis m.

Acromion

Supraspinatus tendon (fused to capsule)

Subdeltoid bursa

Infraspinatus tendon (fused to capsule)

Glenoid fossa

Teres minor tendon (fused to capsule)

Synovial membrane (*cut edge*)

Opening of subtendinous bursa of subscapularis m.

Coracoacromial ligament

Coracoid process

Coracohumeral ligament

Tendon of long head of biceps brachii

Superior glenohumeral ligament

Subscapularis tendon (fused to capsule)

Middle glenohumeral ligament

Inferior glenohumeral ligament

Joint opened: lateral view

Subdeltoid bursa

Supraspinatus tendon

Capsular ligament

Synovial membrane

Acromion

Acromio-clavicular joint

Deltoid m.

Glenoid labrum

Glenoid fossa

Coronal section through shoulder girdle

Plate S–158

Anteroposterior radiograph of right shoulder

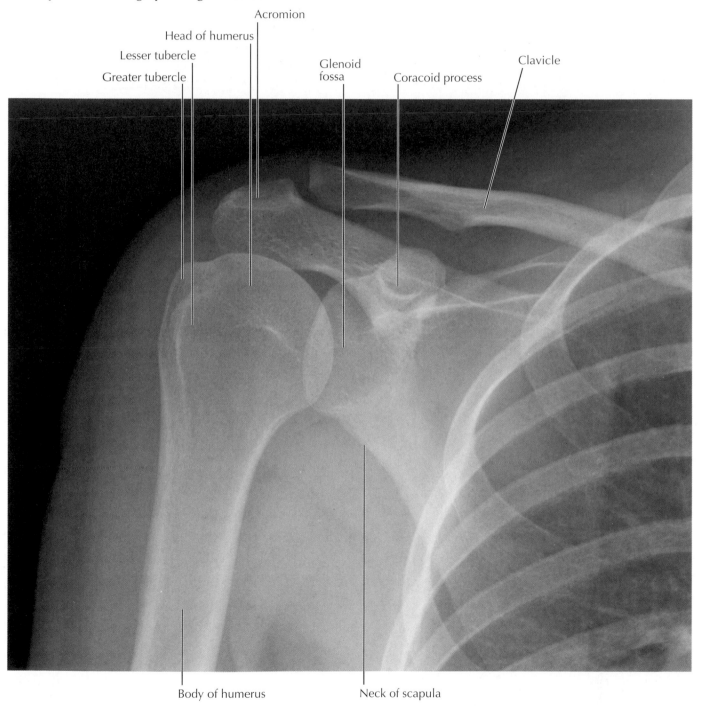

Acromion

Head of humerus

Lesser tubercle

Glenoid
fossa

Greater tubercle

Coracoid process

Clavicle

Body of humerus

Neck of scapula

Right elbow

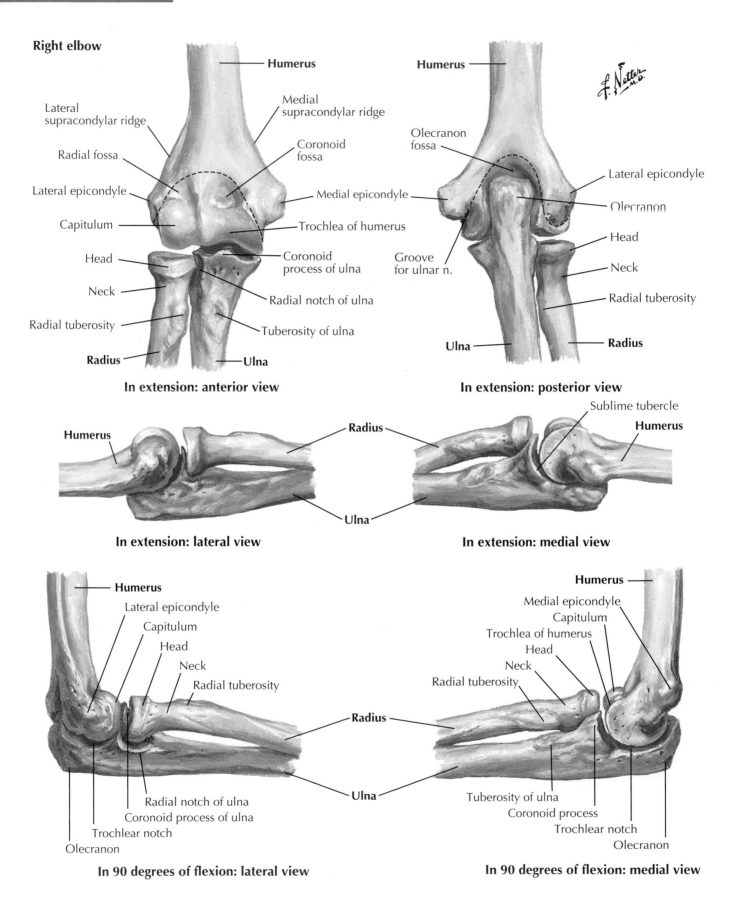

Humerus

Medial supracondylar ridge

Coronoid fossa

Lateral supracondylar ridge

Radial fossa

Lateral epicondyle

Capitulum

Head

Neck

Radial tuberosity

Radius

Medial epicondyle

Trochlea of humerus

Coronoid process of ulna

Radial notch of ulna

Tuberosity of ulna

Ulna

In extension: anterior view

Humerus

Olecranon fossa

Lateral epicondyle

Olecranon

Head

Neck

Radial tuberosity

Groove for ulnar n.

Ulna

Radius

In extension: posterior view

Humerus

Radius

Ulna

In extension: lateral view

Sublime tubercle

Humerus

Radius

Ulna

In extension: medial view

Humerus

Lateral epicondyle

Capitulum

Head

Neck

Radial tuberosity

Radius

Radial notch of ulna

Coronoid process of ulna

Trochlear notch

Olecranon

Ulna

In 90 degrees of flexion: lateral view

Humerus

Medial epicondyle

Capitulum

Trochlea of humerus

Head

Neck

Radial tuberosity

Radius

Tuberosity of ulna

Coronoid process

Trochlear notch

Olecranon

Ulna

In 90 degrees of flexion: medial view

Anteroposterior view

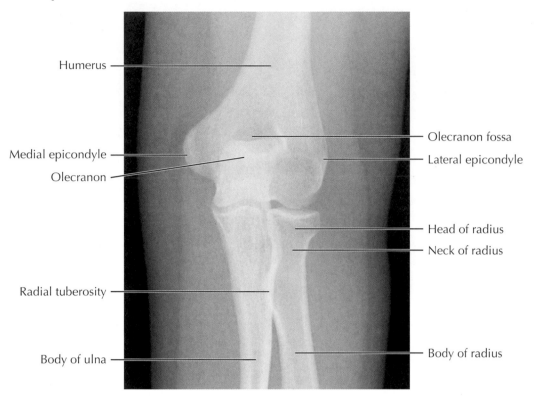

Humerus

Medial epicondyle

Olecranon

Radial tuberosity

Body of ulna

Olecranon fossa

Lateral epicondyle

Head of radius

Neck of radius

Body of radius

Lateral view

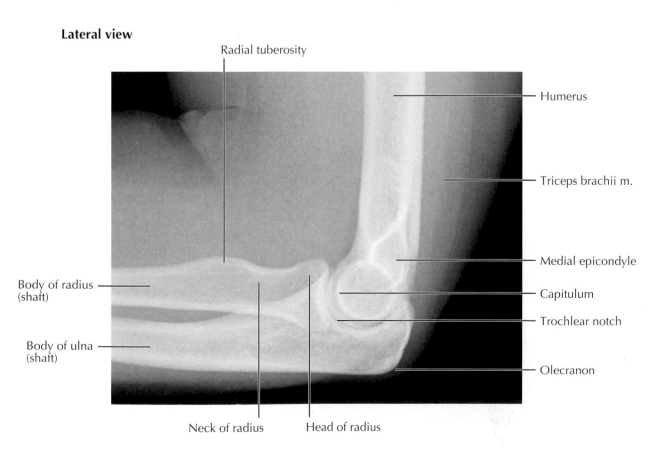

Radial tuberosity

Body of radius
(shaft)

Body of ulna
(shaft)

Neck of radius

Head of radius

Humerus

Triceps brachii m.

Medial epicondyle

Capitulum

Trochlear notch

Olecranon

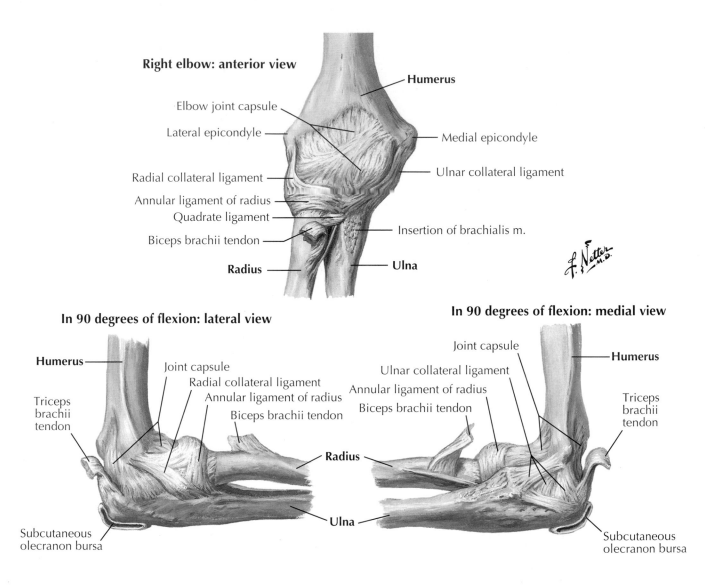

Right elbow: anterior view

Elbow joint capsule

Lateral epicondyle

Radial collateral ligament

Annular ligament of radius

Quadrate ligament

Biceps brachii tendon

Radius

Humerus

Medial epicondyle

Ulnar collateral ligament

Insertion of brachialis m.

Ulna

In 90 degrees of flexion: lateral view

Humerus

Triceps brachii tendon

Joint capsule

Radial collateral ligament

Annular ligament of radius

Biceps brachii tendon

Radius

Ulna

Subcutaneous olecranon bursa

In 90 degrees of flexion: medial view

Joint capsule

Ulnar collateral ligament

Annular ligament of radius

Biceps brachii tendon

Humerus

Triceps brachii tendon

Subcutaneous olecranon bursa

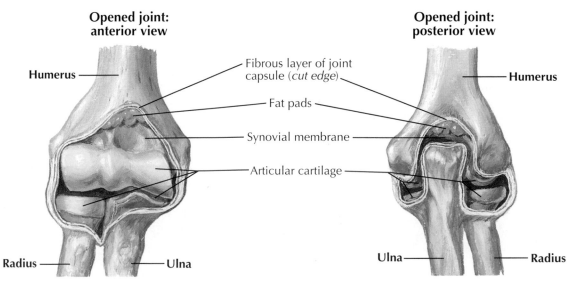

Opened joint: anterior view

Humerus

Radius

Ulna

Fibrous layer of joint capsule (*cut edge*)

Fat pads

Synovial membrane

Articular cartilage

Opened joint: posterior view

Humerus

Ulna

Radius

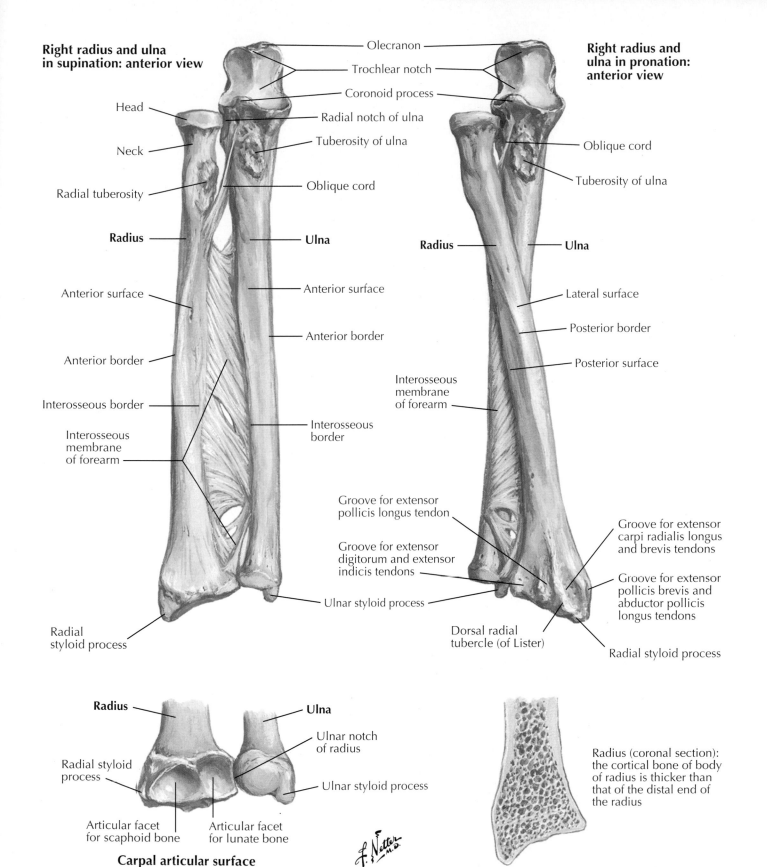

Right radius and ulna in supination: anterior view

- Olecranon
- Trochlear notch
- Coronoid process
- Head
- Radial notch of ulna
- Neck
- Tuberosity of ulna
- Radial tuberosity
- Oblique cord
- **Radius**
- **Ulna**
- Anterior surface
- Anterior surface
- Anterior border
- Anterior border
- Interosseous border
- Interosseous border
- Interosseous membrane of forearm
- Radial styloid process

Right radius and ulna in pronation: anterior view

- Oblique cord
- Tuberosity of ulna
- **Radius**
- **Ulna**
- Lateral surface
- Posterior border
- Posterior surface
- Interosseous membrane of forearm
- Groove for extensor pollicis longus tendon
- Groove for extensor digitorum and extensor indicis tendons
- Ulnar styloid process
- Groove for extensor carpi radialis longus and brevis tendons
- Groove for extensor pollicis brevis and abductor pollicis longus tendons
- Dorsal radial tubercle (of Lister)
- Radial styloid process

Radius
Ulna

- Radial styloid process
- Ulnar notch of radius
- Ulnar styloid process
- Articular facet for scaphoid bone
- Articular facet for lunate bone

Carpal articular surface

Radius (coronal section): the cortical bone of body of radius is thicker than that of the distal end of the radius

f. Netter M.D.

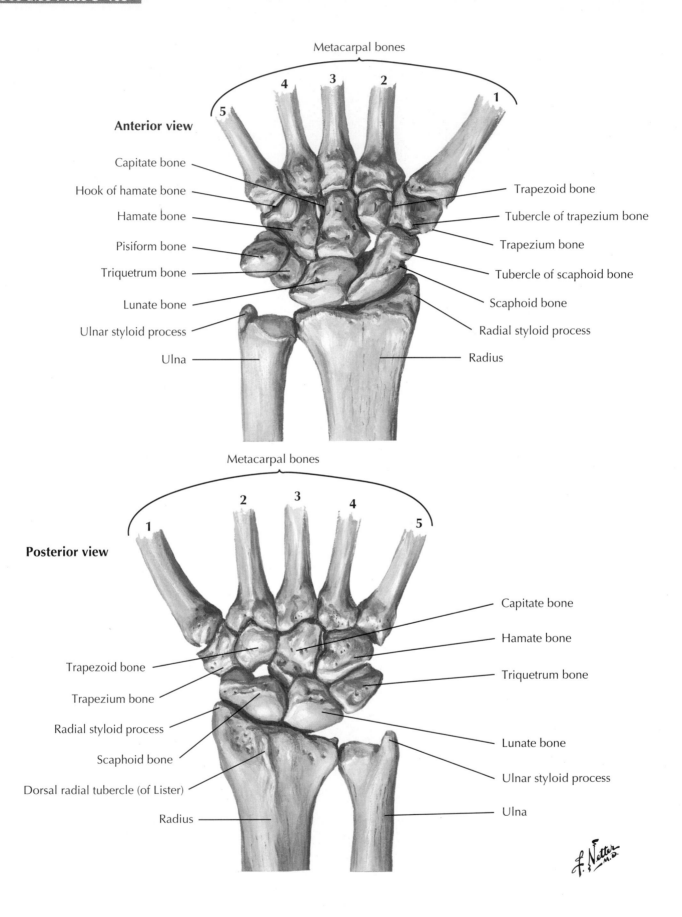

Metacarpal bones

Anterior view

5   4   3   2   1

Capitate bone

Hook of hamate bone

Hamate bone

Pisiform bone

Triquetrum bone

Lunate bone

Ulnar styloid process

Ulna

Trapezoid bone

Tubercle of trapezium bone

Trapezium bone

Tubercle of scaphoid bone

Scaphoid bone

Radial styloid process

Radius

Metacarpal bones

Posterior view

1   2   3   4   5

Trapezoid bone

Trapezium bone

Radial styloid process

Scaphoid bone

Dorsal radial tubercle (of Lister)

Radius

Capitate bone

Hamate bone

Triquetrum bone

Lunate bone

Ulnar styloid process

Ulna

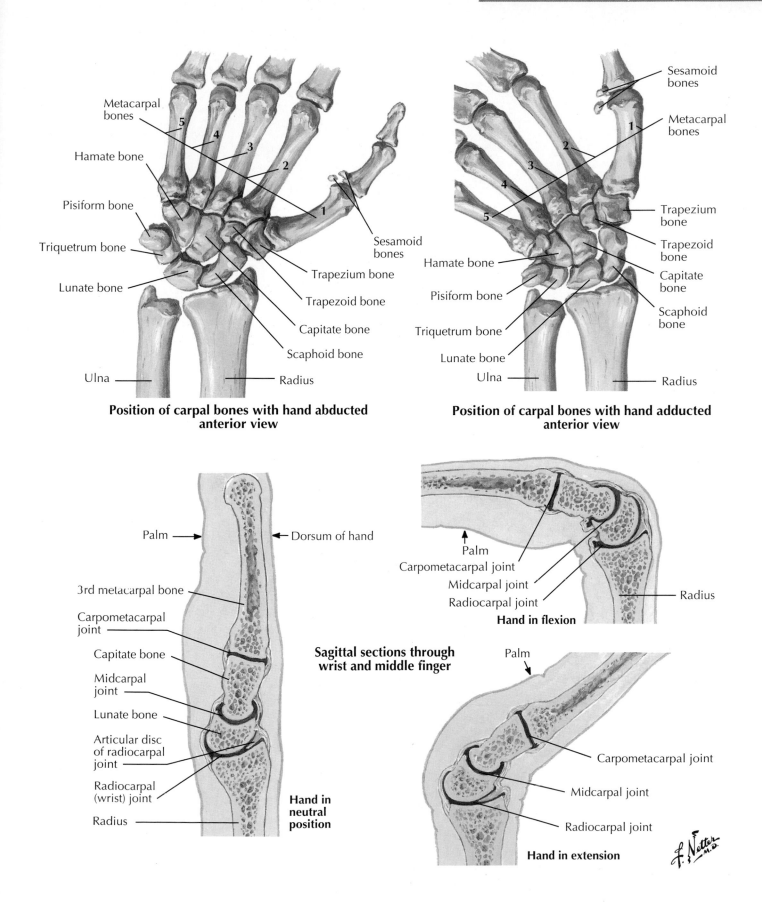

Position of carpal bones with hand abducted anterior view

Metacarpal bones
5
4
3
2
1
Hamate bone
Pisiform bone
Triquetrum bone
Lunate bone
Ulna
Radius
Sesamoid bones
Trapezium bone
Trapezoid bone
Capitate bone
Scaphoid bone

Position of carpal bones with hand adducted anterior view

Sesamoid bones
Metacarpal bones
1
2
3
4
5
Trapezium bone
Trapezoid bone
Capitate bone
Scaphoid bone
Hamate bone
Pisiform bone
Triquetrum bone
Lunate bone
Ulna
Radius

Palm
Dorsum of hand
3rd metacarpal bone
Carpometacarpal joint
Capitate bone
Midcarpal joint
Lunate bone
Articular disc of radiocarpal joint
Radiocarpal (wrist) joint
Radius

Hand in neutral position

Sagittal sections through wrist and middle finger

Palm
Carpometacarpal joint
Midcarpal joint
Radiocarpal joint
Radius

Hand in flexion

Palm
Carpometacarpal joint
Midcarpal joint
Radiocarpal joint

Hand in extension

Deep palm

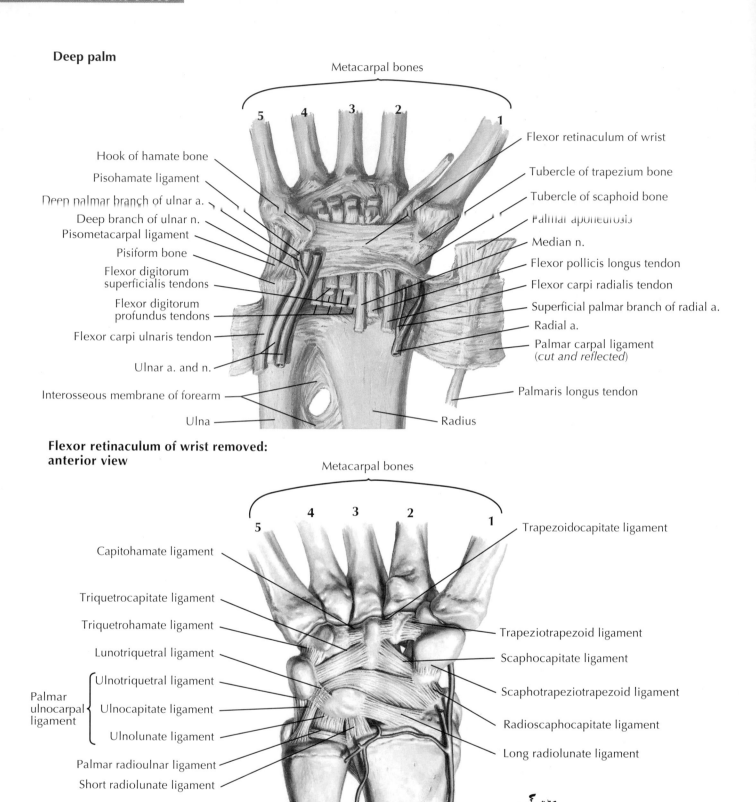

Metacarpal bones

5   4   3   2   1

Hook of hamate bone
Pisohamate ligament
Deep palmar branch of ulnar a.
Deep branch of ulnar n.
Pisometacarpal ligament
Pisiform bone
Flexor digitorum superficialis tendons
Flexor digitorum profundus tendons
Flexor carpi ulnaris tendon
Ulnar a. and n.
Interosseous membrane of forearm
Ulna

Flexor retinaculum of wrist
Tubercle of trapezium bone
Tubercle of scaphoid bone
Palmar aponeurosis
Median n.
Flexor pollicis longus tendon
Flexor carpi radialis tendon
Superficial palmar branch of radial a.
Radial a.
Palmar carpal ligament (cut and reflected)
Palmaris longus tendon
Radius

Flexor retinaculum of wrist removed: anterior view

Metacarpal bones

5   4   3   2   1

Capitohamate ligament
Triquetrocapitate ligament
Triquetrohamate ligament
Lunotriquetral ligament
Palmar ulnocarpal ligament {
  Ulnotriquetral ligament
  Ulnocapitate ligament
  Ulnolunate ligament
}
Palmar radioulnar ligament
Short radiolunate ligament

Trapezoidocapitate ligament
Trapeziotrapezoid ligament
Scaphocapitate ligament
Scaphotrapeziotrapezoid ligament
Radioscaphocapitate ligament
Long radiolunate ligament

Posterior view

Metacarpal bones

1   2   3   4   5

Trapezoidocapitate ligament

Trapeziotrapezoid ligament

Scapholunate ligament

Dorsal radiocarpal ligament

Capitohamate ligament

Dorsal intercarpal ligament

Triquetrohamate ligament

Ulnotriquetral ligament

Dorsal radioulnar ligament

Arcuate dorsal radioulnar ligament
(part of dorsal radioulnar ligament)

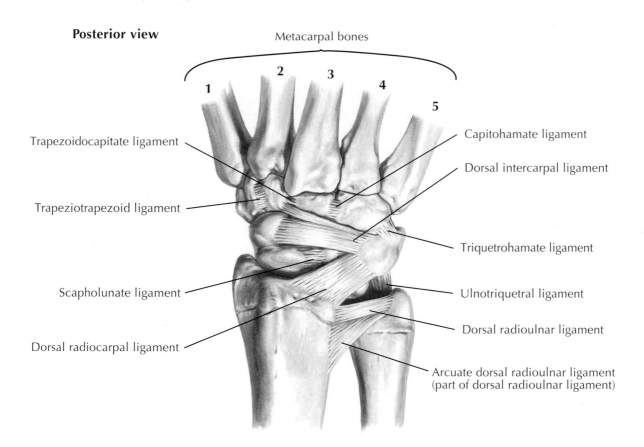

Coronal section: posterior view

Metacarpal bones

1   2   3   4   5

Intermetacarpal joints

Carpometacarpal joint

Trapezium bone

Trapezoid bone

Midcarpal joint

Scaphoid bone

Radiocarpal (wrist) joint

Lunate bone

Radius

Capitate bone

Hamate bone

Triquetrum bone

Interosseous intercarpal ligaments

Pisiform bone

Meniscus

Articular disc of radiocarpal joint

Distal radioulnar joint

Ulna

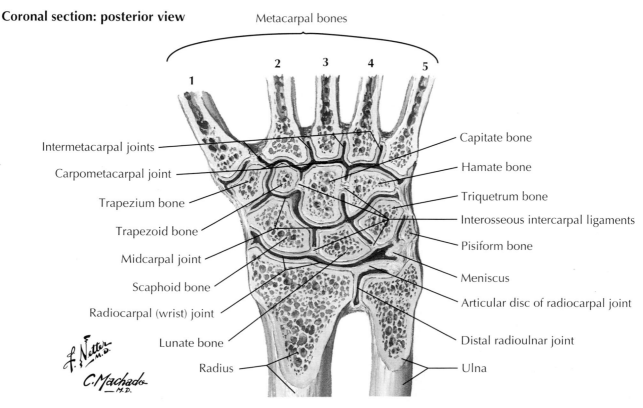

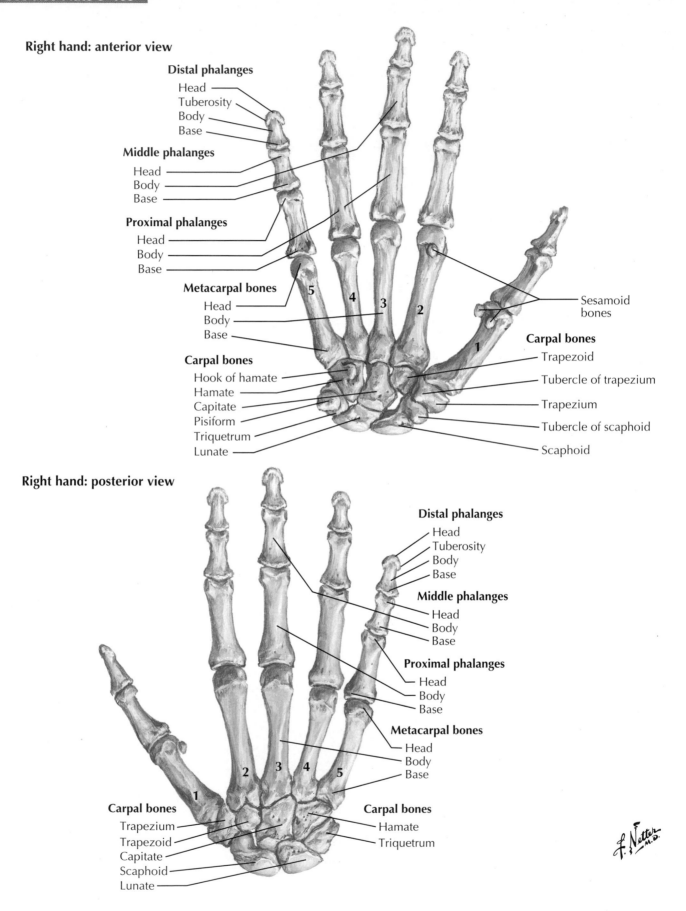

Right hand: anterior view

Distal phalanges
Head
Tuberosity
Body
Base

Middle phalanges
Head
Body
Base

Proximal phalanges
Head
Body
Base

Metacarpal bones
Head
Body
Base

Carpal bones
Hook of hamate
Hamate
Capitate
Pisiform
Triquetrum
Lunate

Sesamoid bones

Carpal bones
Trapezoid
Tubercle of trapezium
Trapezium
Tubercle of scaphoid
Scaphoid

Right hand: posterior view

Distal phalanges
Head
Tuberosity
Body
Base

Middle phalanges
Head
Body
Base

Proximal phalanges
Head
Body
Base

Metacarpal bones
Head
Body
Base

Carpal bones
Trapezium
Trapezoid
Capitate
Scaphoid
Lunate

Carpal bones
Hamate
Triquetrum

Anteroposterior view

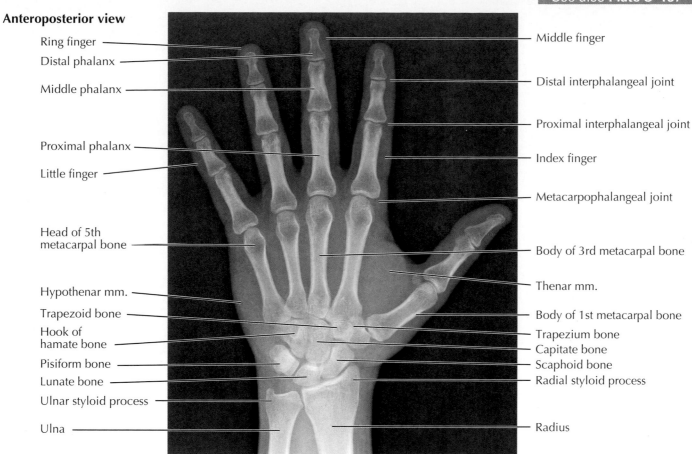

Ring finger

Distal phalanx

Middle phalanx

Proximal phalanx

Little finger

Head of 5th
metacarpal bone

Hypothenar mm.

Trapezoid bone

Hook of
hamate bone

Pisiform bone

Lunate bone

Ulnar styloid process

Ulna

Middle finger

Distal interphalangeal joint

Proximal interphalangeal joint

Index finger

Metacarpophalangeal joint

Body of 3rd metacarpal bone

Thenar mm.

Body of 1st metacarpal bone

Trapezium bone

Capitate bone

Scaphoid bone

Radial styloid process

Radius

Lateral view

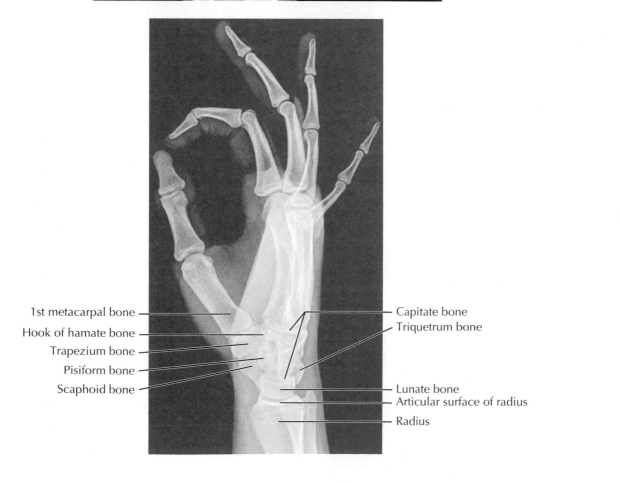

1st metacarpal bone

Hook of hamate bone

Trapezium bone

Pisiform bone

Scaphoid bone

Capitate bone

Triquetrum bone

Lunate bone

Articular surface of radius

Radius

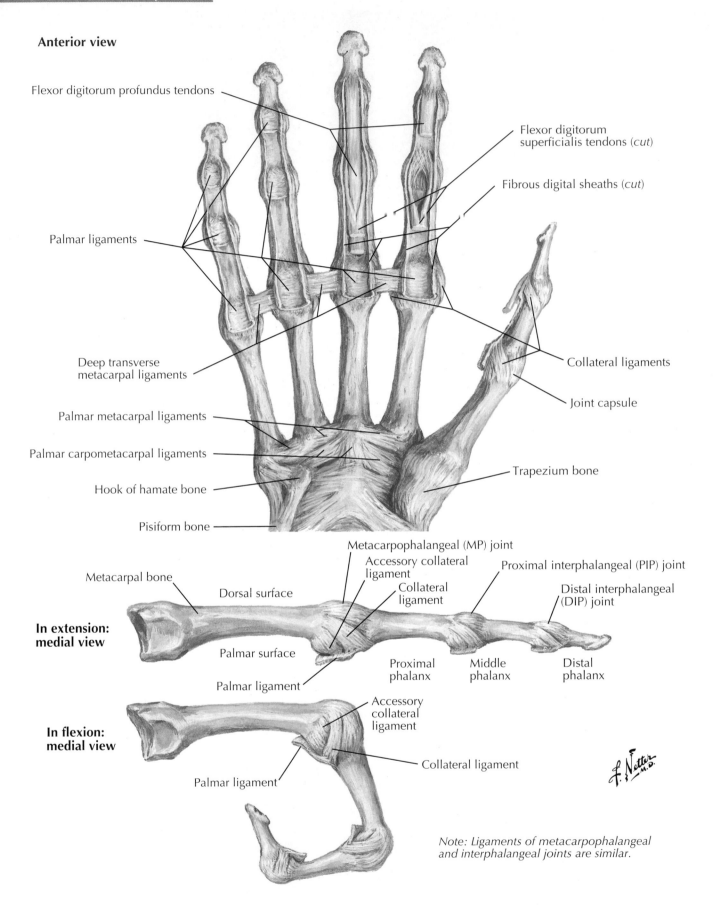

Anterior view

Flexor digitorum profundus tendons

Flexor digitorum superficialis tendons (*cut*)

Fibrous digital sheaths (*cut*)

Palmar ligaments

Collateral ligaments

Joint capsule

Deep transverse metacarpal ligaments

Palmar metacarpal ligaments

Palmar carpometacarpal ligaments

Trapezium bone

Hook of hamate bone

Pisiform bone

Metacarpophalangeal (MP) joint

Accessory collateral ligament

Proximal interphalangeal (PIP) joint

Collateral ligament

Distal interphalangeal (DIP) joint

Metacarpal bone

Dorsal surface

In extension: medial view

Palmar surface

Proximal phalanx

Middle phalanx

Distal phalanx

Palmar ligament

Accessory collateral ligament

In flexion: medial view

Collateral ligament

Palmar ligament

Note: Ligaments of metacarpophalangeal and interphalangeal joints are similar.

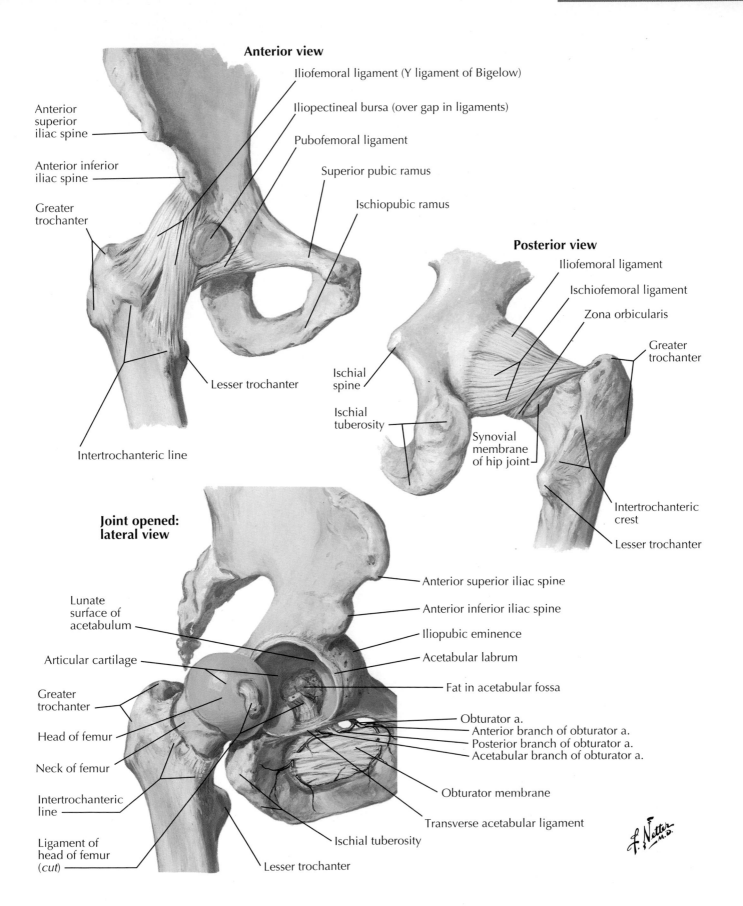

Anterior view

Iliofemoral ligament (Y ligament of Bigelow)

Iliopectineal bursa (over gap in ligaments)

Pubofemoral ligament

Superior pubic ramus

Ischiopubic ramus

Anterior superior iliac spine

Anterior inferior iliac spine

Greater trochanter

Lesser trochanter

Intertrochanteric line

Posterior view

Iliofemoral ligament

Ischiofemoral ligament

Zona orbicularis

Greater trochanter

Ischial spine

Ischial tuberosity

Synovial membrane of hip joint

Intertrochanteric crest

Lesser trochanter

Joint opened: lateral view

Lunate surface of acetabulum

Articular cartilage

Greater trochanter

Head of femur

Neck of femur

Intertrochanteric line

Ligament of head of femur (cut)

Lesser trochanter

Ischial tuberosity

Transverse acetabular ligament

Obturator membrane

Acetabular branch of obturator a.

Posterior branch of obturator a.

Anterior branch of obturator a.

Obturator a.

Fat in acetabular fossa

Acetabular labrum

Iliopubic eminence

Anterior inferior iliac spine

Anterior superior iliac spine

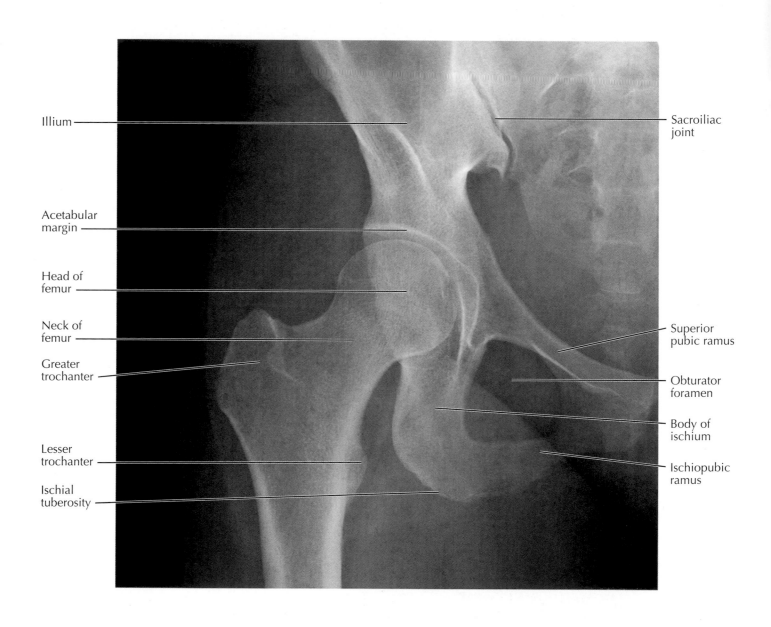

Illium

Acetabular margin

Head of femur

Neck of femur

Greater trochanter

Lesser trochanter

Ischial tuberosity

Sacroiliac joint

Superior pubic ramus

Obturator foramen

Body of ischium

Ischiopubic ramus

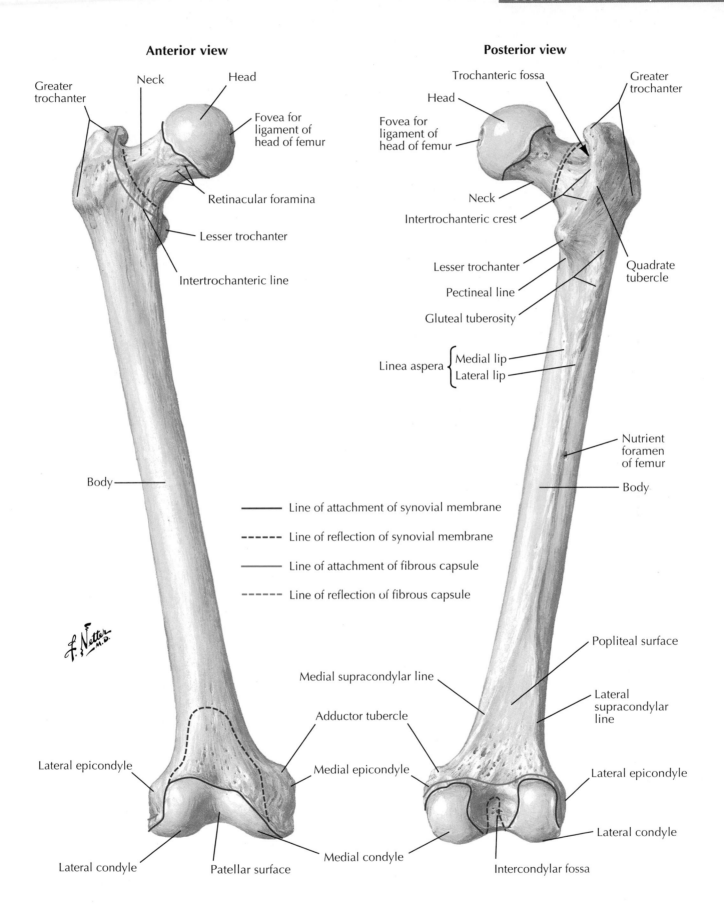

Anterior view

Greater trochanter

Neck

Head

Fovea for ligament of head of femur

Retinacular foramina

Lesser trochanter

Intertrochanteric line

Body

Line of attachment of synovial membrane
- - - - Line of reflection of synovial membrane
Line of attachment of fibrous capsule
- - - - Line of reflection of fibrous capsule

Medial supracondylar line

Adductor tubercle

Medial epicondyle

Lateral epicondyle

Lateral condyle

Patellar surface

Medial condyle

Posterior view

Trochanteric fossa

Head

Greater trochanter

Fovea for ligament of head of femur

Neck

Intertrochanteric crest

Lesser trochanter

Pectineal line

Gluteal tuberosity

Quadrate tubercle

Linea aspera { Medial lip
Lateral lip }

Nutrient foramen of femur

Body

Popliteal surface

Lateral supracondylar line

Lateral epicondyle

Lateral condyle

Intercondylar fossa

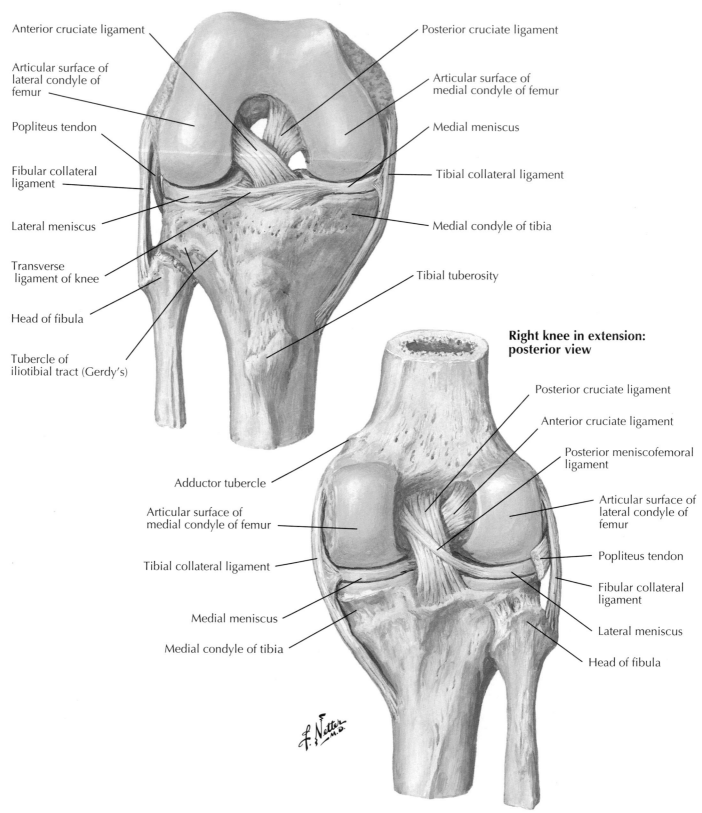

Right knee in flexion: anterior view

Anterior cruciate ligament

Articular surface of lateral condyle of femur

Popliteus tendon

Fibular collateral ligament

Lateral meniscus

Transverse ligament of knee

Head of fibula

Tubercle of iliotibial tract (Gerdy's)

Posterior cruciate ligament

Articular surface of medial condyle of femur

Medial meniscus

Tibial collateral ligament

Medial condyle of tibia

Tibial tuberosity

Right knee in extension: posterior view

Adductor tubercle

Articular surface of medial condyle of femur

Tibial collateral ligament

Medial meniscus

Medial condyle of tibia

Posterior cruciate ligament

Anterior cruciate ligament

Posterior meniscofemoral ligament

Articular surface of lateral condyle of femur

Popliteus tendon

Fibular collateral ligament

Lateral meniscus

Head of fibula

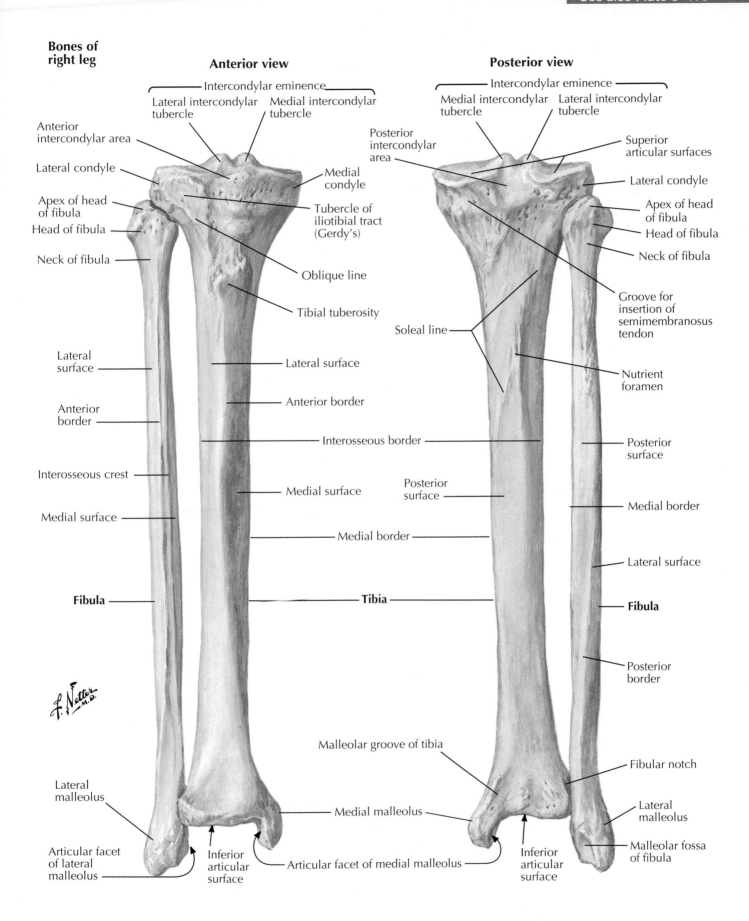

Bones of right leg

Anterior view

Posterior view

Intercondylar eminence

Lateral intercondylar tubercle

Medial intercondylar tubercle

Anterior intercondylar area

Lateral condyle

Apex of head of fibula

Head of fibula

Neck of fibula

Medial condyle

Tubercle of iliotibial tract (Gerdy's)

Oblique line

Tibial tuberosity

Lateral surface

Anterior border

Lateral surface

Anterior border

Interosseous border

Interosseous crest

Medial surface

Medial surface

Medial border

Posterior surface

Fibula

Tibia

Malleolar groove of tibia

Lateral malleolus

Articular facet of lateral malleolus

Inferior articular surface

Medial malleolus

Articular facet of medial malleolus

Intercondylar eminence

Medial intercondylar tubercle

Lateral intercondylar tubercle

Posterior intercondylar area

Superior articular surfaces

Lateral condyle

Apex of head of fibula

Head of fibula

Neck of fibula

Groove for insertion of semimembranosus tendon

Soleal line

Nutrient foramen

Posterior surface

Medial border

Lateral surface

Fibula

Posterior border

Fibular notch

Lateral malleolus

Malleolar fossa of fibula

Inferior articular surface

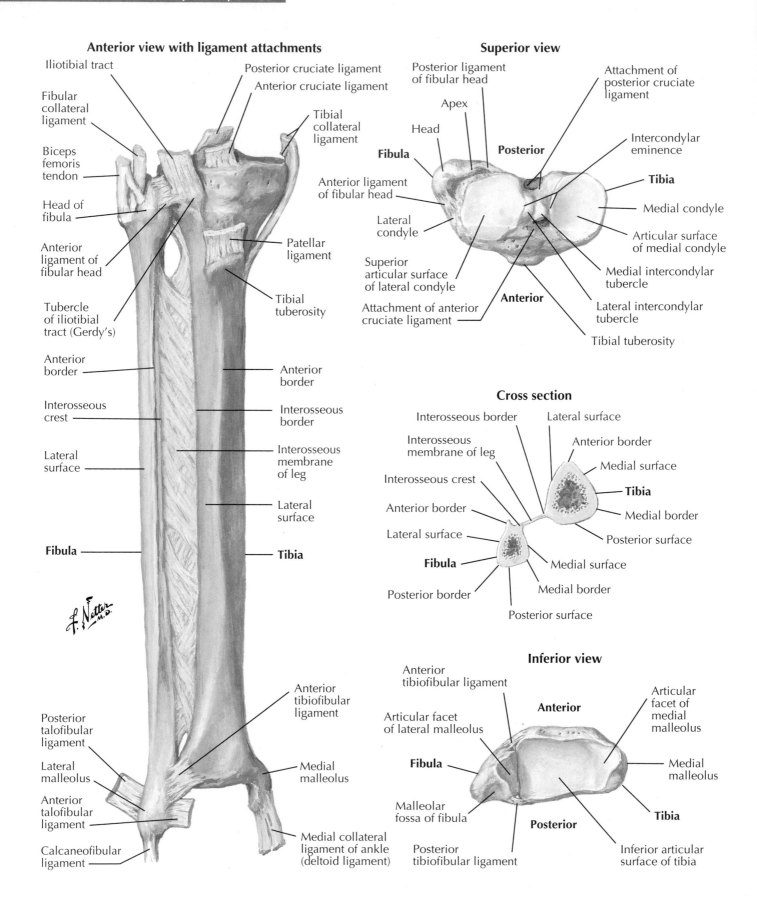

Anterior view with ligament attachments

Iliotibial tract

Fibular collateral ligament

Biceps femoris tendon

Head of fibula

Anterior ligament of fibular head

Tubercle of iliotibial tract (Gerdy's)

Anterior border

Interosseous crest

Lateral surface

Fibula

Posterior talofibular ligament

Lateral malleolus

Anterior talofibular ligament

Calcaneofibular ligament

Posterior cruciate ligament

Anterior cruciate ligament

Tibial collateral ligament

Patellar ligament

Tibial tuberosity

Anterior border

Interosseous border

Interosseous membrane of leg

Lateral surface

Tibia

Anterior tibiofibular ligament

Medial malleolus

Medial collateral ligament of ankle (deltoid ligament)

Superior view

Posterior ligament of fibular head

Apex

Head

Fibula

Posterior

Anterior ligament of fibular head

Lateral condyle

Superior articular surface of lateral condyle

Attachment of anterior cruciate ligament

Anterior

Attachment of posterior cruciate ligament

Intercondylar eminence

Tibia

Medial condyle

Articular surface of medial condyle

Medial intercondylar tubercle

Lateral intercondylar tubercle

Tibial tuberosity

Cross section

Interosseous border

Interosseous membrane of leg

Interosseous crest

Anterior border

Lateral surface

Fibula

Posterior border

Lateral surface

Anterior border

Medial surface

Tibia

Medial border

Posterior surface

Medial surface

Medial border

Posterior surface

Inferior view

Anterior tibiofibular ligament

Articular facet of lateral malleolus

Fibula

Malleolar fossa of fibula

Posterior tibiofibular ligament

Anterior

Articular facet of medial malleolus

Medial malleolus

Tibia

Inferior articular surface of tibia

Posterior

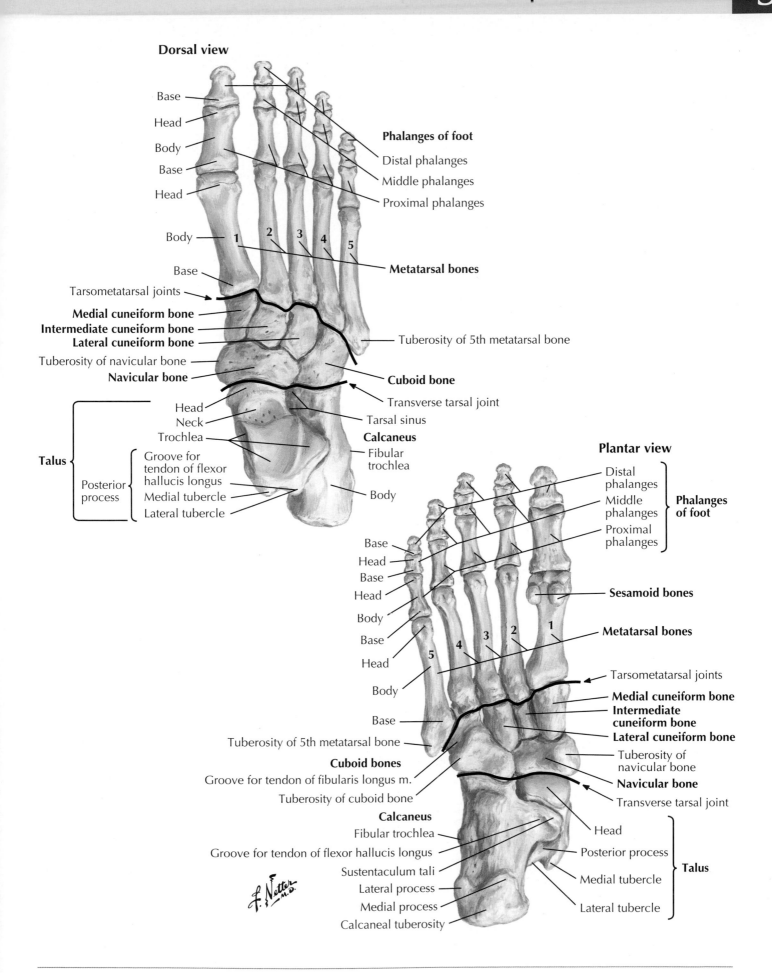

Dorsal view

Base
Head
Body
Base
Head

Phalanges of foot
Distal phalanges
Middle phalanges
Proximal phalanges

Body — 1 2 3 4 5
Metatarsal bones

Base
Tarsometatarsal joints
Medial cuneiform bone
Intermediate cuneiform bone
Lateral cuneiform bone
Tuberosity of navicular bone
Navicular bone

Tuberosity of 5th metatarsal bone

Cuboid bone
Transverse tarsal joint
Tarsal sinus
Calcaneus
Fibular trochlea

Head
Neck
Trochlea
Talus
Posterior process
Groove for tendon of flexor hallucis longus
Medial tubercle
Lateral tubercle

Body

Plantar view

Distal phalanges
Middle phalanges
Proximal phalanges
Phalanges of foot

Base
Head
Base
Head
Body
Base
Head
Body

Sesamoid bones
Metatarsal bones

5 4 3 2 1

Tarsometatarsal joints
Medial cuneiform bone
Intermediate cuneiform bone
Lateral cuneiform bone
Tuberosity of navicular bone
Navicular bone
Transverse tarsal joint

Base
Tuberosity of 5th metatarsal bone
Cuboid bones
Groove for tendon of fibularis longus m.
Tuberosity of cuboid bone
Calcaneus
Fibular trochlea
Groove for tendon of flexor hallucis longus
Sustentaculum tali
Lateral process
Medial process
Calcaneal tuberosity

Head
Posterior process
Medial tubercle
Lateral tubercle
Talus

f. Netter m.d.

Lower Limb

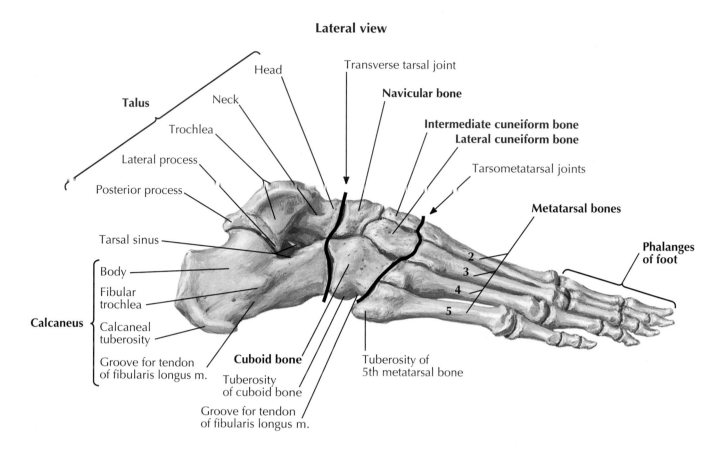

Lateral view

Transverse tarsal joint

Head

Navicular bone

Neck

Talus

Trochlea

Intermediate cuneiform bone
Lateral cuneiform bone

Lateral process

Tarsometatarsal joints

Posterior process

Metatarsal bones

Tarsal sinus

Phalanges of foot

Body

2

Fibular trochlea

3

Calcaneus

Calcaneal tuberosity

4

5

Groove for tendon of fibularis longus m.

Cuboid bone

Tuberosity of 5th metatarsal bone

Tuberosity of cuboid bone

Groove for tendon of fibularis longus m.

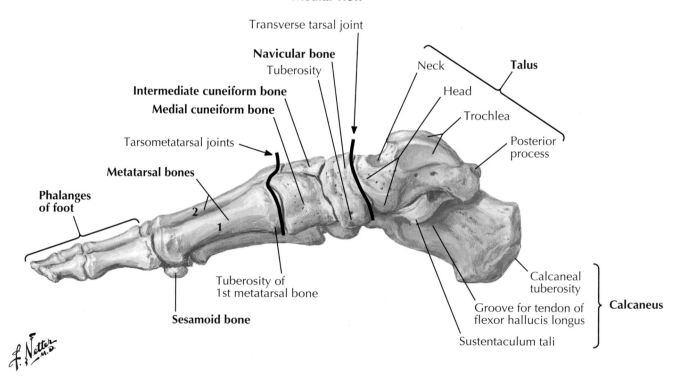

Medial view

Transverse tarsal joint

Navicular bone
Tuberosity

Neck

Talus

Intermediate cuneiform bone

Head

Medial cuneiform bone

Trochlea

Tarsometatarsal joints

Posterior process

Metatarsal bones

Phalanges of foot

2

1

Calcaneal tuberosity

Tuberosity of 1st metatarsal bone

Calcaneus

Groove for tendon of flexor hallucis longus

Sesamoid bone

Sustentaculum tali

Right foot

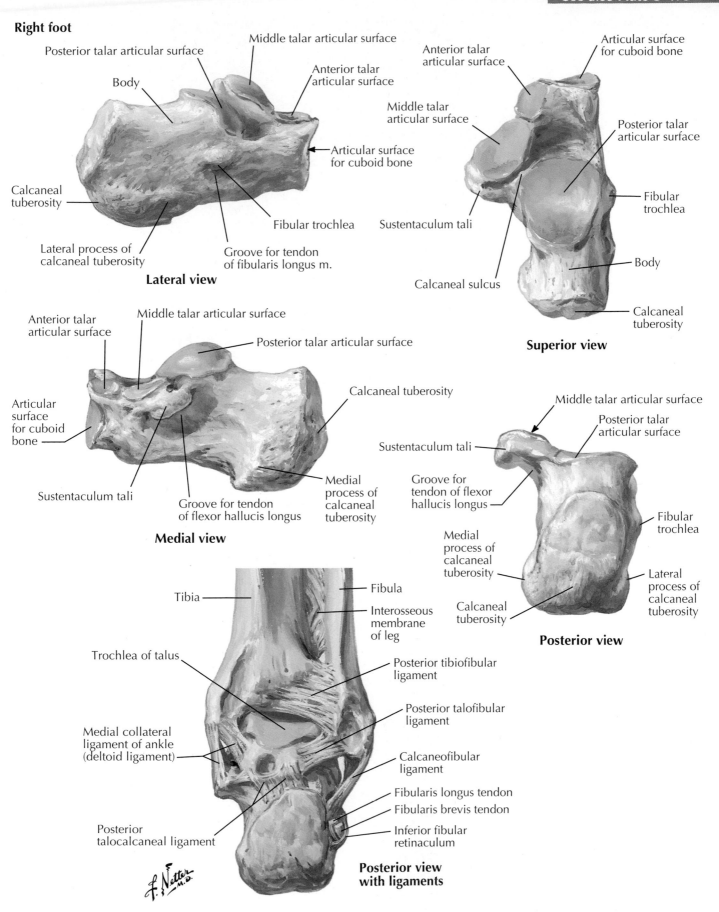

Posterior talar articular surface

Middle talar articular surface

Body

Anterior talar
articular surface

Calcaneal
tuberosity

Articular surface
for cuboid bone

Lateral process of
calcaneal tuberosity

Fibular trochlea

Groove for tendon
of fibularis longus m.

Lateral view

Anterior talar
articular surface

Middle talar articular surface

Posterior talar articular surface

Articular
surface
for cuboid
bone

Calcaneal tuberosity

Sustentaculum tali

Groove for tendon
of flexor hallucis longus

Medial
process of
calcaneal
tuberosity

Medial view

Anterior talar
articular surface

Articular surface
for cuboid bone

Middle talar
articular surface

Posterior talar
articular surface

Fibular
trochlea

Sustentaculum tali

Body

Calcaneal sulcus

Calcaneal
tuberosity

Superior view

Middle talar articular surface

Posterior talar
articular surface

Sustentaculum tali

Groove for
tendon of flexor
hallucis longus

Medial
process of
calcaneal
tuberosity

Calcaneal
tuberosity

Fibular
trochlea

Lateral
process of
calcaneal
tuberosity

Posterior view

Tibia

Trochlea of talus

Medial collateral
ligament of ankle
(deltoid ligament)

Posterior
talocalcaneal ligament

Fibula

Interosseous
membrane
of leg

Posterior tibiofibular
ligament

Posterior talofibular
ligament

Calcaneofibular
ligament

Fibularis longus tendon

Fibularis brevis tendon

Inferior fibular
retinaculum

**Posterior view
with ligaments**

Lateral view

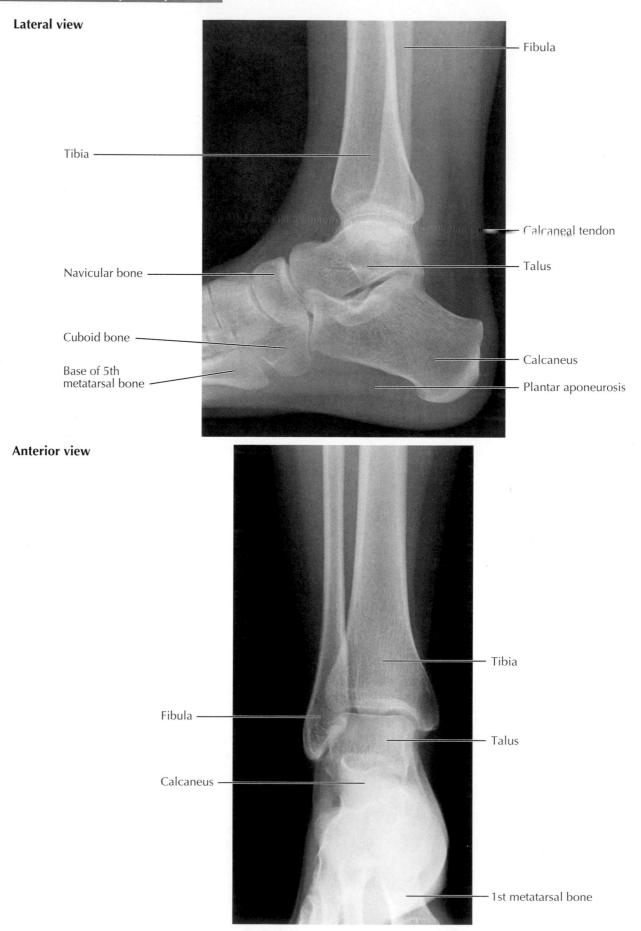

Fibula

Tibia

Calcaneal tendon

Talus

Navicular bone

Cuboid bone

Calcaneus

Base of 5th
metatarsal bone

Plantar aponeurosis

Anterior view

Tibia

Fibula

Talus

Calcaneus

1st metatarsal bone

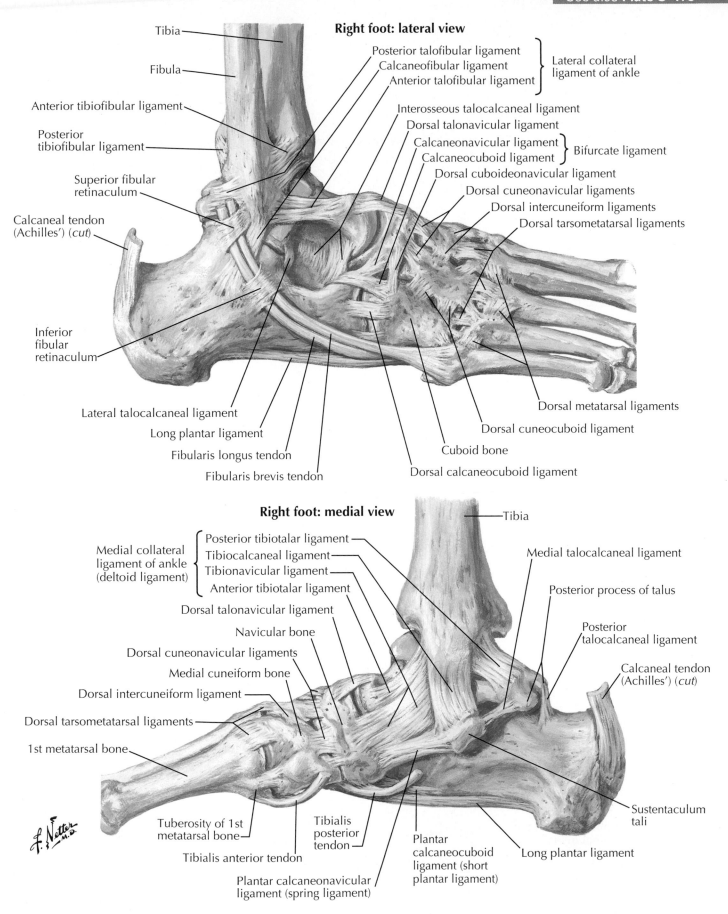

Right foot: lateral view

Tibia

Fibula

Anterior tibiofibular ligament

Posterior tibiofibular ligament

Superior fibular retinaculum

Calcaneal tendon (Achilles') (cut)

Inferior fibular retinaculum

Posterior talofibular ligament
Calcaneofibular ligament
Anterior talofibular ligament
} Lateral collateral ligament of ankle

Interosseous talocalcaneal ligament
Dorsal talonavicular ligament
Calcaneonavicular ligament
Calcaneocuboid ligament
} Bifurcate ligament
Dorsal cuboideonavicular ligament
Dorsal cuneonavicular ligaments
Dorsal intercuneiform ligaments
Dorsal tarsometatarsal ligaments

Dorsal metatarsal ligaments

Dorsal cuneocuboid ligament

Cuboid bone

Dorsal calcaneocuboid ligament

Lateral talocalcaneal ligament

Long plantar ligament

Fibularis longus tendon

Fibularis brevis tendon

Right foot: medial view

Tibia

Medial collateral ligament of ankle (deltoid ligament)
{ Posterior tibiotalar ligament
Tibiocalcaneal ligament
Tibionavicular ligament
Anterior tibiotalar ligament

Dorsal talonavicular ligament

Navicular bone

Dorsal cuneonavicular ligaments

Medial cuneiform bone

Dorsal intercuneiform ligament

Dorsal tarsometatarsal ligaments

1st metatarsal bone

Medial talocalcaneal ligament

Posterior process of talus

Posterior talocalcaneal ligament

Calcaneal tendon (Achilles') (cut)

Sustentaculum tali

Tuberosity of 1st metatarsal bone

Tibialis anterior tendon

Tibialis posterior tendon

Plantar calcaneonavicular ligament (spring ligament)

Plantar calcaneocuboid ligament (short plantar ligament)

Long plantar ligament

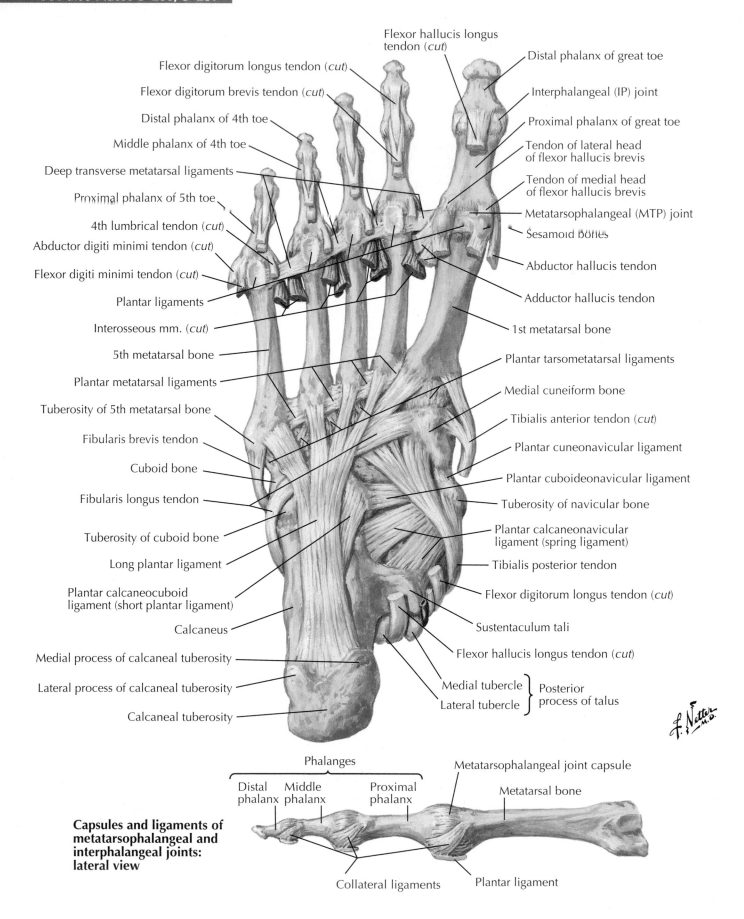

Flexor hallucis longus tendon (*cut*)

Flexor digitorum longus tendon (*cut*)

Flexor digitorum brevis tendon (*cut*)

Distal phalanx of 4th toe

Middle phalanx of 4th toe

Deep transverse metatarsal ligaments

Proximal phalanx of 5th toe

4th lumbrical tendon (*cut*)

Abductor digiti minimi tendon (*cut*)

Flexor digiti minimi tendon (*cut*)

Plantar ligaments

Interosseous mm. (*cut*)

5th metatarsal bone

Plantar metatarsal ligaments

Tuberosity of 5th metatarsal bone

Fibularis brevis tendon

Cuboid bone

Fibularis longus tendon

Tuberosity of cuboid bone

Long plantar ligament

Plantar calcaneocuboid ligament (short plantar ligament)

Calcaneus

Medial process of calcaneal tuberosity

Lateral process of calcaneal tuberosity

Calcaneal tuberosity

Distal phalanx of great toe

Interphalangeal (IP) joint

Proximal phalanx of great toe

Tendon of lateral head of flexor hallucis brevis

Tendon of medial head of flexor hallucis brevis

Metatarsophalangeal (MTP) joint

Sesamoid bones

Abductor hallucis tendon

Adductor hallucis tendon

1st metatarsal bone

Plantar tarsometatarsal ligaments

Medial cuneiform bone

Tibialis anterior tendon (*cut*)

Plantar cuneonavicular ligament

Plantar cuboideonavicular ligament

Tuberosity of navicular bone

Plantar calcaneonavicular ligament (spring ligament)

Tibialis posterior tendon

Flexor digitorum longus tendon (*cut*)

Sustentaculum tali

Flexor hallucis longus tendon (*cut*)

Medial tubercle
Lateral tubercle } Posterior process of talus

Phalanges

Distal phalanx   Middle phalanx   Proximal phalanx

Capsules and ligaments of metatarsophalangeal and interphalangeal joints: lateral view

Metatarsophalangeal joint capsule

Metatarsal bone

Collateral ligaments

Plantar ligament

Axial T2-weighted MRI

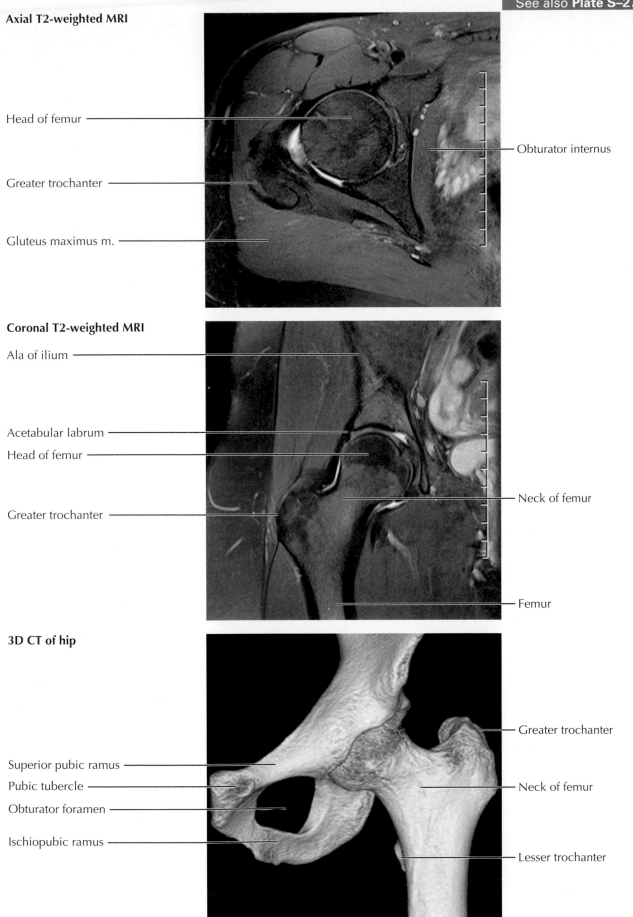

Head of femur

Greater trochanter

Gluteus maximus m.

Obturator internus

Coronal T2-weighted MRI

Ala of ilium

Acetabular labrum

Head of femur

Greater trochanter

Neck of femur

Femur

3D CT of hip

Superior pubic ramus

Pubic tubercle

Obturator foramen

Ischiopubic ramus

Greater trochanter

Neck of femur

Lesser trochanter

ANATOMIC STRUCTURES	CLINICAL IMPORTANCE	PLATE NUMBERS
Head and Neck		
Orbit	Most facial trauma involves orbit; traumatic fractures can occur in rim or walls; a "blowout" fracture involves inferior wall and may injure inferior rectus muscle and/or infraorbital nerve; rim fractures affect contours of orbital rim and occur in zygomaticomaxillary fractures	S–123, S–368
Pterion	Intersection of frontal, parietal, temporal, and sphenoid bones; thin, weak region of skull that is susceptible to fracture; frontal branch of middle meningeal artery lies immediately deep to this region and may be injured	S–125
Asterion	Landmark at posterior end of parietotemporal suture, used in lateral neurosurgical approaches to posterior cranial fossa	S–125
Temporomandibular joint	Temporomandibular joint disorders are common sources of pain and joint dysfunction; poor replacement results to date; dislocations/subluxations can be reduced via retromolar fossa	S–136
Cranial sutures	Premature fusion may result in skull deformity known as craniosynostosis; sagittal suture is most often affected	S–124, S–127
Cervical vertebrae	Degenerative changes causing narrowing of intervertebral foramina may result in cervical radiculopathy; C1–C4 and C5–C7 vertebrae are common areas of pathology for children and adults, respectively; neck hyperextension from abrupt deceleration may cause bilateral pedicle fractures in axis (C2 vertebra) known as hangman's fracture; hyperflexion may cause anterior wedge vertebral fracture; axial loads may cause burst fracture; dens axis (odontoid) fractures may occur with either forceful extension or flexion; all fractures are classified as stable or unstable depending on whether structural integrity of cervical spine has been sufficiently disrupted to permit compression of spinal cord	S–139, S–140, S–143
Laryngeal cartilages	Thyroid and cricoid cartilages are palpable landmarks of anterior neck used for cricothyrotomy, tracheostomy, and cricoid pressure during airway intubation	S–375, S–525
Hyoid bone	Palpable anterior neck landmark at C3 vertebral level that can be fractured during sporting activities, compromising swallowing and speech; fractures may also indicate strangulation	S–192
Auditory ossicles	Pathologic conditions involving ossicles (e.g., otosclerosis) can cause conductive hearing loss	S–93, S–94
Back		
Spinous processes	Palpable landmarks used to assess spinal curvatures and determine location of spinal cord for procedures such as lumbar puncture and injection of spinal anesthesia	S–6, S–199
Spinous process of C7 vertebra (vertebra prominens)	Most prominent spinous process in cervical region; often used to begin counting vertebrae	S–6, S–144
Intervertebral disc	Age-related changes may produce herniation of nucleus pulposus, causing back pain; occurs most commonly in lower lumbar regions of vertebral column	S–17, S–145
Lamina of vertebral arch	Surgically removed in laminectomy to gain access to vertebral canal and spinal cord	S–145
Intervertebral foramen	May become narrowed by age-related changes (e.g., osteophyte formation) or changes in intervertebral disc height, producing compression of its contents	S–145, S–147, S–148
Sacral hiatus	Provides access to epidural space to administer caudal epidural anesthesia	S–146
Fifth lumbar vertebra	Spondylolysis is clinical condition in which vertebral body separates from part of its vertebral arch bearing inferior articulating process (defect is through pars interarticularis); if this occurs bilaterally, L5 vertebral body and transverse process may slide forward over sacrum, giving rise to spondylolisthesis	S–147

Table 3.1   **Structures with High Clinical Significance**

ANATOMIC STRUCTURES	CLINICAL IMPORTANCE	PLATE NUMBERS
L5–S1 intervertebral disc	Most common level of intervertebral disc herniation, which may result in nerve compression and lower back pain associated with pain and weakness in ipsilateral lower limb (sciatica)	S–18, S–147
Vertebral foramen	May be congenitally stenotic in cervical region or narrowed by arthritic changes in lumbar vertebrae; can lead to back pain, sciatica, numbness or tingling, and weakness in lower limbs	S–144, S–145
Thorax		
Ribs	Rib fractures may cause respiratory dysfunction, predispose to pneumonia, and injure underlying structures (e.g., liver and spleen); severe fractures may breach pleural space and cause pneumothorax; flail chest occurs when multiple fractures in adjacent ribs create "floating" area of thorax with paradoxical motion during inspiration	S–149
Costochondral and sternochondral (sternocostal) joints	Frequent site of pain and tenderness after thoracic wall injury or excessive weightlifting (costochondritis); generally reproducible with palpation of joints	S–150
Clavicles	Common site of fracture, often after falling onto outstretched limb or onto shoulder; fractures typically occur in middle third; supraclavicular nerve block relieves pain associated with fracture	S–149
Sternal angle (of Louis)	Surface landmark for counting ribs (2nd pair of ribs articulate between manubrium and body of sternum) and intercostal spaces; divides superior from inferior mediastinum, and marks transition from aortic arch to descending thoracic aorta	S–149
Superior thoracic aperture	Compression of neurovascular structures (inferior trunk of brachial plexus and great vessels) that traverse superior thoracic aperture may produce thoracic outlet syndrome	S–338
Intercostal spaces	Relationship of intercostal neurovascular bundle to ribs is crucial when placing chest tube to relieve pneumothorax or hemothorax, or needles to anesthetize nerves; tubes should be inserted along superior margin of ribs to avoid these bundles; of note, however, intercostal nerve may have collateral branch running along superior border of lower rib, which can result in pain	S–204, S–534
Median cricothyroid ligament	Also known as cricothyroid membrane; site of cricothyrotomy, an emergency procedure to establish surgical airway	S–191, S–525
Abdomen		
Xiphoid process and pubic symphysis	Palpable landmarks used to locate transpyloric plane (of Addison; L1 plane), located halfway between these structures; plane may contain pylorus of stomach, horizontal portion of duodenum, head and neck of pancreas, superior mesenteric artery, and hilum of spleen	S–8
Anterior superior iliac spine (ASIS)	Palpable landmark used to locate McBurney's point; tenderness over McBurney's point is indication of appendicitis	S–151, S–427
Pelvis		
Pubic symphysis	Palpable landmark used to obtain pelvic measurements (e.g., diagonal conjugate) that may be used to assess adequacy of pelvis for childbirth; during prenatal examinations, used for estimating fetal growth (symphysis–fundal height measurement); injury may result in widening on x-ray	S–477
Ischial spine	Palpable landmark used to estimate interspinous diameter for childbirth and to locate pudendal nerve for pudendal nerve block	S–477
Ischial tuberosity	Palpable landmark used to estimate width of pelvic outlet for childbirth; proximal attachment site of hamstring muscles	S–477

Structures with High Clinical Significance   **Table 3.2**

ANATOMIC STRUCTURES	CLINICAL IMPORTANCE	PLATE NUMBERS
Pelvis—Continued		
Superior pubic ramus	Often fractured by lateral compression injury of pelvis in anteroposterior plane by crush injury or falls in elderly with osteoporosis	S–154
Sacroiliac joint	Stiffening, sclerosis, and fusion occur in autoimmune disease known as ankylosing spondylitis; difficult diagnosis to make and known to refer pain to adjacent joints	S–477
Upper Limb		
Clavicle	Most clavicular fractures are caused from a fall on an outstretched arm or direct trauma delivered to lateral side of shoulder; middle third of clavicle is most commonly fractured; supraclavicular nerve block relieves acute pain associated with fracture	S–155, S–156
Humerus	Proximal humerus, especially surgical neck, is fractured due to low-energy falls in elderly persons and high-energy trauma in young persons; axillary nerve and circumflex humeral arteries can be injured; hematoma from anterior/posterior circumflex humeral artery damage as a result of dislocation may complicate reductions; midbody fractures are also relatively common and may affect radial nerve and/or deep brachial artery; distal humerus fractures may affect ulnar nerve medially and radial nerve laterally	S–155, S–158, S–264
Ulna	Subcutaneous location of olecranon makes it vulnerable to fracture by direct trauma, especially when elbow is flexed; ulnar styloid process may also be fractured with distal radial fractures	S–159, S–162
Radius	Fractures of distal radius are most common fracture of upper limb (Colles' fracture), typically caused by fall on outstretched hand (FOOSH)	S–162
Scaphoid bone	Most commonly fractured carpal bone, typically from fall on outstretched hand (FOOSH)	S–163, S–262, S–263
Lower Limb		
Neck of femur	Common fracture in elderly from falls; can lead to avascular necrosis of head of femur due to disrupted blood supply	S–171, S–172, S–278
Body of femur	Midshaft is common fracture site in high-energy trauma (motor vehicle collisions)	S–172
Hip joint	Potential for avascular necrosis of head of femur in hip dislocations or fractures	S–170, S–278
Anterior cruciate ligament	Most commonly injured knee ligament, typically from sudden pivot of knee causing excessive valgus stress coupled with medial rotation of the tibia	S–173, S–280, S–281
Tibial (medial) collateral and anterior cruciate ligaments and medial meniscus	"Unhappy triad of the knee"; damage to these structures can result from blow to lateral aspect of joint in extension	S–173, S–280, S–281
Tibia and fibula	High-energy fractures of shaft (boot-top skiing fracture) from falling forward at high speed	S–174
First metatarsophalangeal joint	Joint misalignment leads to hallux valgus (bunion); wearing narrow shoes can contribute, though there is also a strong genetic component	S–176
Calcaneus	Most common tarsal bone fracture, usually caused by landing forcefully on heel after falling from height	S–178
Ankle joint	Most sprains are inversion injuries that occur when foot is plantar flexed, placing stress on lateral collateral ligaments of ankle; fractures often occur to lateral malleolus of fibula and inferior articular surface of tibia	S–180

*Selections are based largely on clinical data and commonly discussed clinical correlations in macroscopic ("gross") anatomy courses.

MUSCULAR SYSTEM 4

ELECTRONIC BONUS PLATES

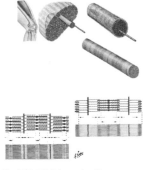

S–BP 34 Muscle Structure

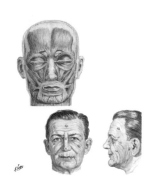

S–BP 35 Muscles of Facial Expression: Anterior View

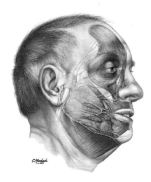

S–BP 36 Musculature of Face

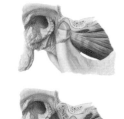

S–BP 37 Opening of the Mandible

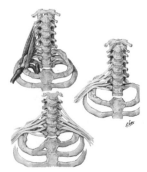

S–BP 38 Cervical Ribs and Related Variations

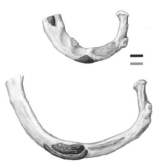

S–BP 39 Muscle Attachments of Ribs

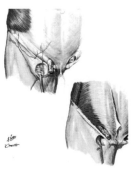

S–BP 40 Inguinal and Femoral Regions

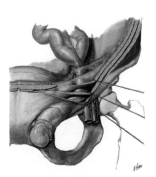

S–BP 41 Indirect Inguinal Hernia

ELECTRONIC BONUS PLATES—*cont'd*

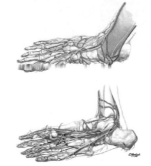

S–BP 42 Flexor and Extensor Zones of Hand

S–BP 43 Foot: Nerves and Arteries

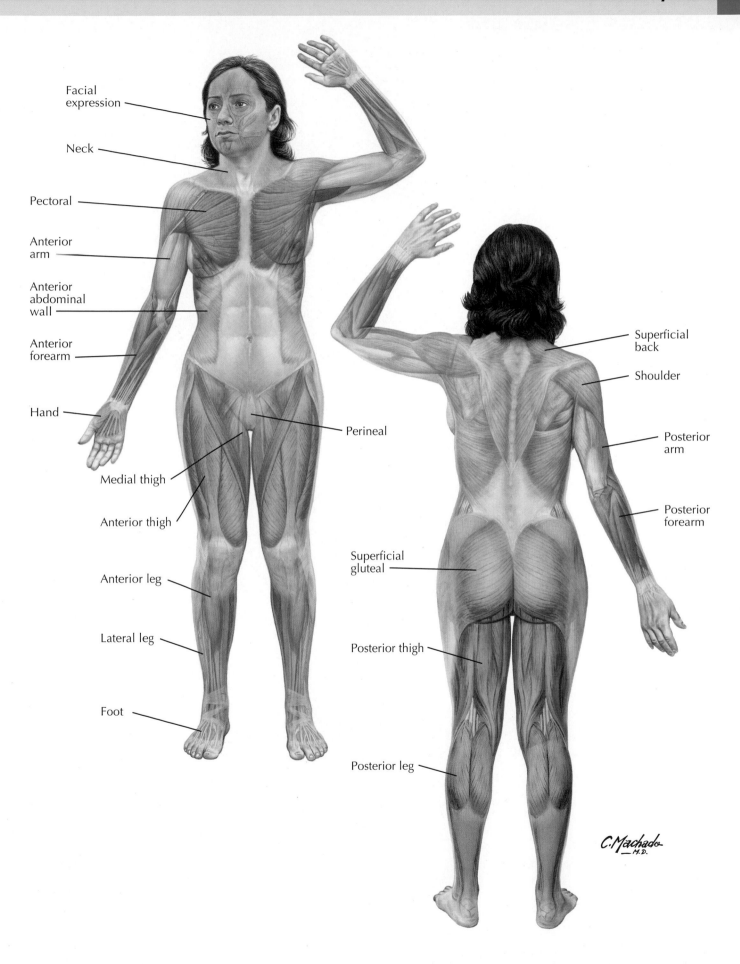

Facial expression

Neck

Pectoral

Anterior arm

Anterior abdominal wall

Anterior forearm

Hand

Medial thigh

Anterior thigh

Anterior leg

Lateral leg

Foot

Perineal

Superficial back

Shoulder

Posterior arm

Posterior forearm

Superficial gluteal

Posterior thigh

Posterior leg

C. Machado
M.D.

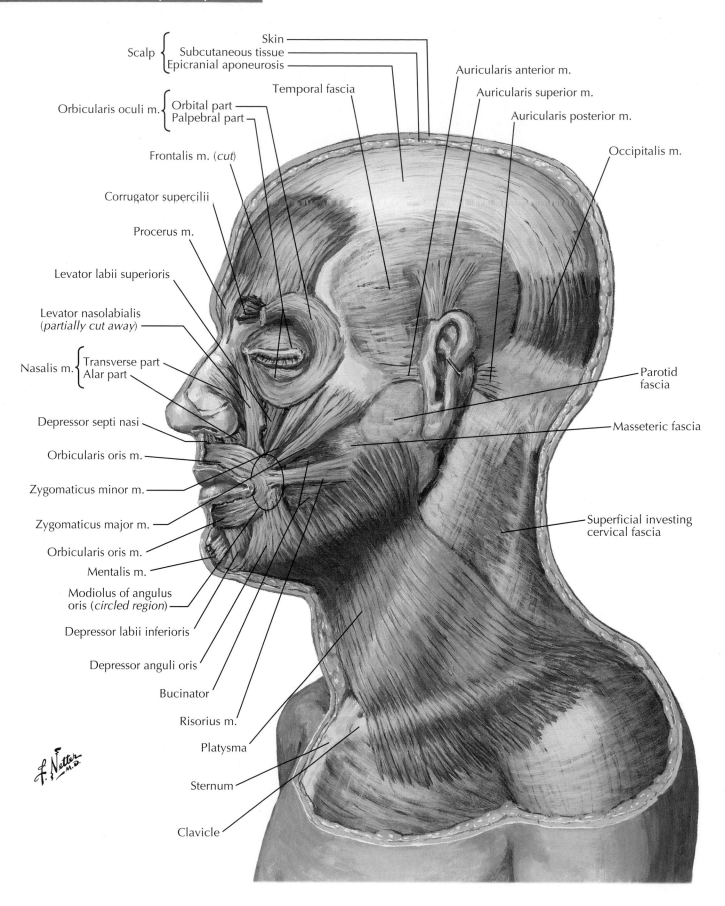

Scalp
- Skin
- Subcutaneous tissue
- Epicranial aponeurosis

Temporal fascia

Auricularis anterior m.

Auricularis superior m.

Auricularis posterior m.

Orbicularis oculi m.
- Orbital part
- Palpebral part

Occipitalis m.

Frontalis m. (*cut*)

Corrugator supercilii

Procerus m.

Levator labii superioris

Levator nasolabialis (*partially cut away*)

Nasalis m.
- Transverse part
- Alar part

Depressor septi nasi

Orbicularis oris m.

Zygomaticus minor m.

Zygomaticus major m.

Orbicularis oris m.

Mentalis m.

Modiolus of angulus oris (*circled region*)

Depressor labii inferioris

Depressor anguli oris

Bucinator

Risorius m.

Platysma

Sternum

Clavicle

Parotid fascia

Masseteric fascia

Superficial investing cervical fascia

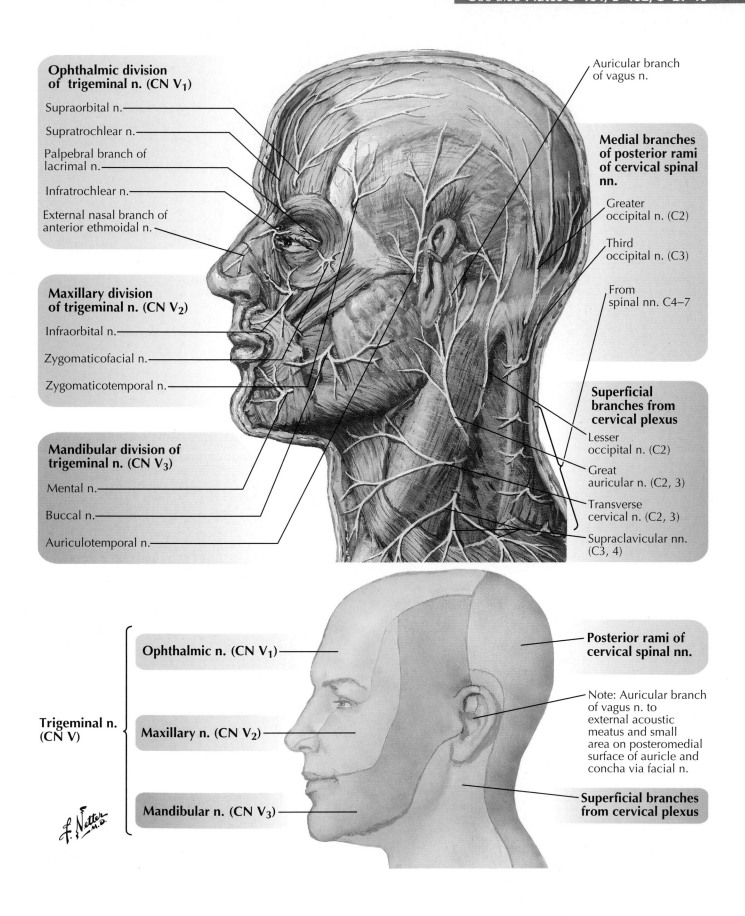

Ophthalmic division of trigeminal n. (CN V₁)

Supraorbital n.

Supratrochlear n.

Palpebral branch of lacrimal n.

Infratrochlear n.

External nasal branch of anterior ethmoidal n.

Maxillary division of trigeminal n. (CN V₂)

Infraorbital n.

Zygomaticofacial n.

Zygomaticotemporal n.

Mandibular division of trigeminal n. (CN V₃)

Mental n.

Buccal n.

Auriculotemporal n.

Auricular branch of vagus n.

Medial branches of posterior rami of cervical spinal nn.

Greater occipital n. (C2)

Third occipital n. (C3)

From spinal nn. C4–7

Superficial branches from cervical plexus

Lesser occipital n. (C2)

Great auricular n. (C2, 3)

Transverse cervical n. (C2, 3)

Supraclavicular nn. (C3, 4)

Trigeminal n. (CN V)

Ophthalmic n. (CN V₁)

Maxillary n. (CN V₂)

Mandibular n. (CN V₃)

Posterior rami of cervical spinal nn.

Note: Auricular branch of vagus n. to external acoustic meatus and small area on posteromedial surface of auricle and concha via facial n.

Superficial branches from cervical plexus

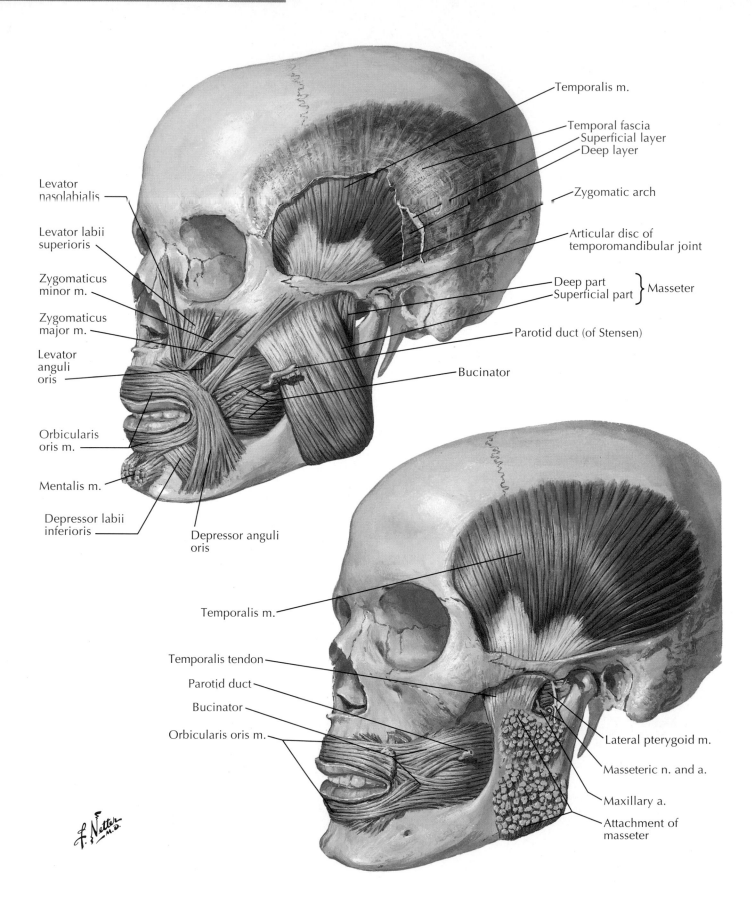

Temporalis m.

Temporal fascia
Superficial layer
Deep layer

Zygomatic arch

Articular disc of
temporomandibular joint

Deep part
Superficial part } Masseter

Parotid duct (of Stensen)

Bucinator

Levator
nasolabialis

Levator labii
superioris

Zygomaticus
minor m.

Zygomaticus
major m.

Levator
anguli
oris

Orbicularis
oris m.

Mentalis m.

Depressor labii
inferioris

Depressor anguli
oris

Temporalis m.

Temporalis tendon

Parotid duct

Bucinator

Orbicularis oris m.

Lateral pterygoid m.

Masseteric n. and a.

Maxillary a.

Attachment of
masseter

Lateral view

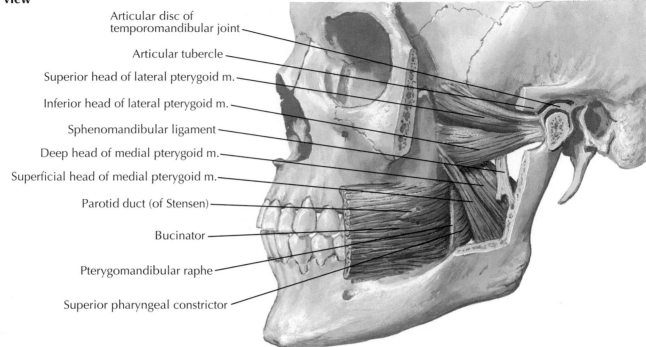

Articular disc of temporomandibular joint

Articular tubercle

Superior head of lateral pterygoid m.

Inferior head of lateral pterygoid m.

Sphenomandibular ligament

Deep head of medial pterygoid m.

Superficial head of medial pterygoid m.

Parotid duct (of Stensen)

Bucinator

Pterygomandibular raphe

Superior pharyngeal constrictor

Posterior view

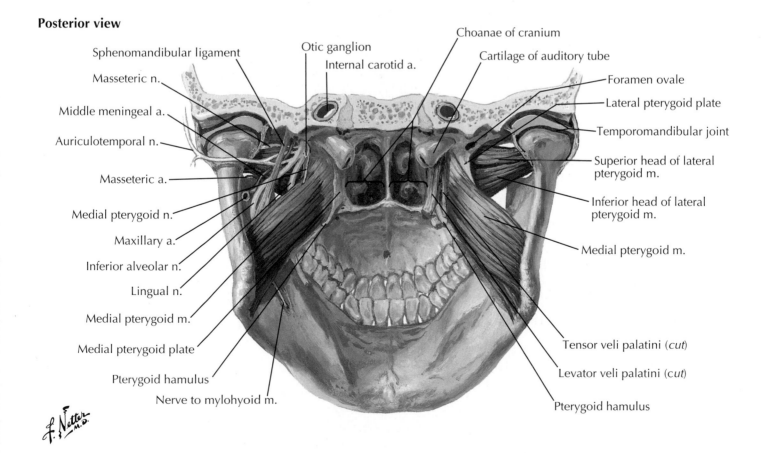

Sphenomandibular ligament

Masseteric n.

Middle meningeal a.

Auriculotemporal n.

Masseteric a.

Medial pterygoid n.

Maxillary a.

Inferior alveolar n.

Lingual n.

Medial pterygoid m.

Medial pterygoid plate

Pterygoid hamulus

Nerve to mylohyoid m.

Otic ganglion

Internal carotid a.

Choanae of cranium

Cartilage of auditory tube

Foramen ovale

Lateral pterygoid plate

Temporomandibular joint

Superior head of lateral pterygoid m.

Inferior head of lateral pterygoid m.

Medial pterygoid m.

Tensor veli palatini (cut)

Levator veli palatini (cut)

Pterygoid hamulus

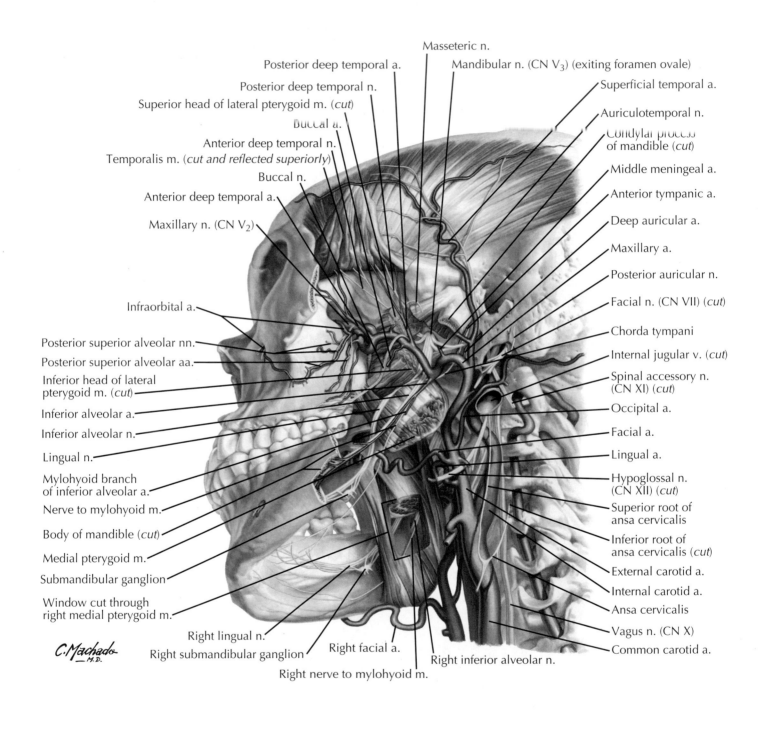

Masseteric n.

Posterior deep temporal a.

Mandibular n. (CN V₃) (exiting foramen ovale)

Posterior deep temporal n.

Superficial temporal a.

Superior head of lateral pterygoid m. (cut)

Buccal a.

Auriculotemporal n.

Anterior deep temporal n.

Condylar process of mandible (cut)

Temporalis m. (cut and reflected superiorly)

Middle meningeal a.

Buccal n.

Anterior tympanic a.

Anterior deep temporal a.

Deep auricular a.

Maxillary n. (CN V₂)

Maxillary a.

Posterior auricular n.

Infraorbital a.

Facial n. (CN VII) (cut)

Posterior superior alveolar nn.

Chorda tympani

Posterior superior alveolar aa.

Internal jugular v. (cut)

Inferior head of lateral pterygoid m. (cut)

Spinal accessory n. (CN XI) (cut)

Inferior alveolar a.

Occipital a.

Inferior alveolar n.

Facial a.

Lingual n.

Lingual a.

Mylohyoid branch of inferior alveolar a.

Hypoglossal n. (CN XII) (cut)

Nerve to mylohyoid m.

Superior root of ansa cervicalis

Body of mandible (cut)

Inferior root of ansa cervicalis (cut)

Medial pterygoid m.

External carotid a.

Submandibular ganglion

Internal carotid a.

Window cut through right medial pterygoid m.

Ansa cervicalis

Vagus n. (CN X)

C. Machado
_M.D._

Right lingual n.

Common carotid a.

Right submandibular ganglion

Right facial a.

Right inferior alveolar n.

Right nerve to mylohyoid m.

See also **Plates** S–129, S–130, S–136, S–194, S–322

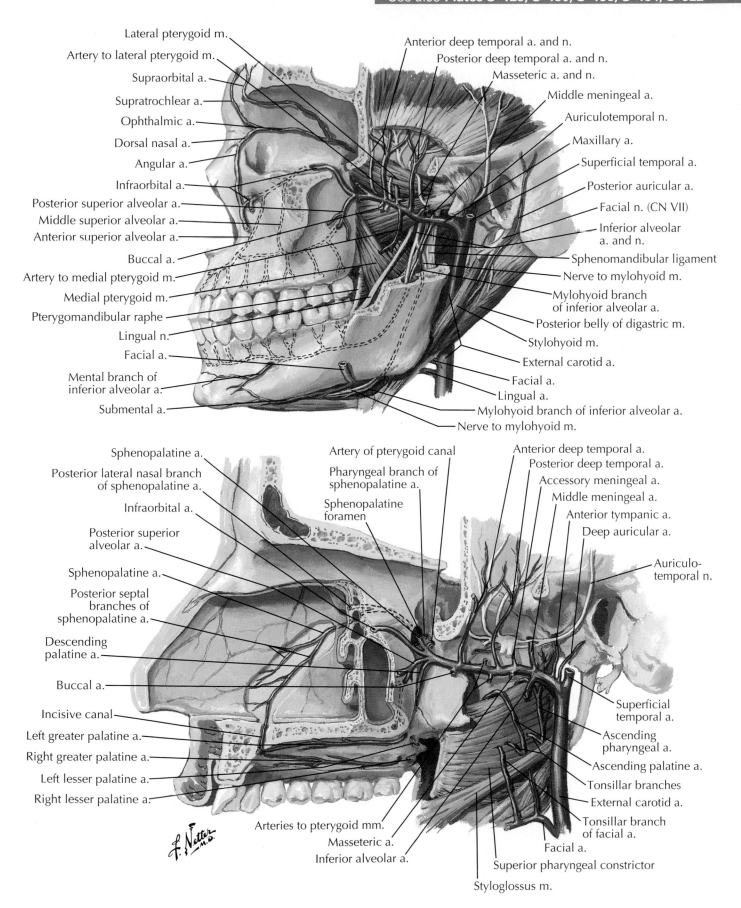

Lateral pterygoid m.

Artery to lateral pterygoid m.

Supraorbital a.

Supratrochlear a.

Ophthalmic a.

Dorsal nasal a.

Angular a.

Infraorbital a.

Posterior superior alveolar a.

Middle superior alveolar a.

Anterior superior alveolar a.

Buccal a.

Artery to medial pterygoid m.

Medial pterygoid m.

Pterygomandibular raphe

Lingual n.

Facial a.

Mental branch of inferior alveolar a.

Submental a.

Anterior deep temporal a. and n.

Posterior deep temporal a. and n.

Masseteric a. and n.

Middle meningeal a.

Auriculotemporal n.

Maxillary a.

Superficial temporal a.

Posterior auricular a.

Facial n. (CN VII)

Inferior alveolar a. and n.

Sphenomandibular ligament

Nerve to mylohyoid m.

Mylohyoid branch of inferior alveolar a.

Posterior belly of digastric m.

Stylohyoid m.

External carotid a.

Facial a.

Lingual a.

Mylohyoid branch of inferior alveolar a.

Nerve to mylohyoid m.

Sphenopalatine a.

Posterior lateral nasal branch of sphenopalatine a.

Infraorbital a.

Posterior superior alveolar a.

Sphenopalatine a.

Posterior septal branches of sphenopalatine a.

Descending palatine a.

Buccal a.

Incisive canal

Left greater palatine a.

Right greater palatine a.

Left lesser palatine a.

Right lesser palatine a.

Artery of pterygoid canal

Pharyngeal branch of sphenopalatine a.

Sphenopalatine foramen

Arteries to pterygoid mm.

Masseteric a.

Inferior alveolar a.

Anterior deep temporal a.

Posterior deep temporal a.

Accessory meningeal a.

Middle meningeal a.

Anterior tympanic a.

Deep auricular a.

Auriculo-temporal n.

Superficial temporal a.

Ascending pharyngeal a.

Ascending palatine a.

Tonsillar branches

External carotid a.

Tonsillar branch of facial a.

Facial a.

Superior pharyngeal constrictor

Styloglossus m.

f. Netter M.D.

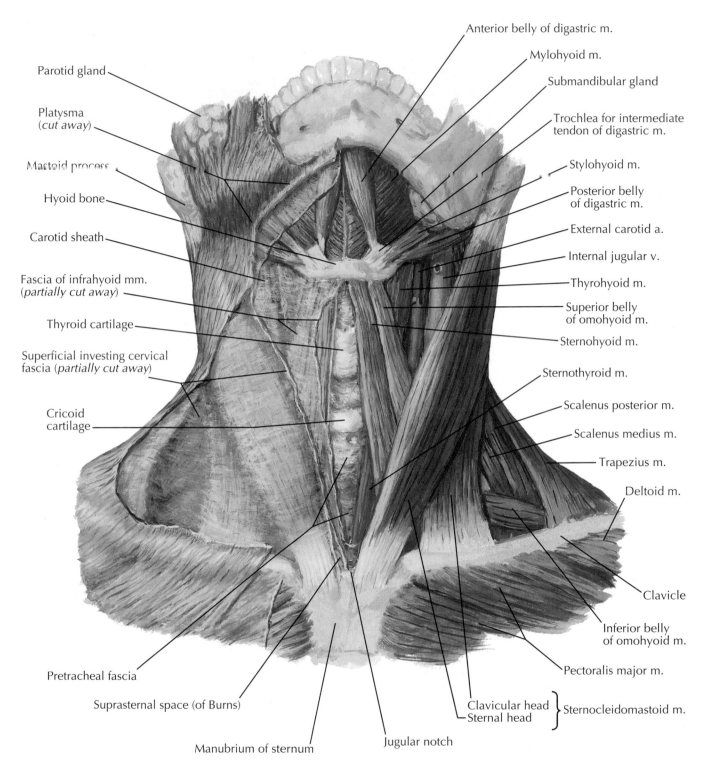

Parotid gland

Platysma
(*cut away*)

Mastoid process

Hyoid bone

Carotid sheath

Fascia of infrahyoid mm.
(*partially cut away*)

Thyroid cartilage

Superficial investing cervical
fascia (*partially cut away*)

Cricoid
cartilage

Pretracheal fascia

Suprasternal space (of Burns)

Manubrium of sternum

Anterior belly of digastric m.

Mylohyoid m.

Submandibular gland

Trochlea for intermediate
tendon of digastric m.

Stylohyoid m.

Posterior belly
of digastric m.

External carotid a.

Internal jugular v.

Thyrohyoid m.

Superior belly
of omohyoid m.

Sternohyoid m.

Sternothyroid m.

Scalenus posterior m.

Scalenus medius m.

Trapezius m.

Deltoid m.

Clavicle

Inferior belly
of omohyoid m.

Pectoralis major m.

Clavicular head
Sternal head } Sternocleidomastoid m.

Jugular notch

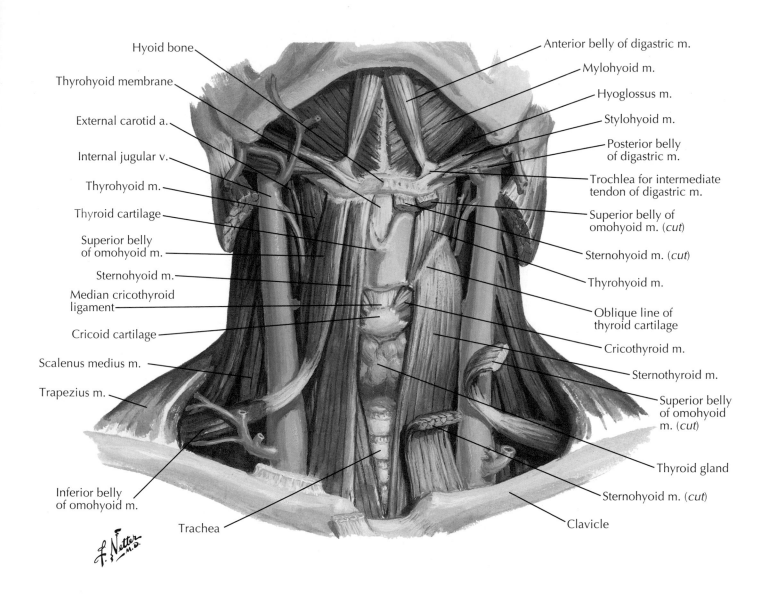

Hyoid bone

Thyrohyoid membrane

External carotid a.

Internal jugular v.

Thyrohyoid m.

Thyroid cartilage

Superior belly of omohyoid m.

Sternohyoid m.

Median cricothyroid ligament

Cricoid cartilage

Scalenus medius m.

Trapezius m.

Inferior belly of omohyoid m.

Trachea

Anterior belly of digastric m.

Mylohyoid m.

Hyoglossus m.

Stylohyoid m.

Posterior belly of digastric m.

Trochlea for intermediate tendon of digastric m.

Superior belly of omohyoid m. (cut)

Sternohyoid m. (cut)

Thyrohyoid m.

Oblique line of thyroid cartilage

Cricothyroid m.

Sternothyroid m.

Superior belly of omohyoid m. (cut)

Thyroid gland

Sternohyoid m. (cut)

Clavicle

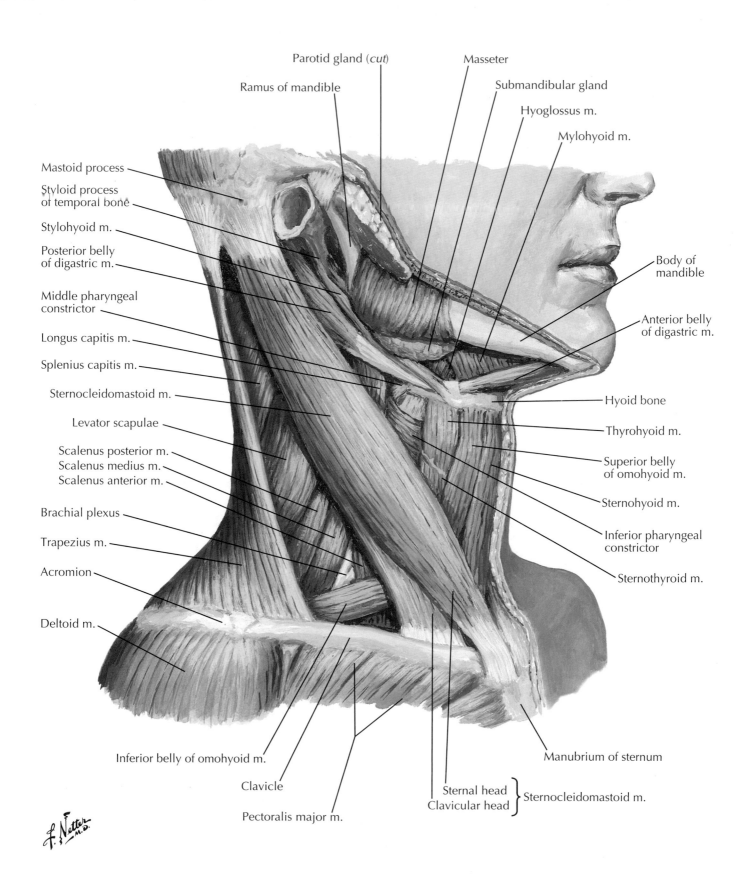

Parotid gland (*cut*)

Masseter

Ramus of mandible

Submandibular gland

Hyoglossus m.

Mylohyoid m.

Mastoid process

Styloid process
of temporal bone

Body of
mandible

Stylohyoid m.

Posterior belly
of digastric m.

Anterior belly
of digastric m.

Middle pharyngeal
constrictor

Longus capitis m.

Splenius capitis m.

Sternocleidomastoid m.

Hyoid bone

Levator scapulae

Thyrohyoid m.

Scalenus posterior m.

Scalenus medius m.

Scalenus anterior m.

Superior belly
of omohyoid m.

Sternohyoid m.

Brachial plexus

Inferior pharyngeal
constrictor

Trapezius m.

Acromion

Sternothyroid m.

Deltoid m.

Inferior belly of omohyoid m.

Manubrium of sternum

Clavicle

Sternal head
Clavicular head

Sternocleidomastoid m.

Pectoralis major m.

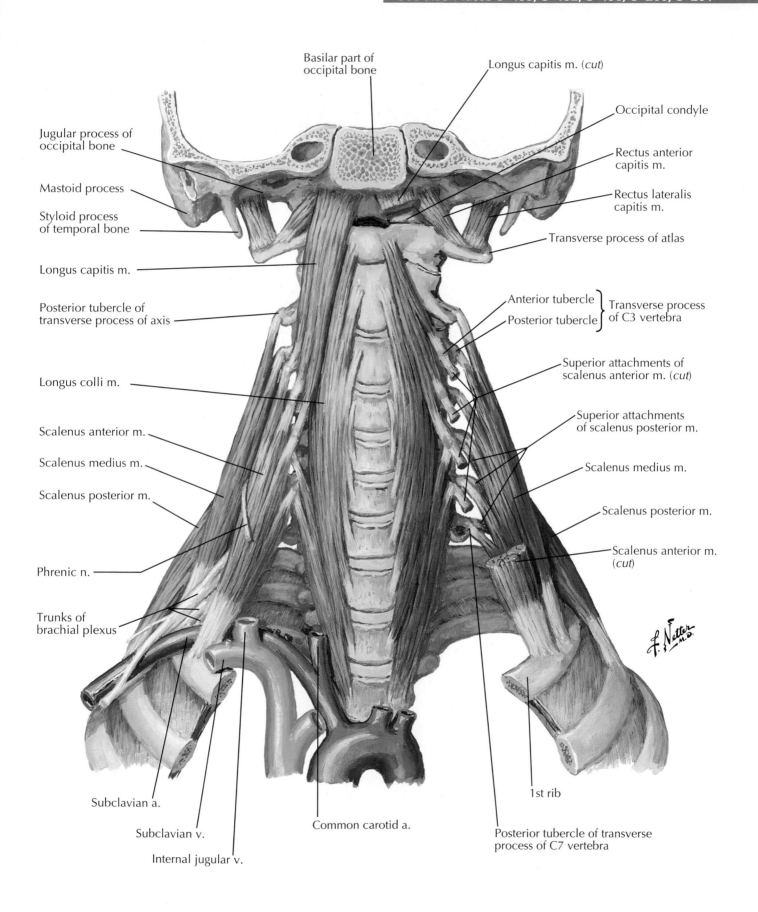

Basilar part of occipital bone

Longus capitis m. (*cut*)

Occipital condyle

Jugular process of occipital bone

Rectus anterior capitis m.

Mastoid process

Rectus lateralis capitis m.

Styloid process of temporal bone

Transverse process of atlas

Longus capitis m.

Anterior tubercle ⎫ Transverse process
Posterior tubercle ⎭ of C3 vertebra

Posterior tubercle of transverse process of axis

Superior attachments of scalenus anterior m. (*cut*)

Longus colli m.

Superior attachments of scalenus posterior m.

Scalenus anterior m.

Scalenus medius m.

Scalenus medius m.

Scalenus posterior m.

Scalenus posterior m.

Scalenus anterior m. (*cut*)

Phrenic n.

Trunks of brachial plexus

f. Netter M.D.

Subclavian a.

1st rib

Subclavian v.

Common carotid a.

Posterior tubercle of transverse process of C7 vertebra

Internal jugular v.

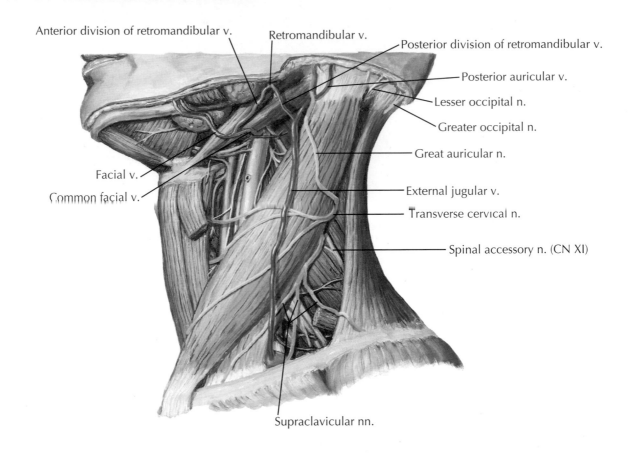

Anterior division of retromandibular v.

Retromandibular v.

Posterior division of retromandibular v.

Posterior auricular v.

Lesser occipital n.

Greater occipital n.

Great auricular n.

Facial v.

Common facial v.

External jugular v.

Transverse cervical n.

Spinal accessory n. (CN XI)

Supraclavicular nn.

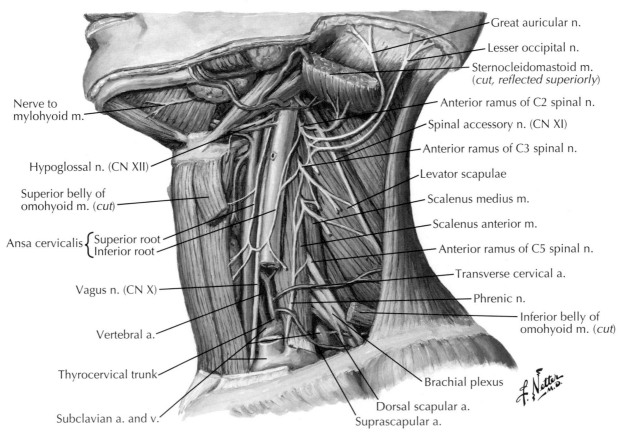

Great auricular n.

Lesser occipital n.

Sternocleidomastoid m. (cut, reflected superiorly)

Nerve to mylohyoid m.

Anterior ramus of C2 spinal n.

Spinal accessory n. (CN XI)

Anterior ramus of C3 spinal n.

Hypoglossal n. (CN XII)

Levator scapulae

Superior belly of omohyoid m. (cut)

Scalenus medius m.

Scalenus anterior m.

Ansa cervicalis { Superior root / Inferior root }

Anterior ramus of C5 spinal n.

Transverse cervical a.

Vagus n. (CN X)

Phrenic n.

Inferior belly of omohyoid m. (cut)

Vertebral a.

Thyrocervical trunk

Brachial plexus

Subclavian a. and v.

Dorsal scapular a.

Suprascapular a.

Cervical plexus: schema

(S = gray ramus communicans from superior cervical ganglion)

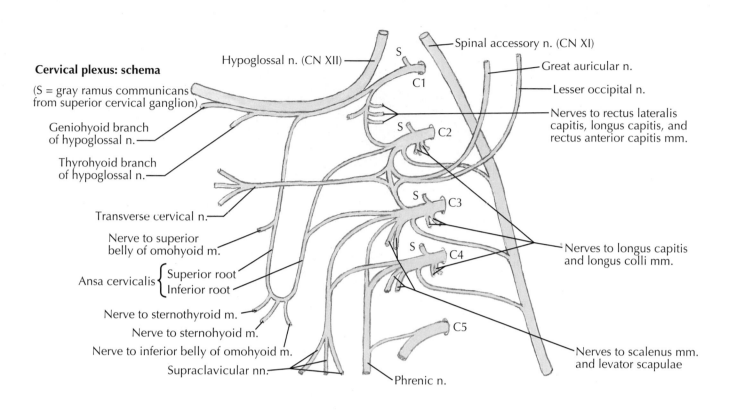

Hypoglossal n. (CN XII)

Spinal accessory n. (CN XI)

Great auricular n.

Lesser occipital n.

Nerves to rectus lateralis capitis, longus capitis, and rectus anterior capitis mm.

Geniohyoid branch of hypoglossal n.

Thyrohyoid branch of hypoglossal n.

Transverse cervical n.

Nerve to superior belly of omohyoid m.

Ansa cervicalis { Superior root, Inferior root }

Nerve to sternothyroid m.

Nerve to sternohyoid m.

Nerve to inferior belly of omohyoid m.

Supraclavicular nn.

Phrenic n.

Nerves to longus capitis and longus colli mm.

Nerves to scalenus mm. and levator scapulae

C1, C2, C3, C4, C5

Right anterior view

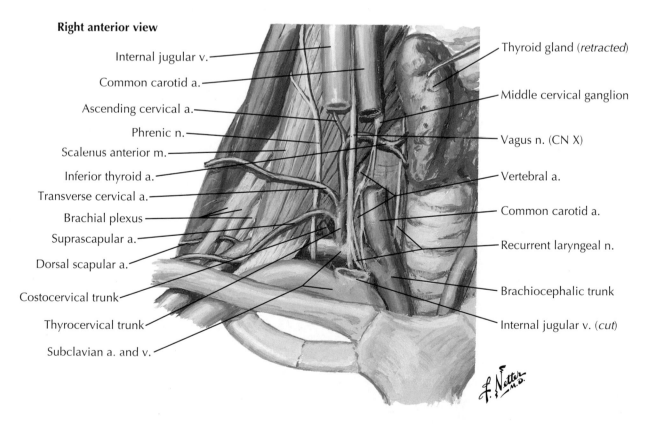

Internal jugular v.

Common carotid a.

Ascending cervical a.

Phrenic n.

Scalenus anterior m.

Inferior thyroid a.

Transverse cervical a.

Brachial plexus

Suprascapular a.

Dorsal scapular a.

Costocervical trunk

Thyrocervical trunk

Subclavian a. and v.

Thyroid gland (*retracted*)

Middle cervical ganglion

Vagus n. (CN X)

Vertebral a.

Common carotid a.

Recurrent laryngeal n.

Brachiocephalic trunk

Internal jugular v. (*cut*)

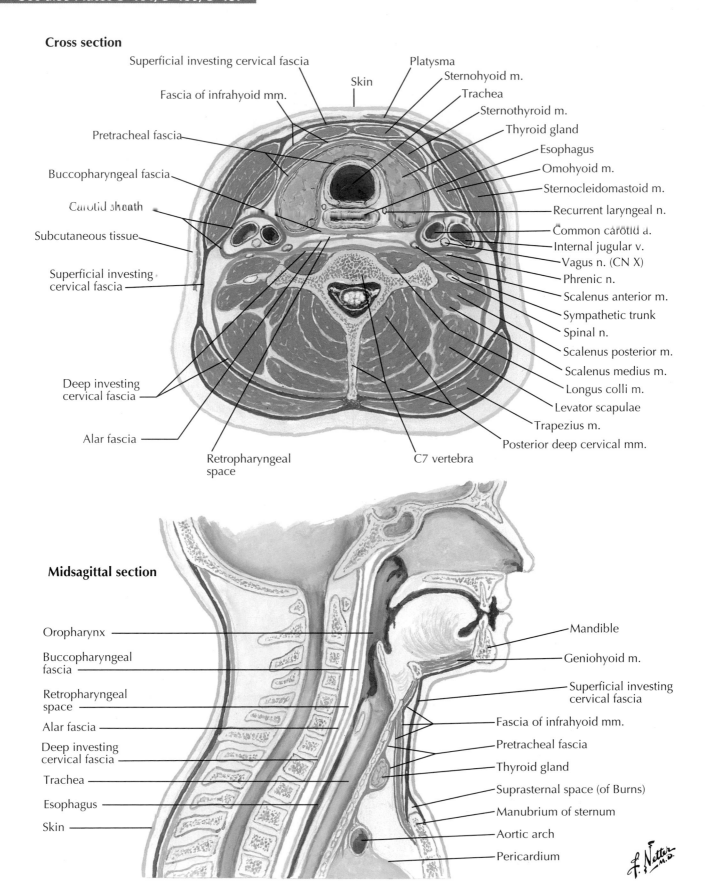

Cross section

Superficial investing cervical fascia

Fascia of infrahyoid mm.

Pretracheal fascia

Buccopharyngeal fascia

Carotid sheath

Subcutaneous tissue

Superficial investing cervical fascia

Deep investing cervical fascia

Alar fascia

Retropharyngeal space

Skin

Platysma

Sternohyoid m.

Trachea

Sternothyroid m.

Thyroid gland

Esophagus

Omohyoid m.

Sternocleidomastoid m.

Recurrent laryngeal n.

Common carotid a.

Internal jugular v.

Vagus n. (CN X)

Phrenic n.

Scalenus anterior m.

Sympathetic trunk

Spinal n.

Scalenus posterior m.

Scalenus medius m.

Longus colli m.

Levator scapulae

Trapezius m.

Posterior deep cervical mm.

C7 vertebra

Midsagittal section

Oropharynx

Buccopharyngeal fascia

Retropharyngeal space

Alar fascia

Deep investing cervical fascia

Trachea

Esophagus

Skin

Mandible

Geniohyoid m.

Superficial investing cervical fascia

Fascia of infrahyoid mm.

Pretracheal fascia

Thyroid gland

Suprasternal space (of Burns)

Manubrium of sternum

Aortic arch

Pericardium

■ Superficial investing cervical fascia
■ Fascia of infrahyoid mm.
■ Pretracheal fascia
■ Buccopharyngeal fascia
■ Carotid sheath
■ Deep investing cervical fascia

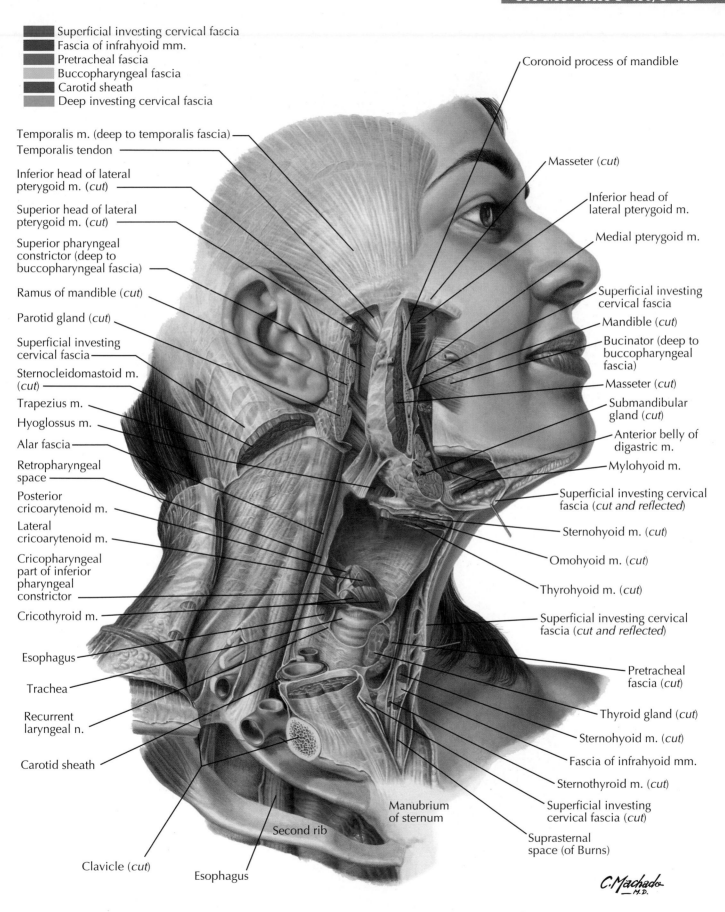

Temporalis m. (deep to temporalis fascia)

Temporalis tendon

Inferior head of lateral pterygoid m. (*cut*)

Superior head of lateral pterygoid m. (*cut*)

Superior pharyngeal constrictor (deep to buccopharyngeal fascia)

Ramus of mandible (*cut*)

Parotid gland (*cut*)

Superficial investing cervical fascia

Sternocleidomastoid m. (*cut*)

Trapezius m.

Hyoglossus m.

Alar fascia

Retropharyngeal space

Posterior cricoarytenoid m.

Lateral cricoarytenoid m.

Cricopharyngeal part of inferior pharyngeal constrictor

Cricothyroid m.

Esophagus

Trachea

Recurrent laryngeal n.

Carotid sheath

Clavicle (*cut*)

Esophagus

Second rib

Manubrium of sternum

Coronoid process of mandible

Masseter (*cut*)

Inferior head of lateral pterygoid m.

Medial pterygoid m.

Superficial investing cervical fascia

Mandible (*cut*)

Bucinator (deep to buccopharyngeal fascia)

Masseter (*cut*)

Submandibular gland (*cut*)

Anterior belly of digastric m.

Mylohyoid m.

Superficial investing cervical fascia (*cut and reflected*)

Sternohyoid m. (*cut*)

Omohyoid m. (*cut*)

Thyrohyoid m. (*cut*)

Superficial investing cervical fascia (*cut and reflected*)

Pretracheal fascia (*cut*)

Thyroid gland (*cut*)

Sternohyoid m. (*cut*)

Fascia of infrahyoid mm.

Sternothyroid m. (*cut*)

Superficial investing cervical fascia (*cut*)

Suprasternal space (of Burns)

C.Machado
_M.D._

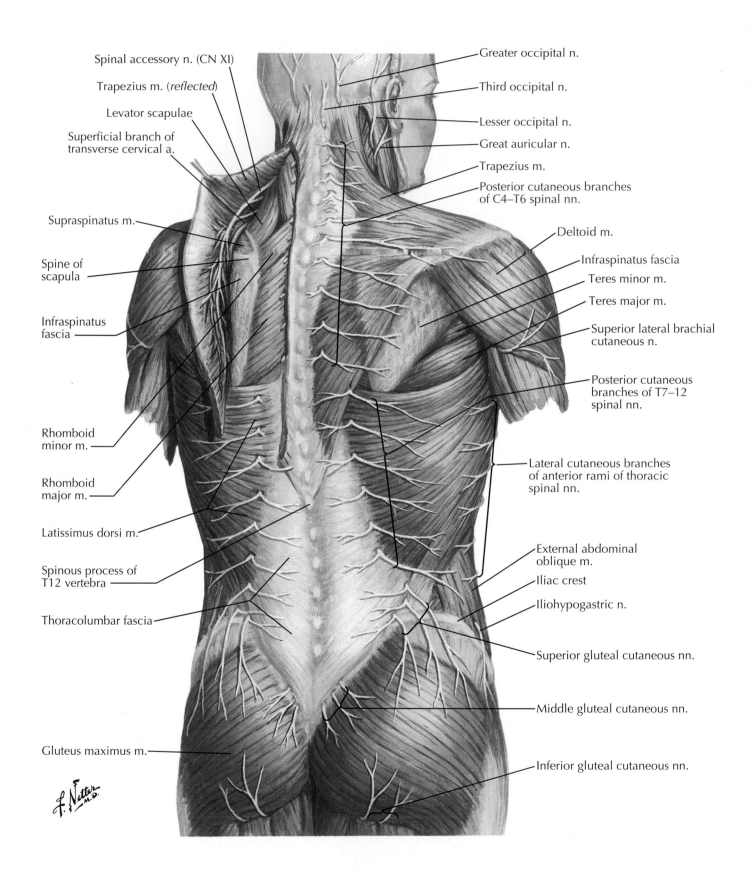

Spinal accessory n. (CN XI)

Trapezius m. (*reflected*)

Levator scapulae

Superficial branch of
transverse cervical a.

Supraspinatus m.

Spine of
scapula

Infraspinatus
fascia

Rhomboid
minor m.

Rhomboid
major m.

Latissimus dorsi m.

Spinous process of
T12 vertebra

Thoracolumbar fascia

Gluteus maximus m.

Greater occipital n.

Third occipital n.

Lesser occipital n.

Great auricular n.

Trapezius m.

Posterior cutaneous branches
of C4–T6 spinal nn.

Deltoid m.

Infraspinatus fascia

Teres minor m.

Teres major m.

Superior lateral brachial
cutaneous n.

Posterior cutaneous
branches of T7–12
spinal nn.

Lateral cutaneous branches
of anterior rami of thoracic
spinal nn.

External abdominal
oblique m.

Iliac crest

Iliohypogastric n.

Superior gluteal cutaneous nn.

Middle gluteal cutaneous nn.

Inferior gluteal cutaneous nn.

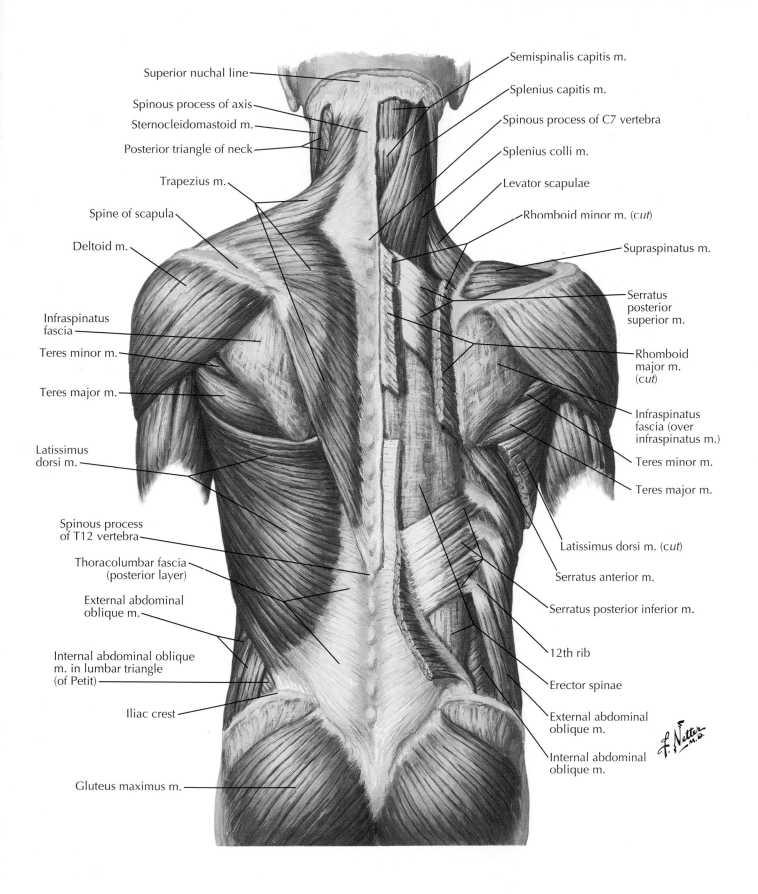

Superior nuchal line

Spinous process of axis

Sternocleidomastoid m.

Posterior triangle of neck

Trapezius m.

Spine of scapula

Deltoid m.

Infraspinatus fascia

Teres minor m.

Teres major m.

Latissimus dorsi m.

Spinous process of T12 vertebra

Thoracolumbar fascia (posterior layer)

External abdominal oblique m.

Internal abdominal oblique m. in lumbar triangle (of Petit)

Iliac crest

Gluteus maximus m.

Semispinalis capitis m.

Splenius capitis m.

Spinous process of C7 vertebra

Splenius colli m.

Levator scapulae

Rhomboid minor m. (*cut*)

Supraspinatus m.

Serratus posterior superior m.

Rhomboid major m. (*cut*)

Infraspinatus fascia (over infraspinatus m.)

Teres minor m.

Teres major m.

Latissimus dorsi m. (*cut*)

Serratus anterior m.

Serratus posterior inferior m.

12th rib

Erector spinae

External abdominal oblique m.

Internal abdominal oblique m.

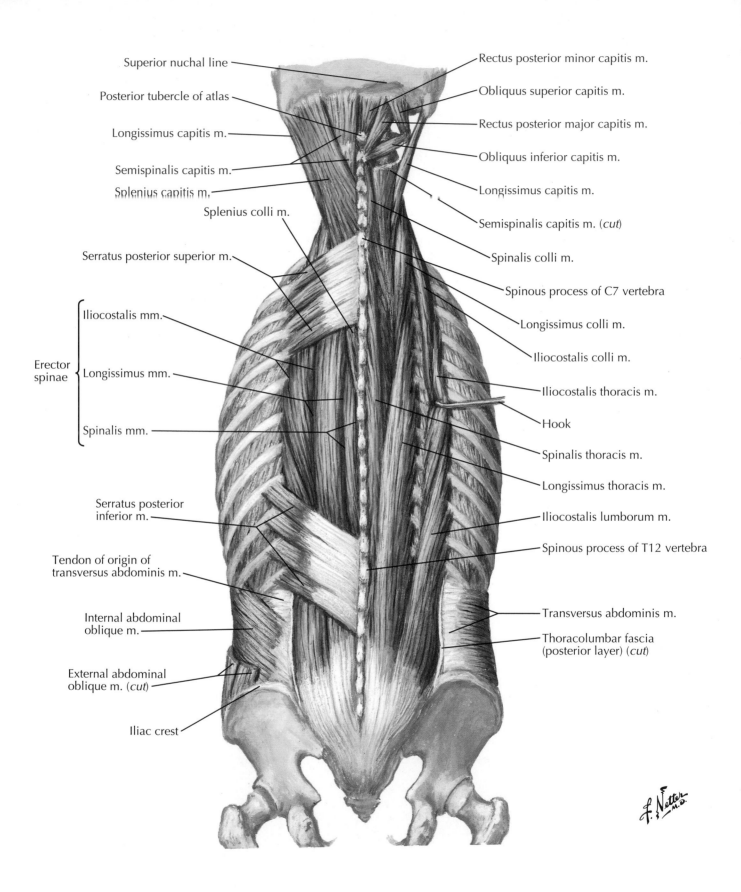

Superior nuchal line

Posterior tubercle of atlas

Longissimus capitis m.

Semispinalis capitis m.

Splenius capitis m.

Splenius colli m.

Serratus posterior superior m.

Iliocostalis mm.

Erector spinae — Longissimus mm.

Spinalis mm.

Serratus posterior inferior m.

Tendon of origin of transversus abdominis m.

Internal abdominal oblique m.

External abdominal oblique m. (*cut*)

Iliac crest

Rectus posterior minor capitis m.

Obliquus superior capitis m.

Rectus posterior major capitis m.

Obliquus inferior capitis m.

Longissimus capitis m.

Semispinalis capitis m. (*cut*)

Spinalis colli m.

Spinous process of C7 vertebra

Longissimus colli m.

Iliocostalis colli m.

Iliocostalis thoracis m.

Hook

Spinalis thoracis m.

Longissimus thoracis m.

Iliocostalis lumborum m.

Spinous process of T12 vertebra

Transversus abdominis m.

Thoracolumbar fascia (posterior layer) (*cut*)

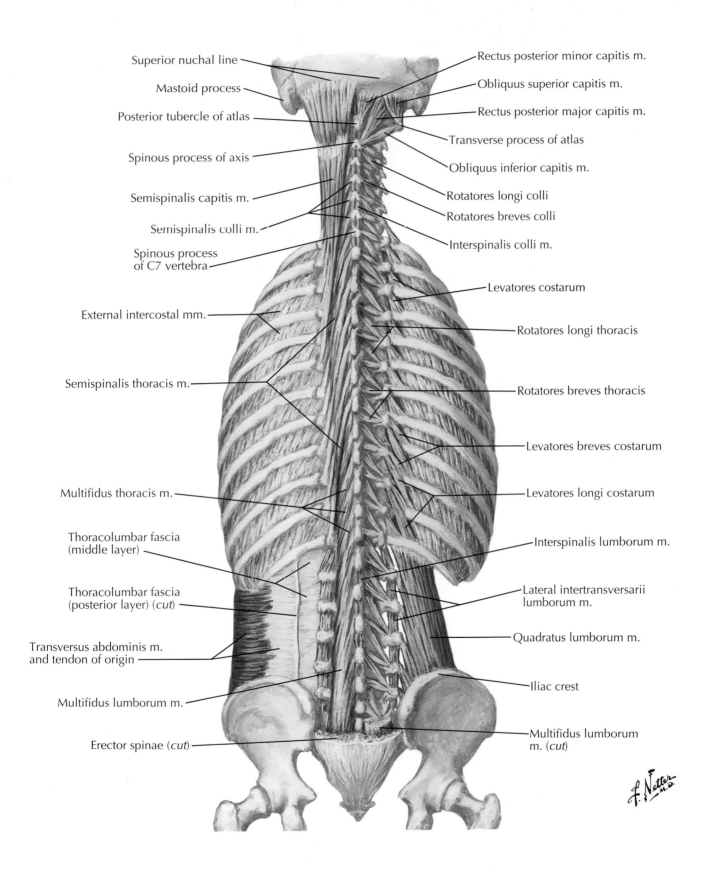

Superior nuchal line

Mastoid process

Posterior tubercle of atlas

Spinous process of axis

Semispinalis capitis m.

Semispinalis colli m.

Spinous process of C7 vertebra

External intercostal mm.

Semispinalis thoracis m.

Multifidus thoracis m.

Thoracolumbar fascia (middle layer)

Thoracolumbar fascia (posterior layer) (*cut*)

Transversus abdominis m. and tendon of origin

Multifidus lumborum m.

Erector spinae (*cut*)

Rectus posterior minor capitis m.

Obliquus superior capitis m.

Rectus posterior major capitis m.

Transverse process of atlas

Obliquus inferior capitis m.

Rotatores longi colli

Rotatores breves colli

Interspinalis colli m.

Levatores costarum

Rotatores longi thoracis

Rotatores breves thoracis

Levatores breves costarum

Levatores longi costarum

Interspinalis lumborum m.

Lateral intertransversarii lumborum m.

Quadratus lumborum m.

Iliac crest

Multifidus lumborum m. (*cut*)

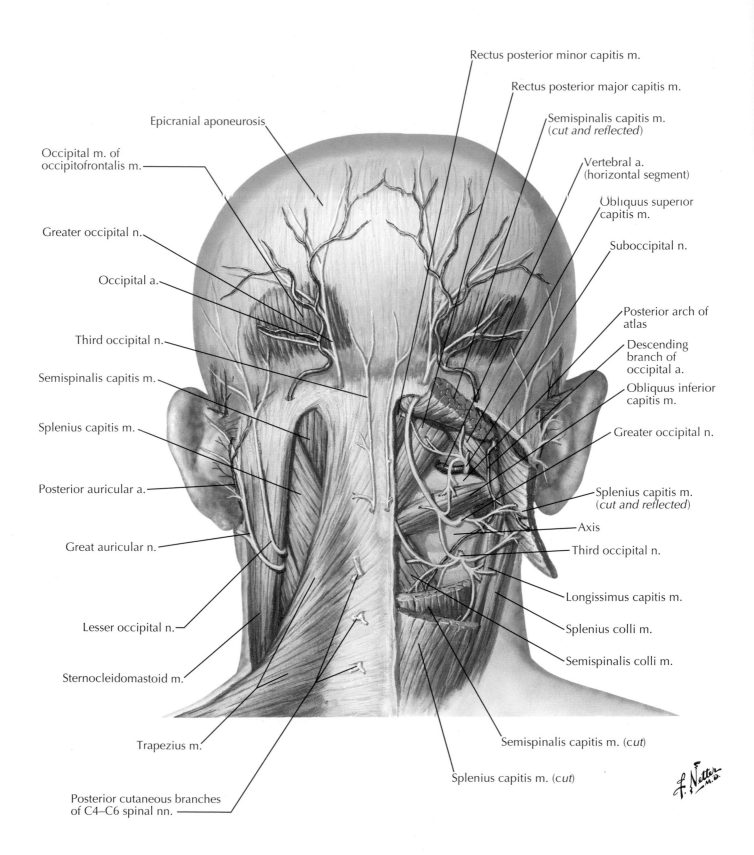

Rectus posterior minor capitis m.

Rectus posterior major capitis m.

Semispinalis capitis m.
(*cut and reflected*)

Vertebral a.
(horizontal segment)

Obliquus superior
capitis m.

Suboccipital n.

Posterior arch of
atlas

Descending
branch of
occipital a.

Obliquus inferior
capitis m.

Greater occipital n.

Splenius capitis m.
(*cut and reflected*)

Axis

Third occipital n.

Longissimus capitis m.

Splenius colli m.

Semispinalis colli m.

Semispinalis capitis m. (c*ut*)

Splenius capitis m. (c*ut*)

Epicranial aponeurosis

Occipital m. of
occipitofrontalis m.

Greater occipital n.

Occipital a.

Third occipital n.

Semispinalis capitis m.

Splenius capitis m.

Posterior auricular a.

Great auricular n.

Lesser occipital n.

Sternocleidomastoid m.

Trapezius m.

Posterior cutaneous branches
of C4–C6 spinal nn.

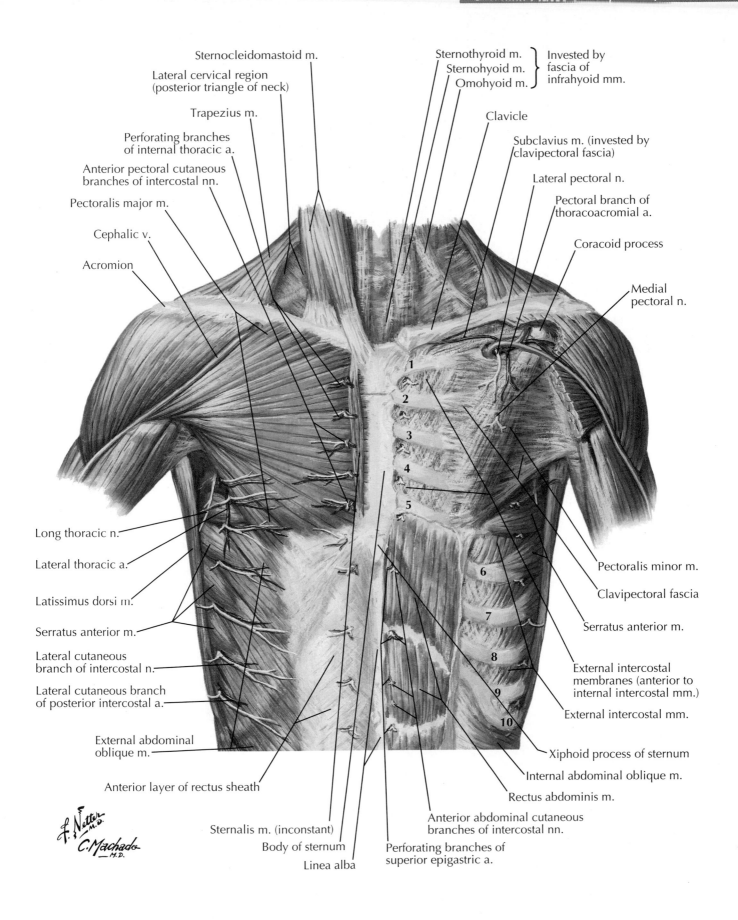

Sternocleidomastoid m.

Lateral cervical region
(posterior triangle of neck)

Trapezius m.

Perforating branches
of internal thoracic a.

Anterior pectoral cutaneous
branches of intercostal nn.

Pectoralis major m.

Cephalic v.

Acromion

Sternothyroid m.
Sternohyoid m. } Invested by
Omohyoid m. } fascia of
infrahyoid mm.

Clavicle

Subclavius m. (invested by
clavipectoral fascia)

Lateral pectoral n.

Pectoral branch of
thoracoacromial a.

Coracoid process

Medial
pectoral n.

Long thoracic n.

Lateral thoracic a.

Latissimus dorsi m.

Serratus anterior m.

Lateral cutaneous
branch of intercostal n.

Lateral cutaneous branch
of posterior intercostal a.

External abdominal
oblique m.

Anterior layer of rectus sheath

Pectoralis minor m.

Clavipectoral fascia

Serratus anterior m.

External intercostal
membranes (anterior to
internal intercostal mm.)

External intercostal mm.

Xiphoid process of sternum

Internal abdominal oblique m.

Rectus abdominis m.

Anterior abdominal cutaneous
branches of intercostal nn.

Sternalis m. (inconstant)

Body of sternum

Linea alba

Perforating branches of
superior epigastric a.

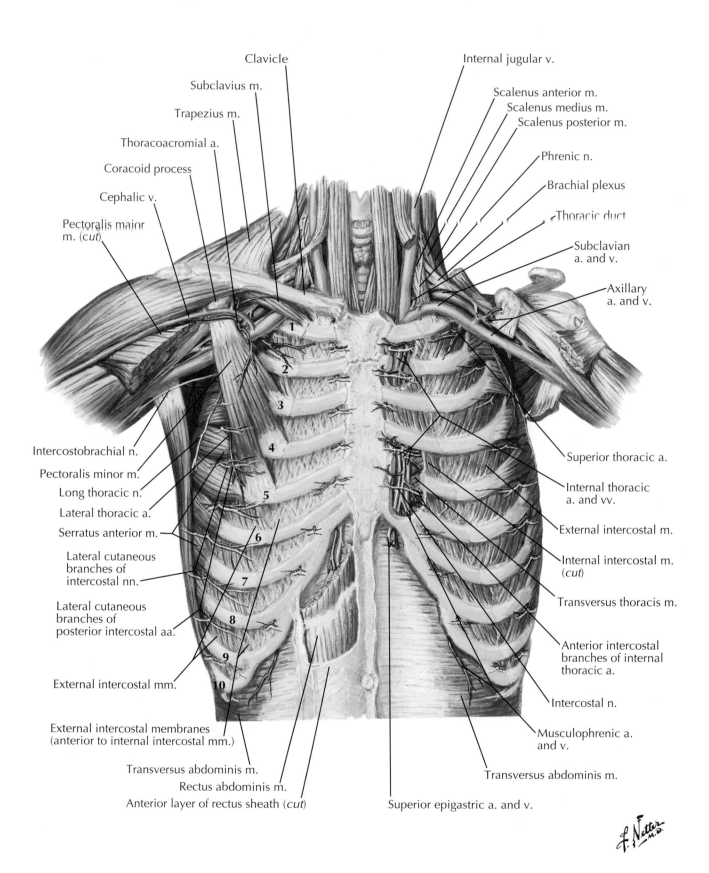

Clavicle

Subclavius m.

Trapezius m.

Thoracoacromial a.

Coracoid process

Cephalic v.

Pectoralis major m. (cut)

Internal jugular v.

Scalenus anterior m.

Scalenus medius m.

Scalenus posterior m.

Phrenic n.

Brachial plexus

Thoracic duct

Subclavian a. and v.

Axillary a. and v.

Intercostobrachial n.

Pectoralis minor m.

Long thoracic n.

Lateral thoracic a.

Serratus anterior m.

Lateral cutaneous branches of intercostal nn.

Lateral cutaneous branches of posterior intercostal aa.

External intercostal mm.

External intercostal membranes (anterior to internal intercostal mm.)

Transversus abdominis m.

Rectus abdominis m.

Anterior layer of rectus sheath (cut)

Superior thoracic a.

Internal thoracic a. and vv.

External intercostal m.

Internal intercostal m. (cut)

Transversus thoracis m.

Anterior intercostal branches of internal thoracic a.

Intercostal n.

Musculophrenic a. and v.

Transversus abdominis m.

Superior epigastric a. and v.

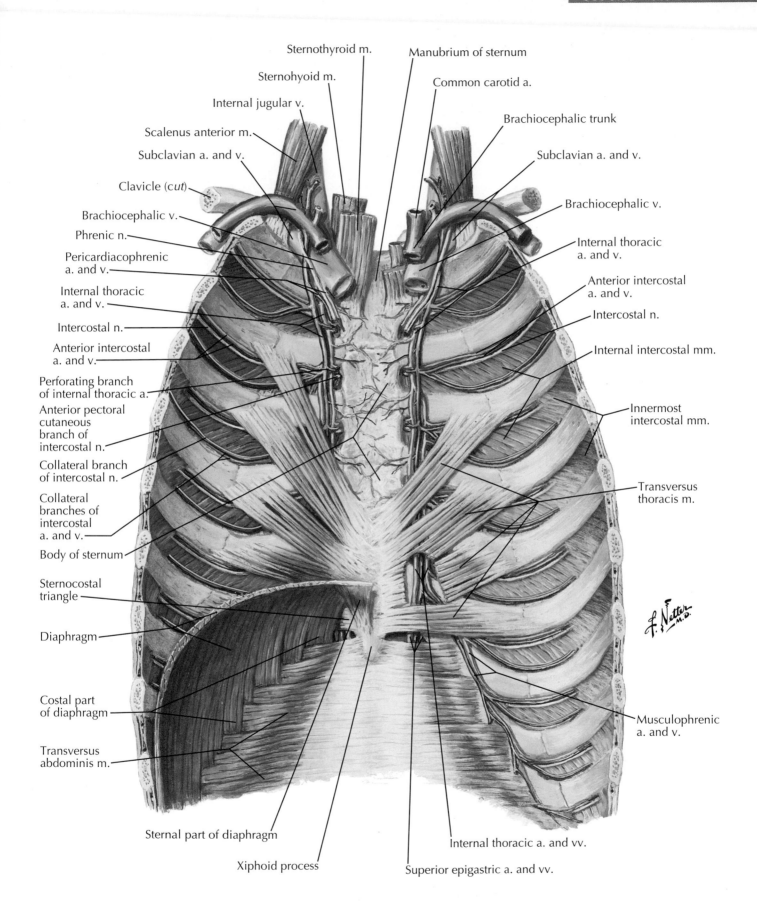

Sternothyroid m.

Sternohyoid m.

Internal jugular v.

Scalenus anterior m.

Subclavian a. and v.

Clavicle (*cut*)

Brachiocephalic v.

Phrenic n.

Pericardiacophrenic a. and v.

Internal thoracic a. and v.

Intercostal n.

Anterior intercostal a. and v.

Perforating branch of internal thoracic a.

Anterior pectoral cutaneous branch of intercostal n.

Collateral branch of intercostal n.

Collateral branches of intercostal a. and v.

Body of sternum

Sternocostal triangle

Diaphragm

Costal part of diaphragm

Transversus abdominis m.

Manubrium of sternum

Common carotid a.

Brachiocephalic trunk

Subclavian a. and v.

Brachiocephalic v.

Internal thoracic a. and v.

Anterior intercostal a. and v.

Intercostal n.

Internal intercostal mm.

Innermost intercostal mm.

Transversus thoracis m.

Musculophrenic a. and v.

Sternal part of diaphragm

Xiphoid process

Internal thoracic a. and vv.

Superior epigastric a. and vv.

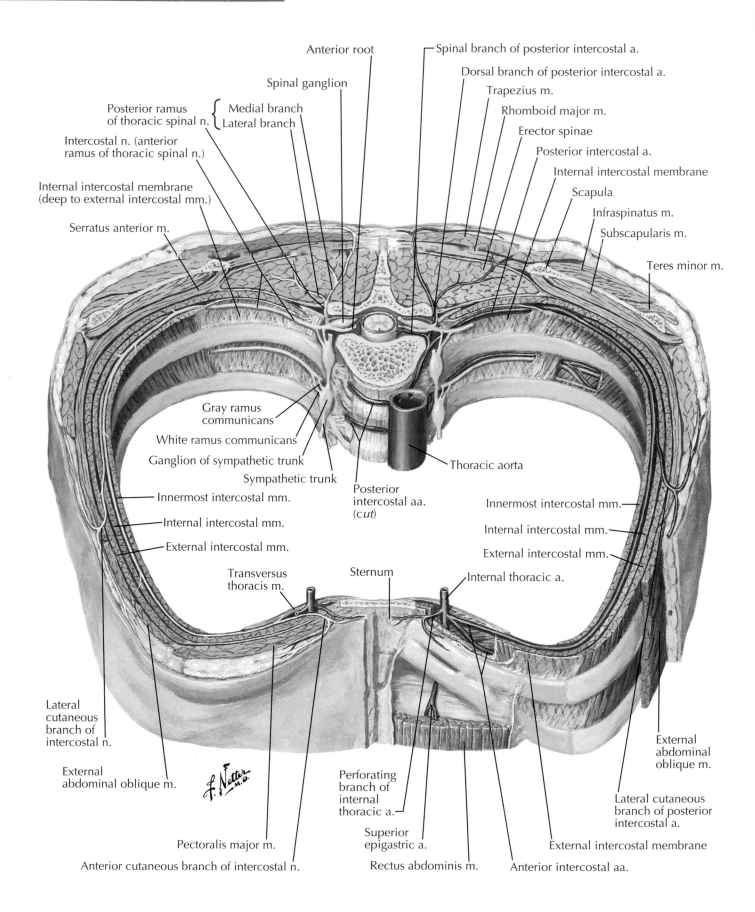

Anterior root

Spinal ganglion

Spinal branch of posterior intercostal a.

Dorsal branch of posterior intercostal a.

Trapezius m.

Rhomboid major m.

Erector spinae

Posterior intercostal a.

Internal intercostal membrane

Scapula

Infraspinatus m.

Subscapularis m.

Teres minor m.

Posterior ramus of thoracic spinal n.
Medial branch
Lateral branch

Intercostal n. (anterior ramus of thoracic spinal n.)

Internal intercostal membrane (deep to external intercostal mm.)

Serratus anterior m.

Gray ramus communicans

White ramus communicans

Ganglion of sympathetic trunk

Sympathetic trunk

Posterior intercostal aa. (cut)

Thoracic aorta

Innermost intercostal mm.

Internal intercostal mm.

External intercostal mm.

Innermost intercostal mm.

Internal intercostal mm.

External intercostal mm.

Transversus thoracis m.

Sternum

Internal thoracic a.

Lateral cutaneous branch of intercostal n.

External abdominal oblique m.

Perforating branch of internal thoracic a.

Superior epigastric a.

Rectus abdominis m.

Anterior intercostal aa.

External abdominal oblique m.

Lateral cutaneous branch of posterior intercostal a.

External intercostal membrane

Pectoralis major m.

Anterior cutaneous branch of intercostal n.

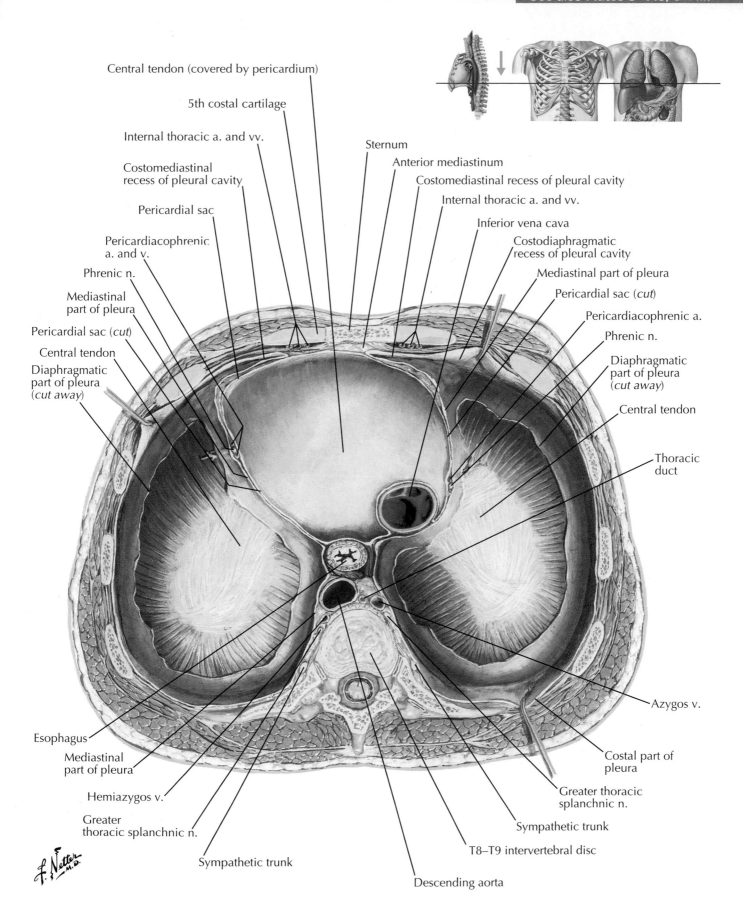

Central tendon (covered by pericardium)

5th costal cartilage

Internal thoracic a. and vv.

Costomediastinal
recess of pleural cavity

Pericardial sac

Pericardiacophrenic
a. and v.

Phrenic n.

Mediastinal
part of pleura

Pericardial sac (*cut*)

Central tendon

Diaphragmatic
part of pleura
(*cut away*)

Sternum

Anterior mediastinum

Costomediastinal recess of pleural cavity

Internal thoracic a. and vv.

Inferior vena cava

Costodiaphragmatic
recess of pleural cavity

Mediastinal part of pleura

Pericardial sac (*cut*)

Pericardiacophrenic a.

Phrenic n.

Diaphragmatic
part of pleura
(*cut away*)

Central tendon

Thoracic
duct

Azygos v.

Costal part of
pleura

Greater thoracic
splanchnic n.

Sympathetic trunk

T8–T9 intervertebral disc

Descending aorta

Sympathetic trunk

Greater
thoracic splanchnic n.

Hemiazygos v.

Mediastinal
part of pleura

Esophagus

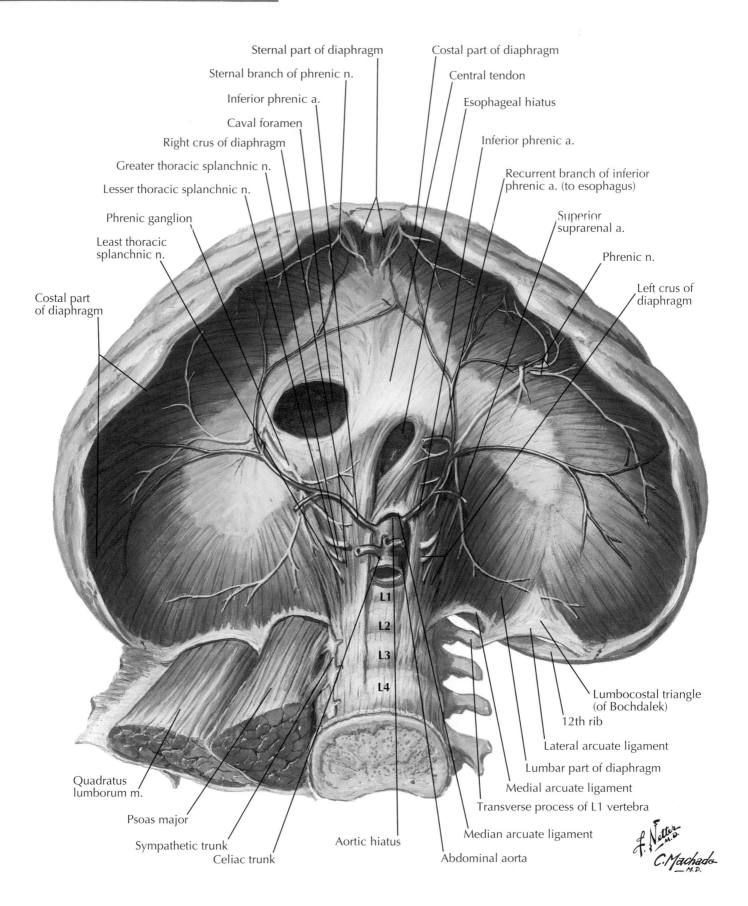

Sternal part of diaphragm

Sternal branch of phrenic n.

Inferior phrenic a.

Caval foramen

Right crus of diaphragm

Greater thoracic splanchnic n.

Lesser thoracic splanchnic n.

Phrenic ganglion

Least thoracic splanchnic n.

Costal part of diaphragm

Costal part of diaphragm

Central tendon

Esophageal hiatus

Inferior phrenic a.

Recurrent branch of inferior phrenic a. (to esophagus)

Superior suprarenal a.

Phrenic n.

Left crus of diaphragm

L1

L2

L3

L4

Lumbocostal triangle (of Bochdalek)

12th rib

Lateral arcuate ligament

Lumbar part of diaphragm

Medial arcuate ligament

Transverse process of L1 vertebra

Median arcuate ligament

Abdominal aorta

Aortic hiatus

Celiac trunk

Sympathetic trunk

Psoas major

Quadratus lumborum m.

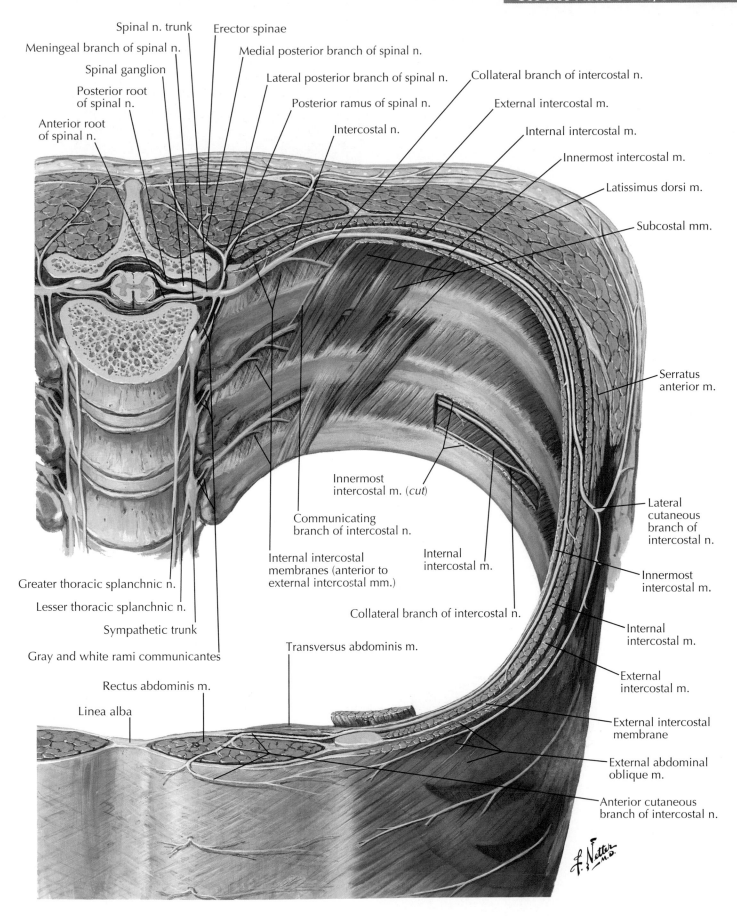

Spinal n. trunk

Meningeal branch of spinal n.

Spinal ganglion

Posterior root of spinal n.

Anterior root of spinal n.

Erector spinae

Medial posterior branch of spinal n.

Lateral posterior branch of spinal n.

Posterior ramus of spinal n.

Intercostal n.

Collateral branch of intercostal n.

External intercostal m.

Internal intercostal m.

Innermost intercostal m.

Latissimus dorsi m.

Subcostal mm.

Serratus anterior m.

Innermost intercostal m. (cut)

Communicating branch of intercostal n.

Internal intercostal membranes (anterior to external intercostal mm.)

Internal intercostal m.

Collateral branch of intercostal n.

Lateral cutaneous branch of intercostal n.

Innermost intercostal m.

Internal intercostal m.

Greater thoracic splanchnic n.

Lesser thoracic splanchnic n.

Sympathetic trunk

Gray and white rami communicantes

Transversus abdominis m.

External intercostal m.

Rectus abdominis m.

Linea alba

External intercostal membrane

External abdominal oblique m.

Anterior cutaneous branch of intercostal n.

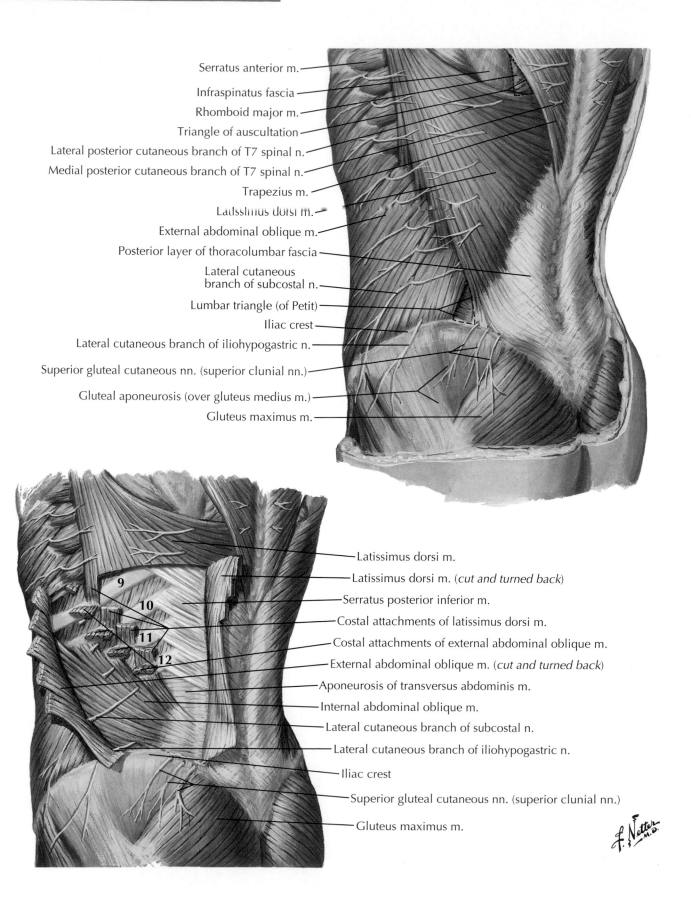

Serratus anterior m.

Infraspinatus fascia

Rhomboid major m.

Triangle of auscultation

Lateral posterior cutaneous branch of T7 spinal n.

Medial posterior cutaneous branch of T7 spinal n.

Trapezius m.

Latissimus dorsi m.

External abdominal oblique m.

Posterior layer of thoracolumbar fascia

Lateral cutaneous branch of subcostal n.

Lumbar triangle (of Petit)

Iliac crest

Lateral cutaneous branch of iliohypogastric n.

Superior gluteal cutaneous nn. (superior clunial nn.)

Gluteal aponeurosis (over gluteus medius m.)

Gluteus maximus m.

9
10
11
12

Latissimus dorsi m.

Latissimus dorsi m. (*cut and turned back*)

Serratus posterior inferior m.

Costal attachments of latissimus dorsi m.

Costal attachments of external abdominal oblique m.

External abdominal oblique m. (*cut and turned back*)

Aponeurosis of transversus abdominis m.

Internal abdominal oblique m.

Lateral cutaneous branch of subcostal n.

Lateral cutaneous branch of iliohypogastric n.

Iliac crest

Superior gluteal cutaneous nn. (superior clunial nn.)

Gluteus maximus m.

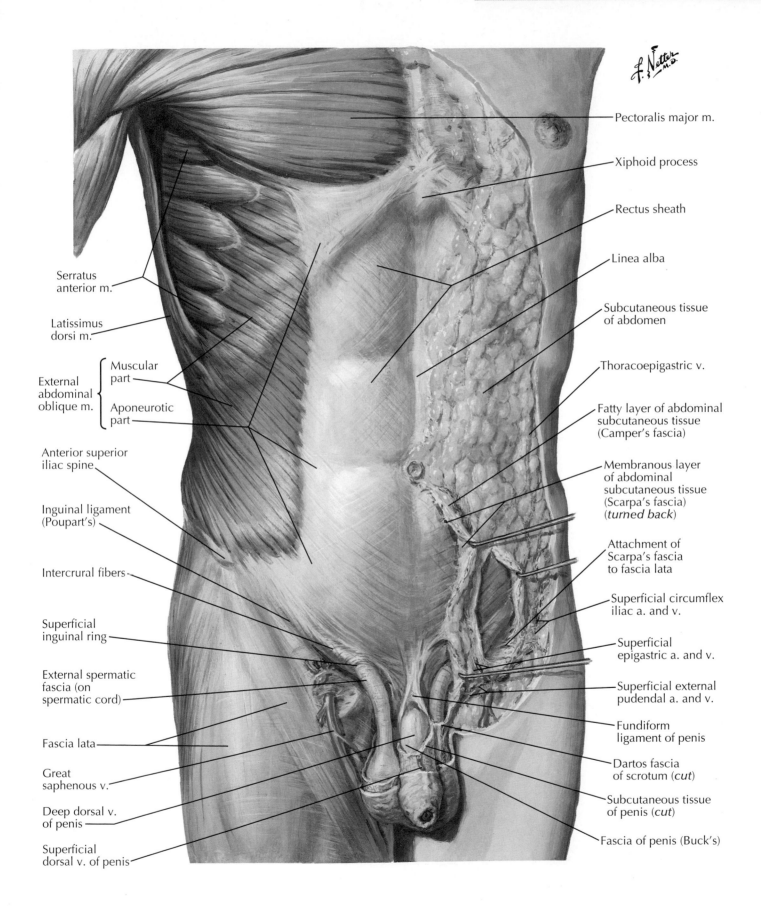

Pectoralis major m.

Xiphoid process

Rectus sheath

Linea alba

Subcutaneous tissue of abdomen

Thoracoepigastric v.

Fatty layer of abdominal subcutaneous tissue (Camper's fascia)

Membranous layer of abdominal subcutaneous tissue (Scarpa's fascia) (*turned back*)

Attachment of Scarpa's fascia to fascia lata

Superficial circumflex iliac a. and v.

Superficial epigastric a. and v.

Superficial external pudendal a. and v.

Fundiform ligament of penis

Dartos fascia of scrotum (*cut*)

Subcutaneous tissue of penis (*cut*)

Fascia of penis (Buck's)

Pectoralis major m.

Serratus anterior m.

Latissimus dorsi m.

External abdominal oblique m.
Muscular part
Aponeurotic part

Anterior superior iliac spine

Inguinal ligament (Poupart's)

Intercrural fibers

Superficial inguinal ring

External spermatic fascia (on spermatic cord)

Fascia lata

Great saphenous v.

Deep dorsal v. of penis

Superficial dorsal v. of penis

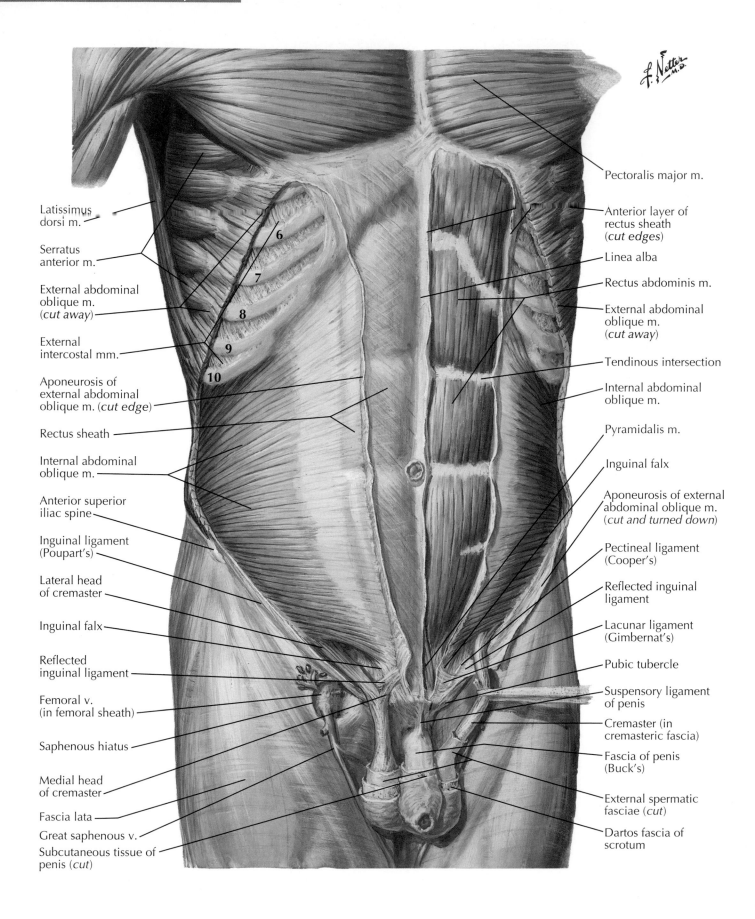

Latissimus dorsi m.

Serratus anterior m.

External abdominal oblique m. (*cut away*)

External intercostal mm.

Aponeurosis of external abdominal oblique m. (*cut edge*)

Rectus sheath

Internal abdominal oblique m.

Anterior superior iliac spine

Inguinal ligament (Poupart's)

Lateral head of cremaster

Inguinal falx

Reflected inguinal ligament

Femoral v. (in femoral sheath)

Saphenous hiatus

Medial head of cremaster

Fascia lata

Great saphenous v.

Subcutaneous tissue of penis (*cut*)

6
7
8
9
10

Pectoralis major m.

Anterior layer of rectus sheath (*cut edges*)

Linea alba

Rectus abdominis m.

External abdominal oblique m. (*cut away*)

Tendinous intersection

Internal abdominal oblique m.

Pyramidalis m.

Inguinal falx

Aponeurosis of external abdominal oblique m. (*cut and turned down*)

Pectineal ligament (Cooper's)

Reflected inguinal ligament

Lacunar ligament (Gimbernat's)

Pubic tubercle

Suspensory ligament of penis

Cremaster (in cremasteric fascia)

Fascia of penis (Buck's)

External spermatic fasciae (*cut*)

Dartos fascia of scrotum

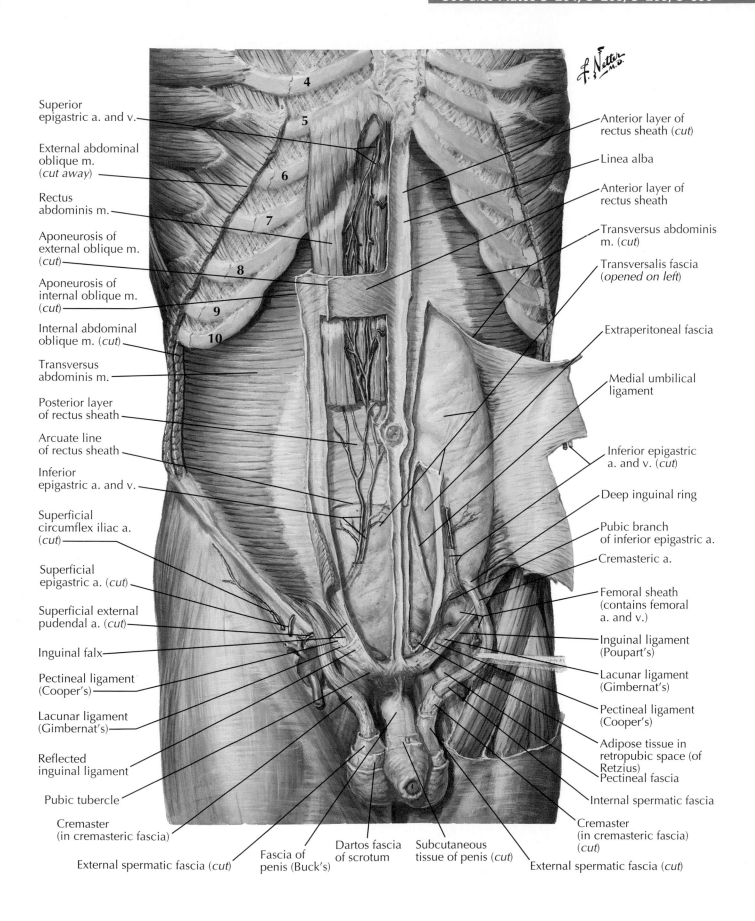

Superior epigastric a. and v.

External abdominal oblique m. (*cut away*)

Rectus abdominis m.

Aponeurosis of external oblique m. (*cut*)

Aponeurosis of internal oblique m. (*cut*)

Internal abdominal oblique m. (*cut*)

Transversus abdominis m.

Posterior layer of rectus sheath

Arcuate line of rectus sheath

Inferior epigastric a. and v.

Superficial circumflex iliac a. (*cut*)

Superficial epigastric a. (*cut*)

Superficial external pudendal a. (*cut*)

Inguinal falx

Pectineal ligament (Cooper's)

Lacunar ligament (Gimbernat's)

Reflected inguinal ligament

Pubic tubercle

Cremaster (in cremasteric fascia)

External spermatic fascia (*cut*)

Fascia of penis (Buck's)

Dartos fascia of scrotum

Subcutaneous tissue of penis (*cut*)

Anterior layer of rectus sheath (*cut*)

Linea alba

Anterior layer of rectus sheath

Transversus abdominis m. (*cut*)

Transversalis fascia (*opened on left*)

Extraperitoneal fascia

Medial umbilical ligament

Inferior epigastric a. and v. (*cut*)

Deep inguinal ring

Pubic branch of inferior epigastric a.

Cremasteric a.

Femoral sheath (contains femoral a. and v.)

Inguinal ligament (Poupart's)

Lacunar ligament (Gimbernat's)

Pectineal ligament (Cooper's)

Adipose tissue in retropubic space (of Retzius)

Pectineal fascia

Internal spermatic fascia

Cremaster (in cremasteric fascia) (*cut*)

External spermatic fascia (*cut*)

Section superior to arcuate line of rectus sheath

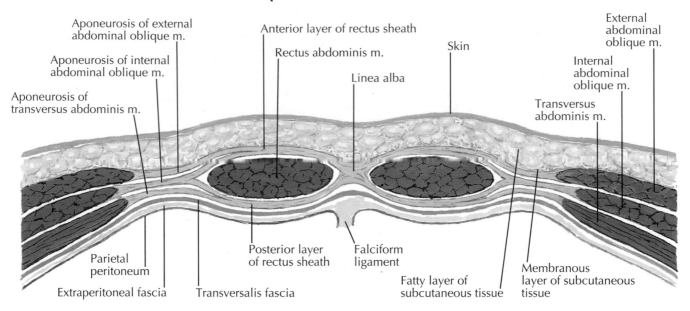

Aponeurosis of external abdominal oblique m.

Aponeurosis of internal abdominal oblique m.

Aponeurosis of transversus abdominis m.

Anterior layer of rectus sheath

Rectus abdominis m.

Linea alba

Skin

External abdominal oblique m.

Internal abdominal oblique m.

Transversus abdominis m.

Parietal peritoneum

Extraperitoneal fascia

Transversalis fascia

Posterior layer of rectus sheath

Falciform ligament

Fatty layer of subcutaneous tissue

Membranous layer of subcutaneous tissue

Aponeurosis of internal abdominal oblique muscle splits to form anterior and posterior layers of rectus sheath. Aponeurosis of external abdominal oblique muscle joins anterior layer; aponeurosis of transversus abdominis muscle joins posterior layer. Anterior and posterior layers of rectus sheath unite medially to form linea alba.

Section inferior to arcuate line of rectus sheath

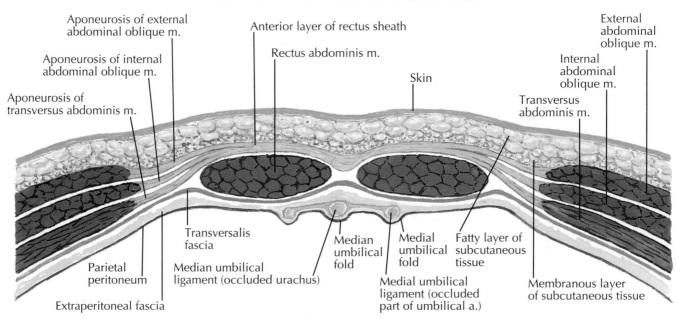

Aponeurosis of external abdominal oblique m.

Aponeurosis of internal abdominal oblique m.

Aponeurosis of transversus abdominis m.

Anterior layer of rectus sheath

Rectus abdominis m.

Skin

External abdominal oblique m.

Internal abdominal oblique m.

Transversus abdominis m.

Transversalis fascia

Median umbilical fold

Medial umbilical fold

Fatty layer of subcutaneous tissue

Parietal peritoneum

Median umbilical ligament (occluded urachus)

Medial umbilical ligament (occluded part of umbilical a.)

Membranous layer of subcutaneous tissue

Extraperitoneal fascia

Aponeurosis of internal abdominal oblique muscle does not split at this level but passes completely anterior to rectus abdominis muscle and is fused there with both the aponeurosis of external abdominal oblique muscle and that of transversus abdominis muscle. Thus, the posterior layer of the rectus sheath is absent below arcuate line, leaving only transversalis fascia.

Abdomen and Pelvis

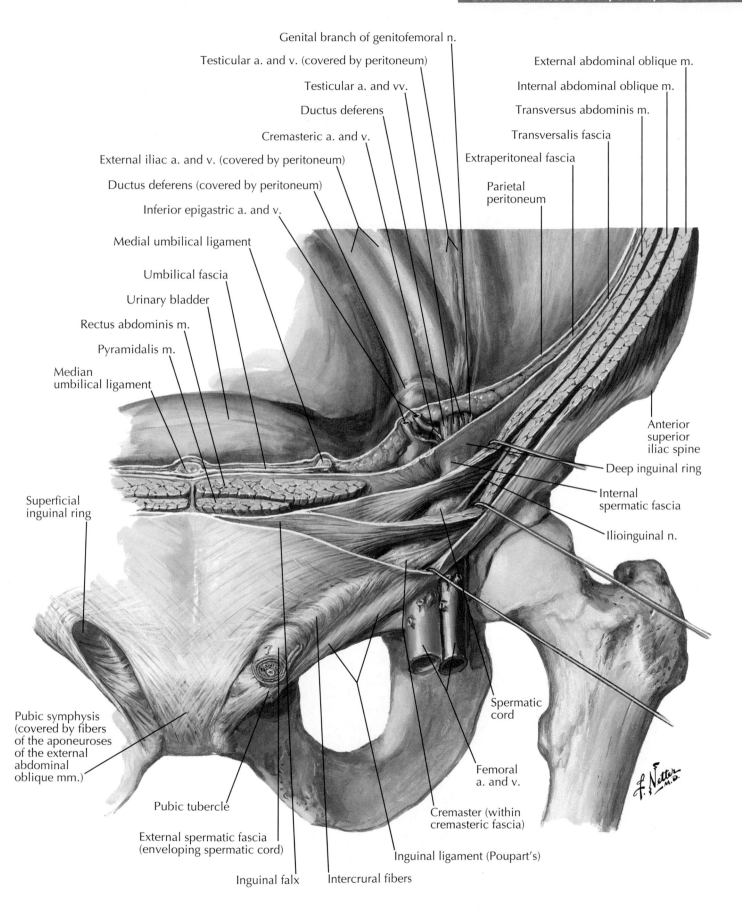

Genital branch of genitofemoral n.

Testicular a. and v. (covered by peritoneum)

Testicular a. and vv.

Ductus deferens

Cremasteric a. and v.

External iliac a. and v. (covered by peritoneum)

Ductus deferens (covered by peritoneum)

Inferior epigastric a. and v.

Medial umbilical ligament

Umbilical fascia

Urinary bladder

Rectus abdominis m.

Pyramidalis m.

Median umbilical ligament

Superficial inguinal ring

Pubic symphysis (covered by fibers of the aponeuroses of the external abdominal oblique mm.)

Pubic tubercle

External spermatic fascia (enveloping spermatic cord)

Inguinal falx

Intercrural fibers

External abdominal oblique m.

Internal abdominal oblique m.

Transversus abdominis m.

Transversalis fascia

Extraperitoneal fascia

Parietal peritoneum

Anterior superior iliac spine

Deep inguinal ring

Internal spermatic fascia

Ilioinguinal n.

Spermatic cord

Femoral a. and v.

Cremaster (within cremasteric fascia)

Inguinal ligament (Poupart's)

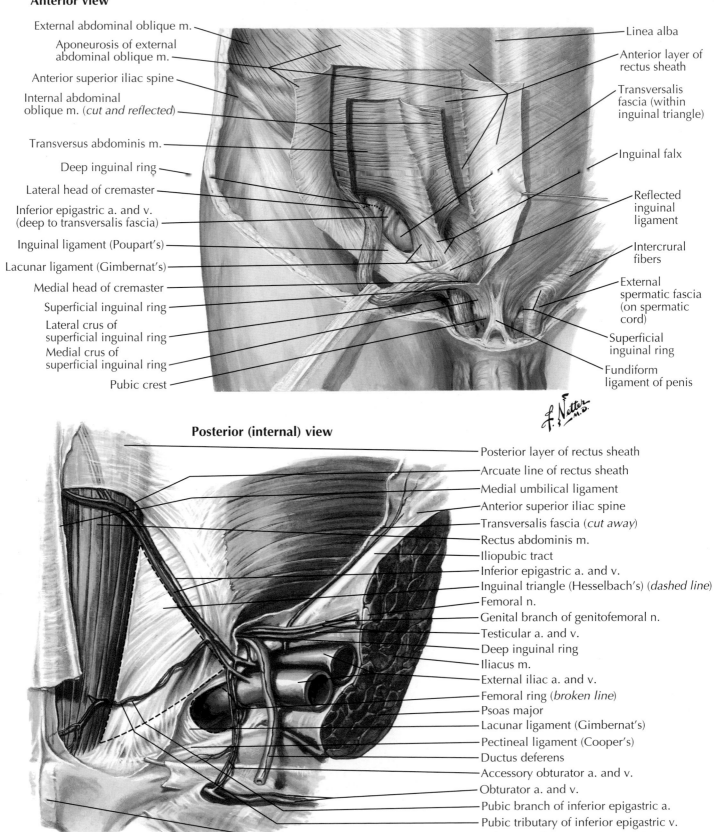

Anterior view

External abdominal oblique m.

Aponeurosis of external
abdominal oblique m.

Anterior superior iliac spine

Internal abdominal
oblique m. (*cut and reflected*)

Transversus abdominis m.

Deep inguinal ring

Lateral head of cremaster

Inferior epigastric a. and v.
(deep to transversalis fascia)

Inguinal ligament (Poupart's)

Lacunar ligament (Gimbernat's)

Medial head of cremaster

Superficial inguinal ring

Lateral crus of
superficial inguinal ring

Medial crus of
superficial inguinal ring

Pubic crest

Linea alba

Anterior layer of
rectus sheath

Transversalis
fascia (within
inguinal triangle)

Inguinal falx

Reflected
inguinal
ligament

Intercrural
fibers

External
spermatic fascia
(on spermatic
cord)

Superficial
inguinal ring

Fundiform
ligament of penis

Posterior (internal) view

Posterior layer of rectus sheath

Arcuate line of rectus sheath

Medial umbilical ligament

Anterior superior iliac spine

Transversalis fascia (*cut away*)

Rectus abdominis m.

Iliopubic tract

Inferior epigastric a. and v.

Inguinal triangle (Hesselbach's) (*dashed line*)

Femoral n.

Genital branch of genitofemoral n.

Testicular a. and v.

Deep inguinal ring

Iliacus m.

External iliac a. and v.

Femoral ring (*broken line*)

Psoas major

Lacunar ligament (Gimbernat's)

Pectineal ligament (Cooper's)

Ductus deferens

Accessory obturator a. and v.

Obturator a. and v.

Pubic branch of inferior epigastric a.

Pubic tributary of inferior epigastric v.

Pubic symphysis

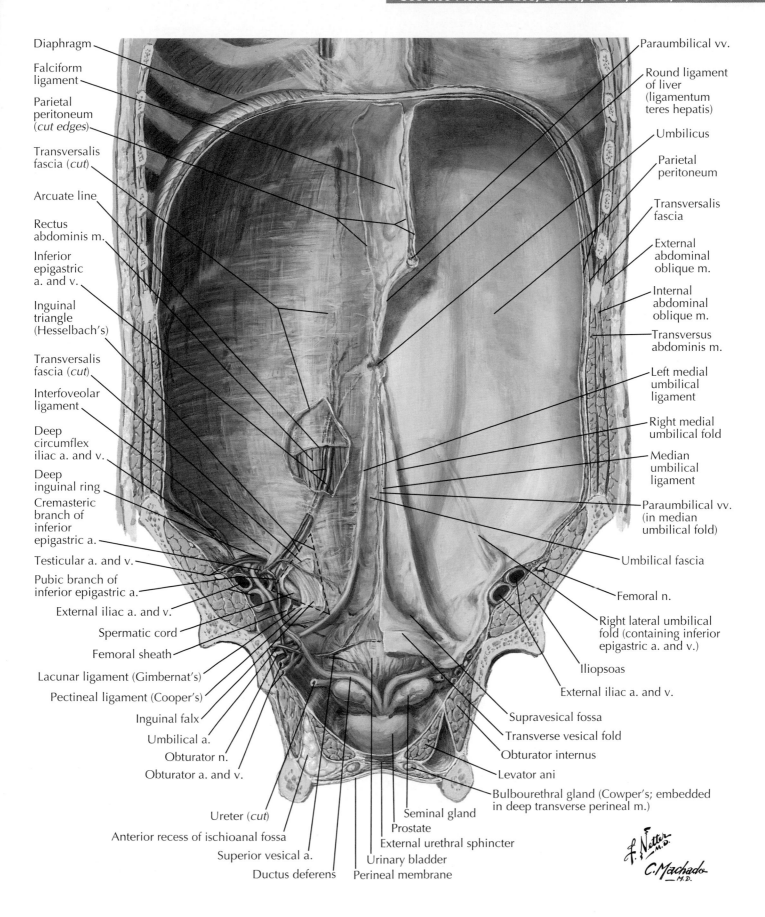

Diaphragm

Falciform ligament

Parietal peritoneum (*cut edges*)

Transversalis fascia (*cut*)

Arcuate line

Rectus abdominis m.

Inferior epigastric a. and v.

Inguinal triangle (Hesselbach's)

Transversalis fascia (*cut*)

Interfoveolar ligament

Deep circumflex iliac a. and v.

Deep inguinal ring

Cremasteric branch of inferior epigastric a.

Testicular a. and v.

Pubic branch of inferior epigastric a.

External iliac a. and v.

Spermatic cord

Femoral sheath

Lacunar ligament (Gimbernat's)

Pectineal ligament (Cooper's)

Inguinal falx

Umbilical a.

Obturator n.

Obturator a. and v.

Ureter (*cut*)

Anterior recess of ischioanal fossa

Superior vesical a.

Ductus deferens

Paraumbilical vv.

Round ligament of liver (ligamentum teres hepatis)

Umbilicus

Parietal peritoneum

Transversalis fascia

External abdominal oblique m.

Internal abdominal oblique m.

Transversus abdominis m.

Left medial umbilical ligament

Right medial umbilical fold

Median umbilical ligament

Paraumbilical vv. (in median umbilical fold)

Umbilical fascia

Femoral n.

Right lateral umbilical fold (containing inferior epigastric a. and v.)

Iliopsoas

External iliac a. and v.

Supravesical fossa

Transverse vesical fold

Obturator internus

Levator ani

Bulbourethral gland (Cowper's; embedded in deep transverse perineal m.)

Seminal gland

Prostate

External urethral sphincter

Urinary bladder

Perineal membrane

f. Netter M.D.

C. Machado M.D.

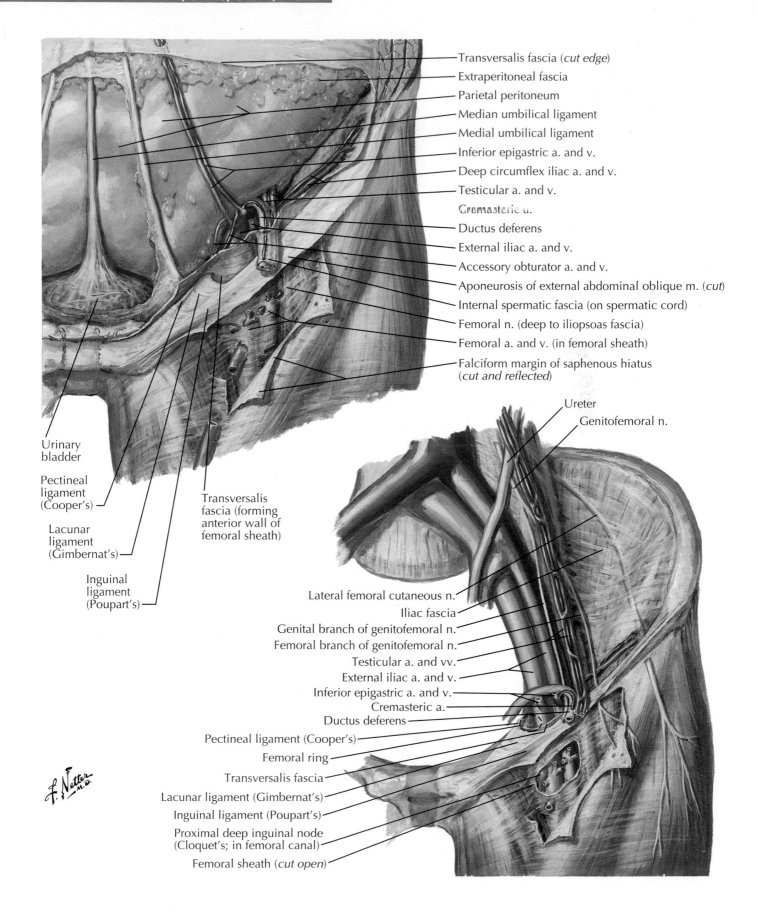

Transversalis fascia (*cut edge*)

Extraperitoneal fascia

Parietal peritoneum

Median umbilical ligament

Medial umbilical ligament

Inferior epigastric a. and v.

Deep circumflex iliac a. and v.

Testicular a. and v.

Cremasteric a.

Ductus deferens

External iliac a. and v.

Accessory obturator a. and v.

Aponeurosis of external abdominal oblique m. (*cut*)

Internal spermatic fascia (on spermatic cord)

Femoral n. (deep to iliopsoas fascia)

Femoral a. and v. (in femoral sheath)

Falciform margin of saphenous hiatus
(*cut and reflected*)

Urinary bladder

Pectineal ligament (Cooper's)

Lacunar ligament (Gimbernat's)

Inguinal ligament (Poupart's)

Transversalis fascia (forming anterior wall of femoral sheath)

Ureter

Genitofemoral n.

Lateral femoral cutaneous n.

Iliac fascia

Genital branch of genitofemoral n.

Femoral branch of genitofemoral n.

Testicular a. and vv.

External iliac a. and v.

Inferior epigastric a. and v.

Cremasteric a.

Ductus deferens

Pectineal ligament (Cooper's)

Femoral ring

Transversalis fascia

Lacunar ligament (Gimbernat's)

Inguinal ligament (Poupart's)

Proximal deep inguinal node
(Cloquet's; in femoral canal)

Femoral sheath (*cut open*)

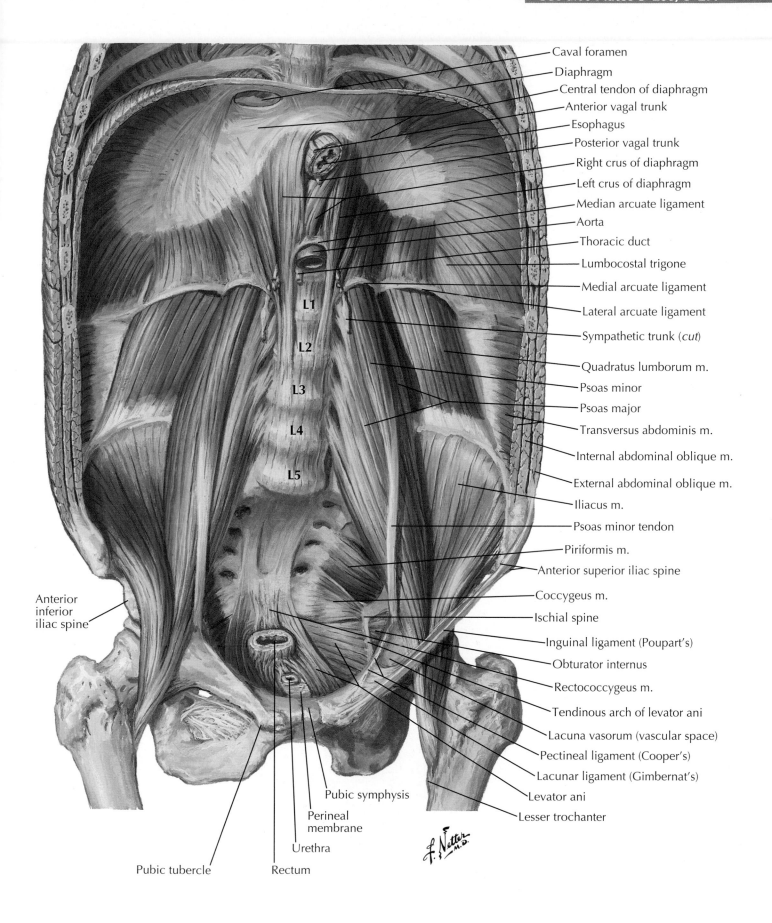

Caval foramen
Diaphragm
Central tendon of diaphragm
Anterior vagal trunk
Esophagus
Posterior vagal trunk
Right crus of diaphragm
Left crus of diaphragm
Median arcuate ligament
Aorta
Thoracic duct
Lumbocostal trigone
Medial arcuate ligament
Lateral arcuate ligament
Sympathetic trunk (*cut*)
Quadratus lumborum m.
Psoas minor
Psoas major
Transversus abdominis m.
Internal abdominal oblique m.
External abdominal oblique m.
Iliacus m.
Psoas minor tendon
Piriformis m.
Anterior superior iliac spine
Coccygeus m.
Ischial spine
Inguinal ligament (Poupart's)
Obturator internus
Rectococcygeus m.
Tendinous arch of levator ani
Lacuna vasorum (vascular space)
Pectineal ligament (Cooper's)
Lacunar ligament (Gimbernat's)
Levator ani
Lesser trochanter

L1
L2
L3
L4
L5

Anterior inferior iliac spine

Pubic symphysis
Perineal membrane
Urethra
Pubic tubercle
Rectum

Medial view

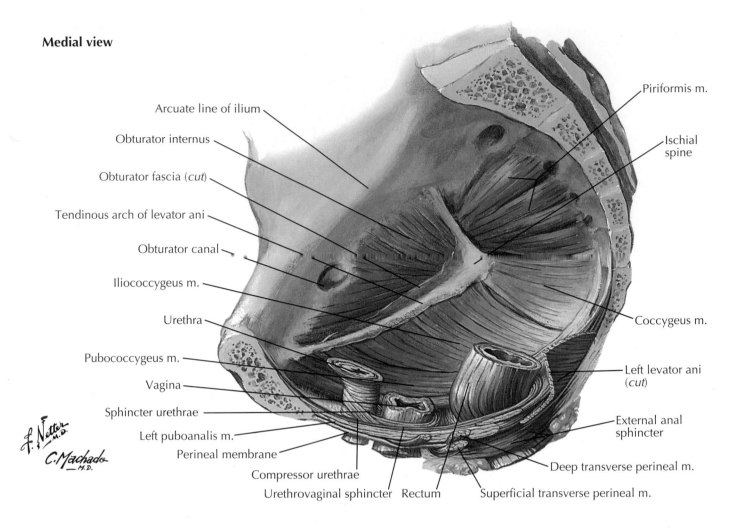

Arcuate line of ilium

Obturator internus

Obturator fascia (cut)

Tendinous arch of levator ani

Obturator canal

Iliococcygeus m.

Urethra

Pubococcygeus m.

Vagina

Sphincter urethrae

Left puboanalis m.

Perineal membrane

Compressor urethrae

Urethrovaginal sphincter   Rectum

Piriformis m.

Ischial spine

Coccygeus m.

Left levator ani (cut)

External anal sphincter

Deep transverse perineal m.

Superficial transverse perineal m.

Lateral view

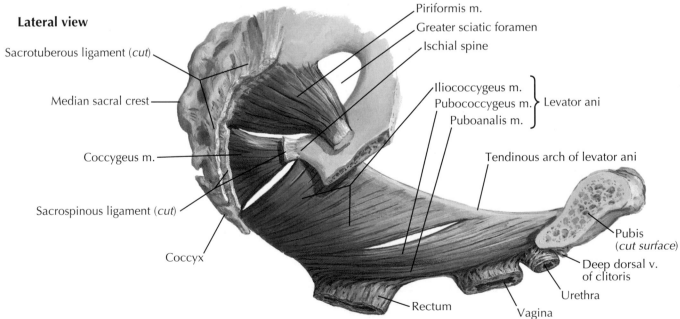

Sacrotuberous ligament (cut)

Median sacral crest

Coccygeus m.

Sacrospinous ligament (cut)

Coccyx

Piriformis m.

Greater sciatic foramen

Ischial spine

Iliococcygeus m.
Pubococcygeus m.   } Levator ani
Puboanalis m.

Tendinous arch of levator ani

Pubis (cut surface)

Deep dorsal v. of clitoris

Urethra

Rectum   Vagina

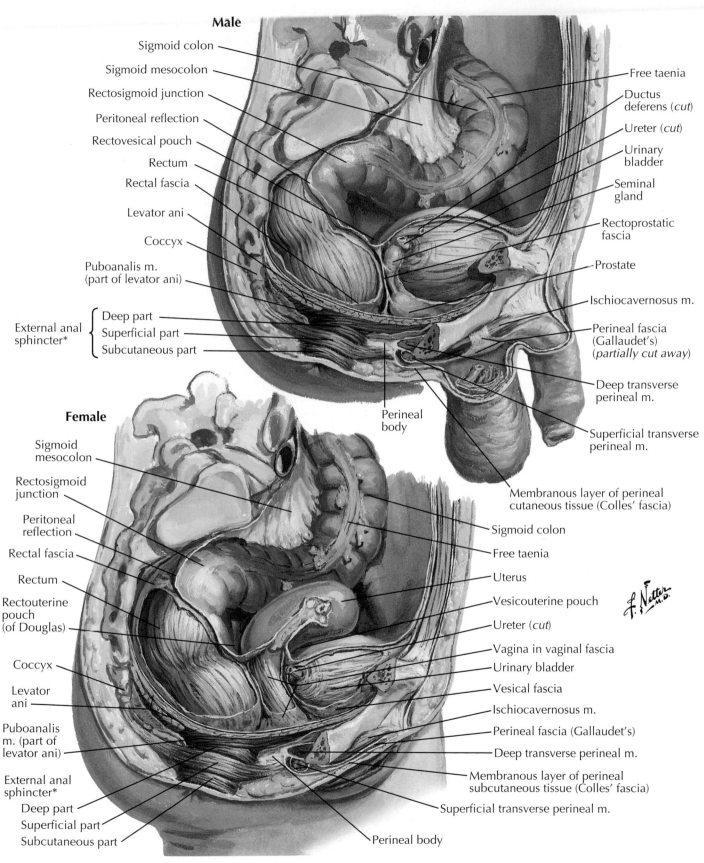

Male

Sigmoid colon

Sigmoid mesocolon

Rectosigmoid junction

Peritoneal reflection

Rectovesical pouch

Rectum

Rectal fascia

Levator ani

Coccyx

Puboanalis m. (part of levator ani)

External anal sphincter*
  Deep part
  Superficial part
  Subcutaneous part

Free taenia

Ductus deferens (cut)

Ureter (cut)

Urinary bladder

Seminal gland

Rectoprostatic fascia

Prostate

Ischiocavernosus m.

Perineal fascia (Gallaudet's) (partially cut away)

Deep transverse perineal m.

Perineal body

Superficial transverse perineal m.

Membranous layer of perineal cutaneous tissue (Colles' fascia)

Female

Sigmoid mesocolon

Rectosigmoid junction

Peritoneal reflection

Rectal fascia

Rectum

Rectouterine pouch (of Douglas)

Coccyx

Levator ani

Puboanalis m. (part of levator ani)

External anal sphincter*
  Deep part
  Superficial part
  Subcutaneous part

Sigmoid colon

Free taenia

Uterus

Vesicouterine pouch

Ureter (cut)

Vagina in vaginal fascia

Urinary bladder

Vesical fascia

Ischiocavernosus m.

Perineal fascia (Gallaudet's)

Deep transverse perineal m.

Membranous layer of perineal subcutaneous tissue (Colles' fascia)

Superficial transverse perineal m.

Perineal body

*Parts variable and often indistinct

Abdomen and Pelvis

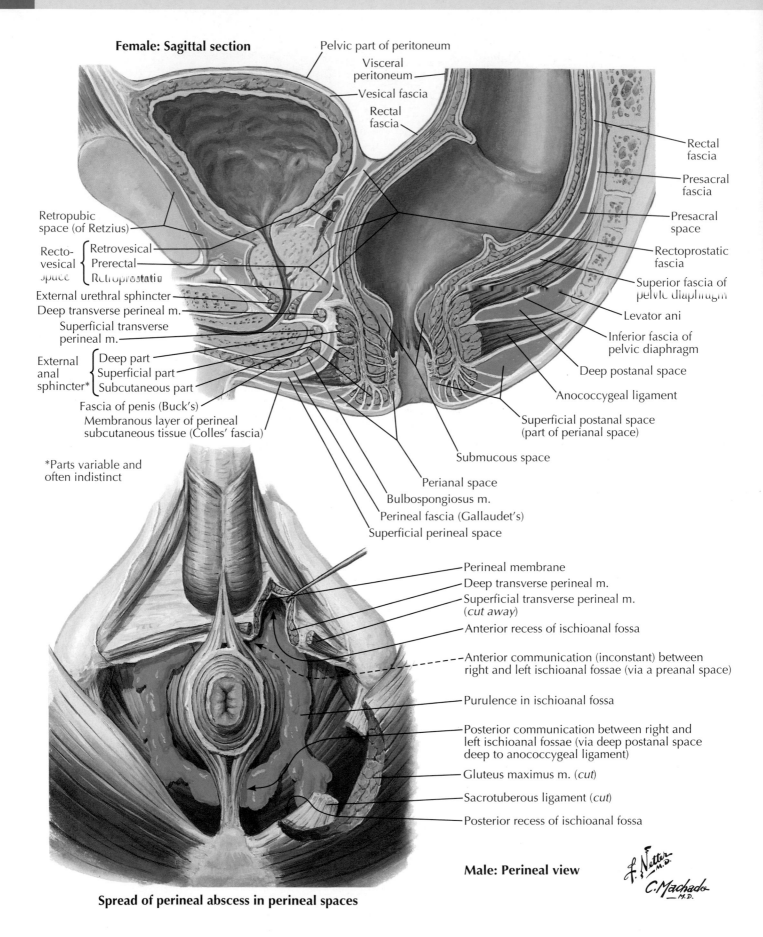

Female: Sagittal section

Pelvic part of peritoneum

Visceral peritoneum

Vesical fascia

Rectal fascia

Rectal fascia

Presacral fascia

Presacral space

Rectoprostatic fascia

Superior fascia of pelvic diaphragm

Levator ani

Inferior fascia of pelvic diaphragm

Deep postanal space

Anococcygeal ligament

Superficial postanal space (part of perianal space)

Submucous space

Perianal space

Bulbospongiosus m.

Perineal fascia (Gallaudet's)

Superficial perineal space

Retropubic space (of Retzius)

Recto-vesical space {
Retrovesical
Prerectal
Retroprostatic
}

External urethral sphincter

Deep transverse perineal m.

Superficial transverse perineal m.

External anal sphincter* {
Deep part
Superficial part
Subcutaneous part
}

Fascia of penis (Buck's)

Membranous layer of perineal subcutaneous tissue (Colles' fascia)

*Parts variable and often indistinct

Perineal membrane

Deep transverse perineal m.

Superficial transverse perineal m. (cut away)

Anterior recess of ischioanal fossa

Anterior communication (inconstant) between right and left ischioanal fossae (via a preanal space)

Purulence in ischioanal fossa

Posterior communication between right and left ischioanal fossae (via deep postanal space deep to anococcygeal ligament)

Gluteus maximus m. (cut)

Sacrotuberous ligament (cut)

Posterior recess of ischioanal fossa

Spread of perineal abscess in perineal spaces

Male: Perineal view

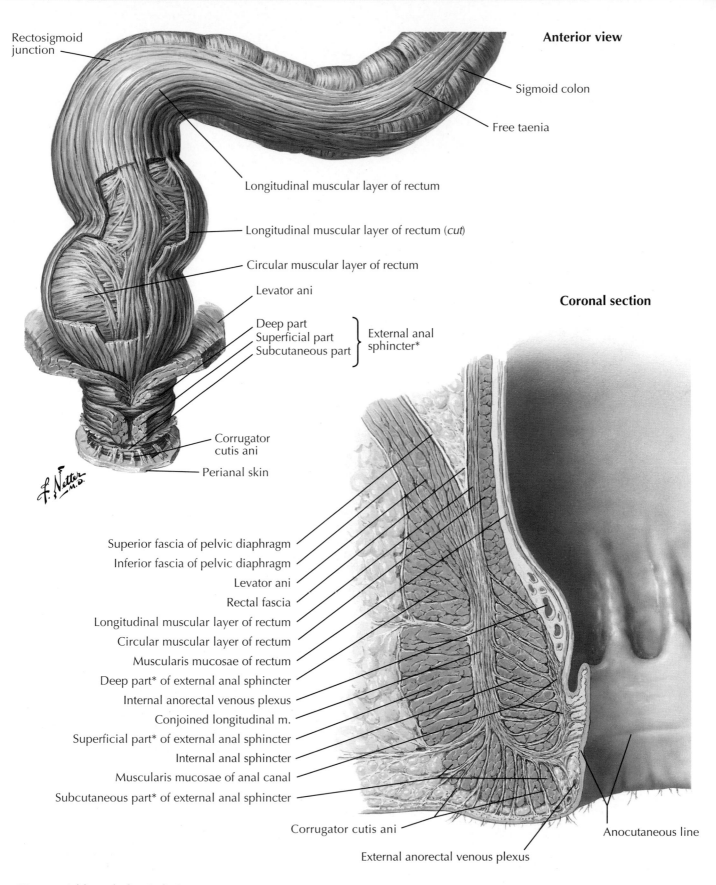

Anterior view

Rectosigmoid junction

Sigmoid colon

Free taenia

Longitudinal muscular layer of rectum

Longitudinal muscular layer of rectum (*cut*)

Circular muscular layer of rectum

Levator ani

Deep part
Superficial part
Subcutaneous part
} External anal sphincter*

Coronal section

Corrugator cutis ani

Perianal skin

Superior fascia of pelvic diaphragm
Inferior fascia of pelvic diaphragm
Levator ani
Rectal fascia
Longitudinal muscular layer of rectum
Circular muscular layer of rectum
Muscularis mucosae of rectum
Deep part* of external anal sphincter
Internal anorectal venous plexus
Conjoined longitudinal m.
Superficial part* of external anal sphincter
Internal anal sphincter
Muscularis mucosae of anal canal
Subcutaneous part* of external anal sphincter

Anocutaneous line

Corrugator cutis ani

External anorectal venous plexus

*Parts variable and often indistinct

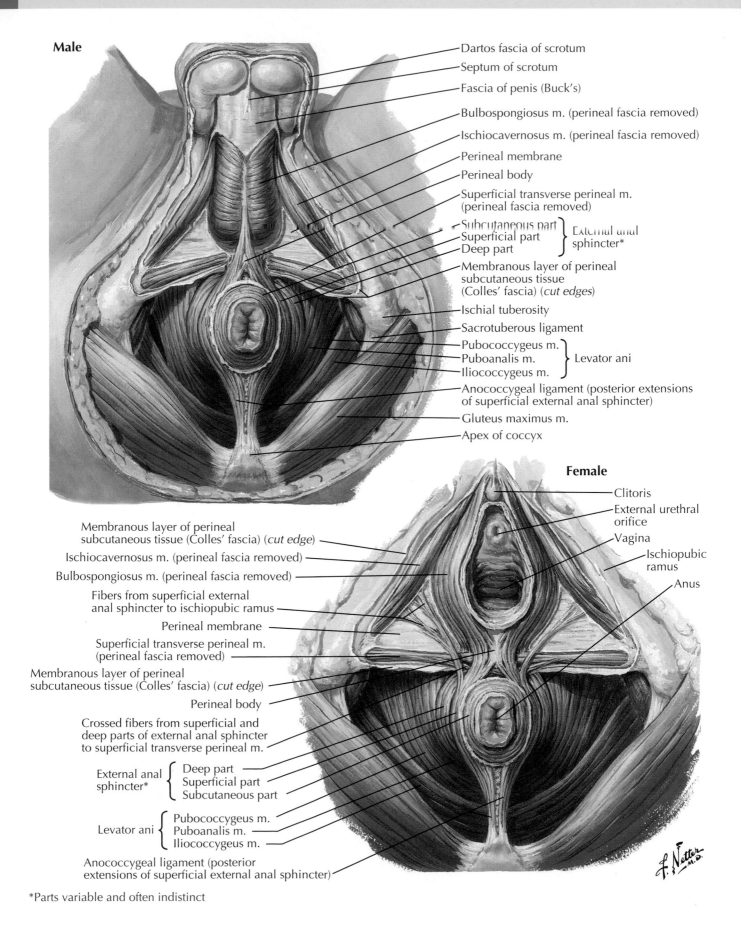

Male

Dartos fascia of scrotum

Septum of scrotum

Fascia of penis (Buck's)

Bulbospongiosus m. (perineal fascia removed)

Ischiocavernosus m. (perineal fascia removed)

Perineal membrane

Perineal body

Superficial transverse perineal m. (perineal fascia removed)

Subcutaneous part ⎤
Superficial part ⎬ External anal sphincter*
Deep part ⎦

Membranous layer of perineal subcutaneous tissue (Colles' fascia) (cut edges)

Ischial tuberosity

Sacrotuberous ligament

Pubococcygeus m. ⎤
Puboanalis m. ⎬ Levator ani
Iliococcygeus m. ⎦

Anococcygeal ligament (posterior extensions of superficial external anal sphincter)

Gluteus maximus m.

Apex of coccyx

Female

Clitoris

External urethral orifice

Vagina

Ischiopubic ramus

Anus

Membranous layer of perineal subcutaneous tissue (Colles' fascia) (cut edge)

Ischiocavernosus m. (perineal fascia removed)

Bulbospongiosus m. (perineal fascia removed)

Fibers from superficial external anal sphincter to ischiopubic ramus

Perineal membrane

Superficial transverse perineal m. (perineal fascia removed)

Membranous layer of perineal subcutaneous tissue (Colles' fascia) (cut edge)

Perineal body

Crossed fibers from superficial and deep parts of external anal sphincter to superficial transverse perineal m.

External anal sphincter* ⎰ Deep part
⎱ Superficial part
Subcutaneous part

Levator ani ⎰ Pubococcygeus m.
⎱ Puboanalis m.
Iliococcygeus m.

Anococcygeal ligament (posterior extensions of superficial external anal sphincter)

*Parts variable and often indistinct

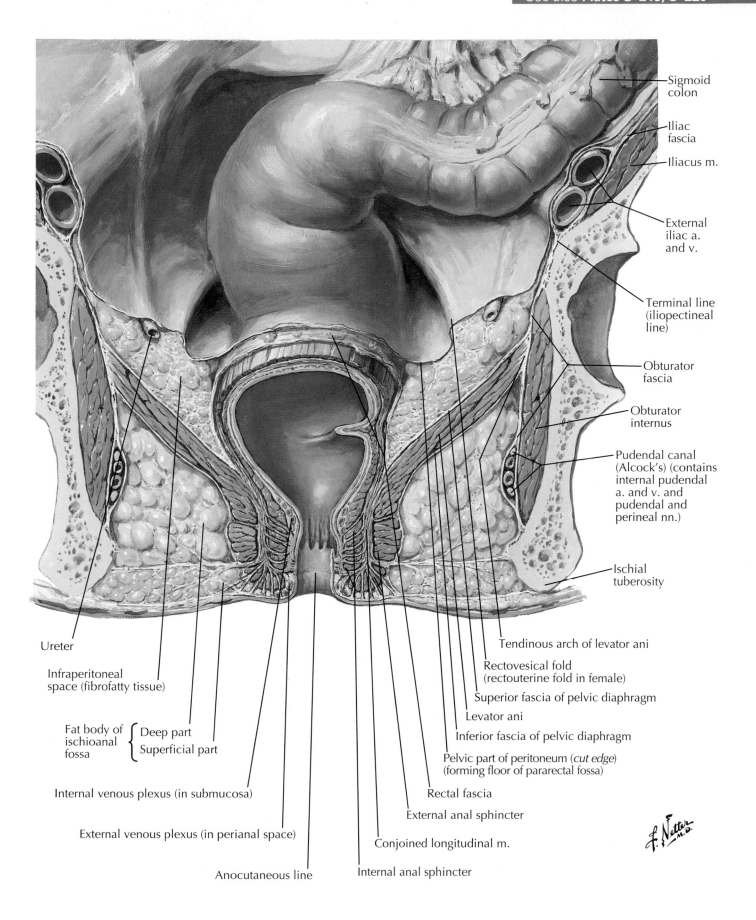

Sigmoid colon

Iliac fascia

Iliacus m.

External iliac a. and v.

Terminal line (iliopectineal line)

Obturator fascia

Obturator internus

Pudendal canal (Alcock's) (contains internal pudendal a. and v. and pudendal and perineal nn.)

Ischial tuberosity

Ureter

Infraperitoneal space (fibrofatty tissue)

Fat body of ischioanal fossa { Deep part / Superficial part

Internal venous plexus (in submucosa)

External venous plexus (in perianal space)

Anocutaneous line

Internal anal sphincter

Conjoined longitudinal m.

External anal sphincter

Rectal fascia

Pelvic part of peritoneum (*cut edge*) (forming floor of pararectal fossa)

Inferior fascia of pelvic diaphragm

Levator ani

Superior fascia of pelvic diaphragm

Rectovesical fold (rectouterine fold in female)

Tendinous arch of levator ani

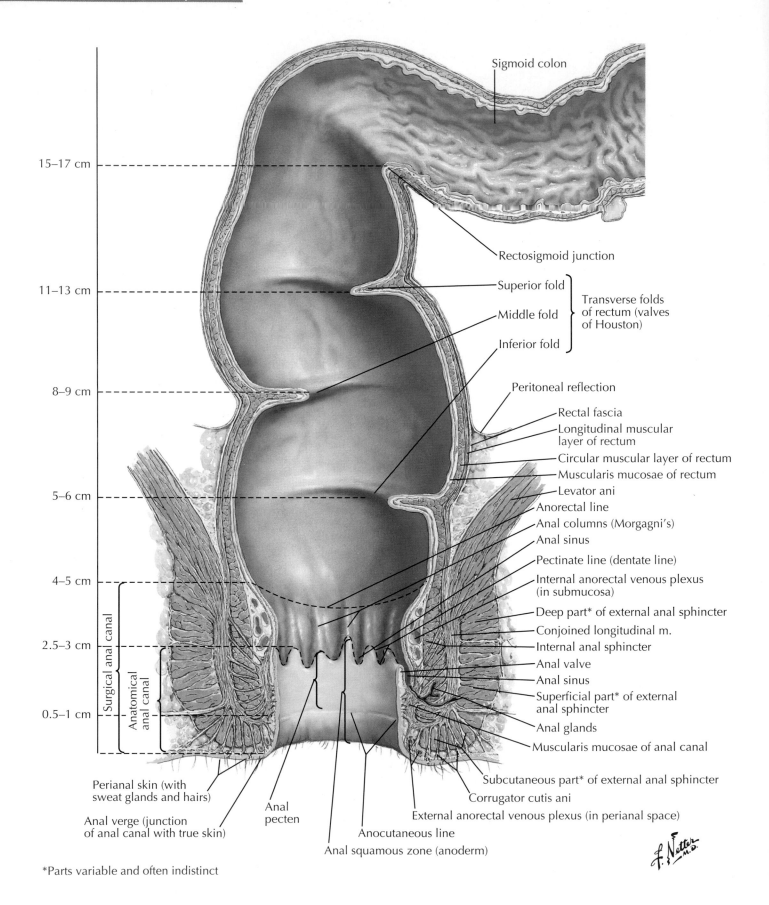

Sigmoid colon

15–17 cm

Rectosigmoid junction

Superior fold

11–13 cm

Middle fold

Transverse folds of rectum (valves of Houston)

Inferior fold

Peritoneal reflection

8–9 cm

Rectal fascia

Longitudinal muscular layer of rectum

Circular muscular layer of rectum

Muscularis mucosae of rectum

Levator ani

5–6 cm

Anorectal line

Anal columns (Morgagni's)

Anal sinus

Pectinate line (dentate line)

Internal anorectal venous plexus (in submucosa)

4–5 cm

Deep part* of external anal sphincter

Conjoined longitudinal m.

Internal anal sphincter

2.5–3 cm

Anal valve

Anal sinus

Superficial part* of external anal sphincter

Surgical anal canal

Anatomical anal canal

Anal glands

0.5–1 cm

Muscularis mucosae of anal canal

Subcutaneous part* of external anal sphincter

Corrugator cutis ani

Perianal skin (with sweat glands and hairs)

Anal pecten

External anorectal venous plexus (in perianal space)

Anal verge (junction of anal canal with true skin)

Anocutaneous line

Anal squamous zone (anoderm)

*Parts variable and often indistinct

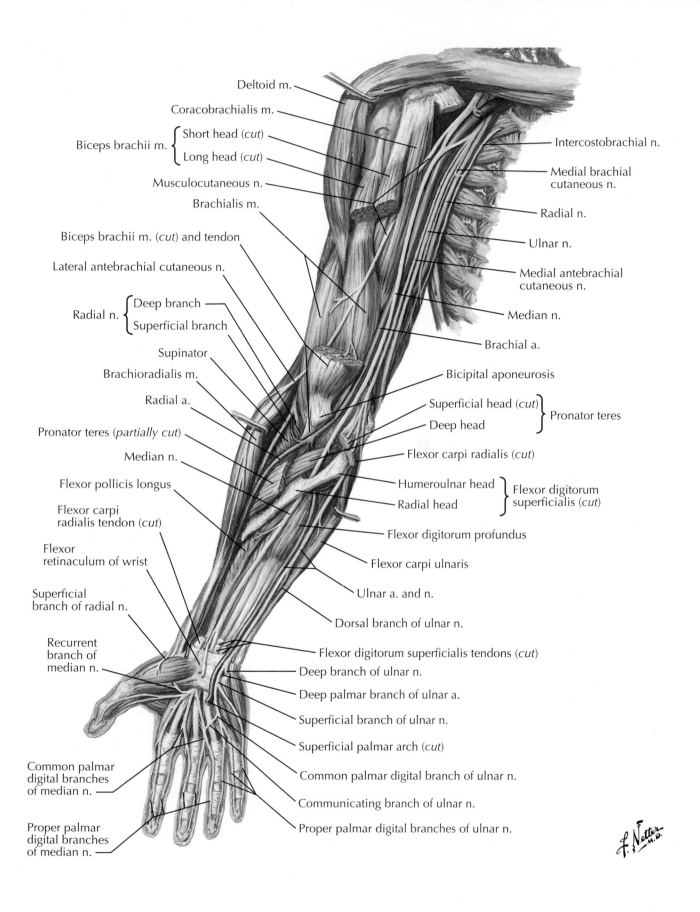

Deltoid m.

Coracobrachialis m.

Biceps brachii m. { Short head (*cut*)
Long head (*cut*)

Musculocutaneous n.

Brachialis m.

Biceps brachii m. (*cut*) and tendon

Lateral antebrachial cutaneous n.

Radial n. { Deep branch
Superficial branch

Supinator

Brachioradialis m.

Radial a.

Pronator teres (*partially cut*)

Median n.

Flexor pollicis longus

Flexor carpi radialis tendon (*cut*)

Flexor retinaculum of wrist

Superficial branch of radial n.

Recurrent branch of median n.

Common palmar digital branches of median n.

Proper palmar digital branches of median n.

Intercostobrachial n.

Medial brachial cutaneous n.

Radial n.

Ulnar n.

Medial antebrachial cutaneous n.

Median n.

Brachial a.

Bicipital aponeurosis

Superficial head (*cut*) } Pronator teres
Deep head

Flexor carpi radialis (*cut*)

Humeroulnar head } Flexor digitorum superficialis (*cut*)
Radial head

Flexor digitorum profundus

Flexor carpi ulnaris

Ulnar a. and n.

Dorsal branch of ulnar n.

Flexor digitorum superficialis tendons (*cut*)

Deep branch of ulnar n.

Deep palmar branch of ulnar a.

Superficial branch of ulnar n.

Superficial palmar arch (*cut*)

Common palmar digital branch of ulnar n.

Communicating branch of ulnar n.

Proper palmar digital branches of ulnar n.

Upper Limb

Anterior view

Posterior view

Medial supraclavicular n.

Intermediate supraclavicular n.

Lateral supraclavicular n.

Acromial tributary of thoracoacromial v.

Medial brachial cutaneous n.

Intercostobrachial n.

Superior lateral brachial cutaneous n.

Cephalic v.

Inferior lateral brachial cutaneous n.

Posterior antebrachial cutaneous n.

Accessory cephalic v.

Branches of medial antebrachial cutaneous n.

Basilic v.

Lateral antebrachial cutaneous n.

Median cubital v.

Median antebrachial v.

Basilic v.

Perforating vv.

Cephalic v.

Lateral supraclavicular n.

Acromial tributary of thoracoacromial v.

Posterior circumflex humeral v.

Intercosto-brachial n.

Medial brachial cutaneous n.

Superior lateral brachial cutaneous n.

Posterior brachial cutaneous n.

Inferior lateral brachial cutaneous n.

Posterior antebrachial cutaneous n.

Branches of medial antebrachial cutaneous n.

Branches of lateral antebrachial cutaneous n.

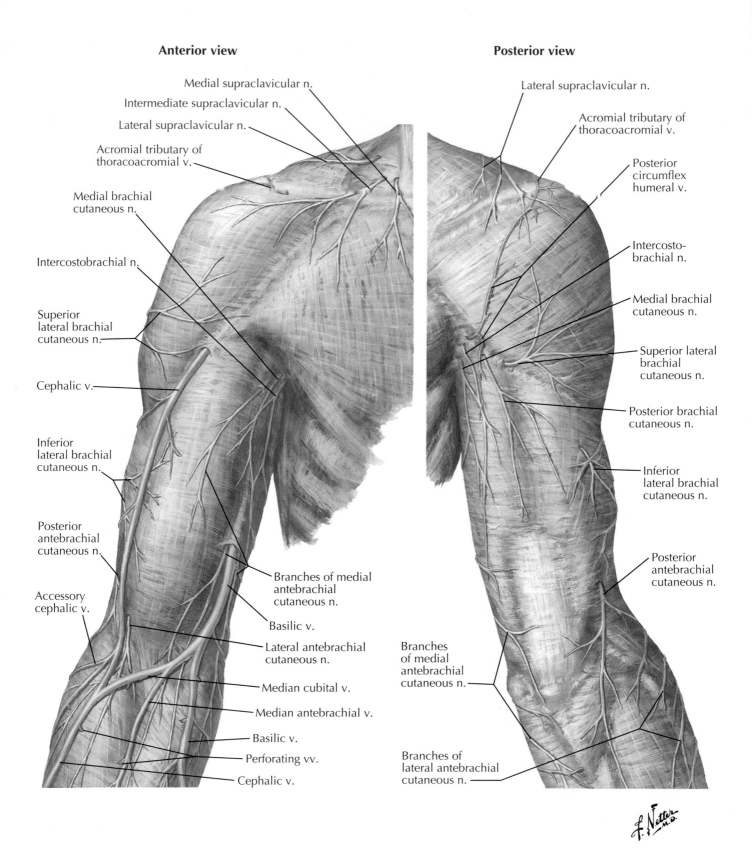

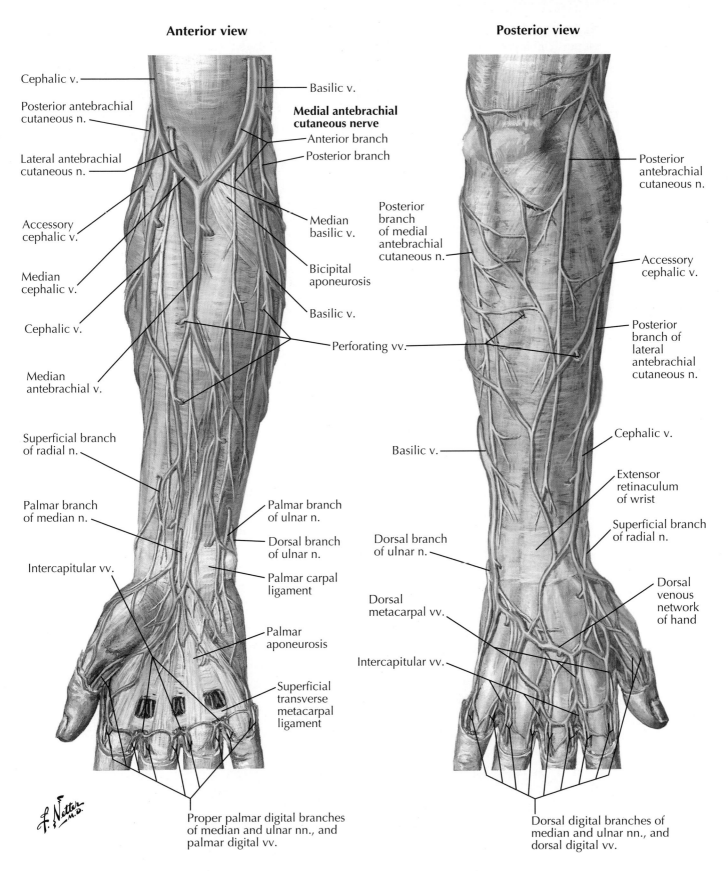

Anterior view

Cephalic v.

Posterior antebrachial cutaneous n.

Lateral antebrachial cutaneous n.

Accessory cephalic v.

Median cephalic v.

Cephalic v.

Median antebrachial v.

Superficial branch of radial n.

Palmar branch of median n.

Intercapitular vv.

Basilic v.

Medial antebrachial cutaneous nerve

Anterior branch

Posterior branch

Median basilic v.

Bicipital aponeurosis

Basilic v.

Perforating vv.

Palmar branch of ulnar n.

Dorsal branch of ulnar n.

Palmar carpal ligament

Palmar aponeurosis

Superficial transverse metacarpal ligament

Proper palmar digital branches of median and ulnar nn., and palmar digital vv.

Posterior view

Posterior branch of medial antebrachial cutaneous n.

Posterior antebrachial cutaneous n.

Accessory cephalic v.

Posterior branch of lateral antebrachial cutaneous n.

Cephalic v.

Extensor retinaculum of wrist

Superficial branch of radial n.

Basilic v.

Dorsal branch of ulnar n.

Dorsal metacarpal vv.

Intercapitular vv.

Dorsal venous network of hand

Dorsal digital branches of median and ulnar nn., and dorsal digital vv.

*Note: In 70% of cases, a median cubital vein (tributary to basilic vein) replaces median cephalic and median basilic veins (see Plate S–228).

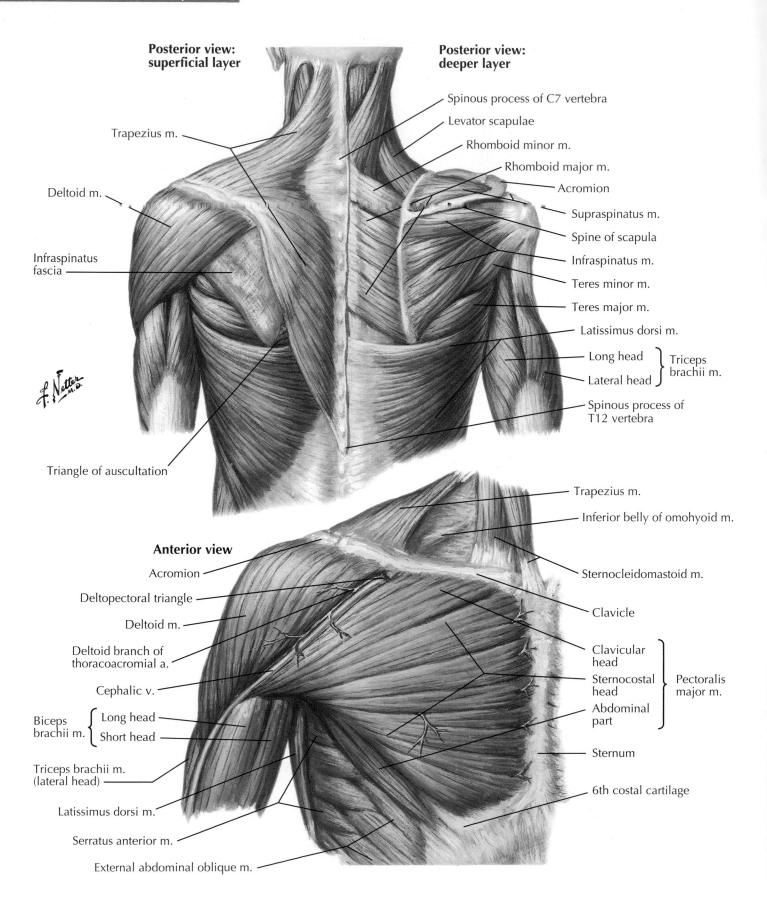

Posterior view: superficial layer

Trapezius m.

Deltoid m.

Infraspinatus fascia

Triangle of auscultation

Posterior view: deeper layer

Spinous process of C7 vertebra

Levator scapulae

Rhomboid minor m.

Rhomboid major m.

Acromion

Supraspinatus m.

Spine of scapula

Infraspinatus m.

Teres minor m.

Teres major m.

Latissimus dorsi m.

Long head
Lateral head } Triceps brachii m.

Spinous process of T12 vertebra

Anterior view

Acromion

Deltopectoral triangle

Deltoid m.

Deltoid branch of thoracoacromial a.

Cephalic v.

Biceps brachii m. { Long head
Short head }

Triceps brachii m. (lateral head)

Latissimus dorsi m.

Serratus anterior m.

External abdominal oblique m.

Trapezius m.

Inferior belly of omohyoid m.

Sternocleidomastoid m.

Clavicle

Clavicular head
Sternocostal head
Abdominal part } Pectoralis major m.

Sternum

6th costal cartilage

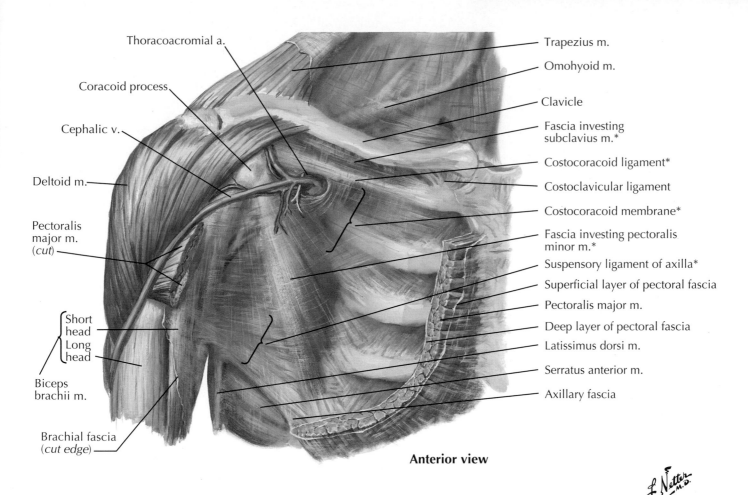

Thoracoacromial a.

Coracoid process

Cephalic v.

Deltoid m.

Pectoralis
major m.
(*cut*)

Short
head
Long
head

Biceps
brachii m.

Brachial fascia
(*cut edge*)

Trapezius m.

Omohyoid m.

Clavicle

Fascia investing
subclavius m.*

Costocoracoid ligament*

Costoclavicular ligament

Costocoracoid membrane*

Fascia investing pectoralis
minor m.*

Suspensory ligament of axilla*

Superficial layer of pectoral fascia

Pectoralis major m.

Deep layer of pectoral fascia

Latissimus dorsi m.

Serratus anterior m.

Axillary fascia

Anterior view

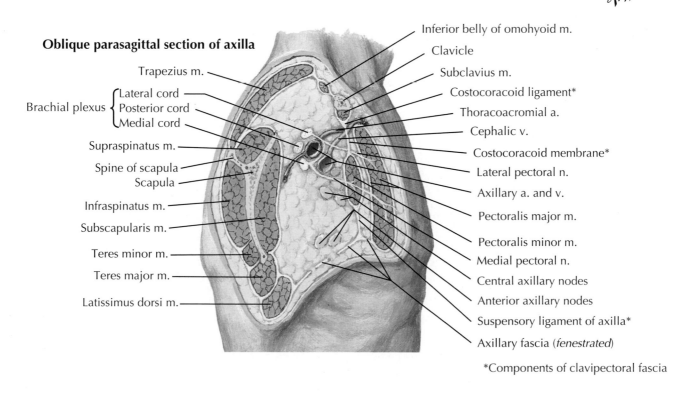

Oblique parasagittal section of axilla

Trapezius m.

Brachial plexus
{ Lateral cord
Posterior cord
Medial cord }

Supraspinatus m.

Spine of scapula

Scapula

Infraspinatus m.

Subscapularis m.

Teres minor m.

Teres major m.

Latissimus dorsi m.

Inferior belly of omohyoid m.

Clavicle

Subclavius m.

Costocoracoid ligament*

Thoracoacromial a.

Cephalic v.

Costocoracoid membrane*

Lateral pectoral n.

Axillary a. and v.

Pectoralis major m.

Pectoralis minor m.

Medial pectoral n.

Central axillary nodes

Anterior axillary nodes

Suspensory ligament of axilla*

Axillary fascia (*fenestrated*)

*Components of clavipectoral fascia

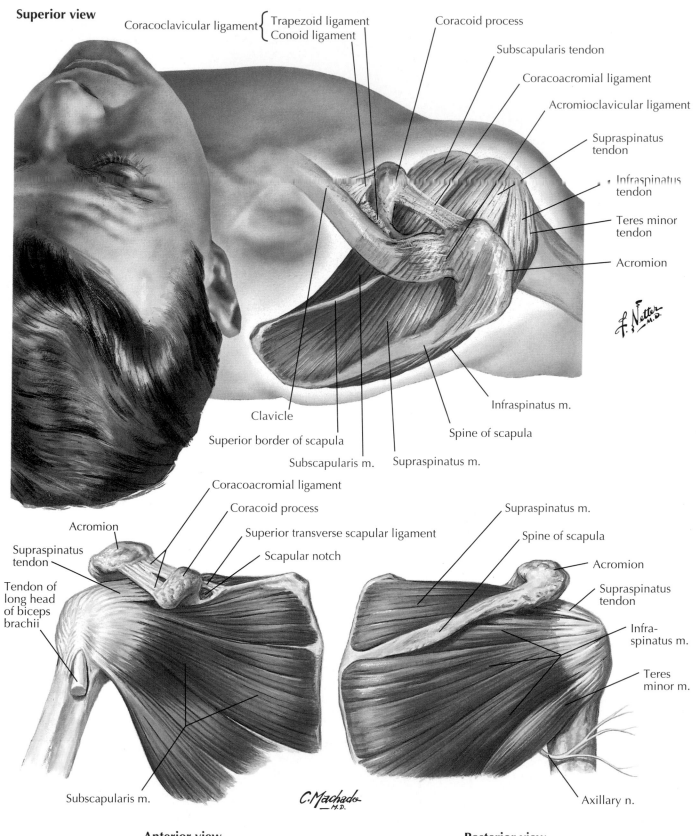

Superior view

Coracoclavicular ligament { Trapezoid ligament
Conoid ligament

Coracoid process

Subscapularis tendon

Coracoacromial ligament

Acromioclavicular ligament

Supraspinatus tendon

Infraspinatus tendon

Teres minor tendon

Acromion

Infraspinatus m.

Spine of scapula

Supraspinatus m.

Subscapularis m.

Superior border of scapula

Clavicle

Coracoacromial ligament

Coracoid process

Acromion

Superior transverse scapular ligament

Scapular notch

Supraspinatus tendon

Tendon of long head of biceps brachii

Subscapularis m.

Anterior view

Supraspinatus m.

Spine of scapula

Acromion

Supraspinatus tendon

Infra-spinatus m.

Teres minor m.

Axillary n.

Posterior view

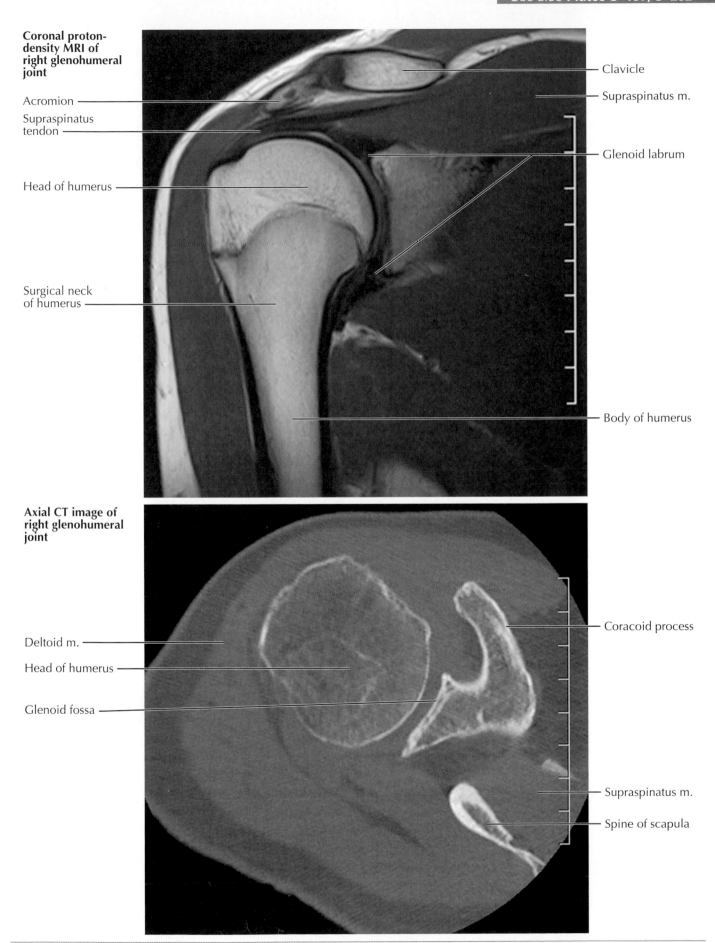

Coronal proton-
density MRI of
right glenohumeral
joint

Acromion

Supraspinatus
tendon

Head of humerus

Surgical neck
of humerus

Clavicle

Supraspinatus m.

Glenoid labrum

Body of humerus

Axial CT image of
right glenohumeral
joint

Deltoid m.

Head of humerus

Glenoid fossa

Coracoid process

Supraspinatus m.

Spine of scapula

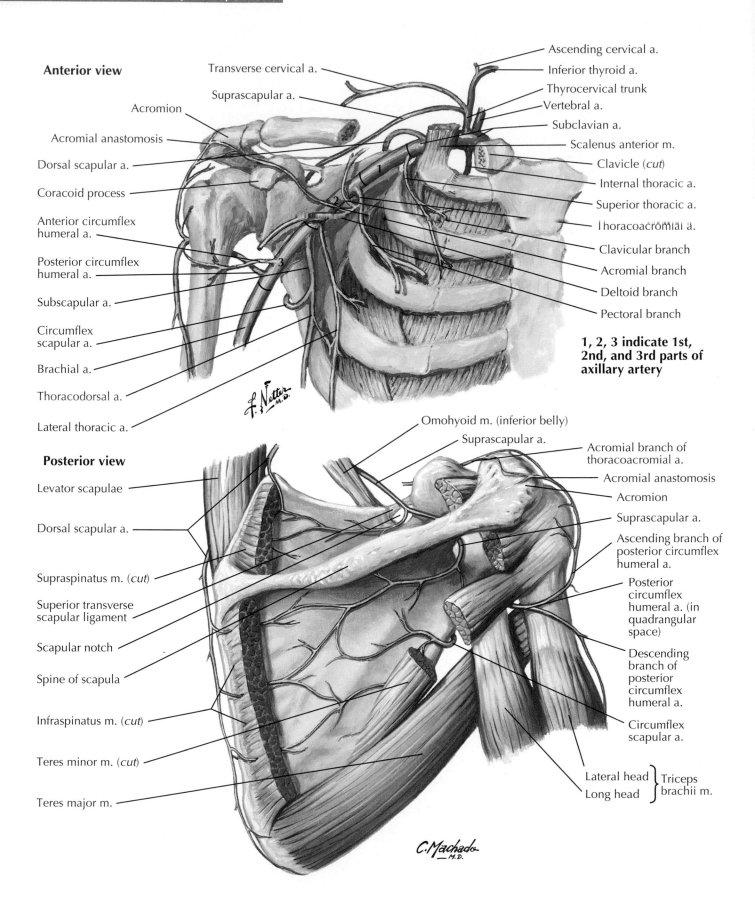

Anterior view

Acromion

Acromial anastomosis

Dorsal scapular a.

Coracoid process

Anterior circumflex humeral a.

Posterior circumflex humeral a.

Subscapular a.

Circumflex scapular a.

Brachial a.

Thoracodorsal a.

Lateral thoracic a.

Transverse cervical a.

Suprascapular a.

Ascending cervical a.

Inferior thyroid a.

Thyrocervical trunk

Vertebral a.

Subclavian a.

Scalenus anterior m.

Clavicle (*cut*)

Internal thoracic a.

Superior thoracic a.

Thoracoacromial a.

Clavicular branch

Acromial branch

Deltoid branch

Pectoral branch

1, 2, 3 indicate 1st, 2nd, and 3rd parts of axillary artery

Posterior view

Levator scapulae

Dorsal scapular a.

Supraspinatus m. (*cut*)

Superior transverse scapular ligament

Scapular notch

Spine of scapula

Infraspinatus m. (*cut*)

Teres minor m. (*cut*)

Teres major m.

Omohyoid m. (inferior belly)

Suprascapular a.

Acromial branch of thoracoacromial a.

Acromial anastomosis

Acromion

Suprascapular a.

Ascending branch of posterior circumflex humeral a.

Posterior circumflex humeral a. (in quadrangular space)

Descending branch of posterior circumflex humeral a.

Circumflex scapular a.

Lateral head } Triceps
Long head } brachii m.

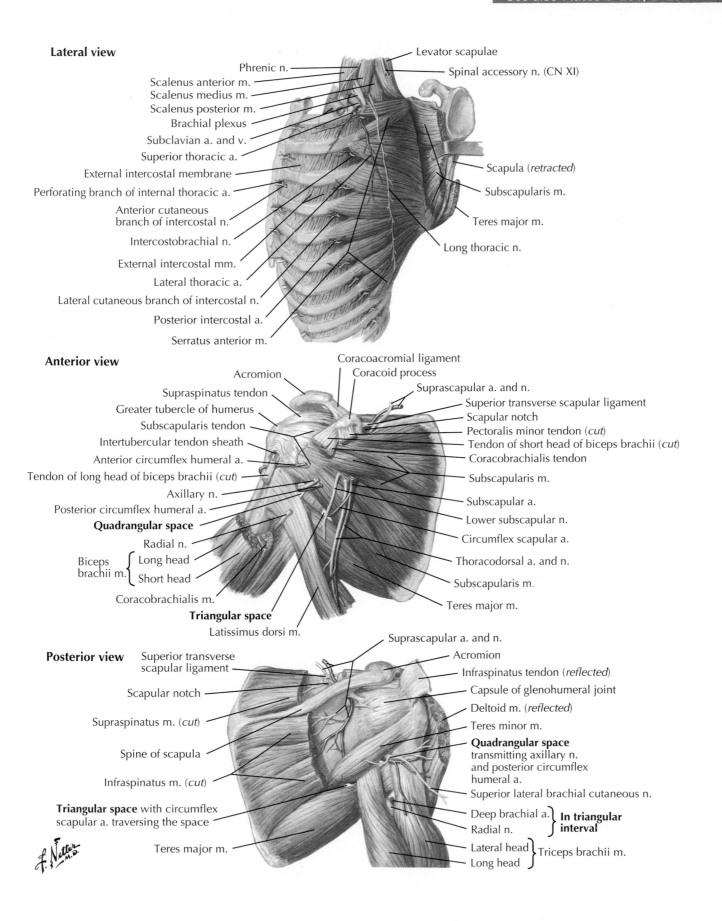

Lateral view

Levator scapulae
Phrenic n.
Spinal accessory n. (CN XI)
Scalenus anterior m.
Scalenus medius m.
Scalenus posterior m.
Brachial plexus
Subclavian a. and v.
Superior thoracic a.
Scapula (*retracted*)
External intercostal membrane
Subscapularis m.
Perforating branch of internal thoracic a.
Anterior cutaneous branch of intercostal n.
Teres major m.
Intercostobrachial n.
Long thoracic n.
External intercostal mm.
Lateral thoracic a.
Lateral cutaneous branch of intercostal n.
Posterior intercostal a.
Serratus anterior m.

Anterior view

Coracoacromial ligament
Coracoid process
Acromion
Suprascapular a. and n.
Supraspinatus tendon
Superior transverse scapular ligament
Greater tubercle of humerus
Scapular notch
Subscapularis tendon
Pectoralis minor tendon (*cut*)
Intertubercular tendon sheath
Tendon of short head of biceps brachii (*cut*)
Anterior circumflex humeral a.
Coracobrachialis tendon
Tendon of long head of biceps brachii (*cut*)
Subscapularis m.
Axillary n.
Subscapular a.
Posterior circumflex humeral a.
Lower subscapular n.
Quadrangular space
Circumflex scapular a.
Radial n.
Thoracodorsal a. and n.
Biceps brachii m. { Long head
Short head
Subscapularis m.
Coracobrachialis m.
Teres major m.
Triangular space
Latissimus dorsi m.

Posterior view

Suprascapular a. and n.
Superior transverse scapular ligament
Acromion
Scapular notch
Infraspinatus tendon (*reflected*)
Capsule of glenohumeral joint
Supraspinatus m. (*cut*)
Deltoid m. (*reflected*)
Spine of scapula
Teres minor m.
Quadrangular space transmitting axillary n. and posterior circumflex humeral a.
Infraspinatus m. (*cut*)
Superior lateral brachial cutaneous n.
Triangular space with circumflex scapular a. traversing the space
Deep brachial a. } **In triangular interval**
Radial n.
Teres major m.
Lateral head } Triceps brachii m.
Long head

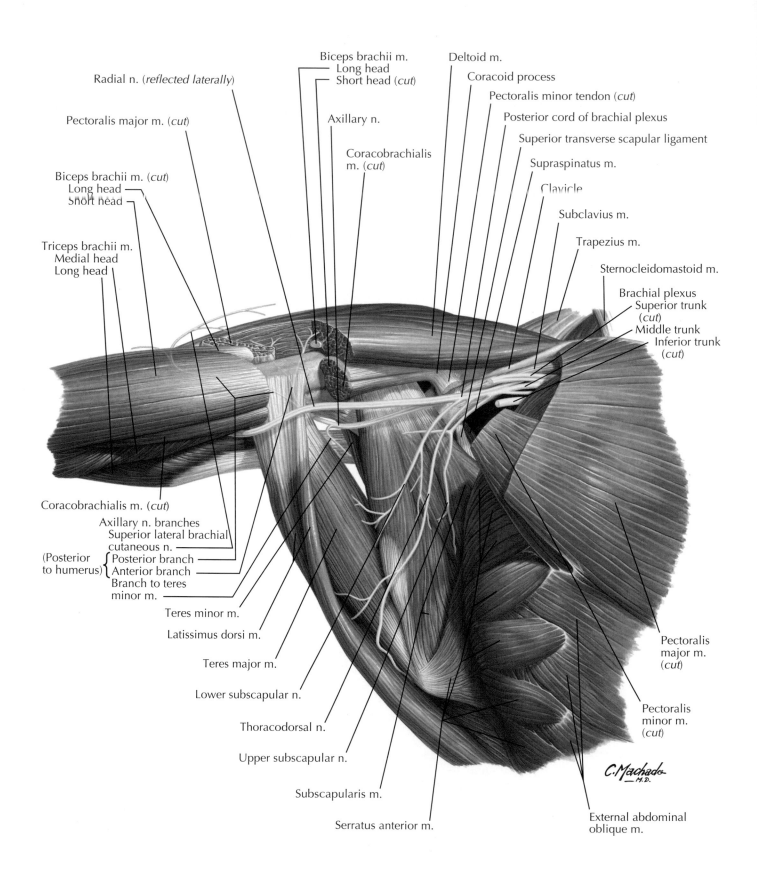

Radial n. (*reflected laterally*)

Pectoralis major m. (*cut*)

Biceps brachii m. (*cut*)
Long head
Short head

Triceps brachii m.
Medial head
Long head

Coracobrachialis m. (*cut*)

Axillary n. branches
Superior lateral brachial
cutaneous n.
(Posterior {Posterior branch
to humerus){Anterior branch
Branch to teres
minor m.

Teres minor m.

Latissimus dorsi m.

Teres major m.

Lower subscapular n.

Thoracodorsal n.

Upper subscapular n.

Subscapularis m.

Serratus anterior m.

Biceps brachii m.
Long head
Short head (*cut*)

Axillary n.

Coracobrachialis m. (*cut*)

Deltoid m.

Coracoid process

Pectoralis minor tendon (*cut*)

Posterior cord of brachial plexus

Superior transverse scapular ligament

Supraspinatus m.

Clavicle

Subclavius m.

Trapezius m.

Sternocleidomastoid m.

Brachial plexus
Superior trunk (*cut*)
Middle trunk
Inferior trunk (*cut*)

Pectoralis major m. (*cut*)

Pectoralis minor m. (*cut*)

External abdominal oblique m.

C. Machado
M.D.

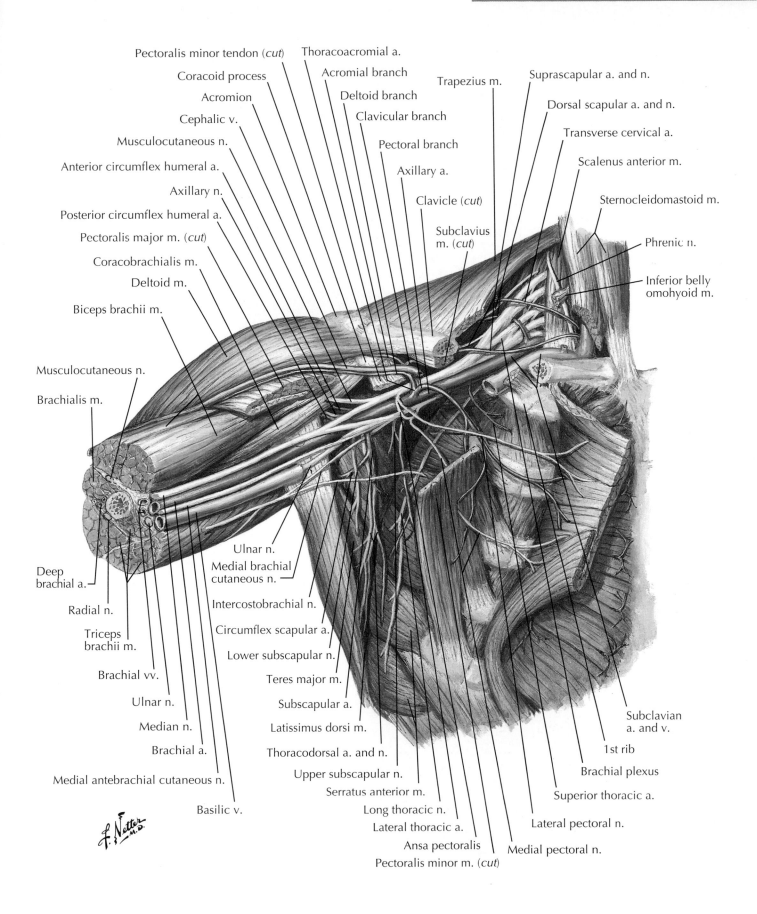

Pectoralis minor tendon (*cut*)
Coracoid process
Acromion
Cephalic v.
Musculocutaneous n.
Anterior circumflex humeral a.
Axillary n.
Posterior circumflex humeral a.
Pectoralis major m. (*cut*)
Coracobrachialis m.
Deltoid m.
Biceps brachii m.

Thoracoacromial a.
Acromial branch
Deltoid branch
Clavicular branch
Pectoral branch
Axillary a.
Clavicle (*cut*)
Subclavius m. (*cut*)

Trapezius m.
Suprascapular a. and n.
Dorsal scapular a. and n.
Transverse cervical a.
Scalenus anterior m.
Sternocleidomastoid m.
Phrenic n.
Inferior belly omohyoid m.

Musculocutaneous n.
Brachialis m.

Deep brachial a.
Radial n.
Triceps brachii m.
Brachial vv.
Ulnar n.
Median n.
Brachial a.
Medial antebrachial cutaneous n.
Basilic v.

Ulnar n.
Medial brachial cutaneous n.
Intercostobrachial n.
Circumflex scapular a.
Lower subscapular n.
Teres major m.
Subscapular a.
Latissimus dorsi m.
Thoracodorsal a. and n.
Upper subscapular n.
Serratus anterior m.
Long thoracic n.
Lateral thoracic a.
Ansa pectoralis
Pectoralis minor m. (*cut*)

Subclavian a. and v.
1st rib
Brachial plexus
Superior thoracic a.
Lateral pectoral n.
Medial pectoral n.

f. Netter M.D.

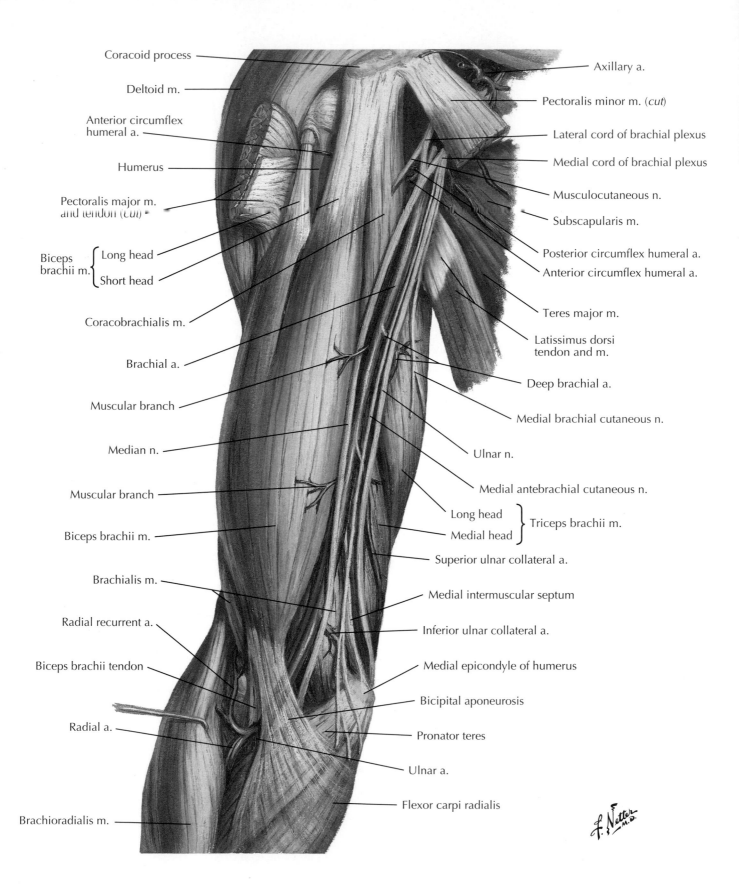

Coracoid process

Deltoid m.

Anterior circumflex humeral a.

Humerus

Pectoralis major m. and tendon (cut)

Biceps brachii m. { Long head / Short head }

Coracobrachialis m.

Brachial a.

Muscular branch

Median n.

Muscular branch

Biceps brachii m.

Brachialis m.

Radial recurrent a.

Biceps brachii tendon

Radial a.

Brachioradialis m.

Axillary a.

Pectoralis minor m. (cut)

Lateral cord of brachial plexus

Medial cord of brachial plexus

Musculocutaneous n.

Subscapularis m.

Posterior circumflex humeral a.

Anterior circumflex humeral a.

Teres major m.

Latissimus dorsi tendon and m.

Deep brachial a.

Medial brachial cutaneous n.

Ulnar n.

Medial antebrachial cutaneous n.

Long head / Medial head } Triceps brachii m.

Superior ulnar collateral a.

Medial intermuscular septum

Inferior ulnar collateral a.

Medial epicondyle of humerus

Bicipital aponeurosis

Pronator teres

Ulnar a.

Flexor carpi radialis

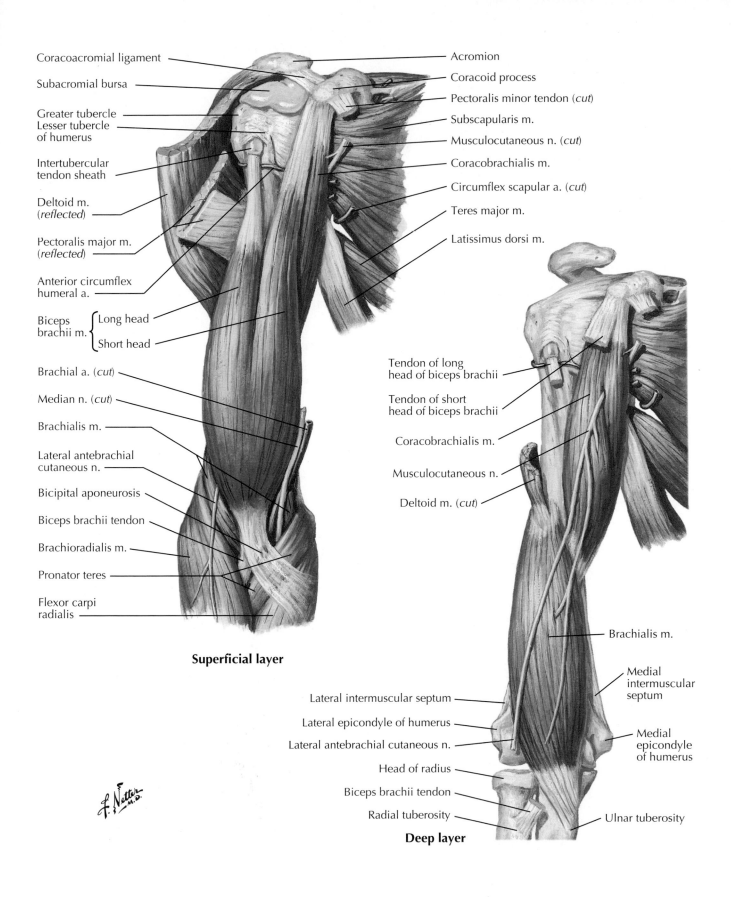

Coracoacromial ligament

Subacromial bursa

Greater tubercle
Lesser tubercle
of humerus

Intertubercular
tendon sheath

Deltoid m.
(*reflected*)

Pectoralis major m.
(*reflected*)

Anterior circumflex
humeral a.

Biceps
brachii m. { Long head
             Short head

Brachial a. (*cut*)

Median n. (*cut*)

Brachialis m.

Lateral antebrachial
cutaneous n.

Bicipital aponeurosis

Biceps brachii tendon

Brachioradialis m.

Pronator teres

Flexor carpi
radialis

Acromion

Coracoid process

Pectoralis minor tendon (*cut*)

Subscapularis m.

Musculocutaneous n. (*cut*)

Coracobrachialis m.

Circumflex scapular a. (*cut*)

Teres major m.

Latissimus dorsi m.

Tendon of long
head of biceps brachii

Tendon of short
head of biceps brachii

Coracobrachialis m.

Musculocutaneous n.

Deltoid m. (*cut*)

Superficial layer

Lateral intermuscular septum

Lateral epicondyle of humerus

Lateral antebrachial cutaneous n.

Head of radius

Biceps brachii tendon

Radial tuberosity

Brachialis m.

Medial
intermuscular
septum

Medial
epicondyle
of humerus

Ulnar tuberosity

Deep layer

Upper Limb

Plate S–239

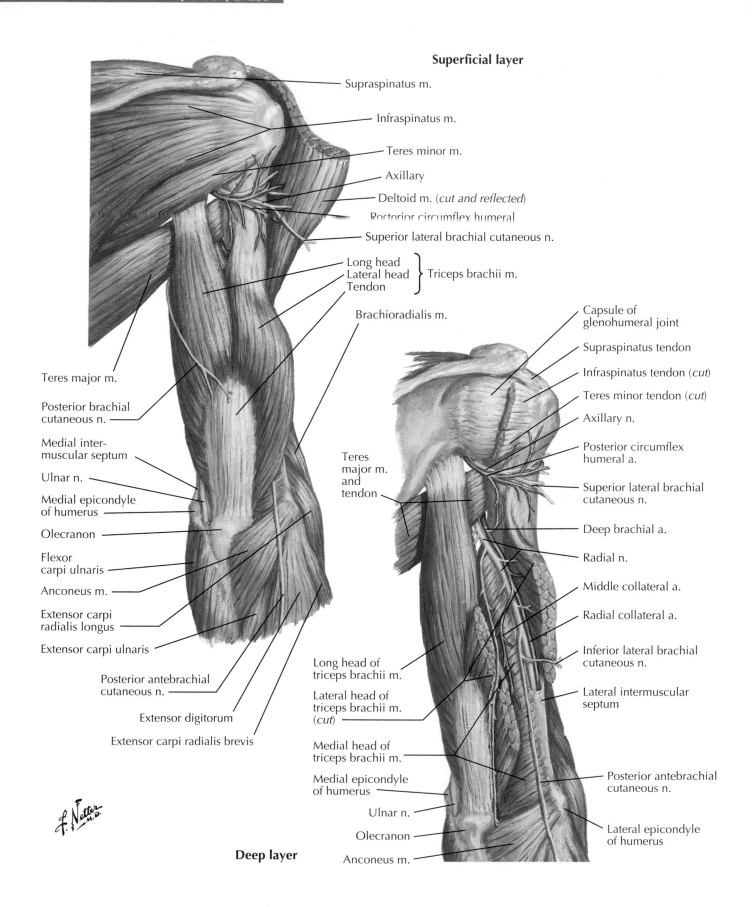

Superficial layer

Supraspinatus m.

Infraspinatus m.

Teres minor m.

Axillary

Deltoid m. (*cut and reflected*)

Posterior circumflex humeral

Superior lateral brachial cutaneous n.

Long head
Lateral head } Triceps brachii m.
Tendon

Brachioradialis m.

Capsule of glenohumeral joint

Supraspinatus tendon

Infraspinatus tendon (*cut*)

Teres minor tendon (*cut*)

Axillary n.

Posterior circumflex humeral a.

Superior lateral brachial cutaneous n.

Teres major m.

Posterior brachial cutaneous n.

Medial inter-muscular septum

Ulnar n.

Medial epicondyle of humerus

Olecranon

Flexor carpi ulnaris

Anconeus m.

Extensor carpi radialis longus

Extensor carpi ulnaris

Posterior antebrachial cutaneous n.

Extensor digitorum

Extensor carpi radialis brevis

Teres major m. and tendon

Deep brachial a.

Radial n.

Middle collateral a.

Radial collateral a.

Inferior lateral brachial cutaneous n.

Lateral intermuscular septum

Long head of triceps brachii m.

Lateral head of triceps brachii m. (*cut*)

Medial head of triceps brachii m.

Medial epicondyle of humerus

Ulnar n.

Olecranon

Anconeus m.

Posterior antebrachial cutaneous n.

Lateral epicondyle of humerus

Deep layer

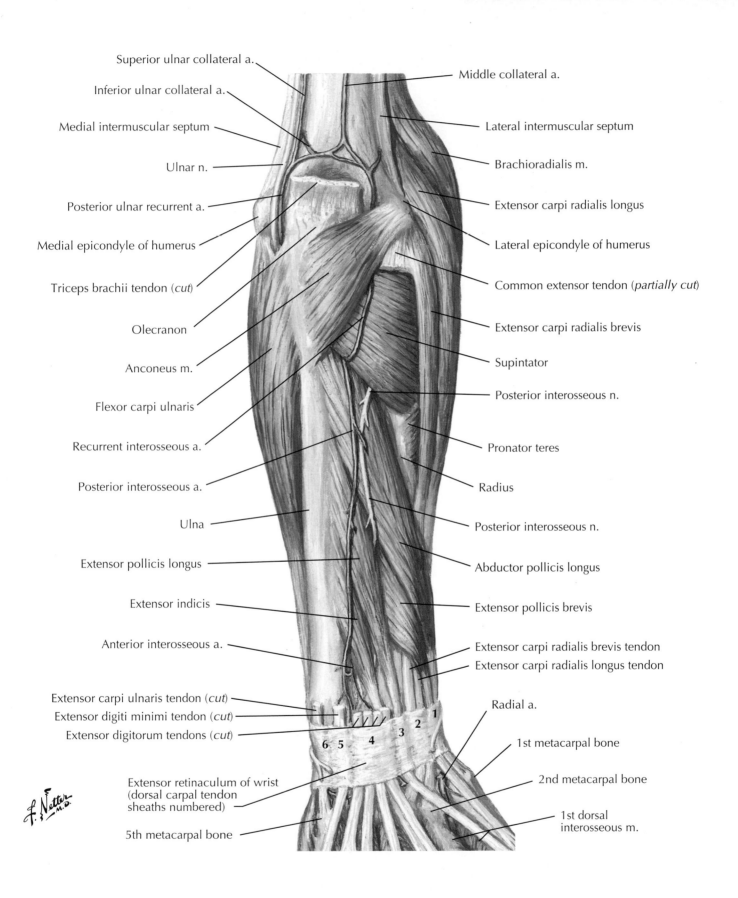

Superior ulnar collateral a.

Inferior ulnar collateral a.

Medial intermuscular septum

Ulnar n.

Posterior ulnar recurrent a.

Medial epicondyle of humerus

Triceps brachii tendon (*cut*)

Olecranon

Anconeus m.

Flexor carpi ulnaris

Recurrent interosseous a.

Posterior interosseous a.

Ulna

Extensor pollicis longus

Extensor indicis

Anterior interosseous a.

Extensor carpi ulnaris tendon (*cut*)

Extensor digiti minimi tendon (*cut*)

Extensor digitorum tendons (*cut*)

Extensor retinaculum of wrist
(dorsal carpal tendon
sheaths numbered)

5th metacarpal bone

Middle collateral a.

Lateral intermuscular septum

Brachioradialis m.

Extensor carpi radialis longus

Lateral epicondyle of humerus

Common extensor tendon (*partially cut*)

Extensor carpi radialis brevis

Supinator

Posterior interosseous n.

Pronator teres

Radius

Posterior interosseous n.

Abductor pollicis longus

Extensor pollicis brevis

Extensor carpi radialis brevis tendon

Extensor carpi radialis longus tendon

Radial a.

1st metacarpal bone

2nd metacarpal bone

1st dorsal
interosseous m.

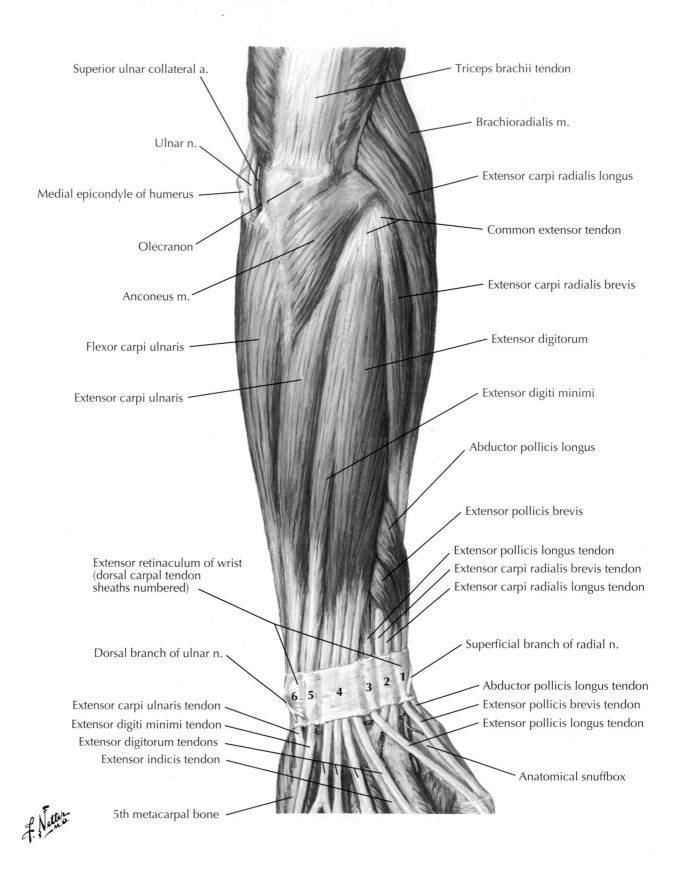

Superior ulnar collateral a.

Ulnar n.

Medial epicondyle of humerus

Olecranon

Anconeus m.

Flexor carpi ulnaris

Extensor carpi ulnaris

Extensor retinaculum of wrist (dorsal carpal tendon sheaths numbered)

Dorsal branch of ulnar n.

Extensor carpi ulnaris tendon

Extensor digiti minimi tendon

Extensor digitorum tendons

Extensor indicis tendon

5th metacarpal bone

Triceps brachii tendon

Brachioradialis m.

Extensor carpi radialis longus

Common extensor tendon

Extensor carpi radialis brevis

Extensor digitorum

Extensor digiti minimi

Abductor pollicis longus

Extensor pollicis brevis

Extensor pollicis longus tendon

Extensor carpi radialis brevis tendon

Extensor carpi radialis longus tendon

Superficial branch of radial n.

Abductor pollicis longus tendon

Extensor pollicis brevis tendon

Extensor pollicis longus tendon

Anatomical snuffbox

6 5   4   3 2 1

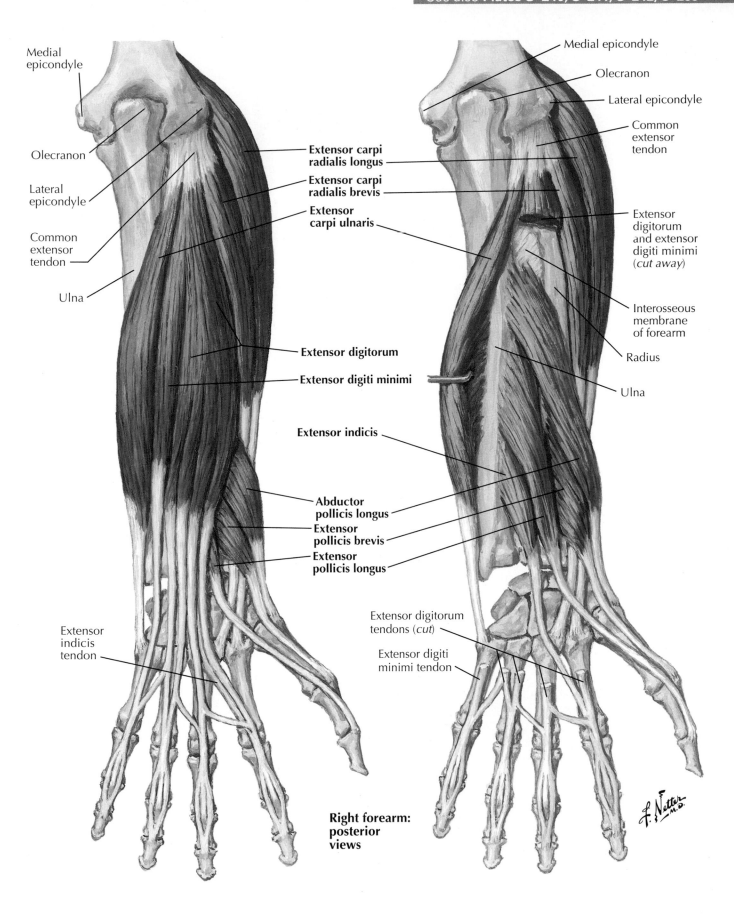

Medial epicondyle

Olecranon

Lateral epicondyle

Common extensor tendon

Ulna

Extensor carpi radialis longus

Extensor carpi radialis brevis

Extensor carpi ulnaris

Extensor digitorum

Extensor digiti minimi

Extensor indicis

Abductor pollicis longus

Extensor pollicis brevis

Extensor pollicis longus

Extensor indicis tendon

Medial epicondyle

Olecranon

Lateral epicondyle

Common extensor tendon

Extensor digitorum and extensor digiti minimi (*cut away*)

Interosseous membrane of forearm

Radius

Ulna

Extensor digitorum tendons (*cut*)

Extensor digiti minimi tendon

Right forearm: posterior views

F. Netter M.D.

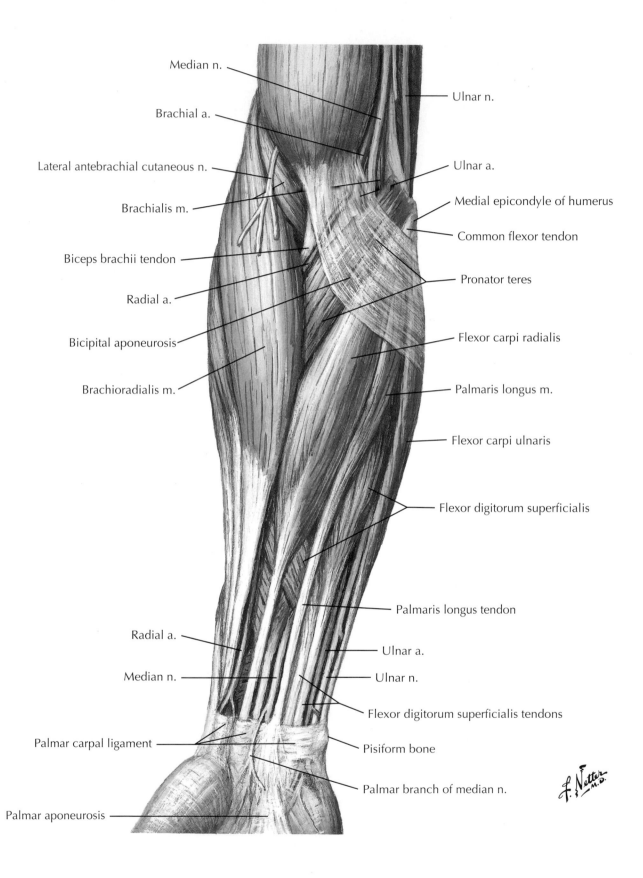

Median n.

Brachial a.

Lateral antebrachial cutaneous n.

Brachialis m.

Biceps brachii tendon

Radial a.

Bicipital aponeurosis

Brachioradialis m.

Radial a.

Median n.

Palmar carpal ligament

Palmar aponeurosis

Ulnar n.

Ulnar a.

Medial epicondyle of humerus

Common flexor tendon

Pronator teres

Flexor carpi radialis

Palmaris longus m.

Flexor carpi ulnaris

Flexor digitorum superficialis

Palmaris longus tendon

Ulnar a.

Ulnar n.

Flexor digitorum superficialis tendons

Pisiform bone

Palmar branch of median n.

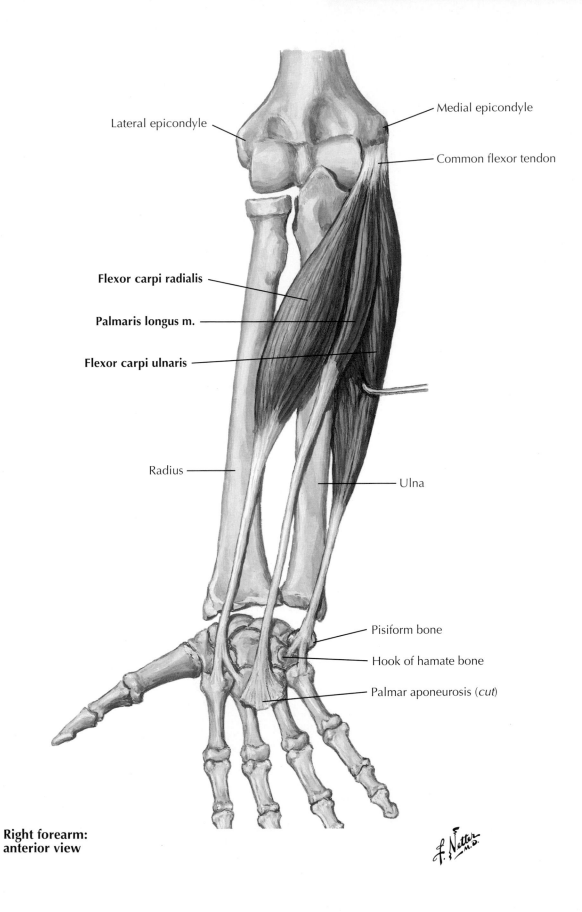

Lateral epicondyle

Medial epicondyle

Common flexor tendon

Flexor carpi radialis

Palmaris longus m.

Flexor carpi ulnaris

Radius

Ulna

Pisiform bone

Hook of hamate bone

Palmar aponeurosis (*cut*)

**Right forearm:
anterior view**

f. Netter M.D.

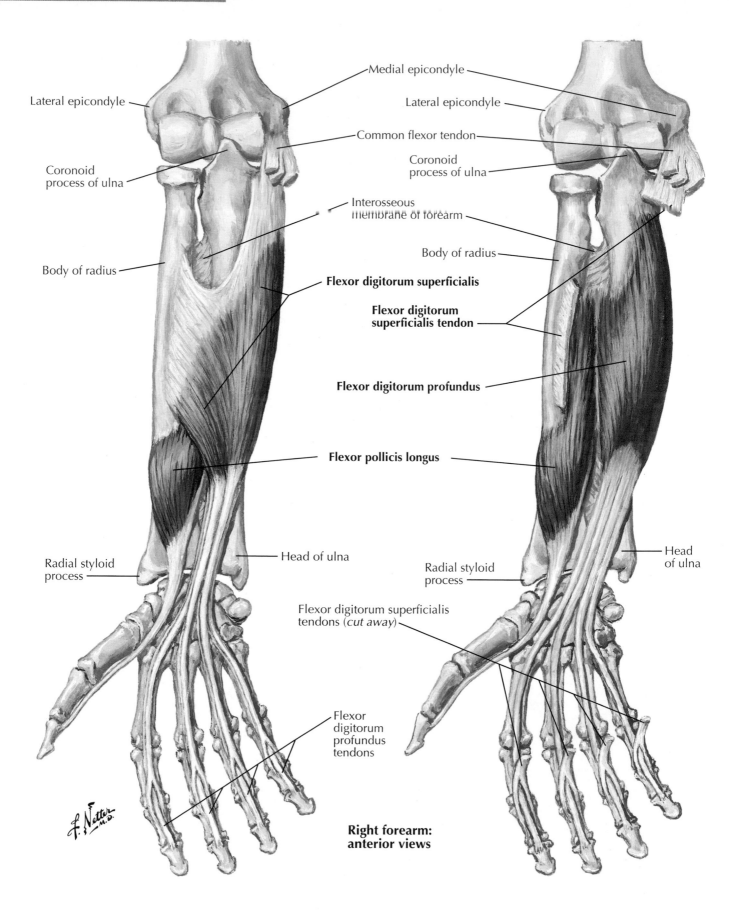

Lateral epicondyle

Medial epicondyle

Lateral epicondyle

Coronoid process of ulna

Common flexor tendon

Coronoid process of ulna

Interosseous membrane of forearm

Body of radius

Body of radius

Flexor digitorum superficialis

Flexor digitorum superficialis tendon

Flexor digitorum profundus

Flexor pollicis longus

Head of ulna

Head of ulna

Radial styloid process

Radial styloid process

Flexor digitorum superficialis tendons (*cut away*)

Flexor digitorum profundus tendons

Right forearm: anterior views

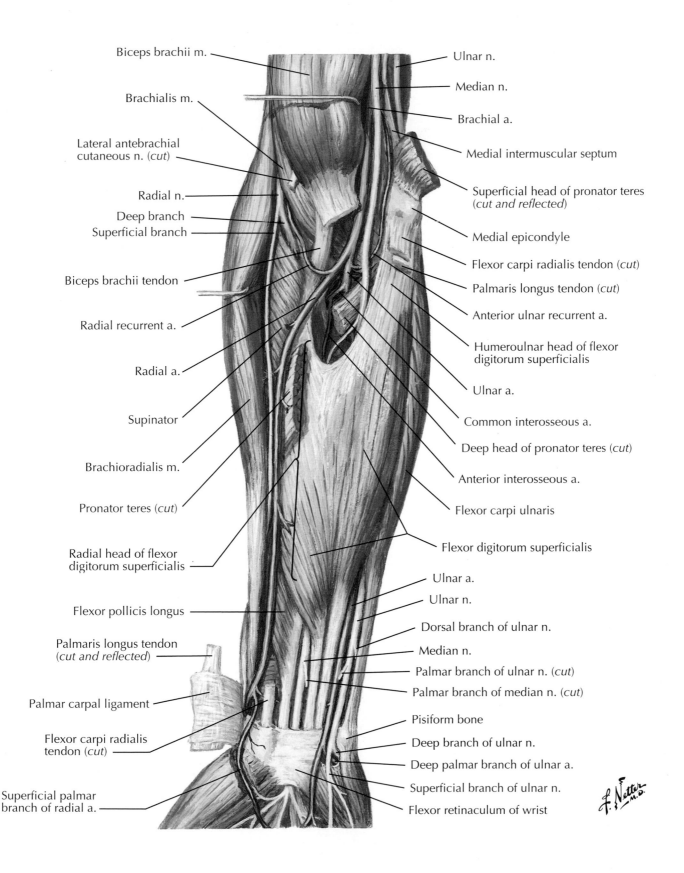

Biceps brachii m.

Brachialis m.

Lateral antebrachial cutaneous n. (cut)

Radial n.

Deep branch

Superficial branch

Biceps brachii tendon

Radial recurrent a.

Radial a.

Supinator

Brachioradialis m.

Pronator teres (cut)

Radial head of flexor digitorum superficialis

Flexor pollicis longus

Palmaris longus tendon (cut and reflected)

Palmar carpal ligament

Flexor carpi radialis tendon (cut)

Superficial palmar branch of radial a.

Ulnar n.

Median n.

Brachial a.

Medial intermuscular septum

Superficial head of pronator teres (cut and reflected)

Medial epicondyle

Flexor carpi radialis tendon (cut)

Palmaris longus tendon (cut)

Anterior ulnar recurrent a.

Humeroulnar head of flexor digitorum superficialis

Ulnar a.

Common interosseous a.

Deep head of pronator teres (cut)

Anterior interosseous a.

Flexor carpi ulnaris

Flexor digitorum superficialis

Ulnar a.

Ulnar n.

Dorsal branch of ulnar n.

Median n.

Palmar branch of ulnar n. (cut)

Palmar branch of median n. (cut)

Pisiform bone

Deep branch of ulnar n.

Deep palmar branch of ulnar a.

Superficial branch of ulnar n.

Flexor retinaculum of wrist

f. Netter M.D.

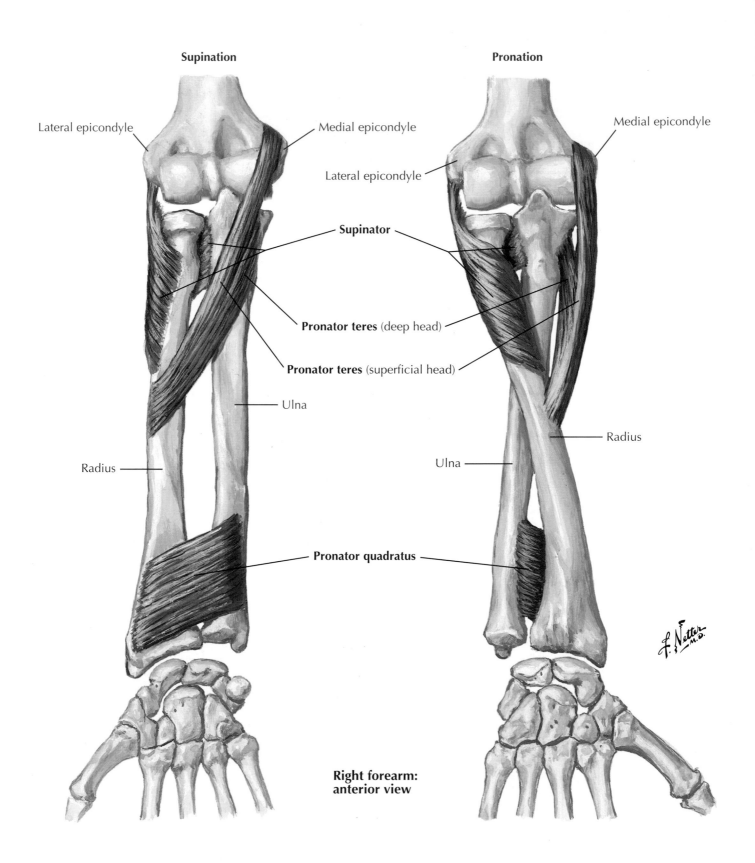

Supination

Pronation

Lateral epicondyle

Medial epicondyle

Medial epicondyle

Lateral epicondyle

Supinator

Pronator teres (deep head)

Pronator teres (superficial head)

Ulna

Radius

Ulna

Radius

Pronator quadratus

**Right forearm:
anterior view**

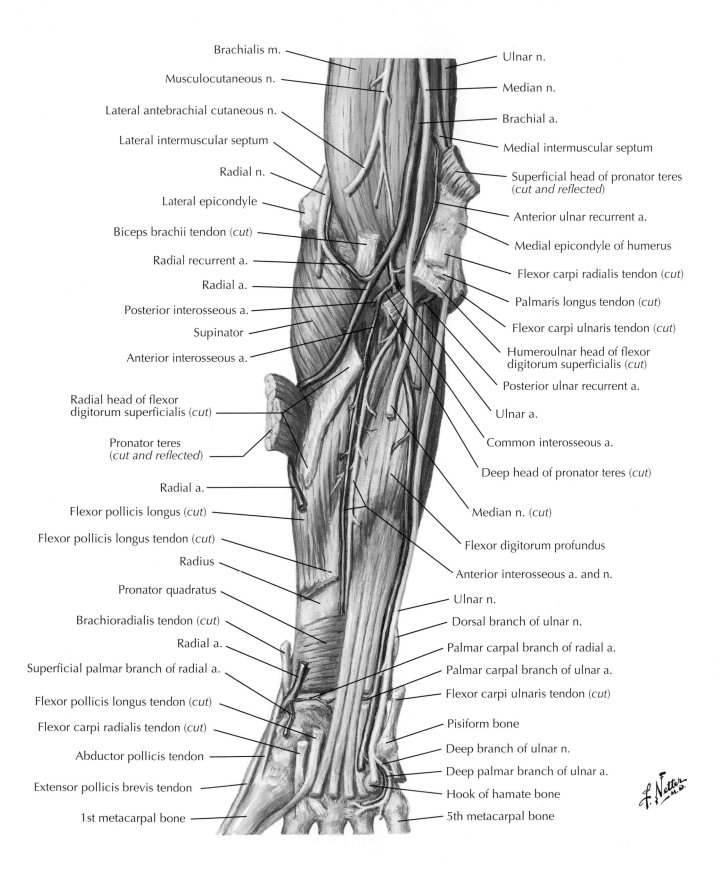

Brachialis m.

Musculocutaneous n.

Lateral antebrachial cutaneous n.

Lateral intermuscular septum

Radial n.

Lateral epicondyle

Biceps brachii tendon (*cut*)

Radial recurrent a.

Radial a.

Posterior interosseous a.

Supinator

Anterior interosseous a.

Radial head of flexor digitorum superficialis (*cut*)

Pronator teres (*cut and reflected*)

Radial a.

Flexor pollicis longus (*cut*)

Flexor pollicis longus tendon (*cut*)

Radius

Pronator quadratus

Brachioradialis tendon (*cut*)

Radial a.

Superficial palmar branch of radial a.

Flexor pollicis longus tendon (*cut*)

Flexor carpi radialis tendon (*cut*)

Abductor pollicis tendon

Extensor pollicis brevis tendon

1st metacarpal bone

Ulnar n.

Median n.

Brachial a.

Medial intermuscular septum

Superficial head of pronator teres (*cut and reflected*)

Anterior ulnar recurrent a.

Medial epicondyle of humerus

Flexor carpi radialis tendon (*cut*)

Palmaris longus tendon (*cut*)

Flexor carpi ulnaris tendon (*cut*)

Humeroulnar head of flexor digitorum superficialis (*cut*)

Posterior ulnar recurrent a.

Ulnar a.

Common interosseous a.

Deep head of pronator teres (*cut*)

Median n. (*cut*)

Flexor digitorum profundus

Anterior interosseous a. and n.

Ulnar n.

Dorsal branch of ulnar n.

Palmar carpal branch of radial a.

Palmar carpal branch of ulnar a.

Flexor carpi ulnaris tendon (*cut*)

Pisiform bone

Deep branch of ulnar n.

Deep palmar branch of ulnar a.

Hook of hamate bone

5th metacarpal bone

Anterior views

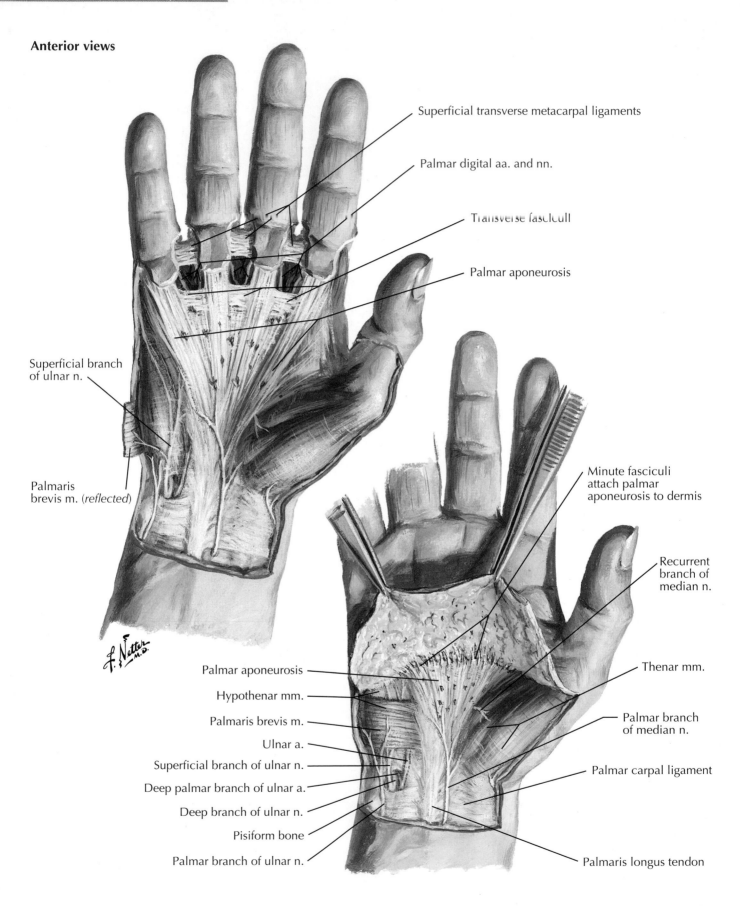

Superficial transverse metacarpal ligaments

Palmar digital aa. and nn.

Transverse fasciculi

Palmar aponeurosis

Superficial branch of ulnar n.

Palmaris brevis m. (*reflected*)

Minute fasciculi attach palmar aponeurosis to dermis

Recurrent branch of median n.

Palmar aponeurosis

Hypothenar mm.

Palmaris brevis m.

Ulnar a.

Superficial branch of ulnar n.

Deep palmar branch of ulnar a.

Deep branch of ulnar n.

Pisiform bone

Palmar branch of ulnar n.

Thenar mm.

Palmar branch of median n.

Palmar carpal ligament

Palmaris longus tendon

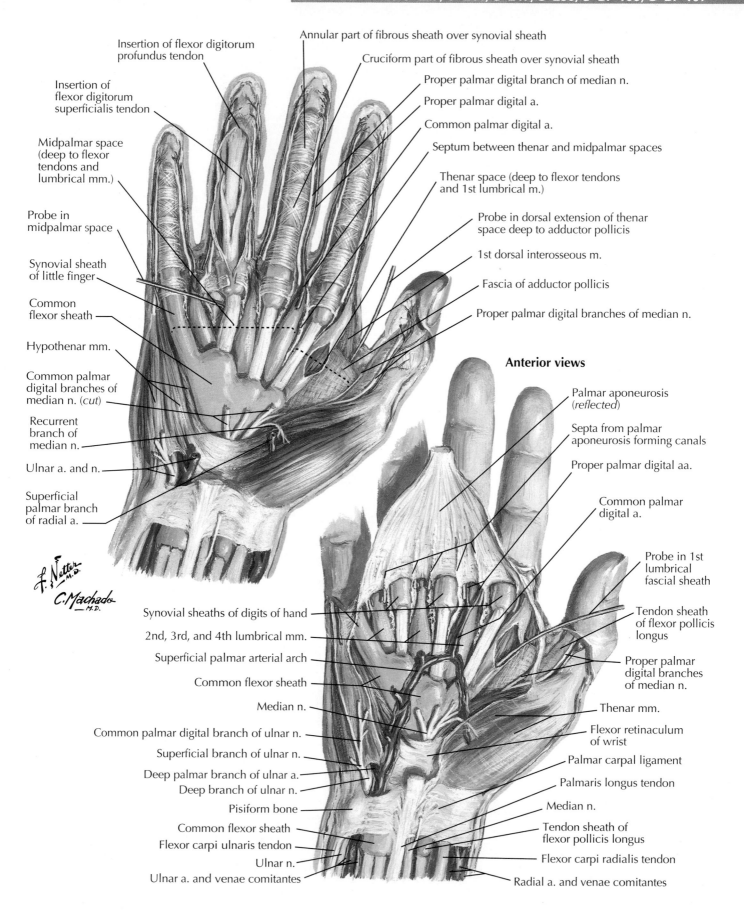

Insertion of flexor digitorum profundus tendon

Insertion of flexor digitorum superficialis tendon

Midpalmar space (deep to flexor tendons and lumbrical mm.)

Probe in midpalmar space

Synovial sheath of little finger

Common flexor sheath

Hypothenar mm.

Common palmar digital branches of median n. (cut)

Recurrent branch of median n.

Ulnar a. and n.

Superficial palmar branch of radial a.

Annular part of fibrous sheath over synovial sheath

Cruciform part of fibrous sheath over synovial sheath

Proper palmar digital branch of median n.

Proper palmar digital a.

Common palmar digital a.

Septum between thenar and midpalmar spaces

Thenar space (deep to flexor tendons and 1st lumbrical m.)

Probe in dorsal extension of thenar space deep to adductor pollicis

1st dorsal interosseous m.

Fascia of adductor pollicis

Proper palmar digital branches of median n.

Anterior views

Palmar aponeurosis (*reflected*)

Septa from palmar aponeurosis forming canals

Proper palmar digital aa.

Common palmar digital a.

Probe in 1st lumbrical fascial sheath

Tendon sheath of flexor pollicis longus

Proper palmar digital branches of median n.

Thenar mm.

Flexor retinaculum of wrist

Palmar carpal ligament

Palmaris longus tendon

Median n.

Tendon sheath of flexor pollicis longus

Flexor carpi radialis tendon

Radial a. and venae comitantes

Synovial sheaths of digits of hand

2nd, 3rd, and 4th lumbrical mm.

Superficial palmar arterial arch

Common flexor sheath

Median n.

Common palmar digital branch of ulnar n.

Superficial branch of ulnar n.

Deep palmar branch of ulnar a.

Deep branch of ulnar n.

Pisiform bone

Common flexor sheath

Flexor carpi ulnaris tendon

Ulnar n.

Ulnar a. and venae comitantes

Common variation

Usual arrangement

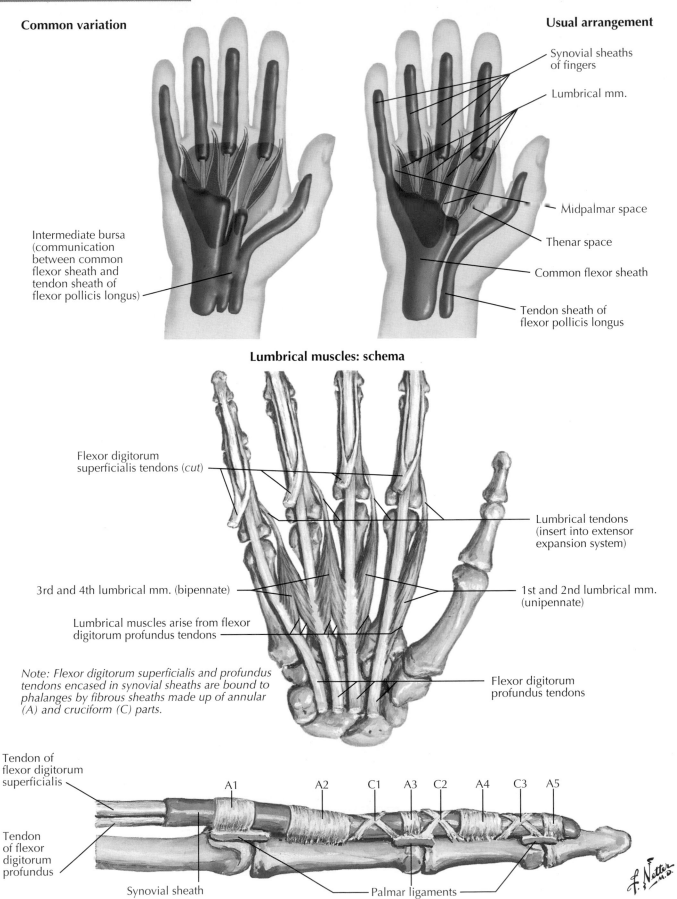

Synovial sheaths
of fingers

Lumbrical mm.

Midpalmar space

Thenar space

Common flexor sheath

Tendon sheath of
flexor pollicis longus

Intermediate bursa
(communication
between common
flexor sheath and
tendon sheath of
flexor pollicis longus)

Lumbrical muscles: schema

Flexor digitorum
superficialis tendons (*cut*)

Lumbrical tendons
(insert into extensor
expansion system)

3rd and 4th lumbrical mm. (bipennate)

1st and 2nd lumbrical mm.
(unipennate)

Lumbrical muscles arise from flexor
digitorum profundus tendons

*Note: Flexor digitorum superficialis and profundus
tendons encased in synovial sheaths are bound to
phalanges by fibrous sheaths made up of annular
(A) and cruciform (C) parts.*

Flexor digitorum
profundus tendons

Tendon of
flexor digitorum
superficialis

A1   A2   C1   A3   C2   A4   C3   A5

Tendon
of flexor
digitorum
profundus

Synovial sheath

Palmar ligaments

Anterior view

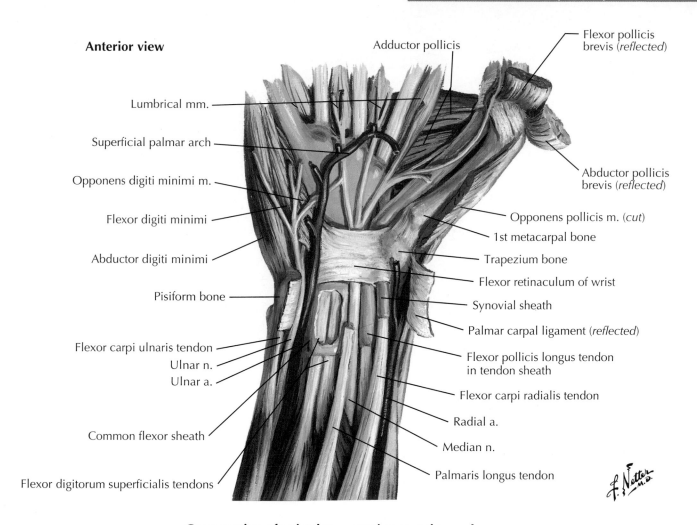

Adductor pollicis

Flexor pollicis brevis (*reflected*)

Lumbrical mm.

Superficial palmar arch

Opponens digiti minimi m.

Flexor digiti minimi

Abductor digiti minimi

Abductor pollicis brevis (*reflected*)

Opponens pollicis m. (*cut*)

1st metacarpal bone

Trapezium bone

Flexor retinaculum of wrist

Synovial sheath

Pisiform bone

Palmar carpal ligament (*reflected*)

Flexor carpi ulnaris tendon

Ulnar n.

Ulnar a.

Flexor pollicis longus tendon in tendon sheath

Flexor carpi radialis tendon

Radial a.

Common flexor sheath

Median n.

Flexor digitorum superficialis tendons

Palmaris longus tendon

Cross section of wrist demonstrating carpal tunnel

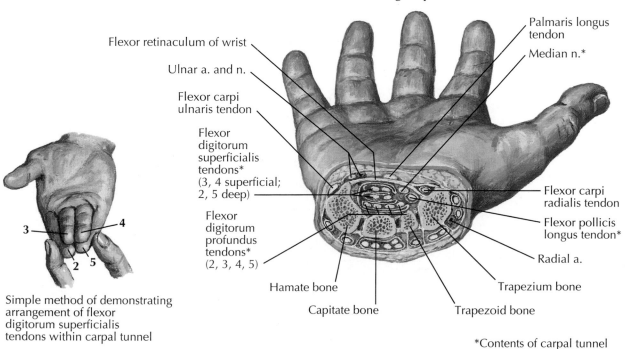

Flexor retinaculum of wrist

Ulnar a. and n.

Flexor carpi ulnaris tendon

Flexor digitorum superficialis tendons* (3, 4 superficial; 2, 5 deep)

Flexor digitorum profundus tendons* (2, 3, 4, 5)

Palmaris longus tendon

Median n.*

Flexor carpi radialis tendon

Flexor pollicis longus tendon*

Radial a.

Hamate bone

Capitate bone

Trapezoid bone

Trapezium bone

Simple method of demonstrating arrangement of flexor digitorum superficialis tendons within carpal tunnel

3   4

2   5

*Contents of carpal tunnel

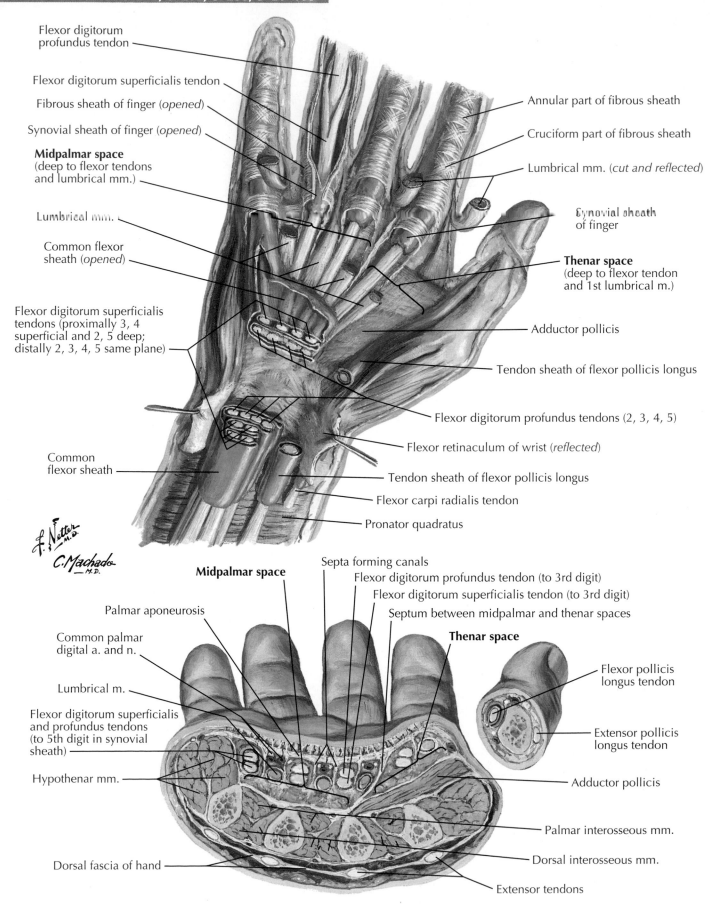

Flexor digitorum profundus tendon

Flexor digitorum superficialis tendon

Fibrous sheath of finger (*opened*)

Synovial sheath of finger (*opened*)

Midpalmar space (deep to flexor tendons and lumbrical mm.)

Lumbrical mm.

Common flexor sheath (*opened*)

Flexor digitorum superficialis tendons (proximally 3, 4 superficial and 2, 5 deep; distally 2, 3, 4, 5 same plane)

Common flexor sheath

Annular part of fibrous sheath

Cruciform part of fibrous sheath

Lumbrical mm. (*cut and reflected*)

Synovial sheath of finger

Thenar space (deep to flexor tendon and 1st lumbrical m.)

Adductor pollicis

Tendon sheath of flexor pollicis longus

Flexor digitorum profundus tendons (2, 3, 4, 5)

Flexor retinaculum of wrist (*reflected*)

Tendon sheath of flexor pollicis longus

Flexor carpi radialis tendon

Pronator quadratus

Septa forming canals

Flexor digitorum profundus tendon (to 3rd digit)

Flexor digitorum superficialis tendon (to 3rd digit)

Septum between midpalmar and thenar spaces

Midpalmar space

Thenar space

Palmar aponeurosis

Common palmar digital a. and n.

Lumbrical m.

Flexor digitorum superficialis and profundus tendons (to 5th digit in synovial sheath)

Hypothenar mm.

Dorsal fascia of hand

Flexor pollicis longus tendon

Extensor pollicis longus tendon

Adductor pollicis

Palmar interosseous mm.

Dorsal interosseous mm.

Extensor tendons

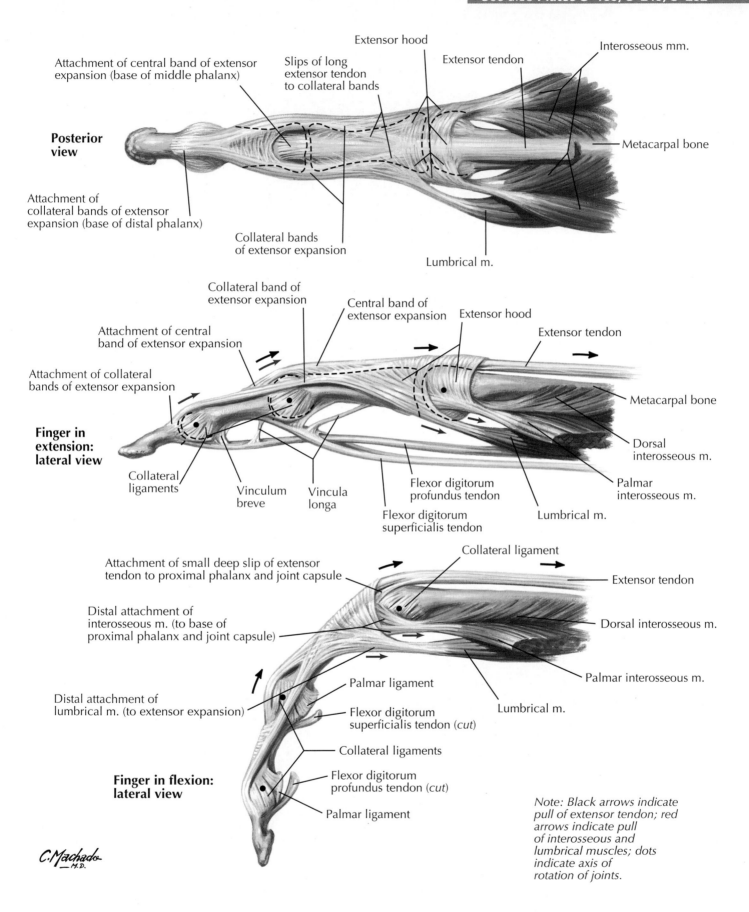

Posterior view

Attachment of central band of extensor expansion (base of middle phalanx)

Slips of long extensor tendon to collateral bands

Extensor hood

Extensor tendon

Interosseous mm.

Metacarpal bone

Attachment of collateral bands of extensor expansion (base of distal phalanx)

Collateral bands of extensor expansion

Lumbrical m.

Collateral band of extensor expansion

Central band of extensor expansion

Extensor hood

Extensor tendon

Attachment of central band of extensor expansion

Attachment of collateral bands of extensor expansion

Metacarpal bone

Finger in extension: lateral view

Collateral ligaments

Vinculum breve

Vincula longa

Flexor digitorum superficialis tendon

Flexor digitorum profundus tendon

Lumbrical m.

Dorsal interosseous m.

Palmar interosseous m.

Attachment of small deep slip of extensor tendon to proximal phalanx and joint capsule

Collateral ligament

Extensor tendon

Distal attachment of interosseous m. (to base of proximal phalanx and joint capsule)

Dorsal interosseous m.

Palmar interosseous m.

Distal attachment of lumbrical m. (to extensor expansion)

Palmar ligament

Flexor digitorum superficialis tendon (*cut*)

Collateral ligaments

Flexor digitorum profundus tendon (*cut*)

Palmar ligament

Lumbrical m.

Finger in flexion: lateral view

Note: Black arrows indicate pull of extensor tendon; red arrows indicate pull of interosseous and lumbrical muscles; dots indicate axis of rotation of joints.

C. Machado —M.D.

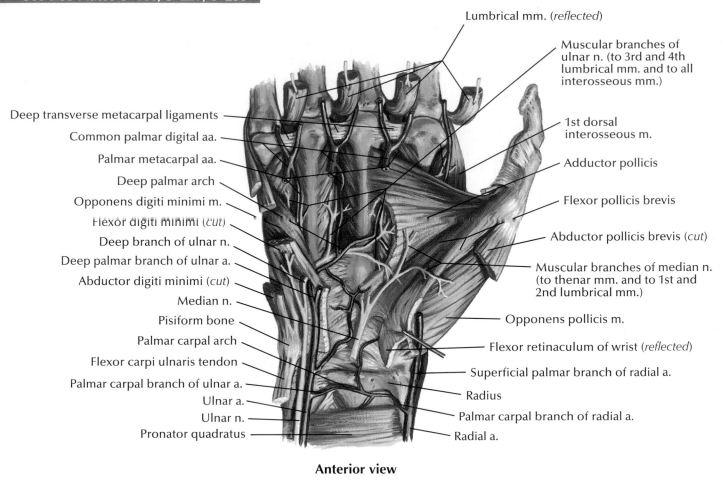

Lumbrical mm. (*reflected*)

Muscular branches of ulnar n. (to 3rd and 4th lumbrical mm. and to all interosseous mm.)

Deep transverse metacarpal ligaments

Common palmar digital aa.

Palmar metacarpal aa.

Deep palmar arch

Opponens digiti minimi m.

Flexor digiti minimi (*cut*)

Deep branch of ulnar n.

Deep palmar branch of ulnar a.

Abductor digiti minimi (*cut*)

Median n.

Pisiform bone

Palmar carpal arch

Flexor carpi ulnaris tendon

Palmar carpal branch of ulnar a.

Ulnar a.

Ulnar n.

Pronator quadratus

1st dorsal interosseous m.

Adductor pollicis

Flexor pollicis brevis

Abductor pollicis brevis (*cut*)

Muscular branches of median n. (to thenar mm. and to 1st and 2nd lumbrical mm.)

Opponens pollicis m.

Flexor retinaculum of wrist (*reflected*)

Superficial palmar branch of radial a.

Radius

Palmar carpal branch of radial a.

Radial a.

Anterior view

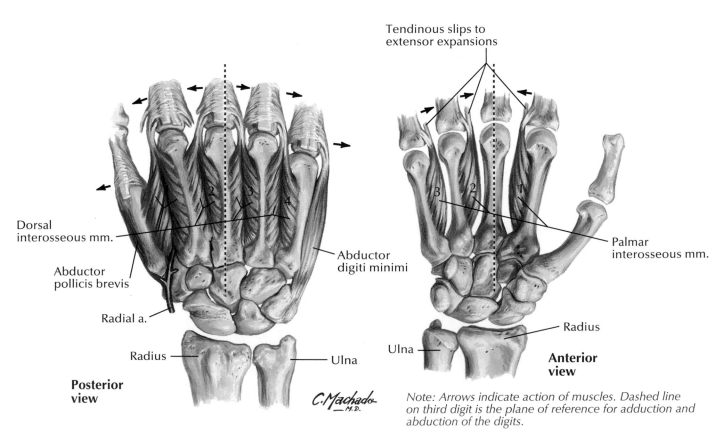

Tendinous slips to extensor expansions

Dorsal interosseous mm.

Abductor pollicis brevis

Radial a.

Radius

Ulna

Abductor digiti minimi

Palmar interosseous mm.

Ulna

Radius

Anterior view

Posterior view

Note: Arrows indicate action of muscles. Dashed line on third digit is the plane of reference for adduction and abduction of the digits.

C. Machado M.D.

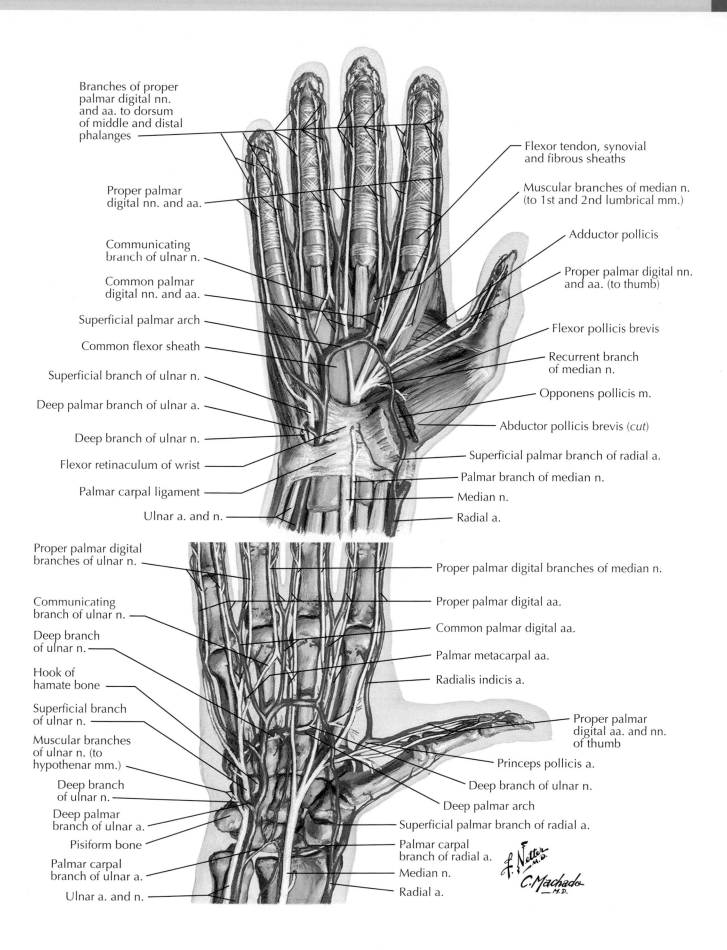

Branches of proper palmar digital nn. and aa. to dorsum of middle and distal phalanges

Proper palmar digital nn. and aa.

Communicating branch of ulnar n.

Common palmar digital nn. and aa.

Superficial palmar arch

Common flexor sheath

Superficial branch of ulnar n.

Deep palmar branch of ulnar a.

Deep branch of ulnar n.

Flexor retinaculum of wrist

Palmar carpal ligament

Ulnar a. and n.

Flexor tendon, synovial and fibrous sheaths

Muscular branches of median n. (to 1st and 2nd lumbrical mm.)

Adductor pollicis

Proper palmar digital nn. and aa. (to thumb)

Flexor pollicis brevis

Recurrent branch of median n.

Opponens pollicis m.

Abductor pollicis brevis (cut)

Superficial palmar branch of radial a.

Palmar branch of median n.

Median n.

Radial a.

Proper palmar digital branches of ulnar n.

Communicating branch of ulnar n.

Deep branch of ulnar n.

Hook of hamate bone

Superficial branch of ulnar n.

Muscular branches of ulnar n. (to hypothenar mm.)

Deep branch of ulnar n.

Deep palmar branch of ulnar a.

Pisiform bone

Palmar carpal branch of ulnar a.

Ulnar a. and n.

Proper palmar digital branches of median n.

Proper palmar digital aa.

Common palmar digital aa.

Palmar metacarpal aa.

Radialis indicis a.

Proper palmar digital aa. and nn. of thumb

Princeps pollicis a.

Deep branch of ulnar n.

Deep palmar arch

Superficial palmar branch of radial a.

Palmar carpal branch of radial a.

Median n.

Radial a.

See also **Plate S–229**

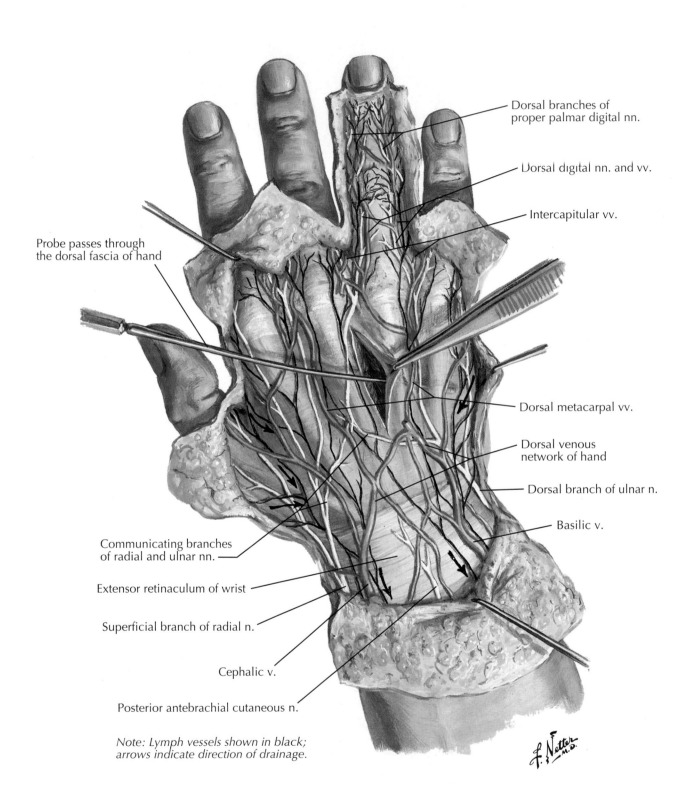

Dorsal branches of proper palmar digital nn.

Dorsal digital nn. and vv.

Intercapitular vv.

Probe passes through the dorsal fascia of hand

Dorsal metacarpal vv.

Dorsal venous network of hand

Dorsal branch of ulnar n.

Basilic v.

Communicating branches of radial and ulnar nn.

Extensor retinaculum of wrist

Superficial branch of radial n.

Cephalic v.

Posterior antebrachial cutaneous n.

Note: Lymph vessels shown in black; arrows indicate direction of drainage.

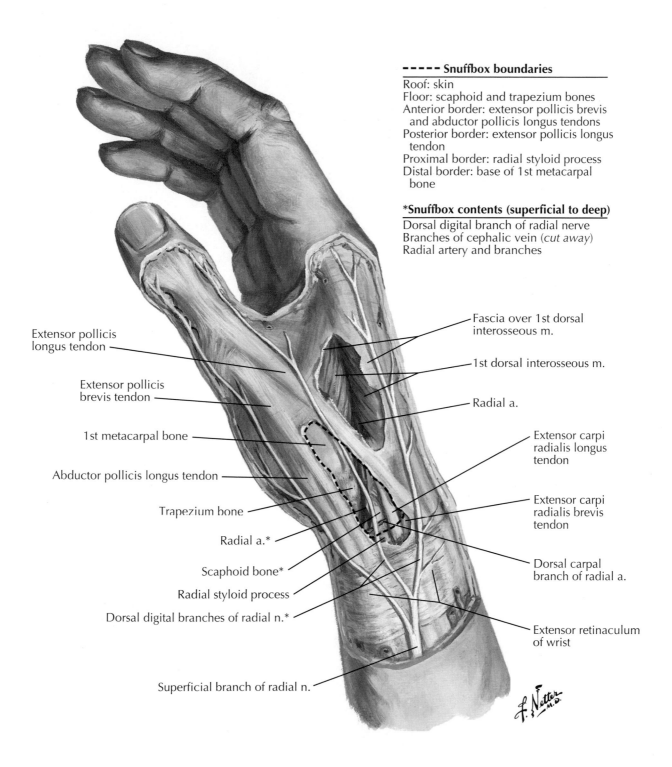

- - - - - Snuffbox boundaries

Roof: skin
Floor: scaphoid and trapezium bones
Anterior border: extensor pollicis brevis and abductor pollicis longus tendons
Posterior border: extensor pollicis longus tendon
Proximal border: radial styloid process
Distal border: base of 1st metacarpal bone

*Snuffbox contents (superficial to deep)

Dorsal digital branch of radial nerve
Branches of cephalic vein (*cut away*)
Radial artery and branches

Fascia over 1st dorsal interosseous m.

1st dorsal interosseous m.

Radial a.

Extensor carpi radialis longus tendon

Extensor carpi radialis brevis tendon

Dorsal carpal branch of radial a.

Extensor retinaculum of wrist

Extensor pollicis longus tendon

Extensor pollicis brevis tendon

1st metacarpal bone

Abductor pollicis longus tendon

Trapezium bone

Radial a.*

Scaphoid bone*

Radial styloid process

Dorsal digital branches of radial n.*

Superficial branch of radial n.

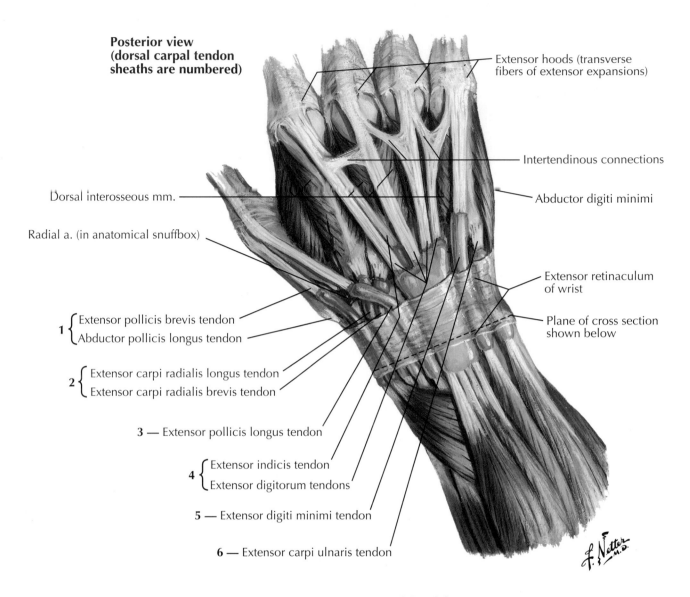

**Posterior view
(dorsal carpal tendon
sheaths are numbered)**

Extensor hoods (transverse
fibers of extensor expansions)

Intertendinous connections

Abductor digiti minimi

Dorsal interosseous mm.

Radial a. (in anatomical snuffbox)

Extensor retinaculum
of wrist

Plane of cross section
shown below

1 { Extensor pollicis brevis tendon
Abductor pollicis longus tendon

2 { Extensor carpi radialis longus tendon
Extensor carpi radialis brevis tendon

3 — Extensor pollicis longus tendon

4 { Extensor indicis tendon
Extensor digitorum tendons

5 — Extensor digiti minimi tendon

6 — Extensor carpi ulnaris tendon

Cross section of distal forearm

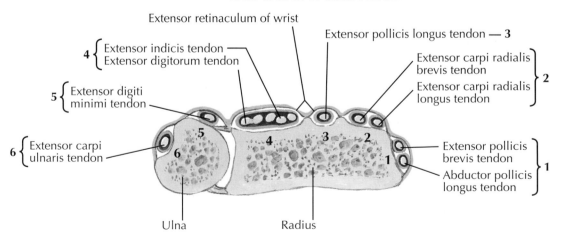

Extensor retinaculum of wrist

Extensor pollicis longus tendon — 3

4 { Extensor indicis tendon
Extensor digitorum tendon

Extensor carpi radialis
brevis tendon

Extensor carpi radialis
longus tendon

5 { Extensor digiti
minimi tendon

6 { Extensor carpi
ulnaris tendon

Extensor pollicis
brevis tendon

Abductor pollicis
longus tendon

Ulna

Radius

Plate S–261

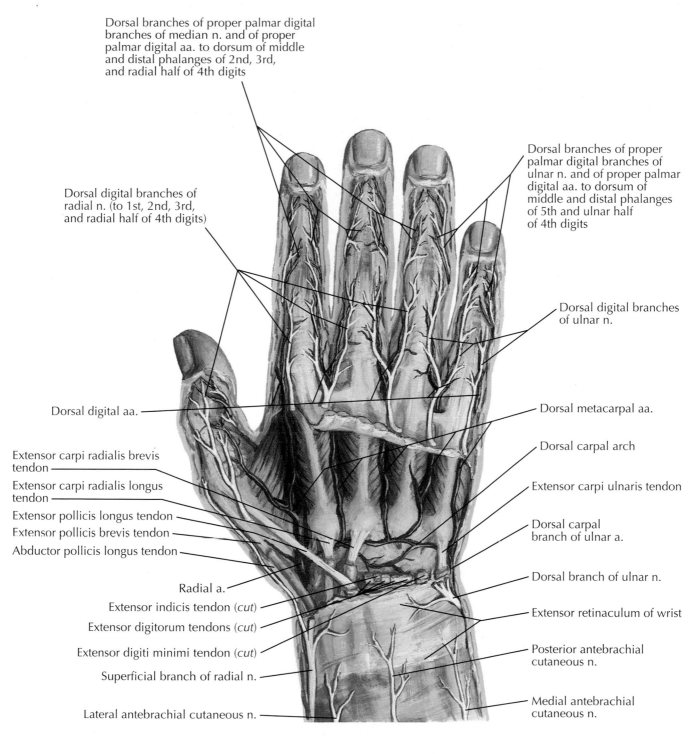

Dorsal branches of proper palmar digital branches of median n. and of proper palmar digital aa. to dorsum of middle and distal phalanges of 2nd, 3rd, and radial half of 4th digits

Dorsal digital branches of radial n. (to 1st, 2nd, 3rd, and radial half of 4th digits)

Dorsal branches of proper palmar digital branches of ulnar n. and of proper palmar digital aa. to dorsum of middle and distal phalanges of 5th and ulnar half of 4th digits

Dorsal digital branches of ulnar n.

Dorsal digital aa.

Dorsal metacarpal aa.

Dorsal carpal arch

Extensor carpi radialis brevis tendon

Extensor carpi radialis longus tendon

Extensor carpi ulnaris tendon

Extensor pollicis longus tendon

Extensor pollicis brevis tendon

Dorsal carpal branch of ulnar a.

Abductor pollicis longus tendon

Radial a.

Dorsal branch of ulnar n.

Extensor indicis tendon (*cut*)

Extensor retinaculum of wrist

Extensor digitorum tendons (*cut*)

Extensor digiti minimi tendon (*cut*)

Posterior antebrachial cutaneous n.

Superficial branch of radial n.

Medial antebrachial cutaneous n.

Lateral antebrachial cutaneous n.

Upper Limb

Plate S–261

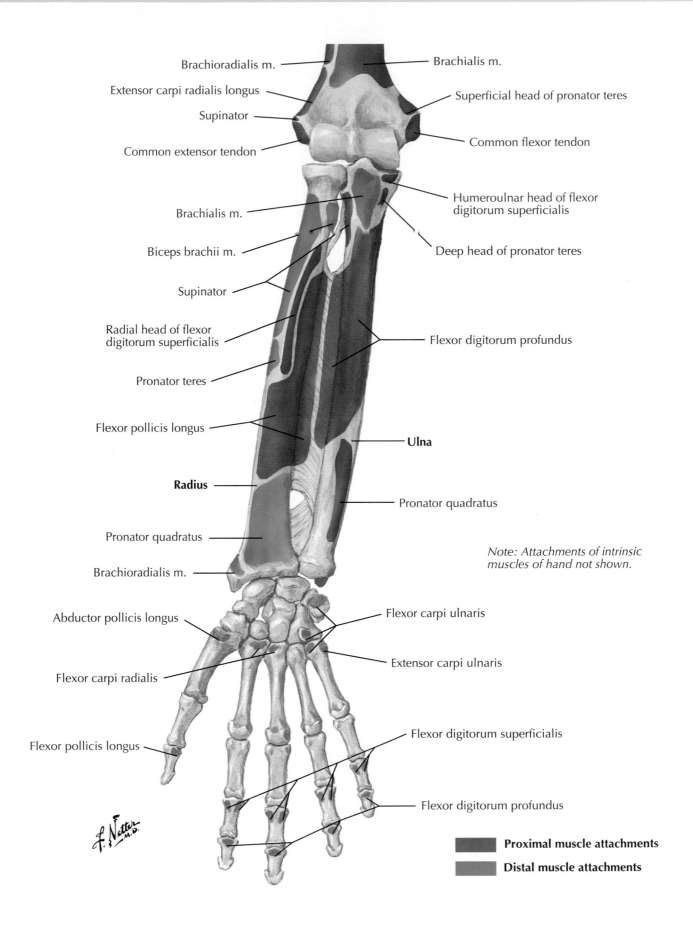

Brachioradialis m.

Brachialis m.

Extensor carpi radialis longus

Superficial head of pronator teres

Supinator

Common extensor tendon

Common flexor tendon

Brachialis m.

Humeroulnar head of flexor digitorum superficialis

Biceps brachii m.

Supinator

Deep head of pronator teres

Radial head of flexor digitorum superficialis

Flexor digitorum profundus

Pronator teres

Flexor pollicis longus

Ulna

Radius

Pronator quadratus

Pronator quadratus

Brachioradialis m.

Note: Attachments of intrinsic muscles of hand not shown.

Abductor pollicis longus

Flexor carpi ulnaris

Extensor carpi ulnaris

Flexor carpi radialis

Flexor pollicis longus

Flexor digitorum superficialis

Flexor digitorum profundus

Proximal muscle attachments

Distal muscle attachments

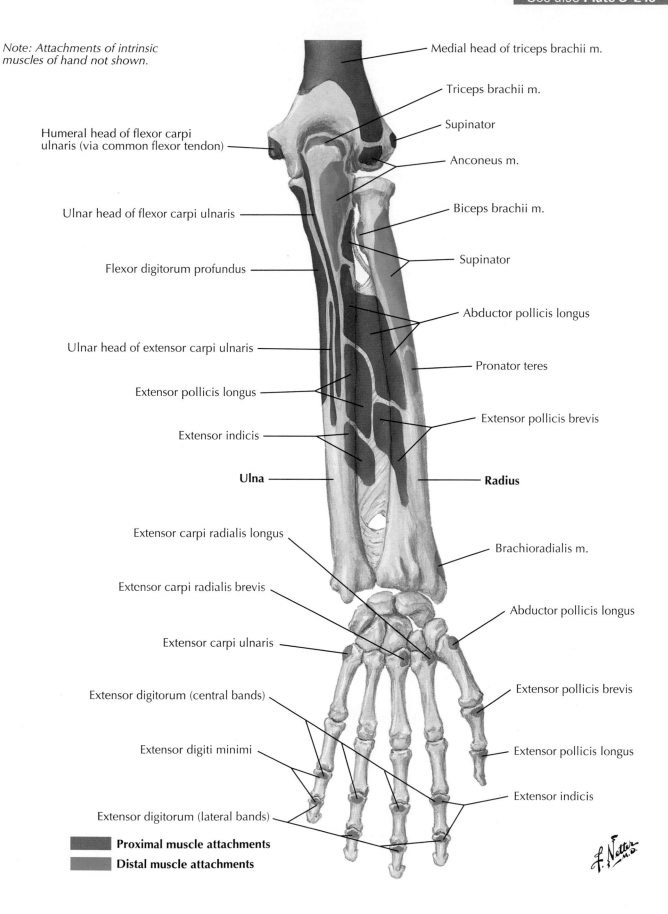

Note: Attachments of intrinsic muscles of hand not shown.

Medial head of triceps brachii m.

Triceps brachii m.

Supinator

Humeral head of flexor carpi ulnaris (via common flexor tendon)

Anconeus m.

Ulnar head of flexor carpi ulnaris

Biceps brachii m.

Supinator

Flexor digitorum profundus

Abductor pollicis longus

Ulnar head of extensor carpi ulnaris

Pronator teres

Extensor pollicis longus

Extensor pollicis brevis

Extensor indicis

Ulna

Radius

Extensor carpi radialis longus

Brachioradialis m.

Extensor carpi radialis brevis

Abductor pollicis longus

Extensor carpi ulnaris

Extensor pollicis brevis

Extensor digitorum (central bands)

Extensor digiti minimi

Extensor pollicis longus

Extensor indicis

Extensor digitorum (lateral bands)

■ **Proximal muscle attachments**
■ **Distal muscle attachments**

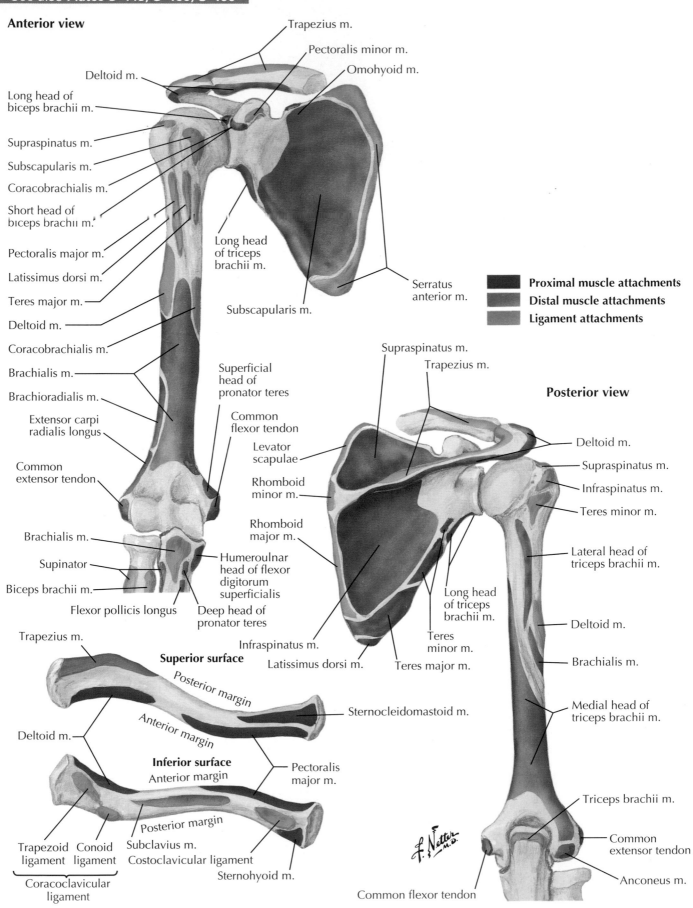

Anterior view

Trapezius m.

Pectoralis minor m.

Omohyoid m.

Deltoid m.

Long head of biceps brachii m.

Supraspinatus m.

Subscapularis m.

Coracobrachialis m.

Short head of biceps brachii m.

Pectoralis major m.

Latissimus dorsi m.

Teres major m.

Deltoid m.

Coracobrachialis m.

Brachialis m.

Brachioradialis m.

Extensor carpi radialis longus

Common extensor tendon

Long head of triceps brachii m.

Subscapularis m.

Serratus anterior m.

| Proximal muscle attachments |
| Distal muscle attachments |
| Ligament attachments |

Superficial head of pronator teres

Common flexor tendon

Levator scapulae

Rhomboid minor m.

Rhomboid major m.

Humeroulnar head of flexor digitorum superficialis

Brachialis m.

Supinator

Biceps brachii m.

Flexor pollicis longus

Deep head of pronator teres

Supraspinatus m.

Trapezius m.

Posterior view

Deltoid m.

Supraspinatus m.

Infraspinatus m.

Teres minor m.

Lateral head of triceps brachii m.

Long head of triceps brachii m.

Teres minor m.

Deltoid m.

Brachialis m.

Medial head of triceps brachii m.

Infraspinatus m.

Latissimus dorsi m.

Teres major m.

Trapezius m.

Superior surface

Posterior margin

Anterior margin

Inferior surface

Anterior margin

Deltoid m.

Sternocleidomastoid m.

Pectoralis major m.

Trapezoid ligament

Conoid ligament

Subclavius m.

Costoclavicular ligament

Sternohyoid m.

Posterior margin

Coracoclavicular ligament

Triceps brachii m.

Common extensor tendon

Anconeus m.

Common flexor tendon

Segmental innervation of upper limb movements

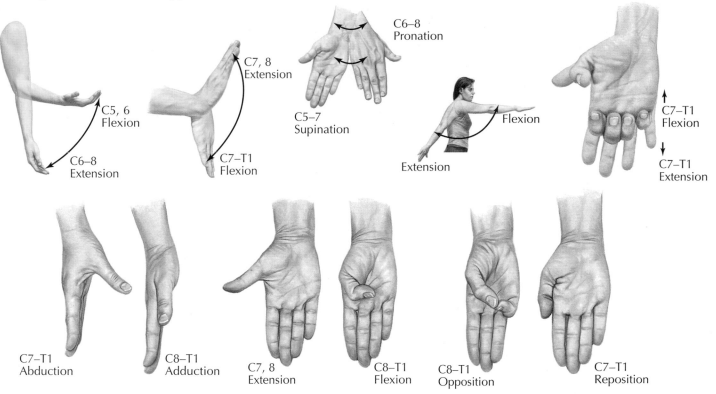

C6–8
Pronation

C7, 8
Extension

C5, 6
Flexion

C6–8
Extension

C7–T1
Flexion

C5–7
Supination

Flexion

Extension

C7–T1
Flexion

C7–T1
Extension

C7–T1
Abduction

C8–T1
Adduction

C7, 8
Extension

C8–T1
Flexion

C8–T1
Opposition

C7–T1
Reposition

Segmental innervation of lower limb movements

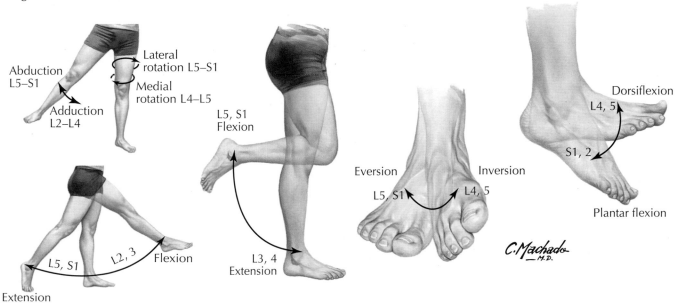

Abduction
L5–S1

Lateral
rotation L5–S1

Medial
rotation L4–L5

Adduction
L2–L4

L5, S1
Flexion

Dorsiflexion
L4, 5

S1, 2

L5, S1

L2, 3
Flexion

L3, 4
Extension

Eversion
L5, S1

Inversion
L4, 5

Plantar flexion

Extension

C. Machado
M.D.

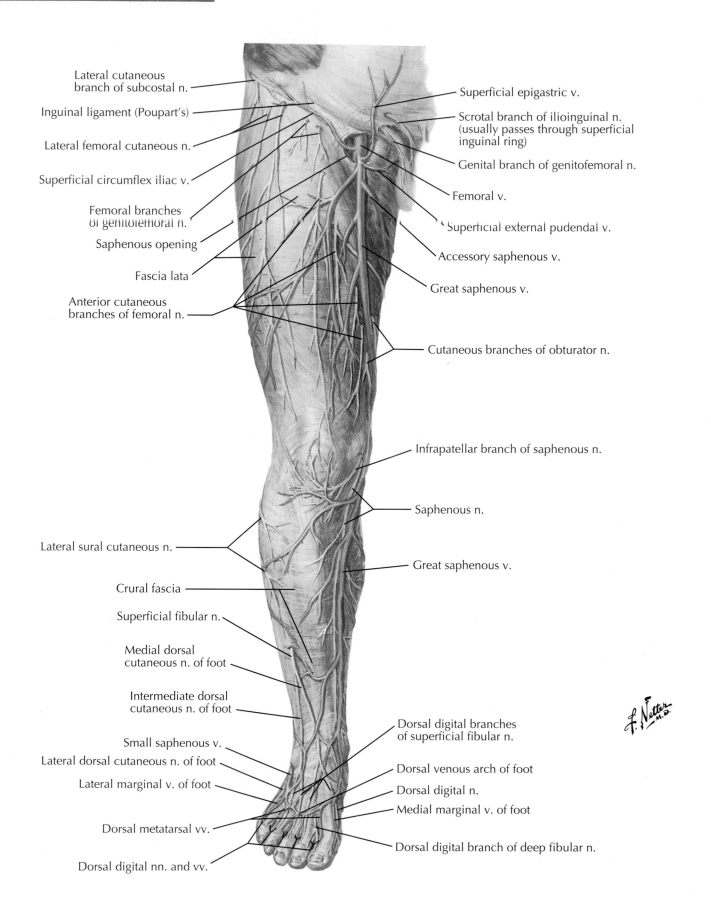

Lateral cutaneous branch of subcostal n.

Inguinal ligament (Poupart's)

Lateral femoral cutaneous n.

Superficial circumflex iliac v.

Femoral branches of genitofemoral n.

Saphenous opening

Fascia lata

Anterior cutaneous branches of femoral n.

Lateral sural cutaneous n.

Crural fascia

Superficial fibular n.

Medial dorsal cutaneous n. of foot

Intermediate dorsal cutaneous n. of foot

Small saphenous v.

Lateral dorsal cutaneous n. of foot

Lateral marginal v. of foot

Dorsal metatarsal vv.

Dorsal digital nn. and vv.

Superficial epigastric v.

Scrotal branch of ilioinguinal n. (usually passes through superficial inguinal ring)

Genital branch of genitofemoral n.

Femoral v.

Superficial external pudendal v.

Accessory saphenous v.

Great saphenous v.

Cutaneous branches of obturator n.

Infrapatellar branch of saphenous n.

Saphenous n.

Great saphenous v.

Dorsal digital branches of superficial fibular n.

Dorsal venous arch of foot

Dorsal digital n.

Medial marginal v. of foot

Dorsal digital branch of deep fibular n.

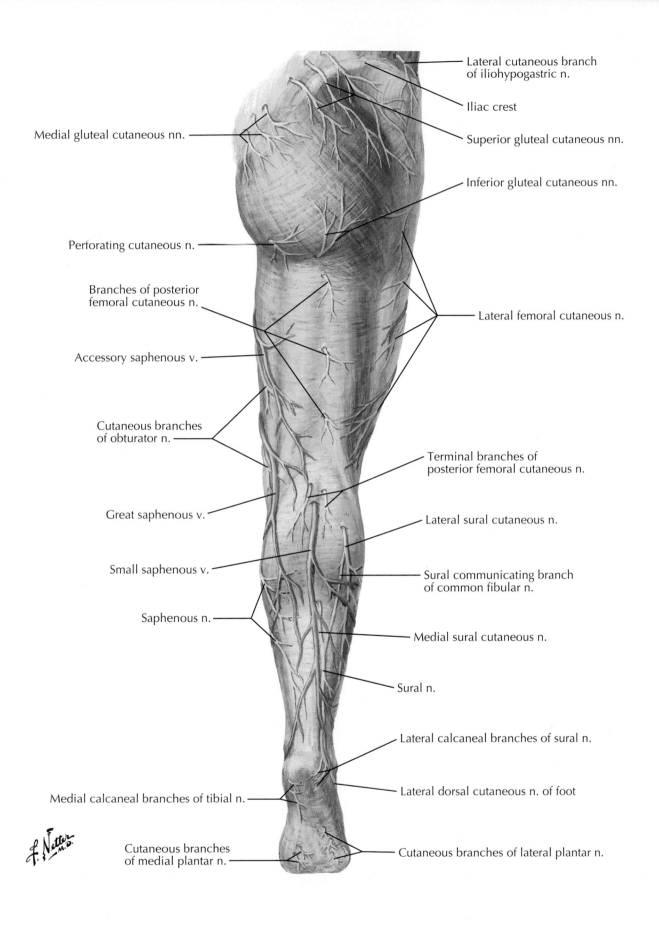

Lateral cutaneous branch of iliohypogastric n.

Iliac crest

Medial gluteal cutaneous nn.

Superior gluteal cutaneous nn.

Inferior gluteal cutaneous nn.

Perforating cutaneous n.

Branches of posterior femoral cutaneous n.

Lateral femoral cutaneous n.

Accessory saphenous v.

Cutaneous branches of obturator n.

Terminal branches of posterior femoral cutaneous n.

Great saphenous v.

Lateral sural cutaneous n.

Small saphenous v.

Sural communicating branch of common fibular n.

Saphenous n.

Medial sural cutaneous n.

Sural n.

Lateral calcaneal branches of sural n.

Medial calcaneal branches of tibial n.

Lateral dorsal cutaneous n. of foot

Cutaneous branches of medial plantar n.

Cutaneous branches of lateral plantar n.

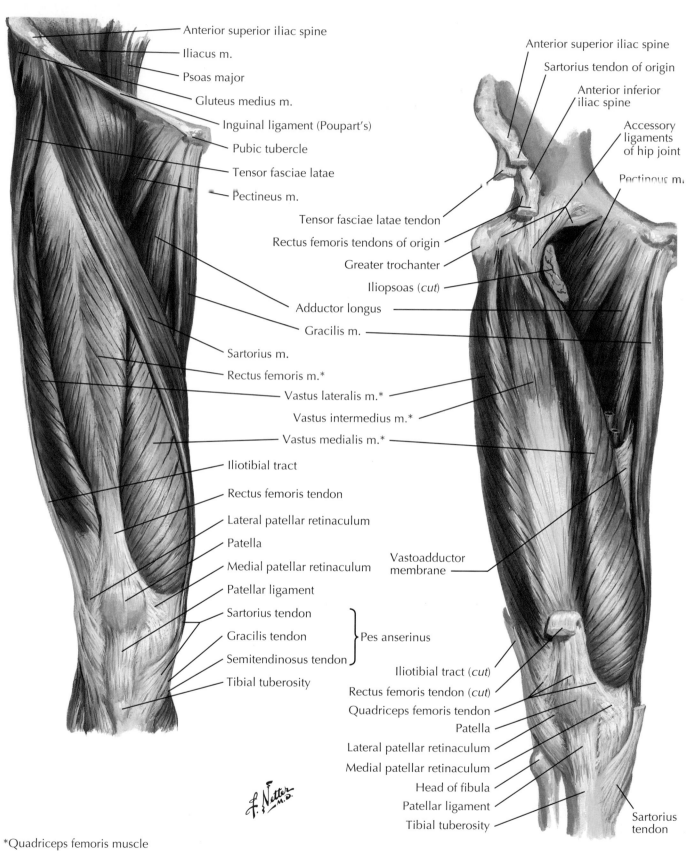

Anterior superior iliac spine
Iliacus m.
Psoas major
Gluteus medius m.
Inguinal ligament (Poupart's)
Pubic tubercle
Tensor fasciae latae
Pectineus m.

Anterior superior iliac spine
Sartorius tendon of origin
Anterior inferior iliac spine
Accessory ligaments of hip joint
Pectineus m.

Tensor fasciae latae tendon
Rectus femoris tendons of origin
Greater trochanter
Iliopsoas (*cut*)
Adductor longus
Gracilis m.
Sartorius m.
Rectus femoris m.*
Vastus lateralis m.*
Vastus intermedius m.*
Vastus medialis m.*
Iliotibial tract
Rectus femoris tendon
Lateral patellar retinaculum
Patella
Medial patellar retinaculum
Patellar ligament
Sartorius tendon
Gracilis tendon
Semitendinosus tendon
Tibial tuberosity

Pes anserinus

Vastoadductor membrane

Iliotibial tract (*cut*)
Rectus femoris tendon (*cut*)
Quadriceps femoris tendon
Patella
Lateral patellar retinaculum
Medial patellar retinaculum
Head of fibula
Patellar ligament
Tibial tuberosity
Sartorius tendon

*Quadriceps femoris muscle

Deep dissection

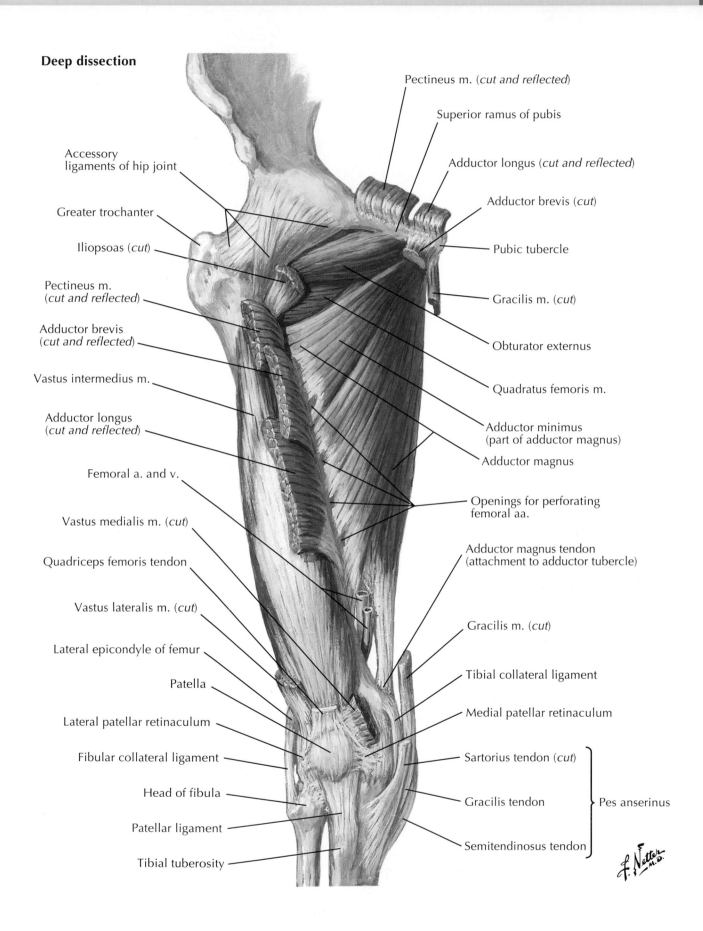

Accessory ligaments of hip joint

Greater trochanter

Iliopsoas (*cut*)

Pectineus m. (*cut and reflected*)

Adductor brevis (*cut and reflected*)

Vastus intermedius m.

Adductor longus (*cut and reflected*)

Femoral a. and v.

Vastus medialis m. (*cut*)

Quadriceps femoris tendon

Vastus lateralis m. (*cut*)

Lateral epicondyle of femur

Patella

Lateral patellar retinaculum

Fibular collateral ligament

Head of fibula

Patellar ligament

Tibial tuberosity

Pectineus m. (*cut and reflected*)

Superior ramus of pubis

Adductor longus (*cut and reflected*)

Adductor brevis (*cut*)

Pubic tubercle

Gracilis m. (*cut*)

Obturator externus

Quadratus femoris m.

Adductor minimus (part of adductor magnus)

Adductor magnus

Openings for perforating femoral aa.

Adductor magnus tendon (attachment to adductor tubercle)

Gracilis m. (*cut*)

Tibial collateral ligament

Medial patellar retinaculum

Sartorius tendon (*cut*)

Gracilis tendon

Semitendinosus tendon

} Pes anserinus

F. Netter M.D.

Superficial dissections

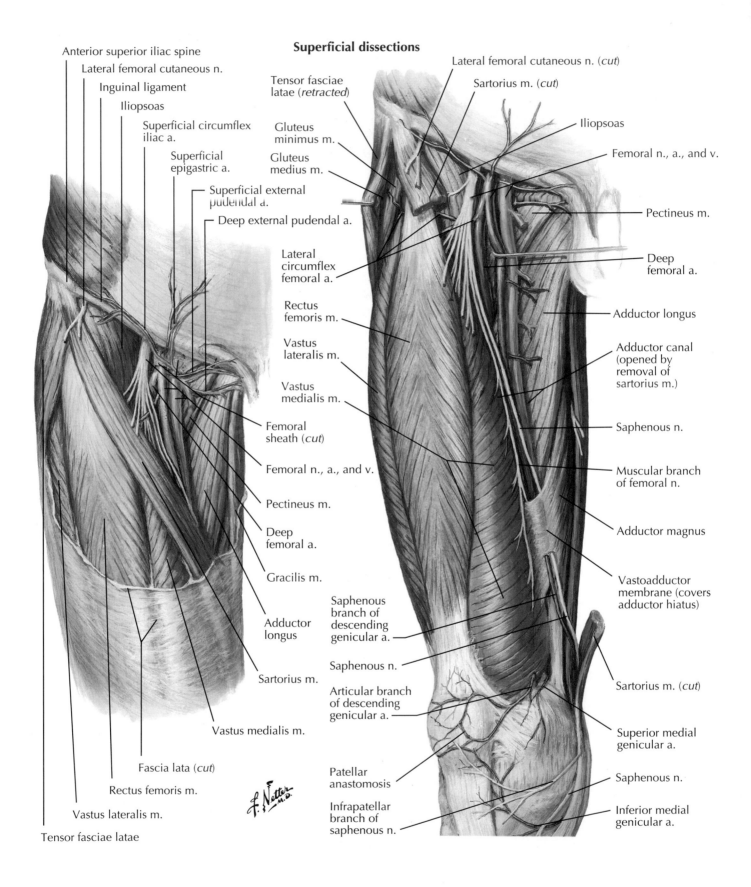

Anterior superior iliac spine

Lateral femoral cutaneous n.

Inguinal ligament

Iliopsoas

Superficial circumflex iliac a.

Superficial epigastric a.

Superficial external pudendal a.

Deep external pudendal a.

Tensor fasciae latae (*retracted*)

Gluteus minimus m.

Gluteus medius m.

Lateral circumflex femoral a.

Rectus femoris m.

Vastus lateralis m.

Vastus medialis m.

Femoral sheath (*cut*)

Femoral n., a., and v.

Pectineus m.

Deep femoral a.

Gracilis m.

Adductor longus

Sartorius m.

Vastus medialis m.

Fascia lata (*cut*)

Rectus femoris m.

Vastus lateralis m.

Tensor fasciae latae

Lateral femoral cutaneous n. (*cut*)

Sartorius m. (*cut*)

Iliopsoas

Femoral n., a., and v.

Pectineus m.

Deep femoral a.

Adductor longus

Adductor canal (opened by removal of sartorius m.)

Saphenous n.

Muscular branch of femoral n.

Adductor magnus

Vastoadductor membrane (covers adductor hiatus)

Sartorius m. (*cut*)

Superior medial genicular a.

Saphenous n.

Inferior medial genicular a.

Saphenous branch of descending genicular a.

Saphenous n.

Articular branch of descending genicular a.

Patellar anastomosis

Infrapatellar branch of saphenous n.

Deep dissection

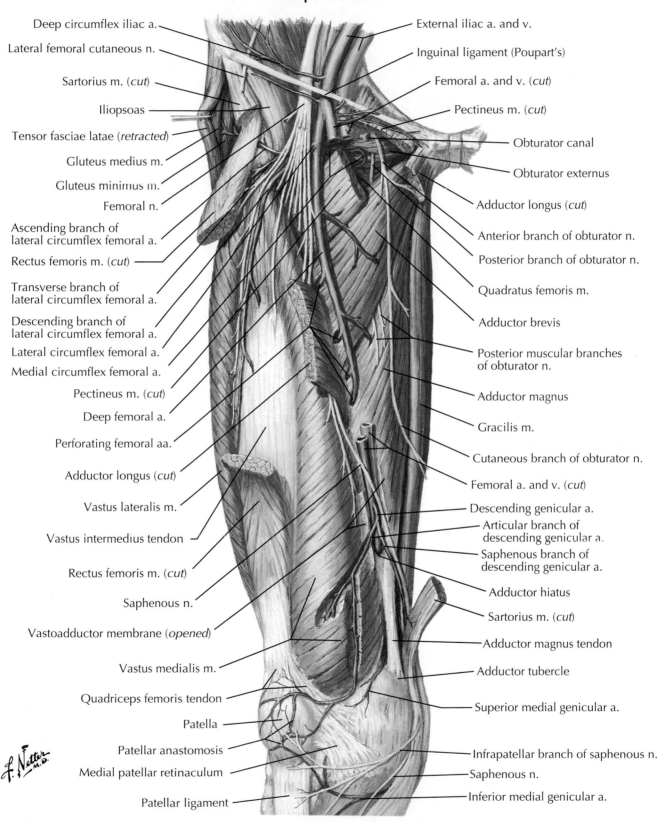

Deep circumflex iliac a.

Lateral femoral cutaneous n.

Sartorius m. (*cut*)

Iliopsoas

Tensor fasciae latae (*retracted*)

Gluteus medius m.

Gluteus minimus m.

Femoral n.

Ascending branch of
lateral circumflex femoral a.

Rectus femoris m. (*cut*)

Transverse branch of
lateral circumflex femoral a.

Descending branch of
lateral circumflex femoral a.

Lateral circumflex femoral a.

Medial circumflex femoral a.

Pectineus m. (*cut*)

Deep femoral a.

Perforating femoral aa.

Adductor longus (*cut*)

Vastus lateralis m.

Vastus intermedius tendon

Rectus femoris m. (*cut*)

Saphenous n.

Vastoadductor membrane (*opened*)

Vastus medialis m.

Quadriceps femoris tendon

Patella

Patellar anastomosis

Medial patellar retinaculum

Patellar ligament

External iliac a. and v.

Inguinal ligament (Poupart's)

Femoral a. and v. (*cut*)

Pectineus m. (*cut*)

Obturator canal

Obturator externus

Adductor longus (*cut*)

Anterior branch of obturator n.

Posterior branch of obturator n.

Quadratus femoris m.

Adductor brevis

Posterior muscular branches
of obturator n.

Adductor magnus

Gracilis m.

Cutaneous branch of obturator n.

Femoral a. and v. (*cut*)

Descending genicular a.

Articular branch of
descending genicular a.

Saphenous branch of
descending genicular a.

Adductor hiatus

Sartorius m. (*cut*)

Adductor magnus tendon

Adductor tubercle

Superior medial genicular a.

Infrapatellar branch of saphenous n.

Saphenous n.

Inferior medial genicular a.

F. Netter
M.D.

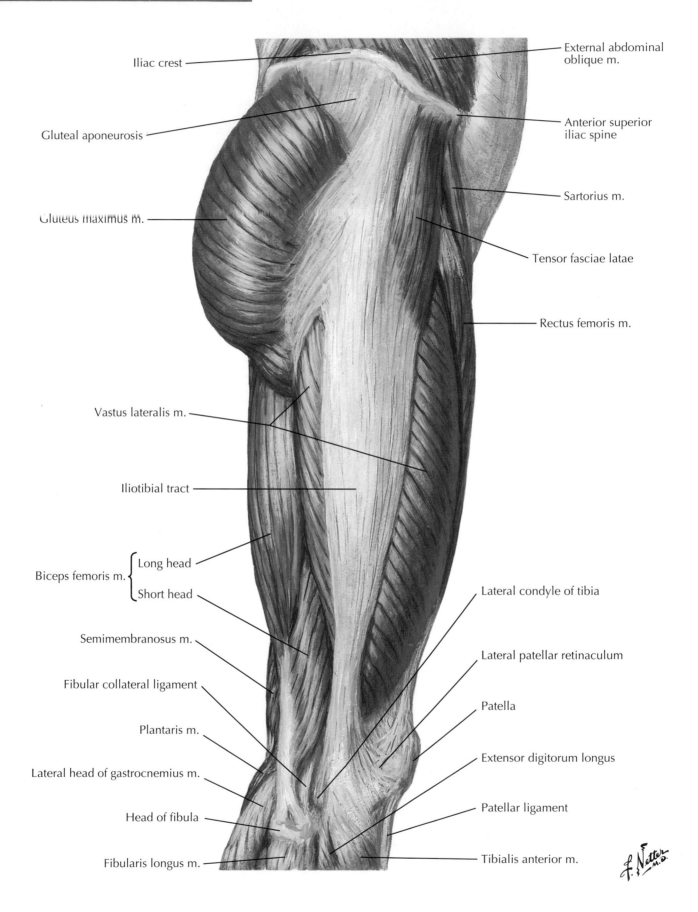

Iliac crest

Gluteal aponeurosis

Gluteus maximus m.

Vastus lateralis m.

Iliotibial tract

Biceps femoris m. { Long head
Short head

Semimembranosus m.

Fibular collateral ligament

Plantaris m.

Lateral head of gastrocnemius m.

Head of fibula

Fibularis longus m.

External abdominal oblique m.

Anterior superior iliac spine

Sartorius m.

Tensor fasciae latae

Rectus femoris m.

Lateral condyle of tibia

Lateral patellar retinaculum

Patella

Extensor digitorum longus

Patellar ligament

Tibialis anterior m.

Superficial dissection

Deeper dissection

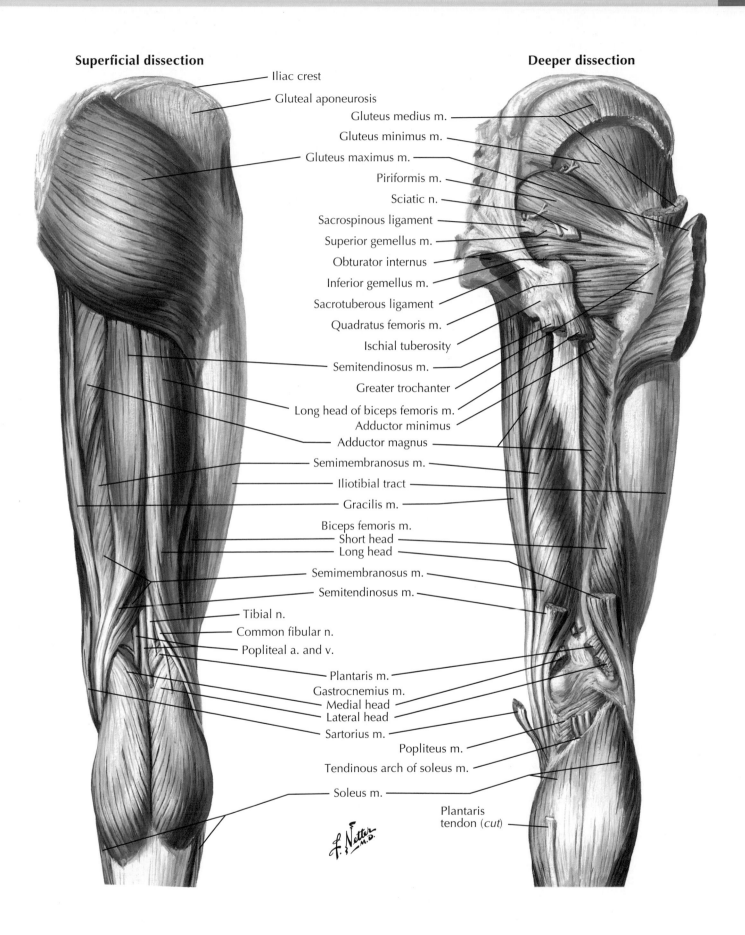

Iliac crest

Gluteal aponeurosis

Gluteus medius m.

Gluteus minimus m.

Gluteus maximus m.

Piriformis m.

Sciatic n.

Sacrospinous ligament

Superior gemellus m.

Obturator internus

Inferior gemellus m.

Sacrotuberous ligament

Quadratus femoris m.

Ischial tuberosity

Semitendinosus m.

Greater trochanter

Long head of biceps femoris m.

Adductor minimus

Adductor magnus

Semimembranosus m.

Iliotibial tract

Gracilis m.

Biceps femoris m.
Short head
Long head

Semimembranosus m.

Semitendinosus m.

Tibial n.

Common fibular n.

Popliteal a. and v.

Plantaris m.

Gastrocnemius m.
Medial head
Lateral head

Sartorius m.

Popliteus m.

Tendinous arch of soleus m.

Soleus m.

Plantaris
tendon (*cut*)

f. Netter, m.d.

Deep dissection

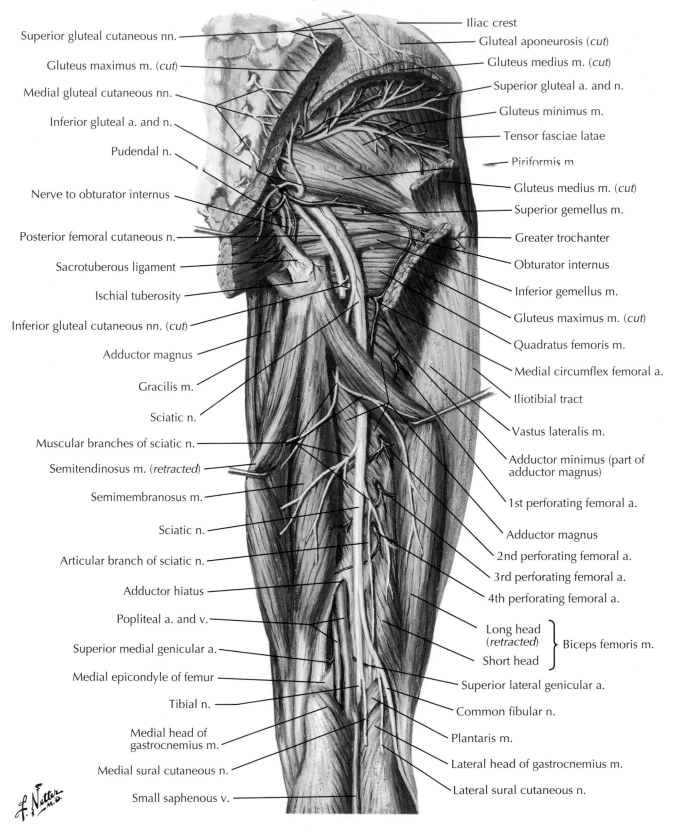

Superior gluteal cutaneous nn.

Gluteus maximus m. (*cut*)

Medial gluteal cutaneous nn.

Inferior gluteal a. and n.

Pudendal n.

Nerve to obturator internus

Posterior femoral cutaneous n.

Sacrotuberous ligament

Ischial tuberosity

Inferior gluteal cutaneous nn. (*cut*)

Adductor magnus

Gracilis m.

Sciatic n.

Muscular branches of sciatic n.

Semitendinosus m. (*retracted*)

Semimembranosus m.

Sciatic n.

Articular branch of sciatic n.

Adductor hiatus

Popliteal a. and v.

Superior medial genicular a.

Medial epicondyle of femur

Tibial n.

Medial head of gastrocnemius m.

Medial sural cutaneous n.

Small saphenous v.

Iliac crest

Gluteal aponeurosis (*cut*)

Gluteus medius m. (*cut*)

Superior gluteal a. and n.

Gluteus minimus m.

Tensor fasciae latae

Piriformis m

Gluteus medius m. (*cut*)

Superior gemellus m.

Greater trochanter

Obturator internus

Inferior gemellus m.

Gluteus maximus m. (*cut*)

Quadratus femoris m.

Medial circumflex femoral a.

Iliotibial tract

Vastus lateralis m.

Adductor minimus (part of adductor magnus)

1st perforating femoral a.

Adductor magnus

2nd perforating femoral a.

3rd perforating femoral a.

4th perforating femoral a.

Long head (*retracted*)

Short head

Biceps femoris m.

Superior lateral genicular a.

Common fibular n.

Plantaris m.

Lateral head of gastrocnemius m.

Lateral sural cutaneous n.

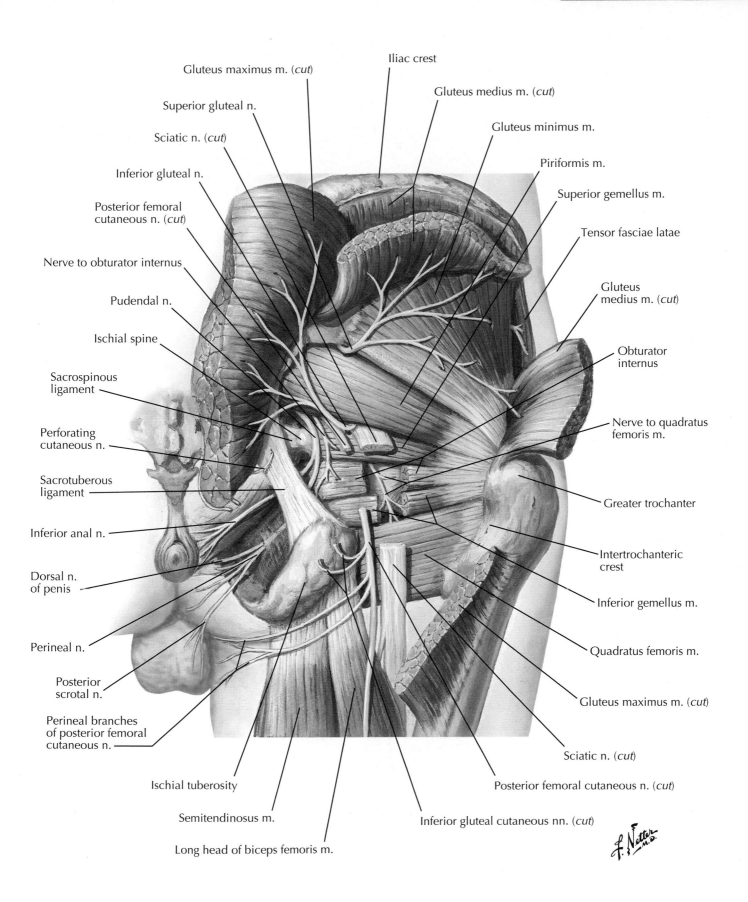

Gluteus maximus m. (*cut*)

Superior gluteal n.

Sciatic n. (*cut*)

Inferior gluteal n.

Posterior femoral cutaneous n. (*cut*)

Nerve to obturator internus

Pudendal n.

Ischial spine

Sacrospinous ligament

Perforating cutaneous n.

Sacrotuberous ligament

Inferior anal n.

Dorsal n. of penis

Perineal n.

Posterior scrotal n.

Perineal branches of posterior femoral cutaneous n.

Iliac crest

Gluteus medius m. (*cut*)

Gluteus minimus m.

Piriformis m.

Superior gemellus m.

Tensor fasciae latae

Gluteus medius m. (*cut*)

Obturator internus

Nerve to quadratus femoris m.

Greater trochanter

Intertrochanteric crest

Inferior gemellus m.

Quadratus femoris m.

Gluteus maximus m. (*cut*)

Sciatic n. (*cut*)

Posterior femoral cutaneous n. (*cut*)

Inferior gluteal cutaneous nn. (*cut*)

Ischial tuberosity

Semitendinosus m.

Long head of biceps femoris m.

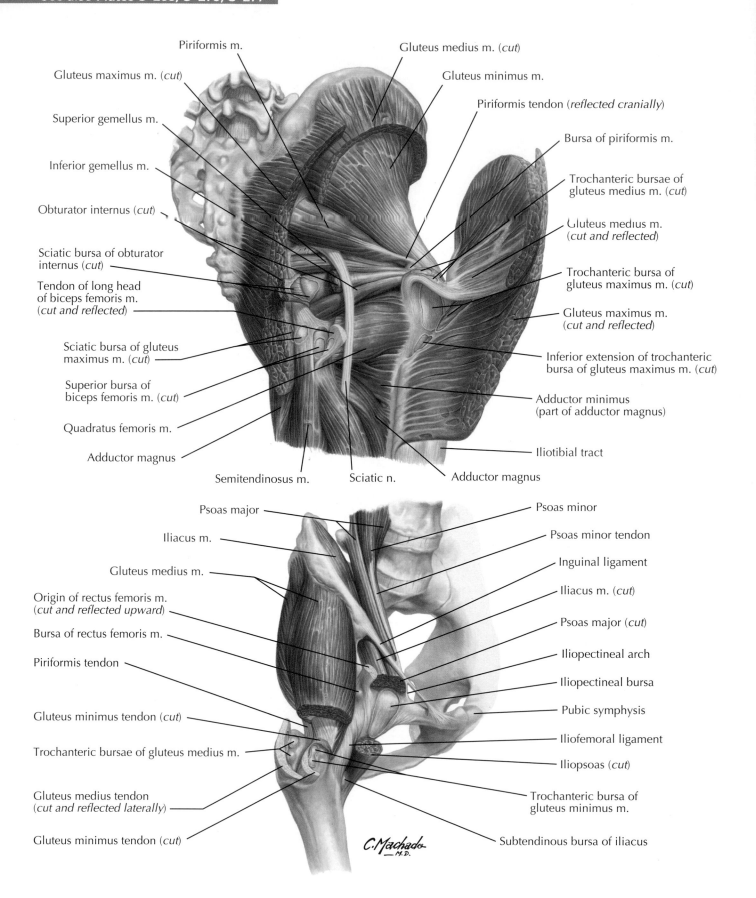

Piriformis m.

Gluteus maximus m. (*cut*)

Superior gemellus m.

Inferior gemellus m.

Obturator internus (*cut*)

Sciatic bursa of obturator internus (*cut*)

Tendon of long head of biceps femoris m. (*cut and reflected*)

Sciatic bursa of gluteus maximus m. (*cut*)

Superior bursa of biceps femoris m. (*cut*)

Quadratus femoris m.

Adductor magnus

Semitendinosus m.

Sciatic n.

Gluteus medius m. (*cut*)

Gluteus minimus m.

Piriformis tendon (*reflected cranially*)

Bursa of piriformis m.

Trochanteric bursae of gluteus medius m. (*cut*)

Gluteus medius m. (*cut and reflected*)

Trochanteric bursa of gluteus maximus m. (*cut*)

Gluteus maximus m. (*cut and reflected*)

Inferior extension of trochanteric bursa of gluteus maximus m. (*cut*)

Adductor minimus (part of adductor magnus)

Iliotibial tract

Adductor magnus

Psoas major

Iliacus m.

Gluteus medius m.

Origin of rectus femoris m. (*cut and reflected upward*)

Bursa of rectus femoris m.

Piriformis tendon

Gluteus minimus tendon (*cut*)

Trochanteric bursae of gluteus medius m.

Gluteus medius tendon (*cut and reflected laterally*)

Gluteus minimus tendon (*cut*)

Psoas minor

Psoas minor tendon

Inguinal ligament

Iliacus m. (*cut*)

Psoas major (*cut*)

Iliopectineal arch

Iliopectineal bursa

Pubic symphysis

Iliofemoral ligament

Iliopsoas (*cut*)

Trochanteric bursa of gluteus minimus m.

Subtendinous bursa of iliacus

C.Machado M.D.

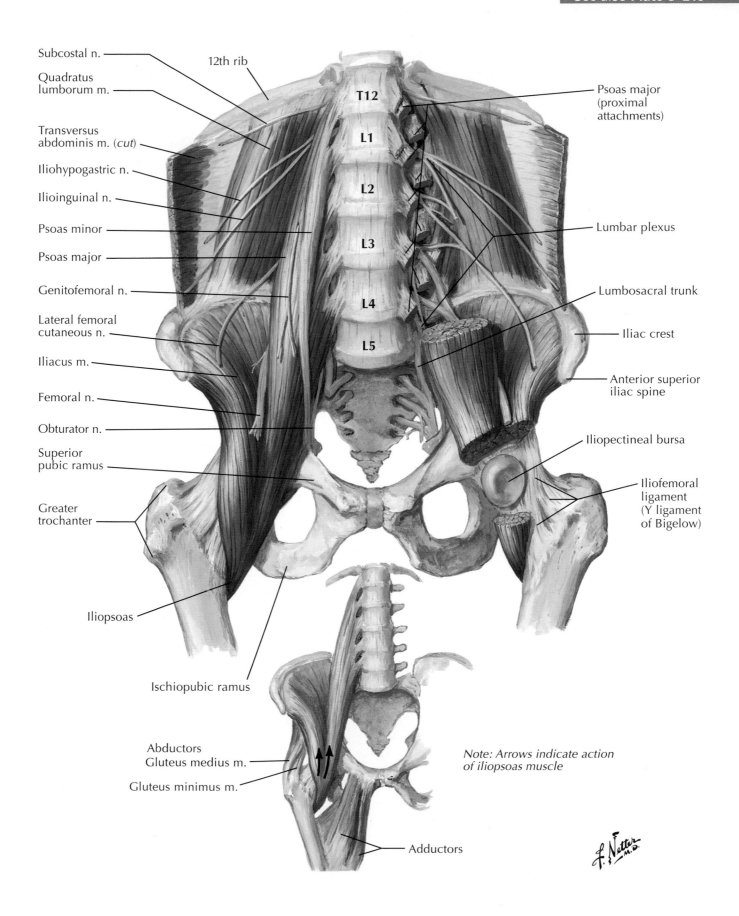

Subcostal n.

12th rib

Quadratus lumborum m.

Transversus abdominis m. (*cut*)

Iliohypogastric n.

Ilioinguinal n.

Psoas minor

Psoas major

Genitofemoral n.

Lateral femoral cutaneous n.

Iliacus m.

Femoral n.

Obturator n.

Superior pubic ramus

Greater trochanter

Iliopsoas

Ischiopubic ramus

Abductors
Gluteus medius m.

Gluteus minimus m.

Adductors

T12

L1

L2

L3

L4

L5

Psoas major (proximal attachments)

Lumbar plexus

Lumbosacral trunk

Iliac crest

Anterior superior iliac spine

Iliopectineal bursa

Iliofemoral ligament (Y ligament of Bigelow)

Note: Arrows indicate action of iliopsoas muscle

F. Netter M.D.

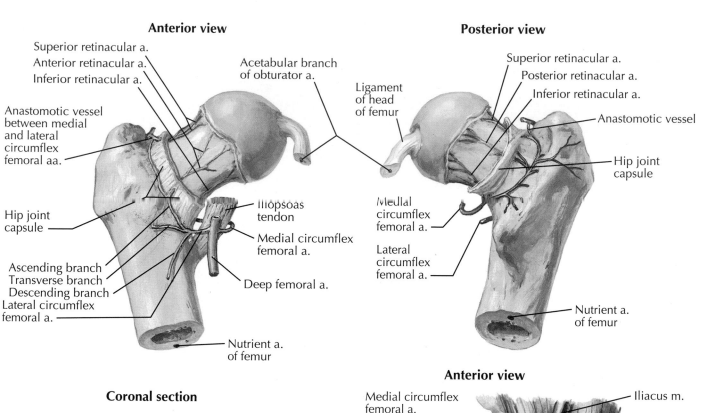

Anterior view

Superior retinacular a.
Anterior retinacular a.
Inferior retinacular a.

Anastomotic vessel between medial and lateral circumflex femoral aa.

Acetabular branch of obturator a.

Ligament of head of femur

Hip joint capsule

Iliopsoas tendon

Medial circumflex femoral a.

Ascending branch
Transverse branch
Descending branch
Lateral circumflex femoral a.

Deep femoral a.

Nutrient a. of femur

Posterior view

Superior retinacular a.
Posterior retinacular a.
Inferior retinacular a.

Anastomotic vessel

Hip joint capsule

Medial circumflex femoral a.

Lateral circumflex femoral a.

Nutrient a. of femur

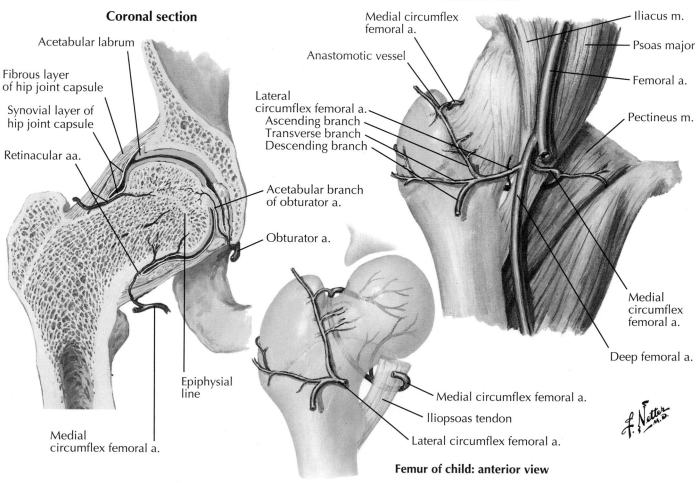

Coronal section

Acetabular labrum

Fibrous layer of hip joint capsule

Synovial layer of hip joint capsule

Retinacular aa.

Acetabular branch of obturator a.

Obturator a.

Epiphysial line

Medial circumflex femoral a.

Anterior view

Medial circumflex femoral a.

Anastomotic vessel

Lateral circumflex femoral a.
Ascending branch
Transverse branch
Descending branch

Iliacus m.

Psoas major

Femoral a.

Pectineus m.

Medial circumflex femoral a.

Deep femoral a.

Medial circumflex femoral a.

Iliopsoas tendon

Lateral circumflex femoral a.

Femur of child: anterior view

Medial view

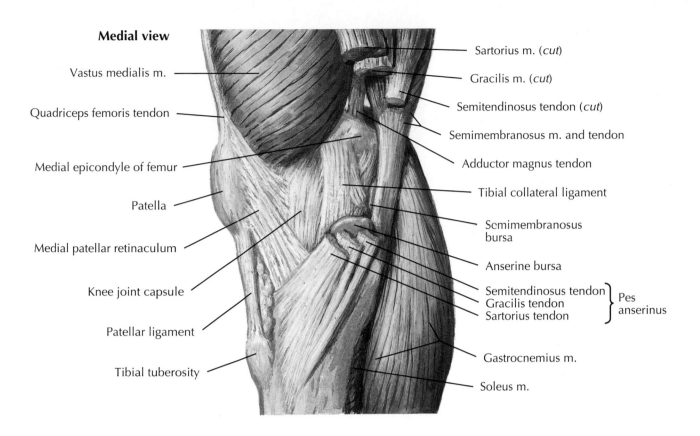

Vastus medialis m.

Quadriceps femoris tendon

Medial epicondyle of femur

Patella

Medial patellar retinaculum

Knee joint capsule

Patellar ligament

Tibial tuberosity

Sartorius m. (*cut*)

Gracilis m. (*cut*)

Semitendinosus tendon (*cut*)

Semimembranosus m. and tendon

Adductor magnus tendon

Tibial collateral ligament

Semimembranosus bursa

Anserine bursa

Semitendinosus tendon
Gracilis tendon
Sartorius tendon
} Pes anserinus

Gastrocnemius m.

Soleus m.

Lateral view

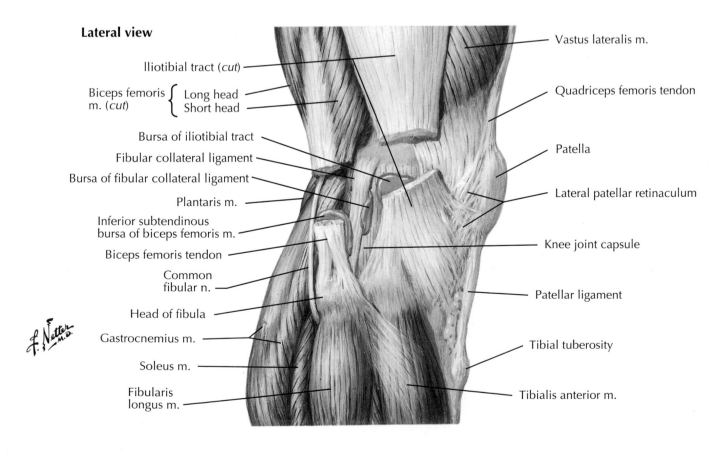

Iliotibial tract (*cut*)

Biceps femoris m. (*cut*) { Long head
Short head

Bursa of iliotibial tract

Fibular collateral ligament

Bursa of fibular collateral ligament

Plantaris m.

Inferior subtendinous bursa of biceps femoris m.

Biceps femoris tendon

Common fibular n.

Head of fibula

Gastrocnemius m.

Soleus m.

Fibularis longus m.

Vastus lateralis m.

Quadriceps femoris tendon

Patella

Lateral patellar retinaculum

Knee joint capsule

Patellar ligament

Tibial tuberosity

Tibialis anterior m.

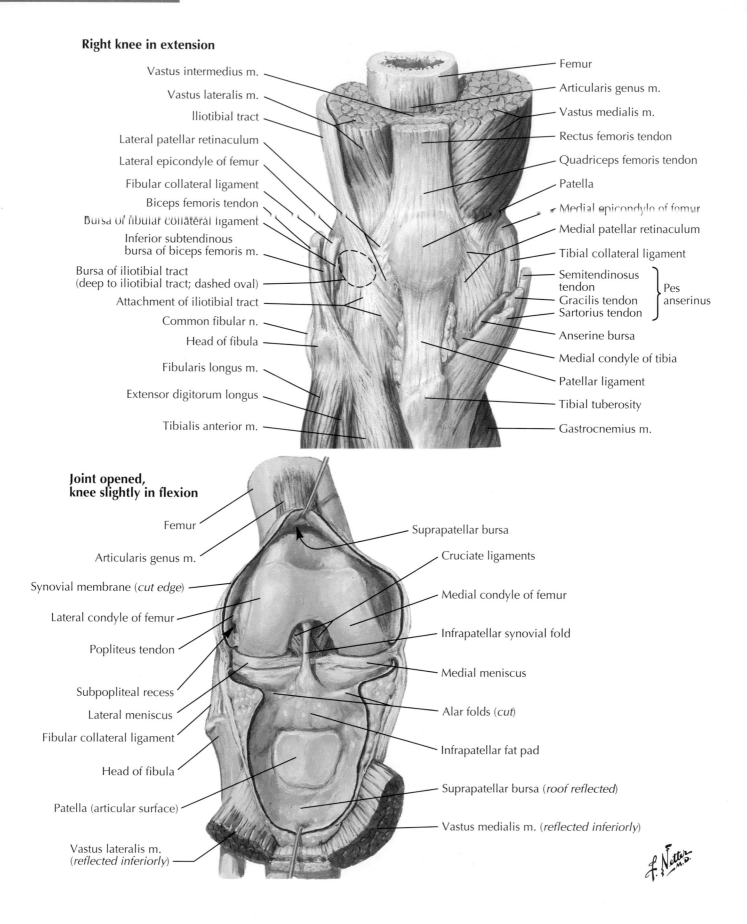

Right knee in extension

Vastus intermedius m.
Vastus lateralis m.
Iliotibial tract
Lateral patellar retinaculum
Lateral epicondyle of femur
Fibular collateral ligament
Biceps femoris tendon
Bursa of fibular collateral ligament
Inferior subtendinous bursa of biceps femoris m.
Bursa of iliotibial tract (deep to iliotibial tract; dashed oval)
Attachment of iliotibial tract
Common fibular n.
Head of fibula
Fibularis longus m.
Extensor digitorum longus
Tibialis anterior m.

Femur
Articularis genus m.
Vastus medialis m.
Rectus femoris tendon
Quadriceps femoris tendon
Patella
Medial epicondyle of femur
Medial patellar retinaculum
Tibial collateral ligament
Semitendinosus tendon
Gracilis tendon
Sartorius tendon
} Pes anserinus
Anserine bursa
Medial condyle of tibia
Patellar ligament
Tibial tuberosity
Gastrocnemius m.

Joint opened, knee slightly in flexion

Femur
Articularis genus m.
Synovial membrane (*cut edge*)
Lateral condyle of femur
Popliteus tendon
Subpopliteal recess
Lateral meniscus
Fibular collateral ligament
Head of fibula
Patella (articular surface)
Vastus lateralis m. (*reflected inferiorly*)

Suprapatellar bursa
Cruciate ligaments
Medial condyle of femur
Infrapatellar synovial fold
Medial meniscus
Alar folds (*cut*)
Infrapatellar fat pad
Suprapatellar bursa (*roof reflected*)
Vastus medialis m. (*reflected inferiorly*)

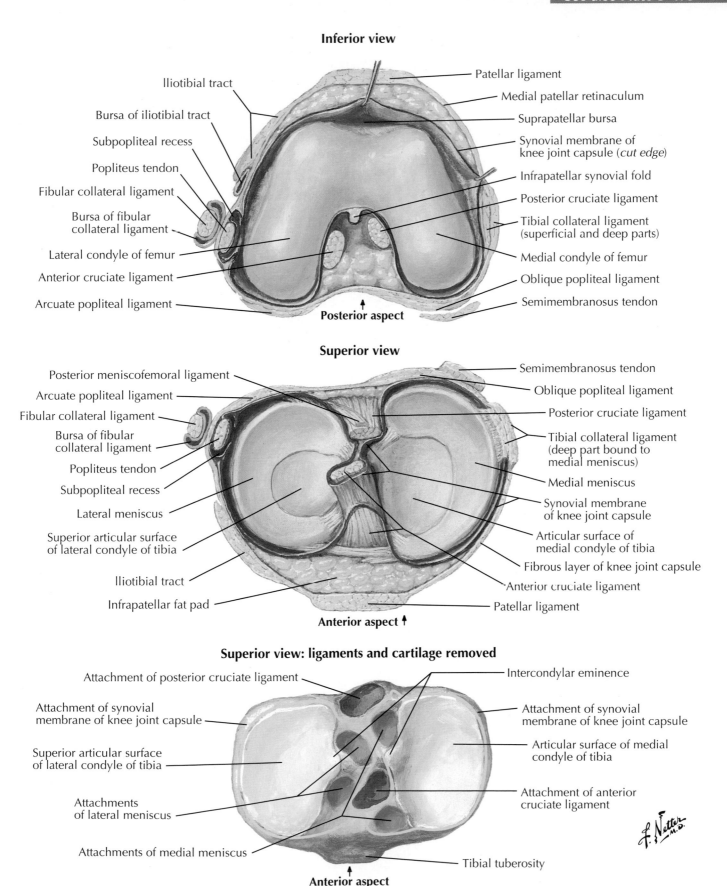

Inferior view

Iliotibial tract

Bursa of iliotibial tract

Subpopliteal recess

Popliteus tendon

Fibular collateral ligament

Bursa of fibular collateral ligament

Lateral condyle of femur

Anterior cruciate ligament

Arcuate popliteal ligament

Patellar ligament

Medial patellar retinaculum

Suprapatellar bursa

Synovial membrane of knee joint capsule (*cut edge*)

Infrapatellar synovial fold

Posterior cruciate ligament

Tibial collateral ligament (superficial and deep parts)

Medial condyle of femur

Oblique popliteal ligament

Semimembranosus tendon

Posterior aspect

Superior view

Posterior meniscofemoral ligament

Arcuate popliteal ligament

Fibular collateral ligament

Bursa of fibular collateral ligament

Popliteus tendon

Subpopliteal recess

Lateral meniscus

Superior articular surface of lateral condyle of tibia

Iliotibial tract

Infrapatellar fat pad

Semimembranosus tendon

Oblique popliteal ligament

Posterior cruciate ligament

Tibial collateral ligament (deep part bound to medial meniscus)

Medial meniscus

Synovial membrane of knee joint capsule

Articular surface of medial condyle of tibia

Fibrous layer of knee joint capsule

Anterior cruciate ligament

Patellar ligament

Anterior aspect

Superior view: ligaments and cartilage removed

Attachment of posterior cruciate ligament

Attachment of synovial membrane of knee joint capsule

Superior articular surface of lateral condyle of tibia

Attachments of lateral meniscus

Attachments of medial meniscus

Intercondylar eminence

Attachment of synovial membrane of knee joint capsule

Articular surface of medial condyle of tibia

Attachment of anterior cruciate ligament

Tibial tuberosity

Anterior aspect

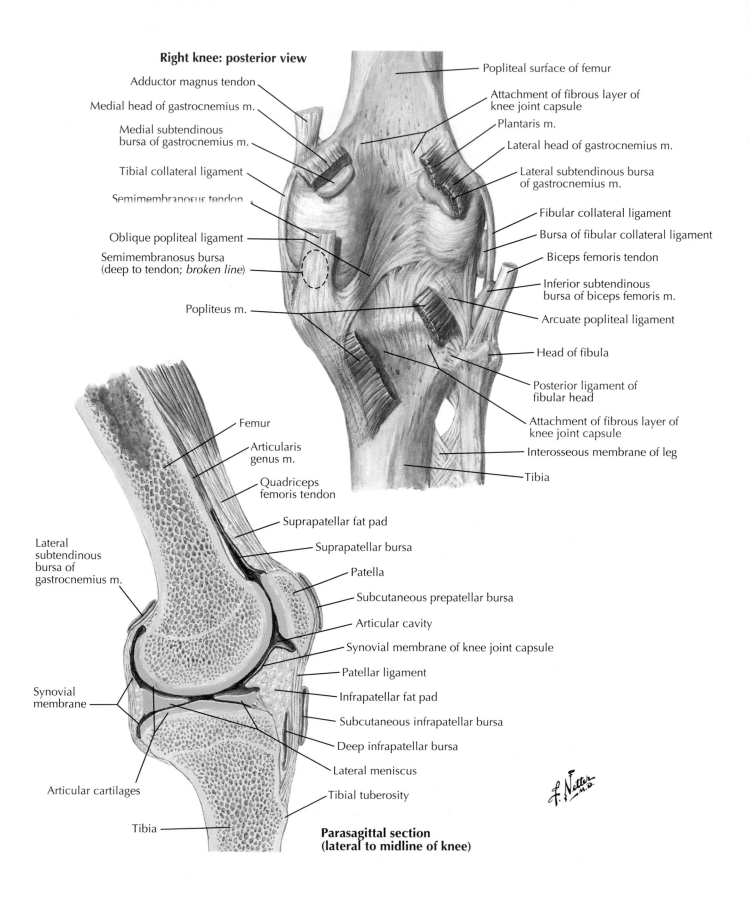

Right knee: posterior view

Adductor magnus tendon

Medial head of gastrocnemius m.

Medial subtendinous bursa of gastrocnemius m.

Tibial collateral ligament

Semimembranosus tendon

Oblique popliteal ligament

Semimembranosus bursa (deep to tendon; *broken line*)

Popliteus m.

Popliteal surface of femur

Attachment of fibrous layer of knee joint capsule

Plantaris m.

Lateral head of gastrocnemius m.

Lateral subtendinous bursa of gastrocnemius m.

Fibular collateral ligament

Bursa of fibular collateral ligament

Biceps femoris tendon

Inferior subtendinous bursa of biceps femoris m.

Arcuate popliteal ligament

Head of fibula

Posterior ligament of fibular head

Attachment of fibrous layer of knee joint capsule

Interosseous membrane of leg

Tibia

Femur

Articularis genus m.

Quadriceps femoris tendon

Suprapatellar fat pad

Suprapatellar bursa

Patella

Subcutaneous prepatellar bursa

Articular cavity

Synovial membrane of knee joint capsule

Patellar ligament

Infrapatellar fat pad

Subcutaneous infrapatellar bursa

Deep infrapatellar bursa

Lateral meniscus

Tibial tuberosity

Lateral subtendinous bursa of gastrocnemius m.

Synovial membrane

Articular cartilages

Tibia

Parasagittal section (lateral to midline of knee)

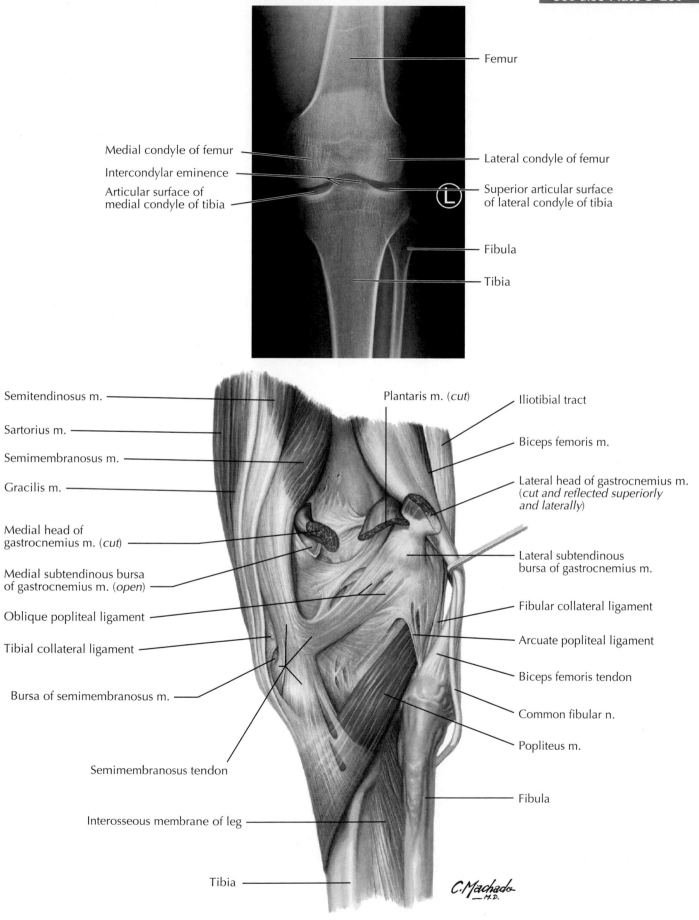

Femur

Medial condyle of femur

Intercondylar eminence

Articular surface of medial condyle of tibia

Lateral condyle of femur

Superior articular surface of lateral condyle of tibia

Fibula

Tibia

Semitendinosus m.

Sartorius m.

Semimembranosus m.

Gracilis m.

Medial head of gastrocnemius m. (*cut*)

Medial subtendinous bursa of gastrocnemius m. (*open*)

Oblique popliteal ligament

Tibial collateral ligament

Bursa of semimembranosus m.

Semimembranosus tendon

Interosseous membrane of leg

Tibia

Plantaris m. (*cut*)

Iliotibial tract

Biceps femoris m.

Lateral head of gastrocnemius m. (*cut and reflected superiorly and laterally*)

Lateral subtendinous bursa of gastrocnemius m.

Fibular collateral ligament

Arcuate popliteal ligament

Biceps femoris tendon

Common fibular n.

Popliteus m.

Fibula

C. Machado _M.D._

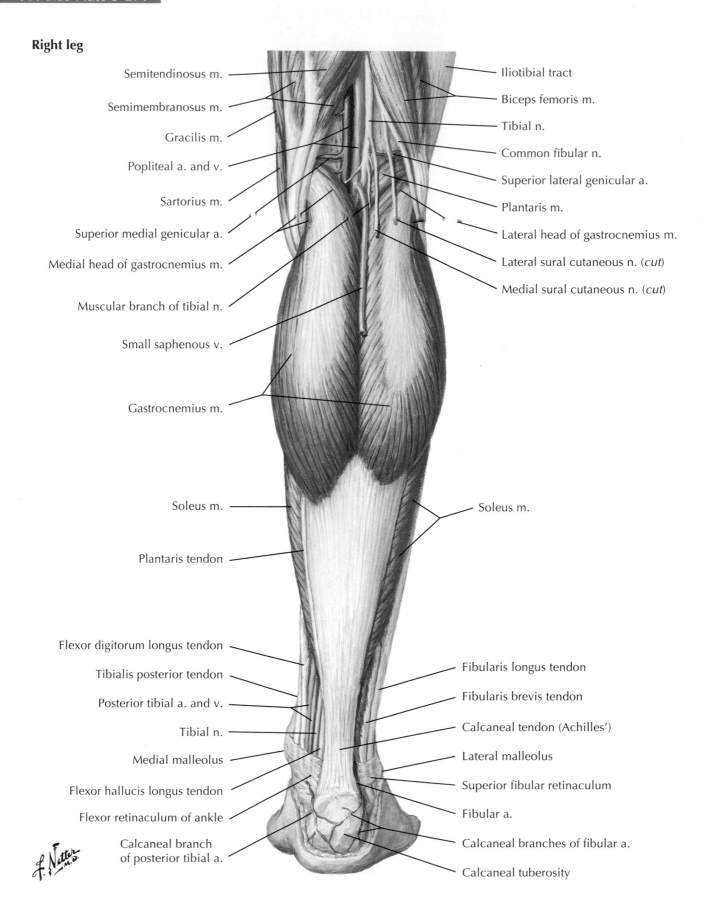

Right leg

Semitendinosus m.

Semimembranosus m.

Gracilis m.

Popliteal a. and v.

Sartorius m.

Superior medial genicular a.

Medial head of gastrocnemius m.

Muscular branch of tibial n.

Small saphenous v.

Gastrocnemius m.

Soleus m.

Plantaris tendon

Flexor digitorum longus tendon

Tibialis posterior tendon

Posterior tibial a. and v.

Tibial n.

Medial malleolus

Flexor hallucis longus tendon

Flexor retinaculum of ankle

Calcaneal branch
of posterior tibial a.

Iliotibial tract

Biceps femoris m.

Tibial n.

Common fibular n.

Superior lateral genicular a.

Plantaris m.

Lateral head of gastrocnemius m.

Lateral sural cutaneous n. (*cut*)

Medial sural cutaneous n. (*cut*)

Soleus m.

Fibularis longus tendon

Fibularis brevis tendon

Calcaneal tendon (Achilles')

Lateral malleolus

Superior fibular retinaculum

Fibular a.

Calcaneal branches of fibular a.

Calcaneal tuberosity

Right leg

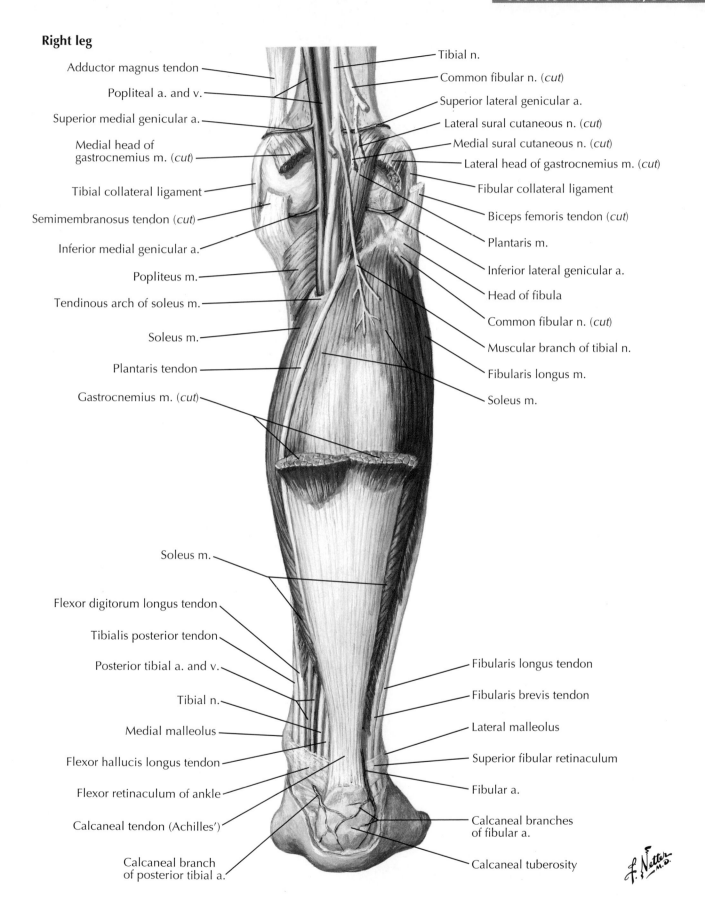

Adductor magnus tendon

Popliteal a. and v.

Superior medial genicular a.

Medial head of gastrocnemius m. (*cut*)

Tibial collateral ligament

Semimembranosus tendon (*cut*)

Inferior medial genicular a.

Popliteus m.

Tendinous arch of soleus m.

Soleus m.

Plantaris tendon

Gastrocnemius m. (*cut*)

Soleus m.

Flexor digitorum longus tendon

Tibialis posterior tendon

Posterior tibial a. and v.

Tibial n.

Medial malleolus

Flexor hallucis longus tendon

Flexor retinaculum of ankle

Calcaneal tendon (Achilles')

Calcaneal branch of posterior tibial a.

Tibial n.

Common fibular n. (*cut*)

Superior lateral genicular a.

Lateral sural cutaneous n. (*cut*)

Medial sural cutaneous n. (*cut*)

Lateral head of gastrocnemius m. (*cut*)

Fibular collateral ligament

Biceps femoris tendon (*cut*)

Plantaris m.

Inferior lateral genicular a.

Head of fibula

Common fibular n. (*cut*)

Muscular branch of tibial n.

Fibularis longus m.

Soleus m.

Fibularis longus tendon

Fibularis brevis tendon

Lateral malleolus

Superior fibular retinaculum

Fibular a.

Calcaneal branches of fibular a.

Calcaneal tuberosity

Right leg

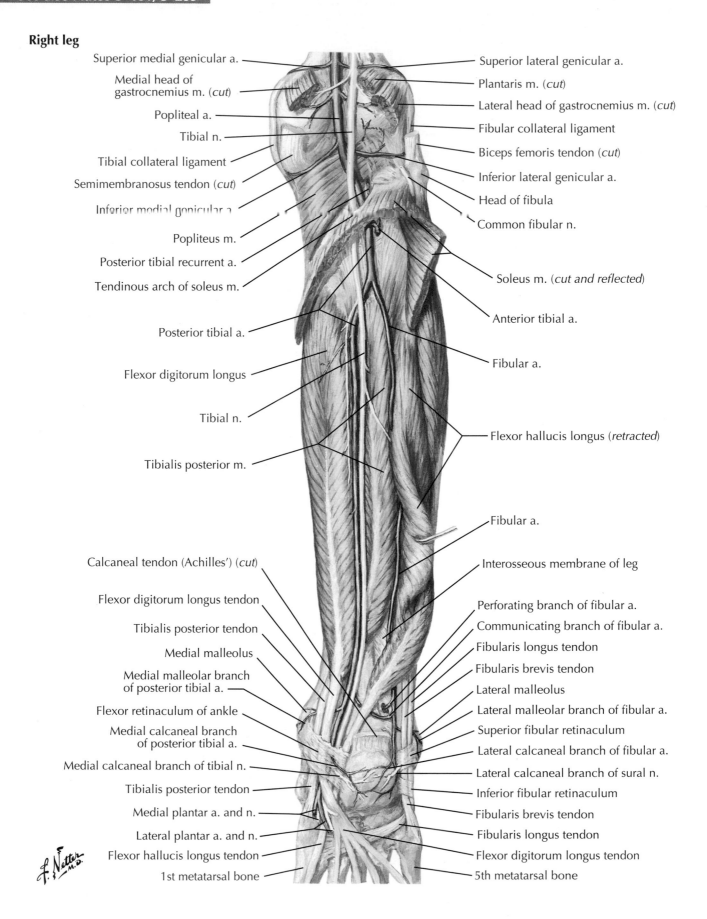

Superior medial genicular a.

Medial head of gastrocnemius m. (*cut*)

Popliteal a.

Tibial n.

Tibial collateral ligament

Semimembranosus tendon (*cut*)

Inferior medial genicular a.

Popliteus m.

Posterior tibial recurrent a.

Tendinous arch of soleus m.

Posterior tibial a.

Flexor digitorum longus

Tibial n.

Tibialis posterior m.

Calcaneal tendon (Achilles') (*cut*)

Flexor digitorum longus tendon

Tibialis posterior tendon

Medial malleolus

Medial malleolar branch of posterior tibial a.

Flexor retinaculum of ankle

Medial calcaneal branch of posterior tibial a.

Medial calcaneal branch of tibial n.

Tibialis posterior tendon

Medial plantar a. and n.

Lateral plantar a. and n.

Flexor hallucis longus tendon

1st metatarsal bone

Superior lateral genicular a.

Plantaris m. (*cut*)

Lateral head of gastrocnemius m. (*cut*)

Fibular collateral ligament

Biceps femoris tendon (*cut*)

Inferior lateral genicular a.

Head of fibula

Common fibular n.

Soleus m. (*cut and reflected*)

Anterior tibial a.

Fibular a.

Flexor hallucis longus (*retracted*)

Fibular a.

Interosseous membrane of leg

Perforating branch of fibular a.

Communicating branch of fibular a.

Fibularis longus tendon

Fibularis brevis tendon

Lateral malleolus

Lateral malleolar branch of fibular a.

Superior fibular retinaculum

Lateral calcaneal branch of fibular a.

Lateral calcaneal branch of sural n.

Inferior fibular retinaculum

Fibularis brevis tendon

Fibularis longus tendon

Flexor digitorum longus tendon

5th metatarsal bone

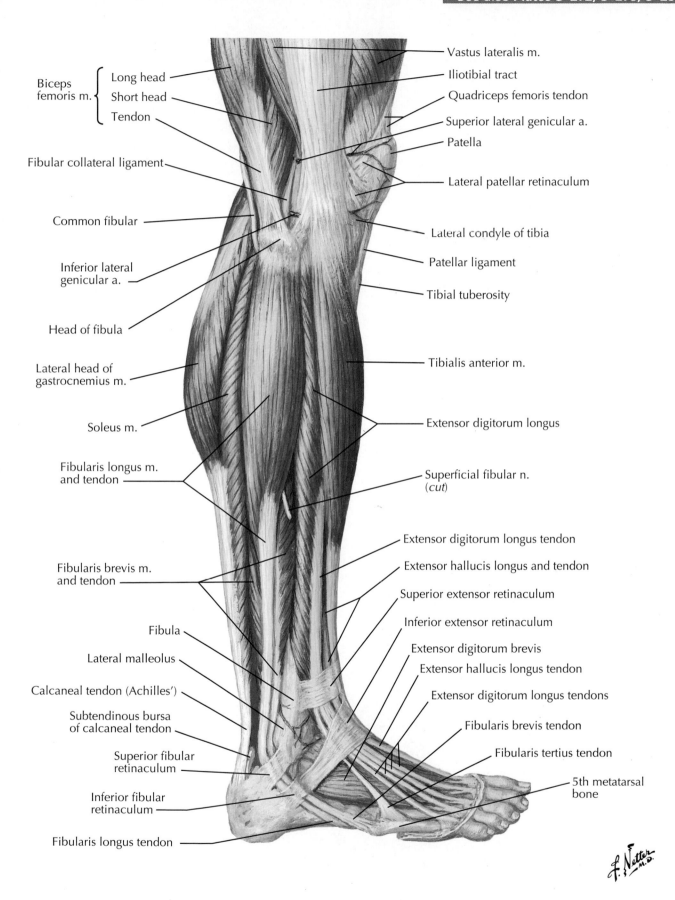

Biceps femoris m. {
- Long head
- Short head
- Tendon
}

Fibular collateral ligament

Common fibular

Inferior lateral genicular a.

Head of fibula

Lateral head of gastrocnemius m.

Soleus m.

Fibularis longus m. and tendon

Fibularis brevis m. and tendon

Fibula

Lateral malleolus

Calcaneal tendon (Achilles')

Subtendinous bursa of calcaneal tendon

Superior fibular retinaculum

Inferior fibular retinaculum

Fibularis longus tendon

Vastus lateralis m.

Iliotibial tract

Quadriceps femoris tendon

Superior lateral genicular a.

Patella

Lateral patellar retinaculum

Lateral condyle of tibia

Patellar ligament

Tibial tuberosity

Tibialis anterior m.

Extensor digitorum longus

Superficial fibular n. (*cut*)

Extensor digitorum longus tendon

Extensor hallucis longus and tendon

Superior extensor retinaculum

Inferior extensor retinaculum

Extensor digitorum brevis

Extensor hallucis longus tendon

Extensor digitorum longus tendons

Fibularis brevis tendon

Fibularis tertius tendon

5th metatarsal bone

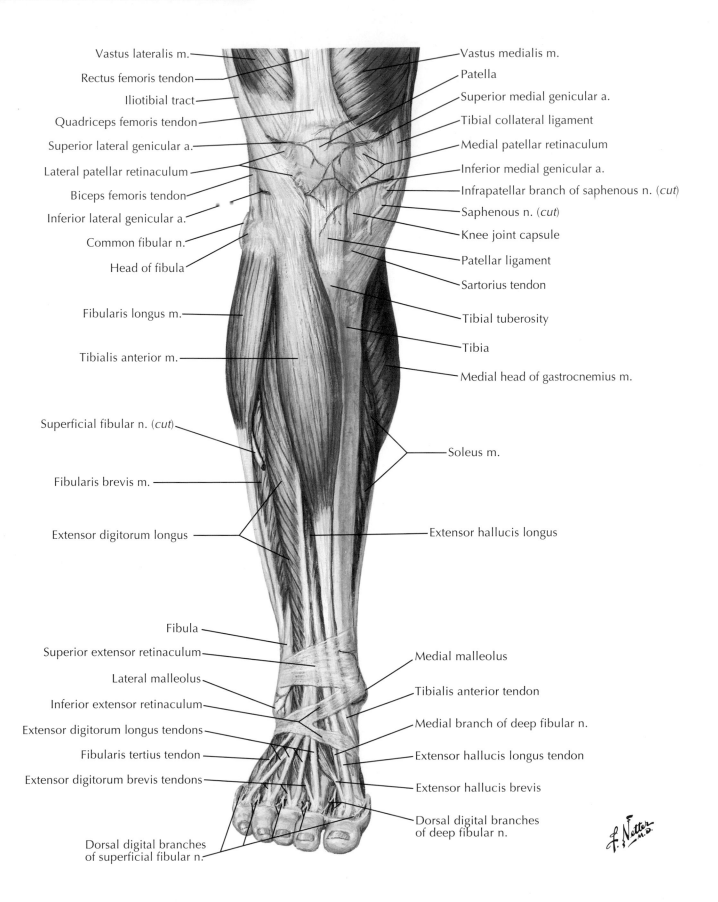

Vastus lateralis m.

Rectus femoris tendon

Iliotibial tract

Quadriceps femoris tendon

Superior lateral genicular a.

Lateral patellar retinaculum

Biceps femoris tendon

Inferior lateral genicular a.

Common fibular n.

Head of fibula

Fibularis longus m.

Tibialis anterior m.

Superficial fibular n. (*cut*)

Fibularis brevis m.

Extensor digitorum longus

Fibula

Superior extensor retinaculum

Lateral malleolus

Inferior extensor retinaculum

Extensor digitorum longus tendons

Fibularis tertius tendon

Extensor digitorum brevis tendons

Dorsal digital branches
of superficial fibular n.

Vastus medialis m.

Patella

Superior medial genicular a.

Tibial collateral ligament

Medial patellar retinaculum

Inferior medial genicular a.

Infrapatellar branch of saphenous n. (*cut*)

Saphenous n. (*cut*)

Knee joint capsule

Patellar ligament

Sartorius tendon

Tibial tuberosity

Tibia

Medial head of gastrocnemius m.

Soleus m.

Extensor hallucis longus

Medial malleolus

Tibialis anterior tendon

Medial branch of deep fibular n.

Extensor hallucis longus tendon

Extensor hallucis brevis

Dorsal digital branches
of deep fibular n.

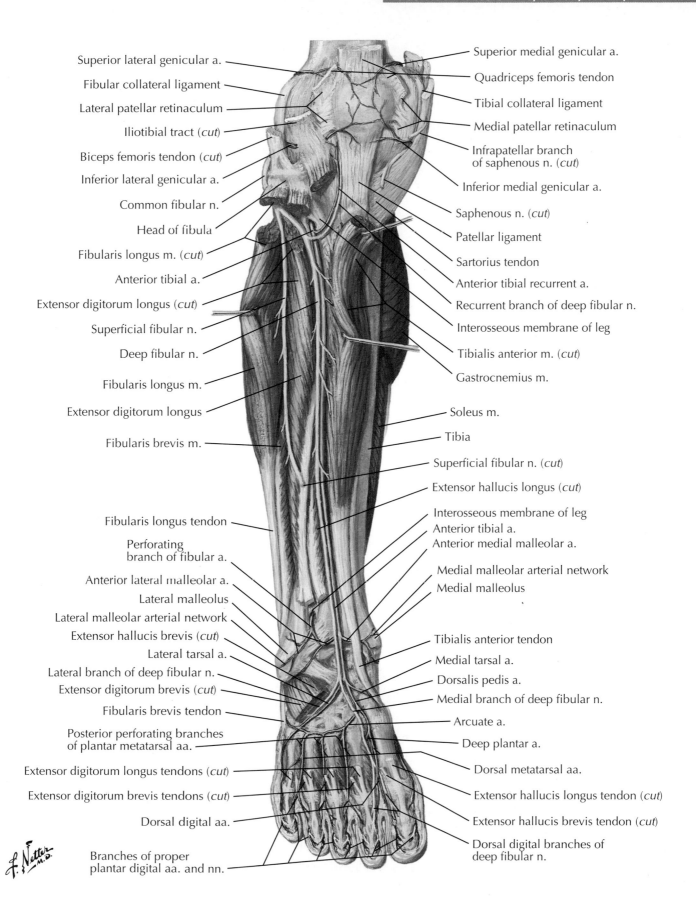

Superior lateral genicular a.

Fibular collateral ligament

Lateral patellar retinaculum

Iliotibial tract (*cut*)

Biceps femoris tendon (*cut*)

Inferior lateral genicular a.

Common fibular n.

Head of fibula

Fibularis longus m. (*cut*)

Anterior tibial a.

Extensor digitorum longus (*cut*)

Superficial fibular n.

Deep fibular n.

Fibularis longus m.

Extensor digitorum longus

Fibularis brevis m.

Fibularis longus tendon

Perforating branch of fibular a.

Anterior lateral malleolar a.

Lateral malleolus

Lateral malleolar arterial network

Extensor hallucis brevis (*cut*)

Lateral tarsal a.

Lateral branch of deep fibular n.

Extensor digitorum brevis (*cut*)

Fibularis brevis tendon

Posterior perforating branches of plantar metatarsal aa.

Extensor digitorum longus tendons (*cut*)

Extensor digitorum brevis tendons (*cut*)

Dorsal digital aa.

Branches of proper plantar digital aa. and nn.

Superior medial genicular a.

Quadriceps femoris tendon

Tibial collateral ligament

Medial patellar retinaculum

Infrapatellar branch of saphenous n. (*cut*)

Inferior medial genicular a.

Saphenous n. (*cut*)

Patellar ligament

Sartorius tendon

Anterior tibial recurrent a.

Recurrent branch of deep fibular n.

Interosseous membrane of leg

Tibialis anterior m. (*cut*)

Gastrocnemius m.

Soleus m.

Tibia

Superficial fibular n. (*cut*)

Extensor hallucis longus (*cut*)

Interosseous membrane of leg
Anterior tibial a.
Anterior medial malleolar a.

Medial malleolar arterial network
Medial malleolus

Tibialis anterior tendon

Medial tarsal a.

Dorsalis pedis a.

Medial branch of deep fibular n.

Arcuate a.

Deep plantar a.

Dorsal metatarsal aa.

Extensor hallucis longus tendon (*cut*)

Extensor hallucis brevis tendon (*cut*)

Dorsal digital branches of deep fibular n.

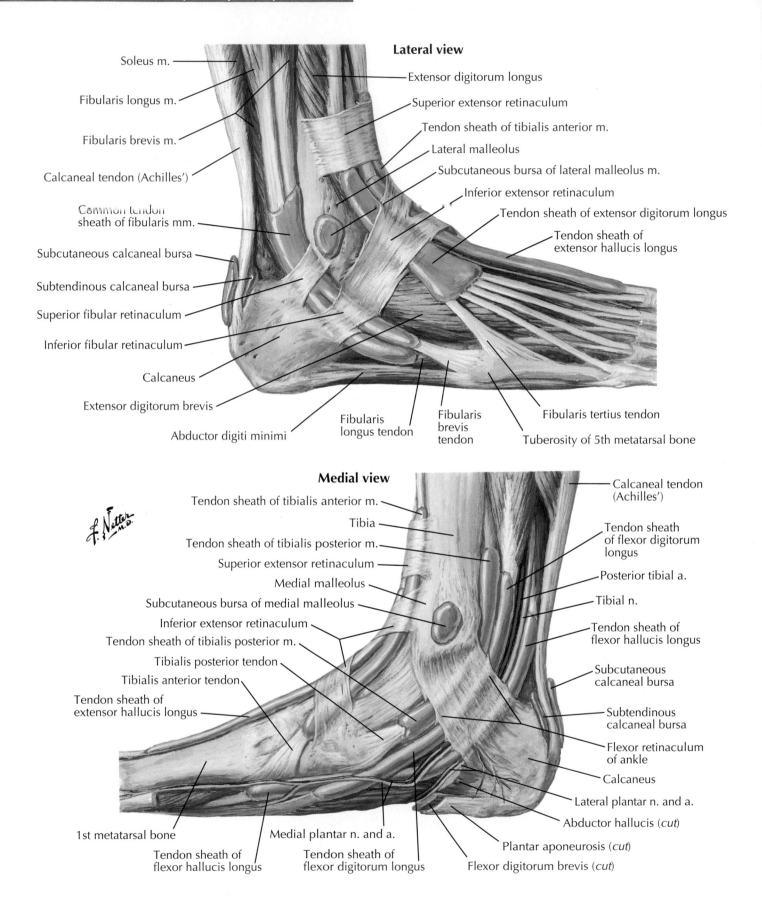

Lateral view

Soleus m.

Fibularis longus m.

Fibularis brevis m.

Calcaneal tendon (Achilles')

Common tendon sheath of fibularis mm.

Subcutaneous calcaneal bursa

Subtendinous calcaneal bursa

Superior fibular retinaculum

Inferior fibular retinaculum

Calcaneus

Extensor digitorum brevis

Abductor digiti minimi

Fibularis longus tendon

Fibularis brevis tendon

Extensor digitorum longus

Superior extensor retinaculum

Tendon sheath of tibialis anterior m.

Lateral malleolus

Subcutaneous bursa of lateral malleolus m.

Inferior extensor retinaculum

Tendon sheath of extensor digitorum longus

Tendon sheath of extensor hallucis longus

Fibularis tertius tendon

Tuberosity of 5th metatarsal bone

Medial view

Tendon sheath of tibialis anterior m.

Tibia

Tendon sheath of tibialis posterior m.

Superior extensor retinaculum

Medial malleolus

Subcutaneous bursa of medial malleolus

Inferior extensor retinaculum

Tendon sheath of tibialis posterior m.

Tibialis posterior tendon

Tibialis anterior tendon

Tendon sheath of extensor hallucis longus

1st metatarsal bone

Tendon sheath of flexor hallucis longus

Medial plantar n. and a.

Tendon sheath of flexor digitorum longus

Calcaneal tendon (Achilles')

Tendon sheath of flexor digitorum longus

Posterior tibial a.

Tibial n.

Tendon sheath of flexor hallucis longus

Subcutaneous calcaneal bursa

Subtendinous calcaneal bursa

Flexor retinaculum of ankle

Calcaneus

Lateral plantar n. and a.

Abductor hallucis (*cut*)

Plantar aponeurosis (*cut*)

Flexor digitorum brevis (*cut*)

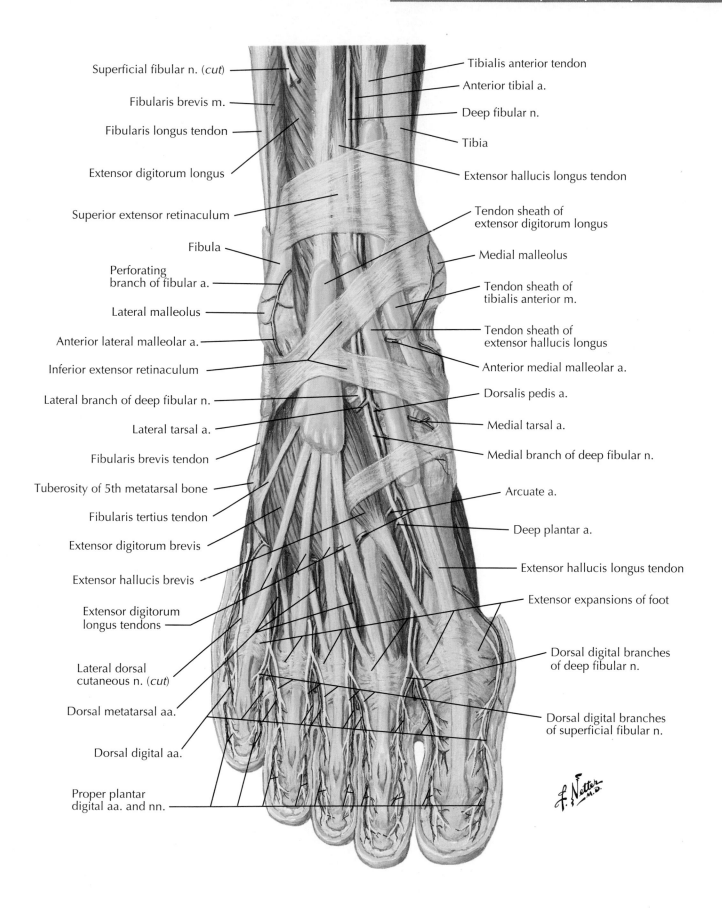

Superficial fibular n. (*cut*)

Fibularis brevis m.

Fibularis longus tendon

Extensor digitorum longus

Superior extensor retinaculum

Fibula

Perforating branch of fibular a.

Lateral malleolus

Anterior lateral malleolar a.

Inferior extensor retinaculum

Lateral branch of deep fibular n.

Lateral tarsal a.

Fibularis brevis tendon

Tuberosity of 5th metatarsal bone

Fibularis tertius tendon

Extensor digitorum brevis

Extensor hallucis brevis

Extensor digitorum longus tendons

Lateral dorsal cutaneous n. (*cut*)

Dorsal metatarsal aa.

Dorsal digital aa.

Proper plantar digital aa. and nn.

Tibialis anterior tendon

Anterior tibial a.

Deep fibular n.

Tibia

Extensor hallucis longus tendon

Tendon sheath of extensor digitorum longus

Medial malleolus

Tendon sheath of tibialis anterior m.

Tendon sheath of extensor hallucis longus

Anterior medial malleolar a.

Dorsalis pedis a.

Medial tarsal a.

Medial branch of deep fibular n.

Arcuate a.

Deep plantar a.

Extensor hallucis longus tendon

Extensor expansions of foot

Dorsal digital branches of deep fibular n.

Dorsal digital branches of superficial fibular n.

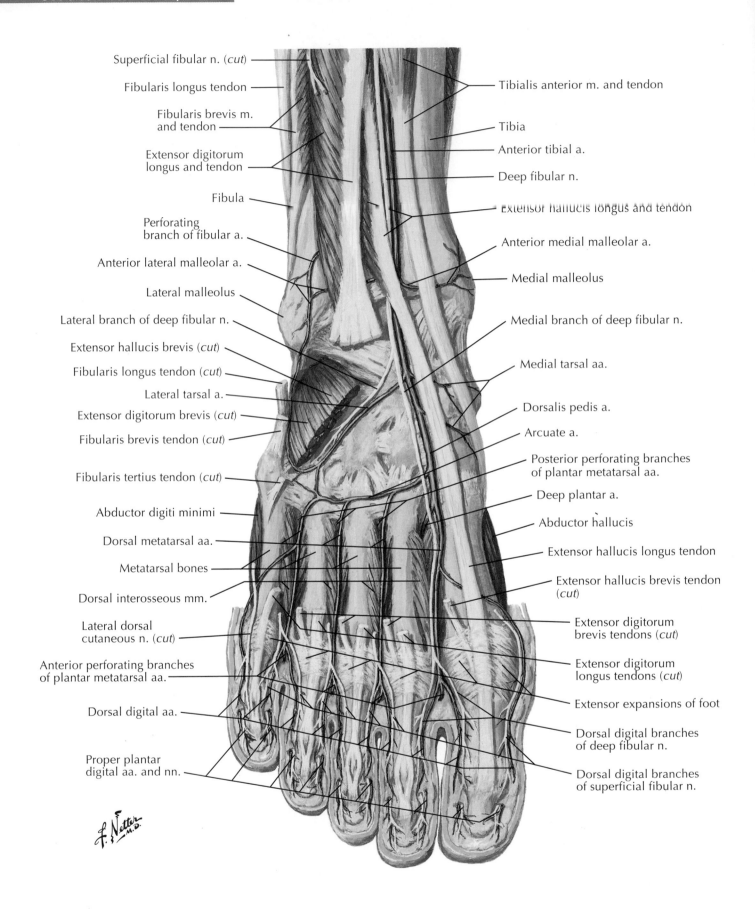

Superficial fibular n. (*cut*)

Fibularis longus tendon

Fibularis brevis m. and tendon

Extensor digitorum longus and tendon

Fibula

Perforating branch of fibular a.

Anterior lateral malleolar a.

Lateral malleolus

Lateral branch of deep fibular n.

Extensor hallucis brevis (*cut*)

Fibularis longus tendon (*cut*)

Lateral tarsal a.

Extensor digitorum brevis (*cut*)

Fibularis brevis tendon (*cut*)

Fibularis tertius tendon (*cut*)

Abductor digiti minimi

Dorsal metatarsal aa.

Metatarsal bones

Dorsal interosseous mm.

Lateral dorsal cutaneous n. (*cut*)

Anterior perforating branches of plantar metatarsal aa.

Dorsal digital aa.

Proper plantar digital aa. and nn.

Tibialis anterior m. and tendon

Tibia

Anterior tibial a.

Deep fibular n.

Extensor hallucis longus and tendon

Anterior medial malleolar a.

Medial malleolus

Medial branch of deep fibular n.

Medial tarsal aa.

Dorsalis pedis a.

Arcuate a.

Posterior perforating branches of plantar metatarsal aa.

Deep plantar a.

Abductor hallucis

Extensor hallucis longus tendon

Extensor hallucis brevis tendon (*cut*)

Extensor digitorum brevis tendons (*cut*)

Extensor digitorum longus tendons (*cut*)

Extensor expansions of foot

Dorsal digital branches of deep fibular n.

Dorsal digital branches of superficial fibular n.

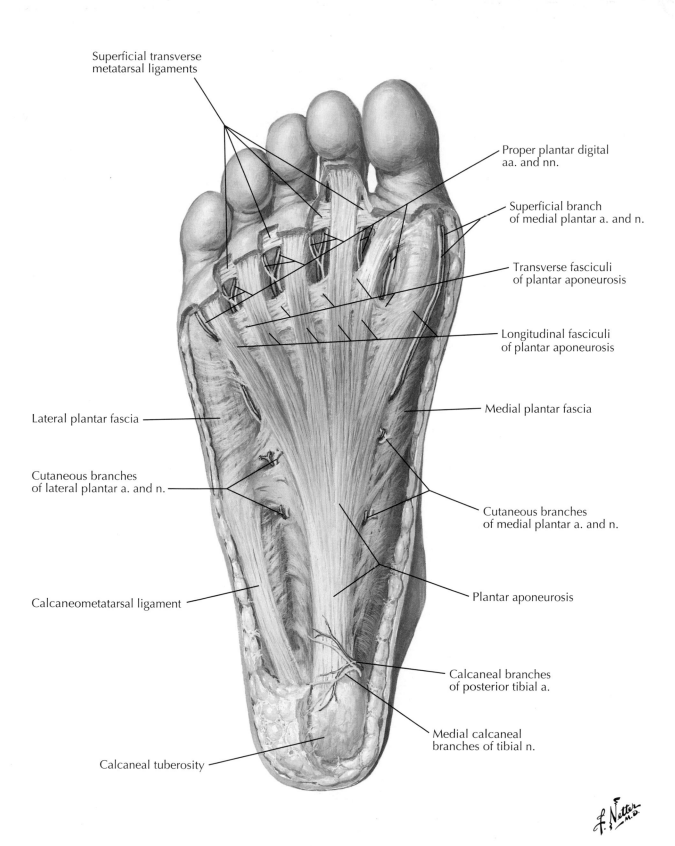

Superficial transverse
metatarsal ligaments

Proper plantar digital
aa. and nn.

Superficial branch
of medial plantar a. and n.

Transverse fasciculi
of plantar aponeurosis

Longitudinal fasciculi
of plantar aponeurosis

Medial plantar fascia

Lateral plantar fascia

Cutaneous branches
of lateral plantar a. and n.

Cutaneous branches
of medial plantar a. and n.

Calcaneometatarsal ligament

Plantar aponeurosis

Calcaneal branches
of posterior tibial a.

Medial calcaneal
branches of tibial n.

Calcaneal tuberosity

First layer of muscles in bold

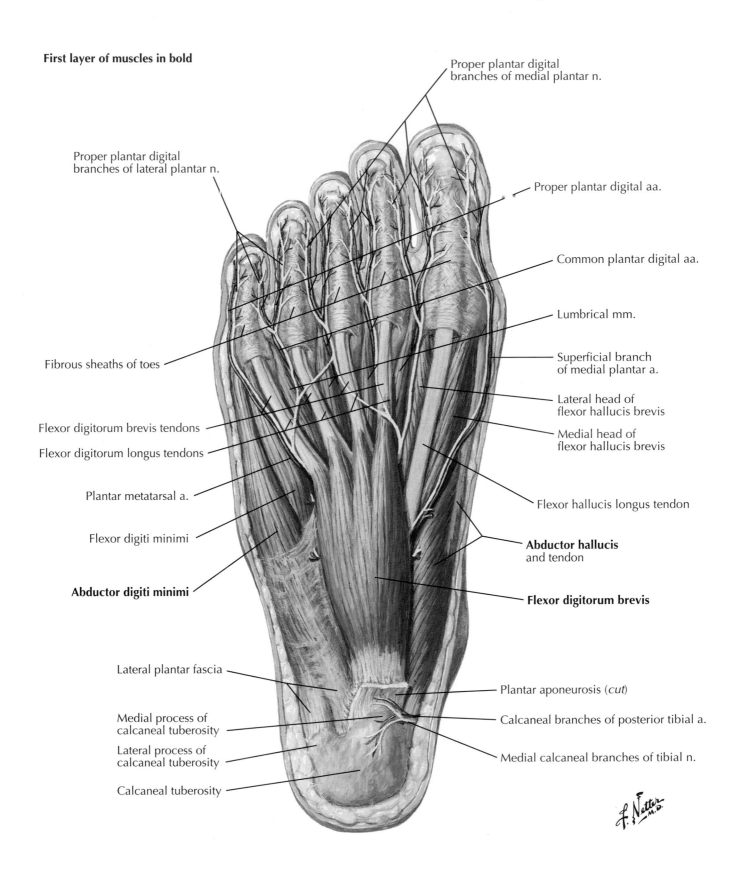

Proper plantar digital
branches of medial plantar n.

Proper plantar digital
branches of lateral plantar n.

Proper plantar digital aa.

Common plantar digital aa.

Lumbrical mm.

Fibrous sheaths of toes

Superficial branch
of medial plantar a.

Lateral head of
flexor hallucis brevis

Flexor digitorum brevis tendons

Medial head of
flexor hallucis brevis

Flexor digitorum longus tendons

Plantar metatarsal a.

Flexor hallucis longus tendon

Flexor digiti minimi

Abductor hallucis
and tendon

Abductor digiti minimi

Flexor digitorum brevis

Lateral plantar fascia

Plantar aponeurosis (cut)

Medial process of
calcaneal tuberosity

Calcaneal branches of posterior tibial a.

Lateral process of
calcaneal tuberosity

Medial calcaneal branches of tibial n.

Calcaneal tuberosity

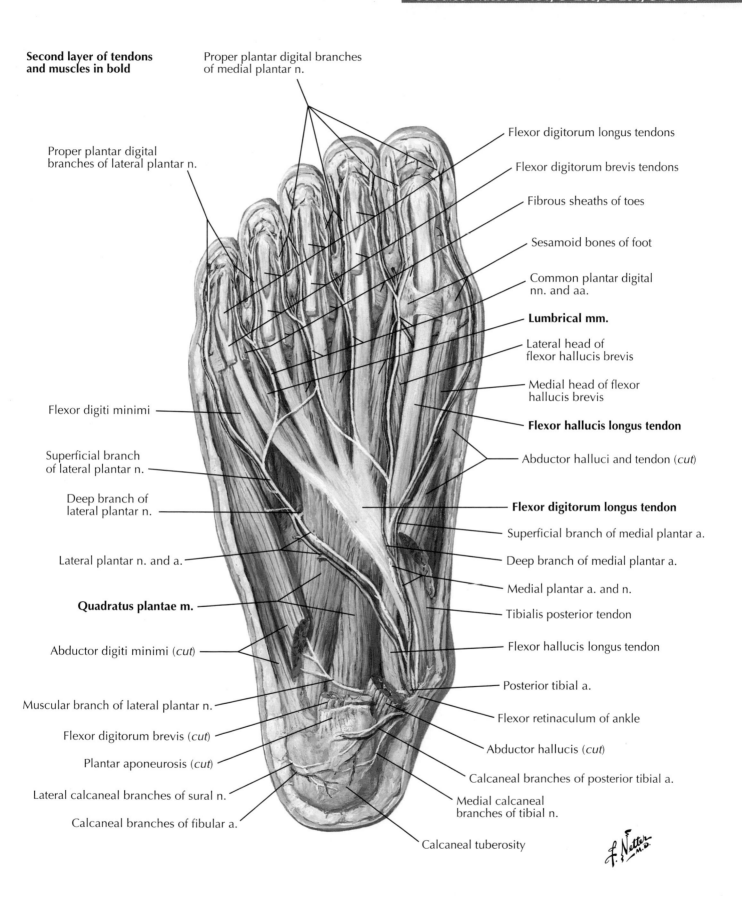

Second layer of tendons and muscles in bold

Proper plantar digital branches of medial plantar n.

Proper plantar digital branches of lateral plantar n.

Flexor digitorum longus tendons

Flexor digitorum brevis tendons

Fibrous sheaths of toes

Sesamoid bones of foot

Common plantar digital nn. and aa.

Lumbrical mm.

Lateral head of flexor hallucis brevis

Medial head of flexor hallucis brevis

Flexor hallucis longus tendon

Abductor halluci and tendon (*cut*)

Flexor digiti minimi

Superficial branch of lateral plantar n.

Deep branch of lateral plantar n.

Flexor digitorum longus tendon

Superficial branch of medial plantar a.

Deep branch of medial plantar a.

Lateral plantar n. and a.

Medial plantar a. and n.

Quadratus plantae m.

Tibialis posterior tendon

Abductor digiti minimi (*cut*)

Flexor hallucis longus tendon

Muscular branch of lateral plantar n.

Posterior tibial a.

Flexor retinaculum of ankle

Flexor digitorum brevis (*cut*)

Abductor hallucis (*cut*)

Plantar aponeurosis (*cut*)

Calcaneal branches of posterior tibial a.

Lateral calcaneal branches of sural n.

Medial calcaneal branches of tibial n.

Calcaneal branches of fibular a.

Calcaneal tuberosity

f. Netter M.D.

Third layer of muscles in bold

Proper plantar digital branches
of medial plantar n.

Proper plantar digital
branches of lateral plantar n.

Proper plantar digital
branch of medial plantar a.

Anterior perforating branches
of plantar metatarsal aa.

Lumbrical tendons (*cut*)

Sesamoid bones of foot

**Transverse head
of adductor hallucis**

Flexor digitorum
longus tendons (*cut*)

**Oblique head of
adductor hallucis**

Flexor digitorum
brevis tendons (*cut*)

**Medial head of
flexor hallucis brevis**

Flexor digiti minimi

**Lateral head of
flexor hallucis brevis**

Superficial branch of
medial plantar a.

Plantar metatarsal aa.

Flexor hallucis
longus tendon (*cut*)

Plantar interosseous mm.

Abductor hallucis (*cut*)

Superficial branch
of lateral plantar n.

Deep branch of
medial plantar a.

Plantar arch

Flexor digitorum
longus tendon (*cut*)

Deep branch of lateral plantar n.

Tibialis posterior tendon

Tuberosity of 5th metatarsal bone

Fibularis brevis tendon

Medial plantar a. and n.

Plantar tendinous
sheath of fibularis longus m.

Flexor hallucis longus tendon

Fibularis longus tendon

Flexor retinaculum of ankle

Quadratus plantae m.
(*cut and slightly retracted*)

Abductor hallucis (*cut*)

Lateral plantar a. and n.

Flexor digitorum brevis (*cut*)

Abductor digiti minimi (*cut*)

Plantar aponeurosis (*cut*)

Medial calcaneal branches of tibial n.

Lateral calcaneal branches of sural n.

Calcaneal branches of fibular a.

Calcaneal branch
of posterior tibial a.

Calcaneal tuberosity

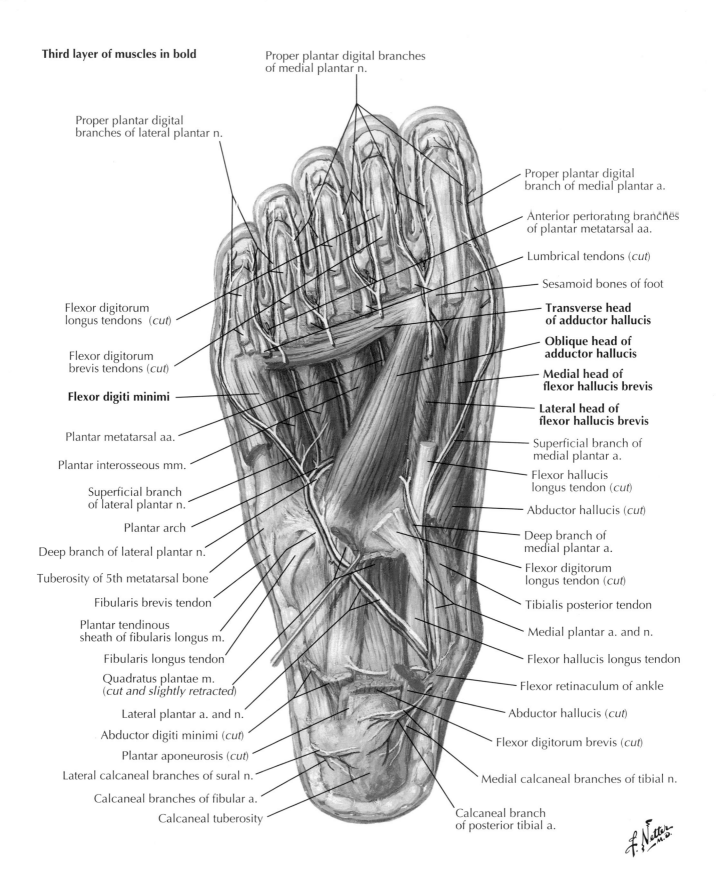

Dorsal view

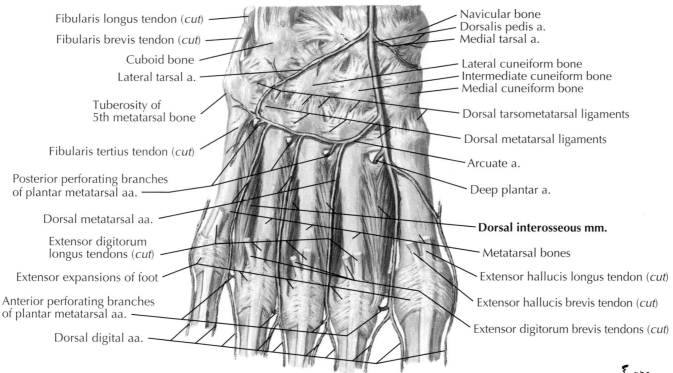

Fibularis longus tendon (*cut*)
Fibularis brevis tendon (*cut*)
Cuboid bone
Lateral tarsal a.
Tuberosity of 5th metatarsal bone
Fibularis tertius tendon (*cut*)
Posterior perforating branches of plantar metatarsal aa.
Dorsal metatarsal aa.
Extensor digitorum longus tendons (*cut*)
Extensor expansions of foot
Anterior perforating branches of plantar metatarsal aa.
Dorsal digital aa.

Navicular bone
Dorsalis pedis a.
Medial tarsal a.
Lateral cuneiform bone
Intermediate cuneiform bone
Medial cuneiform bone
Dorsal tarsometatarsal ligaments
Dorsal metatarsal ligaments
Arcuate a.
Deep plantar a.
Dorsal interosseous mm.
Metatarsal bones
Extensor hallucis longus tendon (*cut*)
Extensor hallucis brevis tendon (*cut*)
Extensor digitorum brevis tendons (*cut*)

Plantar view

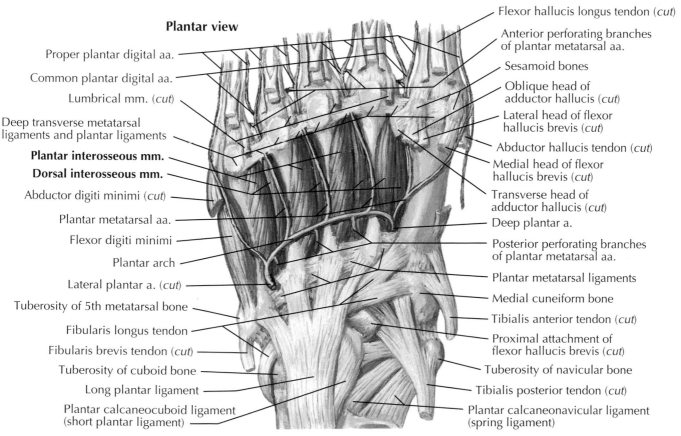

Proper plantar digital aa.
Common plantar digital aa.
Lumbrical mm. (*cut*)
Deep transverse metatarsal ligaments and plantar ligaments
Plantar interosseous mm.
Dorsal interosseous mm.
Abductor digiti minimi (*cut*)
Plantar metatarsal aa.
Flexor digiti minimi
Plantar arch
Lateral plantar a. (*cut*)
Tuberosity of 5th metatarsal bone
Fibularis longus tendon
Fibularis brevis tendon (*cut*)
Tuberosity of cuboid bone
Long plantar ligament
Plantar calcaneocuboid ligament (short plantar ligament)

Flexor hallucis longus tendon (*cut*)
Anterior perforating branches of plantar metatarsal aa.
Sesamoid bones
Oblique head of adductor hallucis (*cut*)
Lateral head of flexor hallucis brevis (*cut*)
Abductor hallucis tendon (*cut*)
Medial head of flexor hallucis brevis (*cut*)
Transverse head of adductor hallucis (*cut*)
Deep plantar a.
Posterior perforating branches of plantar metatarsal aa.
Plantar metatarsal ligaments
Medial cuneiform bone
Tibialis anterior tendon (*cut*)
Proximal attachment of flexor hallucis brevis (*cut*)
Tuberosity of navicular bone
Tibialis posterior tendon (*cut*)
Plantar calcaneonavicular ligament (spring ligament)

Fourth layer of muscles in bold

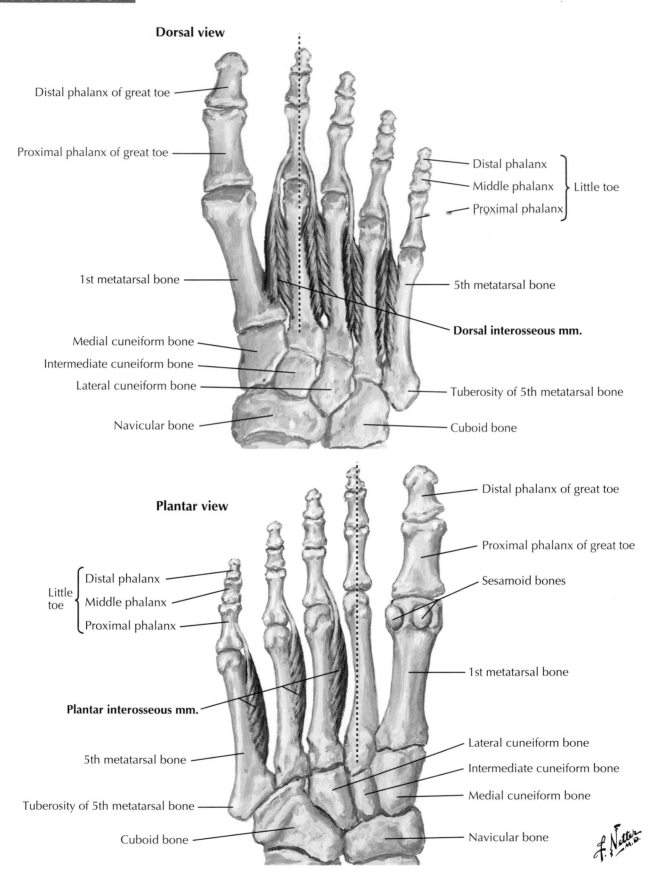

Dorsal view

Distal phalanx of great toe

Proximal phalanx of great toe

Distal phalanx
Middle phalanx } Little toe
Proximal phalanx

1st metatarsal bone

5th metatarsal bone

Dorsal interosseous mm.

Medial cuneiform bone

Intermediate cuneiform bone

Lateral cuneiform bone

Tuberosity of 5th metatarsal bone

Navicular bone

Cuboid bone

Plantar view

Distal phalanx of great toe

Proximal phalanx of great toe

Sesamoid bones

Little toe {
Distal phalanx
Middle phalanx
Proximal phalanx
}

Plantar interosseous mm.

1st metatarsal bone

5th metatarsal bone

Lateral cuneiform bone

Intermediate cuneiform bone

Medial cuneiform bone

Tuberosity of 5th metatarsal bone

Cuboid bone

Navicular bone

Note: dashed line is the line of reference for abduction and adduction of the toes

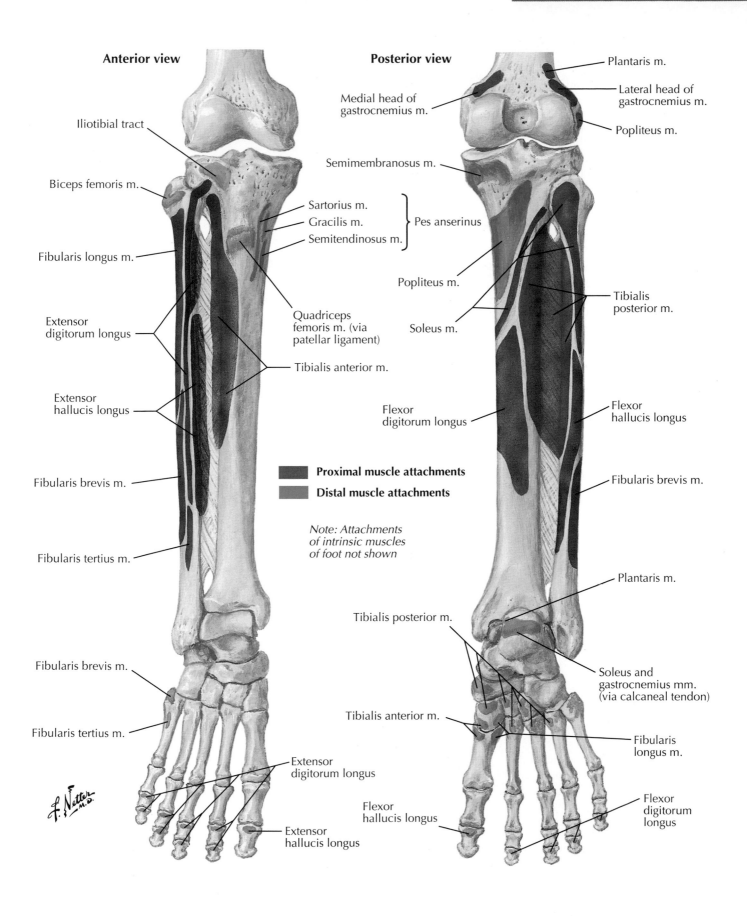

Anterior view

Posterior view

Iliotibial tract

Biceps femoris m.

Fibularis longus m.

Extensor digitorum longus

Extensor hallucis longus

Fibularis brevis m.

Fibularis tertius m.

Fibularis brevis m.

Fibularis tertius m.

Medial head of gastrocnemius m.

Semimembranosus m.

Sartorius m.
Gracilis m.
Semitendinosus m.
} Pes anserinus

Quadriceps femoris m. (via patellar ligament)

Tibialis anterior m.

Plantaris m.

Lateral head of gastrocnemius m.

Popliteus m.

Popliteus m.

Soleus m.

Tibialis posterior m.

Flexor digitorum longus

Flexor hallucis longus

Fibularis brevis m.

Plantaris m.

Tibialis posterior m.

Tibialis anterior m.

Soleus and gastrocnemius mm. (via calcaneal tendon)

Fibularis longus m.

Extensor digitorum longus

Extensor hallucis longus

Flexor hallucis longus

Flexor digitorum longus

Proximal muscle attachments
Distal muscle attachments

Note: Attachments of intrinsic muscles of foot not shown

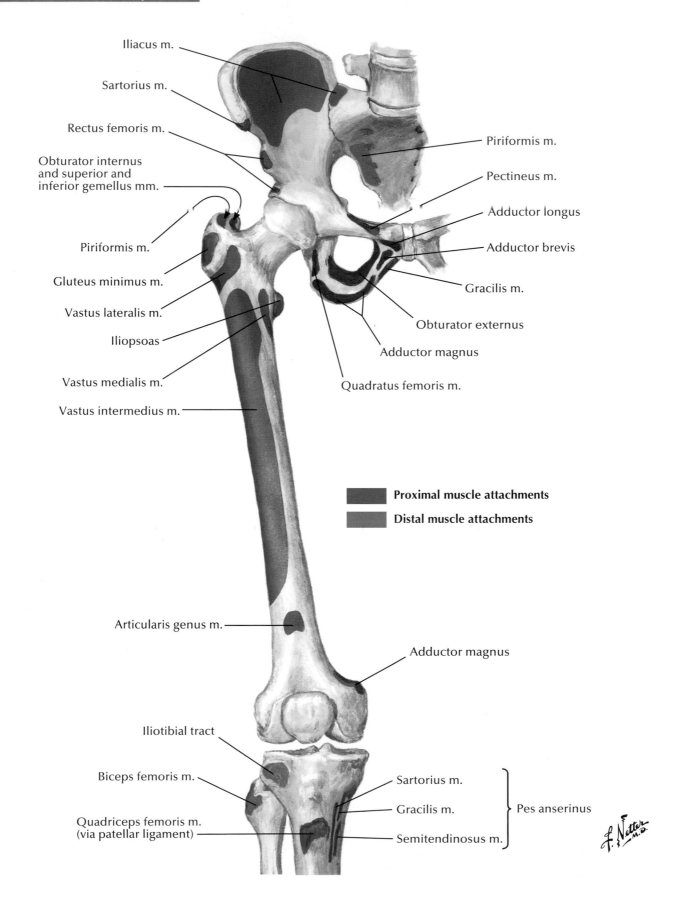

Iliacus m.

Sartorius m.

Rectus femoris m.

Obturator internus
and superior and
inferior gemellus mm.

Piriformis m.

Gluteus minimus m.

Vastus lateralis m.

Iliopsoas

Vastus medialis m.

Vastus intermedius m.

Piriformis m.

Pectineus m.

Adductor longus

Adductor brevis

Gracilis m.

Obturator externus

Adductor magnus

Quadratus femoris m.

Proximal muscle attachments

Distal muscle attachments

Articularis genus m.

Adductor magnus

Iliotibial tract

Biceps femoris m.

Sartorius m.

Gracilis m.

Pes anserinus

Quadriceps femoris m.
(via patellar ligament)

Semitendinosus m.

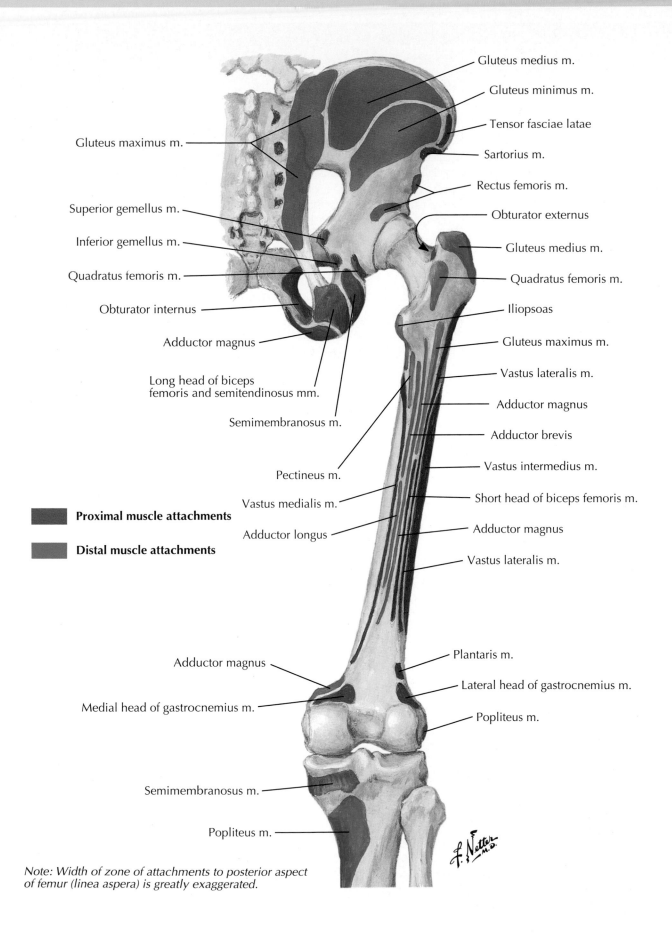

Gluteus medius m.

Gluteus minimus m.

Tensor fasciae latae

Sartorius m.

Rectus femoris m.

Obturator externus

Gluteus medius m.

Quadratus femoris m.

Iliopsoas

Gluteus maximus m.

Vastus lateralis m.

Adductor magnus

Adductor brevis

Vastus intermedius m.

Short head of biceps femoris m.

Adductor magnus

Vastus lateralis m.

Plantaris m.

Lateral head of gastrocnemius m.

Popliteus m.

Gluteus maximus m.

Superior gemellus m.

Inferior gemellus m.

Quadratus femoris m.

Obturator internus

Adductor magnus

Long head of biceps femoris and semitendinosus mm.

Semimembranosus m.

Pectineus m.

Vastus medialis m.

Adductor longus

Adductor magnus

Medial head of gastrocnemius m.

Semimembranosus m.

Popliteus m.

■ **Proximal muscle attachments**

■ **Distal muscle attachments**

Note: Width of zone of attachments to posterior aspect of femur (linea aspera) is greatly exaggerated.

ANATOMIC STRUCTURES	CLINICAL IMPORTANCE	PLATE NUMBERS
Head and Neck		
Muscles of facial expression	Used to assess function of facial nerve (CN VII) during cranial nerve examination; may become weak or paralyzed with CN VII dysfunction (e.g., Bell's palsy); often targeted with botulinum toxin injections for aesthetics (wrinkles), headache, and teeth grinding	S–65, S–184
Sternocleidomastoid muscle	Important landmark that divides neck into anterior and posterior cervical triangles; palpated to identify "nerve point of neck" for administration of anesthesia to cervical plexus and also used as landmark for insertion of central venous catheters; in children, abnormal shortening or fibrosis of sternocleidomastoid muscle results in head tilt, a condition known as torticollis	S–194
Sternocleidomastoid and trapezius muscles	Used to assess function of the spinal accessory nerve (CN XI) during cranial nerve examination	S–69, S–194
Muscles of mastication	Used to assess function of the trigeminal nerve (CN V) during cranial nerve examination; masseter is involved in teeth grinding and associated headaches, which can be treated with botulinum toxin injection	S–186, S–187
Levator veli palatini and musculus uvulae	Used to assess function of vagus nerve (CN X) during cranial nerve examination; contralateral deviation of uvula of soft palate during elevation indicates CN X dysfunction	S–398
Genioglossus muscle	Used to assess hypoglossal nerve (CN XII) function during cranial nerve examination; tongue deviates to side of lesion when protruded following CN XII injury	S–70, S–401
Styloglossus muscle	Posterosuperior movement covers laryngeal inlet, protecting vocal cords and airway, which is especially crucial if epiglottis has been surgically removed because of malignancy	S–372, S–401, S–408
Stapedius muscle	Smallest skeletal muscle in the body, regulates movement of stapes bone to control amplitude of sound	S–95
Levator palpebrae superioris and superior tarsal muscle	Responsible for elevating eyelid; ptosis indicates pathologic change in oculomotor nerve (CN III) or sympathetic fibers (if only superior tarsal muscle is affected)	S–82, S–84
Extraocular muscles	Used to assess function of oculomotor (CN III), trochlear (CN IV), and abducens (CN VI) nerves during cranial nerve examination; anormalities in tone result in disconjugate eye movements, a condition known as strabismus	S–84, S–86
Dilator pupillae	Important in assessment of sympathetic function in head; lack of dilation indicates interruption in sympathetic outflow (e.g., Horner's syndrome)	S–58, S–89
Sphincter pupillae	Involved in pupillary light reflex and accommodation reflex	S–58, S–89
Pterygomandibular raphe	Valuable intraoral landmark for inferior alveolar nerve blocks to anesthetize mandibular teeth	S–73, S–323, S–407, S–408
Modiolus of angulus oris	Fibrous intersection of facial muscles, approximately 1 cm lateral to corner of mouth; valuable landmark when considering reconstruction of injured facial muscles associated with oral region	S–184
Retromolar fossa	Important clinical landmark for reducing dislocated temporomandibular joints; buccal and lingual nerves cross retromolar fossa and may be harmed in dental molar implant surgery	S–135
Back		
Trapezius muscle	Responsible for holding scapula against thoracic wall against gravity; drooping of shoulder may indicate injury to spinal accessory nerve	S–199
Deep (or intrinsic) back muscles	Microscopic stretching or tearing of muscle fibers produces back strain, a common cause of low back pain	S–200, S–201

Table 4.1   **Structures with High Clinical Significance**

ANATOMIC STRUCTURES	CLINICAL IMPORTANCE	PLATE NUMBERS
Thorax		
Diaphragm	Widening of esophageal hiatus at T8 vertebral level or congenital defect allows for protrusion of stomach into thorax (hiatal hernia), which increases incidence of gastroesophageal reflux	S–208, S–413
Abdomen		
Linea alba	Site used for abdominal wall incisions as location provides access to many organs during exploratory surgery and is unlikely to have significant vessels crossing it	S–8, S–213
Inguinal ligament	Surface landmark from ASIS to pubic tubercle that marks division between abdomen and lower limb; formed from external abdominal oblique aponeurosis	S–8, S–212
Inguinal (Hesselbach's) triangle	Important region on interior surface of anterior abdominal wall, bounded by inferior epigastric vessels, inguinal ligament, and rectus abdominis muscle, through which abdominal contents may herniate to produce direct inguinal hernias	S–216, S–217
Deep inguinal ring	Slit-like opening in transversalis fascia just above midpoint of inguinal ligament and lateral to inferior epigastric artery, through which abdominal contents may herniate to produce indirect inguinal hernias	S–215, S–216
Superficial inguinal ring	Triangular opening in external abdominal oblique aponeurosis superior and lateral to pubic tubercle and medial to inferior epigastric artery, through which abdominal contents may herniate to produce indirect inguinal hernias	S–211, S–215, S–216
Femoral ring	Superior opening of femoral canal bounded by medial part of inguinal ligament, femoral vein, and lacunar ligament; abdominal contents may herniate through femoral ring into upper femoral triangle situated inferolateral to pubic tubercle, producing femoral hernia	S–218
Esophageal hiatus	Widening of opening through diaphragm allows stomach to protrude into mediastinum, which can increase incidence of gastroesophageal reflux disease (GERD)	S–103, S–422
Rectus abdominis muscle	Separation (abdominal diastasis) commonly caused from multiple pregnancies, abdominal surgeries, and excessive weight gain; bleeding of inferior epigastric artery may cause blood to accumulate in rectus abdominis muscle (rectus sheath hematoma), which may be mistaken for acute abdominal pathologies, such as appendicitis	S–212
Pelvis		
Pelvic diaphragm (levator ani and coccygeus muscle)	Provides support to urethrovesical angle, helping to maintain urinary continence; weakness or injury during childbirth can lead to stress urinary incontinence in women	S–466, S–479
Pelvic extraperitoneal (endopelvic) fascia	Weakness or tearing of endopelvic fascial ligaments (e.g., pubovesical or cardinal ligaments) may result in prolapse of pelvic organs	S–464, S–488
Perineal body	Tearing of perineal body may occur during childbirth; prophylactic incision into or lateral to perineal body, known as episiotomy, may be performed to facilitate vaginal delivery in some circumstances	S–493
Upper Limb		
Palmar aponeurosis	Progressive fibrosis may result in nodules and eventually a palpable cord that limits finger extension (Dupuytren's contracture)	S–250
Rotator cuff muscles	Injuries to this group of muscles can result from acute injury or chronic overuse and are a common cause of shoulder pain and disability	S–157, S–232, S–240
Supraspinatus tendon	Most commonly torn rotator cuff tendon	S–231, S–232, S–235, S–240

Structures with High Clinical Significance

Table 4.2

ANATOMIC STRUCTURES	CLINICAL IMPORTANCE	PLATE NUMBERS
Upper Limb—Continued		
Biceps brachii tendon	Can rupture from sudden load on muscle when contracting; used in flexor compartment reflex assessing C5 and C6 spinal nerves	S–238, S–239
Long head of biceps brachii muscle	Tendon of long head of biceps brachii muscle can cause shoulder pain from tendinosis of intraarticular portion and can rupture in elderly persons from falls on outstretched arm; when long head has been ruptured, it usually tears from supraglenoid tubercle and retracts down into arm; muscle commonly bulges (Popeye deformity) at midbody of humerus; spontaneous rupture may occur in amyloidosis, an infiltrative disease that also causes cardiomyopathy	S–239
Muscles of posterior compartment of forearm	Repetitive use of muscles arising from common extensor origin can damage tendons and produce pain over lateral epicondyle region (epicondylitis); activities such as swinging tennis racquet and poor technique with hammer can result in "tennis elbow"; muscle most likely involved is extensor carpi radialis brevis	S–243
Muscles of anterior compartment of forearm	Repetitive use of muscles arising from common flexor origin can damage tendons and produce pain over medial epicondyle region (golfer's elbow)	S–245, S–246
Lower Limb		
Muscles of medial compartment of thigh	Excessive stretching or tearing of adductor and gracilis muscles is common in sports that require repeated sprints or quick changes in direction (e.g., soccer, hockey)	S–268, S–269
Patellar ligament	Striking patellar ligament with reflex hammer elicits patellar (knee jerk) reflex to test L3–L4 spinal cord levels (innervation of quadriceps femoris muscle by femoral nerve)	S–268, S–280
Iliotibial tract	Can cause lateral knee pain in runners when tight iliotibial tract rubs repetitively across lateral epicondyle of femur (iliotibial band syndrome)	S–272, S–279
Semitendinosus and semimembranosus muscles and long head of biceps femoris muscle	Excessive stretching or tearing of hamstring muscles occurs most often during high-speed running or activities with high kicks	S–273
Piriformis muscle	Piriformis muscle strain or structural variations (e.g., split piriformis muscle) may produce compression of sciatic nerve	S–274
Gluteus medius and minimus muscles	Paralysis results in contralateral pelvic dip due to weakened hip abduction when standing on affected limb (Trendelenburg sign or gait)	S–275
Calcaneal (Achilles') tendon	Inflammation results from repetitive stress on tendon, often from running on uneven surfaces; extreme stress may cause tendon to rupture; striking calcaneal tendon with reflex hammer elicits ankle (ankle jerk) reflex to test S1–S2 spinal cord levels (innervation of superficial calf muscles by tibial nerve)	S–180, S–284, S–285
Compartments of leg	Acute compartment syndrome may occur after trauma, such as long bone fracture, that increases compartment pressure and thereby compromises vascular flow; symptoms include pain, paresthesias, pallor, pulselessness, and paralysis (five P's); fasciotomy often required to relieve pressure	S–555
Plantar aponeurosis	Inflammation results from increased tension, weight, or overuse, causing heel and foot pain (plantar fasciitis)	S–293

*Selections are based largely on clinical data and commonly discussed clinical correlations in macroscopic ("gross") anatomy courses.

CARDIOVASCULAR SYSTEM 5

ELECTRONIC BONUS PLATES

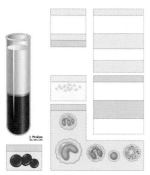

S–BP 44 Cardiovascular System: Composition of Blood

S–BP 45 Arterial Wall

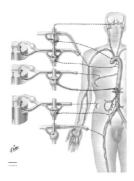

S–BP 46 Innervation of Blood Vessels: Schema

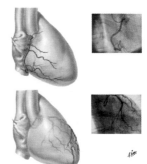

S–BP 47 Coronary Arteries: Right Anterolateral Views with Arteriograms

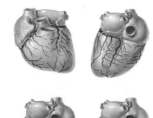

S–BP 48 Coronary Arteries and Cardiac Veins: Variations

S–BP 49 Subclavian Artery

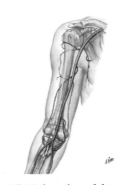

S–BP 50 Arteries of Arm and Proximal Forearm

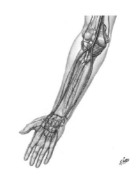

S–BP 51 Arteries of Forearm and Hand

ELECTRONIC BONUS PLATES—*cont'd*

S–BP 52 Arteries of Knee
and Foot

S–BP 53 Arteries of Thigh
and Knee

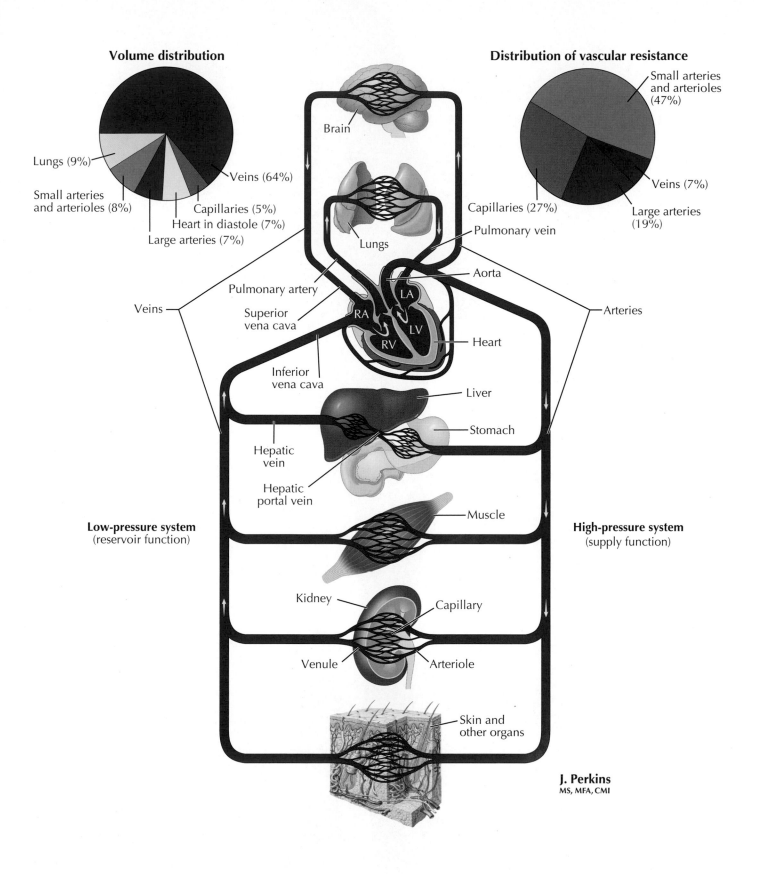

Volume distribution

Lungs (9%)

Small arteries
and arterioles (8%)

Capillaries (5%)

Heart in diastole (7%)

Large arteries (7%)

Veins (64%)

Distribution of vascular resistance

Small arteries
and arterioles
(47%)

Capillaries (27%)

Veins (7%)

Large arteries
(19%)

Brain

Lungs

Pulmonary artery

Superior
vena cava

Veins

Inferior
vena cava

Hepatic
vein

Hepatic
portal vein

RA

LA

RV

LV

Pulmonary vein

Aorta

Arteries

Heart

Liver

Stomach

Muscle

Kidney

Capillary

Venule

Arteriole

Skin and
other organs

Low-pressure system
(reservoir function)

High-pressure system
(supply function)

J. Perkins
MS, MFA, CMI

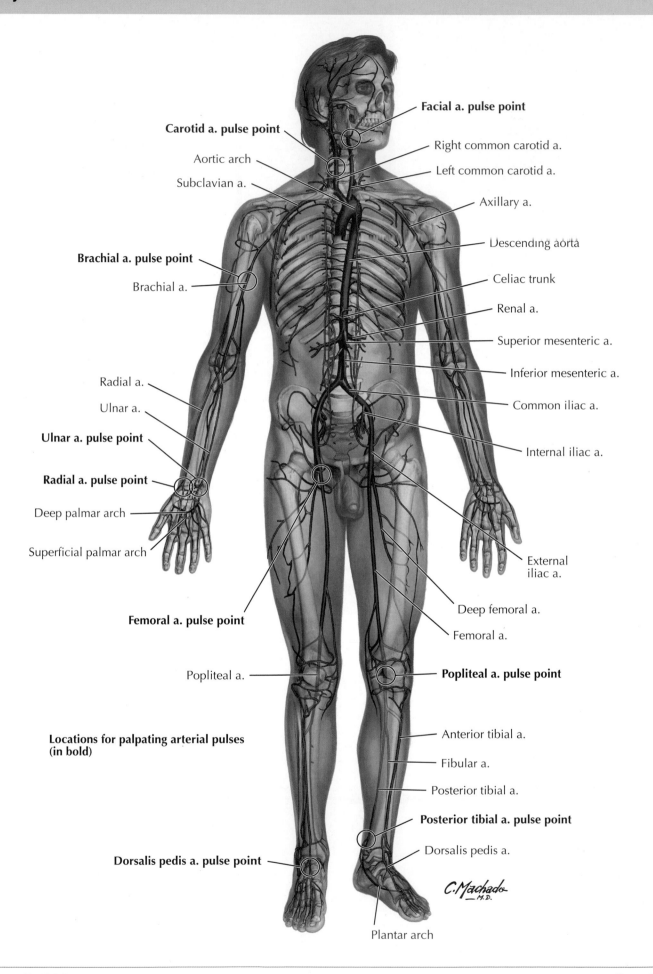

Carotid a. pulse point

Facial a. pulse point

Aortic arch

Right common carotid a.

Subclavian a.

Left common carotid a.

Axillary a.

Descending aorta

Brachial a. pulse point

Celiac trunk

Brachial a.

Renal a.

Superior mesenteric a.

Inferior mesenteric a.

Radial a.

Common iliac a.

Ulnar a.

Ulnar a. pulse point

Internal iliac a.

Radial a. pulse point

Deep palmar arch

Superficial palmar arch

External
iliac a.

Deep femoral a.

Femoral a. pulse point

Femoral a.

Popliteal a.

Popliteal a. pulse point

Locations for palpating arterial pulses
(in bold)

Anterior tibial a.

Fibular a.

Posterior tibial a.

Posterior tibial a. pulse point

Dorsalis pedis a. pulse point

Dorsalis pedis a.

Plantar arch

Major veins

Superficial veins
Deep veins

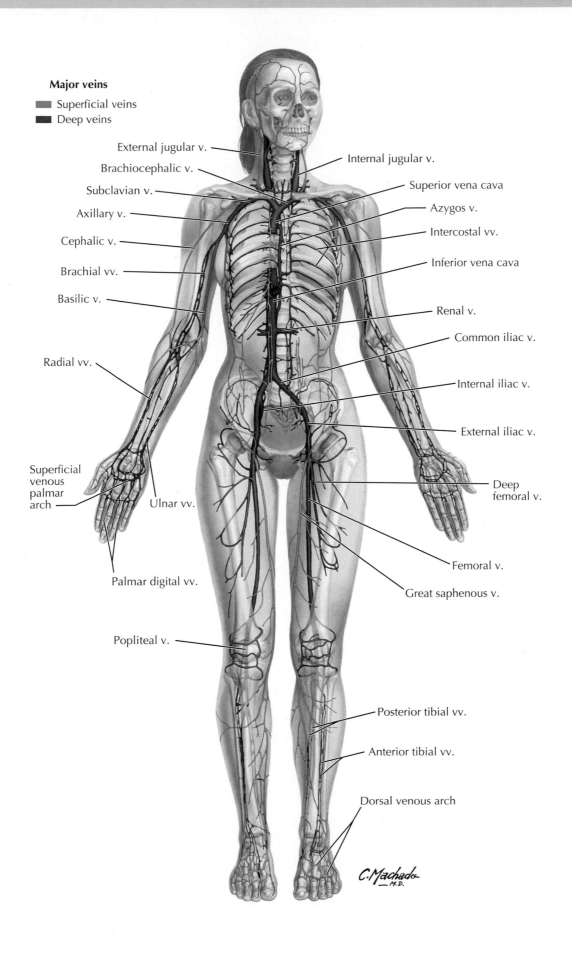

External jugular v.
Brachiocephalic v.
Subclavian v.
Axillary v.
Cephalic v.
Brachial vv.
Basilic v.

Radial vv.

Superficial venous palmar arch

Ulnar vv.

Palmar digital vv.

Popliteal v.

Internal jugular v.
Superior vena cava
Azygos v.
Intercostal vv.
Inferior vena cava
Renal v.
Common iliac v.
Internal iliac v.
External iliac v.
Deep femoral v.
Femoral v.
Great saphenous v.

Posterior tibial vv.
Anterior tibial vv.
Dorsal venous arch

C. Machado
M.D.

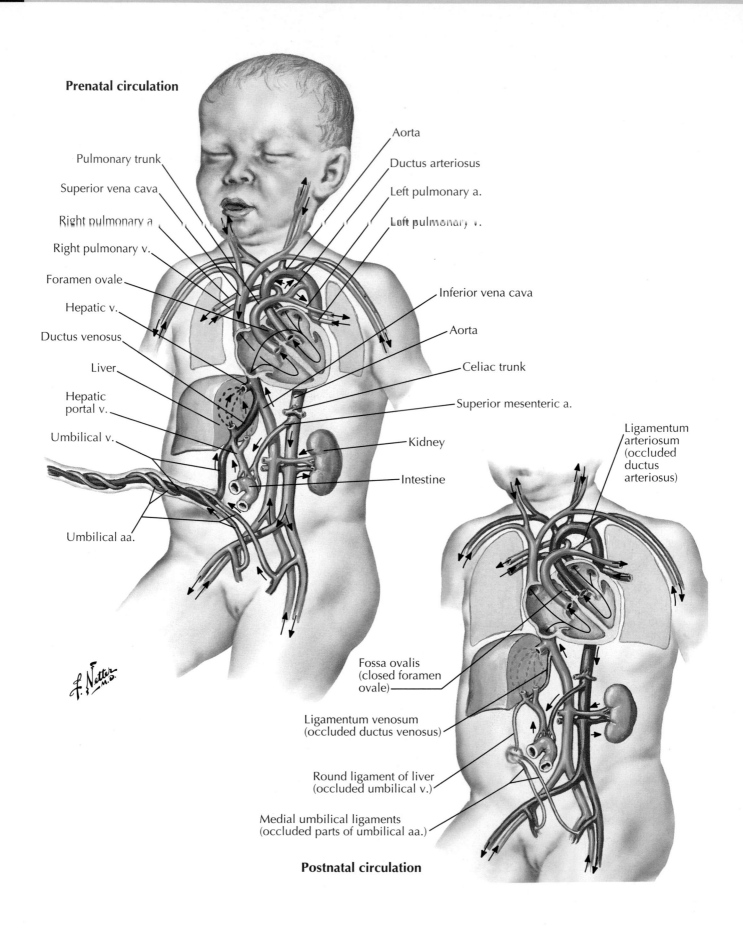

Prenatal circulation

Pulmonary trunk

Superior vena cava

Right pulmonary a.

Right pulmonary v.

Foramen ovale

Hepatic v.

Ductus venosus

Liver

Hepatic portal v.

Umbilical v.

Umbilical aa.

Aorta

Ductus arteriosus

Left pulmonary a.

Left pulmonary v.

Inferior vena cava

Aorta

Celiac trunk

Superior mesenteric a.

Kidney

Intestine

Ligamentum arteriosum (occluded ductus arteriosus)

Fossa ovalis (closed foramen ovale)

Ligamentum venosum (occluded ductus venosus)

Round ligament of liver (occluded umbilical v.)

Medial umbilical ligaments (occluded parts of umbilical aa.)

Postnatal circulation

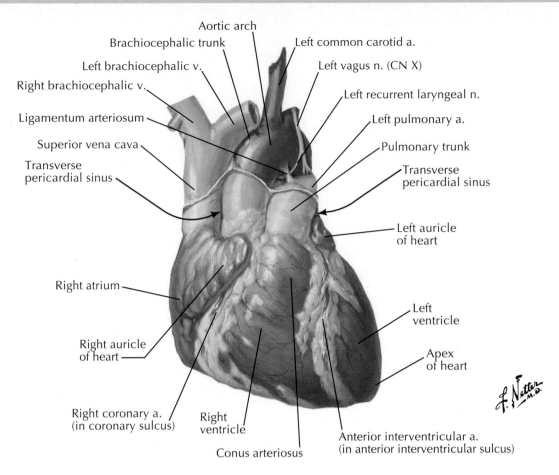

Aortic arch

Brachiocephalic trunk

Left common carotid a.

Left brachiocephalic v.

Left vagus n. (CN X)

Right brachiocephalic v.

Left recurrent laryngeal n.

Ligamentum arteriosum

Left pulmonary a.

Superior vena cava

Pulmonary trunk

Transverse pericardial sinus

Transverse pericardial sinus

Left auricle of heart

Right atrium

Left ventricle

Right auricle of heart

Apex of heart

Right coronary a. (in coronary sulcus)

Right ventricle

Conus arteriosus

Anterior interventricular a. (in anterior interventricular sulcus)

Precordial areas of auscultation:
One listens to the closing of a heart valve downstream from the heart valve, that is, in the right and left ventricles for the tricuspid and mitral valves, respectively, and over the pulmonary trunk and ascending aorta for the pulmonic and aortic valves, respectively.

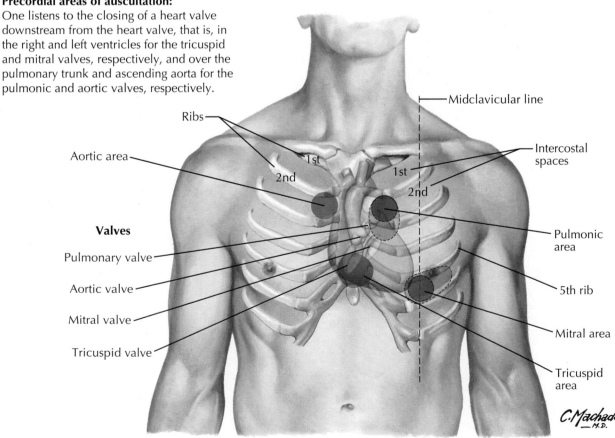

Ribs

Midclavicular line

Aortic area

Intercostal spaces

1st

1st

2nd

2nd

Pulmonic area

Valves

Pulmonary valve

Aortic valve

5th rib

Mitral valve

Tricuspid valve

Mitral area

Tricuspid area

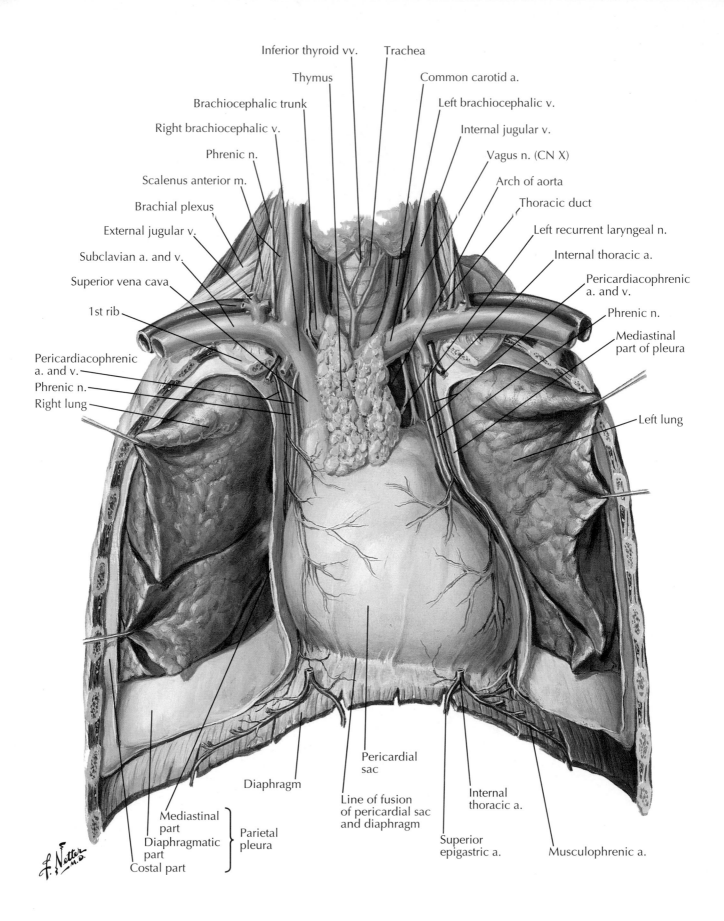

Inferior thyroid vv.

Trachea

Thymus

Common carotid a.

Brachiocephalic trunk

Left brachiocephalic v.

Right brachiocephalic v.

Internal jugular v.

Phrenic n.

Vagus n. (CN X)

Scalenus anterior m.

Arch of aorta

Brachial plexus

Thoracic duct

External jugular v.

Left recurrent laryngeal n.

Subclavian a. and v.

Internal thoracic a.

Superior vena cava

Pericardiacophrenic a. and v.

1st rib

Phrenic n.

Pericardiacophrenic a. and v.

Mediastinal part of pleura

Phrenic n.

Right lung

Left lung

Pericardial sac

Diaphragm

Line of fusion of pericardial sac and diaphragm

Internal thoracic a.

Mediastinal part

Parietal pleura

Diaphragmatic part

Costal part

Superior epigastric a.

Musculophrenic a.

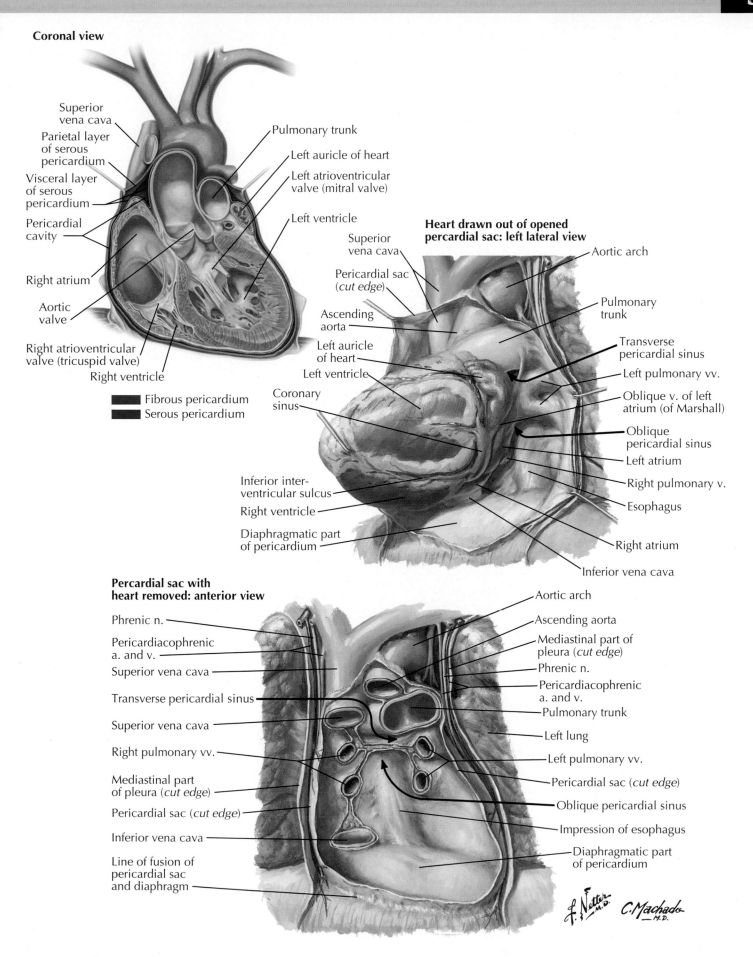

Coronal view

Superior vena cava

Parietal layer of serous pericardium

Visceral layer of serous pericardium

Pericardial cavity

Right atrium

Aortic valve

Right atrioventricular valve (tricuspid valve)

Right ventricle

Pulmonary trunk

Left auricle of heart

Left atrioventricular valve (mitral valve)

Left ventricle

Fibrous pericardium
Serous pericardium

Heart drawn out of opened percardial sac: left lateral view

Superior vena cava

Pericardial sac (*cut edge*)

Ascending aorta

Left auricle of heart

Left ventricle

Coronary sinus

Inferior interventricular sulcus

Right ventricle

Diaphragmatic part of pericardium

Aortic arch

Pulmonary trunk

Transverse pericardial sinus

Left pulmonary vv.

Oblique v. of left atrium (of Marshall)

Oblique pericardial sinus

Left atrium

Right pulmonary v.

Esophagus

Right atrium

Inferior vena cava

Percardial sac with heart removed: anterior view

Phrenic n.

Pericardiacophrenic a. and v.

Superior vena cava

Transverse pericardial sinus

Superior vena cava

Right pulmonary vv.

Mediastinal part of pleura (*cut edge*)

Pericardial sac (*cut edge*)

Inferior vena cava

Line of fusion of pericardial sac and diaphragm

Aortic arch

Ascending aorta

Mediastinal part of pleura (*cut edge*)

Phrenic n.

Pericardiacophrenic a. and v.

Pulmonary trunk

Left lung

Left pulmonary vv.

Pericardial sac (*cut edge*)

Oblique pericardial sinus

Impression of esophagus

Diaphragmatic part of pericardium

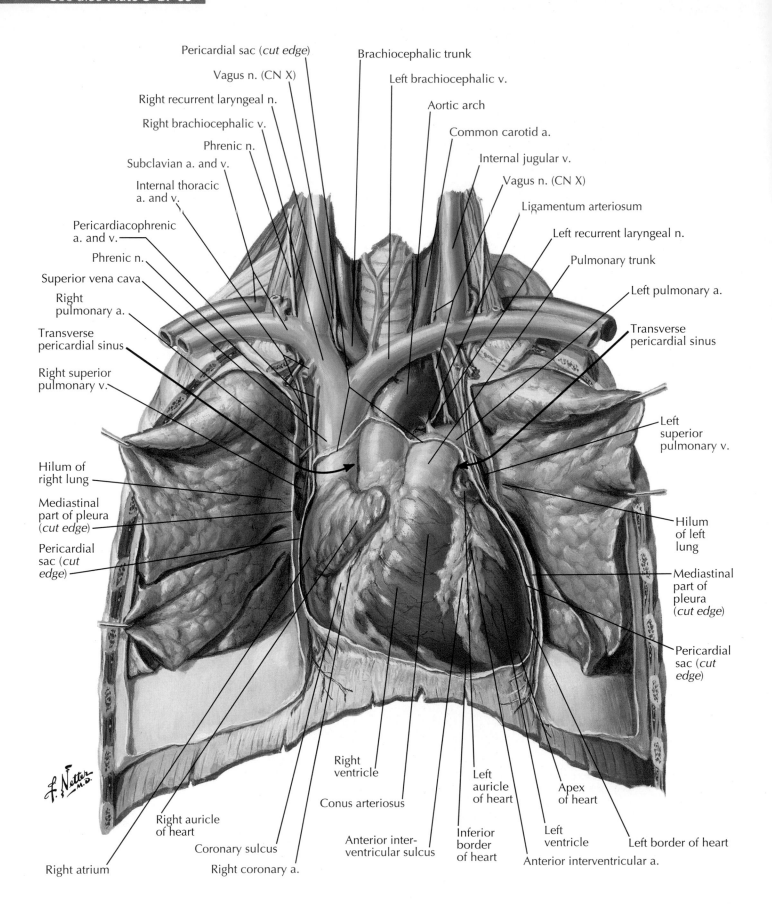

Pericardial sac (*cut edge*)

Vagus n. (CN X)

Right recurrent laryngeal n.

Right brachiocephalic v.

Phrenic n.

Subclavian a. and v.

Internal thoracic a. and v.

Pericardiacophrenic a. and v.

Phrenic n.

Superior vena cava

Right pulmonary a.

Transverse pericardial sinus

Right superior pulmonary v.

Hilum of right lung

Mediastinal part of pleura (*cut edge*)

Pericardial sac (*cut edge*)

Brachiocephalic trunk

Left brachiocephalic v.

Aortic arch

Common carotid a.

Internal jugular v.

Vagus n. (CN X)

Ligamentum arteriosum

Left recurrent laryngeal n.

Pulmonary trunk

Left pulmonary a.

Transverse pericardial sinus

Left superior pulmonary v.

Hilum of left lung

Mediastinal part of pleura (*cut edge*)

Pericardial sac (*cut edge*)

Right ventricle

Conus arteriosus

Right auricle of heart

Coronary sulcus

Right coronary a.

Right atrium

Anterior interventricular sulcus

Left auricle of heart

Inferior border of heart

Apex of heart

Left ventricle

Left border of heart

Anterior interventricular a.

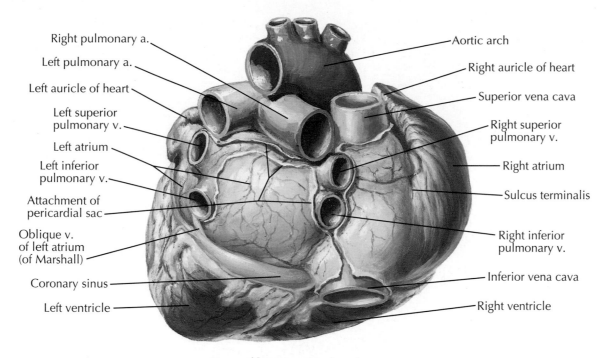

Right pulmonary a.

Left pulmonary a.

Left auricle of heart

Left superior
pulmonary v.

Left atrium

Left inferior
pulmonary v.

Attachment of
pericardial sac

Oblique v.
of left atrium
(of Marshall)

Coronary sinus

Left ventricle

Aortic arch

Right auricle of heart

Superior vena cava

Right superior
pulmonary v.

Right atrium

Sulcus terminalis

Right inferior
pulmonary v.

Inferior vena cava

Right ventricle

Base of heart: posterior view

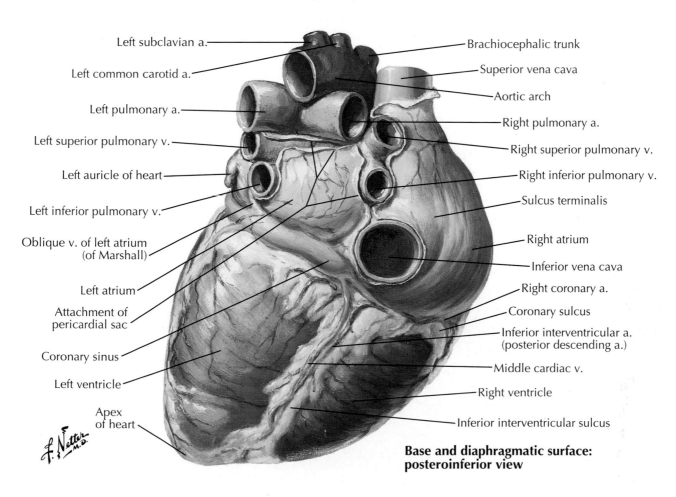

Left subclavian a.

Left common carotid a.

Left pulmonary a.

Left superior pulmonary v.

Left auricle of heart

Left inferior pulmonary v.

Oblique v. of left atrium
(of Marshall)

Left atrium

Attachment of
pericardial sac

Coronary sinus

Left ventricle

Apex
of heart

Brachiocephalic trunk

Superior vena cava

Aortic arch

Right pulmonary a.

Right superior pulmonary v.

Right inferior pulmonary v.

Sulcus terminalis

Right atrium

Inferior vena cava

Right coronary a.

Coronary sulcus

Inferior interventricular a.
(posterior descending a.)

Middle cardiac v.

Right ventricle

Inferior interventricular sulcus

**Base and diaphragmatic surface:
posteroinferior view**

Heart

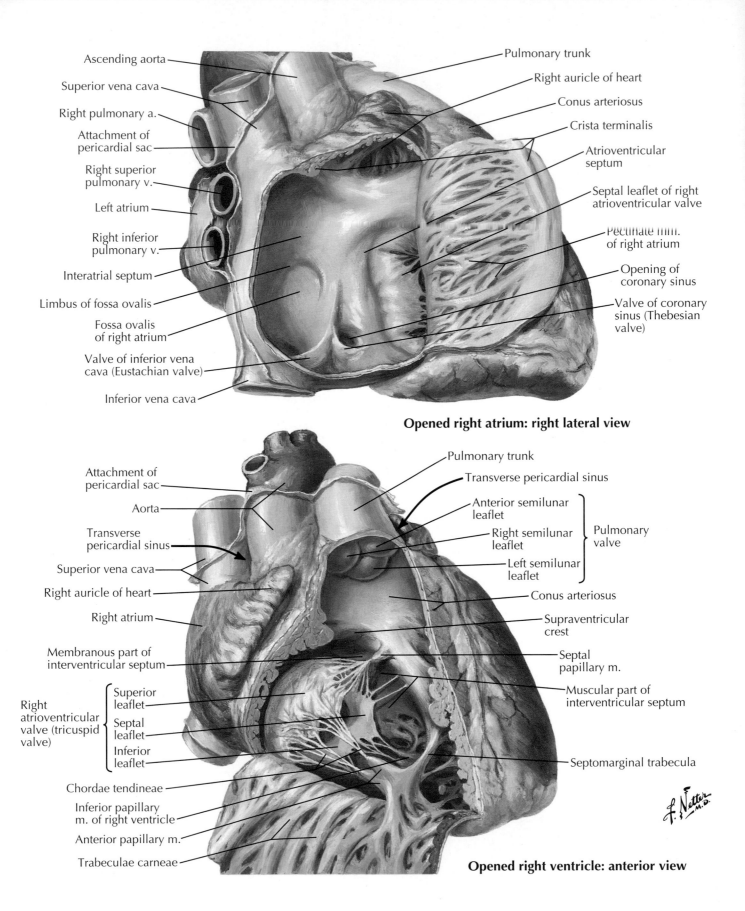

Ascending aorta

Superior vena cava

Right pulmonary a.

Attachment of
pericardial sac

Right superior
pulmonary v.

Left atrium

Right inferior
pulmonary v.

Interatrial septum

Limbus of fossa ovalis

Fossa ovalis
of right atrium

Valve of inferior vena
cava (Eustachian valve)

Inferior vena cava

Pulmonary trunk

Right auricle of heart

Conus arteriosus

Crista terminalis

Atrioventricular
septum

Septal leaflet of right
atrioventricular valve

Pectinate mm.
of right atrium

Opening of
coronary sinus

Valve of coronary
sinus (Thebesian
valve)

Opened right atrium: right lateral view

Attachment of
pericardial sac

Aorta

Transverse
pericardial sinus

Superior vena cava

Right auricle of heart

Right atrium

Membranous part of
interventricular septum

Right
atrioventricular
valve (tricuspid
valve)
{ Superior
leaflet

Septal
leaflet

Inferior
leaflet

Chordae tendineae

Inferior papillary
m. of right ventricle

Anterior papillary m.

Trabeculae carneae

Pulmonary trunk

Transverse pericardial sinus

Anterior semilunar
leaflet

Right semilunar
leaflet

Left semilunar
leaflet

Pulmonary
valve

Conus arteriosus

Supraventricular
crest

Septal
papillary m.

Muscular part of
interventricular septum

Septomarginal trabecula

Opened right ventricle: anterior view

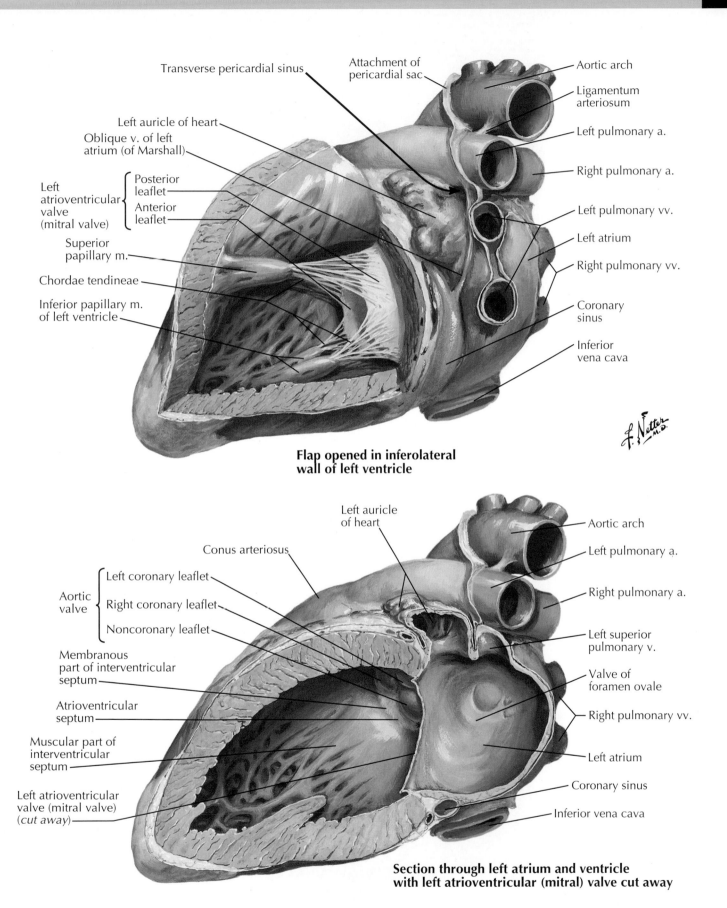

Transverse pericardial sinus

Attachment of pericardial sac

Aortic arch

Ligamentum arteriosum

Left pulmonary a.

Right pulmonary a.

Left auricle of heart

Oblique v. of left atrium (of Marshall)

Left atrioventricular valve (mitral valve)
- Posterior leaflet
- Anterior leaflet

Left pulmonary vv.

Left atrium

Right pulmonary vv.

Superior papillary m.

Chordae tendineae

Inferior papillary m. of left ventricle

Coronary sinus

Inferior vena cava

Flap opened in inferolateral wall of left ventricle

Left auricle of heart

Conus arteriosus

Aortic arch

Left pulmonary a.

Right pulmonary a.

Aortic valve
- Left coronary leaflet
- Right coronary leaflet
- Noncoronary leaflet

Membranous part of interventricular septum

Atrioventricular septum

Muscular part of interventricular septum

Left atrioventricular valve (mitral valve) (cut away)

Left superior pulmonary v.

Valve of foramen ovale

Right pulmonary vv.

Left atrium

Coronary sinus

Inferior vena cava

Section through left atrium and ventricle with left atrioventricular (mitral) valve cut away

Heart

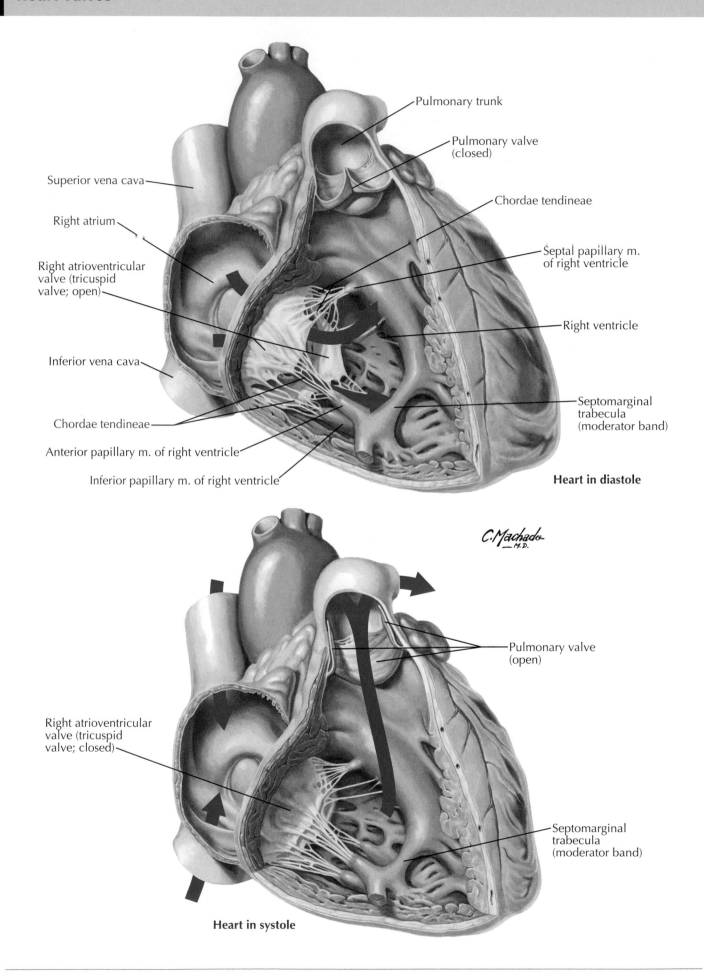

Pulmonary trunk

Pulmonary valve (closed)

Superior vena cava

Chordae tendineae

Right atrium

Septal papillary m. of right ventricle

Right atrioventricular valve (tricuspid valve; open)

Right ventricle

Inferior vena cava

Septomarginal trabecula (moderator band)

Chordae tendineae

Anterior papillary m. of right ventricle

Inferior papillary m. of right ventricle

Heart in diastole

C.Machado ─M.D.

Pulmonary valve (open)

Right atrioventricular valve (tricuspid valve; closed)

Septomarginal trabecula (moderator band)

Heart in systole

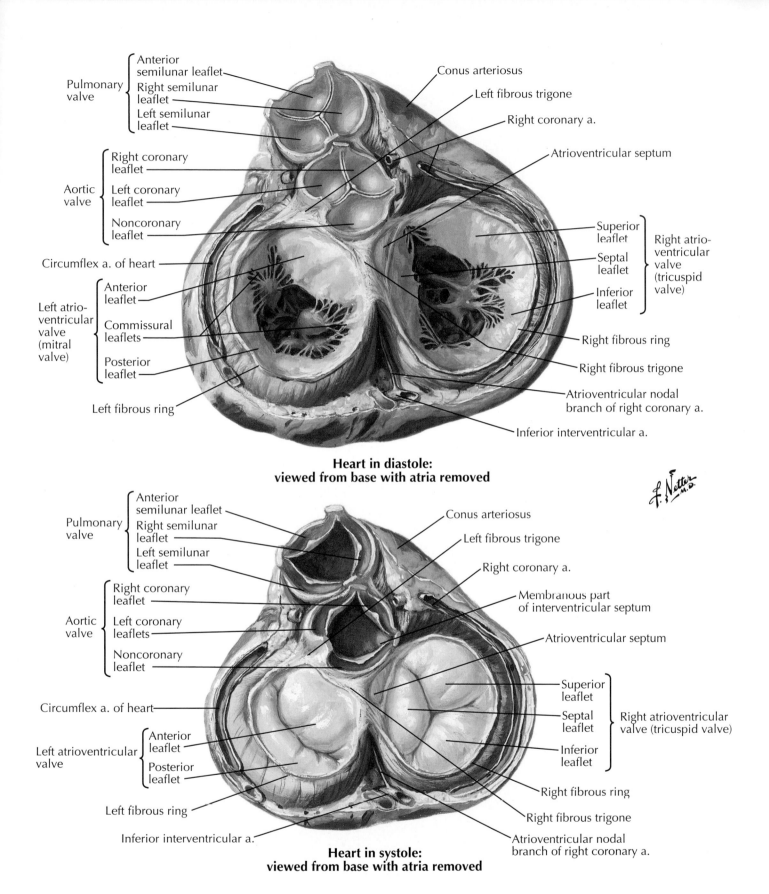

Pulmonary valve {
Anterior semilunar leaflet
Right semilunar leaflet
Left semilunar leaflet
}

Conus arteriosus

Left fibrous trigone

Right coronary a.

Aortic valve {
Right coronary leaflet
Left coronary leaflet
Noncoronary leaflet
}

Atrioventricular septum

Superior leaflet
Septal leaflet
Inferior leaflet
} Right atrio-ventricular valve (tricuspid valve)

Circumflex a. of heart

Left atrio-ventricular valve (mitral valve) {
Anterior leaflet
Commissural leaflets
Posterior leaflet
}

Right fibrous ring

Right fibrous trigone

Atrioventricular nodal branch of right coronary a.

Left fibrous ring

Inferior interventricular a.

Heart in diastole:
viewed from base with atria removed

Pulmonary valve {
Anterior semilunar leaflet
Right semilunar leaflet
Left semilunar leaflet
}

Conus arteriosus

Left fibrous trigone

Right coronary a.

Aortic valve {
Right coronary leaflet
Left coronary leaflets
Noncoronary leaflet
}

Membranous part of interventricular septum

Atrioventricular septum

Superior leaflet
Septal leaflet
Inferior leaflet
} Right atrioventricular valve (tricuspid valve)

Circumflex a. of heart

Left atrioventricular valve {
Anterior leaflet
Posterior leaflet
}

Right fibrous ring

Right fibrous trigone

Left fibrous ring

Inferior interventricular a.

Atrioventricular nodal branch of right coronary a.

Heart in systole:
viewed from base with atria removed

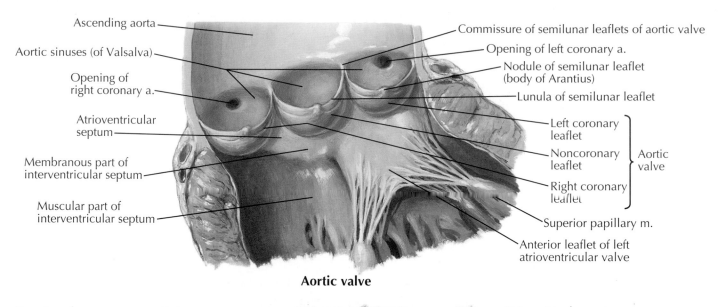

Ascending aorta

Aortic sinuses (of Valsalva)

Opening of
right coronary a.

Atrioventricular
septum

Membranous part of
interventricular septum

Muscular part of
interventricular septum

Commissure of semilunar leaflets of aortic valve

Opening of left coronary a.

Nodule of semilunar leaflet
(body of Arantius)

Lunula of semilunar leaflet

Left coronary
leaflet

Noncoronary
leaflet

Right coronary
leaflet

Aortic
valve

Superior papillary m.

Anterior leaflet of left
atrioventricular valve

Aortic valve

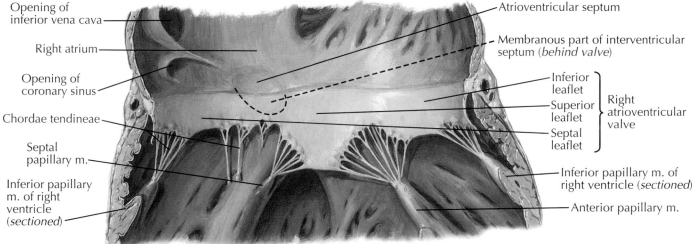

Opening of
inferior vena cava

Right atrium

Opening of
coronary sinus

Chordae tendineae

Septal
papillary m.

Inferior papillary
m. of right
ventricle
(*sectioned*)

Atrioventricular septum

Membranous part of interventricular
septum (*behind valve*)

Inferior
leaflet

Superior
leaflet

Septal
leaflet

Right
atrioventricular
valve

Inferior papillary m. of
right ventricle (*sectioned*)

Anterior papillary m.

Right atrioventricular valve (tricuspid valve)

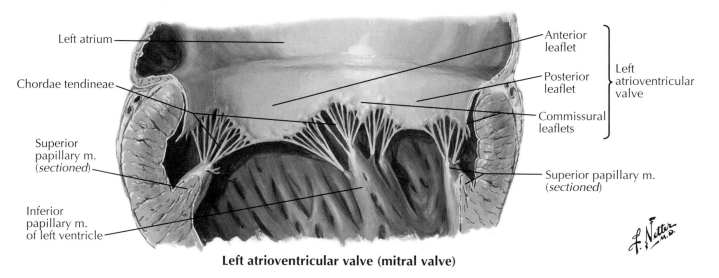

Left atrium

Chordae tendineae

Superior
papillary m.
(*sectioned*)

Inferior
papillary m.
of left ventricle

Anterior
leaflet

Posterior
leaflet

Commissural
leaflets

Left
atrioventricular
valve

Superior papillary m.
(*sectioned*)

Left atrioventricular valve (mitral valve)

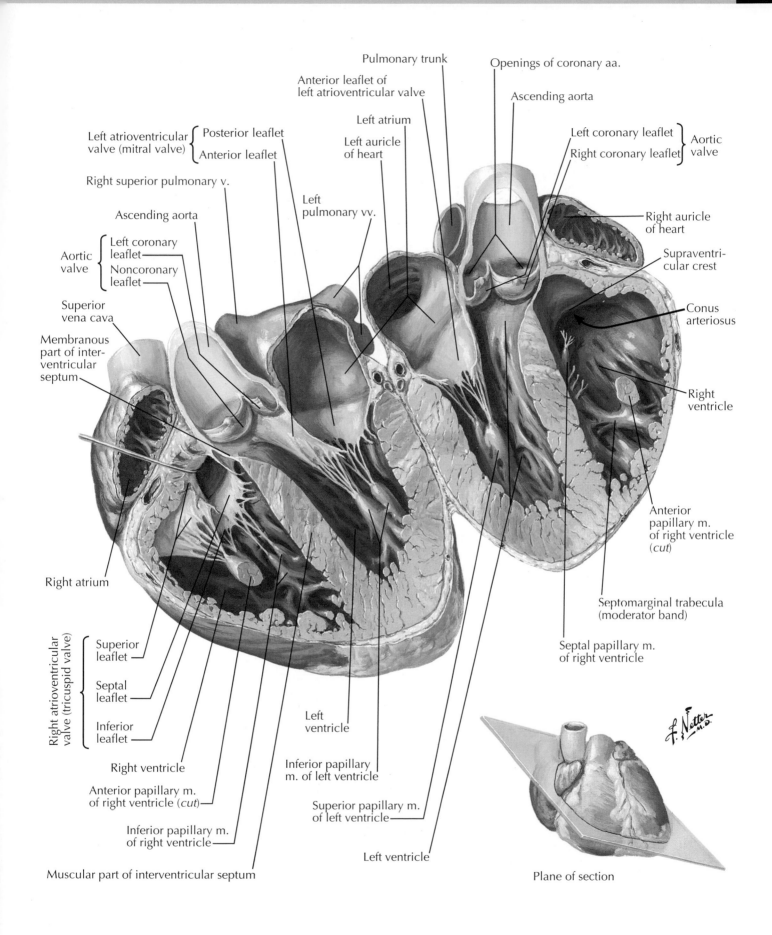

Pulmonary trunk

Anterior leaflet of
left atrioventricular valve

Left atrium

Openings of coronary aa.

Ascending aorta

Left auricle
of heart

Left coronary leaflet
Right coronary leaflet
} Aortic
valve

Left atrioventricular
valve (mitral valve) {
Posterior leaflet
Anterior leaflet

Right superior pulmonary v.

Left
pulmonary vv.

Right auricle
of heart

Ascending aorta

Supraventri-
cular crest

Aortic
valve {
Left coronary
leaflet
Noncoronary
leaflet

Conus
arteriosus

Superior
vena cava

Right
ventricle

Membranous
part of inter-
ventricular
septum

Right atrium

Anterior
papillary m.
of right ventricle
(cut)

Right atrioventricular
valve (tricuspid valve) {
Superior
leaflet

Septal
leaflet

Inferior
leaflet

Septomarginal trabecula
(moderator band)

Septal papillary m.
of right ventricle

Right ventricle

Left
ventricle

Anterior papillary m.
of right ventricle (cut)

Inferior papillary
m. of left ventricle

Inferior papillary m.
of right ventricle

Superior papillary m.
of left ventricle

Muscular part of interventricular septum

Left ventricle

Plane of section

Sternocostal surface

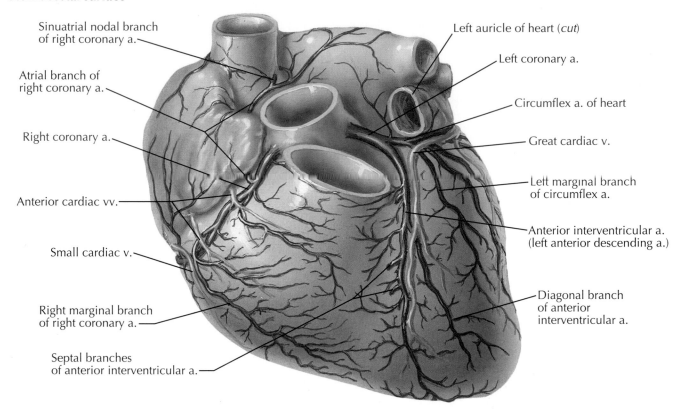

Sinuatrial nodal branch of right coronary a.

Atrial branch of right coronary a.

Right coronary a.

Anterior cardiac vv.

Small cardiac v.

Right marginal branch of right coronary a.

Septal branches of anterior interventricular a.

Left auricle of heart (*cut*)

Left coronary a.

Circumflex a. of heart

Great cardiac v.

Left marginal branch of circumflex a.

Anterior interventricular a. (left anterior descending a.)

Diagonal branch of anterior interventricular a.

Diaphragmatic surface

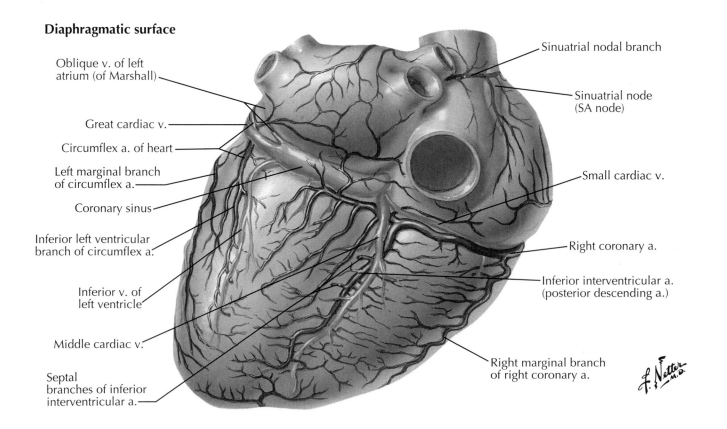

Oblique v. of left atrium (of Marshall)

Great cardiac v.

Circumflex a. of heart

Left marginal branch of circumflex a.

Coronary sinus

Inferior left ventricular branch of circumflex a.

Inferior v. of left ventricle

Middle cardiac v.

Septal branches of inferior interventricular a.

Sinuatrial nodal branch

Sinuatrial node (SA node)

Small cardiac v.

Right coronary a.

Inferior interventricular a. (posterior descending a.)

Right marginal branch of right coronary a.

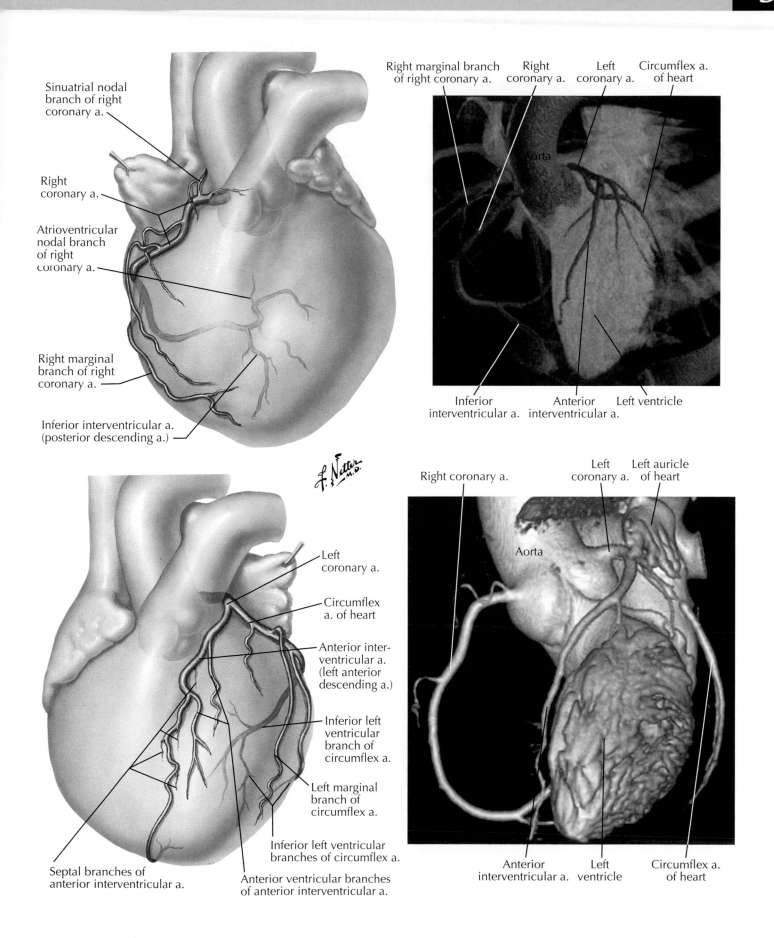

Sinuatrial nodal branch of right coronary a.

Right coronary a.

Atrioventricular nodal branch of right coronary a.

Right marginal branch of right coronary a.

Inferior interventricular a. (posterior descending a.)

Right marginal branch of right coronary a.

Right coronary a.

Left coronary a.

Circumflex a. of heart

Aorta

Inferior interventricular a.

Anterior interventricular a.

Left ventricle

Left coronary a.

Circumflex a. of heart

Anterior interventricular a. (left anterior descending a.)

Inferior left ventricular branch of circumflex a.

Left marginal branch of circumflex a.

Inferior left ventricular branches of circumflex a.

Anterior ventricular branches of anterior interventricular a.

Septal branches of anterior interventricular a.

Right coronary a.

Left coronary a.

Left auricle of heart

Aorta

Anterior interventricular a.

Left ventricle

Circumflex a. of heart

f. Netter
_M.D._

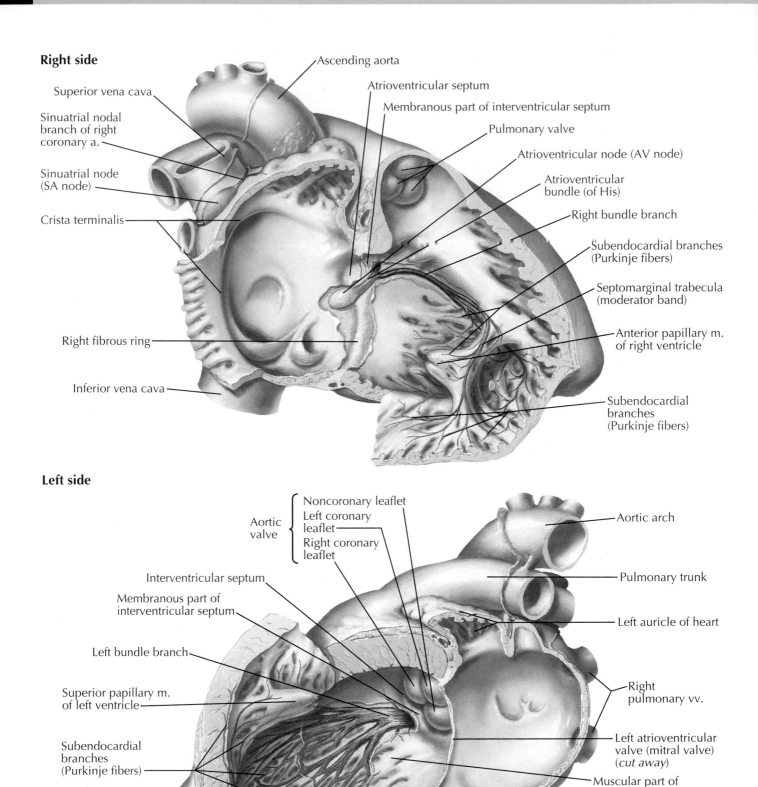

Right side

Superior vena cava

Sinuatrial nodal branch of right coronary a.

Sinuatrial node (SA node)

Crista terminalis

Right fibrous ring

Inferior vena cava

Ascending aorta

Atrioventricular septum

Membranous part of interventricular septum

Pulmonary valve

Atrioventricular node (AV node)

Atrioventricular bundle (of His)

Right bundle branch

Subendocardial branches (Purkinje fibers)

Septomarginal trabecula (moderator band)

Anterior papillary m. of right ventricle

Subendocardial branches (Purkinje fibers)

Left side

Aortic valve

Noncoronary leaflet

Left coronary leaflet

Right coronary leaflet

Interventricular septum

Membranous part of interventricular septum

Left bundle branch

Superior papillary m. of left ventricle

Subendocardial branches (Purkinje fibers)

Inferior papillary m. of left ventricle

Aortic arch

Pulmonary trunk

Left auricle of heart

Right pulmonary vv.

Left atrioventricular valve (mitral valve) (*cut away*)

Muscular part of interventricular septum

Inferior vena cava

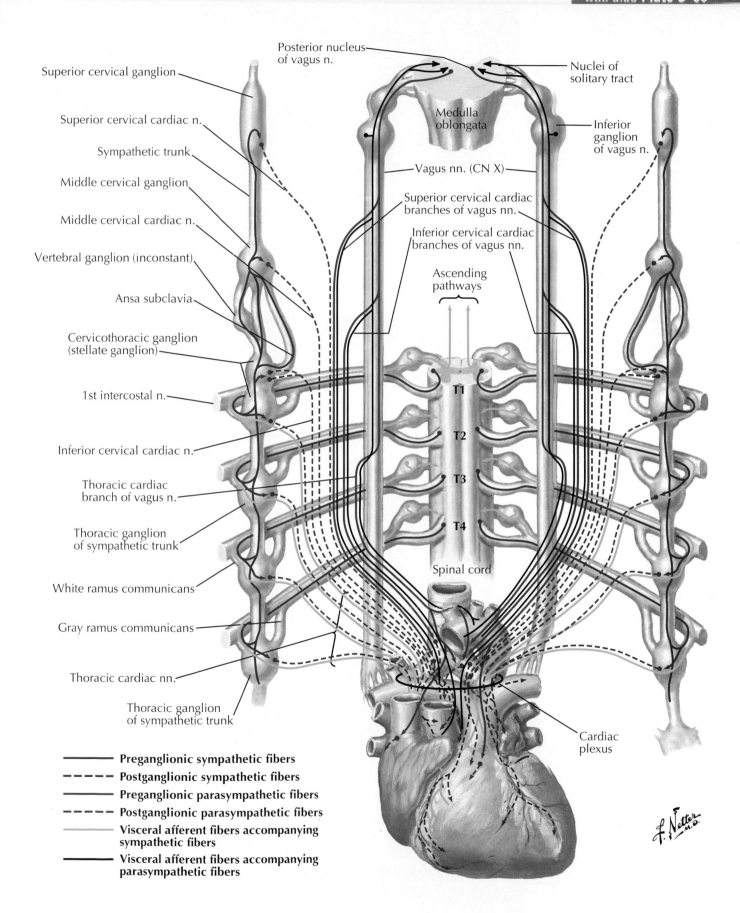

Posterior nucleus
of vagus n.

Superior cervical ganglion

Nuclei of
solitary tract

Superior cervical cardiac n.

Medulla
oblongata

Inferior
ganglion
of vagus n.

Sympathetic trunk

Middle cervical ganglion

Vagus nn. (CN X)

Middle cervical cardiac n.

Superior cervical cardiac
branches of vagus nn.

Vertebral ganglion (inconstant)

Inferior cervical cardiac
branches of vagus nn.

Ansa subclavia

Ascending
pathways

Cervicothoracic ganglion
(stellate ganglion)

1st intercostal n.

T1

Inferior cervical cardiac n.

T2

Thoracic cardiac
branch of vagus n.

T3

Thoracic ganglion
of sympathetic trunk

T4

White ramus communicans

Spinal cord

Gray ramus communicans

Thoracic cardiac nn.

Thoracic ganglion
of sympathetic trunk

Cardiac
plexus

——————— Preganglionic sympathetic fibers
- - - - - - - Postganglionic sympathetic fibers
——————— Preganglionic parasympathetic fibers
- - - - - - - Postganglionic parasympathetic fibers
——————— Visceral afferent fibers accompanying
sympathetic fibers
——————— Visceral afferent fibers accompanying
parasympathetic fibers

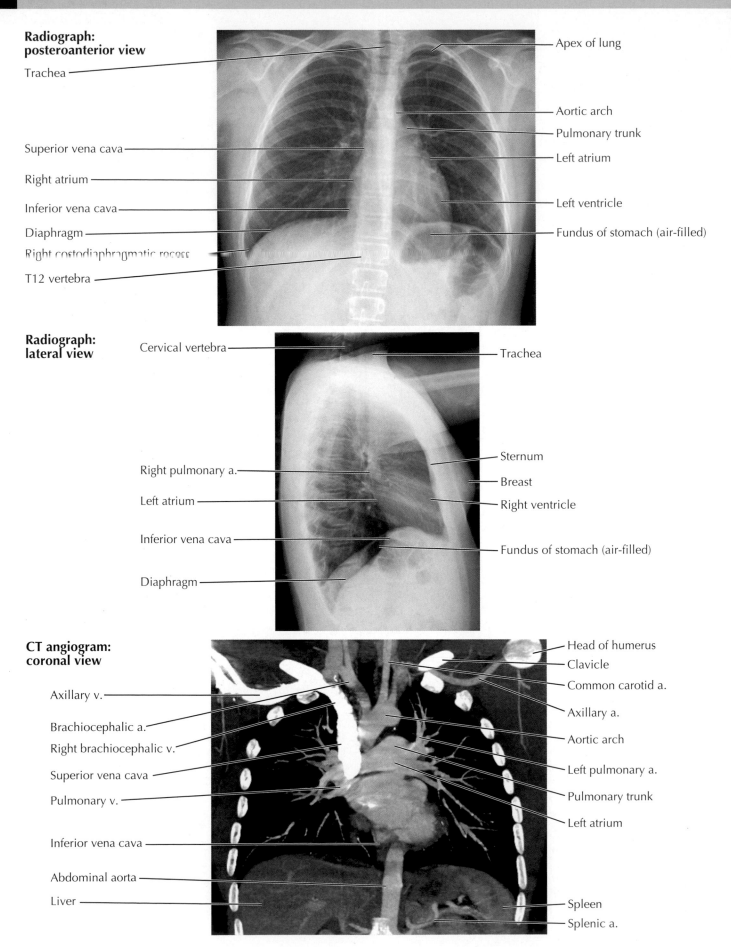

Radiograph: posteroanterior view

Trachea

Superior vena cava

Right atrium

Inferior vena cava

Diaphragm

Right costodiaphragmatic recess

T12 vertebra

Apex of lung

Aortic arch

Pulmonary trunk

Left atrium

Left ventricle

Fundus of stomach (air-filled)

Radiograph: lateral view

Cervical vertebra

Right pulmonary a.

Left atrium

Inferior vena cava

Diaphragm

Trachea

Sternum

Breast

Right ventricle

Fundus of stomach (air-filled)

CT angiogram: coronal view

Axillary v.

Brachiocephalic a.

Right brachiocephalic v.

Superior vena cava

Pulmonary v.

Inferior vena cava

Abdominal aorta

Liver

Head of humerus

Clavicle

Common carotid a.

Axillary a.

Aortic arch

Left pulmonary a.

Pulmonary trunk

Left atrium

Spleen

Splenic a.

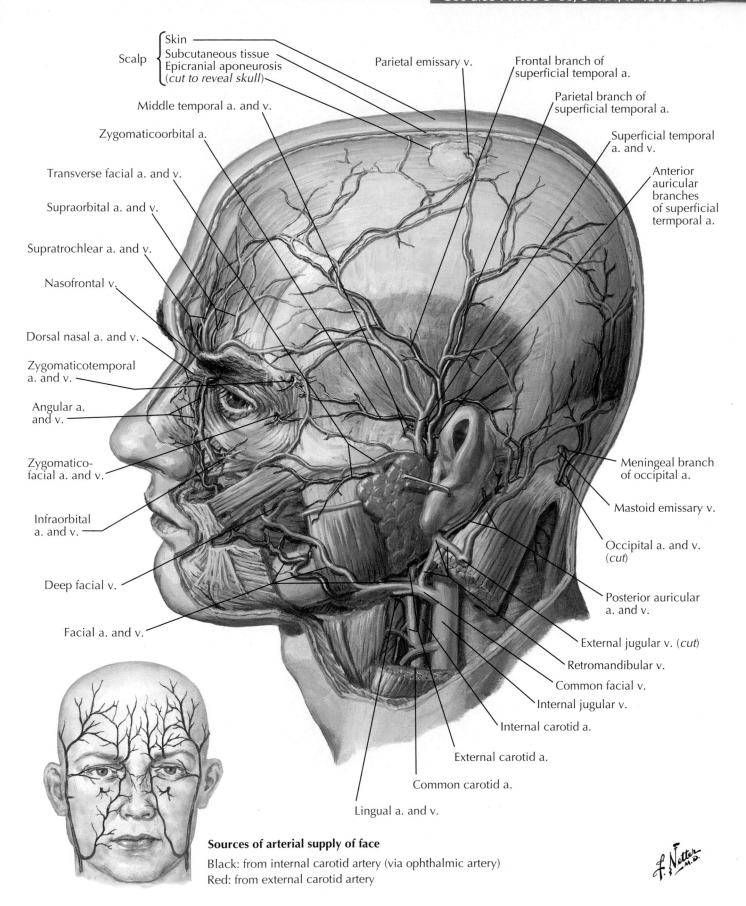

Scalp {
Skin
Subcutaneous tissue
Epicranial aponeurosis
(cut to reveal skull)

Middle temporal a. and v.

Zygomaticoorbital a.

Transverse facial a. and v.

Supraorbital a. and v.

Supratrochlear a. and v.

Nasofrontal v.

Dorsal nasal a. and v.

Zygomaticotemporal a. and v.

Angular a. and v.

Zygomatico-facial a. and v.

Infraorbital a. and v.

Deep facial v.

Facial a. and v.

Parietal emissary v.

Frontal branch of superficial temporal a.

Parietal branch of superficial temporal a.

Superficial temporal a. and v.

Anterior auricular branches of superficial termporal a.

Meningeal branch of occipital a.

Mastoid emissary v.

Occipital a. and v. (cut)

Posterior auricular a. and v.

External jugular v. (cut)

Retromandibular v.

Common facial v.

Internal jugular v.

Internal carotid a.

External carotid a.

Common carotid a.

Lingual a. and v.

Sources of arterial supply of face

Black: from internal carotid artery (via ophthalmic artery)
Red: from external carotid artery

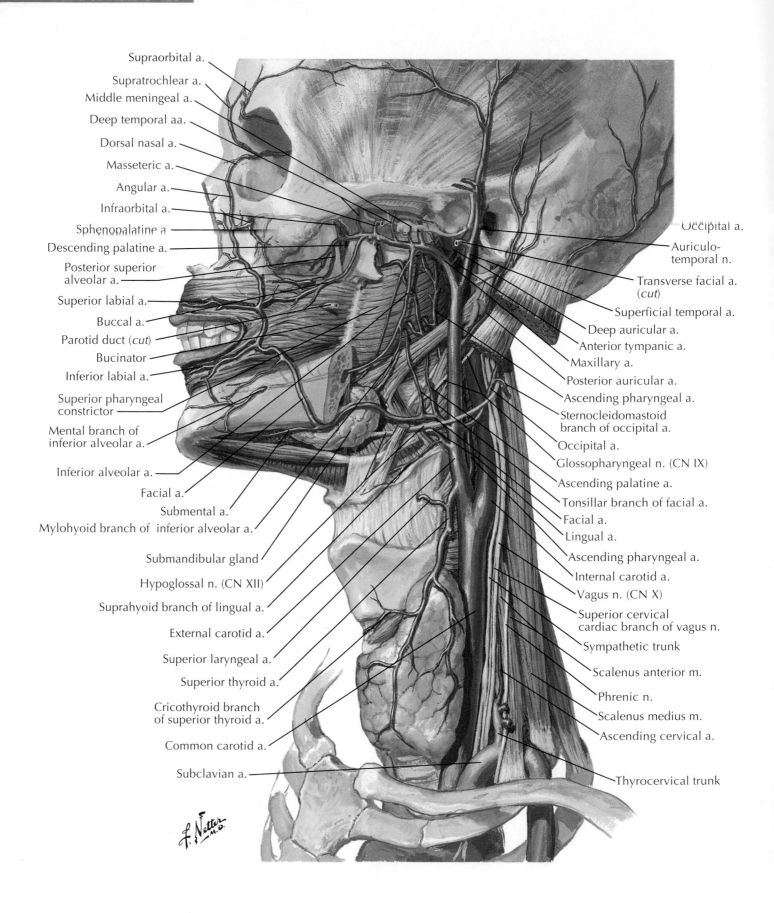

Supraorbital a.

Supratrochlear a.

Middle meningeal a.

Deep temporal aa.

Dorsal nasal a.

Masseteric a.

Angular a.

Infraorbital a.

Sphenopalatine a

Descending palatine a.

Posterior superior alveolar a.

Superior labial a.

Buccal a.

Parotid duct (cut)

Bucinator

Inferior labial a.

Superior pharyngeal constrictor

Mental branch of inferior alveolar a.

Inferior alveolar a.

Facial a.

Submental a.

Mylohyoid branch of inferior alveolar a.

Submandibular gland

Hypoglossal n. (CN XII)

Suprahyoid branch of lingual a.

External carotid a.

Superior laryngeal a.

Superior thyroid a.

Cricothyroid branch of superior thyroid a.

Common carotid a.

Subclavian a.

Occipital a.

Auriculo-temporal n.

Transverse facial a. (cut)

Superficial temporal a.

Deep auricular a.

Anterior tympanic a.

Maxillary a.

Posterior auricular a.

Ascending pharyngeal a.

Sternocleidomastoid branch of occipital a.

Occipital a.

Glossopharyngeal n. (CN IX)

Ascending palatine a.

Tonsillar branch of facial a.

Facial a.

Lingual a.

Ascending pharyngeal a.

Internal carotid a.

Vagus n. (CN X)

Superior cervical cardiac branch of vagus n.

Sympathetic trunk

Scalenus anterior m.

Phrenic n.

Scalenus medius m.

Ascending cervical a.

Thyrocervical trunk

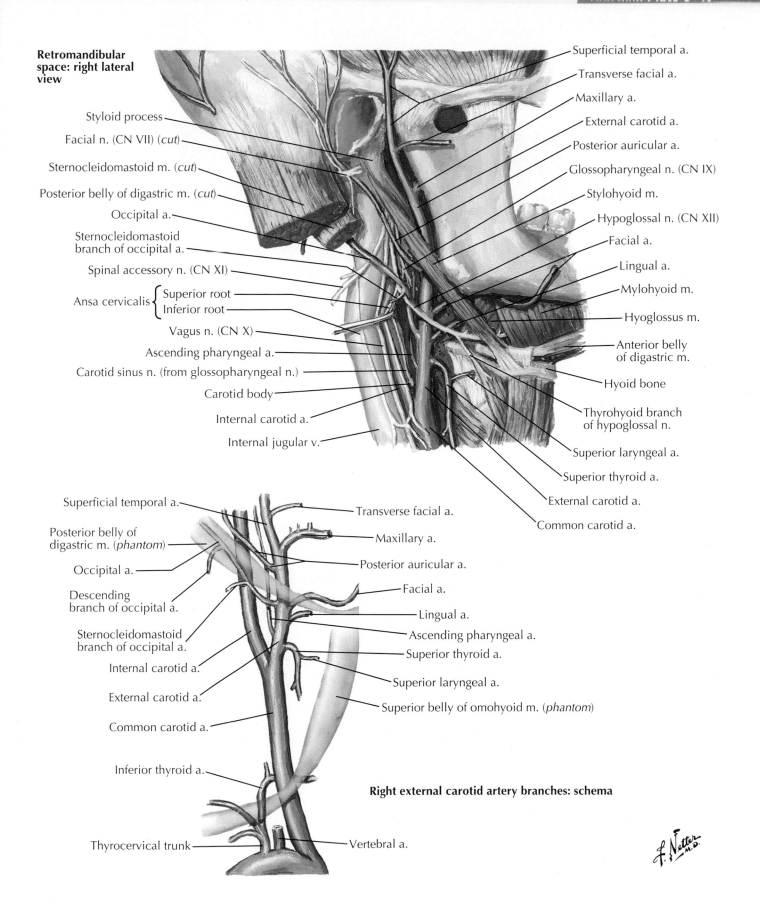

Retromandibular space: right lateral view

Styloid process

Facial n. (CN VII) (*cut*)

Sternocleidomastoid m. (*cut*)

Posterior belly of digastric m. (*cut*)

Occipital a.

Sternocleidomastoid branch of occipital a.

Spinal accessory n. (CN XI)

Ansa cervicalis { Superior root

Inferior root

Vagus n. (CN X)

Ascending pharyngeal a.

Carotid sinus n. (from glossopharyngeal n.)

Carotid body

Internal carotid a.

Internal jugular v.

Superficial temporal a.

Transverse facial a.

Maxillary a.

External carotid a.

Posterior auricular a.

Glossopharyngeal n. (CN IX)

Stylohyoid m.

Hypoglossal n. (CN XII)

Facial a.

Lingual a.

Mylohyoid m.

Hyoglossus m.

Anterior belly of digastric m.

Hyoid bone

Thyrohyoid branch of hypoglossal n.

Superior laryngeal a.

Superior thyroid a.

External carotid a.

Common carotid a.

Superficial temporal a.

Posterior belly of digastric m. (*phantom*)

Occipital a.

Descending branch of occipital a.

Sternocleidomastoid branch of occipital a.

Internal carotid a.

External carotid a.

Common carotid a.

Inferior thyroid a.

Thyrocervical trunk

Transverse facial a.

Maxillary a.

Posterior auricular a.

Facial a.

Lingual a.

Ascending pharyngeal a.

Superior thyroid a.

Superior laryngeal a.

Superior belly of omohyoid m. (*phantom*)

Right external carotid artery branches: schema

Vertebral a.

Blood Vessels of the Head and Neck

Plate S–324

Lateral view

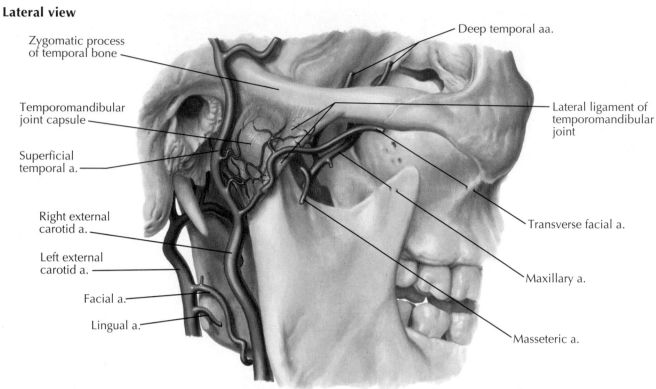

Zygomatic process of temporal bone

Temporomandibular joint capsule

Superficial temporal a.

Right external carotid a.

Left external carotid a.

Facial a.

Lingual a.

Deep temporal aa.

Lateral ligament of temporomandibular joint

Transverse facial a.

Maxillary a.

Masseteric a.

Medial view

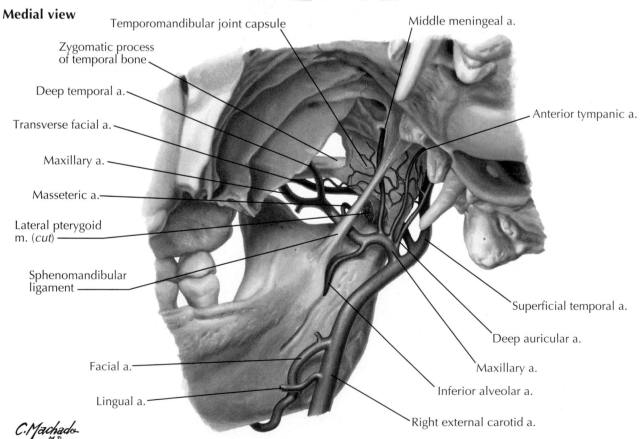

Temporomandibular joint capsule

Middle meningeal a.

Zygomatic process of temporal bone

Deep temporal a.

Transverse facial a.

Maxillary a.

Masseteric a.

Lateral pterygoid m. (cut)

Sphenomandibular ligament

Facial a.

Lingual a.

Anterior tympanic a.

Superficial temporal a.

Deep auricular a.

Maxillary a.

Inferior alveolar a.

Right external carotid a.

C. Machado
M.D.

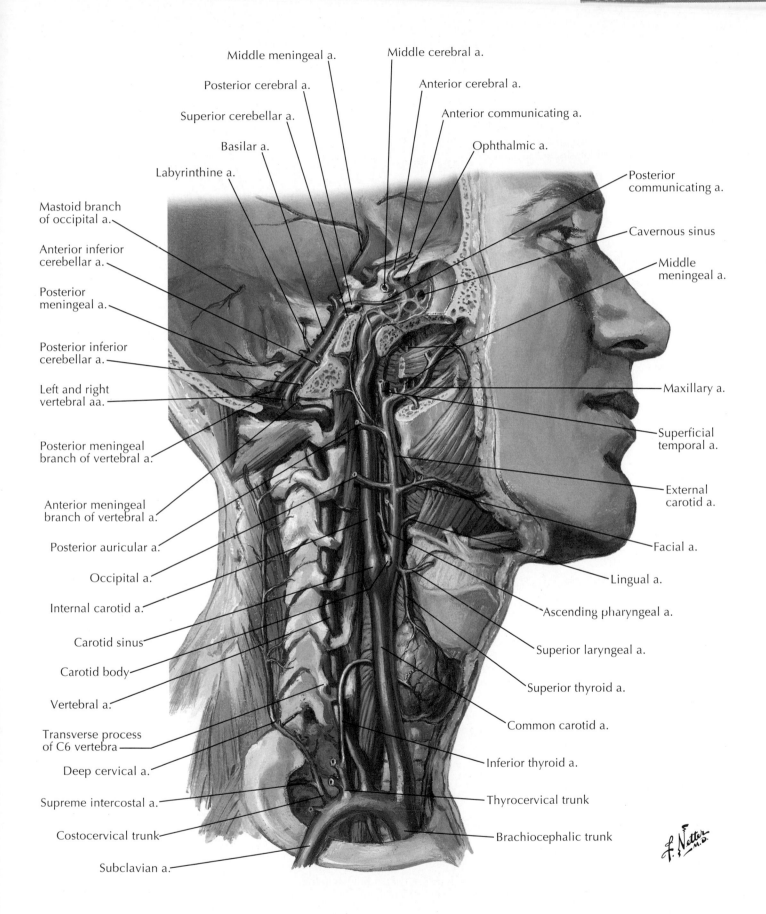

Middle meningeal a.

Middle cerebral a.

Posterior cerebral a.

Anterior cerebral a.

Superior cerebellar a.

Anterior communicating a.

Basilar a.

Ophthalmic a.

Labyrinthine a.

Posterior communicating a.

Mastoid branch of occipital a.

Cavernous sinus

Anterior inferior cerebellar a.

Middle meningeal a.

Posterior meningeal a.

Posterior inferior cerebellar a.

Maxillary a.

Left and right vertebral aa.

Superficial temporal a.

Posterior meningeal branch of vertebral a.

Anterior meningeal branch of vertebral a.

External carotid a.

Posterior auricular a.

Facial a.

Occipital a.

Lingual a.

Internal carotid a.

Ascending pharyngeal a.

Carotid sinus

Superior laryngeal a.

Carotid body

Superior thyroid a.

Vertebral a.

Common carotid a.

Transverse process of C6 vertebra

Inferior thyroid a.

Deep cervical a.

Thyrocervical trunk

Supreme intercostal a.

Costocervical trunk

Brachiocephalic trunk

Subclavian a.

f. Netter M.D.

Blood Vessels of the Head and Neck

Plate S–326

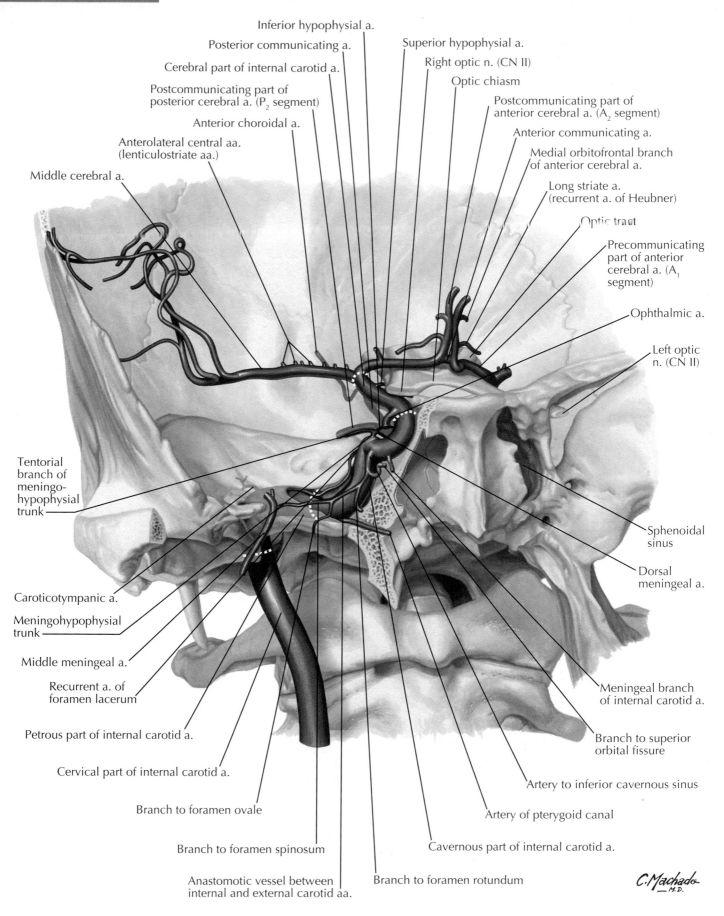

Inferior hypophysial a.

Posterior communicating a.

Cerebral part of internal carotid a.

Postcommunicating part of posterior cerebral a. (P$_2$ segment)

Anterior choroidal a.

Anterolateral central aa. (lenticulostriate aa.)

Middle cerebral a.

Superior hypophysial a.

Right optic n. (CN II)

Optic chiasm

Postcommunicating part of anterior cerebral a. (A$_2$ segment)

Anterior communicating a.

Medial orbitofrontal branch of anterior cerebral a.

Long striate a. (recurrent a. of Heubner)

Optic tract

Precommunicating part of anterior cerebral a. (A$_1$ segment)

Ophthalmic a.

Left optic n. (CN II)

Tentorial branch of meningo-hypophysial trunk

Caroticotympanic a.

Meningohypophysial trunk

Middle meningeal a.

Recurrent a. of foramen lacerum

Petrous part of internal carotid a.

Cervical part of internal carotid a.

Branch to foramen ovale

Branch to foramen spinosum

Anastomotic vessel between internal and external carotid aa.

Branch to foramen rotundum

Cavernous part of internal carotid a.

Artery of pterygoid canal

Artery to inferior cavernous sinus

Branch to superior orbital fissure

Meningeal branch of internal carotid a.

Dorsal meningeal a.

Sphenoidal sinus

C.Machado
M.D.

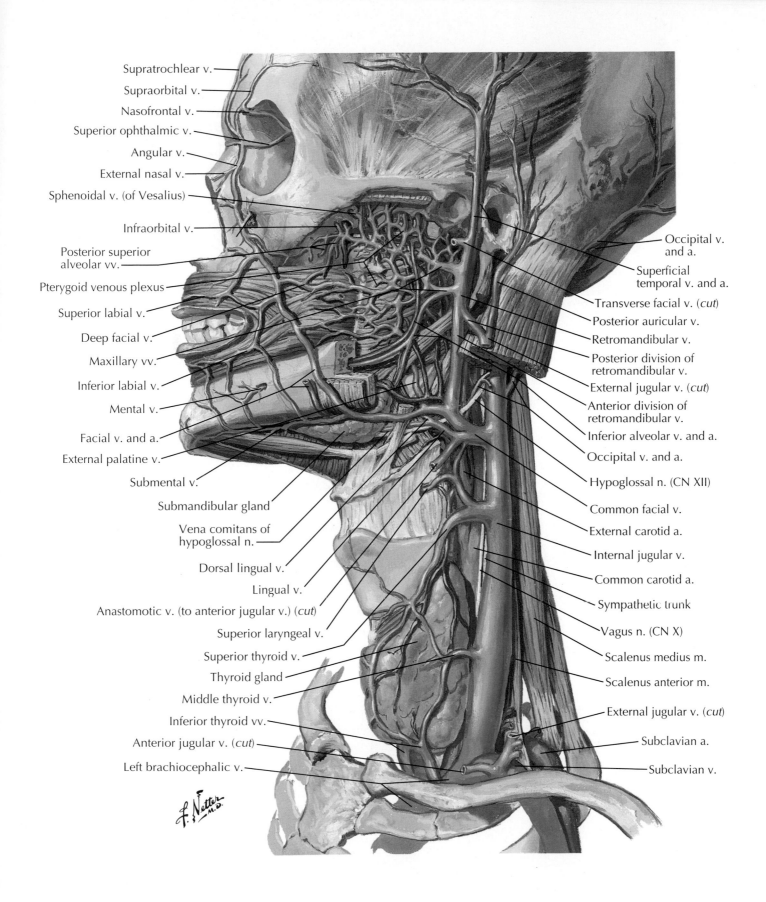

Supratrochlear v.
Supraorbital v.
Nasofrontal v.
Superior ophthalmic v.
Angular v.
External nasal v.
Sphenoidal v. (of Vesalius)
Infraorbital v.
Posterior superior alveolar vv.
Pterygoid venous plexus
Superior labial v.
Deep facial v.
Maxillary vv.
Inferior labial v.
Mental v.
Facial v. and a.
External palatine v.
Submental v.
Submandibular gland
Vena comitans of hypoglossal n.
Dorsal lingual v.
Lingual v.
Anastomotic v. (to anterior jugular v.) (cut)
Superior laryngeal v.
Superior thyroid v.
Thyroid gland
Middle thyroid v.
Inferior thyroid vv.
Anterior jugular v. (cut)
Left brachiocephalic v.

Occipital v. and a.
Superficial temporal v. and a.
Transverse facial v. (cut)
Posterior auricular v.
Retromandibular v.
Posterior division of retromandibular v.
External jugular v. (cut)
Anterior division of retromandibular v.
Inferior alveolar v. and a.
Occipital v. and a.
Hypoglossal n. (CN XII)
Common facial v.
External carotid a.
Internal jugular v.
Common carotid a.
Sympathetic trunk
Vagus n. (CN X)
Scalenus medius m.
Scalenus anterior m.
External jugular v. (cut)
Subclavian a.
Subclavian v.

F. Netter M.D.

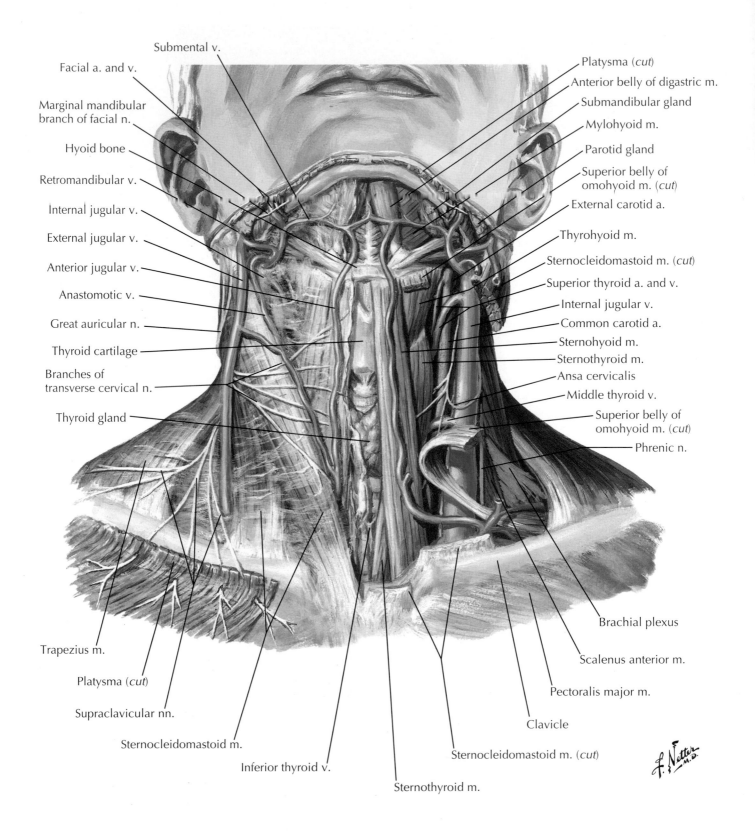

Submental v.

Facial a. and v.

Marginal mandibular branch of facial n.

Hyoid bone

Retromandibular v.

Internal jugular v.

External jugular v.

Anterior jugular v.

Anastomotic v.

Great auricular n.

Thyroid cartilage

Branches of transverse cervical n.

Thyroid gland

Platysma (*cut*)

Anterior belly of digastric m.

Submandibular gland

Mylohyoid m.

Parotid gland

Superior belly of omohyoid m. (*cut*)

External carotid a.

Thyrohyoid m.

Sternocleidomastoid m. (*cut*)

Superior thyroid a. and v.

Internal jugular v.

Common carotid a.

Sternohyoid m.

Sternothyroid m.

Ansa cervicalis

Middle thyroid v.

Superior belly of omohyoid m. (*cut*)

Phrenic n.

Trapezius m.

Platysma (*cut*)

Supraclavicular nn.

Sternocleidomastoid m.

Inferior thyroid v.

Sternothyroid m.

Brachial plexus

Scalenus anterior m.

Pectoralis major m.

Clavicle

Sternocleidomastoid m. (*cut*)

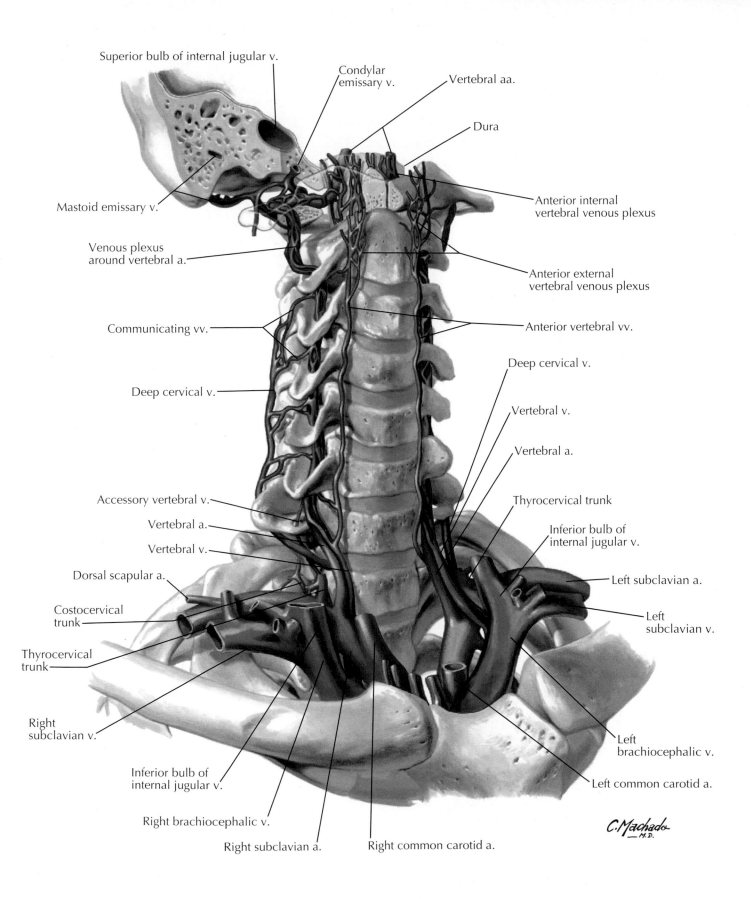

Superior bulb of internal jugular v.

Condylar emissary v.

Vertebral aa.

Dura

Mastoid emissary v.

Anterior internal vertebral venous plexus

Venous plexus around vertebral a.

Anterior external vertebral venous plexus

Communicating vv.

Anterior vertebral vv.

Deep cervical v.

Deep cervical v.

Vertebral v.

Vertebral a.

Accessory vertebral v.

Thyrocervical trunk

Vertebral a.

Inferior bulb of internal jugular v.

Vertebral v.

Left subclavian a.

Dorsal scapular a.

Costocervical trunk

Left subclavian v.

Thyrocervical trunk

Right subclavian v.

Left brachiocephalic v.

Inferior bulb of internal jugular v.

Left common carotid a.

Right brachiocephalic v.

C. Machado
—M.D.

Right subclavian a.

Right common carotid a.

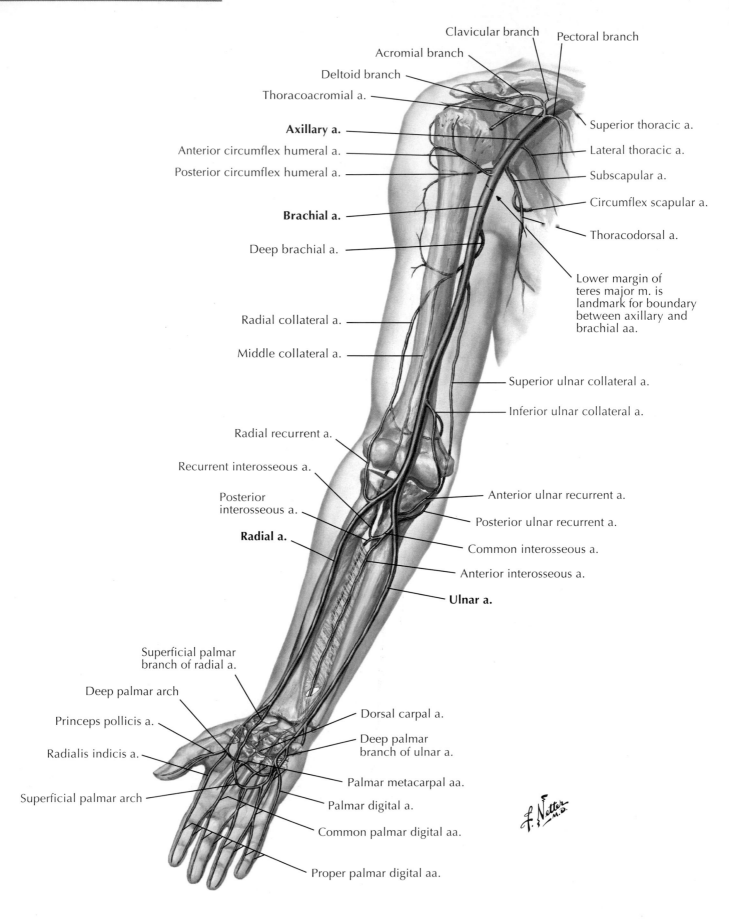

Clavicular branch
Pectoral branch
Acromial branch
Deltoid branch
Thoracoacromial a.
Axillary a.
Anterior circumflex humeral a.
Posterior circumflex humeral a.
Brachial a.
Deep brachial a.
Radial collateral a.
Middle collateral a.
Radial recurrent a.
Recurrent interosseous a.
Posterior interosseous a.
Radial a.
Superficial palmar branch of radial a.
Deep palmar arch
Princeps pollicis a.
Radialis indicis a.
Superficial palmar arch

Superior thoracic a.
Lateral thoracic a.
Subscapular a.
Circumflex scapular a.
Thoracodorsal a.
Lower margin of teres major m. is landmark for boundary between axillary and brachial aa.
Superior ulnar collateral a.
Inferior ulnar collateral a.
Anterior ulnar recurrent a.
Posterior ulnar recurrent a.
Common interosseous a.
Anterior interosseous a.
Ulnar a.
Dorsal carpal a.
Deep palmar branch of ulnar a.
Palmar metacarpal aa.
Palmar digital a.
Common palmar digital aa.
Proper palmar digital aa.

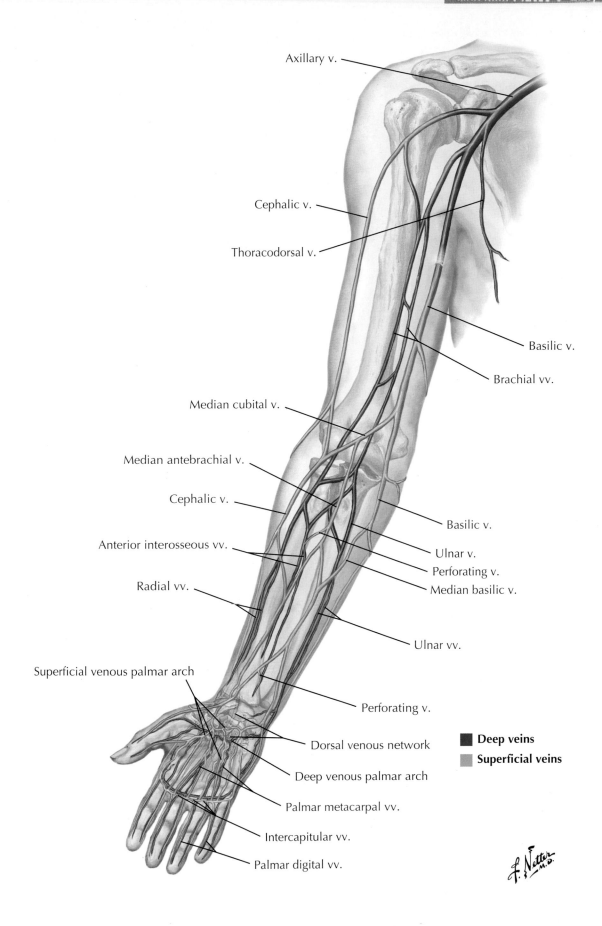

Axillary v.

Cephalic v.

Thoracodorsal v.

Basilic v.

Brachial vv.

Median cubital v.

Median antebrachial v.

Cephalic v.

Basilic v.

Anterior interosseous vv.

Ulnar v.

Perforating v.

Median basilic v.

Radial vv.

Ulnar vv.

Superficial venous palmar arch

Perforating v.

Dorsal venous network

Deep venous palmar arch

Palmar metacarpal vv.

Intercapitular vv.

Palmar digital vv.

■ Deep veins
■ Superficial veins

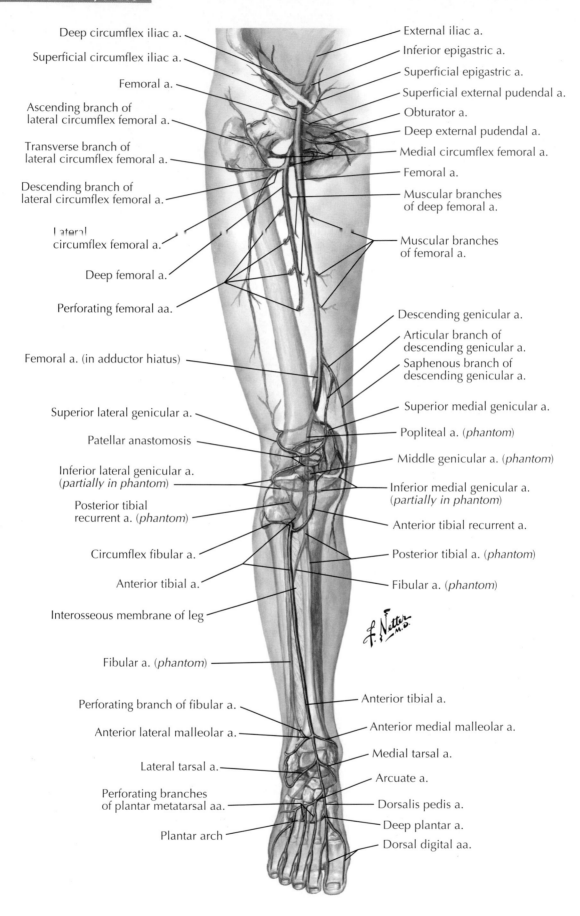

Deep circumflex iliac a.

Superficial circumflex iliac a.

Femoral a.

Ascending branch of
lateral circumflex femoral a.

Transverse branch of
lateral circumflex femoral a.

Descending branch of
lateral circumflex femoral a.

Lateral
circumflex femoral a.

Deep femoral a.

Perforating femoral aa.

Femoral a. (in adductor hiatus)

Superior lateral genicular a.

Patellar anastomosis

Inferior lateral genicular a.
(*partially in phantom*)

Posterior tibial
recurrent a. (*phantom*)

Circumflex fibular a.

Anterior tibial a.

Interosseous membrane of leg

Fibular a. (*phantom*)

Perforating branch of fibular a.

Anterior lateral malleolar a.

Lateral tarsal a.

Perforating branches
of plantar metatarsal aa.

Plantar arch

External iliac a.

Inferior epigastric a.

Superficial epigastric a.

Superficial external pudendal a.

Obturator a.

Deep external pudendal a.

Medial circumflex femoral a.

Femoral a.

Muscular branches
of deep femoral a.

Muscular branches
of femoral a.

Descending genicular a.

Articular branch of
descending genicular a.

Saphenous branch of
descending genicular a.

Superior medial genicular a.

Popliteal a. (*phantom*)

Middle genicular a. (*phantom*)

Inferior medial genicular a.
(*partially in phantom*)

Anterior tibial recurrent a.

Posterior tibial a. (*phantom*)

Fibular a. (*phantom*)

Anterior tibial a.

Anterior medial malleolar a.

Medial tarsal a.

Arcuate a.

Dorsalis pedis a.

Deep plantar a.

Dorsal digital aa.

See also **Plates S–266, S–267**

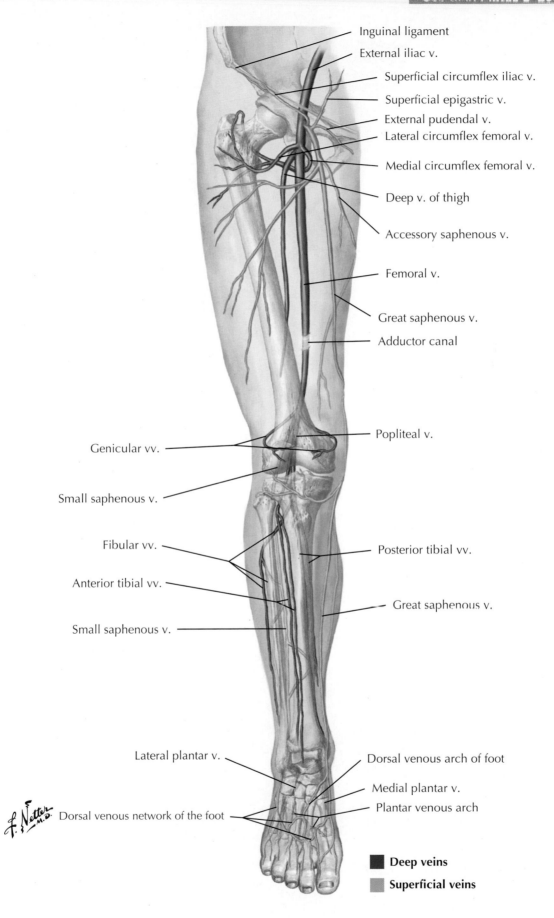

Inguinal ligament

External iliac v.

Superficial circumflex iliac v.

Superficial epigastric v.

External pudendal v.

Lateral circumflex femoral v.

Medial circumflex femoral v.

Deep v. of thigh

Accessory saphenous v.

Femoral v.

Great saphenous v.

Adductor canal

Genicular vv.

Popliteal v.

Small saphenous v.

Fibular vv.

Posterior tibial vv.

Anterior tibial vv.

Great saphenous v.

Small saphenous v.

Lateral plantar v.

Dorsal venous arch of foot

Medial plantar v.

Plantar venous arch

Dorsal venous network of the foot

■ Deep veins

■ Superficial veins

f. Netter M.D.

Blood Vessels of the Limbs

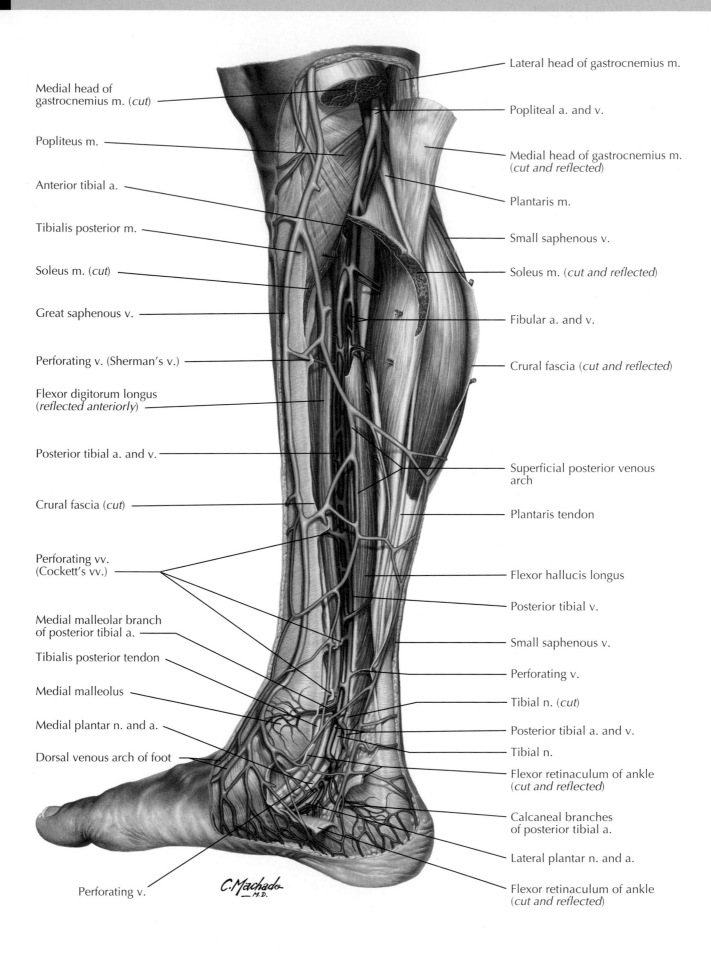

Medial head of gastrocnemius m. (*cut*)

Popliteus m.

Anterior tibial a.

Tibialis posterior m.

Soleus m. (*cut*)

Great saphenous v.

Perforating v. (Sherman's v.)

Flexor digitorum longus (*reflected anteriorly*)

Posterior tibial a. and v.

Crural fascia (*cut*)

Perforating vv. (Cockett's vv.)

Medial malleolar branch of posterior tibial a.

Tibialis posterior tendon

Medial malleolus

Medial plantar n. and a.

Dorsal venous arch of foot

Perforating v.

Lateral head of gastrocnemius m.

Popliteal a. and v.

Medial head of gastrocnemius m. (*cut and reflected*)

Plantaris m.

Small saphenous v.

Soleus m. (*cut and reflected*)

Fibular a. and v.

Crural fascia (*cut and reflected*)

Superficial posterior venous arch

Plantaris tendon

Flexor hallucis longus

Posterior tibial v.

Small saphenous v.

Perforating v.

Tibial n. (*cut*)

Posterior tibial a. and v.

Tibial n.

Flexor retinaculum of ankle (*cut and reflected*)

Calcaneal branches of posterior tibial a.

Lateral plantar n. and a.

Flexor retinaculum of ankle (*cut and reflected*)

C. Machado _M.D._

See also Plates S–204, S–206, S 213, S–270

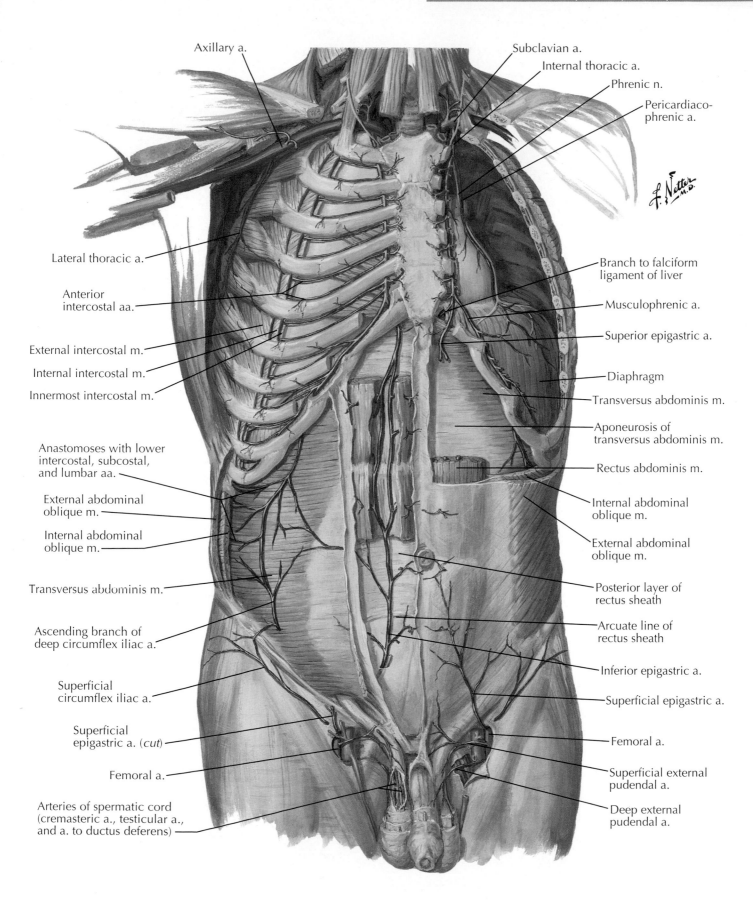

Axillary a.

Subclavian a.

Internal thoracic a.

Phrenic n.

Pericardiaco-phrenic a.

Lateral thoracic a.

Anterior intercostal aa.

External intercostal m.

Internal intercostal m.

Innermost intercostal m.

Anastomoses with lower intercostal, subcostal, and lumbar aa.

External abdominal oblique m.

Internal abdominal oblique m.

Transversus abdominis m.

Ascending branch of deep circumflex iliac a.

Superficial circumflex iliac a.

Superficial epigastric a. (cut)

Femoral a.

Arteries of spermatic cord (cremasteric a., testicular a., and a. to ductus deferens)

Branch to falciform ligament of liver

Musculophrenic a.

Superior epigastric a.

Diaphragm

Transversus abdominis m.

Aponeurosis of transversus abdominis m.

Rectus abdominis m.

Internal abdominal oblique m.

External abdominal oblique m.

Posterior layer of rectus sheath

Arcuate line of rectus sheath

Inferior epigastric a.

Superficial epigastric a.

Femoral a.

Superficial external pudendal a.

Deep external pudendal a.

Blood Vessels of the Trunk

Plate S–336

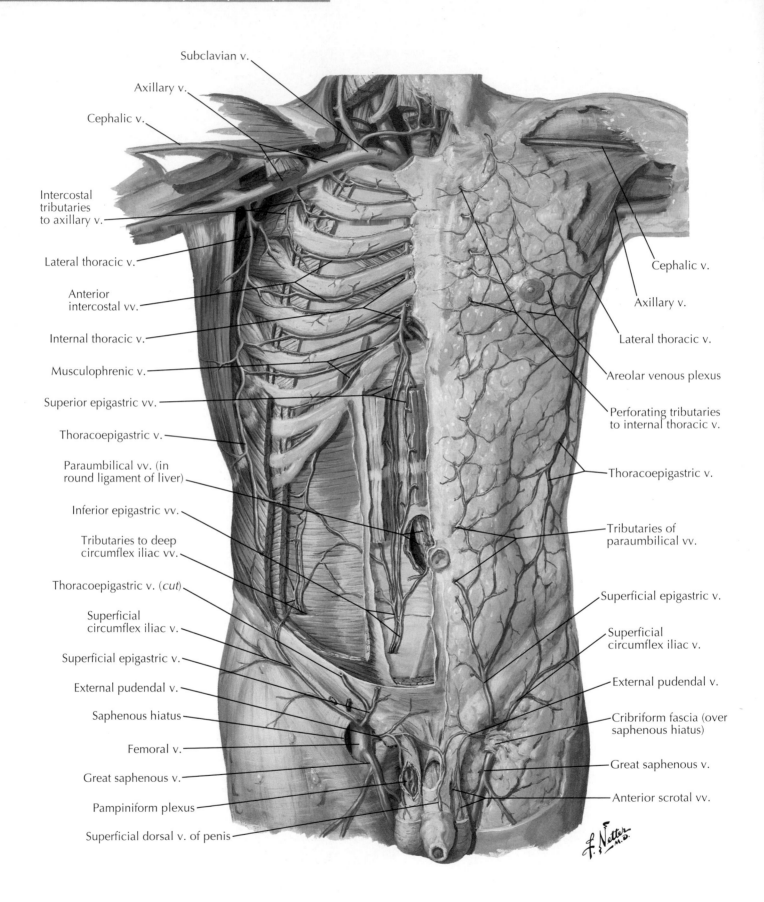

Subclavian v.

Axillary v.

Cephalic v.

Intercostal tributaries to axillary v.

Lateral thoracic v.

Anterior intercostal vv.

Internal thoracic v.

Musculophrenic v.

Superior epigastric vv.

Thoracoepigastric v.

Paraumbilical vv. (in round ligament of liver)

Inferior epigastric vv.

Tributaries to deep circumflex iliac vv.

Thoracoepigastric v. (cut)

Superficial circumflex iliac v.

Superficial epigastric v.

External pudendal v.

Saphenous hiatus

Femoral v.

Great saphenous v.

Pampiniform plexus

Superficial dorsal v. of penis

Cephalic v.

Axillary v.

Lateral thoracic v.

Areolar venous plexus

Perforating tributaries to internal thoracic v.

Thoracoepigastric v.

Tributaries of paraumbilical vv.

Superficial epigastric v.

Superficial circumflex iliac v.

External pudendal v.

Cribriform fascia (over saphenous hiatus)

Great saphenous v.

Anterior scrotal vv.

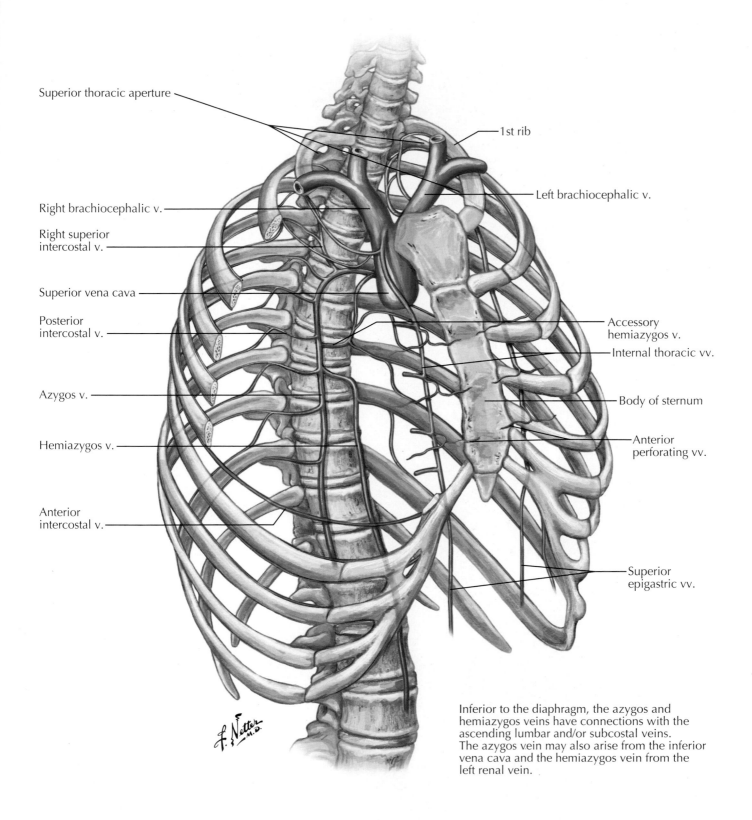

Superior thoracic aperture

1st rib

Right brachiocephalic v.

Left brachiocephalic v.

Right superior intercostal v.

Superior vena cava

Posterior intercostal v.

Accessory hemiazygos v.

Internal thoracic vv.

Azygos v.

Body of sternum

Hemiazygos v.

Anterior perforating vv.

Anterior intercostal v.

Superior epigastric vv.

Inferior to the diaphragm, the azygos and hemiazygos veins have connections with the ascending lumbar and/or subcostal veins. The azygos vein may also arise from the inferior vena cava and the hemiazygos vein from the left renal vein.

F. Netter M.D.

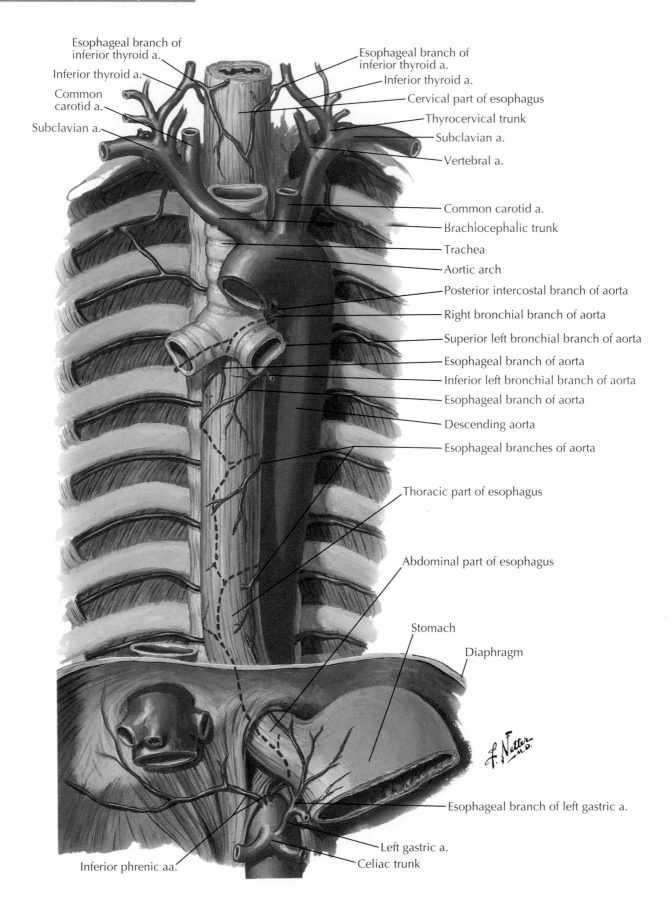

Esophageal branch of inferior thyroid a.

Inferior thyroid a.

Common carotid a.

Subclavian a.

Esophageal branch of inferior thyroid a.

Inferior thyroid a.

Cervical part of esophagus

Thyrocervical trunk

Subclavian a.

Vertebral a.

Common carotid a.

Brachiocephalic trunk

Trachea

Aortic arch

Posterior intercostal branch of aorta

Right bronchial branch of aorta

Superior left bronchial branch of aorta

Esophageal branch of aorta

Inferior left bronchial branch of aorta

Esophageal branch of aorta

Descending aorta

Esophageal branches of aorta

Thoracic part of esophagus

Abdominal part of esophagus

Stomach

Diaphragm

Esophageal branch of left gastric a.

Left gastric a.

Celiac trunk

Inferior phrenic aa.

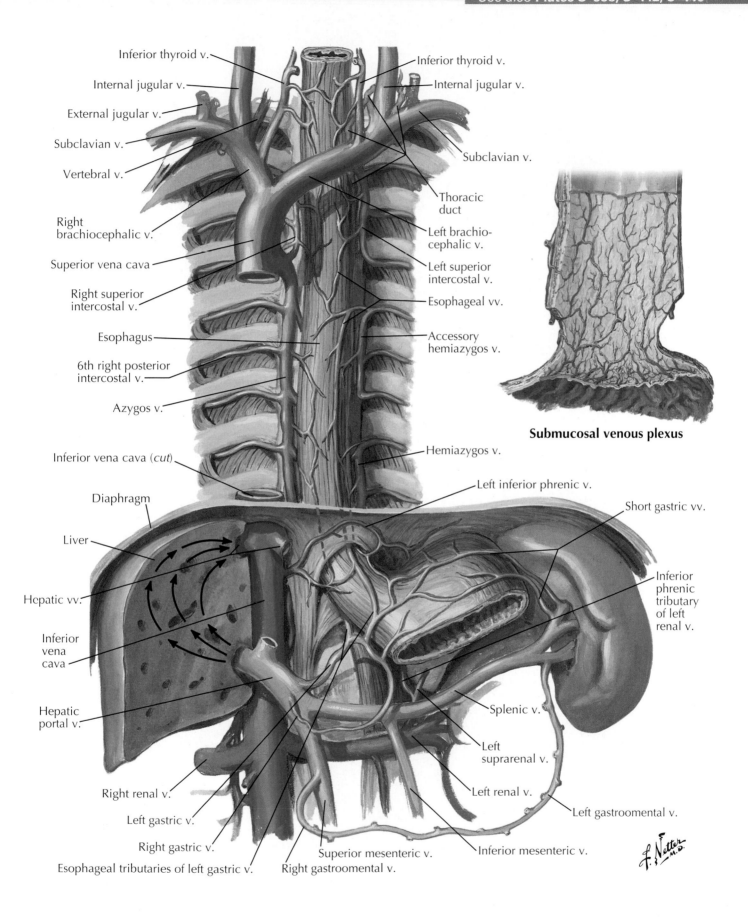

Inferior thyroid v.

Internal jugular v.

External jugular v.

Subclavian v.

Vertebral v.

Right brachiocephalic v.

Superior vena cava

Right superior intercostal v.

Esophagus

6th right posterior intercostal v.

Azygos v.

Inferior vena cava (*cut*)

Diaphragm

Liver

Hepatic vv.

Inferior vena cava

Hepatic portal v.

Right renal v.

Left gastric v.

Right gastric v.

Esophageal tributaries of left gastric v.

Inferior thyroid v.

Internal jugular v.

Subclavian v.

Thoracic duct

Left brachio-cephalic v.

Left superior intercostal v.

Esophageal vv.

Accessory hemiazygos v.

Hemiazygos v.

Left inferior phrenic v.

Short gastric vv.

Inferior phrenic tributary of left renal v.

Splenic v.

Left suprarenal v.

Left renal v.

Left gastroomental v.

Superior mesenteric v.

Right gastroomental v.

Inferior mesenteric v.

Submucosal venous plexus

Blood Vessels of the Trunk

Plate S–340

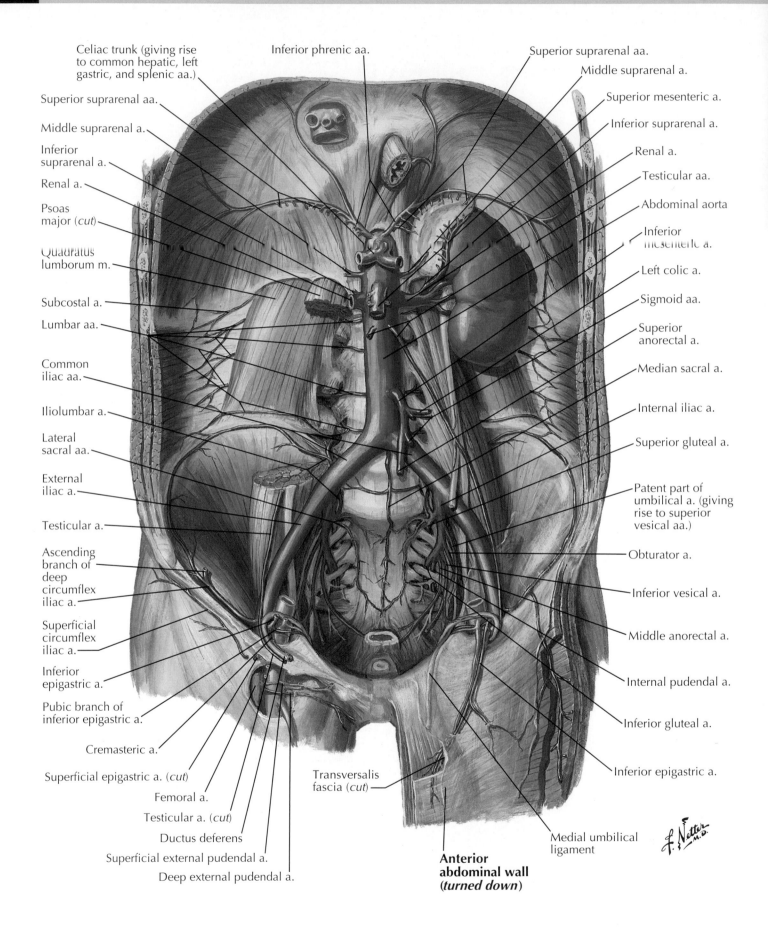

Celiac trunk (giving rise to common hepatic, left gastric, and splenic aa.)

Inferior phrenic aa.

Superior suprarenal aa.

Middle suprarenal a.

Superior suprarenal aa.

Superior mesenteric a.

Middle suprarenal a.

Inferior suprarenal a.

Inferior suprarenal a.

Renal a.

Renal a.

Testicular aa.

Psoas major (cut)

Abdominal aorta

Quadratus lumborum m.

Inferior mesenteric a.

Left colic a.

Subcostal a.

Sigmoid aa.

Lumbar aa.

Superior anorectal a.

Common iliac aa.

Median sacral a.

Iliolumbar a.

Internal iliac a.

Lateral sacral aa.

Superior gluteal a.

External iliac a.

Patent part of umbilical a. (giving rise to superior vesical aa.)

Testicular a.

Obturator a.

Ascending branch of deep circumflex iliac a.

Inferior vesical a.

Superficial circumflex iliac a.

Middle anorectal a.

Inferior epigastric a.

Internal pudendal a.

Pubic branch of inferior epigastric a.

Inferior gluteal a.

Cremasteric a.

Inferior epigastric a.

Superficial epigastric a. (cut)

Transversalis fascia (cut)

Femoral a.

Testicular a. (cut)

Ductus deferens

Superficial external pudendal a.

Medial umbilical ligament

Deep external pudendal a.

Anterior abdominal wall (turned down)

f. Netter M.D.

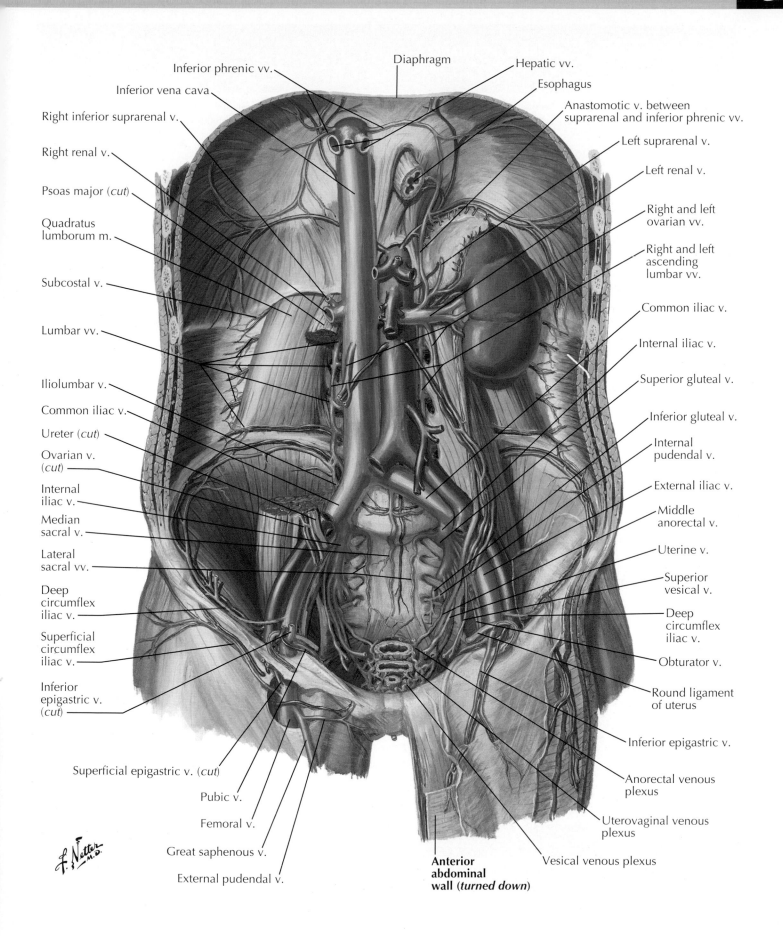

Inferior phrenic vv.

Diaphragm

Hepatic vv.

Esophagus

Inferior vena cava

Anastomotic v. between suprarenal and inferior phrenic vv.

Right inferior suprarenal v.

Left suprarenal v.

Right renal v.

Left renal v.

Psoas major (cut)

Right and left ovarian vv.

Quadratus lumborum m.

Right and left ascending lumbar vv.

Subcostal v.

Common iliac v.

Lumbar vv.

Internal iliac v.

Superior gluteal v.

Iliolumbar v.

Inferior gluteal v.

Common iliac v.

Internal pudendal v.

Ureter (cut)

External iliac v.

Ovarian v. (cut)

Middle anorectal v.

Internal iliac v.

Median sacral v.

Uterine v.

Lateral sacral vv.

Superior vesical v.

Deep circumflex iliac v.

Deep circumflex iliac v.

Superficial circumflex iliac v.

Obturator v.

Inferior epigastric v. (cut)

Round ligament of uterus

Inferior epigastric v.

Superficial epigastric v. (cut)

Anorectal venous plexus

Pubic v.

Uterovaginal venous plexus

Femoral v.

Great saphenous v.

Anterior abdominal wall (turned down)

Vesical venous plexus

External pudendal v.

ANATOMIC STRUCTURES	CLINICAL IMPORTANCE	PLATE NUMBERS
Head and Neck		
Right internal and external jugular veins	Examined to assess right atrial pressure, estimated as height of jugular pulsation above sternal angle (in centimeters) plus 5; right internal jugular vein is preferred because it is in line with superior vena cava	S–329
Internal jugular vein	Thrombosis may occur secondary to local extension of inflammation from severe pharyngitis, a condition known as Lemierre's syndrome	S–329
Internal jugular and subclavian veins	Used to obtain venous access via insertion of central venous catheter	S–328, S–329
Inferior thyroid artery	At risk during thyroidectomy; must be preserved to maintain blood supply to parathyroid glands; identified by its redundant loop shape and is landmark to identify recurrent laryngeal nerve	S–195, S–526
Common carotid artery	Palpate in neck to assess carotid pulse; bifurcation typically at C4 vertebral level	S–43, S–324
Internal carotid artery	Common site for atherosclerosis, which may be treated with stent or endarterectomy for stroke prevention; carotid sinus is sensitive to changes in circulating blood volume and may be massaged to induce vagal reaction; internal carotid artery has no branches in neck; classified into seven parts: C1, cervical; C2, petrous; C3, lacerum; C4, cavernous; C5, clinoid; C6, ophthalmic; and C7, communicating	S–326, S–327
Anterior ethmoidal, sphenopalatine, and facial arteries	Anastomosis site of branches of these vessels in nasal vestibule, known as Kiesselbach's plexus or Little's area, is common site of anterior nosebleeds (epistaxis); sphenopalatine artery injury causes posterior nosebleeds	S–363
Pterygoid venous plexus	Common route for spread of infection due to connections between face, orbit, and venous sinuses; valveless veins allow retrograde flow	S–87, S–328
Ophthalmic artery	Primary source of blood to retina; blindness may occur if artery is occluded	S–87, S–91
Arteries of scalp	Scalp lacerations bleed profusely owing to rich blood supply	S–26, S–322
Superior cerebral veins	May be torn from their junction with superior sagittal sinus, producing subdural hematoma	S–28, S–29, S–35
Middle meningeal artery	Trauma to pterion can tear middle meningeal artery (frontal branch), often causing epidural hematoma	S–27
Dural venous sinuses	Infections in head may spread to sinuses, causing dural venous sinus thrombosis; cavernous sinus is most common site	S–29, S–30
Cavernous sinus	Fistula (anastomosis) between internal carotid artery and cavernous sinus may form, especially following trauma	S–30, S–45
Carotid sinus	Compressed during carotid sinus massage, which may result in bradycardia and/or hypertension; carotid sinus hypersensitivity, most common among older adults, may cause syncope	S–76
Cerebral arterial circle of Willis	Common site of aneurysms and important site of collateral cerebral circulation; aneurysmal rupture produces subarachnoid hemorrhage	S–44
Emissary veins	Valveless veins, which can convey infection extracranially to intracranially and allow alternative route for drainage when dural venous sinuses are obstructed	S–28
Back		
Arteries of spinal cord	Narrowing or damage caused by atherosclerosis, vertebral fractures, or vertebral dislocations may cause ischemia of spinal cord	S–23
Vertebral venous plexuses	Mostly valveless veins along vertebral column allow retrograde flow and can act as conduits for metastasis of cancer cells to spine, lungs, and brain	S–330, S–BP 27

ANATOMIC STRUCTURES	CLINICAL IMPORTANCE	PLATE NUMBERS
Thorax		
Internal thoracic artery	Commonly used for coronary artery bypass grafts, most often for anterior interventricular artery (left anterior descending artery)	S–205, S–206
Pulmonary arteries	Thromboemboli, most often from pelvic and femoral sources, may obstruct pulmonary arteries (pulmonary embolus), leading to hypoxemia, hemodynamic compromise, and pulmonary infarction	S–389, S–392
Pericardium	Pericardial space can contain small amounts of physiologic fluid (15–50 mL); effusion may compromise heart function (cardiac tamponade); pericardial sac can expand and become quite large with slow, progressive fluid accumulation	S–307, S–308
Coronary arteries	Fixed atherosclerotic disease may cause myocardial ischemia to manifest as thoracic pain; rupture and thrombosis of atherosclerotic plaque is main cause of acute myocardial infarction; severity and outcomes depend on amount of myocardium that vessel subtends, with proximal lesions of large vessels being most morbid	S–317
Pulmonary veins	Atrial fibrillation, a common arrhythmia, is believed to originate from within pulmonary veins; electrical ablation of this arrhythmia creates rings of fibrosis around pulmonary veins as they enter left atrium, thereby preventing propagation of electrical signals into heart	S–312
Foramen ovale	Provides channel for interatrial flow (right-to-left shunt) during fetal development; remains patent in approximately one in four adults and may provide route for venous microthromboses to enter left side of heart and cause ischemic stroke	S–305, S–312
Interventricular septum	Ventricular septal defect is common congenital cardiac defect, most often involving membranous portion of septum; myocardial infarction in territory of anterior interventricular artery (left anterior descending artery), especially if not promptly treated, may generate sufficient ischemia of septum to cause perforation	S–312, S–316
Heart valves	Valvular disease (e.g., aortic stenosis, mitral insufficiency) is common, especially among older adults, and may cause progressive heart failure	S–314
Aortic valve	1% of population has bicuspid aortic valve (i.e., containing two rather than three leaflets), which predisposes to aortic stenosis and insufficiency and is also associated with aortic aneurysm	S–314, S–315
Sinuatrial node	Primary cardiac pacemaker, which generates action potentials that propagate through cardiac conduction system; aging, infiltrative diseases, and heart surgery may cause sinus node dysfunction, resulting in bradycardia	S–319
Atrioventricular node	Conducts action potentials from atria to ventricles; intrinsic refractory period prevents rapid atrial rhythms from causing equivalent tachycardia of ventricles; dysfunction secondary to fibrosis, medications, or cardiac surgery may result in heart block; when block is complete, atria and ventricles have independent rhythms	S–319
Ligamentum arteriosum	Remnant of ductus arteriosus, which connects pulmonary and systemic circulations during fetal development; lack of ductus closure after birth may cause exertional dyspnea, pulmonary vascular disease, or heart failure; acts as landmark to identify left recurrent laryngeal nerve looping inferior to ligament	S–305, S–309

ANATOMIC STRUCTURES	CLINICAL IMPORTANCE	PLATE NUMBERS
Thorax—Continued		
Thoracic aorta	Lies naturally to left of vertebral column in thorax and commences at T4 vertebral level where aortic arch terminates; congenital coarctation (narrowing) of aorta may cause hypertension in children and young adults; significant difference in blood pressure between upper and lower extremities is suggestive	S–339
Thoracic aorta	Aneurysm (enlargement) may occur secondary to age and atherosclerotic risk factors (e.g., tobacco abuse and hypertension), connective tissue disorders, in association with a bicuspid aortic valve, or from infection (e.g., syphilis). Large aneurysm can rupture or dissect; latter occurs when tear in intimal layer allows blood to propagate into false lumen between intima and media	S–339
Azygos vein	Drains posterior thorax and provides important collateral channel between inferior vena cava and superior vena cava	S–340
Abdomen		
Paraumbilical veins	May become dilated in patients with portal hypertension and in late-term pregnancy	S–337, S–446
Cystic artery	Ligated during cholecystectomy; can have multiple origins; found classically in cystohepatic triangle (of Calot)	S–436
Superior mesenteric artery	May compress horizontal (third) part of duodenum in thin patient or patient who has recently lost a large amount of weight; can shear or tear in sudden deceleration injuries	S–436, S–439
Intestinal arteries	Areas without significant collateral circulation between major vessels (watershed areas) are at risk for ischemia, which can occur secondary to atherosclerosis or thromboembolism of mesenteric arteries	S–439, S–440
Marginal artery anastomosis	Marginal artery connects right, middle, and left colic arteries, providing important anastomosis for collateral circulation	S–440
Esophageal veins	May become dilated in portal hypertension, resulting in esophageal varices; variceal hemorrhage can be life-threatening and often requires emergent endoscopic intervention	S–442, S–446
Hepatic portal vein	Increased resistance to blood flow through liver (e.g., due to cirrhosis) may produce portal hypertension and dilation of tributaries of hepatic portal vein; blood may return to heart at sites of portosystemic anastomosis	S–444, S–446
Superior anorectal (rectal) vein	Anastomoses with systemic middle and inferior anorectal veins may become dilated in portal hypertension	S–444, S–446
Abdominal aorta	Common site for aneurysm in abdomen, especially inferior to renal arteries; assessed routinely with ultrasound to exclude aneurysms	S–468
Renal artery	Stenosis may occur secondary to atherosclerosis or fibromuscular dysplasia, resulting in difficult-to-control hypertension; can be affected in abdominal aortic aneurysms	S–303, S–531
Pelvis		
Pampiniform plexus	Dilation can cause testicular varicocele, affecting testicular temperature regulation and potentially contributing to infertility; most commonly occurs on left side due to longer course of left gonadal vein and angle of drainage into renal vein	S–497, S–510

Table 5.3   **Structures with High Clinical Significance**

ANATOMIC STRUCTURES	CLINICAL IMPORTANCE	PLATE NUMBERS
Uterine artery	Ligated or cauterized during hysterectomy; selective embolization of branches may be performed to treat uterine fibroids	S–508, S–511
Deep and dorsal arteries of penis, and cavernous tissue	Blockage or loss of vascular smooth muscle function can lead to erectile dysfunction, treated with vasodilators	S–512
Internal iliac veins	Provide communication between prostatic venous plexus and veins draining vertebral column, which is route of spread for prostate cancer	S–509
Anorectal (rectal) veins	Portal hypertension may cause dilated anorectal veins if portosystemic anastomoses develop between superior anorectal veins (portal drainage) and middle and/or inferior anorectal veins (systemic drainage)	S–445, S–446
Internal and external anorectal (rectal) venous plexuses	Enlargement may result in painful condition known as hemorrhoids	S–223, S–226
Upper Limb		
Median cubital vein	Accessed in cubital fossa for venipuncture	S–228
Suprascapular, dorsal scapular, and circumflex scapular arteries	Provide collateral circulation around scapula, allowing blood to reach distal part of upper limb if axillary artery is blocked or compressed	S–234
Brachial artery	During deflation of sphygmomanometer on upper arm, brachial artery is auscultated for Korotkoff sounds to measure systolic and diastolic blood pressure; identified medial to biceps brachii tendon and deep to bicipital aponeurosis in cubital fossa	S–238, S–331
Radial artery	Palpated at lateral aspect of wrist to assess radial pulse; common site of vascular access for percutaneous cardiac procedures, such as angioplasty, and for sampling of arterial blood	S–303, S–331
Ulnar artery	Provides important collateral circulation to hand via palmar arch during catheterization of radial artery; patency is assessed prior to procedure using Allen test, in which both radial and ulnar arteries are compressed, then pressure over ulnar artery is released; return of color to wrist within a few seconds indicates patent ulnar artery	S–257, S–331
Lower Limb		
Femoral vein	Common site of vascular access for central venous catheters; however, risk of catheter infection is greater than with jugular or subclavian vein access	S–270
Deep veins of lower limb	Venous thrombosis of deep leg veins is due to venous stasis, vessel injury, and/or coagulation disorders (Virchow's triad); can lead to thrombus formation and thromboemboli, such as to the lungs	S–335
Great saphenous vein	Often used as graft in coronary artery bypass surgery	S–266
Superficial veins of lower limb	Varicose veins are dilated, tortuous superficial veins, often associated with reflux of superficial and/or deep veins; progressive venous disease may cause edema, pain, and ulceration	S–266, S–335
Femoral artery	Common site of vascular access for percutaneous cardiac and vascular procedures; target is segment between origins of inferior epigastric and deep femoral arteries, generally identified using fluoroscopy by its location alongside the femoral head; "high stick" (too cephalad) may result in retroperitoneal hemorrhage	S–333

Structures with High Clinical Significance **Table 5.4**

ANATOMIC STRUCTURES	CLINICAL IMPORTANCE	PLATE NUMBERS
Lower Limb—Continued		
Femoral, popliteal, tibial, and fibular arteries	Peripheral arterial disease due to atherosclerosis may occur in major arteries of lower limb, resulting in reduced blood flow; patients experience claudication (cramping pain in thigh or calf) upon exertion	S–333
Arteries of lower limb	Pulse points: femoral artery in femoral triangle, popliteal artery in deep popliteal region of knee, anterior tibial artery between extensor hallucis longus and extensor digitorum longus at ankle joint, dorsalis pedis artery on dorsum of foot, and posterior tibial artery in tarsal tunnel posterior to medial malleolus	S–270, S–284, S–290, S–291, S–303

*Selections are based largely on clinical data and commonly discussed clinical correlations in macroscopic ("gross") anatomy courses.

Table 5.5   **Structures with High Clinical Significance**

LYMPH VESSELS AND LYMPHOID ORGANS

6

ELECTRONIC BONUS PLATES

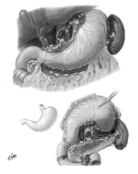

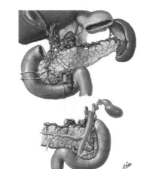

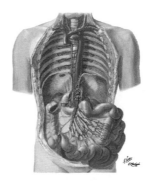

S–BP 54 Lymph Vessels and Lymph Nodes of Stomach

S–BP 55 Lymph Vessels and Lymph Nodes of Pancreas

S–BP 56 Lymph Vessels and Lymph Nodes of Small Intestine

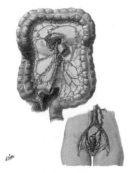

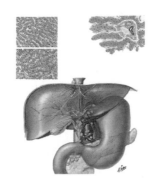

S–BP 57 Lymph Vessels and Lymph Nodes of Large Intestine

S–BP 58 Lymph Vessels and Lymph Nodes of Liver

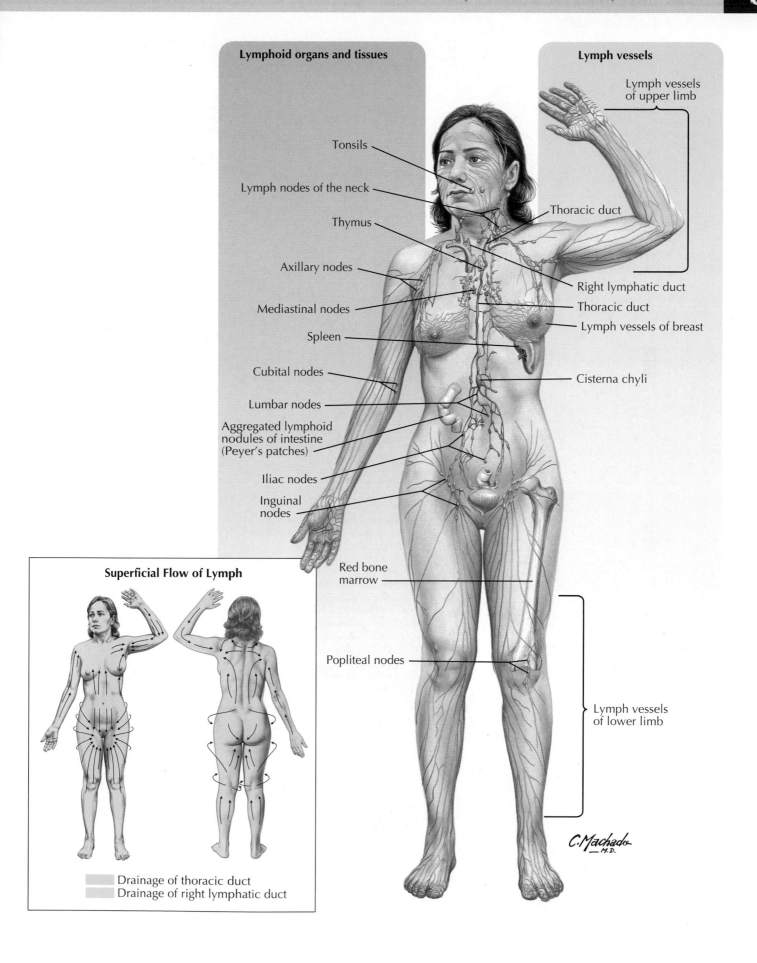

Lymphoid organs and tissues

Lymph vessels

Lymph vessels
of upper limb

Tonsils

Lymph nodes of the neck

Thoracic duct

Thymus

Axillary nodes

Right lymphatic duct

Mediastinal nodes

Thoracic duct

Lymph vessels of breast

Spleen

Cubital nodes

Cisterna chyli

Lumbar nodes

Aggregated lymphoid
nodules of intestine
(Peyer's patches)

Iliac nodes

Inguinal
nodes

Red bone
marrow

Popliteal nodes

Lymph vessels
of lower limb

Superficial Flow of Lymph

Drainage of thoracic duct
Drainage of right lymphatic duct

C. Machado
M.D.

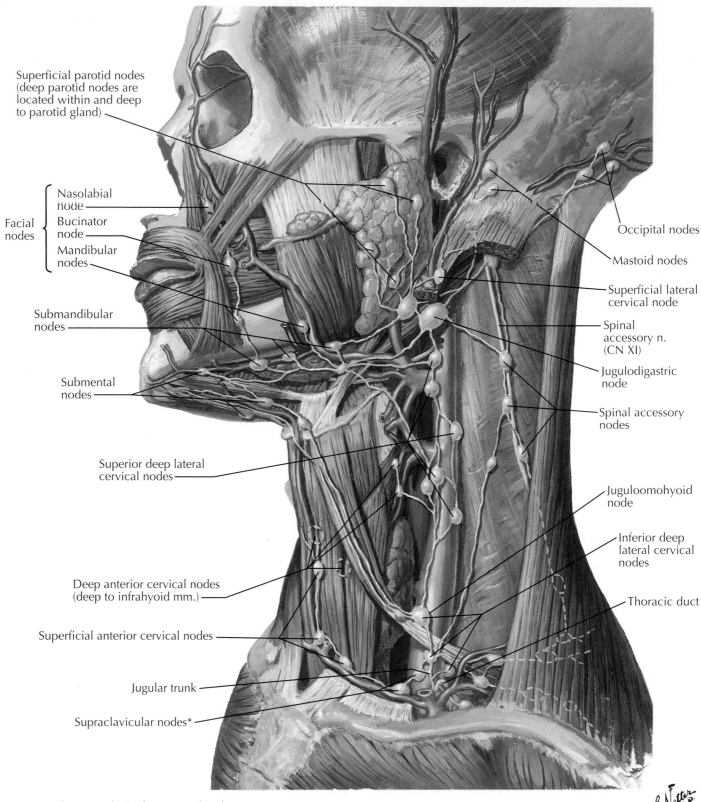

Superficial parotid nodes
(deep parotid nodes are
located within and deep
to parotid gland)

Facial nodes
- Nasolabial node
- Bucinator node
- Mandibular nodes

Submandibular nodes

Submental nodes

Superior deep lateral cervical nodes

Deep anterior cervical nodes
(deep to infrahyoid mm.)

Superficial anterior cervical nodes

Jugular trunk

Supraclavicular nodes*

Occipital nodes

Mastoid nodes

Superficial lateral cervical node

Spinal accessory n. (CN XI)

Jugulodigastric node

Spinal accessory nodes

Juguloomohyoid node

Inferior deep lateral cervical nodes

Thoracic duct

*The supraclavicular group of nodes,
especially on the left, are also sometimes
referred to as the signal or sentinel lymph
nodes of Virchow or Troisier, especially when
sufficiently enlarged and palpable. These
nodes (or a single node) are so termed because
they may be the first recognized presumptive
evidence of malignant disease in the viscera.

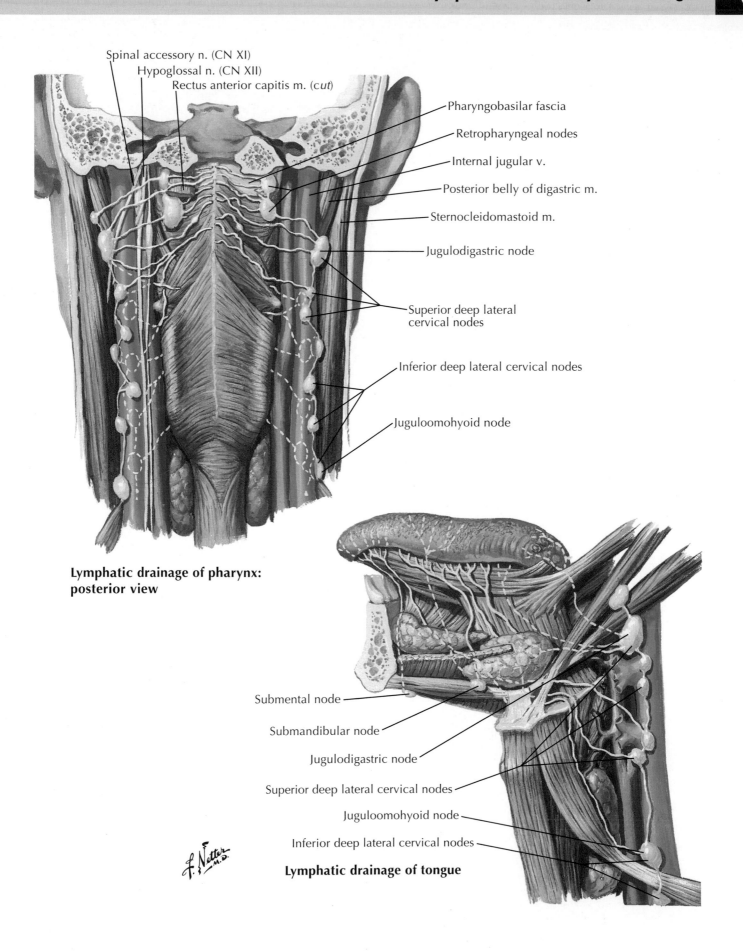

Spinal accessory n. (CN XI)
Hypoglossal n. (CN XII)
Rectus anterior capitis m. (*cut*)

Pharyngobasilar fascia

Retropharyngeal nodes

Internal jugular v.

Posterior belly of digastric m.

Sternocleidomastoid m.

Jugulodigastric node

Superior deep lateral cervical nodes

Inferior deep lateral cervical nodes

Juguloomohyoid node

Lymphatic drainage of pharynx: posterior view

Submental node

Submandibular node

Jugulodigastric node

Superior deep lateral cervical nodes

Juguloomohyoid node

Inferior deep lateral cervical nodes

Lymphatic drainage of tongue

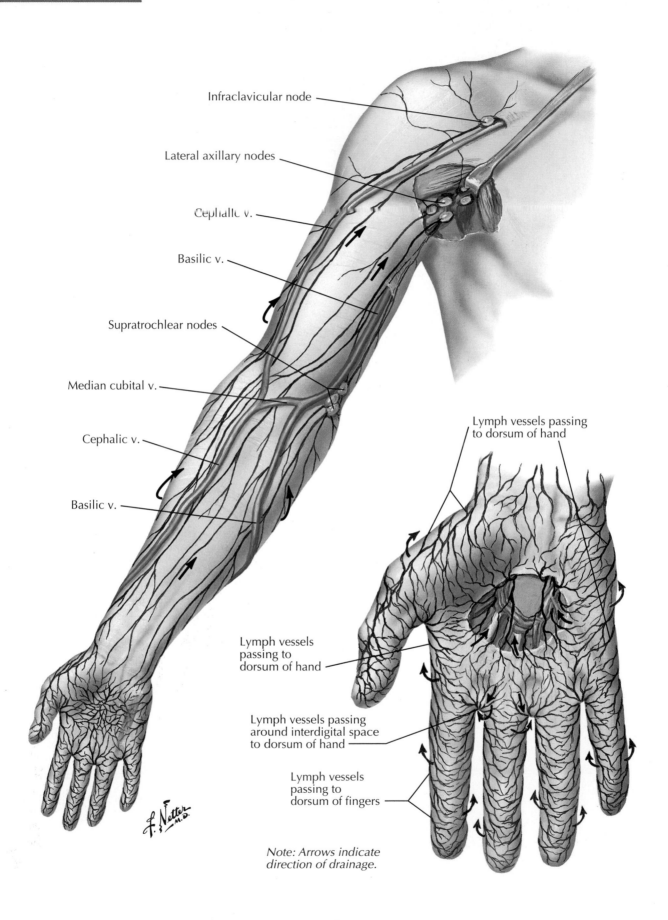

Infraclavicular node

Lateral axillary nodes

Cephalic v.

Basilic v.

Supratrochlear nodes

Median cubital v.

Cephalic v.

Basilic v.

Lymph vessels passing to dorsum of hand

Lymph vessels passing to dorsum of hand

Lymph vessels passing around interdigital space to dorsum of hand

Lymph vessels passing to dorsum of fingers

Note: Arrows indicate direction of drainage.

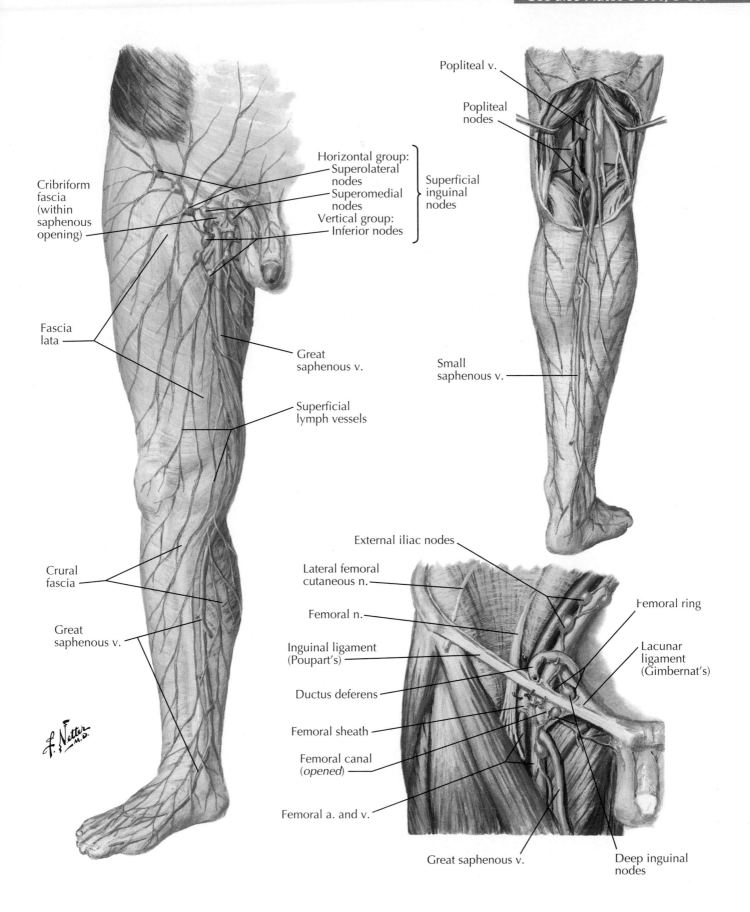

Cribriform fascia (within saphenous opening)

Horizontal group:
Superolateral nodes
Superomedial nodes
Vertical group:
Inferior nodes

Superficial inguinal nodes

Popliteal v.

Popliteal nodes

Fascia lata

Great saphenous v.

Small saphenous v.

Superficial lymph vessels

Crural fascia

Great saphenous v.

External iliac nodes

Lateral femoral cutaneous n.

Femoral n.

Inguinal ligament (Poupart's)

Ductus deferens

Femoral sheath

Femoral canal (opened)

Femoral a. and v.

Great saphenous v.

Femoral ring

Lacunar ligament (Gimbernat's)

Deep inguinal nodes

f. Netter M.D.

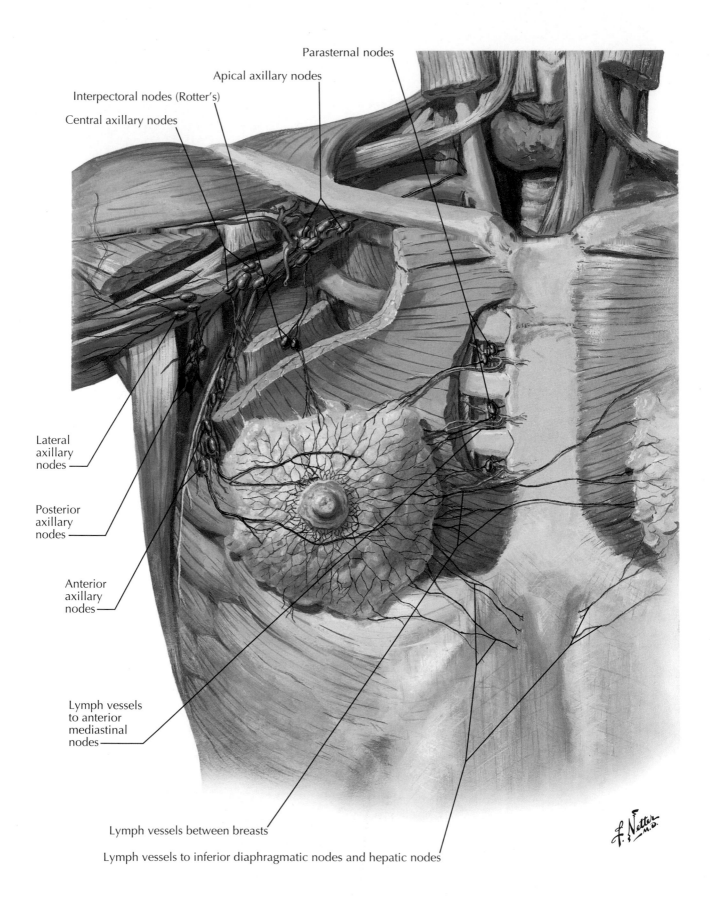

Parasternal nodes

Apical axillary nodes

Interpectoral nodes (Rotter's)

Central axillary nodes

Lateral axillary nodes

Posterior axillary nodes

Anterior axillary nodes

Lymph vessels to anterior mediastinal nodes

Lymph vessels between breasts

Lymph vessels to inferior diaphragmatic nodes and hepatic nodes

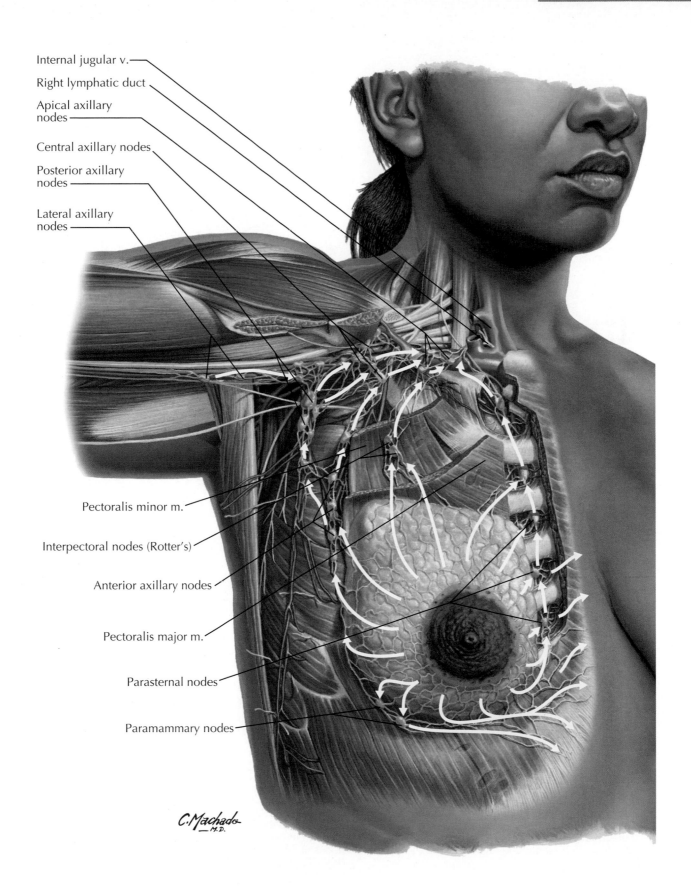

Internal jugular v.

Right lymphatic duct

Apical axillary nodes

Central axillary nodes

Posterior axillary nodes

Lateral axillary nodes

Pectoralis minor m.

Interpectoral nodes (Rotter's)

Anterior axillary nodes

Pectoralis major m.

Parasternal nodes

Paramammary nodes

C. Machado
M.D.

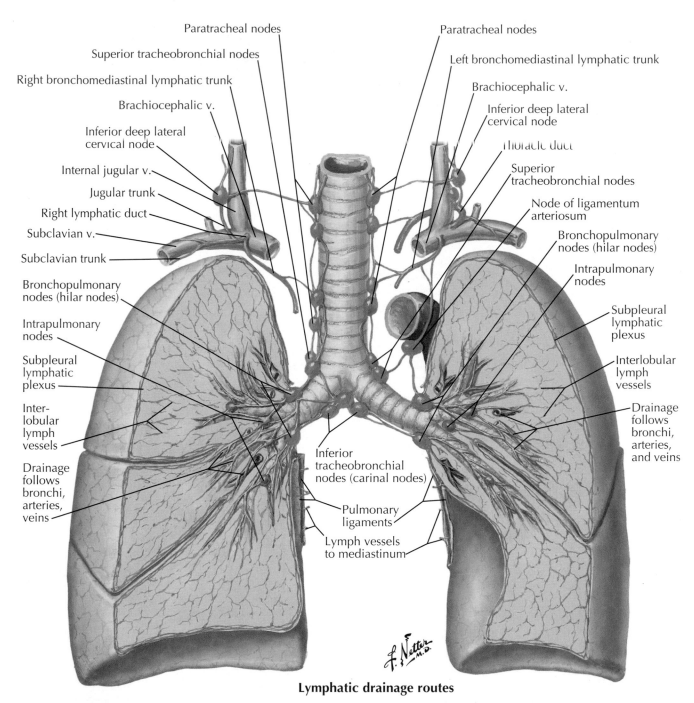

Paratracheal nodes

Superior tracheobronchial nodes

Right bronchomediastinal lymphatic trunk

Brachiocephalic v.

Inferior deep lateral cervical node

Internal jugular v.

Jugular trunk

Right lymphatic duct

Subclavian v.

Subclavian trunk

Bronchopulmonary nodes (hilar nodes)

Intrapulmonary nodes

Subpleural lymphatic plexus

Interlobular lymph vessels

Drainage follows bronchi, arteries, veins

Paratracheal nodes

Left bronchomediastinal lymphatic trunk

Brachiocephalic v.

Inferior deep lateral cervical node

Thoracic duct

Superior tracheobronchial nodes

Node of ligamentum arteriosum

Bronchopulmonary nodes (hilar nodes)

Intrapulmonary nodes

Subpleural lymphatic plexus

Interlobular lymph vessels

Drainage follows bronchi, arteries, and veins

Inferior tracheobronchial nodes (carinal nodes)

Pulmonary ligaments

Lymph vessels to mediastinum

Lymphatic drainage routes

Right lung: All lobes drain to intrapulmonary and bronchopulmonary nodes, then to inferior tracheobronchial nodes, right superior tracheobronchial nodes, and right paratracheal nodes on the way to the brachiocephalic vein via the right bronchomediastinal and jugular trunks.

Left lung: The superior lobe drains to intrapulmonary and bronchopulmonary nodes, then to inferior tracheobronchial nodes, left superior tracheobronchial nodes, left paratracheal nodes and the node of the ligamentum arteriosum on the way to the brachiocephalic vein via the left bronchomediastinal trunk and thoracic duct. The intrapulmonary and bronchopulmonary nodes of the left lung also drain to right superior tracheobronchial nodes, where the lymph follows the same route as lymph from the right lung.

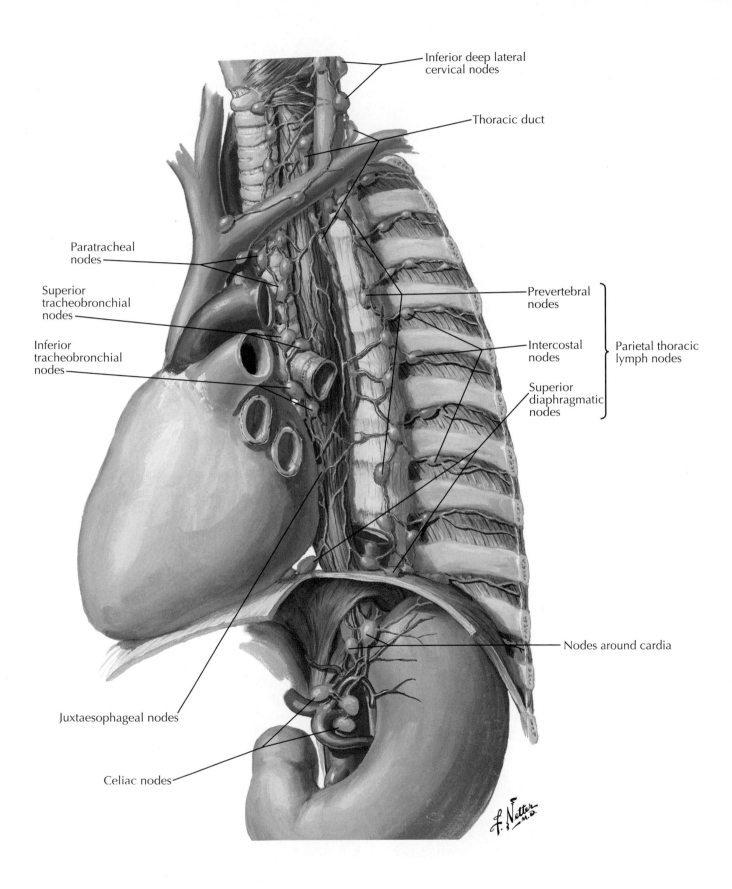

Inferior deep lateral cervical nodes

Thoracic duct

Paratracheal nodes

Superior tracheobronchial nodes

Inferior tracheobronchial nodes

Prevertebral nodes

Intercostal nodes

Parietal thoracic lymph nodes

Superior diaphragmatic nodes

Nodes around cardia

Juxtaesophageal nodes

Celiac nodes

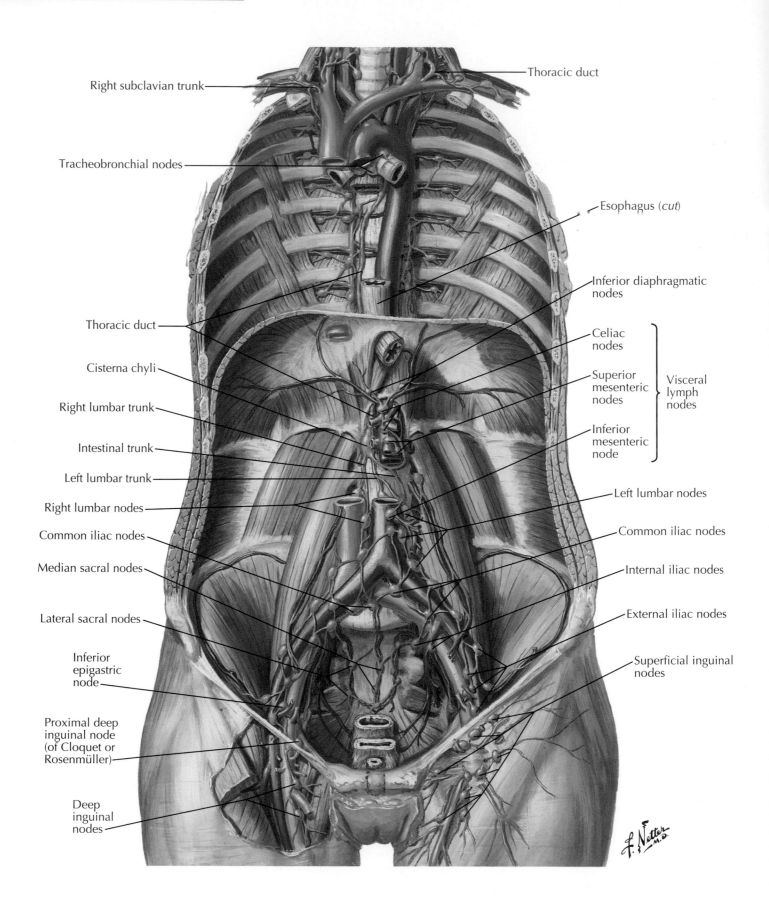

Right subclavian trunk

Tracheobronchial nodes

Thoracic duct

Cisterna chyli

Right lumbar trunk

Intestinal trunk

Left lumbar trunk

Right lumbar nodes

Common iliac nodes

Median sacral nodes

Lateral sacral nodes

Inferior epigastric node

Proximal deep inguinal node (of Cloquet or Rosenmüller)

Deep inguinal nodes

Thoracic duct

Esophagus (*cut*)

Inferior diaphragmatic nodes

Celiac nodes

Superior mesenteric nodes

Inferior mesenteric node

Visceral lymph nodes

Left lumbar nodes

Common iliac nodes

Internal iliac nodes

External iliac nodes

Superficial inguinal nodes

F. Netter M.D.

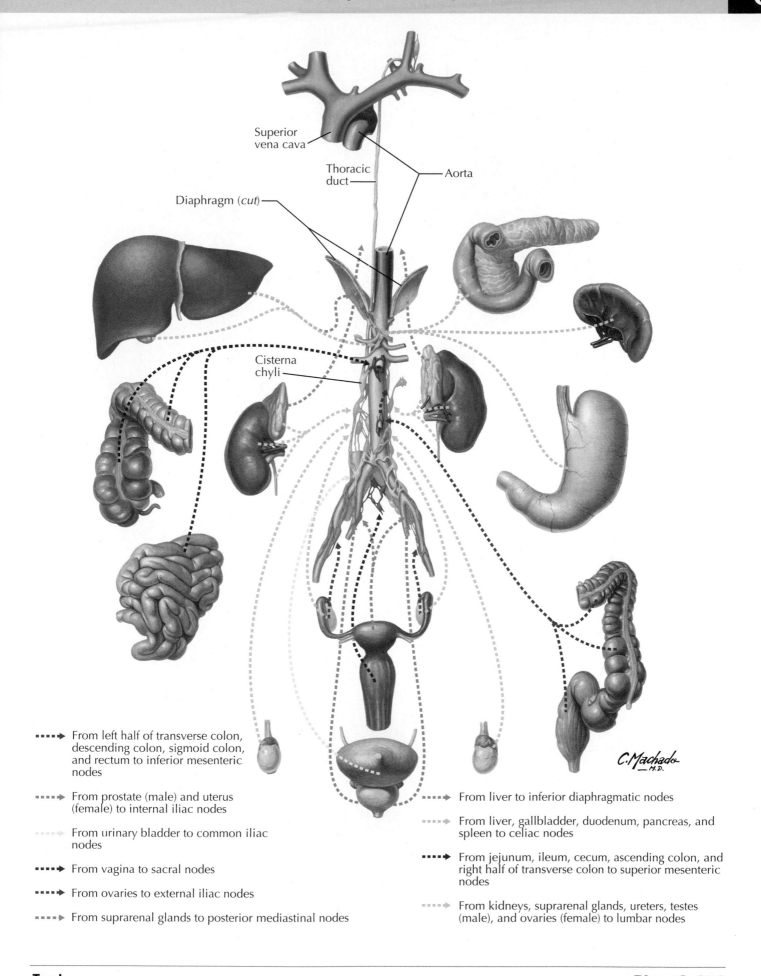

Superior vena cava

Thoracic duct

Aorta

Diaphragm (*cut*)

Cisterna chyli

C. Machado
—M.D.

- ┅► From left half of transverse colon, descending colon, sigmoid colon, and rectum to inferior mesenteric nodes

- ┅► From prostate (male) and uterus (female) to internal iliac nodes

- ┅► From urinary bladder to common iliac nodes

- ┅► From vagina to sacral nodes

- ┅► From ovaries to external iliac nodes

- ┅► From suprarenal glands to posterior mediastinal nodes

- ┅► From liver to inferior diaphragmatic nodes

- ┅► From liver, gallbladder, duodenum, pancreas, and spleen to celiac nodes

- ┅► From jejunum, ileum, cecum, ascending colon, and right half of transverse colon to superior mesenteric nodes

- ┅► From kidneys, suprarenal glands, ureters, testes (male), and ovaries (female) to lumbar nodes

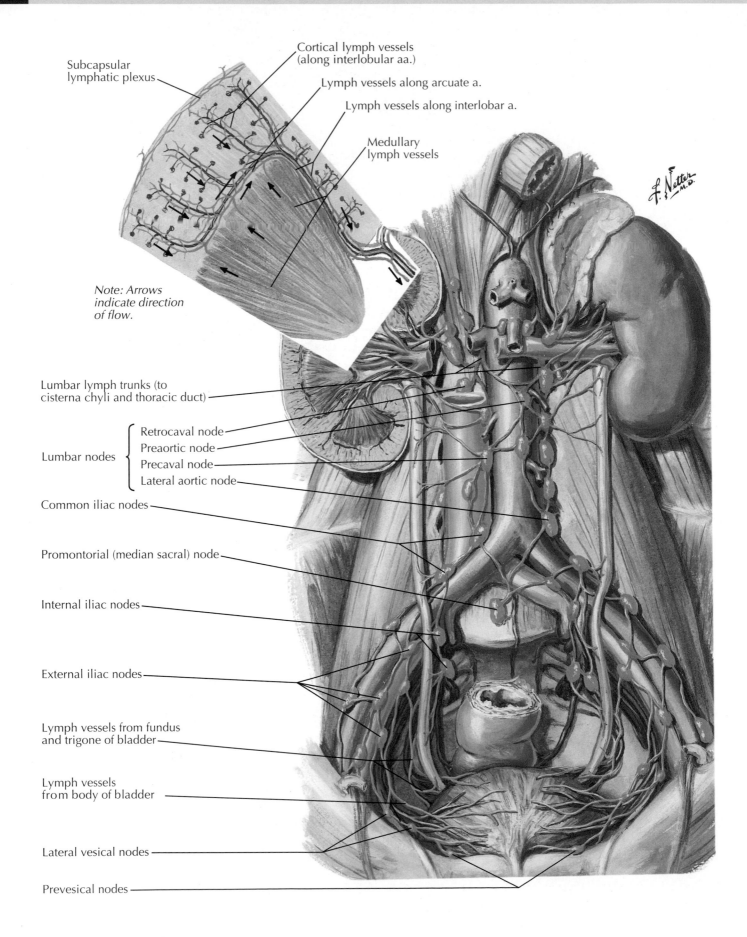

Subcapsular
lymphatic plexus

Cortical lymph vessels
(along interlobular aa.)

Lymph vessels along arcuate a.

Lymph vessels along interlobar a.

Medullary
lymph vessels

Note: Arrows
indicate direction
of flow.

Lumbar lymph trunks (to
cisterna chyli and thoracic duct)

Lumbar nodes
{
Retrocaval node
Preaortic node
Precaval node
Lateral aortic node

Common iliac nodes

Promontorial (median sacral) node

Internal iliac nodes

External iliac nodes

Lymph vessels from fundus
and trigone of bladder

Lymph vessels
from body of bladder

Lateral vesical nodes

Prevesical nodes

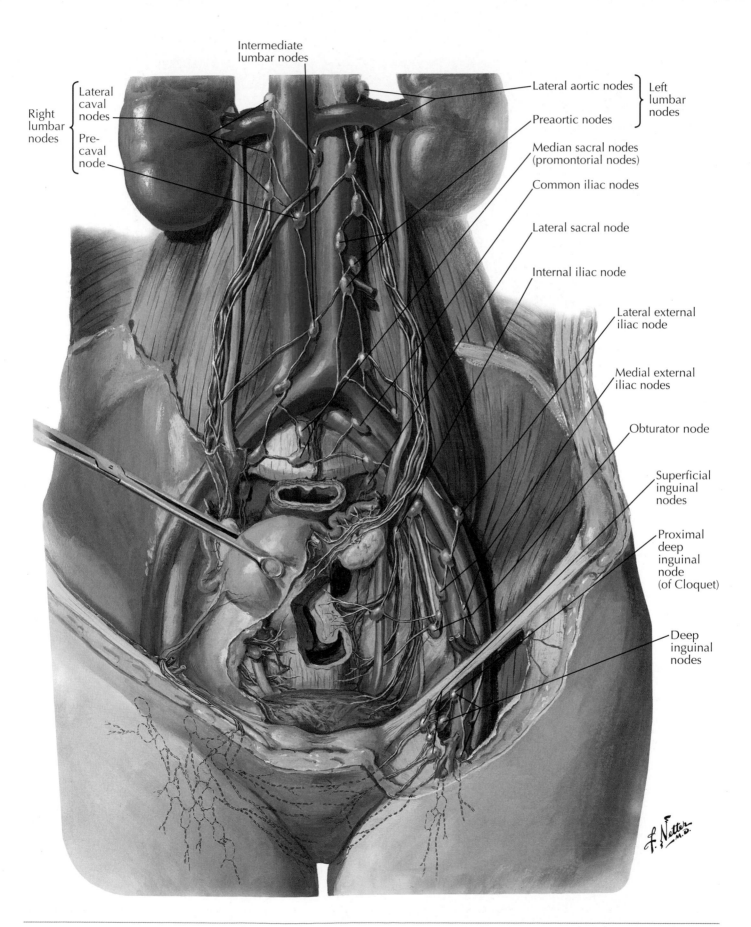

Intermediate lumbar nodes

Lateral caval nodes

Right lumbar nodes

Pre-caval node

Lateral aortic nodes

Preaortic nodes

Left lumbar nodes

Median sacral nodes (promontorial nodes)

Common iliac nodes

Lateral sacral node

Internal iliac node

Lateral external iliac node

Medial external iliac nodes

Obturator node

Superficial inguinal nodes

Proximal deep inguinal node (of Cloquet)

Deep inguinal nodes

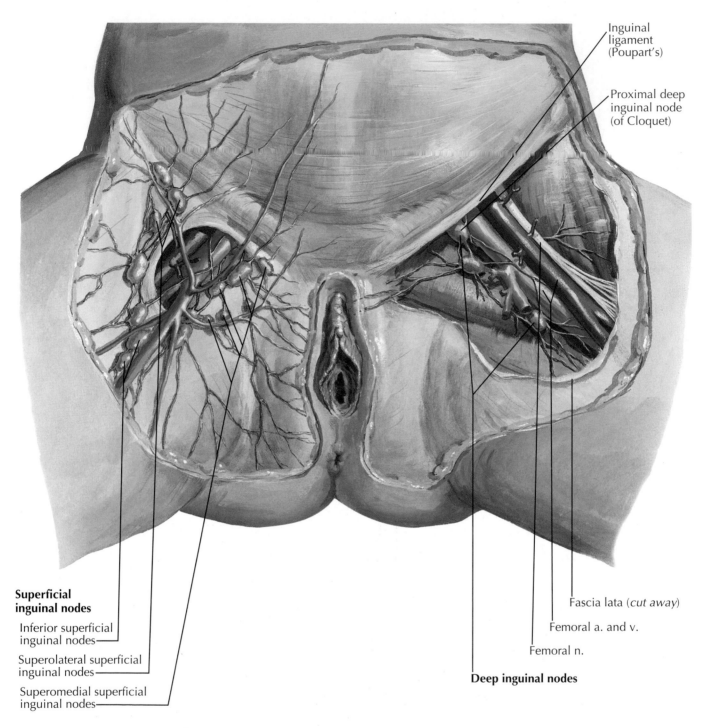

Inguinal
ligament
(Poupart's)

Proximal deep
inguinal node
(of Cloquet)

**Superficial
inguinal nodes**

Inferior superficial
inguinal nodes

Superolateral superficial
inguinal nodes

Superomedial superficial
inguinal nodes

Fascia lata (*cut away*)

Femoral a. and v.

Femoral n.

Deep inguinal nodes

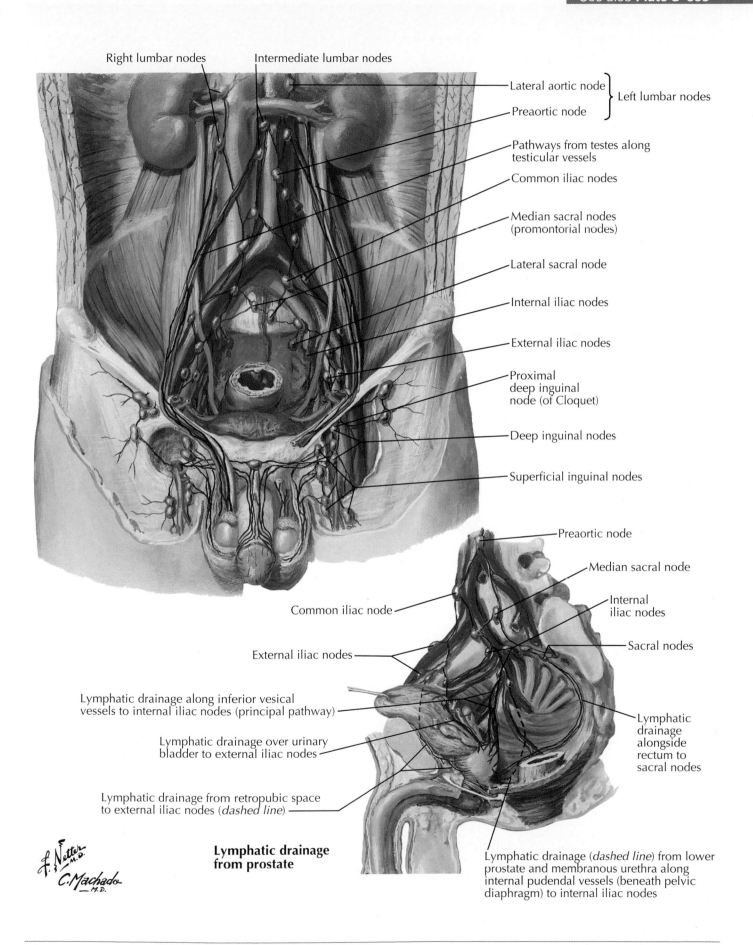

Right lumbar nodes

Intermediate lumbar nodes

Lateral aortic node ⎫
Preaortic node ⎭ Left lumbar nodes

Pathways from testes along testicular vessels

Common iliac nodes

Median sacral nodes (promontorial nodes)

Lateral sacral node

Internal iliac nodes

External iliac nodes

Proximal deep inguinal node (of Cloquet)

Deep inguinal nodes

Superficial inguinal nodes

Preaortic node

Median sacral node

Common iliac node

Internal iliac nodes

External iliac nodes

Sacral nodes

Lymphatic drainage along inferior vesical vessels to internal iliac nodes (principal pathway)

Lymphatic drainage alongside rectum to sacral nodes

Lymphatic drainage over urinary bladder to external iliac nodes

Lymphatic drainage from retropubic space to external iliac nodes (*dashed line*)

Lymphatic drainage from prostate

Lymphatic drainage (*dashed line*) from lower prostate and membranous urethra along internal pudendal vessels (beneath pelvic diaphragm) to internal iliac nodes

ANATOMIC STRUCTURES	CLINICAL IMPORTANCE	PLATE NUMBERS
Head and Neck		
Thoracic duct	Thoracic duct may be injured during surgeries to neck and thorax due to multiple and frequent variants; injury in lower neck region is often at junction of left internal jugular and subclavian veins; increased risk during esophageal surgery and left central venous line placement	S–344, S–351
Superior and inferior deep lateral cervical nodes	Palpated during neck examination to assess size and shape; if coalescent, malignancy must be excluded	S–344, S–345
Palatine and pharyngeal tonsils	Palatine tonsils are commonly involved with viral and bacterial infection; exudative lesions combined with fever, lymphadenopathy, and lack of cough suggestive of streptococcal infection (strep throat); enlarged pharyngeal tonsils (adenoids) cause snoring and can cover auditory (Eustachian) tube opening, increasing risk of middle ear infections	S–403
Thorax		
Lymph vessels of breast	Metastatic spread of cancer cells from mammary gland to axilla and thorax via lymphatics draining breast	S–349
Axillary nodes	Primary nodes that receive lymphatic drainage from upper limb, thoracic wall, and breast; commonly enlarged in patients with breast cancer	S–348, S–349
Abdomen		
Spleen	May be ruptured by fracture of ribs 10 to 12; enlargement (splenomegaly) may occur in cirrhosis, viral infections, and hematologic malignancies; if palpable, the spleen is enlarged	S–416, S–434
Pelvis		
Pelvic and lumbar nodes	Spread of ovarian cancer cells via venous drainage to inferior vena cava and lungs or via lymphatics	S–355
Lumbar and tracheobronchial nodes	Prostate cancer cells may spread via lymphatics to retroperitoneum and mediastinum	S–351, S–357
Lumbar (e.g., lateral aortic, preaortic, lateral caval) nodes	Receive lymphatic drainage from ovary, uterine tube, and fundus of uterus in women and from testis in men; cancers in these organs may therefore spread to retroperitoneum	S–355, S–357
Pelvic nodes	Sampling or dissection is performed to assess spread of gynecologic malignancies	S–355
Lower Limb		
Superficial inguinal nodes	Superficial inguinal nodes drain the lower limb, gluteal region, lower abdominal region, and perineum; are palpable when enlarged	S–347
Lymph vessels of lower limb	Lymphedema (stasis of lymph flow in lymph vessels obstructed by inflammation, fibrosis, tumor, or abnormally small diameter)	S–347

*Selections are based largely on clinical data and commonly discussed clinical correlations in macroscopic ("gross") anatomy courses.

Table 6.1   **Structures with High Clinical Significance**

RESPIRATORY SYSTEM 7

ELECTRONIC BONUS PLATES

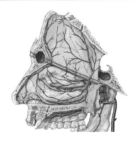

S–BP 59 Arteries of Nasal Cavity: Bony Nasal Septum Turned Up

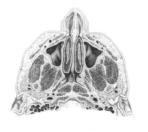

S–BP 60 Nose and Maxillary Sinus: Transverse Section

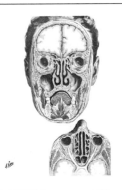

S–BP 61 Paranasal Sinuses

S–BP 62 Intrapulmonary Airways: Schema

S–BP 63 Anatomy of Ventilation and Respiration

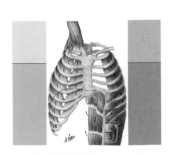

S–BP 64 Muscles of Respiration

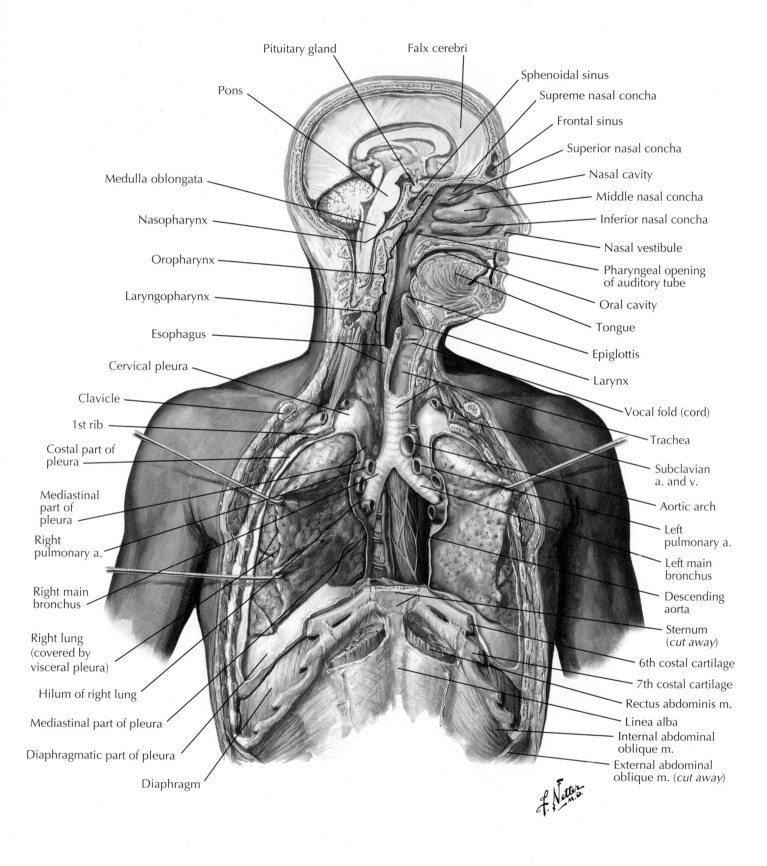

Pituitary gland

Falx cerebri

Sphenoidal sinus

Supreme nasal concha

Frontal sinus

Superior nasal concha

Nasal cavity

Middle nasal concha

Inferior nasal concha

Nasal vestibule

Pharyngeal opening of auditory tube

Oral cavity

Tongue

Epiglottis

Larynx

Vocal fold (cord)

Trachea

Subclavian a. and v.

Aortic arch

Left pulmonary a.

Left main bronchus

Descending aorta

Sternum (*cut away*)

6th costal cartilage

7th costal cartilage

Rectus abdominis m.

Linea alba

Internal abdominal oblique m.

External abdominal oblique m. (*cut away*)

Pons

Medulla oblongata

Nasopharynx

Oropharynx

Laryngopharynx

Esophagus

Cervical pleura

Clavicle

1st rib

Costal part of pleura

Mediastinal part of pleura

Right pulmonary a.

Right main bronchus

Right lung (covered by visceral pleura)

Hilum of right lung

Mediastinal part of pleura

Diaphragmatic part of pleura

Diaphragm

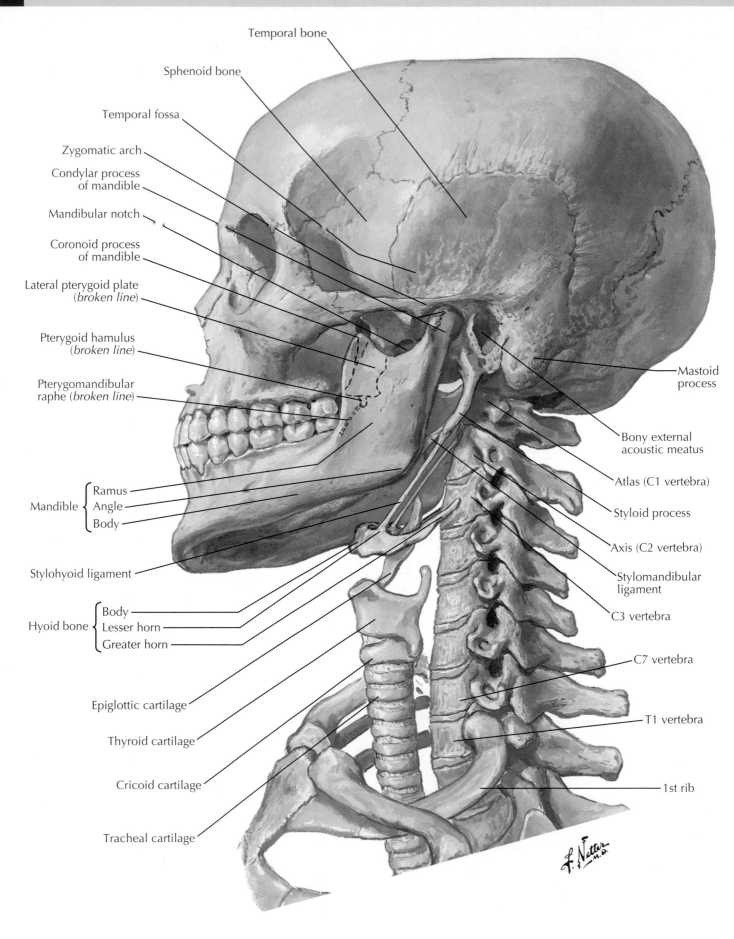

Temporal bone

Sphenoid bone

Temporal fossa

Zygomatic arch

Condylar process
of mandible

Mandibular notch

Coronoid process
of mandible

Lateral pterygoid plate
(*broken line*)

Pterygoid hamulus
(*broken line*)

Pterygomandibular
raphe (*broken line*)

Mandible { Ramus
Angle
Body

Stylohyoid ligament

Hyoid bone { Body
Lesser horn
Greater horn

Epiglottic cartilage

Thyroid cartilage

Cricoid cartilage

Tracheal cartilage

Mastoid
process

Bony external
acoustic meatus

Atlas (C1 vertebra)

Styloid process

Axis (C2 vertebra)

Stylomandibular
ligament

C3 vertebra

C7 vertebra

T1 vertebra

1st rib

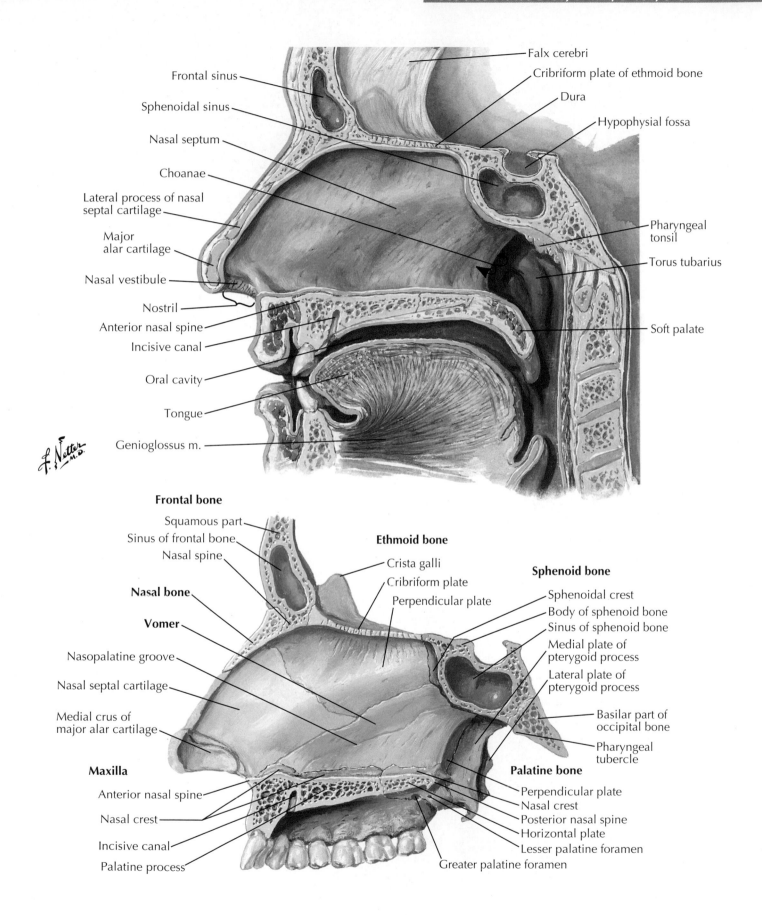

Falx cerebri

Frontal sinus

Cribriform plate of ethmoid bone

Sphenoidal sinus

Dura

Nasal septum

Hypophysial fossa

Choanae

Lateral process of nasal septal cartilage

Pharyngeal tonsil

Major alar cartilage

Torus tubarius

Nasal vestibule

Nostril

Soft palate

Anterior nasal spine

Incisive canal

Oral cavity

Tongue

Genioglossus m.

Frontal bone

Squamous part

Ethmoid bone

Sinus of frontal bone

Nasal spine

Crista galli

Sphenoid bone

Cribriform plate

Nasal bone

Perpendicular plate

Sphenoidal crest

Body of sphenoid bone

Vomer

Sinus of sphenoid bone

Nasopalatine groove

Medial plate of pterygoid process

Nasal septal cartilage

Lateral plate of pterygoid process

Medial crus of major alar cartilage

Basilar part of occipital bone

Pharyngeal tubercle

Maxilla

Palatine bone

Anterior nasal spine

Perpendicular plate

Nasal crest

Nasal crest

Posterior nasal spine

Incisive canal

Horizontal plate

Palatine process

Lesser palatine foramen

Greater palatine foramen

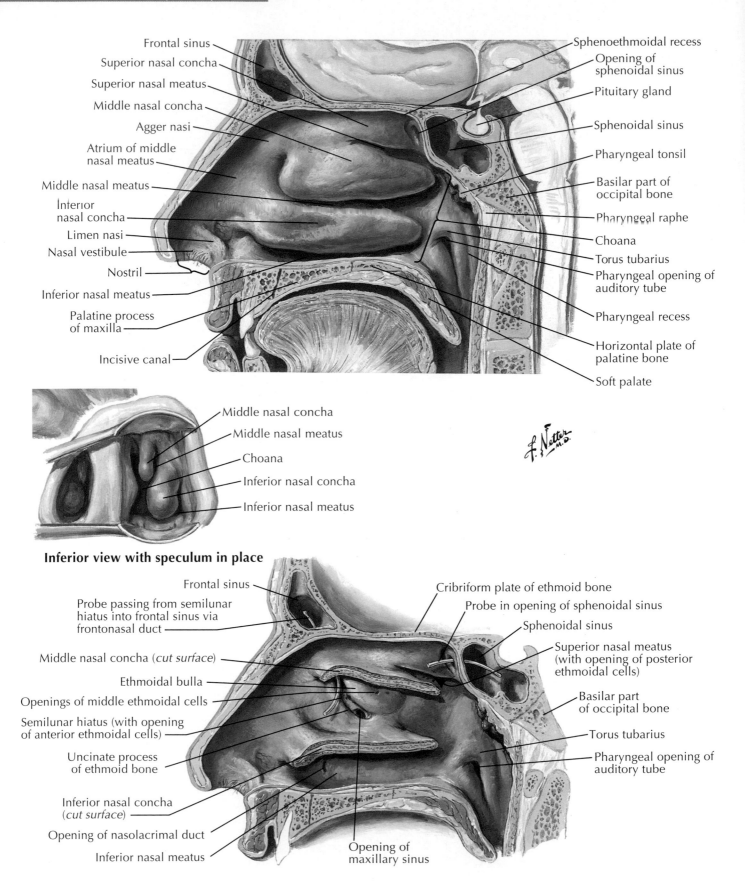

Frontal sinus

Superior nasal concha

Superior nasal meatus

Middle nasal concha

Agger nasi

Atrium of middle nasal meatus

Middle nasal meatus

Inferior nasal concha

Limen nasi

Nasal vestibule

Nostril

Inferior nasal meatus

Palatine process of maxilla

Incisive canal

Sphenoethmoidal recess

Opening of sphenoidal sinus

Pituitary gland

Sphenoidal sinus

Pharyngeal tonsil

Basilar part of occipital bone

Pharyngeal raphe

Choana

Torus tubarius

Pharyngeal opening of auditory tube

Pharyngeal recess

Horizontal plate of palatine bone

Soft palate

Middle nasal concha

Middle nasal meatus

Choana

Inferior nasal concha

Inferior nasal meatus

Inferior view with speculum in place

Frontal sinus

Probe passing from semilunar hiatus into frontal sinus via frontonasal duct

Middle nasal concha (*cut surface*)

Ethmoidal bulla

Openings of middle ethmoidal cells

Semilunar hiatus (with opening of anterior ethmoidal cells)

Uncinate process of ethmoid bone

Inferior nasal concha (*cut surface*)

Opening of nasolacrimal duct

Inferior nasal meatus

Cribriform plate of ethmoid bone

Probe in opening of sphenoidal sinus

Sphenoidal sinus

Superior nasal meatus (with opening of posterior ethmoidal cells)

Basilar part of occipital bone

Torus tubarius

Pharyngeal opening of auditory tube

Opening of maxillary sinus

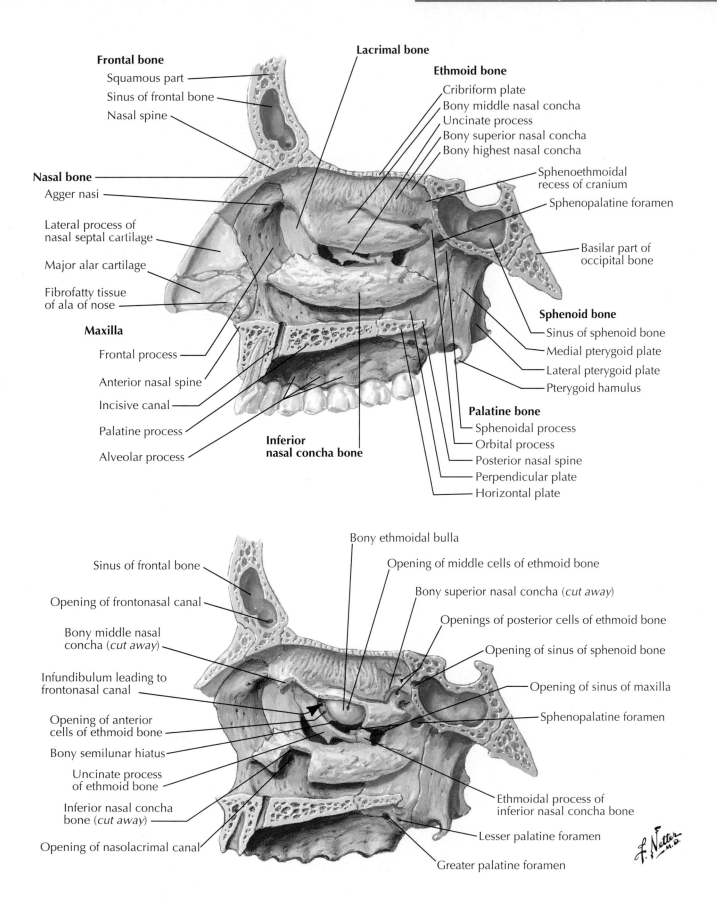

Frontal bone
Squamous part
Sinus of frontal bone
Nasal spine

Nasal bone
Agger nasi

Lateral process of
nasal septal cartilage

Major alar cartilage

Fibrofatty tissue
of ala of nose

Maxilla
Frontal process
Anterior nasal spine
Incisive canal
Palatine process
Alveolar process

Lacrimal bone

Ethmoid bone
Cribriform plate
Bony middle nasal concha
Uncinate process
Bony superior nasal concha
Bony highest nasal concha
Sphenoethmoidal
recess of cranium
Sphenopalatine foramen

Basilar part of
occipital bone

Sphenoid bone
Sinus of sphenoid bone
Medial pterygoid plate
Lateral pterygoid plate
Pterygoid hamulus

Palatine bone
Sphenoidal process
Orbital process
Posterior nasal spine
Perpendicular plate
Horizontal plate

**Inferior
nasal concha bone**

Bony ethmoidal bulla
Opening of middle cells of ethmoid bone
Bony superior nasal concha (*cut away*)
Openings of posterior cells of ethmoid bone
Opening of sinus of sphenoid bone
Opening of sinus of maxilla
Sphenopalatine foramen

Sinus of frontal bone
Opening of frontonasal canal
Bony middle nasal
concha (*cut away*)
Infundibulum leading to
frontonasal canal
Opening of anterior
cells of ethmoid bone
Bony semilunar hiatus
Uncinate process
of ethmoid bone
Inferior nasal concha
bone (*cut away*)
Opening of nasolacrimal canal

Ethmoidal process of
inferior nasal concha bone
Lesser palatine foramen
Greater palatine foramen

Lateral wall of nasal cavity

Anterior lateral nasal branch of anterior ethmoidal a.

Posterior lateral nasal branches of sphenopalatine a.

Alar branches of lateral nasal branch of facial a.

Greater palatine a.

Lesser palatine a.

Anterior ethmoidal a.

Posterior ethmoidal a.

Ophthalmic a.

Internal carotid a.

Sphenopalatine a.

Descending palatine a.

Maxillary a.

Internal carotid a.

External carotid a.

Nasal septum

Anterior septal branches of anterior ethmoidal a.

Kiesselbach's plexus

Posterior septal branch of sphenopalatine a.

Greater palatine a.

Nasal septal branches of superior labial a.

Anastomosis between posterior septal branch of sphenopalatine a. and greater palatine a. in incisive canal

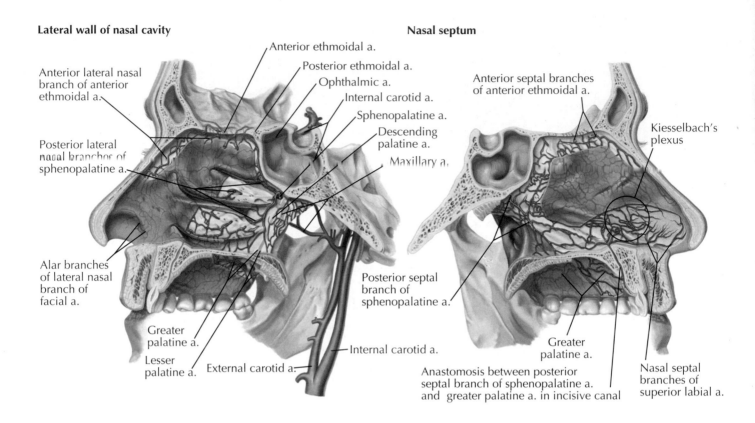

Lateral wall of nasal cavity

Nasofrontal v.

Anterior lateral nasal tributary of anterior ethmoidal v.

External nasal v.

Posterior lateral nasal tributaries of sphenopalatine v.

Facial v.

Anterior ethmoidal v.

Posterior ethmoidal v.

Superior ophthalmic v.

Inferior ophthalmic v.

Cavernous sinus

Sphenopalatine v.

Pterygoid venous plexus

Descending palatine v.

Posterior septal tributary of sphenopalatine v.

Retromandibular v.

Nasal septum

Communicating vv. between oral and nasal cavities in incisive canal

C. Machado M.D.

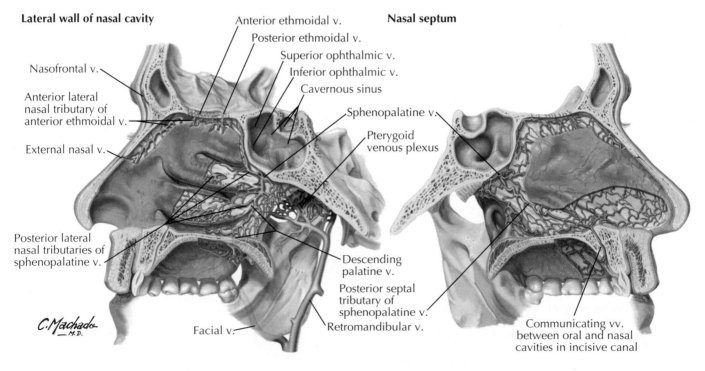

Vasculature and Innervation of Nasal Cavity

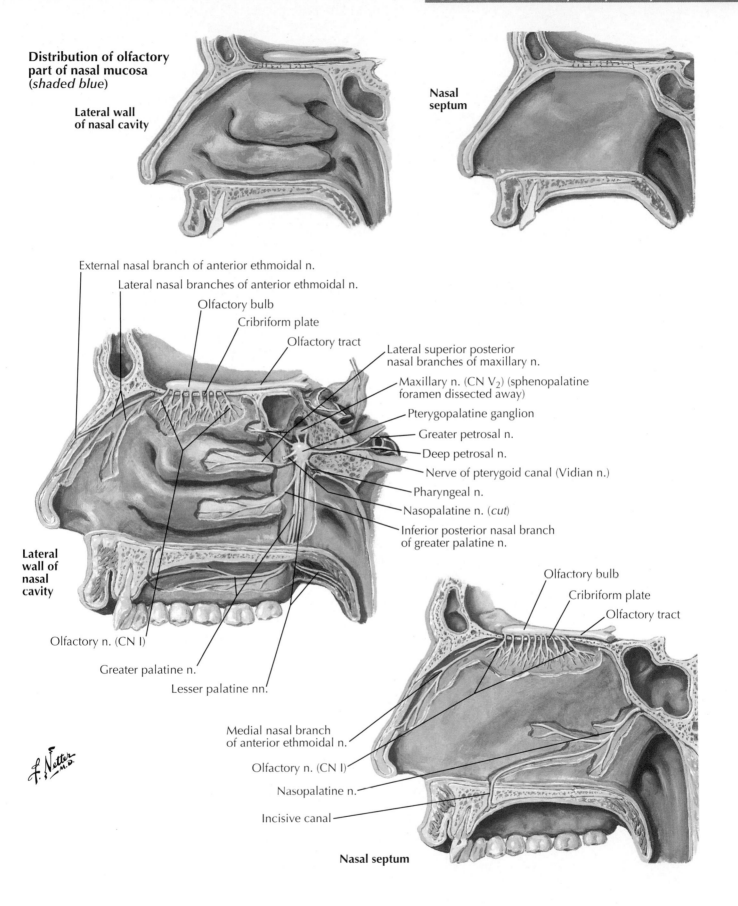

Distribution of olfactory part of nasal mucosa (*shaded blue*)

Lateral wall of nasal cavity

Nasal septum

External nasal branch of anterior ethmoidal n.

Lateral nasal branches of anterior ethmoidal n.

Olfactory bulb

Cribriform plate

Olfactory tract

Lateral superior posterior nasal branches of maxillary n.

Maxillary n. (CN V$_2$) (sphenopalatine foramen dissected away)

Pterygopalatine ganglion

Greater petrosal n.

Deep petrosal n.

Nerve of pterygoid canal (Vidian n.)

Pharyngeal n.

Nasopalatine n. (*cut*)

Inferior posterior nasal branch of greater palatine n.

Lateral wall of nasal cavity

Olfactory n. (CN I)

Greater palatine n.

Lesser palatine nn.

Olfactory bulb

Cribriform plate

Olfactory tract

Medial nasal branch of anterior ethmoidal n.

Olfactory n. (CN I)

Nasopalatine n.

Incisive canal

Nasal septum

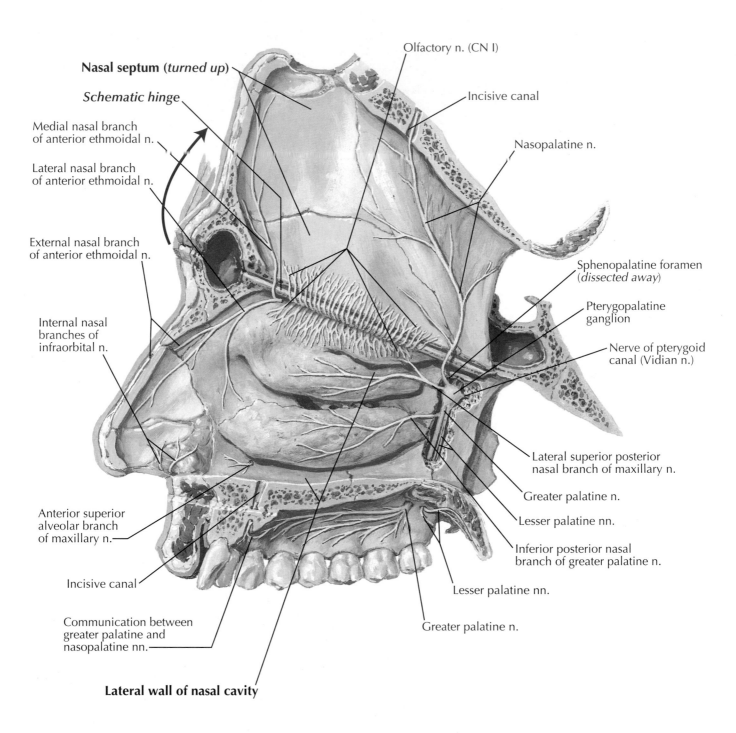

Nasal septum (*turned up*)

Schematic hinge

Medial nasal branch
of anterior ethmoidal n.

Lateral nasal branch
of anterior ethmoidal n.

External nasal branch
of anterior ethmoidal n.

Internal nasal
branches of
infraorbital n.

Anterior superior
alveolar branch
of maxillary n.

Incisive canal

Communication between
greater palatine and
nasopalatine nn.

Lateral wall of nasal cavity

Olfactory n. (CN I)

Incisive canal

Nasopalatine n.

Sphenopalatine foramen
(*dissected away*)

Pterygopalatine
ganglion

Nerve of pterygoid
canal (Vidian n.)

Lateral superior posterior
nasal branch of maxillary n.

Greater palatine n.

Lesser palatine nn.

Inferior posterior nasal
branch of greater palatine n.

Lesser palatine nn.

Greater palatine n.

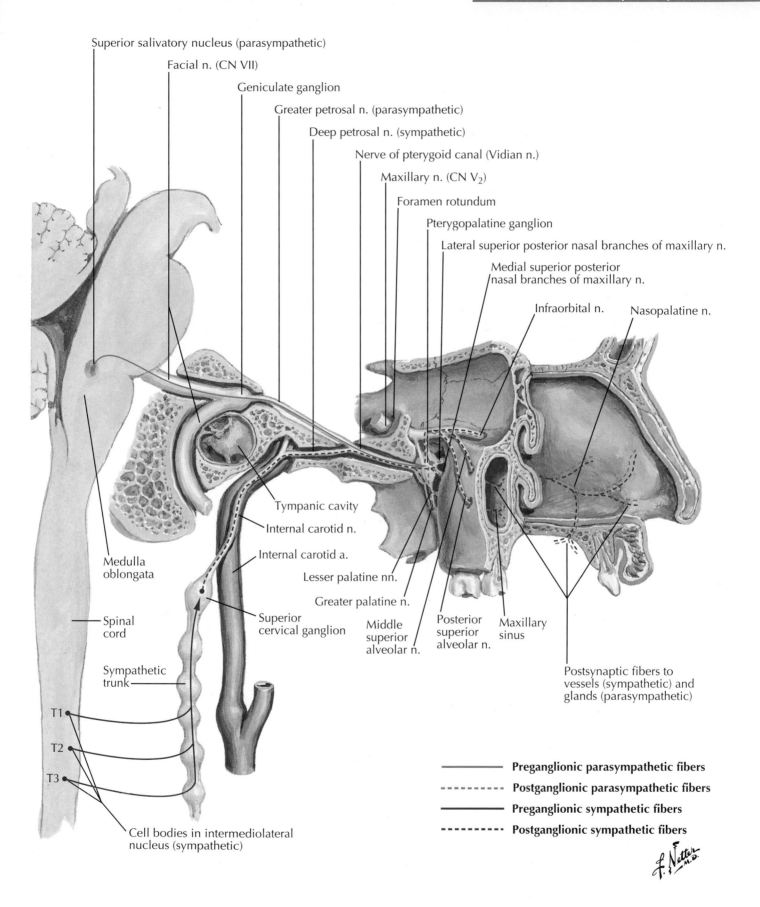

Superior salivatory nucleus (parasympathetic)

Facial n. (CN VII)

Geniculate ganglion

Greater petrosal n. (parasympathetic)

Deep petrosal n. (sympathetic)

Nerve of pterygoid canal (Vidian n.)

Maxillary n. (CN V$_2$)

Foramen rotundum

Pterygopalatine ganglion

Lateral superior posterior nasal branches of maxillary n.

Medial superior posterior nasal branches of maxillary n.

Infraorbital n.

Nasopalatine n.

Tympanic cavity

Internal carotid n.

Internal carotid a.

Lesser palatine nn.

Greater palatine n.

Medulla oblongata

Spinal cord

Superior cervical ganglion

Middle superior alveolar n.

Posterior superior alveolar n.

Maxillary sinus

Postsynaptic fibers to vessels (sympathetic) and glands (parasympathetic)

Sympathetic trunk

T1

T2

T3

Cell bodies in intermediolateral nucleus (sympathetic)

—————— **Preganglionic parasympathetic fibers**

- - - - - - - - **Postganglionic parasympathetic fibers**

—————— **Preganglionic sympathetic fibers**

- - - - - - - - **Postganglionic sympathetic fibers**

Plane of section
1 Ethmoid bone
2 Inferior nasal concha bone

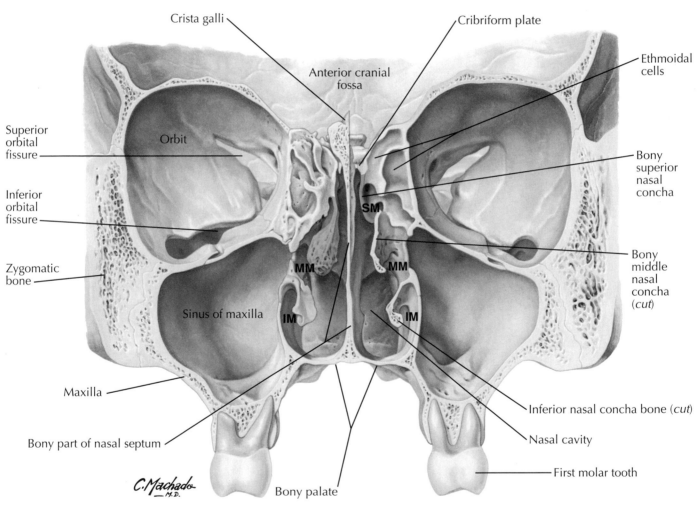

Crista galli

Cribriform plate

Anterior cranial fossa

Ethmoidal cells

Superior orbital fissure

Orbit

SM

Bony superior nasal concha

Inferior orbital fissure

Bony middle nasal concha (cut)

Zygomatic bone

MM

MM

Sinus of maxilla

IM

IM

Inferior nasal concha bone (cut)

Maxilla

Nasal cavity

Bony part of nasal septum

First molar tooth

Bony palate

C. Machado M.D.

Bony superior nasal concha   **SM** Superior nasal meatus

Bony middle nasal concha   **MM** Middle nasal meatus

Inferior nasal concha bone   **IM** Inferior nasal meatus

Coronal section

Falx cerebri

Brain

Nasal septum

Middle nasal concha

Middle nasal meatus

Infraorbital n., a., and v.

Maxillary sinus

Inferior nasal concha

Inferior nasal meatus

Palatine process of maxilla

Oral cavity

Genioglossus m.

Mylohyoid m.

Geniohyoid m.

Olfactory bulbs

Frontal sinus

Ethmoidal cells

Lacrimal gland

Opening of maxillary sinus

Infraorbital recess
Zygomatic recess
Alveolar recess

} Maxillary sinus

Alveolar process of maxilla

Bucinator

Tongue

Sublingual gland

Mandible

Anterior belly of digastric m.

Transverse section

Nasal bone

Lacrimal bone

Ethmoidal cells

Eyeball

Zygomatic bone

Orbital fat body

Sphenoidal sinuses

Optic chiasm

Temporal lobe

Infundibular recess of third ventricle

Mammillary bodies

Crus cerebri

Aqueduct of midbrain

Nasal septum

Medial wall of orbit

Medial rectus m.

Lateral rectus m.

Optic n. (CN II)

Temporalis m.

Optic tract

Interpeduncular fossa

Vermis

C. Machado M.D.

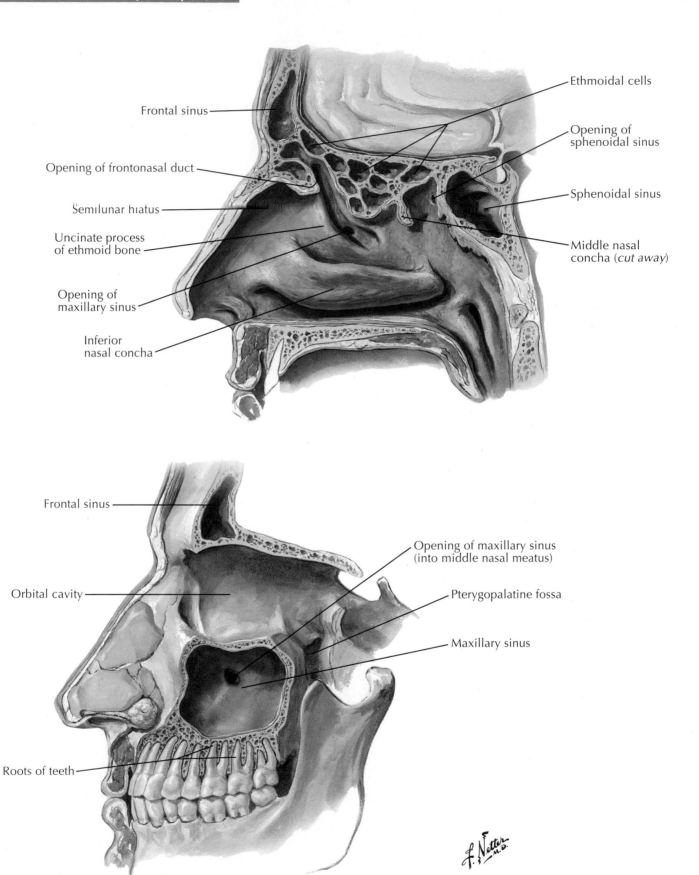

Frontal sinus

Ethmoidal cells

Opening of frontonasal duct

Opening of sphenoidal sinus

Semilunar hiatus

Sphenoidal sinus

Uncinate process of ethmoid bone

Middle nasal concha (*cut away*)

Opening of maxillary sinus

Inferior nasal concha

Frontal sinus

Opening of maxillary sinus (into middle nasal meatus)

Orbital cavity

Pterygopalatine fossa

Maxillary sinus

Roots of teeth

Bones of nasal cavity and paranasal sinuses at birth

Sinus represents one or more anterior ethmoidal cells opening into semilunar hiatus of middle nasal meatus

Sinus represents one or more middle ethmoidal cells opening into middle nasal meatus

Sinuses represent two or more posterior ethmoidal cells opening into superior nasal meatus

Part of nasolacrimal duct that formed in depths of nasolacrimal canal

Sphenoidal sinus within bony shell (sphenoidal concha) located anterior and lateral to body of sphenoid bone (broken line indicates sinus lateral to body of sphenoid bone)

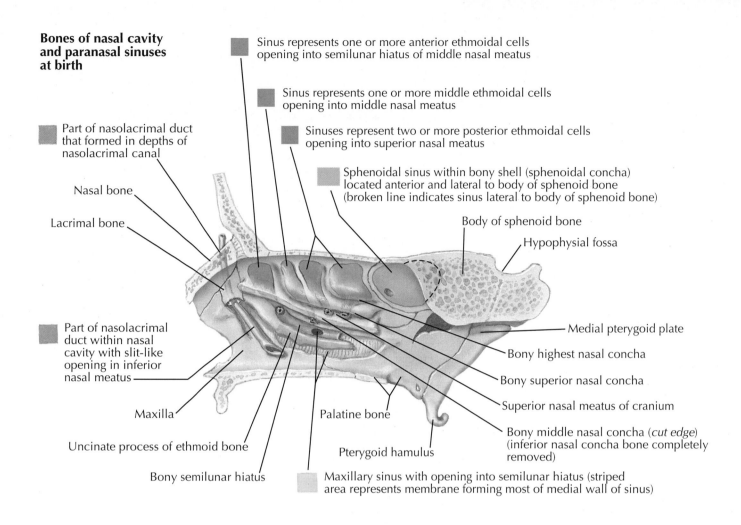

Nasal bone

Lacrimal bone

Body of sphenoid bone

Hypophysial fossa

Medial pterygoid plate

Part of nasolacrimal duct within nasal cavity with slit-like opening in inferior nasal meatus

Bony highest nasal concha

Bony superior nasal concha

Superior nasal meatus of cranium

Maxilla

Uncinate process of ethmoid bone

Palatine bone

Pterygoid hamulus

Bony middle nasal concha (*cut edge*) (inferior nasal concha bone completely removed)

Bony semilunar hiatus

Maxillary sinus with opening into semilunar hiatus (striped area represents membrane forming most of medial wall of sinus)

Growth of sinuses in frontal bone and maxilla throughout life

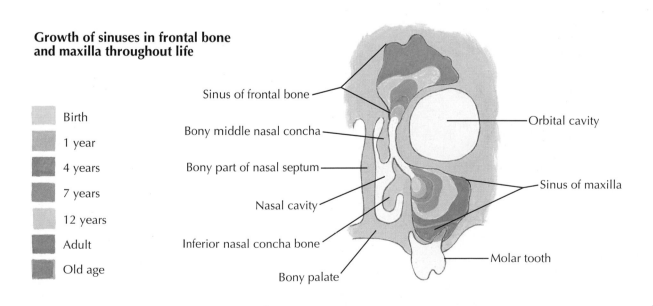

Birth

1 year

4 years

7 years

12 years

Adult

Old age

Sinus of frontal bone

Bony middle nasal concha

Bony part of nasal septum

Nasal cavity

Inferior nasal concha bone

Bony palate

Orbital cavity

Sinus of maxilla

Molar tooth

Paranasal Sinuses

Plate S–370

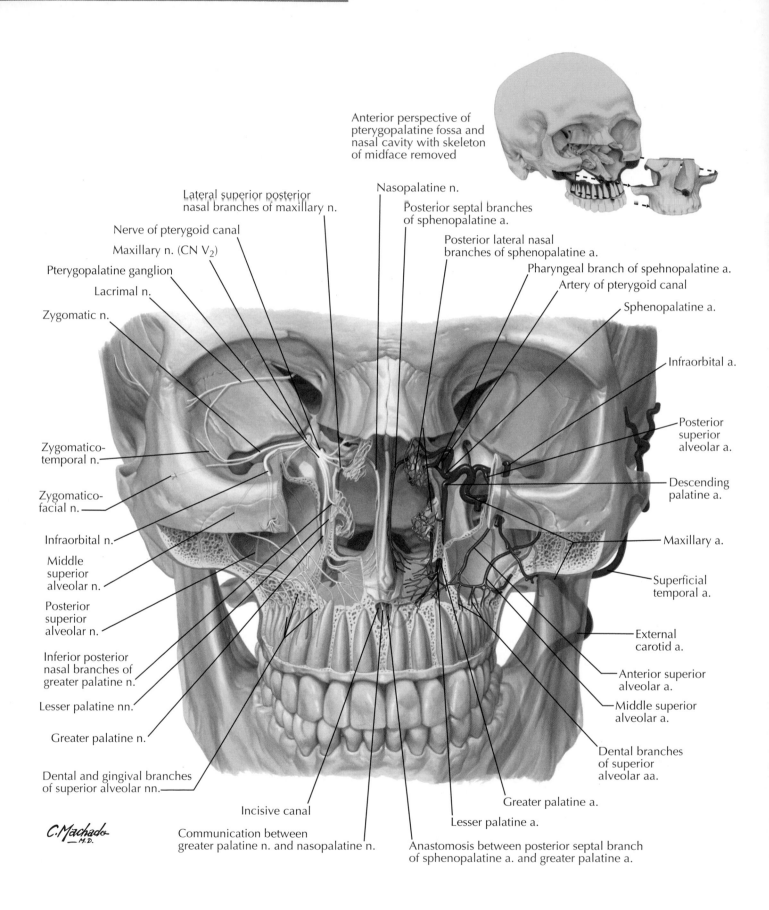

Anterior perspective of pterygopalatine fossa and nasal cavity with skeleton of midface removed

Nasopalatine n.

Posterior septal branches of sphenopalatine a.

Lateral superior posterior nasal branches of maxillary n.

Nerve of pterygoid canal

Maxillary n. (CN V₂)

Pterygopalatine ganglion

Lacrimal n.

Zygomatic n.

Posterior lateral nasal branches of sphenopalatine a.

Pharyngeal branch of spehnopalatine a.

Artery of pterygoid canal

Sphenopalatine a.

Infraorbital a.

Posterior superior alveolar a.

Zygomatico-temporal n.

Zygomatico-facial n.

Infraorbital n.

Middle superior alveolar n.

Posterior superior alveolar n.

Inferior posterior nasal branches of greater palatine n.

Lesser palatine nn.

Greater palatine n.

Dental and gingival branches of superior alveolar nn.

Descending palatine a.

Maxillary a.

Superficial temporal a.

External carotid a.

Anterior superior alveolar a.

Middle superior alveolar a.

Dental branches of superior alveolar aa.

Greater palatine a.

Lesser palatine a.

Incisive canal

Communication between greater palatine n. and nasopalatine n.

Anastomosis between posterior septal branch of sphenopalatine a. and greater palatine a.

C. Machado
—M.D.

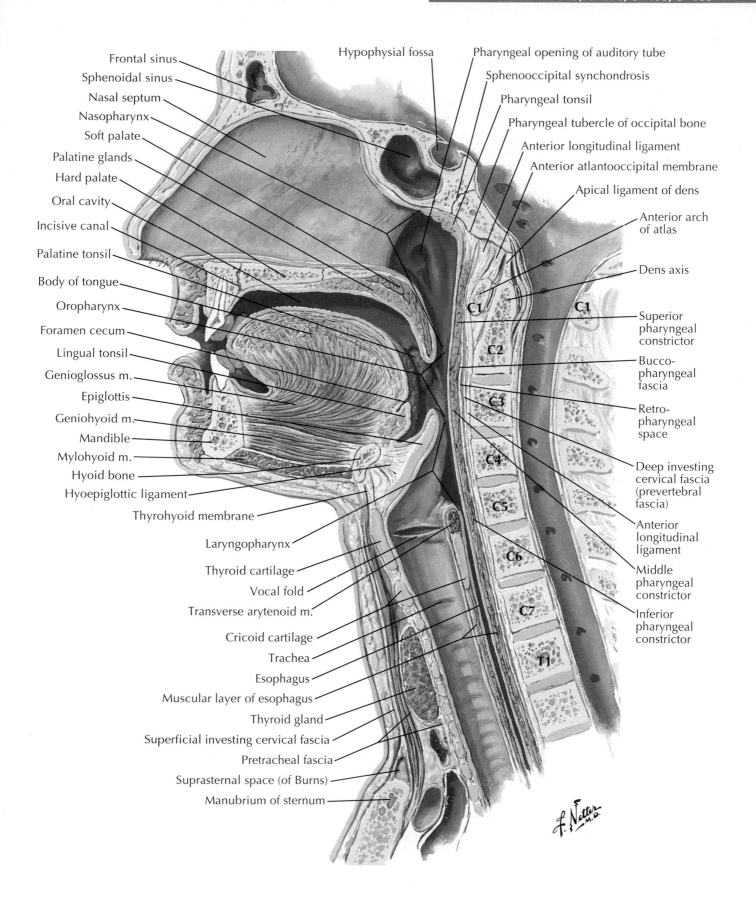

Frontal sinus

Sphenoidal sinus

Nasal septum

Nasopharynx

Soft palate

Palatine glands

Hard palate

Oral cavity

Incisive canal

Palatine tonsil

Body of tongue

Oropharynx

Foramen cecum

Lingual tonsil

Genioglossus m.

Epiglottis

Geniohyoid m.

Mandible

Mylohyoid m.

Hyoid bone

Hyoepiglottic ligament

Thyrohyoid membrane

Laryngopharynx

Thyroid cartilage

Vocal fold

Transverse arytenoid m.

Cricoid cartilage

Trachea

Esophagus

Muscular layer of esophagus

Thyroid gland

Superficial investing cervical fascia

Pretracheal fascia

Suprasternal space (of Burns)

Manubrium of sternum

Hypophysial fossa

Pharyngeal opening of auditory tube

Sphenooccipital synchondrosis

Pharyngeal tonsil

Pharyngeal tubercle of occipital bone

Anterior longitudinal ligament

Anterior atlantooccipital membrane

Apical ligament of dens

Anterior arch of atlas

Dens axis

Superior pharyngeal constrictor

Bucco-pharyngeal fascia

Retro-pharyngeal space

Deep investing cervical fascia (prevertebral fascia)

Anterior longitudinal ligament

Middle pharyngeal constrictor

Inferior pharyngeal constrictor

C1

C1

C2

C3

C4

C5

C6

C7

T1

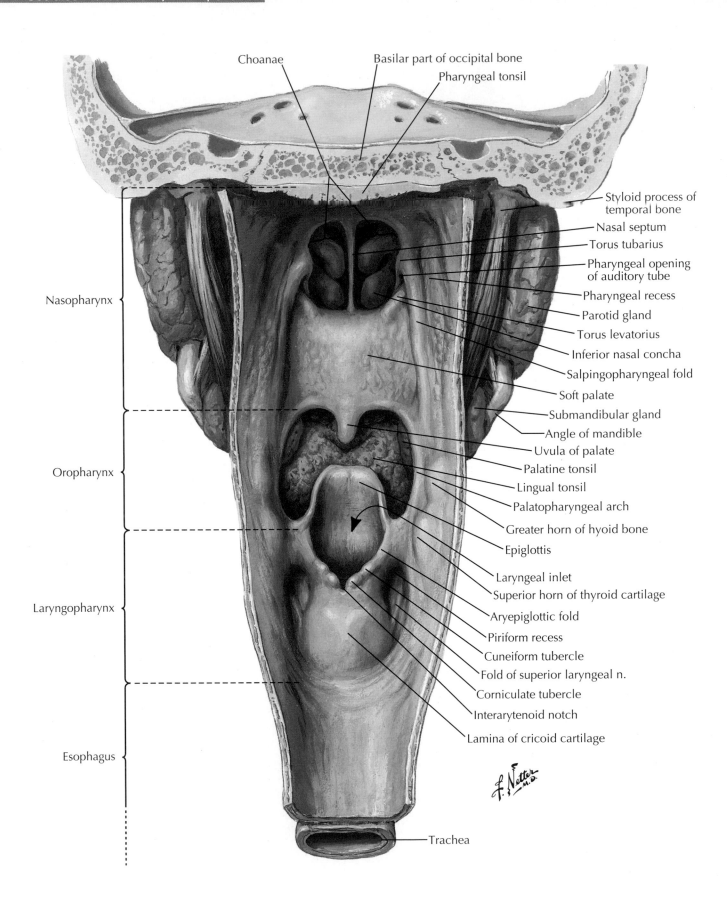

Choanae

Basilar part of occipital bone

Pharyngeal tonsil

Styloid process of temporal bone

Nasal septum

Torus tubarius

Pharyngeal opening of auditory tube

Pharyngeal recess

Parotid gland

Torus levatorius

Inferior nasal concha

Salpingopharyngeal fold

Soft palate

Submandibular gland

Angle of mandible

Uvula of palate

Palatine tonsil

Lingual tonsil

Palatopharyngeal arch

Greater horn of hyoid bone

Epiglottis

Laryngeal inlet

Superior horn of thyroid cartilage

Aryepiglottic fold

Piriform recess

Cuneiform tubercle

Fold of superior laryngeal n.

Corniculate tubercle

Interarytenoid notch

Lamina of cricoid cartilage

Nasopharynx

Oropharynx

Laryngopharynx

Esophagus

Trachea

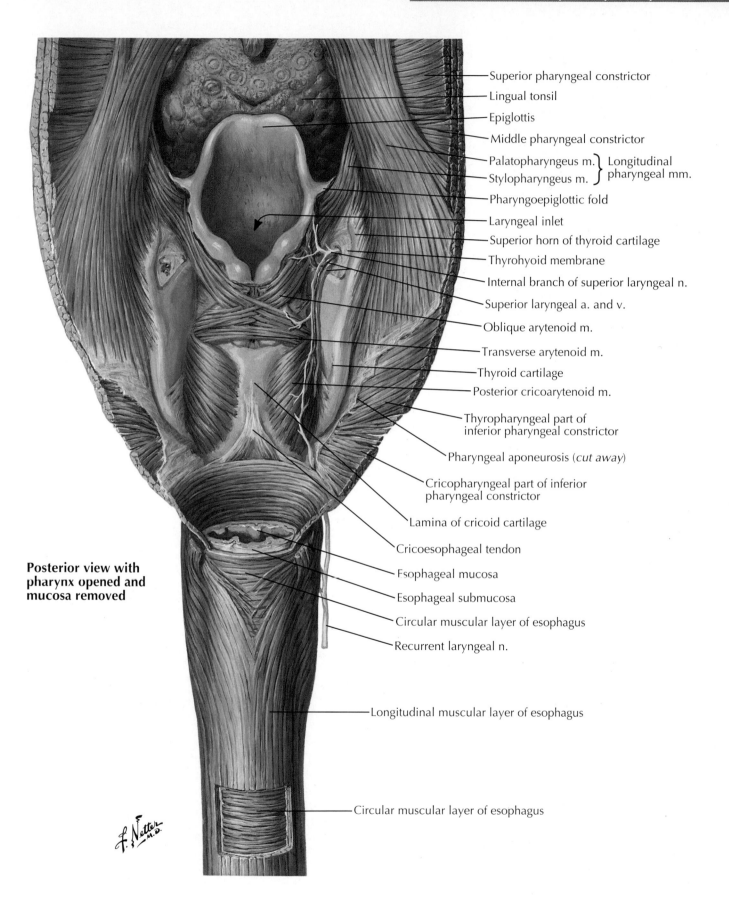

Superior pharyngeal constrictor

Lingual tonsil

Epiglottis

Middle pharyngeal constrictor

Palatopharyngeus m. } Longitudinal
Stylopharyngeus m. } pharyngeal mm.

Pharyngoepiglottic fold

Laryngeal inlet

Superior horn of thyroid cartilage

Thyrohyoid membrane

Internal branch of superior laryngeal n.

Superior laryngeal a. and v.

Oblique arytenoid m.

Transverse arytenoid m.

Thyroid cartilage

Posterior cricoarytenoid m.

Thyropharyngeal part of
inferior pharyngeal constrictor

Pharyngeal aponeurosis (cut away)

Cricopharyngeal part of inferior
pharyngeal constrictor

Lamina of cricoid cartilage

Cricoesophageal tendon

Esophageal mucosa

Esophageal submucosa

Circular muscular layer of esophagus

Recurrent laryngeal n.

Longitudinal muscular layer of esophagus

Circular muscular layer of esophagus

**Posterior view with
pharynx opened and
mucosa removed**

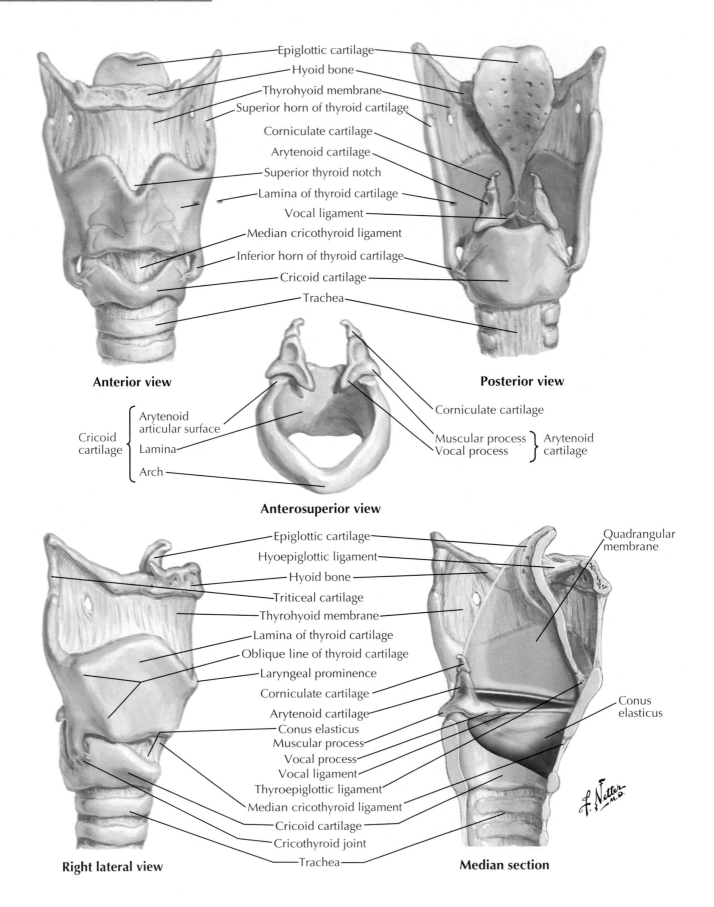

Epiglottic cartilage

Hyoid bone

Thyrohyoid membrane

Superior horn of thyroid cartilage

Corniculate cartilage

Arytenoid cartilage

Superior thyroid notch

Lamina of thyroid cartilage

Vocal ligament

Median cricothyroid ligament

Inferior horn of thyroid cartilage

Cricoid cartilage

Trachea

Anterior view

Posterior view

Cricoid cartilage {
Arytenoid articular surface
Lamina
Arch
}

Corniculate cartilage

Muscular process
Vocal process
} Arytenoid cartilage

Anterosuperior view

Epiglottic cartilage

Hyoepiglottic ligament

Hyoid bone

Triticeal cartilage

Thyrohyoid membrane

Lamina of thyroid cartilage

Oblique line of thyroid cartilage

Laryngeal prominence

Corniculate cartilage

Arytenoid cartilage

Conus elasticus

Muscular process

Vocal process

Vocal ligament

Thyroepiglottic ligament

Median cricothyroid ligament

Cricoid cartilage

Cricothyroid joint

Trachea

Quadrangular membrane

Conus elasticus

Right lateral view

Median section

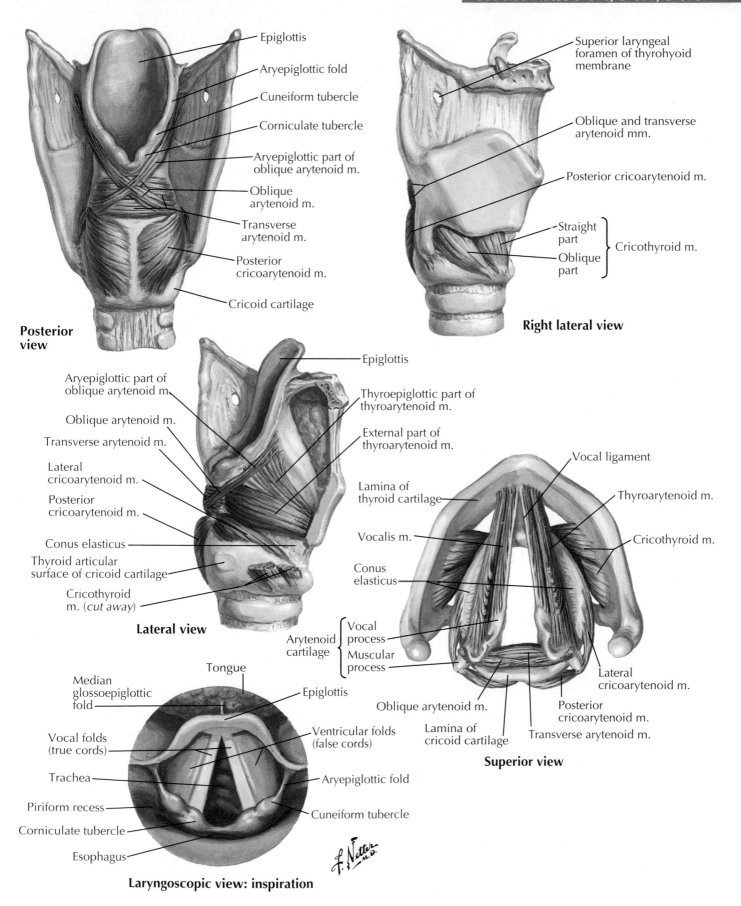

Epiglottis

Aryepiglottic fold

Cuneiform tubercle

Corniculate tubercle

Aryepiglottic part of oblique arytenoid m.

Oblique arytenoid m.

Transverse arytenoid m.

Posterior cricoarytenoid m.

Cricoid cartilage

Posterior view

Superior laryngeal foramen of thyrohyoid membrane

Oblique and transverse arytenoid mm.

Posterior cricoarytenoid m.

Straight part
Oblique part } Cricothyroid m.

Right lateral view

Aryepiglottic part of oblique arytenoid m.

Oblique arytenoid m.

Transverse arytenoid m.

Lateral cricoarytenoid m.

Posterior cricoarytenoid m.

Conus elasticus

Thyroid articular surface of cricoid cartilage

Cricothyroid m. (*cut away*)

Lateral view

Epiglottis

Thyroepiglottic part of thyroarytenoid m.

External part of thyroarytenoid m.

Lamina of thyroid cartilage

Vocal ligament

Thyroarytenoid m.

Vocalis m.

Cricothyroid m.

Conus elasticus

Arytenoid cartilage { Vocal process
Muscular process

Oblique arytenoid m.

Lamina of cricoid cartilage

Posterior cricoarytenoid m.

Transverse arytenoid m.

Lateral cricoarytenoid m.

Superior view

Tongue

Median glossoepiglottic fold

Epiglottis

Vocal folds (true cords)

Ventricular folds (false cords)

Trachea

Piriform recess

Aryepiglottic fold

Corniculate tubercle

Cuneiform tubercle

Esophagus

Laryngoscopic view: inspiration

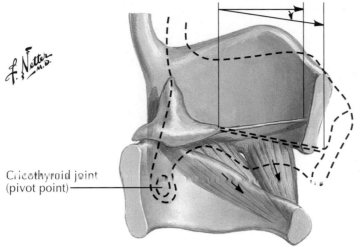

Cricothyroid joint
(pivot point)

Action of cricothyroid muscles
Stretching of vocal ligaments

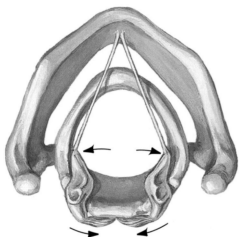

Action of posterior cricoarytenoid muscles
Abduction of vocal ligaments

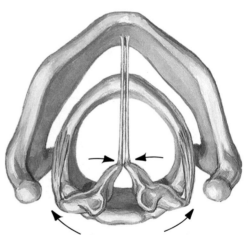

Action of lateral cricoarytenoid muscles
Adduction of vocal ligaments

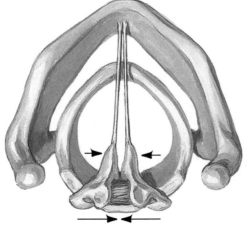

Action of transverse and oblique arytenoid muscles
Adduction of vocal ligaments

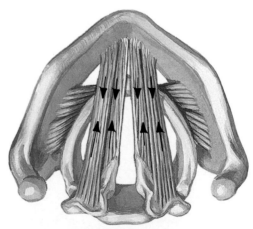

Action of vocalis and thyroarytenoid muscles
Shortening (relaxation) of vocal ligaments

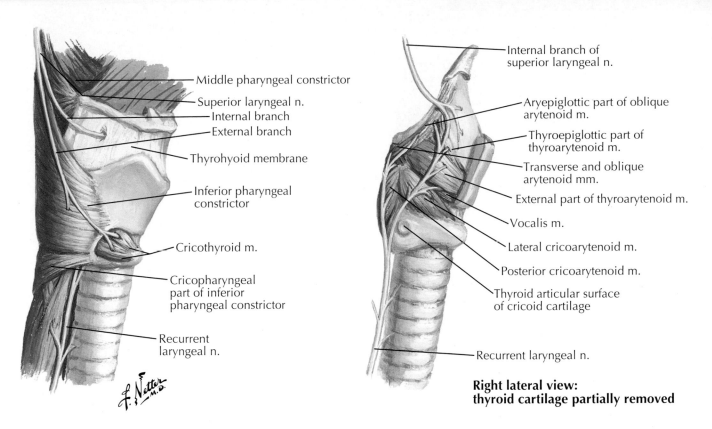

Middle pharyngeal constrictor

Superior laryngeal n.

Internal branch

External branch

Thyrohyoid membrane

Inferior pharyngeal constrictor

Cricothyroid m.

Cricopharyngeal part of inferior pharyngeal constrictor

Recurrent laryngeal n.

Internal branch of superior laryngeal n.

Aryepiglottic part of oblique arytenoid m.

Thyroepiglottic part of thyroarytenoid m.

Transverse and oblique arytenoid mm.

External part of thyroarytenoid m.

Vocalis m.

Lateral cricoarytenoid m.

Posterior cricoarytenoid m.

Thyroid articular surface of cricoid cartilage

Recurrent laryngeal n.

Right lateral view: thyroid cartilage partially removed

Coronal section through larynx

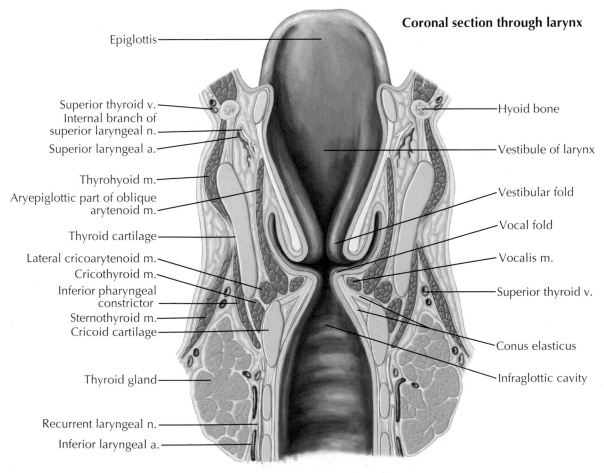

Epiglottis

Superior thyroid v.
Internal branch of superior laryngeal n.
Superior laryngeal a.

Thyrohyoid m.

Aryepiglottic part of oblique arytenoid m.

Thyroid cartilage

Lateral cricoarytenoid m.
Cricothyroid m.
Inferior pharyngeal constrictor
Sternothyroid m.
Cricoid cartilage

Thyroid gland

Recurrent laryngeal n.

Inferior laryngeal a.

Hyoid bone

Vestibule of larynx

Vestibular fold

Vocal fold

Vocalis m.

Superior thyroid v.

Conus elasticus

Infraglottic cavity

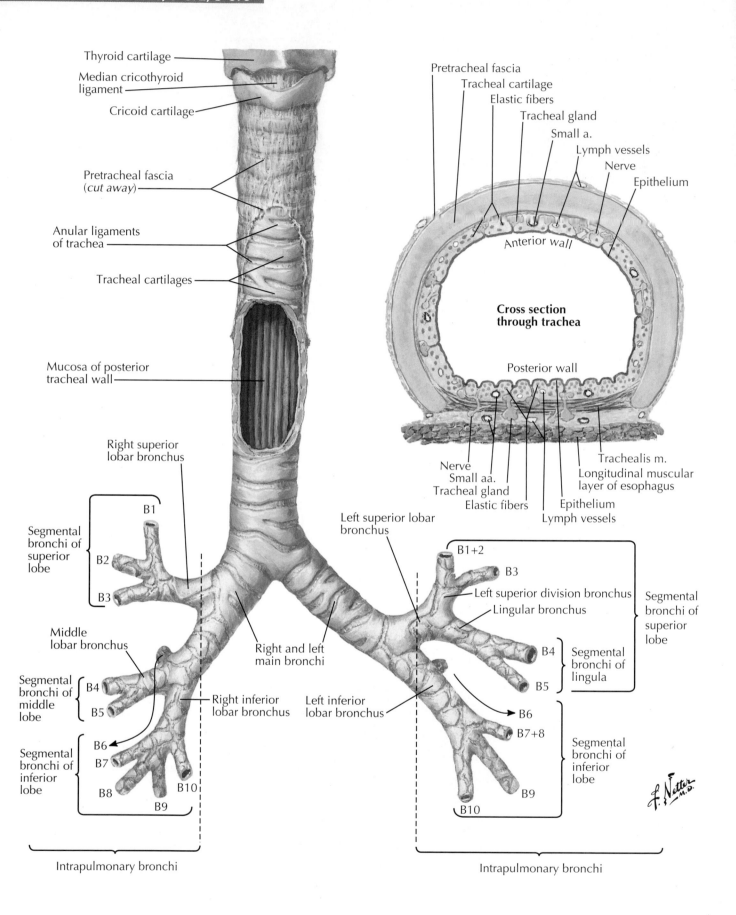

Thyroid cartilage

Median cricothyroid
ligament

Cricoid cartilage

Pretracheal fascia
(*cut away*)

Anular ligaments
of trachea

Tracheal cartilages

Mucosa of posterior
tracheal wall

Right superior
lobar bronchus

Segmental
bronchi of
superior
lobe

B1

B2

B3

Middle
lobar bronchus

Segmental
bronchi of
middle
lobe

B4

B5

Segmental
bronchi of
inferior
lobe

B6

B7

B8   B10

B9

Intrapulmonary bronchi

Pretracheal fascia

Tracheal cartilage

Elastic fibers

Tracheal gland

Small a.

Lymph vessels

Nerve

Epithelium

Anterior *wall*

**Cross section
through trachea**

Posterior wall

Nerve

Small aa.

Tracheal gland

Elastic fibers

Trachealis m.

Longitudinal muscular
layer of esophagus

Epithelium

Lymph vessels

Left superior lobar
bronchus

B1+2

B3

Left superior division bronchus

Lingular bronchus

Segmental
bronchi of
superior
lobe

B4

B5

Segmental
bronchi of
lingula

Right and left
main bronchi

Right inferior
lobar bronchus

Left inferior
lobar bronchus

B6

B7+8

Segmental
bronchi of
inferior
lobe

B9

B10

Intrapulmonary bronchi

See also **Plates S–375, S–379, S–382**

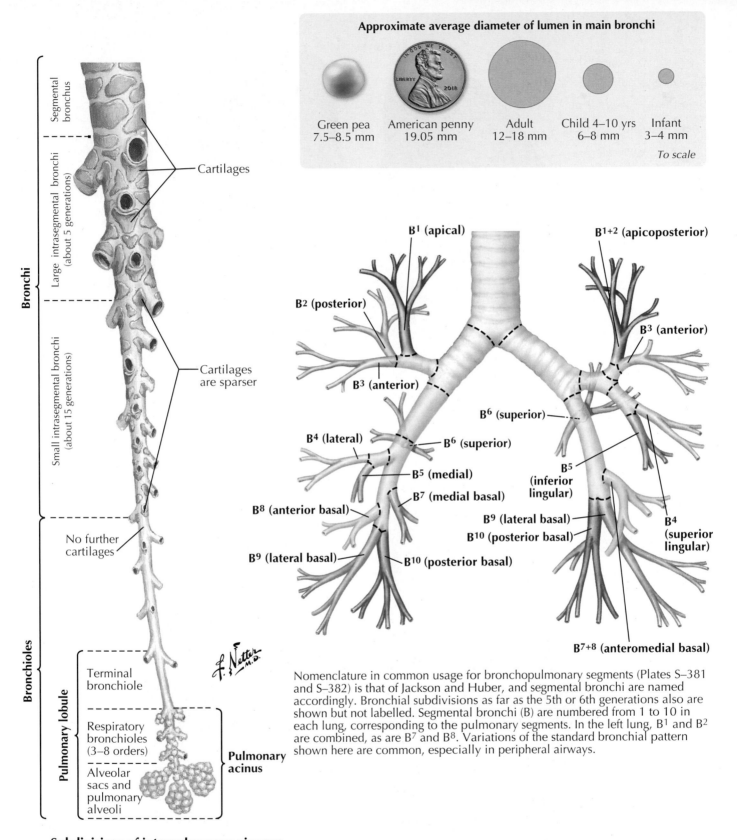

Approximate average diameter of lumen in main bronchi

Green pea
7.5–8.5 mm

American penny
19.05 mm

Adult
12–18 mm

Child 4–10 yrs
6–8 mm

Infant
3–4 mm

To scale

Segmental bronchus

Large intrasegmental bronchi (about 5 generations)

Small intrasegmental bronchi (about 15 generations)

Bronchi

Cartilages

Cartilages are sparser

No further cartilages

Bronchioles

Pulmonary lobule

Terminal bronchiole

Respiratory bronchioles (3–8 orders)

Alveolar sacs and pulmonary alveoli

Pulmonary acinus

Subdivisions of intrapulmonary airways

B^1 (apical)

B^{1+2} (apicoposterior)

B^2 (posterior)

B^3 (anterior)

B^3 (anterior)

B^6 (superior)

B^4 (lateral)

B^6 (superior)

B^5 (medial)

B^5 (inferior lingular)

B^7 (medial basal)

B^8 (anterior basal)

B^9 (lateral basal)

B^{10} (posterior basal)

B^4 (superior lingular)

B^9 (lateral basal)

B^{10} (posterior basal)

B^{7+8} (anteromedial basal)

Nomenclature in common usage for bronchopulmonary segments (Plates S–381 and S–382) is that of Jackson and Huber, and segmental bronchi are named accordingly. Bronchial subdivisions as far as the 5th or 6th generations also are shown but not labelled. Segmental bronchi (B) are numbered from 1 to 10 in each lung, corresponding to the pulmonary segments. In the left lung, B^1 and B^2 are combined, as are B^7 and B^8. Variations of the standard bronchial pattern shown here are common, especially in peripheral airways.

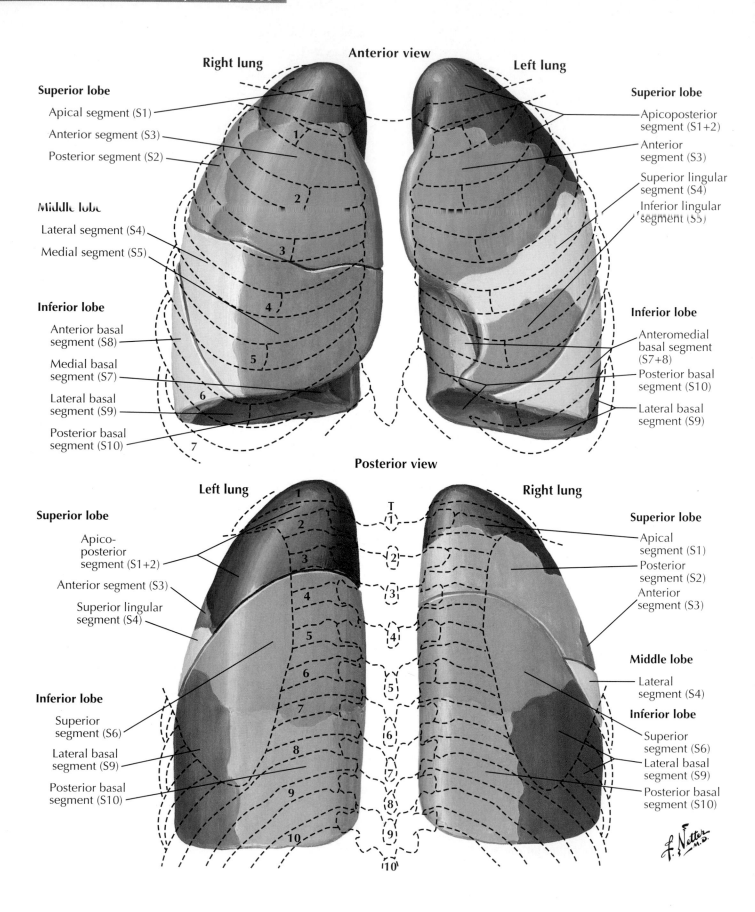

Anterior view

Right lung

Superior lobe

Apical segment (S1)

Anterior segment (S3)

Posterior segment (S2)

Middle lobe

Lateral segment (S4)

Medial segment (S5)

Inferior lobe

Anterior basal segment (S8)

Medial basal segment (S7)

Lateral basal segment (S9)

Posterior basal segment (S10)

Left lung

Superior lobe

Apicoposterior segment (S1+2)

Anterior segment (S3)

Superior lingular segment (S4)

Inferior lingular segment (S5)

Inferior lobe

Anteromedial basal segment (S7+8)

Posterior basal segment (S10)

Lateral basal segment (S9)

Posterior view

Left lung

Superior lobe

Apico-posterior segment (S1+2)

Anterior segment (S3)

Superior lingular segment (S4)

Inferior lobe

Superior segment (S6)

Lateral basal segment (S9)

Posterior basal segment (S10)

Right lung

Superior lobe

Apical segment (S1)

Posterior segment (S2)

Anterior segment (S3)

Middle lobe

Lateral segment (S4)

Inferior lobe

Superior segment (S6)

Lateral basal segment (S9)

Posterior basal segment (S10)

See also **Plates S–380, S–381, S–388**

Lateral views

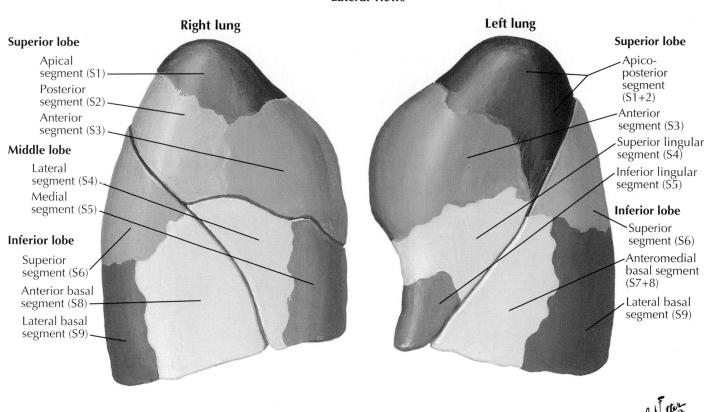

Right lung

Left lung

Superior lobe
Apical segment (S1)
Posterior segment (S2)
Anterior segment (S3)

Middle lobe
Lateral segment (S4)
Medial segment (S5)

Inferior lobe
Superior segment (S6)
Anterior basal segment (S8)
Lateral basal segment (S9)

Superior lobe
Apico-posterior segment (S1+2)
Anterior segment (S3)
Superior lingular segment (S4)
Inferior lingular segment (S5)

Inferior lobe
Superior segment (S6)
Anteromedial basal segment (S7+8)
Lateral basal segment (S9)

Medial views

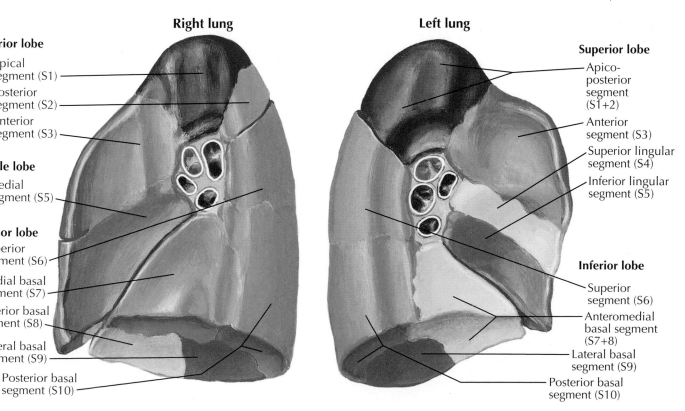

Right lung

Left lung

Superior lobe
Apical segment (S1)
Posterior segment (S2)
Anterior segment (S3)

Middle lobe
Medial segment (S5)

Inferior lobe
Superior segment (S6)
Medial basal segment (S7)
Anterior basal segment (S8)
Lateral basal segment (S9)
Posterior basal segment (S10)

Superior lobe
Apico-posterior segment (S1+2)
Anterior segment (S3)
Superior lingular segment (S4)
Inferior lingular segment (S5)

Inferior lobe
Superior segment (S6)
Anteromedial basal segment (S7+8)
Lateral basal segment (S9)
Posterior basal segment (S10)

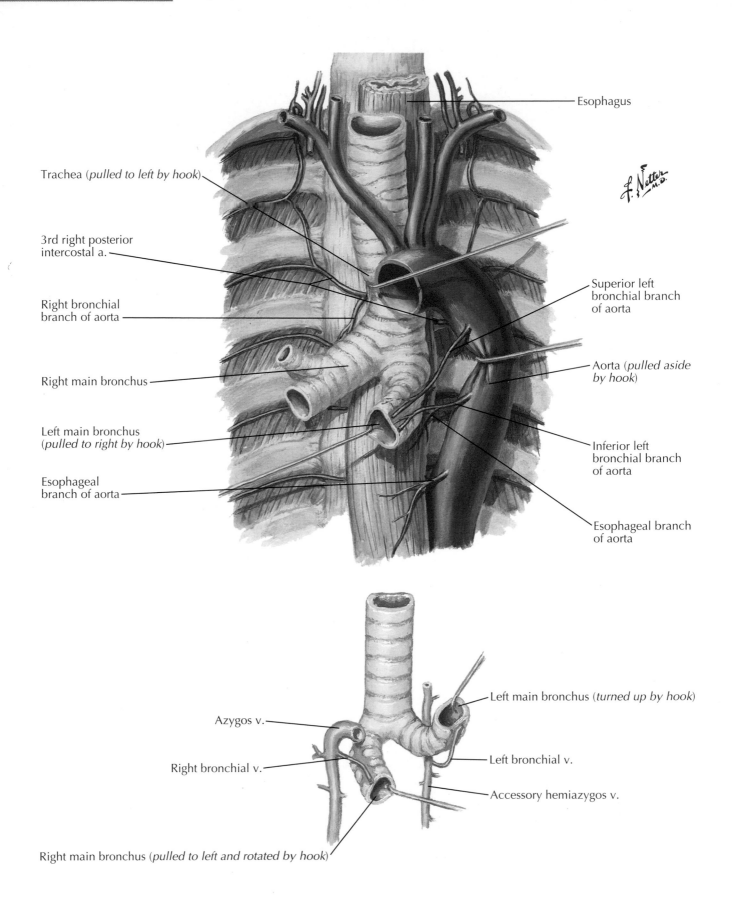

Esophagus

Trachea (*pulled to left by hook*)

3rd right posterior intercostal a.

Right bronchial branch of aorta

Right main bronchus

Left main bronchus (*pulled to right by hook*)

Esophageal branch of aorta

Superior left bronchial branch of aorta

Aorta (*pulled aside by hook*)

Inferior left bronchial branch of aorta

Esophageal branch of aorta

Azygos v.

Right bronchial v.

Right main bronchus (*pulled to left and rotated by hook*)

Left main bronchus (*turned up by hook*)

Left bronchial v.

Accessory hemiazygos v.

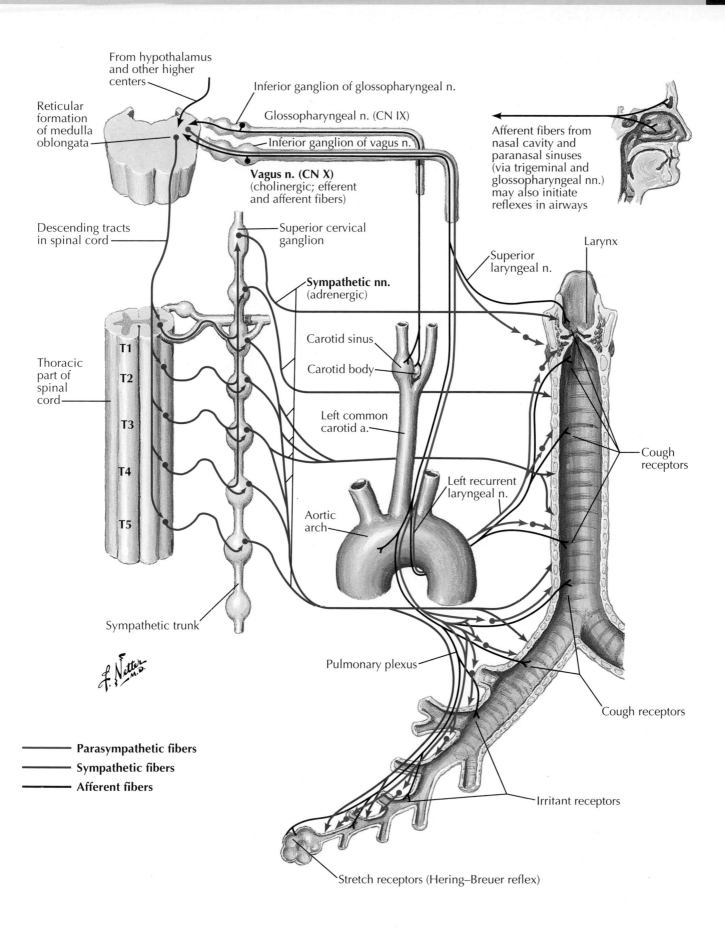

From hypothalamus and other higher centers

Inferior ganglion of glossopharyngeal n.

Glossopharyngeal n. (CN IX)

Reticular formation of medulla oblongata

Inferior ganglion of vagus n.

Vagus n. (CN X) (cholinergic; efferent and afferent fibers)

Afferent fibers from nasal cavity and paranasal sinuses (via trigeminal and glossopharyngeal nn.) may also initiate reflexes in airways

Descending tracts in spinal cord

Superior cervical ganglion

Larynx

Superior laryngeal n.

Sympathetic nn. (adrenergic)

T1

Carotid sinus

Thoracic part of spinal cord

T2

Carotid body

T3

Left common carotid a.

Cough receptors

T4

Left recurrent laryngeal n.

T5

Aortic arch

Sympathetic trunk

Cough receptors

Pulmonary plexus

Irritant receptors

——— Parasympathetic fibers

——— Sympathetic fibers

——— Afferent fibers

Stretch receptors (Hering–Breuer reflex)

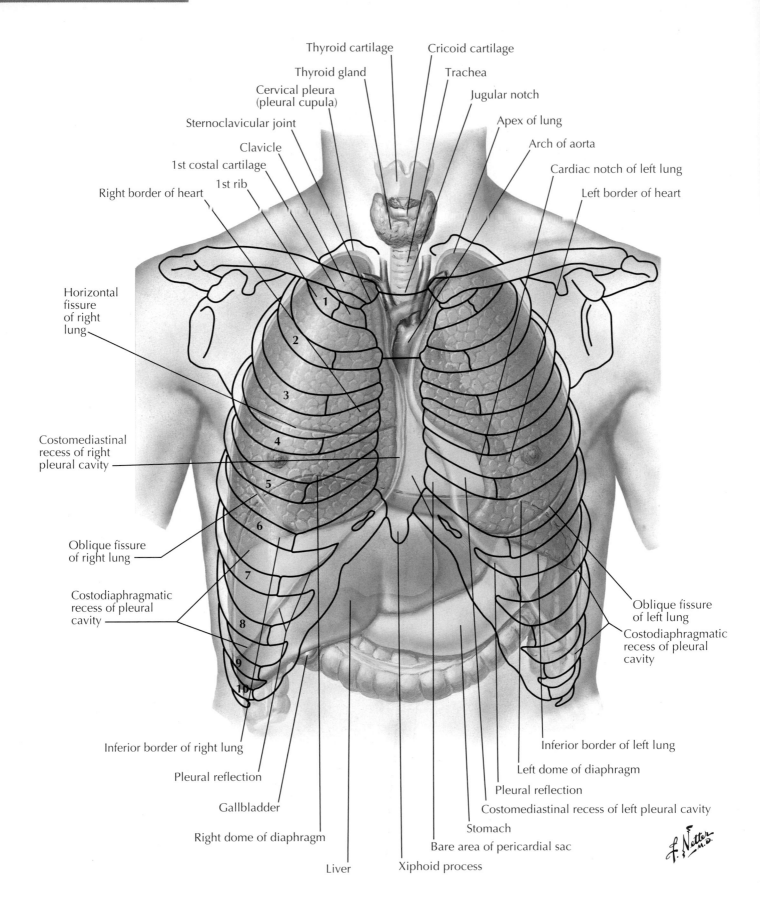

Thyroid cartilage

Cricoid cartilage

Thyroid gland

Trachea

Cervical pleura
(pleural cupula)

Jugular notch

Sternoclavicular joint

Apex of lung

Clavicle

Arch of aorta

1st costal cartilage

Cardiac notch of left lung

1st rib

Left border of heart

Right border of heart

Horizontal
fissure
of right
lung

Costomediastinal
recess of right
pleural cavity

Oblique fissure
of right lung

Oblique fissure
of left lung

Costodiaphragmatic
recess of pleural
cavity

Costodiaphragmatic
recess of pleural
cavity

Inferior border of right lung

Inferior border of left lung

Pleural reflection

Left dome of diaphragm

Gallbladder

Pleural reflection

Costomediastinal recess of left pleural cavity

Stomach

Right dome of diaphragm

Bare area of pericardial sac

Liver

Xiphoid process

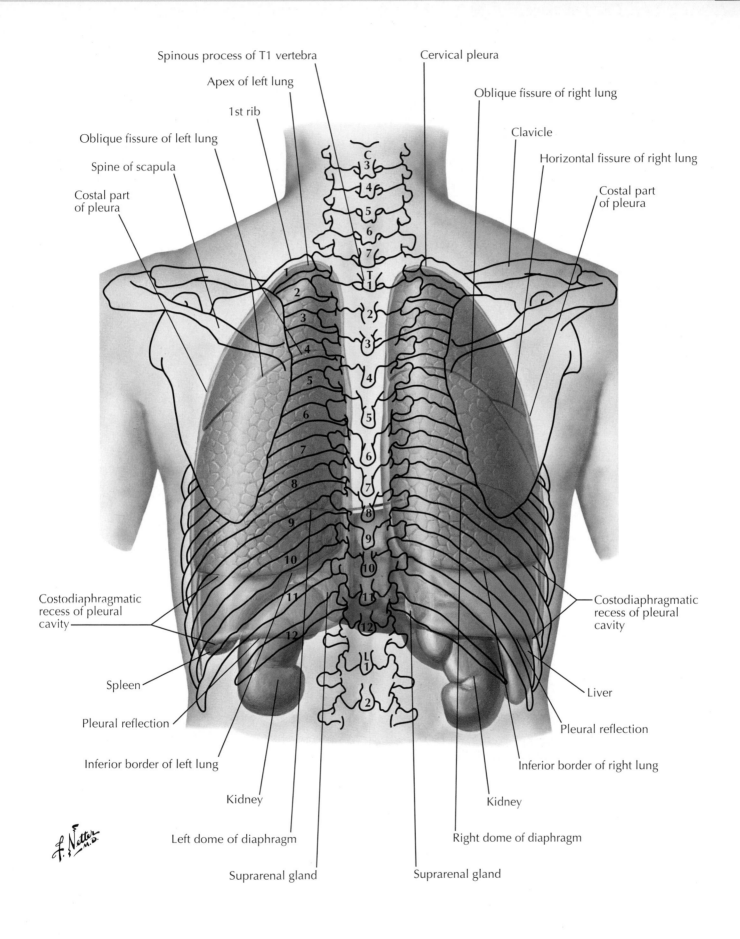

Spinous process of T1 vertebra

Apex of left lung

1st rib

Oblique fissure of left lung

Spine of scapula

Costal part of pleura

Cervical pleura

Oblique fissure of right lung

Clavicle

Horizontal fissure of right lung

Costal part of pleura

Costodiaphragmatic recess of pleural cavity

Costodiaphragmatic recess of pleural cavity

Spleen

Pleural reflection

Inferior border of left lung

Kidney

Left dome of diaphragm

Suprarenal gland

Liver

Pleural reflection

Inferior border of right lung

Kidney

Right dome of diaphragm

Suprarenal gland

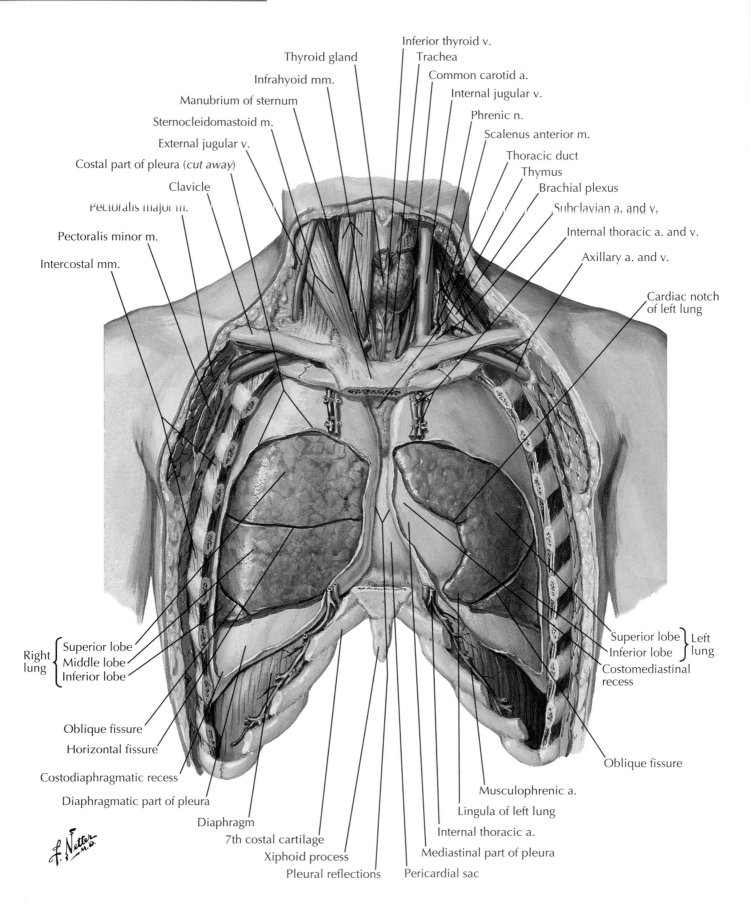

Inferior thyroid v.

Thyroid gland

Trachea

Common carotid a.

Infrahyoid mm.

Internal jugular v.

Manubrium of sternum

Phrenic n.

Sternocleidomastoid m.

Scalenus anterior m.

External jugular v.

Thoracic duct

Costal part of pleura (cut away)

Thymus

Clavicle

Brachial plexus

Pectoralis major m.

Subclavian a. and v.

Pectoralis minor m.

Internal thoracic a. and v.

Intercostal mm.

Axillary a. and v.

Cardiac notch
of left lung

Right
lung { Superior lobe
Middle lobe
Inferior lobe }

Superior lobe } Left
Inferior lobe } lung

Costomediastinal
recess

Oblique fissure

Horizontal fissure

Oblique fissure

Costodiaphragmatic recess

Diaphragmatic part of pleura

Musculophrenic a.

Diaphragm

Lingula of left lung

7th costal cartilage

Internal thoracic a.

Xiphoid process

Mediastinal part of pleura

Pleural reflections

Pericardial sac

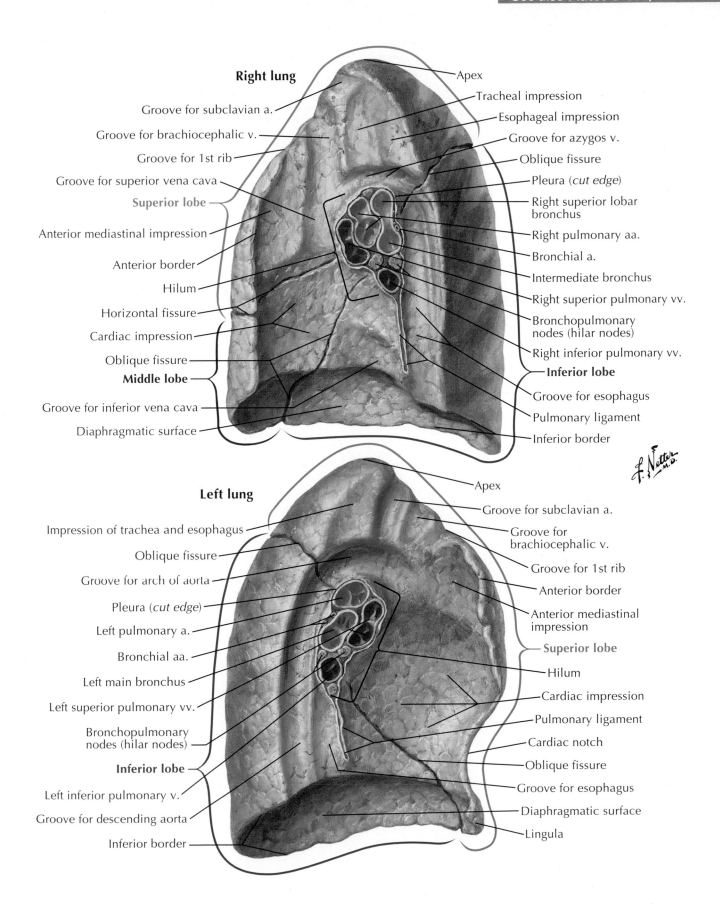

Right lung

Groove for subclavian a.

Groove for brachiocephalic v.

Groove for 1st rib

Groove for superior vena cava

Superior lobe

Anterior mediastinal impression

Anterior border

Hilum

Horizontal fissure

Cardiac impression

Oblique fissure

Middle lobe

Groove for inferior vena cava

Diaphragmatic surface

Apex

Tracheal impression

Esophageal impression

Groove for azygos v.

Oblique fissure

Pleura (cut edge)

Right superior lobar bronchus

Right pulmonary aa.

Bronchial a.

Intermediate bronchus

Right superior pulmonary vv.

Bronchopulmonary nodes (hilar nodes)

Right inferior pulmonary vv.

Inferior lobe

Groove for esophagus

Pulmonary ligament

Inferior border

Left lung

Impression of trachea and esophagus

Oblique fissure

Groove for arch of aorta

Pleura (cut edge)

Left pulmonary a.

Bronchial aa.

Left main bronchus

Left superior pulmonary vv.

Bronchopulmonary nodes (hilar nodes)

Inferior lobe

Left inferior pulmonary v.

Groove for descending aorta

Inferior border

Apex

Groove for subclavian a.

Groove for brachiocephalic v.

Groove for 1st rib

Anterior border

Anterior mediastinal impression

Superior lobe

Hilum

Cardiac impression

Pulmonary ligament

Cardiac notch

Oblique fissure

Groove for esophagus

Diaphragmatic surface

Lingula

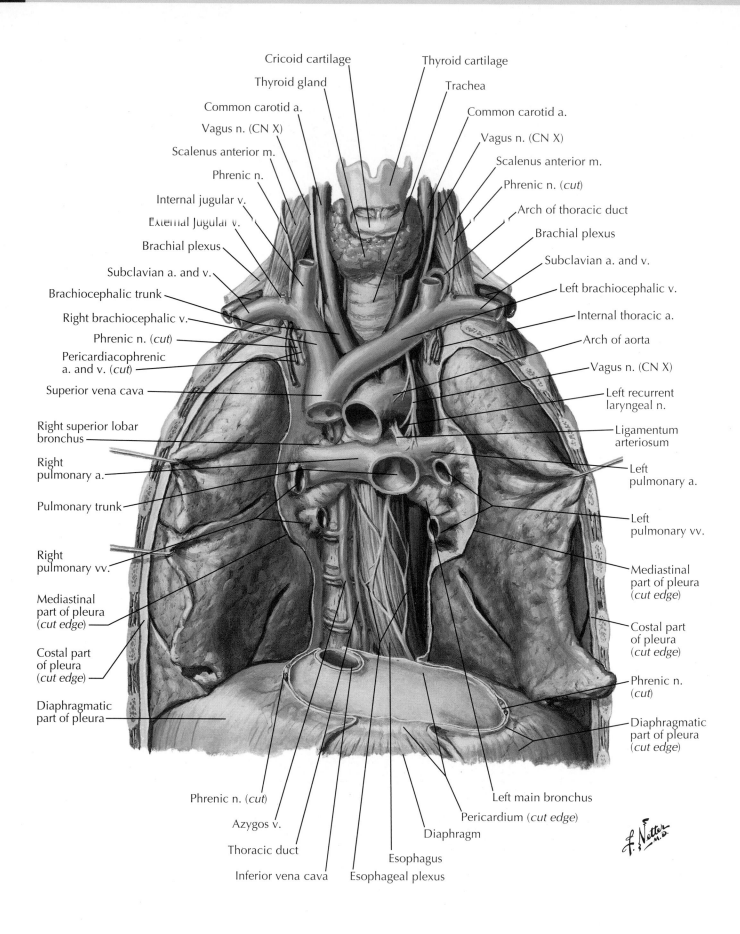

Cricoid cartilage

Thyroid cartilage

Thyroid gland

Trachea

Common carotid a.

Common carotid a.

Vagus n. (CN X)

Vagus n. (CN X)

Scalenus anterior m.

Scalenus anterior m.

Phrenic n.

Phrenic n. (*cut*)

Internal jugular v.

Arch of thoracic duct

External jugular v.

Brachial plexus

Brachial plexus

Subclavian a. and v.

Subclavian a. and v.

Brachiocephalic trunk

Left brachiocephalic v.

Right brachiocephalic v.

Internal thoracic a.

Phrenic n. (*cut*)

Arch of aorta

Pericardiacophrenic
a. and v. (*cut*)

Vagus n. (CN X)

Superior vena cava

Left recurrent
laryngeal n.

Right superior lobar
bronchus

Ligamentum
arteriosum

Right
pulmonary a.

Left
pulmonary a.

Pulmonary trunk

Left
pulmonary vv.

Right
pulmonary vv.

Mediastinal
part of pleura
(*cut edge*)

Mediastinal
part of pleura
(*cut edge*)

Costal part
of pleura
(*cut edge*)

Costal part
of pleura
(*cut edge*)

Phrenic n.
(*cut*)

Diaphragmatic
part of pleura

Diaphragmatic
part of pleura
(*cut edge*)

Phrenic n. (*cut*)

Left main bronchus

Azygos v.

Pericardium (*cut edge*)

Thoracic duct

Diaphragm

Esophagus

Inferior vena cava

Esophageal plexus

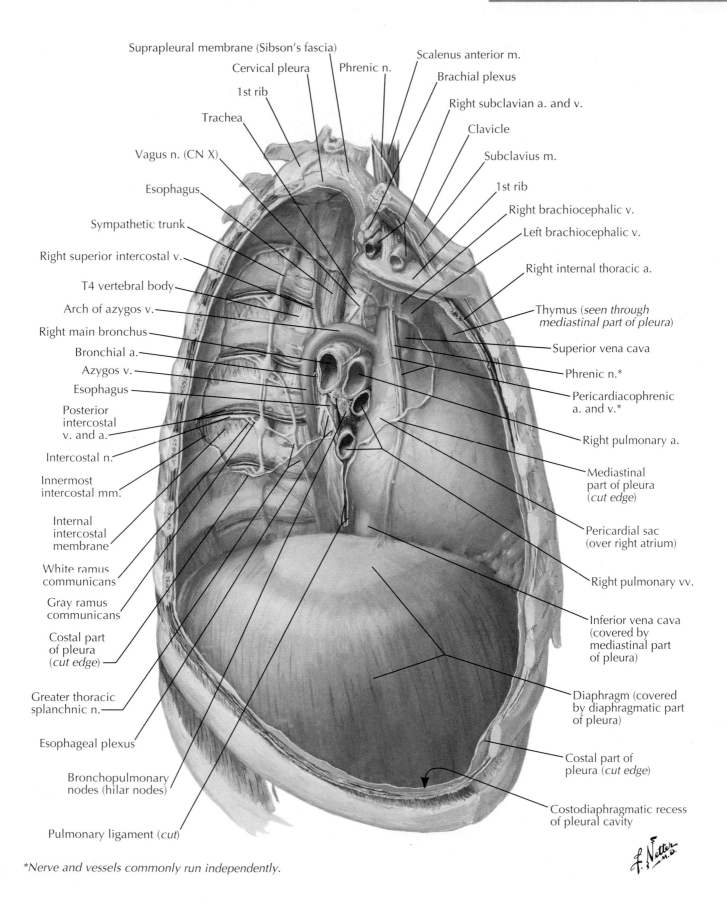

Suprapleural membrane (Sibson's fascia)

Scalenus anterior m.

Cervical pleura

Phrenic n.

Brachial plexus

1st rib

Right subclavian a. and v.

Trachea

Clavicle

Vagus n. (CN X)

Subclavius m.

Esophagus

1st rib

Sympathetic trunk

Right brachiocephalic v.

Right superior intercostal v.

Left brachiocephalic v.

T4 vertebral body

Right internal thoracic a.

Arch of azygos v.

Thymus (*seen through mediastinal part of pleura*)

Right main bronchus

Superior vena cava

Bronchial a.

Phrenic n.*

Azygos v.

Pericardiacophrenic a. and v.*

Esophagus

Posterior intercostal v. and a.

Right pulmonary a.

Intercostal n.

Mediastinal part of pleura (*cut edge*)

Innermost intercostal mm.

Internal intercostal membrane

Pericardial sac (over right atrium)

White ramus communicans

Right pulmonary vv.

Gray ramus communicans

Inferior vena cava (covered by mediastinal part of pleura)

Costal part of pleura (*cut edge*)

Greater thoracic splanchnic n.

Diaphragm (covered by diaphragmatic part of pleura)

Esophageal plexus

Costal part of pleura (*cut edge*)

Bronchopulmonary nodes (hilar nodes)

Costodiaphragmatic recess of pleural cavity

Pulmonary ligament (*cut*)

*Nerve and vessels commonly run independently.

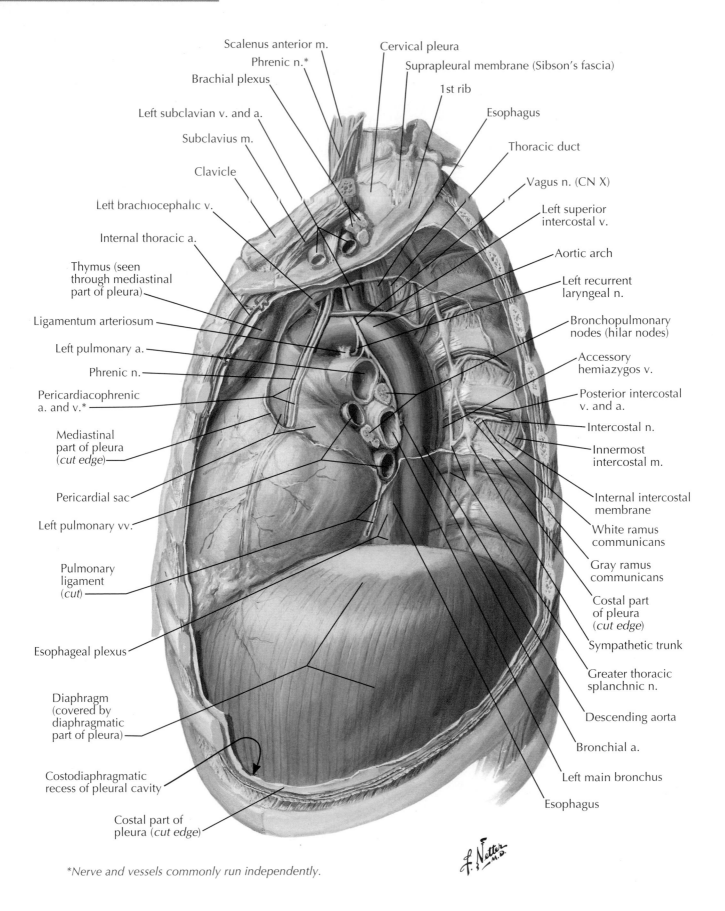

Scalenus anterior m.

Phrenic n.*

Brachial plexus

Left subclavian v. and a.

Subclavius m.

Clavicle

Left brachiocephalic v.

Internal thoracic a.

Thymus (seen through mediastinal part of pleura)

Ligamentum arteriosum

Left pulmonary a.

Phrenic n.

Pericardiacophrenic a. and v.*

Mediastinal part of pleura (cut edge)

Pericardial sac

Left pulmonary vv.

Pulmonary ligament (cut)

Esophageal plexus

Diaphragm (covered by diaphragmatic part of pleura)

Costodiaphragmatic recess of pleural cavity

Costal part of pleura (cut edge)

Cervical pleura

Suprapleural membrane (Sibson's fascia)

1st rib

Esophagus

Thoracic duct

Vagus n. (CN X)

Left superior intercostal v.

Aortic arch

Left recurrent laryngeal n.

Bronchopulmonary nodes (hilar nodes)

Accessory hemiazygos v.

Posterior intercostal v. and a.

Intercostal n.

Innermost intercostal m.

Internal intercostal membrane

White ramus communicans

Gray ramus communicans

Costal part of pleura (cut edge)

Sympathetic trunk

Greater thoracic splanchnic n.

Descending aorta

Bronchial a.

Left main bronchus

Esophagus

Nerve and vessels commonly run independently.

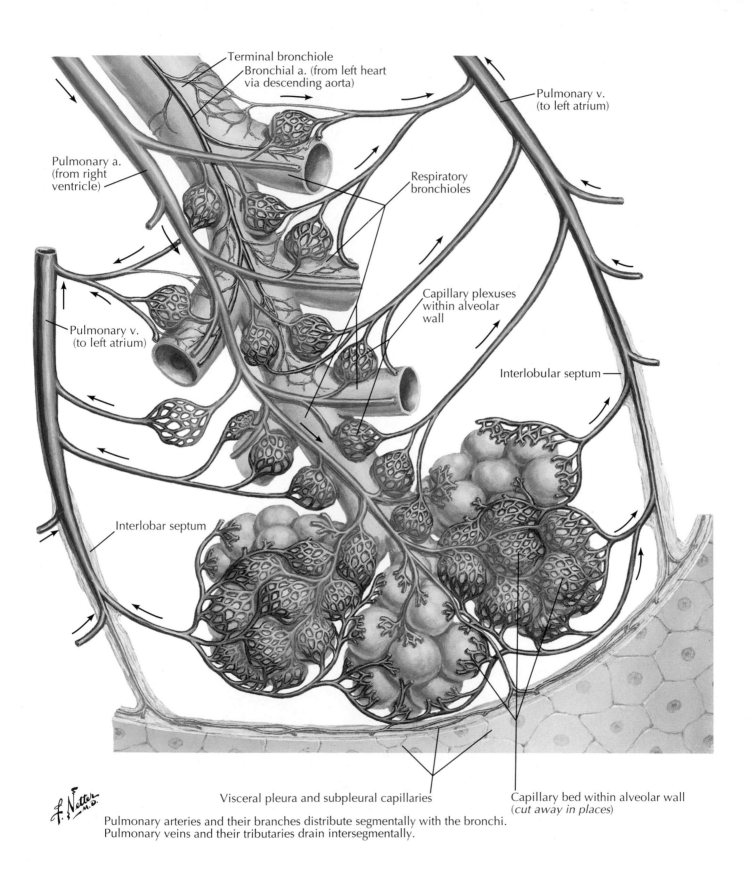

Terminal bronchiole

Bronchial a. (from left heart via descending aorta)

Pulmonary v. (to left atrium)

Pulmonary a. (from right ventricle)

Respiratory bronchioles

Pulmonary v. (to left atrium)

Capillary plexuses within alveolar wall

Interlobular septum

Interlobar septum

Visceral pleura and subpleural capillaries

Capillary bed within alveolar wall (*cut away in places*)

Pulmonary arteries and their branches distribute segmentally with the bronchi.
Pulmonary veins and their tributaries drain intersegmentally.

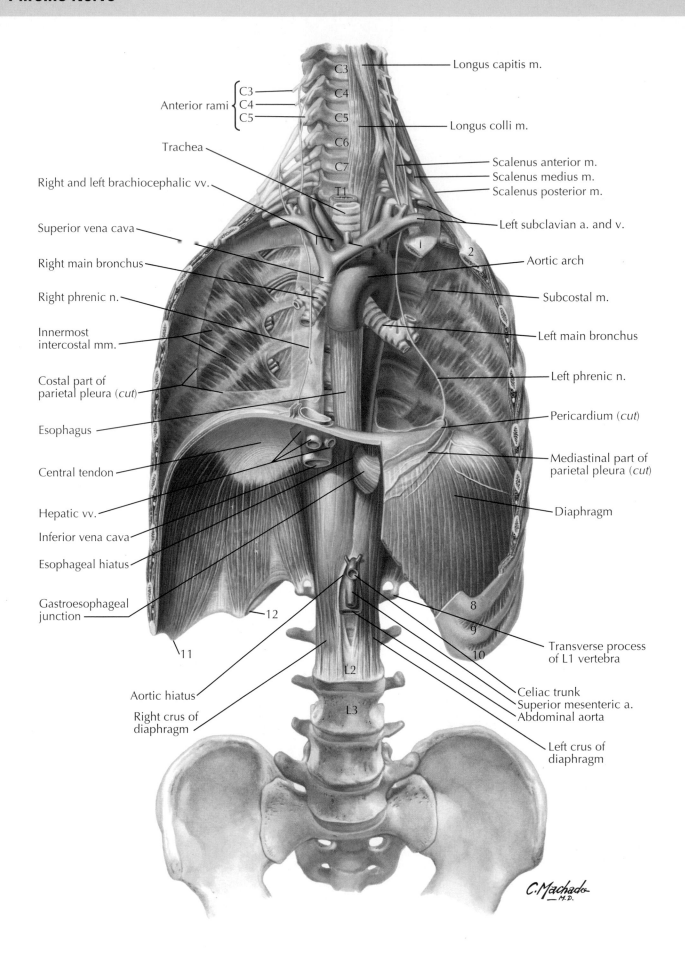

Longus capitis m.

Anterior rami { C3, C4, C5

C3
C4
C5
C6
C7
T1

Longus colli m.

Trachea

Scalenus anterior m.
Scalenus medius m.
Scalenus posterior m.

Right and left brachiocephalic vv.

Left subclavian a. and v.

Superior vena cava

Aortic arch

Right main bronchus

Subcostal m.

Right phrenic n.

Left main bronchus

Innermost intercostal mm.

Left phrenic n.

Costal part of parietal pleura (cut)

Pericardium (cut)

Esophagus

Mediastinal part of parietal pleura (cut)

Central tendon

Diaphragm

Hepatic vv.

Inferior vena cava

Transverse process of L1 vertebra

Esophageal hiatus

Gastroesophageal junction

L2

L3

Celiac trunk
Superior mesenteric a.
Abdominal aorta

Aortic hiatus

Right crus of diaphragm

Left crus of diaphragm

Vasculature and Innervation of Mediastinum

ANATOMIC STRUCTURES	CLINICAL IMPORTANCE	PLATE NUMBERS
Head and Neck		
Epiglottis	Crucial landmark during endotracheal intubation; bacterial or viral infection (e.g., with *Hemophilus influenzae*) may cause epiglottitis, which presents with respiratory distress, sore throat, and hoarse voice	S–372 to S–374
Paranasal sinuses (maxillary, ethmoid, frontal, and sphenoid)	Cavities in skull prone to mucosal inflammation due to bacterial or viral infection	S–368, S–369
Nasal septum	Congenital or acquired deviation may lead to nasal obstruction, treated with septoplasty; perforations may occur with cocaine use or in granulomatosis with polyangiitis	S–367, S–368
Thorax		
Pleura	Air or gas (spontaneous or traumatic) can leak into pleural space between visceral pleura and parietal pleura and compress lung, a condition known as pneumothorax; if severe enough to compromise venous return to heart, resulting in hypotension and dyspnea, condition is known as tension pneumothorax	S–385 to S–387, S–389
Cervical pleura	Extends into neck superior to anterior aspect of angled 1st rib or to the level of 1st rib posteriorly; it may therefore be punctured during neck procedures, producing pneumothorax	S–368, S–385
Tracheal bifurcation	Important landmark when assessing position of endotracheal tube, which should terminate 4–5 cm above; often at T4–T5 vertebral level; right main bronchus is shorter, more vertical, and wider; therefore, aspirated objects more often enter right side; left main bronchus lies over esophagus and posterior to left atrium of heart	S–379
Apex of lung	Pancoast tumor (bronchogenic carcinoma of apex of lung) may compress sympathetic trunk, resulting in Horner's syndrome (ipsilateral miosis, ptosis, anhidrosis, facial flushing); apex is susceptible to pneumothorax from needles introduced in lower neck region	S–385, S–390

*Selections are based largely on clinical data and commonly discussed clinical correlations in macroscopic ("gross") anatomy courses.

Structures with High Clinical Significance   **Table 7.1**

DIGESTIVE SYSTEM 8

ELECTRONIC BONUS PLATES

S–BP 65 Intrinsic Nerves and Variations in Nerves of Esophagus

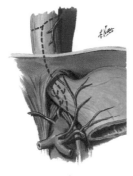

S–BP 66 Arteries of Esophagus: Variations

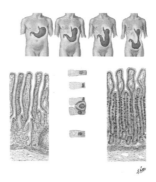

S–BP 67 Variations in Position and Contour of Stomach in Relation to Body Habitus

S–BP 68 Layers of Duodenal Wall

S–BP 69 CT and MRCP Showing Vermiform Appendix, Gallbladder, and Ducts; Nerve Branches on Hepatic Artery

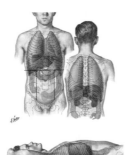

S–BP 70 Topography of Liver

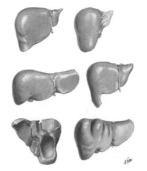

S–BP 71 Variations in Form of Liver

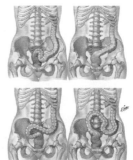

S–BP 72 Sigmoid Colon: Variations in Position

ELECTRONIC BONUS PLATES—*cont'd*

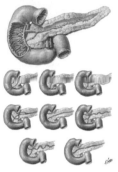

S–BP 73 Variations in Pancreatic Duct

S–BP 74 Variations in Cystic, Hepatic, and Pancreatic Ducts

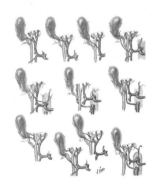

S–BP 75 Variations in Cystic Arteries

S–BP 76 Variations in Hepatic Arteries

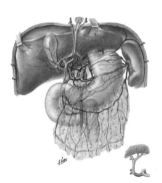

S–BP 77 Arterial Variations and Collateral Supply of Liver and Gallbladder

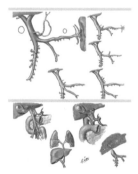

S–BP 78 Variations and Anomalies of Hepatic Portal Vein

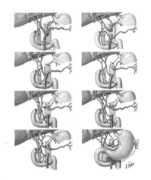

S–BP 79 Variations in Celiac Trunk

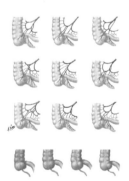

S–BP 80 Variations in Arterial Supply to Cecum and Posterior Peritoneal Attachment of Cecum

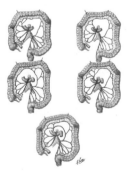

S–BP 81 Variations in Colic Arteries

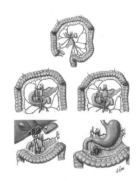

S–BP 82 Variations in Colic Arteries (Continued)

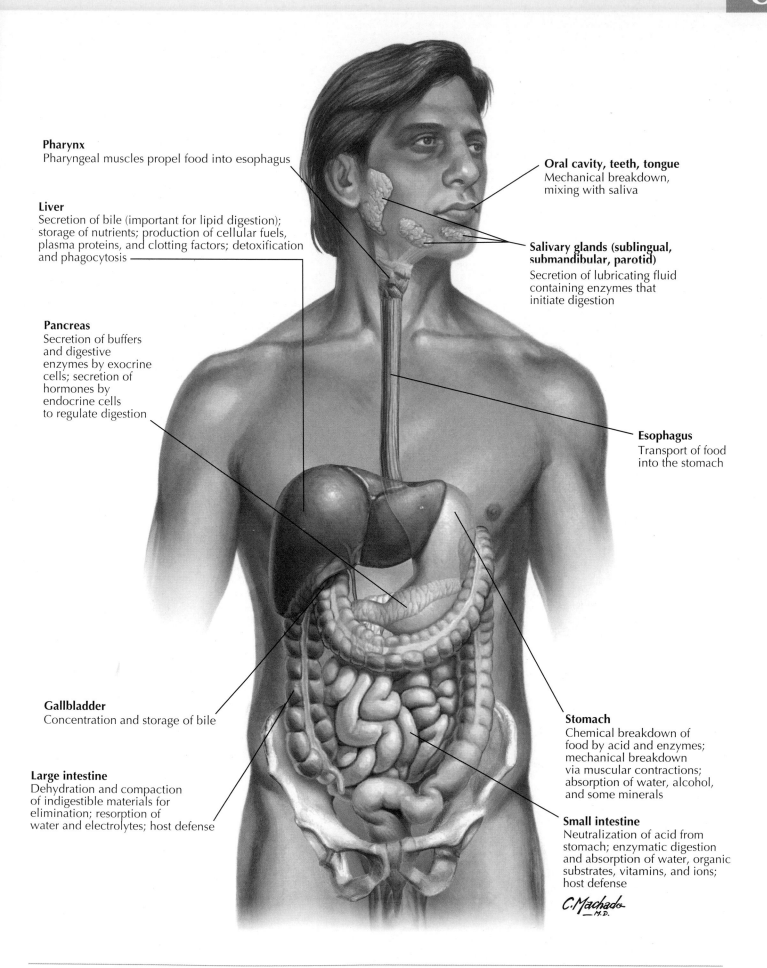

Pharynx
Pharyngeal muscles propel food into esophagus

Liver
Secretion of bile (important for lipid digestion); storage of nutrients; production of cellular fuels, plasma proteins, and clotting factors; detoxification and phagocytosis

Pancreas
Secretion of buffers and digestive enzymes by exocrine cells; secretion of hormones by endocrine cells to regulate digestion

Gallbladder
Concentration and storage of bile

Large intestine
Dehydration and compaction of indigestible materials for elimination; resorption of water and electrolytes; host defense

Oral cavity, teeth, tongue
Mechanical breakdown, mixing with saliva

Salivary glands (sublingual, submandibular, parotid)
Secretion of lubricating fluid containing enzymes that initiate digestion

Esophagus
Transport of food into the stomach

Stomach
Chemical breakdown of food by acid and enzymes; mechanical breakdown via muscular contractions; absorption of water, alcohol, and some minerals

Small intestine
Neutralization of acid from stomach; enzymatic digestion and absorption of water, organic substrates, vitamins, and ions; host defense

C. Machado
_M.D._

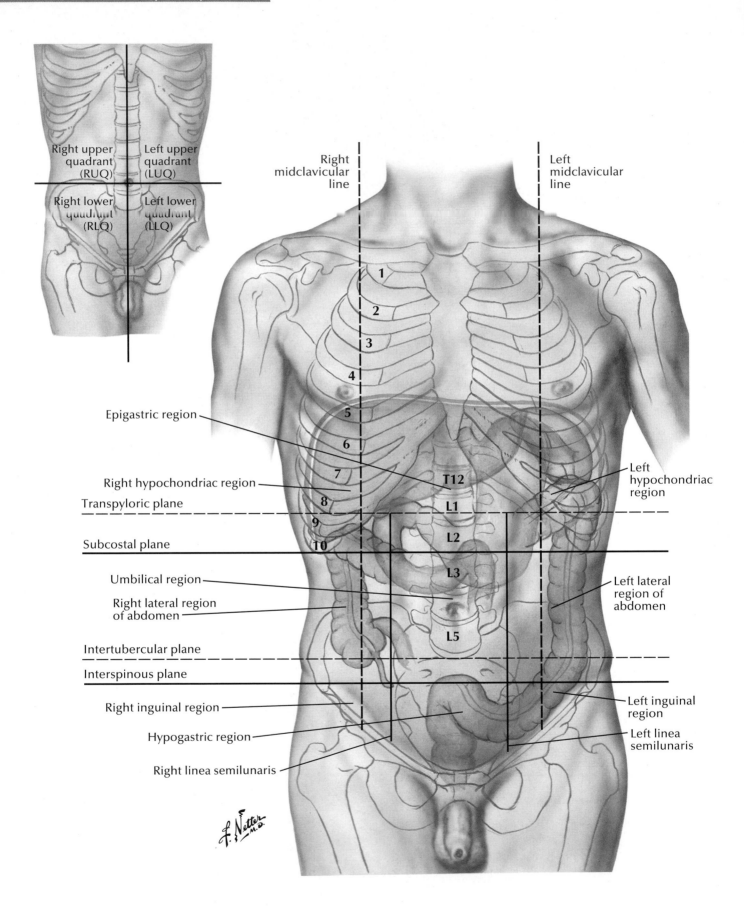

Right upper quadrant (RUQ)

Left upper quadrant (LUQ)

Right lower quadrant (RLQ)

Left lower quadrant (LLQ)

Right midclavicular line

Left midclavicular line

1

2

3

4

5

6

7

T12

8

L1

9

L2

10

L3

L5

Epigastric region

Left hypochondriac region

Right hypochondriac region

Transpyloric plane

Subcostal plane

Umbilical region

Right lateral region of abdomen

Left lateral region of abdomen

Intertubercular plane

Interspinous plane

Right inguinal region

Left inguinal region

Hypogastric region

Left linea semilunaris

Right linea semilunaris

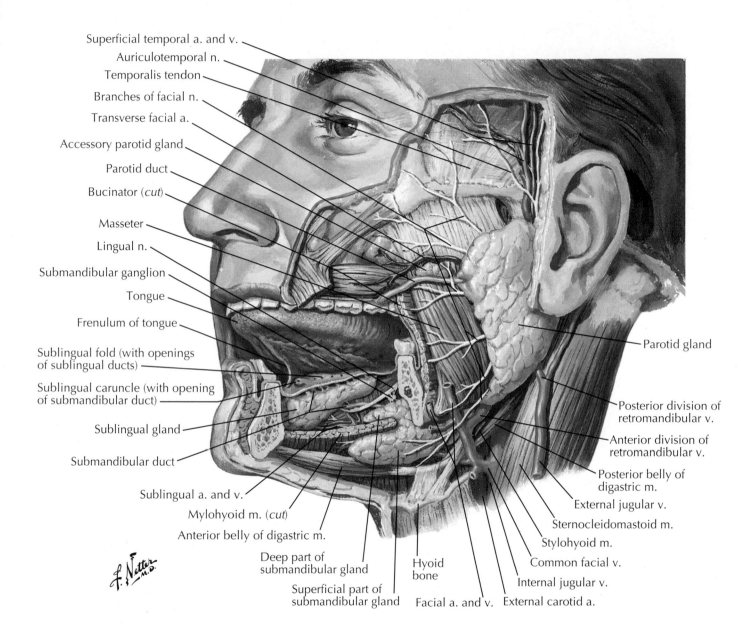

Superficial temporal a. and v.

Auriculotemporal n.

Temporalis tendon

Branches of facial n.

Transverse facial a.

Accessory parotid gland

Parotid duct

Bucinator (*cut*)

Masseter

Lingual n.

Submandibular ganglion

Tongue

Frenulum of tongue

Sublingual fold (with openings of sublingual ducts)

Sublingual caruncle (with opening of submandibular duct)

Sublingual gland

Submandibular duct

Sublingual a. and v.

Mylohyoid m. (*cut*)

Anterior belly of digastric m.

Deep part of submandibular gland

Superficial part of submandibular gland

Hyoid bone

Facial a. and v.

External carotid a.

Internal jugular v.

Common facial v.

Stylohyoid m.

Sternocleidomastoid m.

External jugular v.

Posterior belly of digastric m.

Anterior division of retromandibular v.

Posterior division of retromandibular v.

Parotid gland

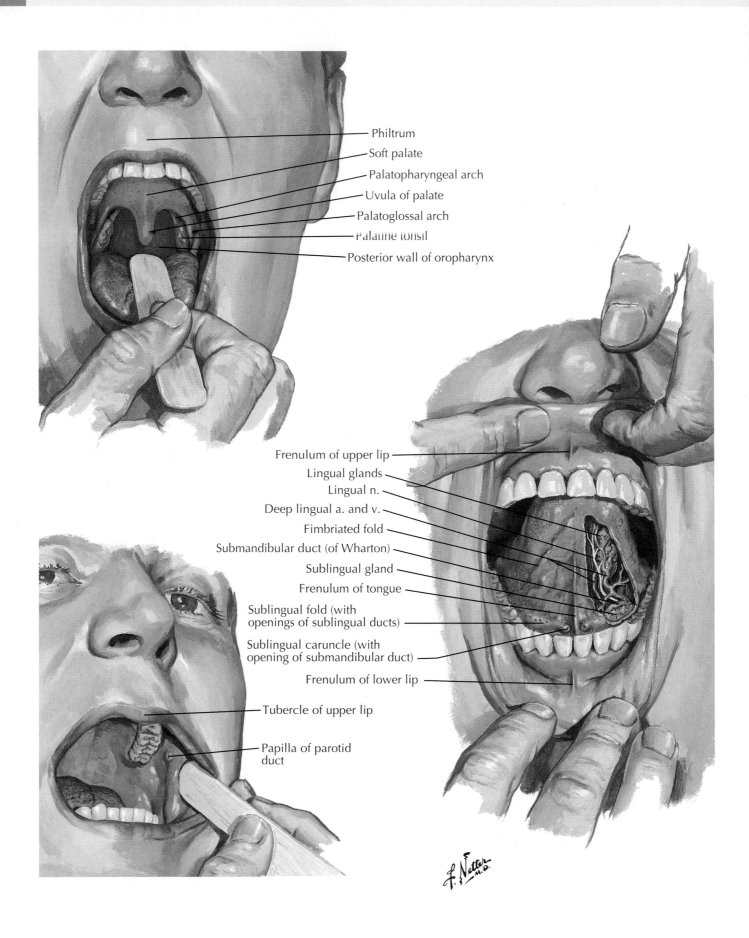

Philtrum

Soft palate

Palatopharyngeal arch

Uvula of palate

Palatoglossal arch

Palatine tonsil

Posterior wall of oropharynx

Frenulum of upper lip

Lingual glands

Lingual n.

Deep lingual a. and v.

Fimbriated fold

Submandibular duct (of Wharton)

Sublingual gland

Frenulum of tongue

Sublingual fold (with openings of sublingual ducts)

Sublingual caruncle (with opening of submandibular duct)

Frenulum of lower lip

Tubercle of upper lip

Papilla of parotid duct

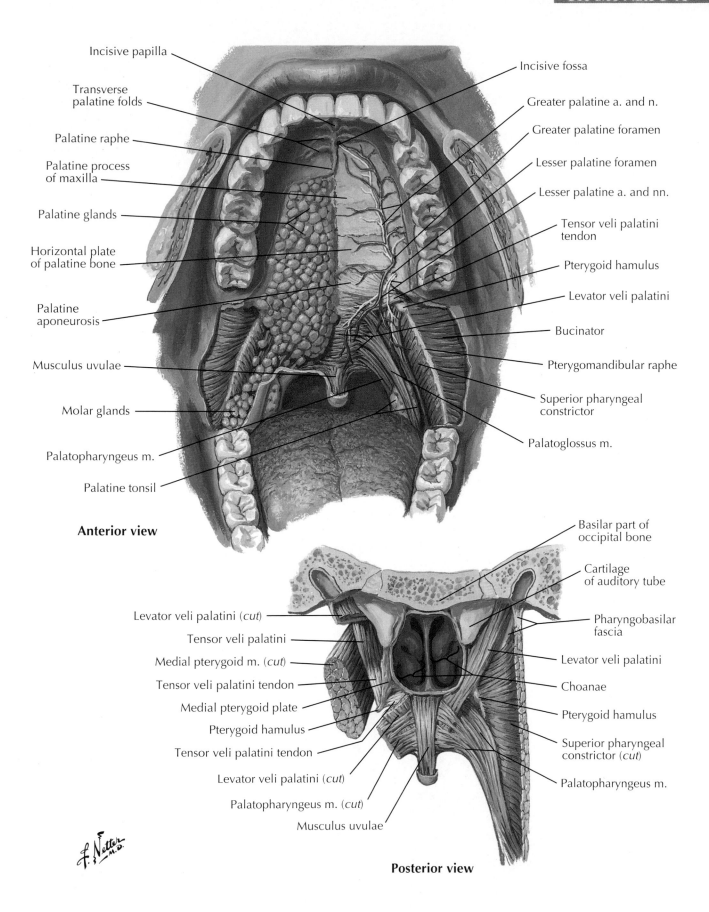

Incisive papilla

Transverse palatine folds

Palatine raphe

Palatine process of maxilla

Palatine glands

Horizontal plate of palatine bone

Palatine aponeurosis

Musculus uvulae

Molar glands

Palatopharyngeus m.

Palatine tonsil

Incisive fossa

Greater palatine a. and n.

Greater palatine foramen

Lesser palatine foramen

Lesser palatine a. and nn.

Tensor veli palatini tendon

Pterygoid hamulus

Levator veli palatini

Bucinator

Pterygomandibular raphe

Superior pharyngeal constrictor

Palatoglossus m.

Anterior view

Levator veli palatini (*cut*)

Tensor veli palatini

Medial pterygoid m. (*cut*)

Tensor veli palatini tendon

Medial pterygoid plate

Pterygoid hamulus

Tensor veli palatini tendon

Levator veli palatini (*cut*)

Palatopharyngeus m. (*cut*)

Musculus uvulae

Basilar part of occipital bone

Cartilage of auditory tube

Pharyngobasilar fascia

Levator veli palatini

Choanae

Pterygoid hamulus

Superior pharyngeal constrictor (*cut*)

Palatopharyngeus m.

Posterior view

Mouth

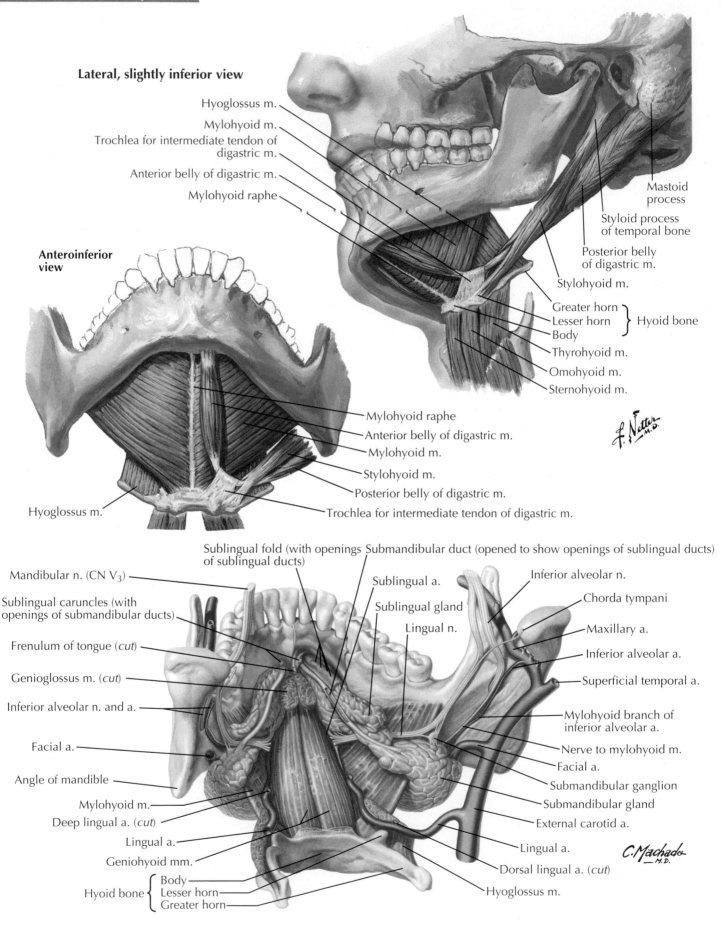

Lateral, slightly inferior view

Hyoglossus m.

Mylohyoid m.

Trochlea for intermediate tendon of digastric m.

Anterior belly of digastric m.

Mylohyoid raphe

Mastoid process

Styloid process of temporal bone

Posterior belly of digastric m.

Stylohyoid m.

Greater horn
Lesser horn } Hyoid bone
Body

Thyrohyoid m.

Omohyoid m.

Sternohyoid m.

Anteroinferior view

Mylohyoid raphe

Anterior belly of digastric m.

Mylohyoid m.

Stylohyoid m.

Posterior belly of digastric m.

Trochlea for intermediate tendon of digastric m.

Hyoglossus m.

Mandibular n. (CN V₃)

Sublingual fold (with openings of sublingual ducts)

Submandibular duct (opened to show openings of sublingual ducts)

Sublingual a.

Sublingual caruncles (with openings of submandibular ducts)

Sublingual gland

Lingual n.

Frenulum of tongue (cut)

Genioglossus m. (cut)

Inferior alveolar n. and a.

Facial a.

Angle of mandible

Mylohyoid m.

Deep lingual a. (cut)

Lingual a.

Geniohyoid mm.

Hyoid bone { Body
Lesser horn
Greater horn

Inferior alveolar n.

Chorda tympani

Maxillary a.

Inferior alveolar a.

Superficial temporal a.

Mylohyoid branch of inferior alveolar a.

Nerve to mylohyoid m.

Facial a.

Submandibular ganglion

Submandibular gland

External carotid a.

Lingual a.

Dorsal lingual a. (cut)

Hyoglossus m.

Horizontal section below lingula of mandible (superior view)

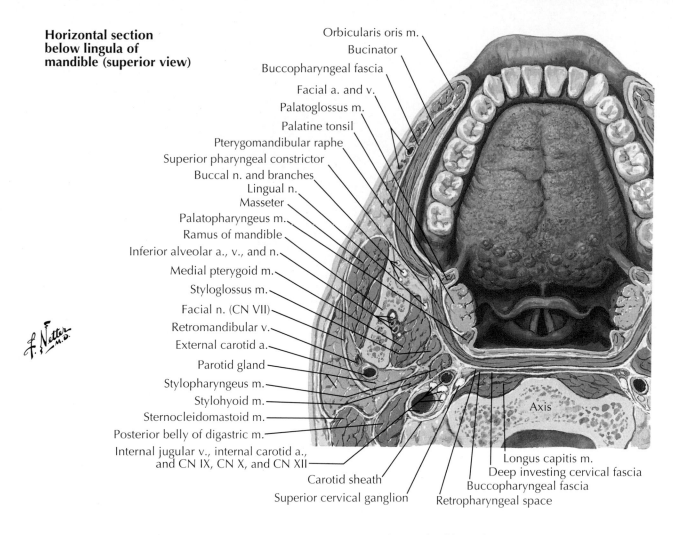

Orbicularis oris m.
Bucinator
Buccopharyngeal fascia
Facial a. and v.
Palatoglossus m.
Palatine tonsil
Pterygomandibular raphe
Superior pharyngeal constrictor
Buccal n. and branches
Lingual n.
Masseter
Palatopharyngeus m.
Ramus of mandible
Inferior alveolar a., v., and n.
Medial pterygoid m.
Styloglossus m.
Facial n. (CN VII)
Retromandibular v.
External carotid a.
Parotid gland
Stylopharyngeus m.
Stylohyoid m.
Sternocleidomastoid m.
Posterior belly of digastric m.
Internal jugular v., internal carotid a., and CN IX, CN X, and CN XII
Carotid sheath
Superior cervical ganglion

Axis
Longus capitis m.
Deep investing cervical fascia
Buccopharyngeal fascia
Retropharyngeal space

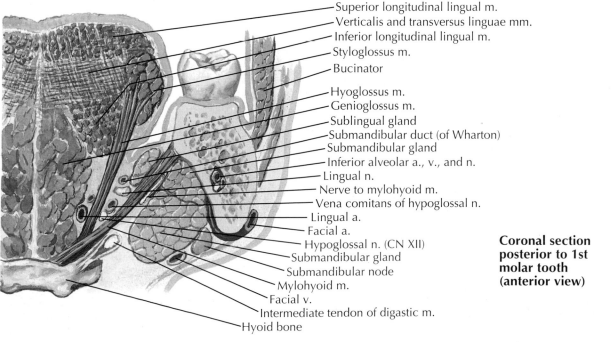

Superior longitudinal lingual m.
Verticalis and transversus linguae mm.
Inferior longitudinal lingual m.
Styloglossus m.
Bucinator
Hyoglossus m.
Genioglossus m.
Sublingual gland
Submandibular duct (of Wharton)
Submandibular gland
Inferior alveolar a., v., and n.
Lingual n.
Nerve to mylohyoid m.
Vena comitans of hypoglossal n.
Lingual a.
Facial a.
Hypoglossal n. (CN XII)
Submandibular gland
Submandibular node
Mylohyoid m.
Facial v.
Intermediate tendon of digastic m.
Hyoid bone

Coronal section posterior to 1st molar tooth (anterior view)

Mouth

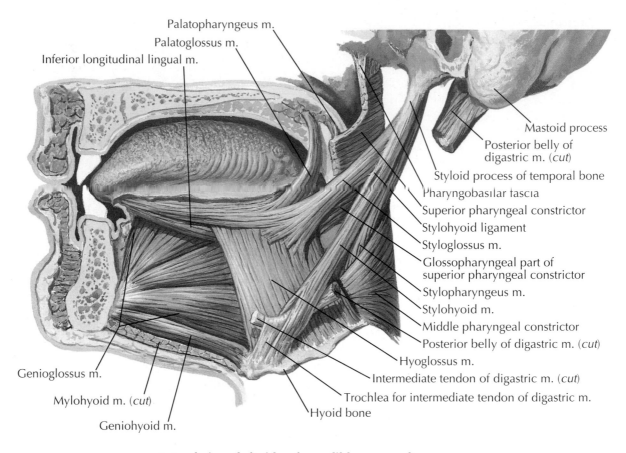

Palatopharyngeus m.

Palatoglossus m.

Inferior longitudinal lingual m.

Mastoid process

Posterior belly of digastric m. (*cut*)

Styloid process of temporal bone

Pharyngobasilar fascia

Superior pharyngeal constrictor

Stylohyoid ligament

Styloglossus m.

Glossopharyngeal part of superior pharyngeal constrictor

Stylopharyngeus m.

Stylohyoid m.

Middle pharyngeal constrictor

Posterior belly of digastric m. (*cut*)

Intermediate tendon of digastric m. (*cut*)

Trochlea for intermediate tendon of digastric m.

Hyoid bone

Genioglossus m.

Mylohyoid m. (*cut*)

Geniohyoid m.

Lateral view (left side of mandible removed)

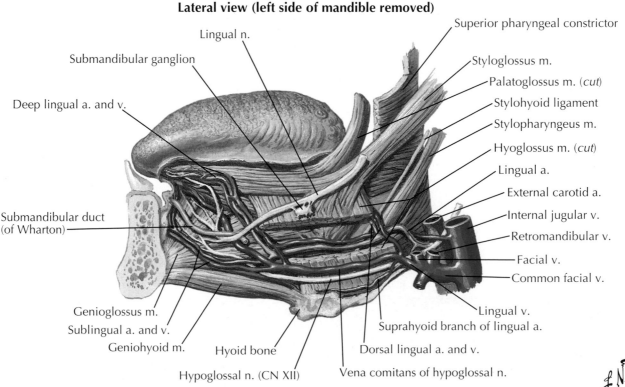

Lingual n.

Submandibular ganglion

Deep lingual a. and v.

Superior pharyngeal constrictor

Styloglossus m.

Palatoglossus m. (*cut*)

Stylohyoid ligament

Stylopharyngeus m.

Hyoglossus m. (*cut*)

Lingual a.

External carotid a.

Internal jugular v.

Retromandibular v.

Facial v.

Common facial v.

Submandibular duct (of Wharton)

Genioglossus m.

Sublingual a. and v.

Geniohyoid m.

Hyoid bone

Hypoglossal n. (CN XII)

Dorsal lingual a. and v.

Vena comitans of hypoglossal n.

Suprahyoid branch of lingual a.

Lingual v.

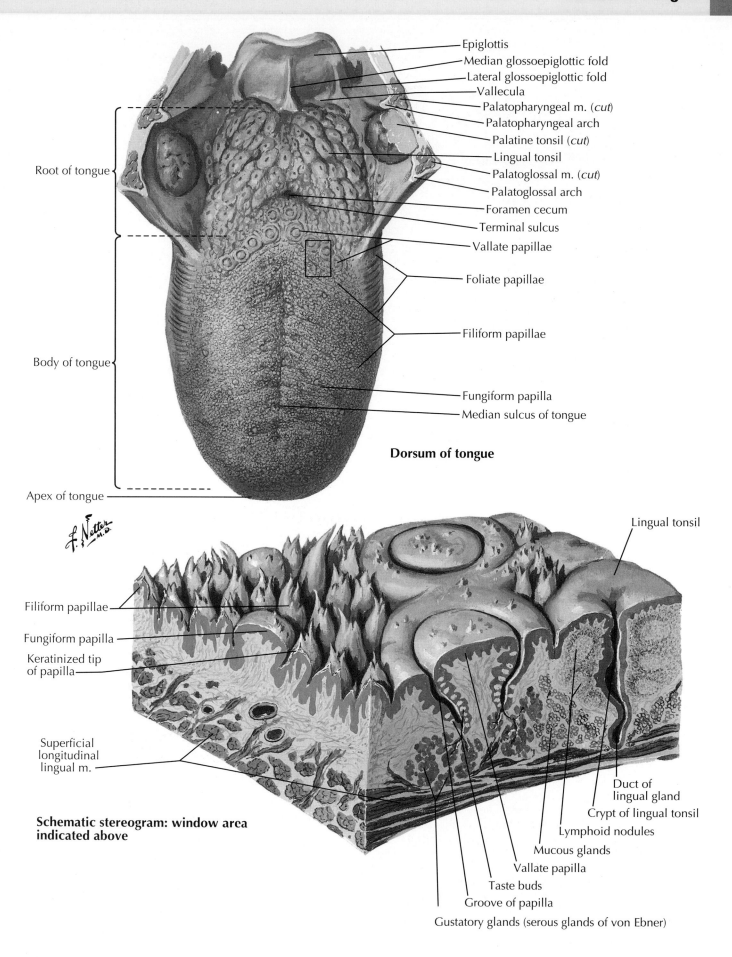

Epiglottis
Median glossoepiglottic fold
Lateral glossoepiglottic fold
Vallecula
Palatopharyngeal m. (*cut*)
Palatopharyngeal arch
Palatine tonsil (*cut*)
Lingual tonsil
Palatoglossal m. (*cut*)
Palatoglossal arch
Foramen cecum
Terminal sulcus
Vallate papillae
Foliate papillae
Filiform papillae
Fungiform papilla
Median sulcus of tongue

Root of tongue
Body of tongue
Apex of tongue

Dorsum of tongue

Lingual tonsil
Filiform papillae
Fungiform papilla
Keratinized tip of papilla
Superficial longitudinal lingual m.

Duct of lingual gland
Crypt of lingual tonsil
Lymphoid nodules
Mucous glands
Vallate papilla
Taste buds
Groove of papilla
Gustatory glands (serous glands of von Ebner)

Schematic stereogram: window area indicated above

Mouth

**Medial view
sagittal section**

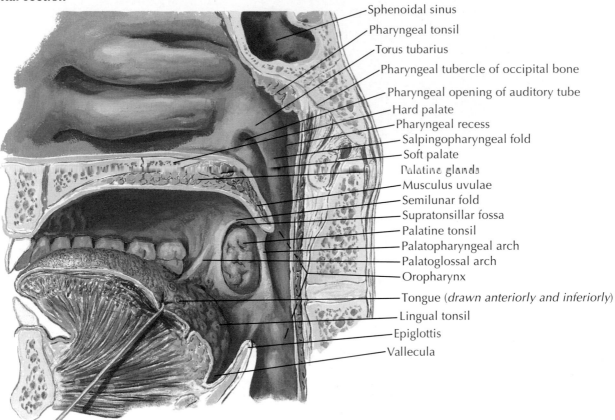

Sphenoidal sinus

Pharyngeal tonsil

Torus tubarius

Pharyngeal tubercle of occipital bone

Pharyngeal opening of auditory tube

Hard palate

Pharyngeal recess

Salpingopharyngeal fold

Soft palate

Palatine glands

Musculus uvulae

Semilunar fold

Supratonsillar fossa

Palatine tonsil

Palatopharyngeal arch

Palatoglossal arch

Oropharynx

Tongue (*drawn anteriorly and inferiorly*)

Lingual tonsil

Epiglottis

Vallecula

Pharyngeal mucosa removed

Pharyngeal tonsil

Cartilage of auditory tube

Medial pterygoid plate

Tensor veli palatini

Levator veli palatini

Ascending palatine a.

Pharyngeal branch of ascending pharyngeal a.

Tensor veli palatini tendon

Lesser palatine a.

Salpingopharyngeus m.

Pterygoid hamulus

Pterygomandibular raphe

Tonsillar branch of lesser palatine a.

Superior pharyngeal constrictor

Tonsillar branch of ascending pharyngeal a.

Palatoglossus m.

Palatopharyngeus m.

Tonsillar branch of ascending palatine a.

Tonsillar branch of facial a.

Dorsal lingual branch of lingual a.

Tonsillar branch of glossopharyngeal n.

Glossopharyngeal n. (CN IX)

Stylohyoid ligament

Hyoglossus m.

Middle pharyngeal constrictor

Stylopharyngeus m.

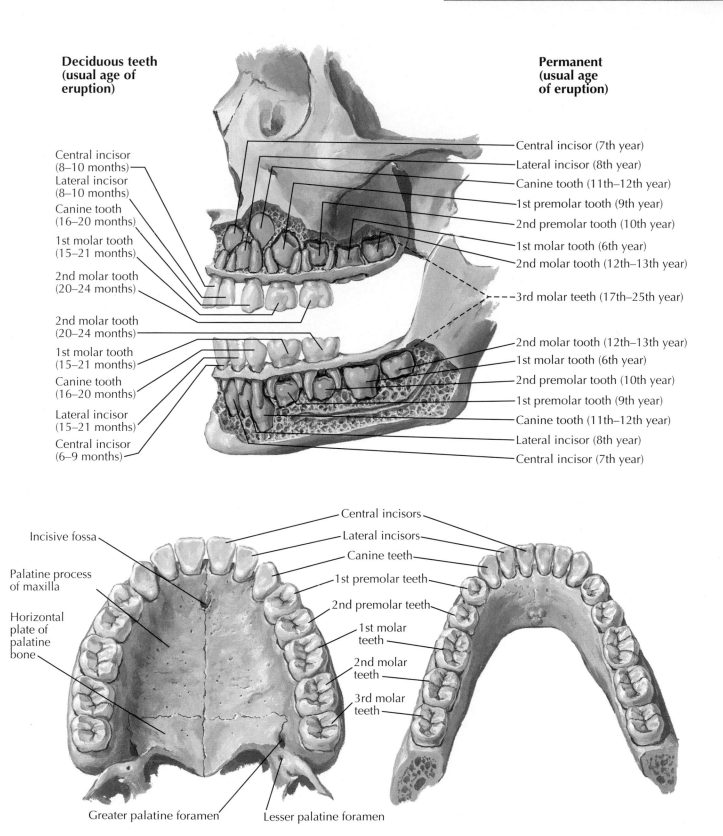

Deciduous teeth (usual age of eruption)

Central incisor (8–10 months)
Lateral incisor (8–10 months)
Canine tooth (16–20 months)
1st molar tooth (15–21 months)
2nd molar tooth (20–24 months)

2nd molar tooth (20–24 months)
1st molar tooth (15–21 months)
Canine tooth (16–20 months)
Lateral incisor (15–21 months)
Central incisor (6–9 months)

Permanent (usual age of eruption)

Central incisor (7th year)
Lateral incisor (8th year)
Canine tooth (11th–12th year)
1st premolar tooth (9th year)
2nd premolar tooth (10th year)
1st molar tooth (6th year)
2nd molar tooth (12th–13th year)
3rd molar teeth (17th–25th year)
2nd molar tooth (12th–13th year)
1st molar tooth (6th year)
2nd premolar tooth (10th year)
1st premolar tooth (9th year)
Canine tooth (11th–12th year)
Lateral incisor (8th year)
Central incisor (7th year)

Incisive fossa
Palatine process of maxilla
Horizontal plate of palatine bone

Central incisors
Lateral incisors
Canine teeth
1st premolar teeth
2nd premolar teeth
1st molar teeth
2nd molar teeth
3rd molar teeth

Greater palatine foramen
Lesser palatine foramen

Upper permanent teeth

Lower permanent teeth

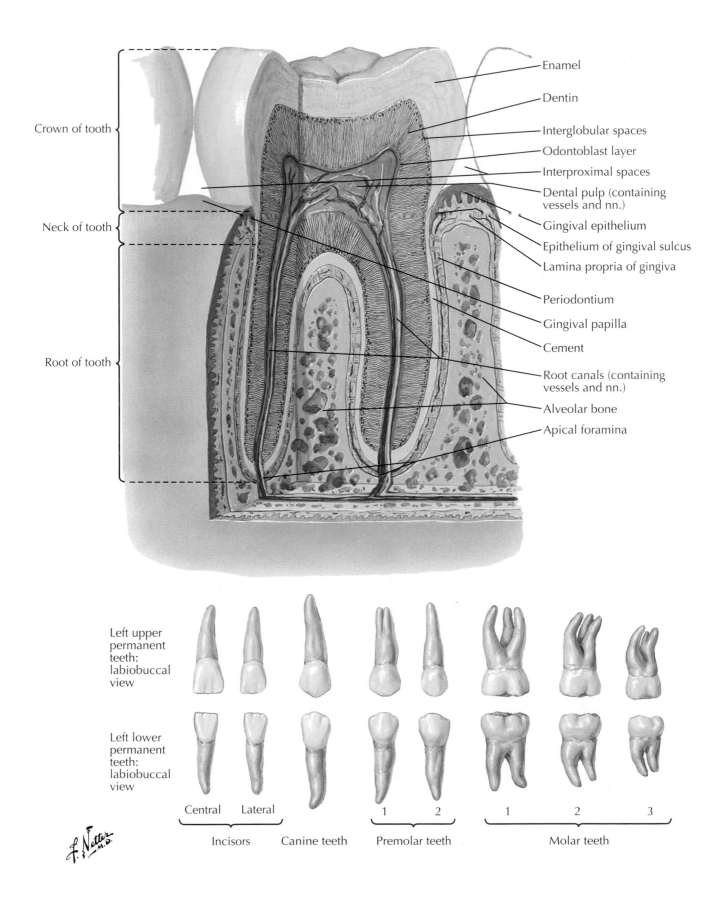

Crown of tooth

Neck of tooth

Root of tooth

Enamel

Dentin

Interglobular spaces

Odontoblast layer

Interproximal spaces

Dental pulp (containing vessels and nn.)

Gingival epithelium

Epithelium of gingival sulcus

Lamina propria of gingiva

Periodontium

Gingival papilla

Cement

Root canals (containing vessels and nn.)

Alveolar bone

Apical foramina

Left upper permanent teeth: labiobuccal view

Left lower permanent teeth: labiobuccal view

Central   Lateral

Incisors

Canine teeth

1   2

Premolar teeth

1   2   3

Molar teeth

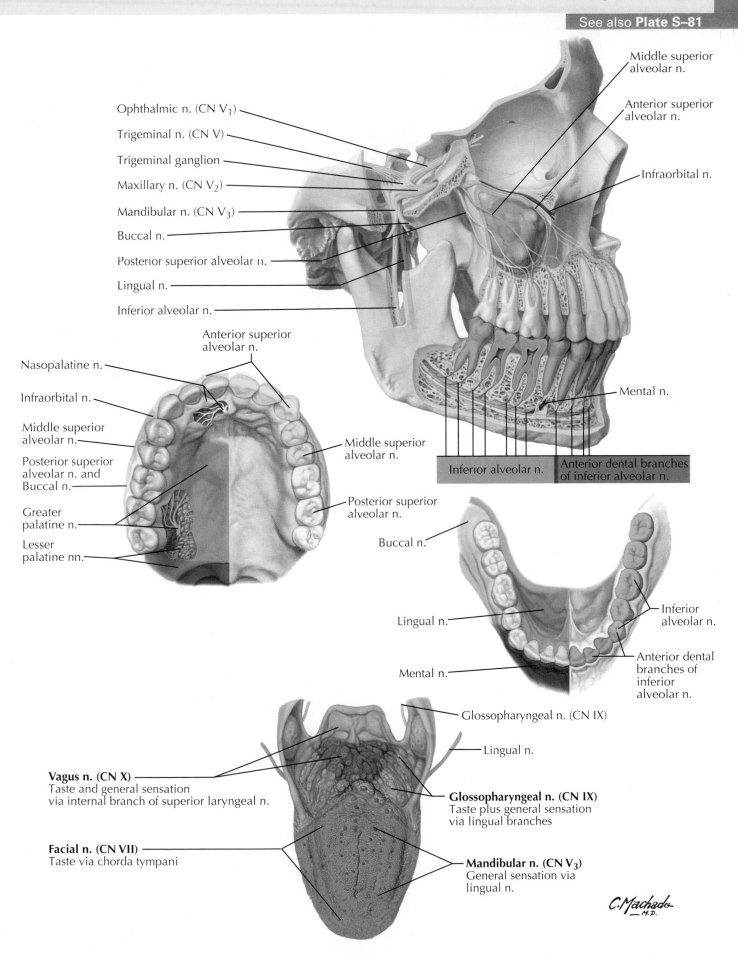

Middle superior alveolar n.

Anterior superior alveolar n.

Ophthalmic n. (CN V₁)

Trigeminal n. (CN V)

Trigeminal ganglion

Infraorbital n.

Maxillary n. (CN V₂)

Mandibular n. (CN V₃)

Buccal n.

Posterior superior alveolar n.

Lingual n.

Inferior alveolar n.

Mental n.

Anterior superior alveolar n.

Nasopalatine n.

Infraorbital n.

Middle superior alveolar n.

Middle superior alveolar n.

Posterior superior alveolar n. and Buccal n.

Posterior superior alveolar n.

Greater palatine n.

Lesser palatine nn.

Inferior alveolar n.

Anterior dental branches of inferior alveolar n.

Buccal n.

Lingual n.

Inferior alveolar n.

Mental n.

Anterior dental branches of inferior alveolar n.

Glossopharyngeal n. (CN IX)

Lingual n.

Vagus n. (CN X)
Taste and general sensation
via internal branch of superior laryngeal n.

Glossopharyngeal n. (CN IX)
Taste plus general sensation
via lingual branches

Facial n. (CN VII)
Taste via chorda tympani

Mandibular n. (CN V₃)
General sensation via
lingual n.

C.Machado
M.D.

Mouth

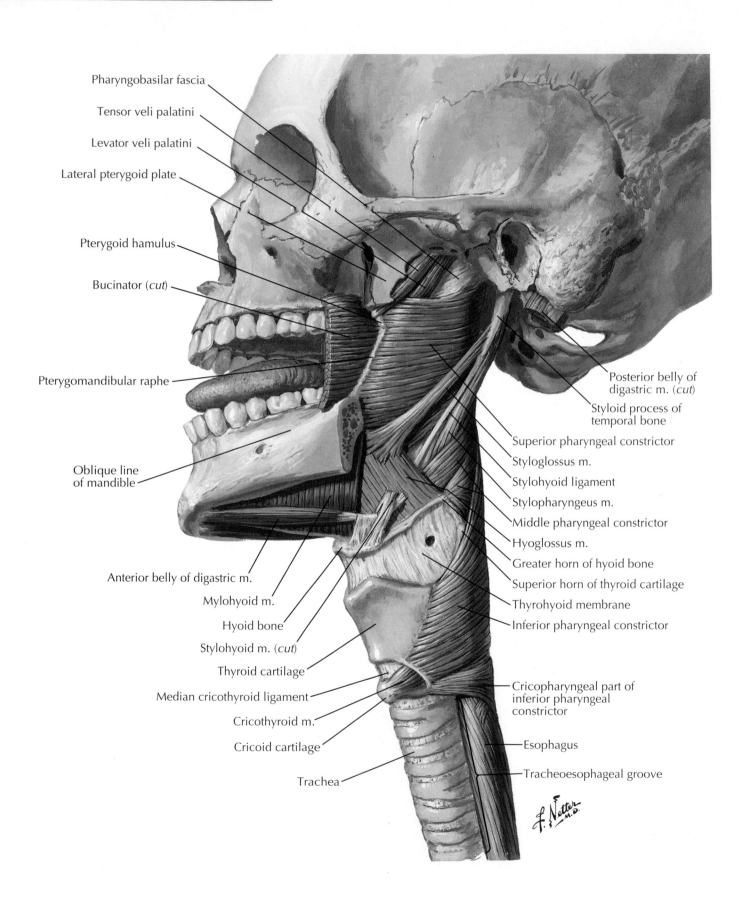

Pharyngobasilar fascia

Tensor veli palatini

Levator veli palatini

Lateral pterygoid plate

Pterygoid hamulus

Bucinator (*cut*)

Pterygomandibular raphe

Oblique line
of mandible

Anterior belly of digastric m.

Mylohyoid m.

Hyoid bone

Stylohyoid m. (*cut*)

Thyroid cartilage

Median cricothyroid ligament

Cricothyroid m.

Cricoid cartilage

Trachea

Posterior belly of
digastric m. (*cut*)

Styloid process of
temporal bone

Superior pharyngeal constrictor

Styloglossus m.

Stylohyoid ligament

Stylopharyngeus m.

Middle pharyngeal constrictor

Hyoglossus m.

Greater horn of hyoid bone

Superior horn of thyroid cartilage

Thyrohyoid membrane

Inferior pharyngeal constrictor

Cricopharyngeal part of
inferior pharyngeal
constrictor

Esophagus

Tracheoesophageal groove

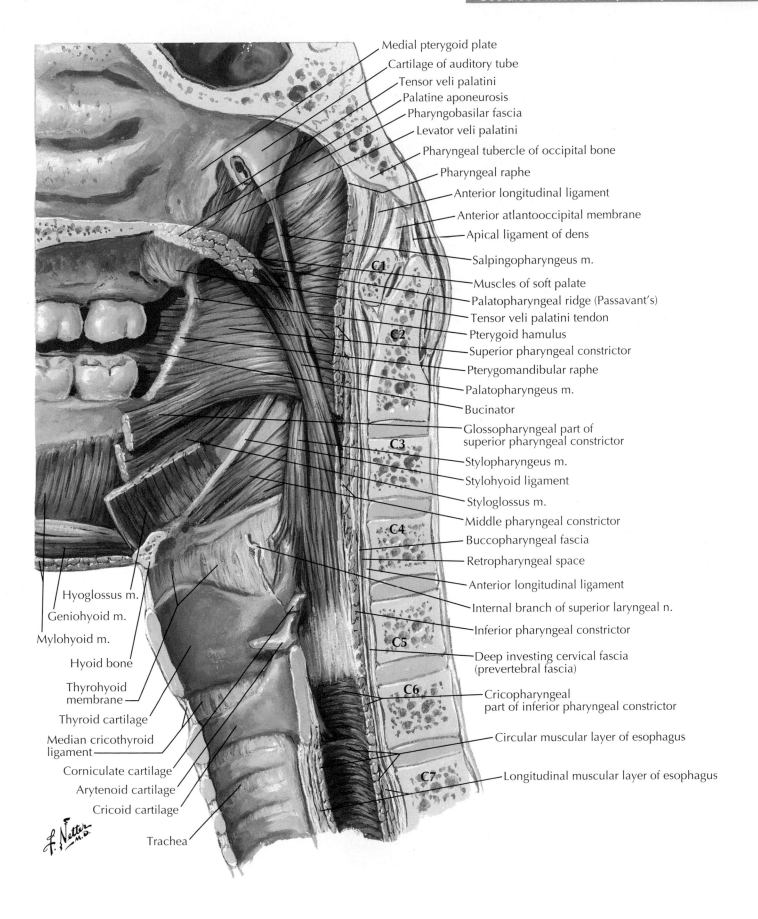

Medial pterygoid plate
Cartilage of auditory tube
Tensor veli palatini
Palatine aponeurosis
Pharyngobasilar fascia
Levator veli palatini
Pharyngeal tubercle of occipital bone
Pharyngeal raphe
Anterior longitudinal ligament
Anterior atlantooccipital membrane
Apical ligament of dens
Salpingopharyngeus m.
Muscles of soft palate
Palatopharyngeal ridge (Passavant's)
Tensor veli palatini tendon
Pterygoid hamulus
Superior pharyngeal constrictor
Pterygomandibular raphe
Palatopharyngeus m.
Bucinator
Glossopharyngeal part of superior pharyngeal constrictor
Stylopharyngeus m.
Stylohyoid ligament
Styloglossus m.
Middle pharyngeal constrictor
Buccopharyngeal fascia
Retropharyngeal space
Anterior longitudinal ligament
Internal branch of superior laryngeal n.
Inferior pharyngeal constrictor
Deep investing cervical fascia (prevertebral fascia)
Cricopharyngeal part of inferior pharyngeal constrictor
Circular muscular layer of esophagus
Longitudinal muscular layer of esophagus

C1
C2
C3
C4
C5
C6
C7

Hyoglossus m.
Geniohyoid m.
Mylohyoid m.
Hyoid bone
Thyrohyoid membrane
Thyroid cartilage
Median cricothyroid ligament
Corniculate cartilage
Arytenoid cartilage
Cricoid cartilage
Trachea

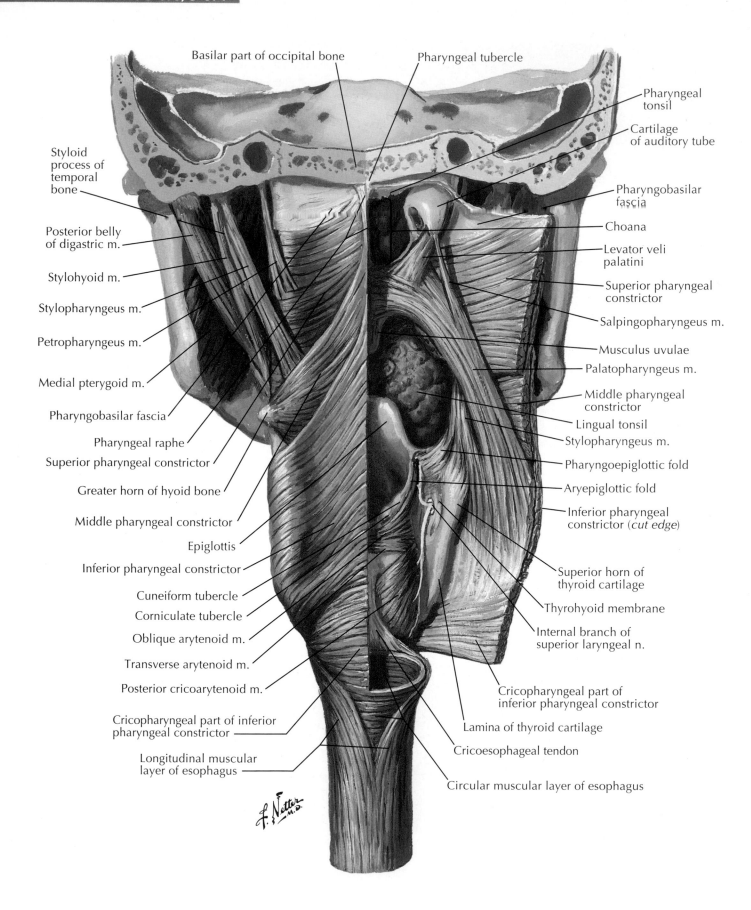

Basilar part of occipital bone

Pharyngeal tubercle

Pharyngeal tonsil

Cartilage of auditory tube

Styloid process of temporal bone

Pharyngobasilar fascia

Choana

Posterior belly of digastric m.

Levator veli palatini

Stylohyoid m.

Superior pharyngeal constrictor

Stylopharyngeus m.

Salpingopharyngeus m.

Petropharyngeus m.

Musculus uvulae

Medial pterygoid m.

Palatopharyngeus m.

Pharyngobasilar fascia

Middle pharyngeal constrictor

Lingual tonsil

Pharyngeal raphe

Stylopharyngeus m.

Superior pharyngeal constrictor

Pharyngoepiglottic fold

Greater horn of hyoid bone

Aryepiglottic fold

Middle pharyngeal constrictor

Inferior pharyngeal constrictor (*cut edge*)

Epiglottis

Inferior pharyngeal constrictor

Superior horn of thyroid cartilage

Cuneiform tubercle

Thyrohyoid membrane

Corniculate tubercle

Oblique arytenoid m.

Internal branch of superior laryngeal n.

Transverse arytenoid m.

Posterior cricoarytenoid m.

Cricopharyngeal part of inferior pharyngeal constrictor

Cricopharyngeal part of inferior pharyngeal constrictor

Lamina of thyroid cartilage

Longitudinal muscular layer of esophagus

Cricoesophageal tendon

Circular muscular layer of esophagus

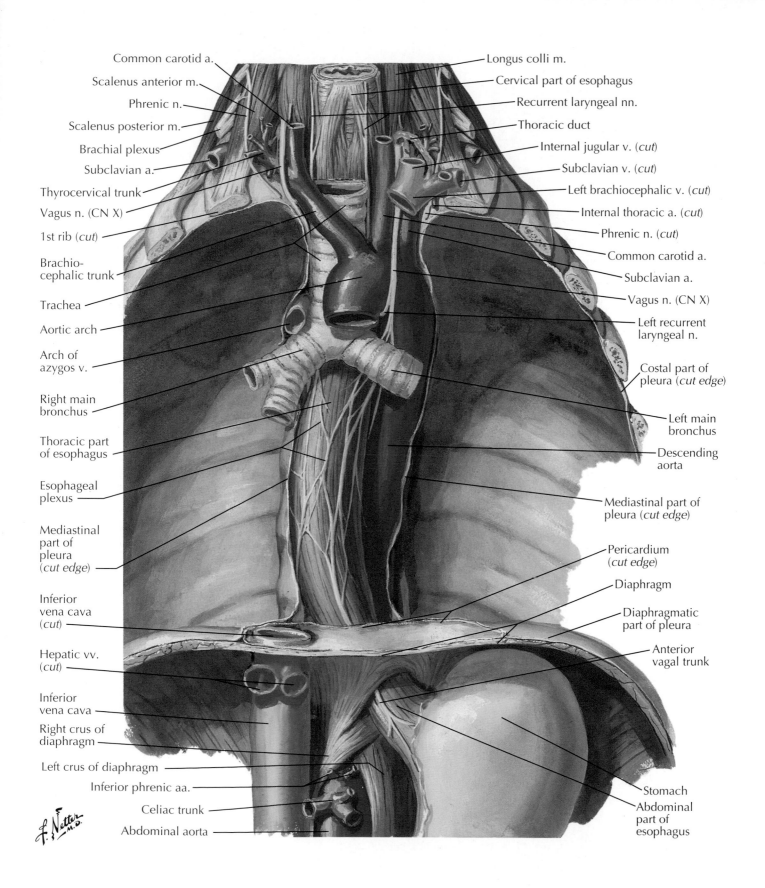

Common carotid a.

Scalenus anterior m.

Phrenic n.

Scalenus posterior m.

Brachial plexus

Subclavian a.

Thyrocervical trunk

Vagus n. (CN X)

1st rib (*cut*)

Brachio-cephalic trunk

Trachea

Aortic arch

Arch of azygos v.

Right main bronchus

Thoracic part of esophagus

Esophageal plexus

Mediastinal part of pleura (*cut edge*)

Inferior vena cava (*cut*)

Hepatic vv. (*cut*)

Inferior vena cava

Right crus of diaphragm

Left crus of diaphragm

Inferior phrenic aa.

Celiac trunk

Abdominal aorta

Longus colli m.

Cervical part of esophagus

Recurrent laryngeal nn.

Thoracic duct

Internal jugular v. (*cut*)

Subclavian v. (*cut*)

Left brachiocephalic v. (*cut*)

Internal thoracic a. (*cut*)

Phrenic n. (*cut*)

Common carotid a.

Subclavian a.

Vagus n. (CN X)

Left recurrent laryngeal n.

Costal part of pleura (*cut edge*)

Left main bronchus

Descending aorta

Mediastinal part of pleura (*cut edge*)

Pericardium (*cut edge*)

Diaphragm

Diaphragmatic part of pleura

Anterior vagal trunk

Stomach

Abdominal part of esophagus

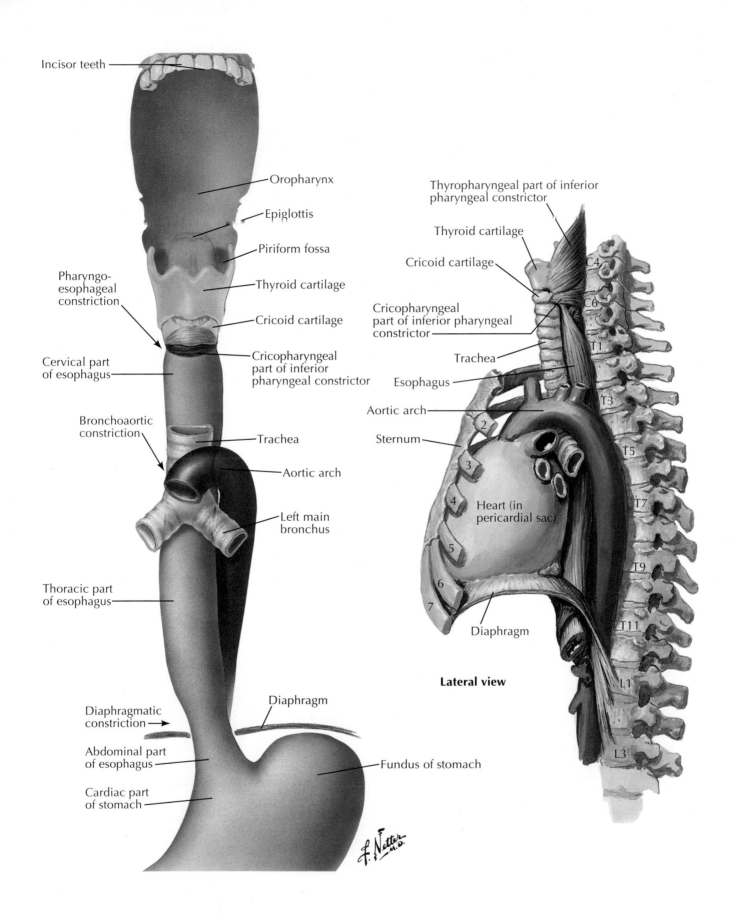

Incisor teeth

Oropharynx

Epiglottis

Piriform fossa

Pharyngo-
esophageal
constriction

Thyroid cartilage

Cricoid cartilage

Cervical part
of esophagus

Cricopharyngeal
part of inferior
pharyngeal constrictor

Bronchoaortic
constriction

Trachea

Aortic arch

Left main
bronchus

Thoracic part
of esophagus

Diaphragmatic
constriction

Diaphragm

Abdominal part
of esophagus

Fundus of stomach

Cardiac part
of stomach

Thyropharyngeal part of inferior
pharyngeal constrictor

Thyroid cartilage

Cricoid cartilage

Cricopharyngeal
part of inferior pharyngeal
constrictor

Trachea

Esophagus

Aortic arch

Sternum

Heart (in
pericardial sac)

Diaphragm

Lateral view

C4

C6

T1

T3

T5

T7

T9

T11

L1

L3

2

3

4

5

6

7

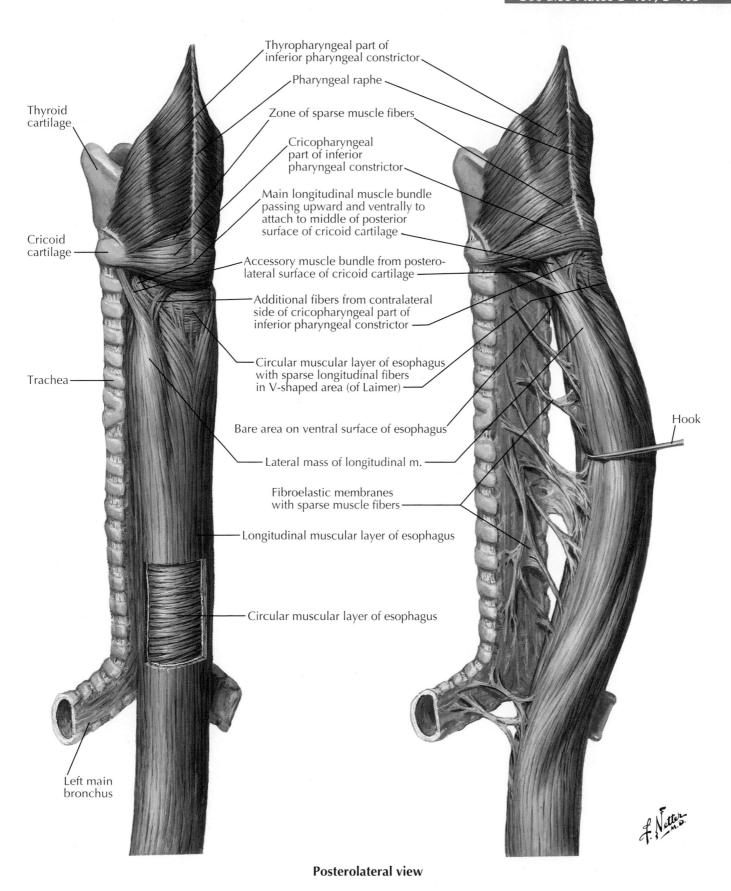

Thyropharyngeal part of inferior pharyngeal constrictor

Pharyngeal raphe

Zone of sparse muscle fibers

Thyroid cartilage

Cricopharyngeal part of inferior pharyngeal constrictor

Main longitudinal muscle bundle passing upward and ventrally to attach to middle of posterior surface of cricoid cartilage

Cricoid cartilage

Accessory muscle bundle from postero-lateral surface of cricoid cartilage

Additional fibers from contralateral side of cricopharyngeal part of inferior pharyngeal constrictor

Trachea

Circular muscular layer of esophagus with sparse longitudinal fibers in V-shaped area (of Laimer)

Bare area on ventral surface of esophagus

Lateral mass of longitudinal m.

Hook

Fibroelastic membranes with sparse muscle fibers

Longitudinal muscular layer of esophagus

Circular muscular layer of esophagus

Left main bronchus

Posterolateral view

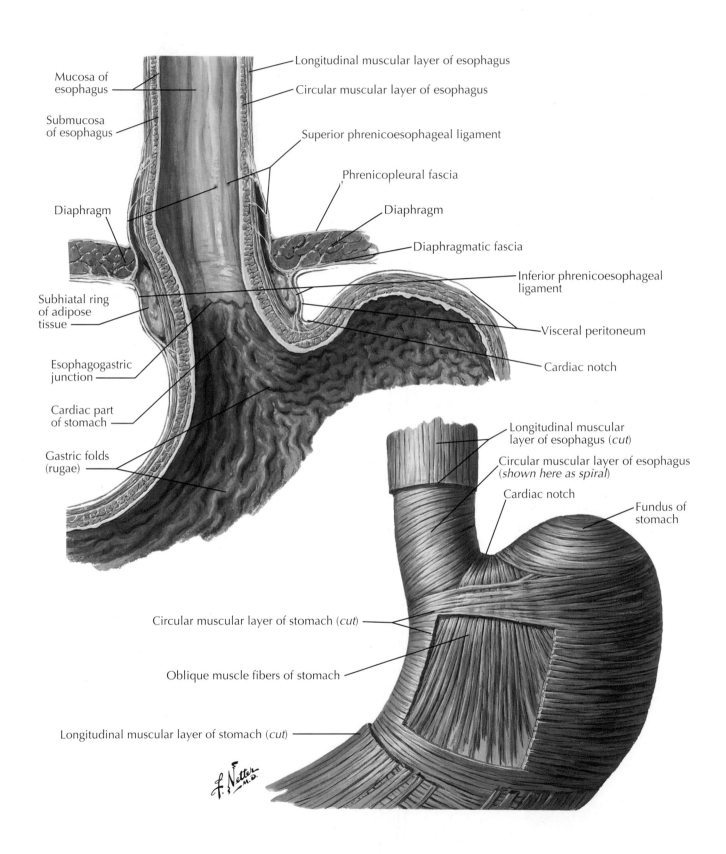

Mucosa of esophagus

Submucosa of esophagus

Diaphragm

Subhiatal ring of adipose tissue

Esophagogastric junction

Cardiac part of stomach

Gastric folds (rugae)

Longitudinal muscular layer of esophagus

Circular muscular layer of esophagus

Superior phrenicoesophageal ligament

Phrenicopleural fascia

Diaphragm

Diaphragmatic fascia

Inferior phrenicoesophageal ligament

Visceral peritoneum

Cardiac notch

Longitudinal muscular layer of esophagus (*cut*)

Circular muscular layer of esophagus (*shown here as spiral*)

Cardiac notch

Fundus of stomach

Circular muscular layer of stomach (*cut*)

Oblique muscle fibers of stomach

Longitudinal muscular layer of stomach (*cut*)

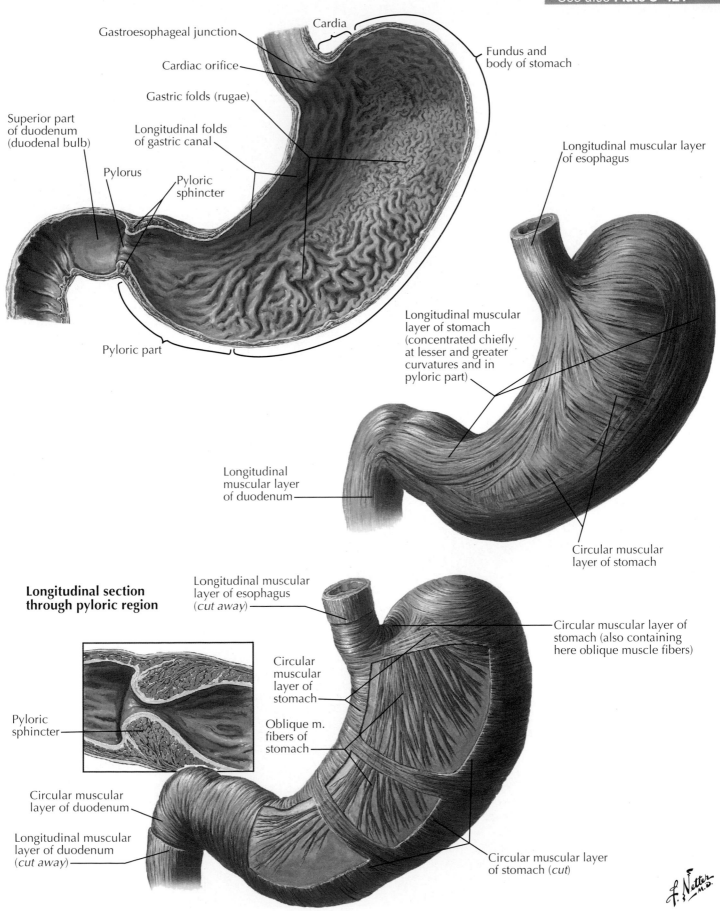

Gastroesophageal junction

Cardia

Cardiac orifice

Fundus and body of stomach

Gastric folds (rugae)

Superior part of duodenum (duodenal bulb)

Longitudinal folds of gastric canal

Longitudinal muscular layer of esophagus

Pylorus

Pyloric sphincter

Longitudinal muscular layer of stomach (concentrated chiefly at lesser and greater curvatures and in pyloric part)

Pyloric part

Longitudinal muscular layer of duodenum

Circular muscular layer of stomach

Longitudinal section through pyloric region

Longitudinal muscular layer of esophagus (*cut away*)

Circular muscular layer of stomach (also containing here oblique muscle fibers)

Circular muscular layer of stomach

Pyloric sphincter

Oblique m. fibers of stomach

Circular muscular layer of duodenum

Longitudinal muscular layer of duodenum (*cut away*)

Circular muscular layer of stomach (*cut*)

Viscera (Esophagus, Stomach, Intestines, Liver, Pancreas)

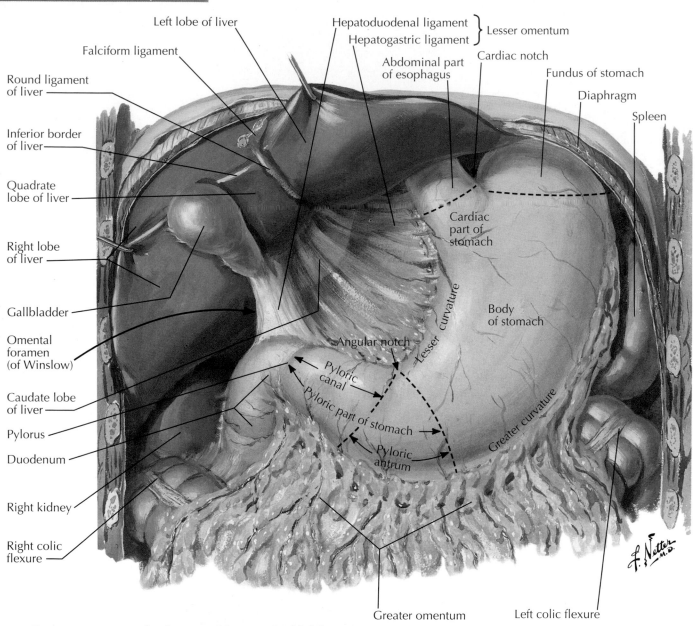

Left lobe of liver

Falciform ligament

Round ligament of liver

Inferior border of liver

Quadrate lobe of liver

Right lobe of liver

Gallbladder

Omental foramen (of Winslow)

Caudate lobe of liver

Pylorus

Duodenum

Right kidney

Right colic flexure

Hepatoduodenal ligament ⎫
Hepatogastric ligament ⎭ Lesser omentum

Abdominal part of esophagus

Cardiac notch

Fundus of stomach

Diaphragm

Spleen

Cardiac part of stomach

Body of stomach

Lesser curvature

Angular notch

Pyloric canal

Pyloric part of stomach

Pyloric antrum

Greater curvature

Greater omentum

Left colic flexure

Transverse gray-scale ultrasound image of midabdomen

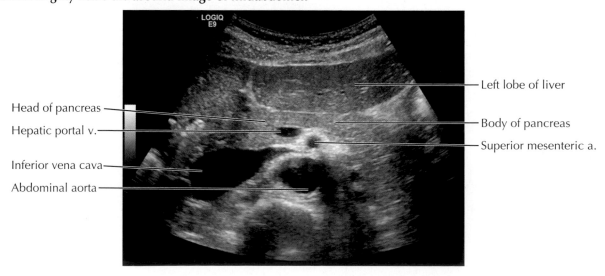

Head of pancreas

Hepatic portal v.

Inferior vena cava

Abdominal aorta

Left lobe of liver

Body of pancreas

Superior mesenteric a.

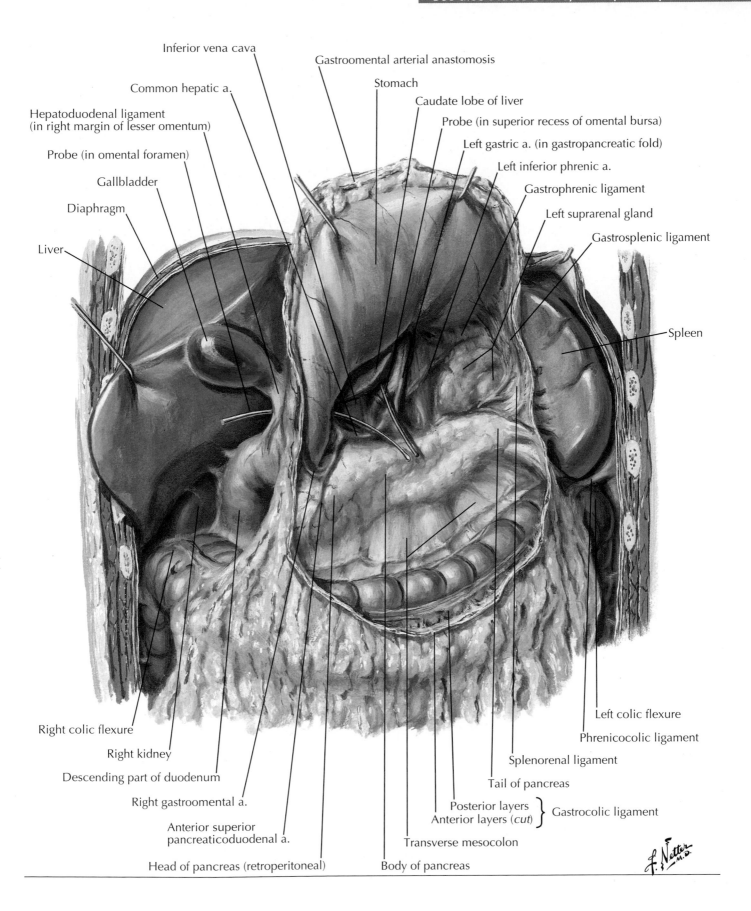

Inferior vena cava

Gastroomental arterial anastomosis

Common hepatic a.

Stomach

Caudate lobe of liver

Hepatoduodenal ligament
(in right margin of lesser omentum)

Probe (in superior recess of omental bursa)

Left gastric a. (in gastropancreatic fold)

Probe (in omental foramen)

Left inferior phrenic a.

Gallbladder

Gastrophrenic ligament

Diaphragm

Left suprarenal gland

Gastrosplenic ligament

Liver

Spleen

Right colic flexure

Left colic flexure

Right kidney

Phrenicocolic ligament

Descending part of duodenum

Splenorenal ligament

Right gastroomental a.

Tail of pancreas

Anterior superior
pancreaticoduodenal a.

Posterior layers
Anterior layers (*cut*) } Gastrocolic ligament

Head of pancreas (retroperitoneal)

Transverse mesocolon

Body of pancreas

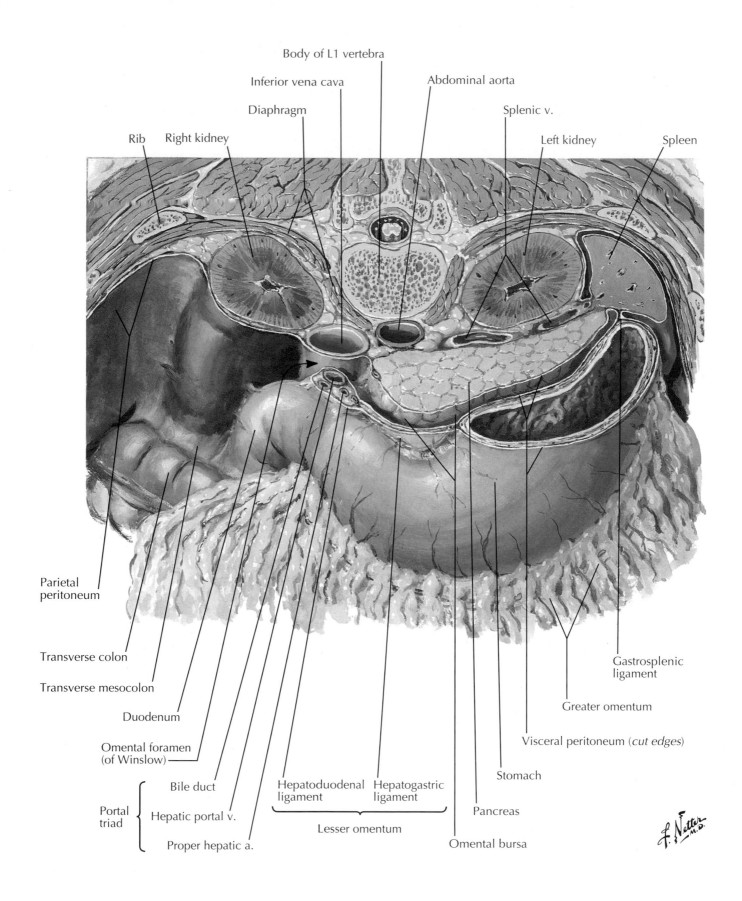

Body of L1 vertebra

Inferior vena cava

Abdominal aorta

Diaphragm

Splenic v.

Rib   Right kidney

Left kidney

Spleen

Parietal
peritoneum

Transverse colon

Transverse mesocolon

Duodenum

Omental foramen
(of Winslow)

Portal
triad

Bile duct

Hepatic portal v.

Proper hepatic a.

Hepatoduodenal
ligament

Hepatogastric
ligament

Lesser omentum

Stomach

Pancreas

Omental bursa

Gastrosplenic
ligament

Greater omentum

Visceral peritoneum (*cut edges*)

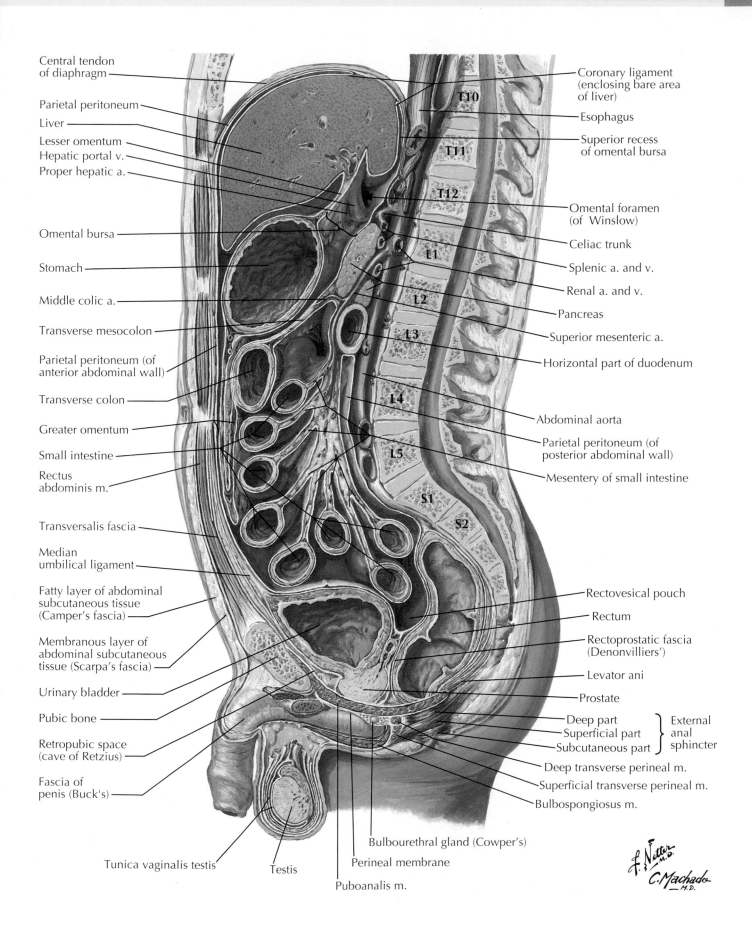

Central tendon of diaphragm

Parietal peritoneum

Liver

Lesser omentum

Hepatic portal v.

Proper hepatic a.

Omental bursa

Stomach

Middle colic a.

Transverse mesocolon

Parietal peritoneum (of anterior abdominal wall)

Transverse colon

Greater omentum

Small intestine

Rectus abdominis m.

Transversalis fascia

Median umbilical ligament

Fatty layer of abdominal subcutaneous tissue (Camper's fascia)

Membranous layer of abdominal subcutaneous tissue (Scarpa's fascia)

Urinary bladder

Pubic bone

Retropubic space (cave of Retzius)

Fascia of penis (Buck's)

Tunica vaginalis testis

Testis

Puboanalis m.

Perineal membrane

Bulbourethral gland (Cowper's)

Coronary ligament (enclosing bare area of liver)

Esophagus

Superior recess of omental bursa

Omental foramen (of Winslow)

Celiac trunk

Splenic a. and v.

Renal a. and v.

Pancreas

Superior mesenteric a.

Horizontal part of duodenum

Abdominal aorta

Parietal peritoneum (of posterior abdominal wall)

Mesentery of small intestine

Rectovesical pouch

Rectum

Rectoprostatic fascia (Denonvilliers')

Levator ani

Prostate

Deep part

Superficial part } External anal sphincter

Subcutaneous part

Deep transverse perineal m.

Superficial transverse perineal m.

Bulbospongiosus m.

T10

T11

T12

L1

L2

L3

L4

L5

S1

S2

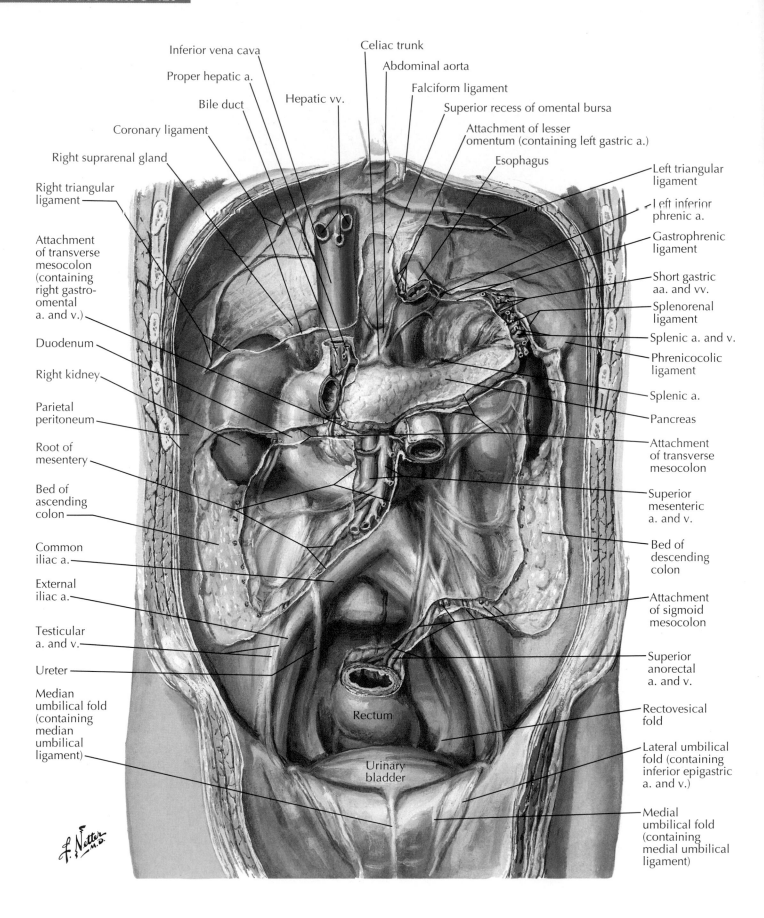

Inferior vena cava

Proper hepatic a.

Bile duct

Coronary ligament

Right suprarenal gland

Right triangular ligament

Attachment of transverse mesocolon (containing right gastro-omental a. and v.)

Duodenum

Right kidney

Parietal peritoneum

Root of mesentery

Bed of ascending colon

Common iliac a.

External iliac a.

Testicular a. and v.

Ureter

Median umbilical fold (containing median umbilical ligament)

Hepatic vv.

Celiac trunk

Abdominal aorta

Falciform ligament

Superior recess of omental bursa

Attachment of lesser omentum (containing left gastric a.)

Esophagus

Left triangular ligament

Left inferior phrenic a.

Gastrophrenic ligament

Short gastric aa. and vv.

Splenorenal ligament

Splenic a. and v.

Phrenicocolic ligament

Splenic a.

Pancreas

Attachment of transverse mesocolon

Superior mesenteric a. and v.

Bed of descending colon

Attachment of sigmoid mesocolon

Superior anorectal a. and v.

Rectovesical fold

Lateral umbilical fold (containing inferior epigastric a. and v.)

Medial umbilical fold (containing medial umbilical ligament)

Rectum

Urinary bladder

See also **Plates S–415, S–422**

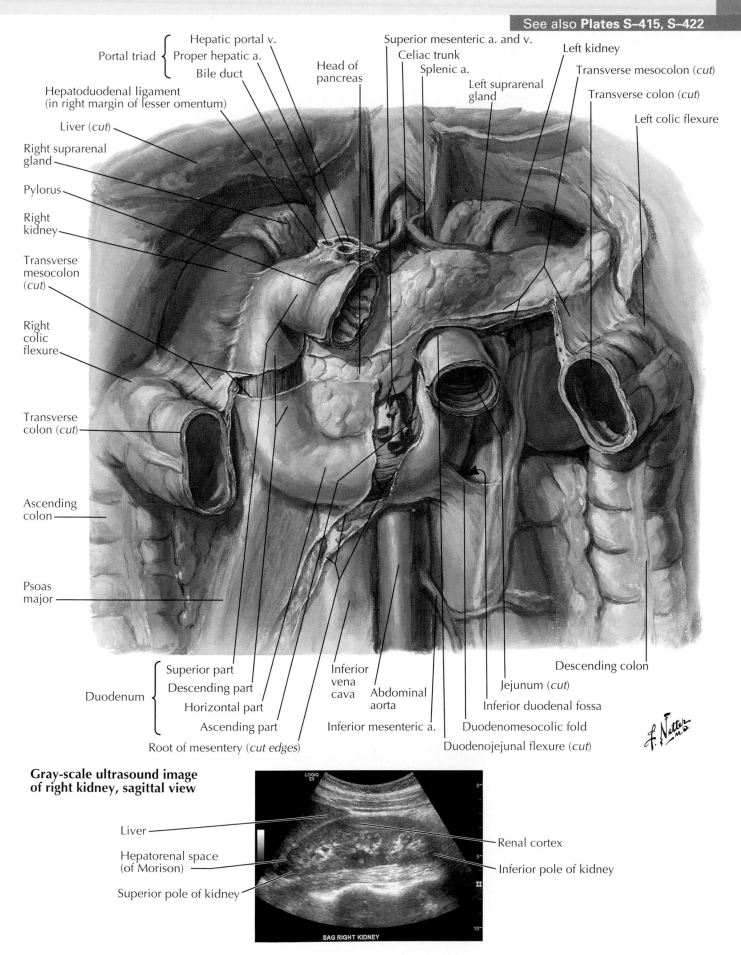

Portal triad {
Hepatic portal v.
Proper hepatic a.
Bile duct

Hepatoduodenal ligament
(in right margin of lesser omentum)

Liver (cut)

Right suprarenal gland

Pylorus

Right kidney

Transverse mesocolon (cut)

Right colic flexure

Transverse colon (cut)

Ascending colon

Psoas major

Head of pancreas

Superior mesenteric a. and v.

Celiac trunk

Splenic a.

Left suprarenal gland

Left kidney

Transverse mesocolon (cut)

Transverse colon (cut)

Left colic flexure

Descending colon

Jejunum (cut)

Inferior duodenal fossa

Duodenomesocolic fold

Duodenojejunal flexure (cut)

Duodenum {
Superior part
Descending part
Horizontal part
Ascending part

Root of mesentery (cut edges)

Inferior vena cava

Abdominal aorta

Inferior mesenteric a.

**Gray-scale ultrasound image
of right kidney, sagittal view**

Liver

Hepatorenal space
(of Morison)

Superior pole of kidney

Renal cortex

Inferior pole of kidney

LOGIQ E9

SAG RIGHT KIDNEY

Viscera (Esophagus, Stomach, Intestines, Liver, Pancreas)

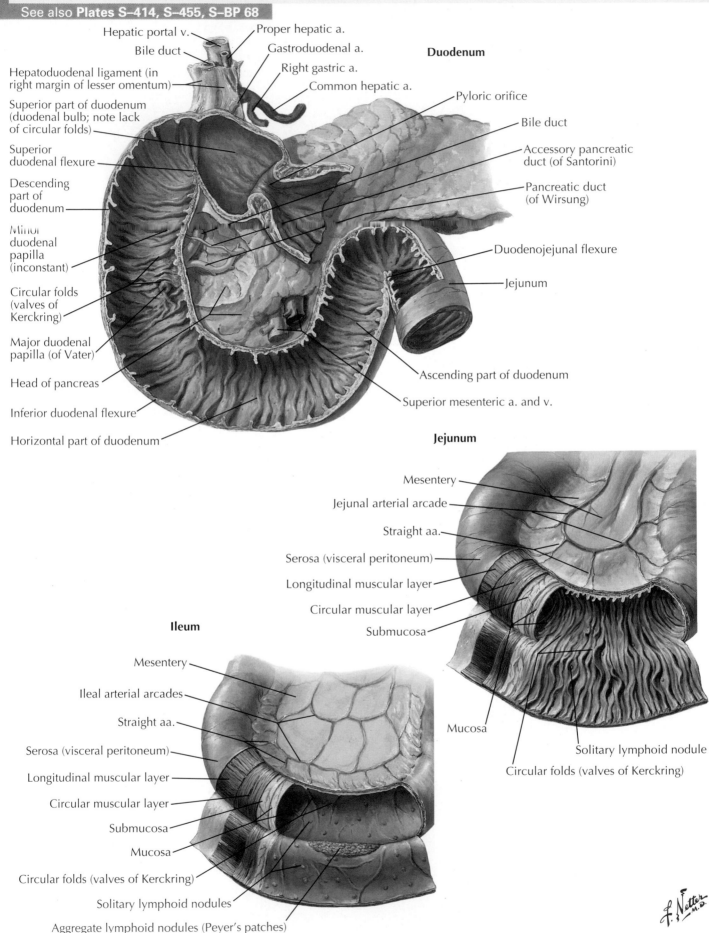

Duodenum

Hepatic portal v.
Bile duct
Hepatoduodenal ligament (in right margin of lesser omentum)
Superior part of duodenum (duodenal bulb; note lack of circular folds)
Superior duodenal flexure
Descending part of duodenum
Minor duodenal papilla (inconstant)
Circular folds (valves of Kerckring)
Major duodenal papilla (of Vater)
Head of pancreas
Inferior duodenal flexure
Horizontal part of duodenum

Proper hepatic a.
Gastroduodenal a.
Right gastric a.
Common hepatic a.

Pyloric orifice
Bile duct
Accessory pancreatic duct (of Santorini)
Pancreatic duct (of Wirsung)
Duodenojejunal flexure
Jejunum
Ascending part of duodenum
Superior mesenteric a. and v.

Jejunum

Mesentery
Jejunal arterial arcade
Straight aa.
Serosa (visceral peritoneum)
Longitudinal muscular layer
Circular muscular layer
Submucosa
Mucosa
Solitary lymphoid nodule
Circular folds (valves of Kerckring)

Ileum

Mesentery
Ileal arterial arcades
Straight aa.
Serosa (visceral peritoneum)
Longitudinal muscular layer
Circular muscular layer
Submucosa
Mucosa
Circular folds (valves of Kerckring)
Solitary lymphoid nodules
Aggregate lymphoid nodules (Peyer's patches)

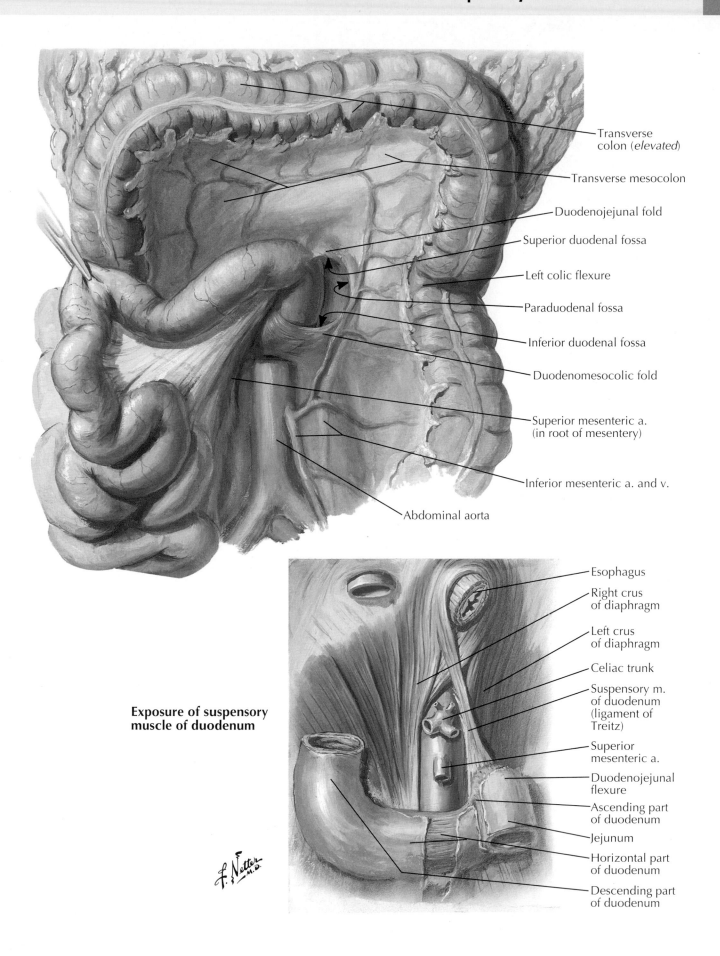

Transverse colon (*elevated*)

Transverse mesocolon

Duodenojejunal fold

Superior duodenal fossa

Left colic flexure

Paraduodenal fossa

Inferior duodenal fossa

Duodenomesocolic fold

Superior mesenteric a. (in root of mesentery)

Inferior mesenteric a. and v.

Abdominal aorta

Exposure of suspensory muscle of duodenum

Esophagus

Right crus of diaphragm

Left crus of diaphragm

Celiac trunk

Suspensory m. of duodenum (ligament of Treitz)

Superior mesenteric a.

Duodenojejunal flexure

Ascending part of duodenum

Jejunum

Horizontal part of duodenum

Descending part of duodenum

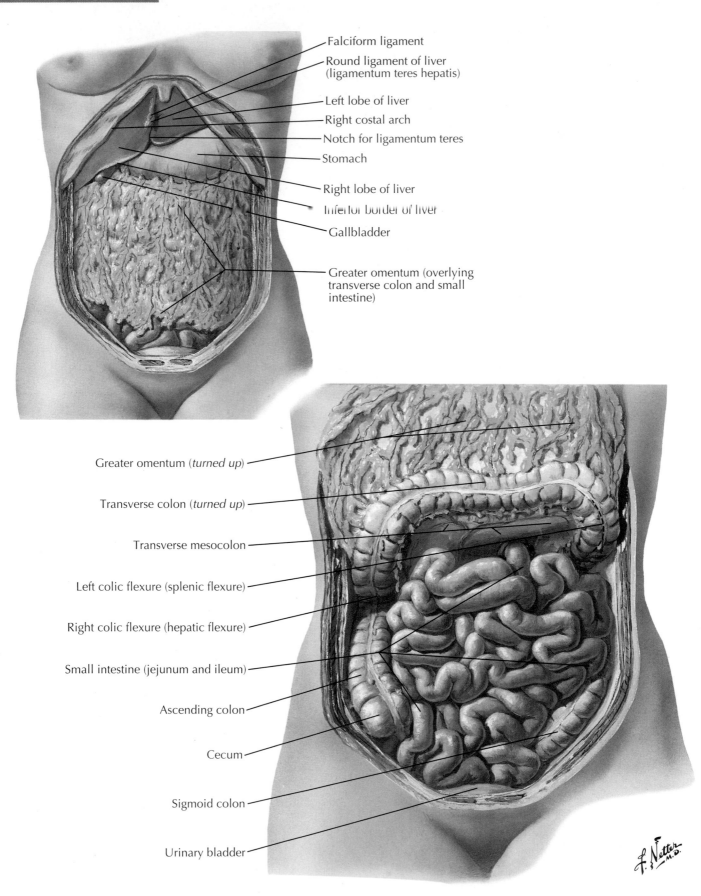

Falciform ligament

Round ligament of liver
(ligamentum teres hepatis)

Left lobe of liver

Right costal arch

Notch for ligamentum teres

Stomach

Right lobe of liver

Inferior border of liver

Gallbladder

Greater omentum (overlying
transverse colon and small
intestine)

Greater omentum (*turned up*)

Transverse colon (*turned up*)

Transverse mesocolon

Left colic flexure (splenic flexure)

Right colic flexure (hepatic flexure)

Small intestine (jejunum and ileum)

Ascending colon

Cecum

Sigmoid colon

Urinary bladder

Viscera (Esophagus, Stomach, Intestines, Liver, Pancreas)

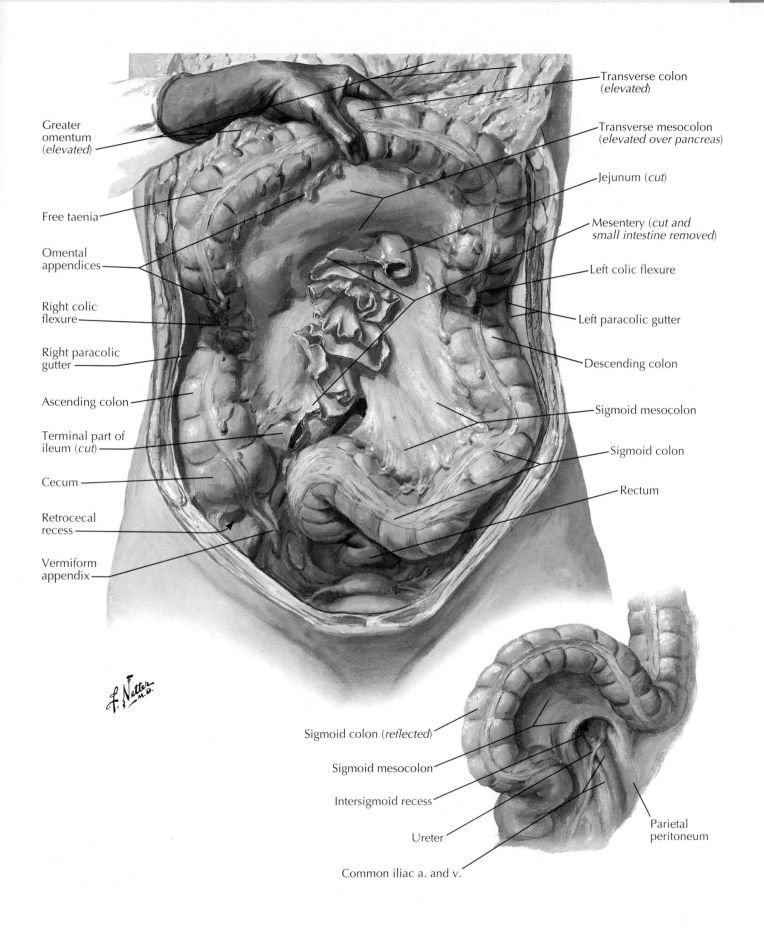

Transverse colon (*elevated*)

Greater omentum (*elevated*)

Transverse mesocolon (*elevated over pancreas*)

Jejunum (*cut*)

Free taenia

Mesentery (*cut and small intestine removed*)

Omental appendices

Left colic flexure

Right colic flexure

Left paracolic gutter

Right paracolic gutter

Descending colon

Ascending colon

Sigmoid mesocolon

Terminal part of ileum (*cut*)

Sigmoid colon

Cecum

Rectum

Retrocecal recess

Vermiform appendix

Sigmoid colon (*reflected*)

Sigmoid mesocolon

Intersigmoid recess

Ureter

Parietal peritoneum

Common iliac a. and v.

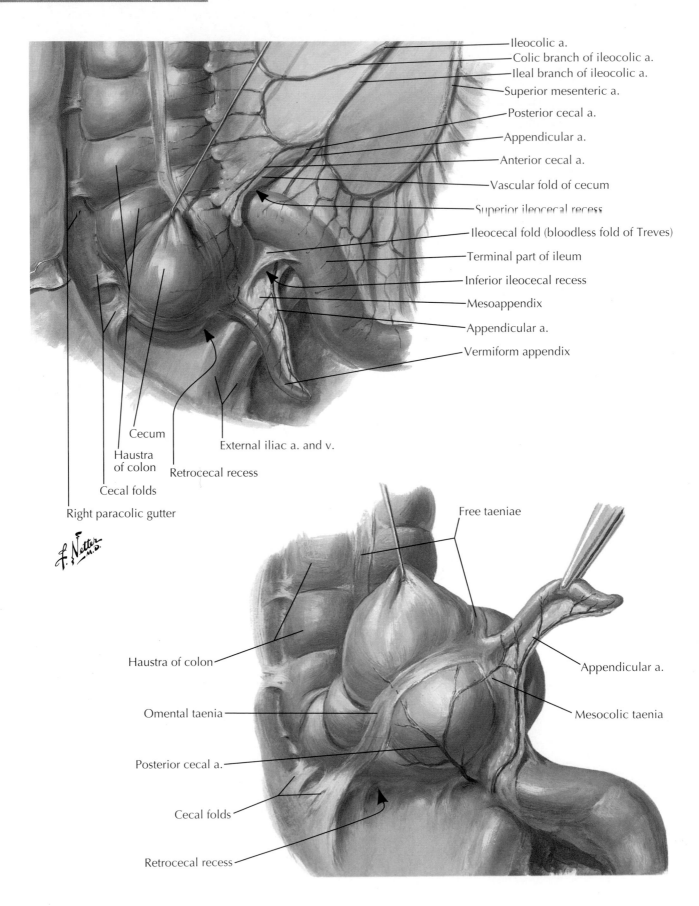

Ileocolic a.

Colic branch of ileocolic a.

Ileal branch of ileocolic a.

Superior mesenteric a.

Posterior cecal a.

Appendicular a.

Anterior cecal a.

Vascular fold of cecum

Superior ileocecal recess

Ileocecal fold (bloodless fold of Treves)

Terminal part of ileum

Inferior ileocecal recess

Mesoappendix

Appendicular a.

Vermiform appendix

Cecum

Haustra of colon

Cecal folds

External iliac a. and v.

Retrocecal recess

Right paracolic gutter

Free taeniae

Haustra of colon

Appendicular a.

Omental taenia

Mesocolic taenia

Posterior cecal a.

Cecal folds

Retrocecal recess

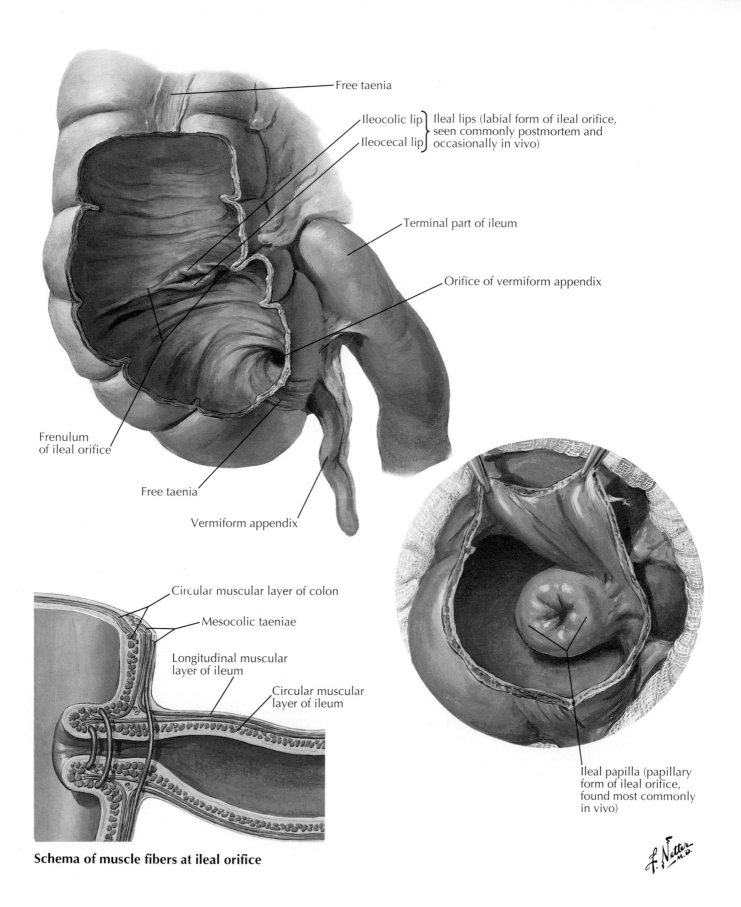

Free taenia

Ileocolic lip ⎫
Ileocecal lip ⎭ Ileal lips (labial form of ileal orifice, seen commonly postmortem and occasionally in vivo)

Terminal part of ileum

Orifice of vermiform appendix

Frenulum of ileal orifice

Free taenia

Vermiform appendix

Circular muscular layer of colon

Mesocolic taeniae

Longitudinal muscular layer of ileum

Circular muscular layer of ileum

Ileal papilla (papillary form of ileal orifice, found most commonly in vivo)

Schema of muscle fibers at ileal orifice

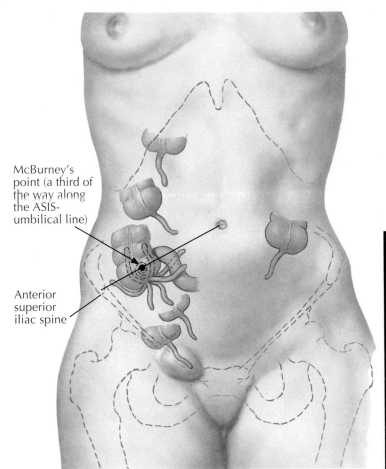

McBurney's point (a third of the way along the ASIS-umbilical line)

Anterior superior iliac spine

Variations in position of vermiform appendix

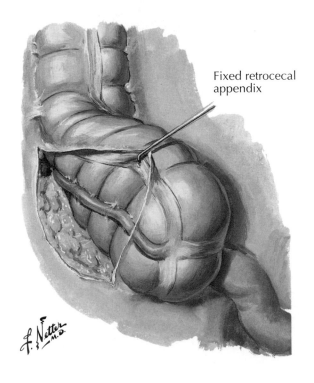

Fixed retrocecal appendix

Coronal CT image with oral and intravenous contrast

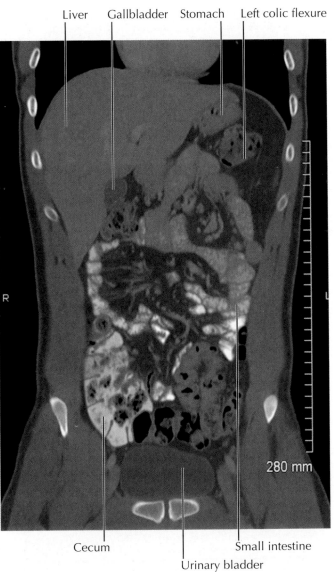

Liver Gallbladder Stomach Left colic flexure

280 mm

Cecum

Urinary bladder

Small intestine

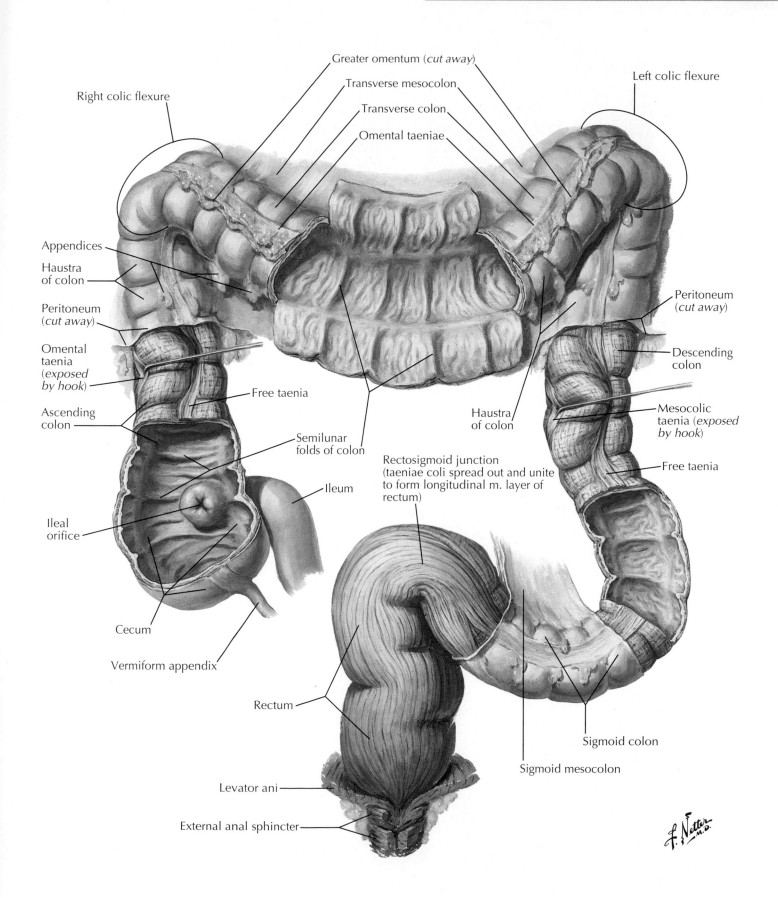

Greater omentum (*cut away*)

Transverse mesocolon

Transverse colon

Omental taeniae

Left colic flexure

Right colic flexure

Appendices

Haustra of colon

Peritoneum (*cut away*)

Omental taenia (*exposed by hook*)

Ascending colon

Free taenia

Ileal orifice

Cecum

Vermiform appendix

Rectum

Levator ani

External anal sphincter

Semilunar folds of colon

Ileum

Haustra of colon

Peritoneum (*cut away*)

Descending colon

Mesocolic taenia (*exposed by hook*)

Free taenia

Rectosigmoid junction (taeniae coli spread out and unite to form longitudinal m. layer of rectum)

Sigmoid colon

Sigmoid mesocolon

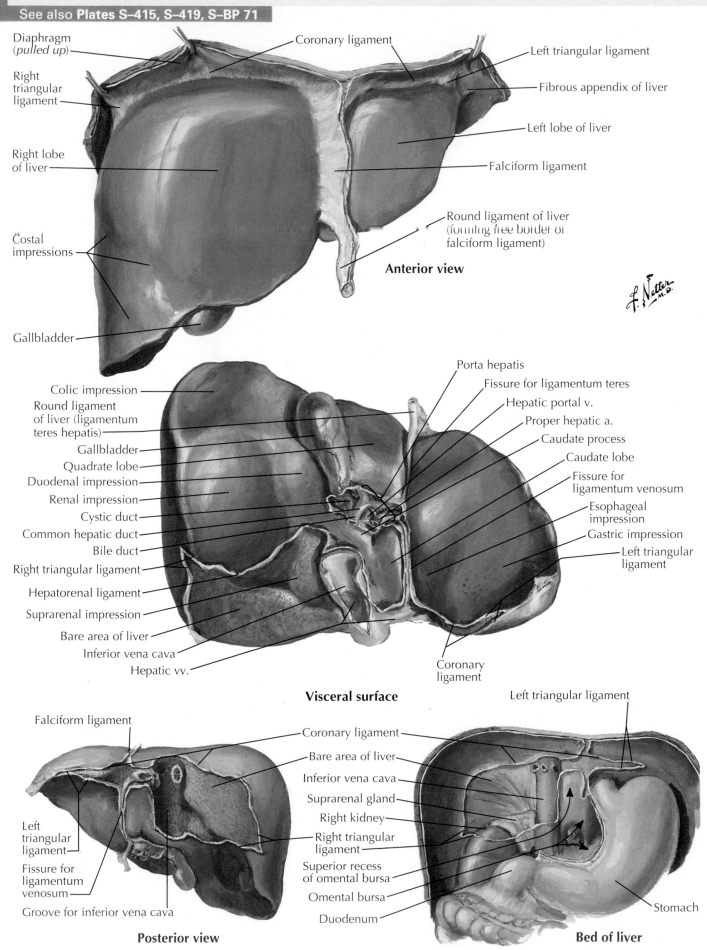

Diaphragm (*pulled up*)

Coronary ligament

Left triangular ligament

Fibrous appendix of liver

Right triangular ligament

Left lobe of liver

Right lobe of liver

Falciform ligament

Round ligament of liver (forming free border of falciform ligament)

Costal impressions

Gallbladder

Anterior view

Colic impression

Porta hepatis

Fissure for ligamentum teres

Round ligament of liver (ligamentum teres hepatis)

Hepatic portal v.

Proper hepatic a.

Gallbladder

Caudate process

Quadrate lobe

Caudate lobe

Duodenal impression

Fissure for ligamentum venosum

Renal impression

Esophageal impression

Cystic duct

Gastric impression

Common hepatic duct

Left triangular ligament

Bile duct

Right triangular ligament

Hepatorenal ligament

Suprarenal impression

Bare area of liver

Inferior vena cava

Hepatic vv.

Coronary ligament

Visceral surface

Falciform ligament

Left triangular ligament

Coronary ligament

Bare area of liver

Inferior vena cava

Suprarenal gland

Right kidney

Left triangular ligament

Right triangular ligament

Fissure for ligamentum venosum

Superior recess of omental bursa

Omental bursa

Groove for inferior vena cava

Duodenum

Stomach

Posterior view

Bed of liver

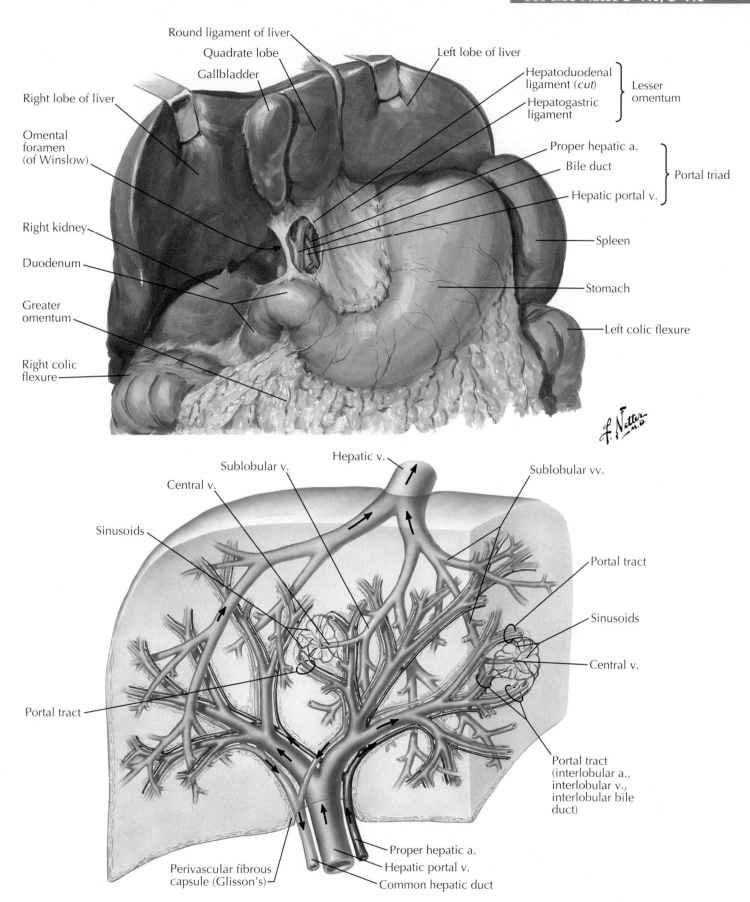

Round ligament of liver

Quadrate lobe

Gallbladder

Right lobe of liver

Omental foramen (of Winslow)

Right kidney

Duodenum

Greater omentum

Right colic flexure

Left lobe of liver

Hepatoduodenal ligament (*cut*)

Hepatogastric ligament

Lesser omentum

Proper hepatic a.

Bile duct

Hepatic portal v.

Portal triad

Spleen

Stomach

Left colic flexure

Hepatic v.

Sublobular v.

Central v.

Sublobular vv.

Sinusoids

Portal tract

Sinusoids

Central v.

Portal tract

Portal tract (interlobular a., interlobular v., interlobular bile duct)

Perivascular fibrous capsule (Glisson's)

Proper hepatic a.

Hepatic portal v.

Common hepatic duct

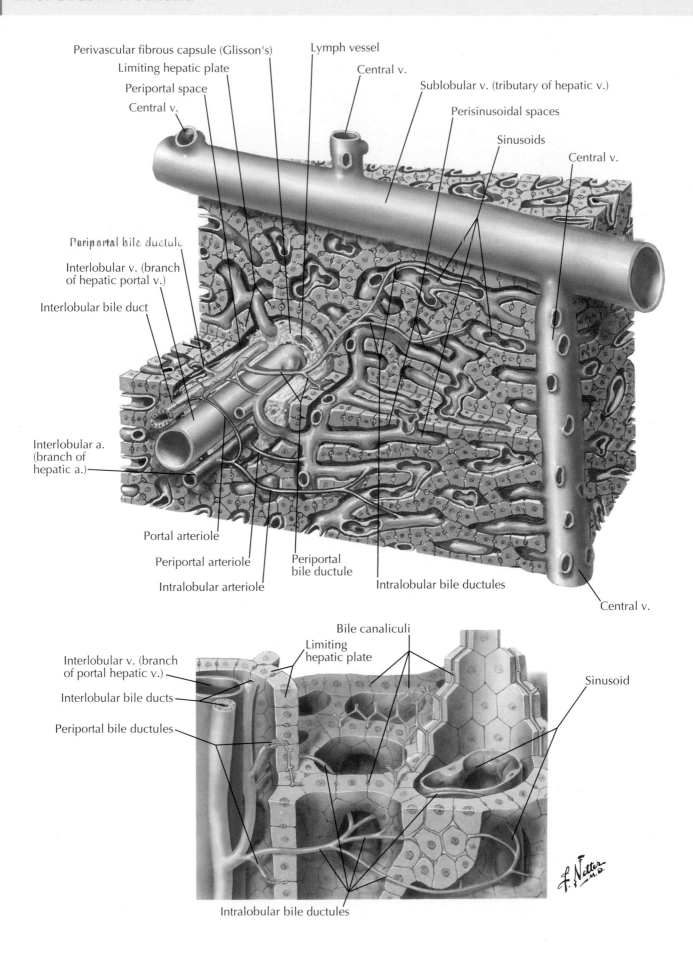

Perivascular fibrous capsule (Glisson's)

Limiting hepatic plate

Periportal space

Central v.

Lymph vessel

Central v.

Sublobular v. (tributary of hepatic v.)

Perisinusoidal spaces

Sinusoids

Central v.

Periportal bile ductule

Interlobular v. (branch of hepatic portal v.)

Interlobular bile duct

Interlobular a. (branch of hepatic a.)

Portal arteriole

Periportal arteriole

Intralobular arteriole

Periportal bile ductule

Intralobular bile ductules

Central v.

Bile canaliculi

Limiting hepatic plate

Interlobular v. (branch of portal hepatic v.)

Interlobular bile ducts

Periportal bile ductules

Sinusoid

Intralobular bile ductules

See also Plates S–415, S–421, S–BP 73, S–BP 74

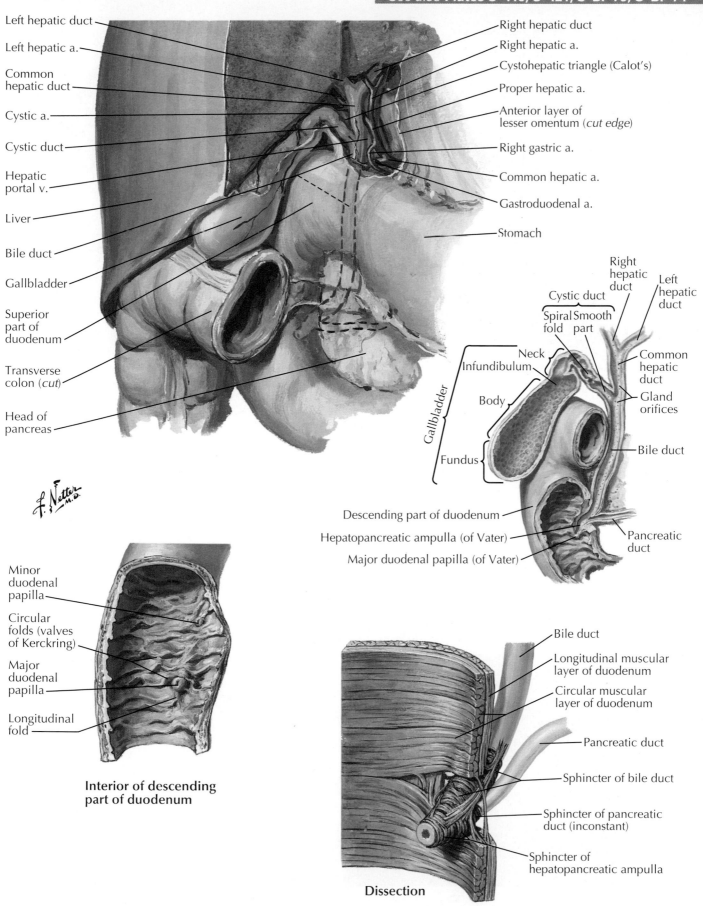

Left hepatic duct

Left hepatic a.

Common hepatic duct

Cystic a.

Cystic duct

Hepatic portal v.

Liver

Bile duct

Gallbladder

Superior part of duodenum

Transverse colon (cut)

Head of pancreas

Right hepatic duct

Right hepatic a.

Cystohepatic triangle (Calot's)

Proper hepatic a.

Anterior layer of lesser omentum (cut edge)

Right gastric a.

Common hepatic a.

Gastroduodenal a.

Stomach

Cystic duct

Spiral fold   Smooth part

Neck

Infundibulum

Body

Gallbladder

Fundus

Right hepatic duct

Left hepatic duct

Common hepatic duct

Gland orifices

Bile duct

Descending part of duodenum

Hepatopancreatic ampulla (of Vater)

Major duodenal papilla (of Vater)

Pancreatic duct

Minor duodenal papilla

Circular folds (valves of Kerckring)

Major duodenal papilla

Longitudinal fold

Interior of descending part of duodenum

Bile duct

Longitudinal muscular layer of duodenum

Circular muscular layer of duodenum

Pancreatic duct

Sphincter of bile duct

Sphincter of pancreatic duct (inconstant)

Sphincter of hepatopancreatic ampulla

Dissection

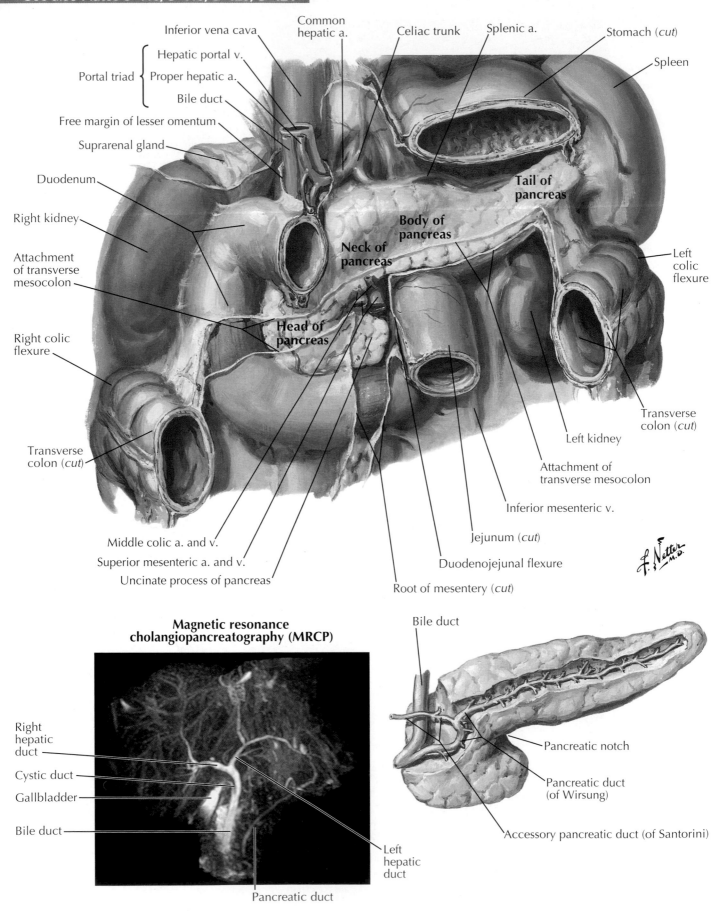

Inferior vena cava

Common hepatic a.

Celiac trunk

Splenic a.

Stomach (*cut*)

Spleen

Portal triad { Hepatic portal v.

Proper hepatic a.

Bile duct

Free margin of lesser omentum

Suprarenal gland

Duodenum

Right kidney

Attachment of transverse mesocolon

Right colic flexure

Transverse colon (*cut*)

Middle colic a. and v.

Superior mesenteric a. and v.

Uncinate process of pancreas

Tail of pancreas

Body of pancreas

Neck of pancreas

Head of pancreas

Left colic flexure

Transverse colon (*cut*)

Left kidney

Attachment of transverse mesocolon

Inferior mesenteric v.

Jejunum (*cut*)

Duodenojejunal flexure

Root of mesentery (*cut*)

Magnetic resonance cholangiopancreatography (MRCP)

Right hepatic duct

Cystic duct

Gallbladder

Bile duct

Left hepatic duct

Pancreatic duct

Bile duct

Pancreatic notch

Pancreatic duct (of Wirsung)

Accessory pancreatic duct (of Santorini)

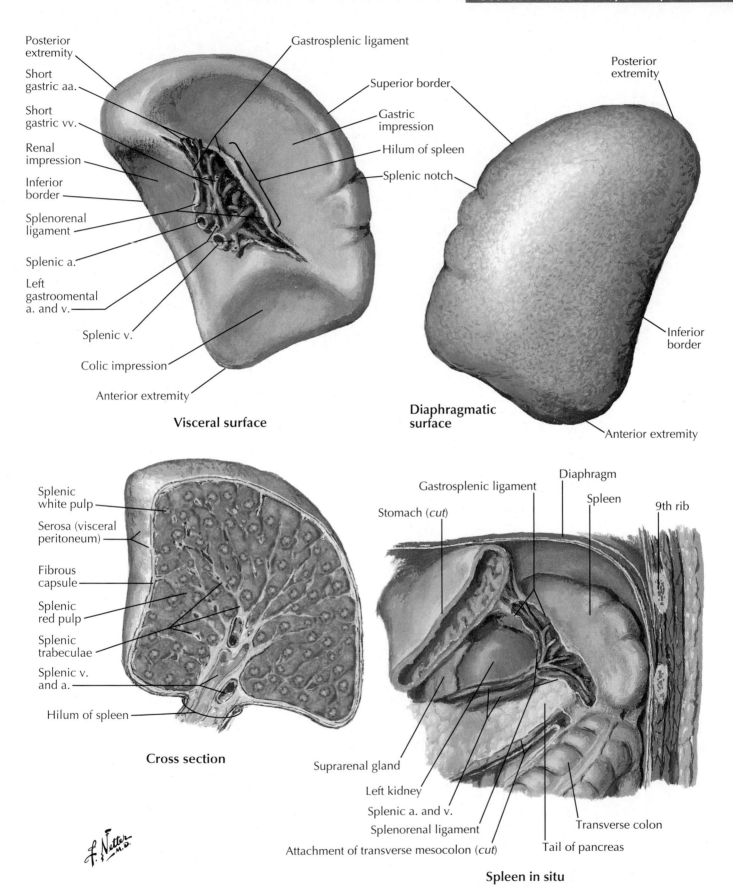

Posterior extremity

Short gastric aa.

Short gastric vv.

Renal impression

Inferior border

Splenorenal ligament

Splenic a.

Left gastroomental a. and v.

Splenic v.

Colic impression

Anterior extremity

Gastrosplenic ligament

Superior border

Gastric impression

Hilum of spleen

Splenic notch

Visceral surface

Posterior extremity

Inferior border

Anterior extremity

Diaphragmatic surface

Splenic white pulp

Serosa (visceral peritoneum)

Fibrous capsule

Splenic red pulp

Splenic trabeculae

Splenic v. and a.

Hilum of spleen

Cross section

Gastrosplenic ligament

Diaphragm

Stomach (*cut*)

Spleen

9th rib

Suprarenal gland

Left kidney

Splenic a. and v.

Splenorenal ligament

Attachment of transverse mesocolon (*cut*)

Tail of pancreas

Transverse colon

Spleen in situ

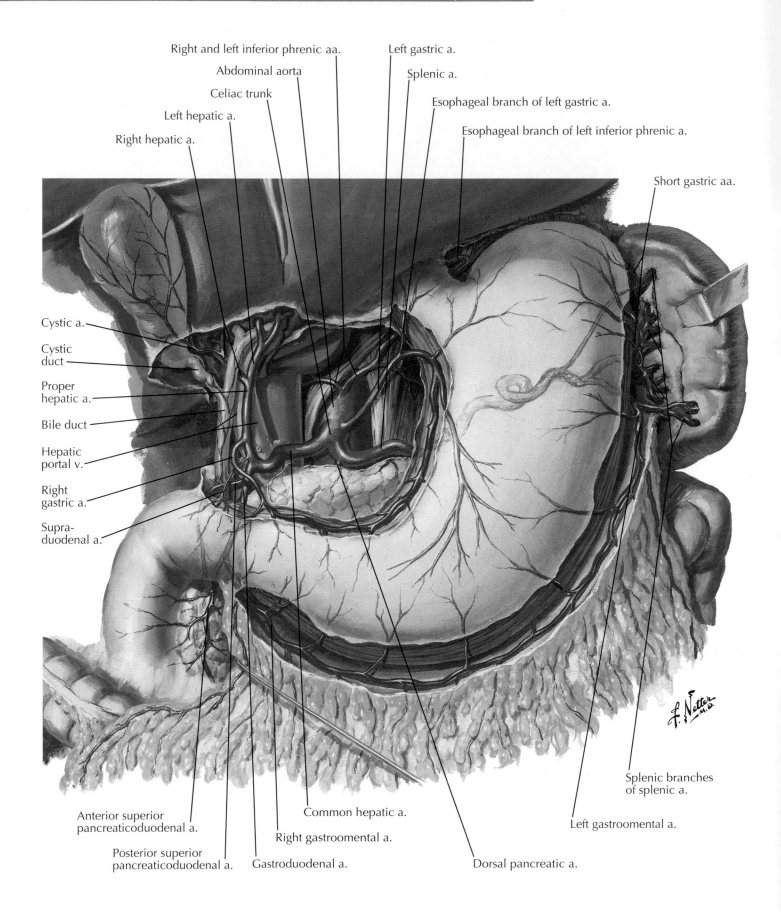

Right and left inferior phrenic aa.

Abdominal aorta

Celiac trunk

Left hepatic a.

Right hepatic a.

Left gastric a.

Splenic a.

Esophageal branch of left gastric a.

Esophageal branch of left inferior phrenic a.

Short gastric aa.

Cystic a.

Cystic duct

Proper hepatic a.

Bile duct

Hepatic portal v.

Right gastric a.

Supra-duodenal a.

Anterior superior pancreaticoduodenal a.

Posterior superior pancreaticoduodenal a.

Gastroduodenal a.

Right gastroomental a.

Common hepatic a.

Dorsal pancreatic a.

Left gastroomental a.

Splenic branches of splenic a.

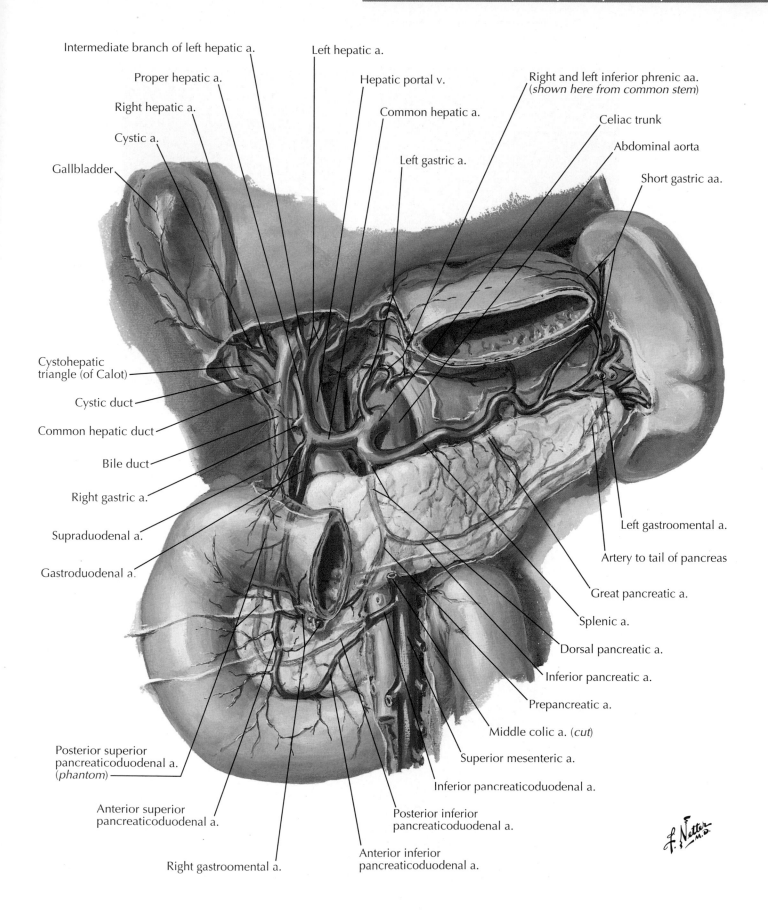

Intermediate branch of left hepatic a.

Proper hepatic a.

Right hepatic a.

Cystic a.

Gallbladder

Left hepatic a.

Hepatic portal v.

Common hepatic a.

Left gastric a.

Right and left inferior phrenic aa. (*shown here from common stem*)

Celiac trunk

Abdominal aorta

Short gastric aa.

Cystohepatic triangle (of Calot)

Cystic duct

Common hepatic duct

Bile duct

Right gastric a.

Supraduodenal a.

Gastroduodenal a.

Posterior superior pancreaticoduodenal a. (*phantom*)

Anterior superior pancreaticoduodenal a.

Right gastroomental a.

Left gastroomental a.

Artery to tail of pancreas

Great pancreatic a.

Splenic a.

Dorsal pancreatic a.

Inferior pancreatic a.

Prepancreatic a.

Middle colic a. (*cut*)

Superior mesenteric a.

Inferior pancreaticoduodenal a.

Posterior inferior pancreaticoduodenal a.

Anterior inferior pancreaticoduodenal a.

3D volume-rendered CT image with intravenous contrast enhancement

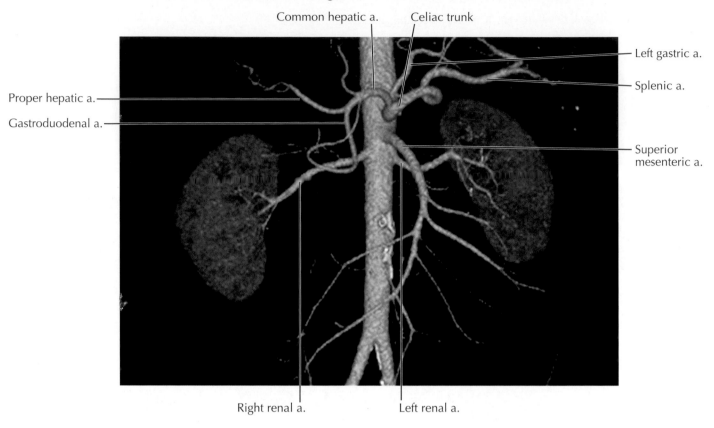

Common hepatic a.   Celiac trunk

Left gastric a.

Splenic a.

Proper hepatic a.

Gastroduodenal a.

Superior mesenteric a.

Right renal a.   Left renal a.

Selective digital subtraction angiogram, celiac trunk

Left gastric a.

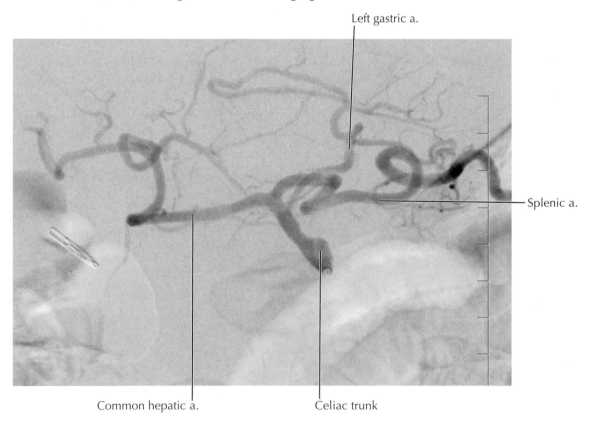

Splenic a.

Common hepatic a.   Celiac trunk

Duodenum and head of pancreas reflected to left

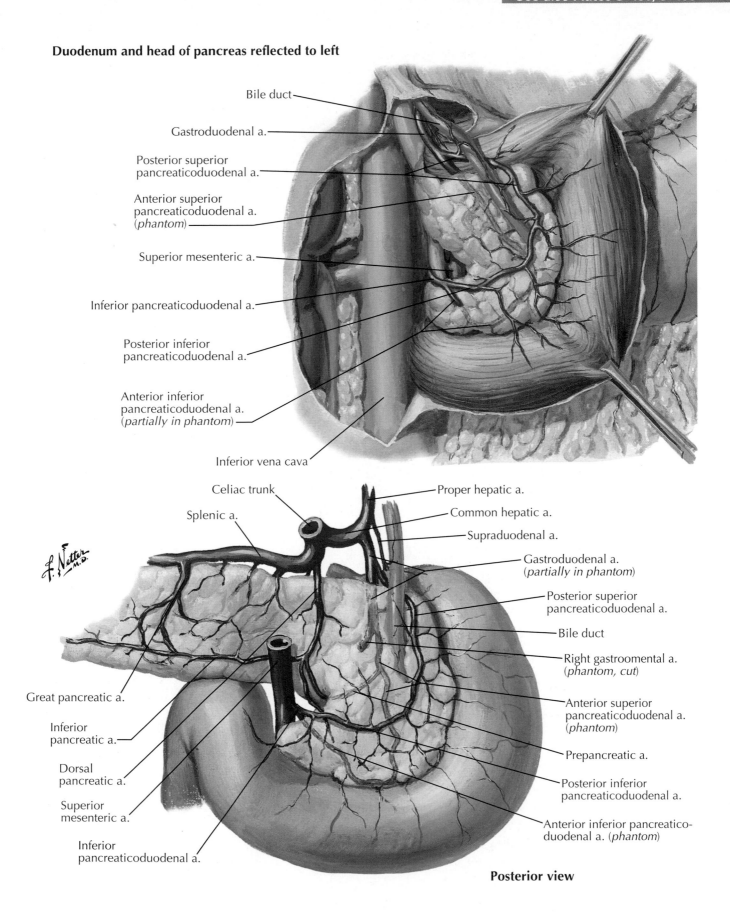

Bile duct

Gastroduodenal a.

Posterior superior pancreaticoduodenal a.

Anterior superior pancreaticoduodenal a. (*phantom*)

Superior mesenteric a.

Inferior pancreaticoduodenal a.

Posterior inferior pancreaticoduodenal a.

Anterior inferior pancreaticoduodenal a. (*partially in phantom*)

Inferior vena cava

Celiac trunk

Splenic a.

Proper hepatic a.

Common hepatic a.

Supraduodenal a.

Gastroduodenal a. (*partially in phantom*)

Posterior superior pancreaticoduodenal a.

Bile duct

Right gastroomental a. (*phantom, cut*)

Anterior superior pancreaticoduodenal a. (*phantom*)

Prepancreatic a.

Posterior inferior pancreaticoduodenal a.

Anterior inferior pancreaticoduodenal a. (*phantom*)

Great pancreatic a.

Inferior pancreatic a.

Dorsal pancreatic a.

Superior mesenteric a.

Inferior pancreaticoduodenal a.

Posterior view

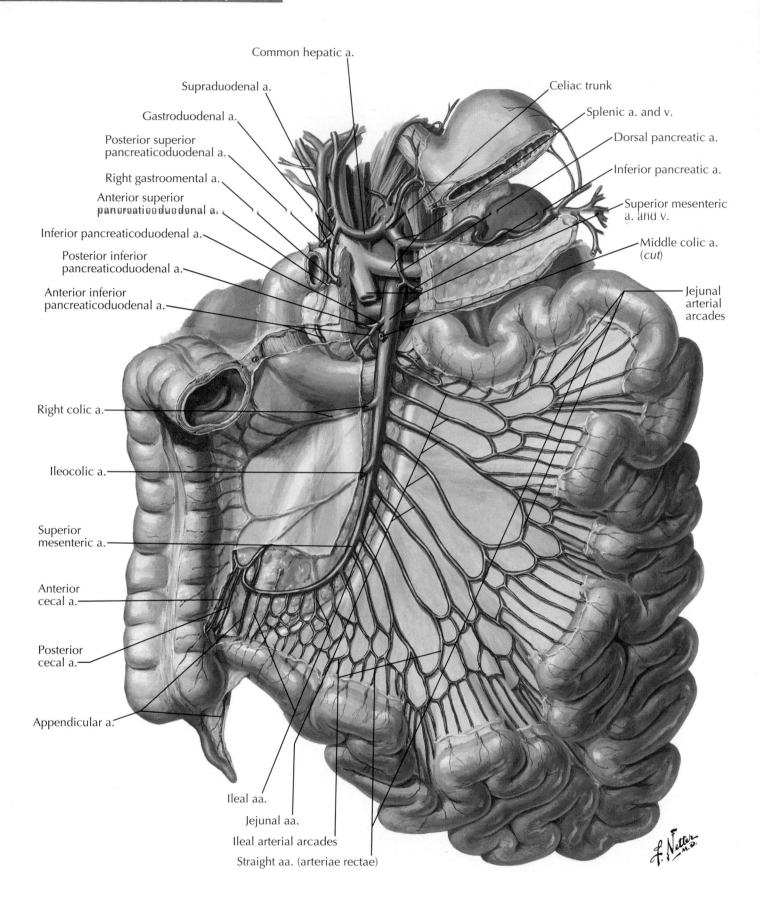

Common hepatic a.

Supraduodenal a.

Gastroduodenal a.

Posterior superior
pancreaticoduodenal a.

Right gastroomental a.

Anterior superior
pancreaticoduodenal a.

Inferior pancreaticoduodenal a.

Posterior inferior
pancreaticoduodenal a.

Anterior inferior
pancreaticoduodenal a.

Right colic a.

Ileocolic a.

Superior
mesenteric a.

Anterior
cecal a.

Posterior
cecal a.

Appendicular a.

Celiac trunk

Splenic a. and v.

Dorsal pancreatic a.

Inferior pancreatic a.

Superior mesenteric
a. and v.

Middle colic a.
(cut)

Jejunal
arterial
arcades

Ileal aa.

Jejunal aa.

Ileal arterial arcades

Straight aa. (arteriae rectae)

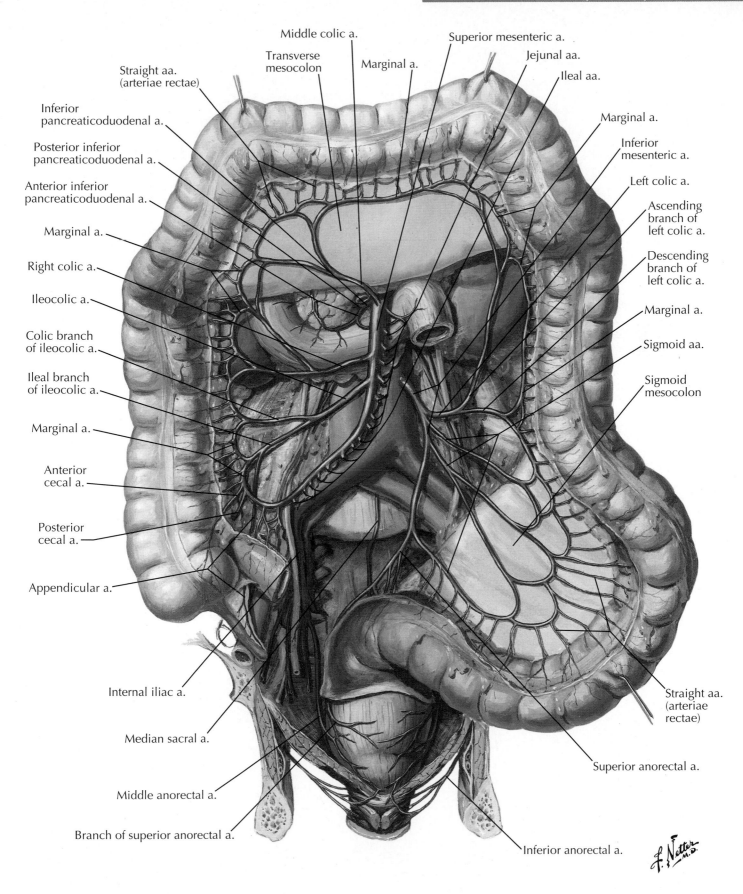

Straight aa. (arteriae rectae)

Inferior pancreaticoduodenal a.

Posterior inferior pancreaticoduodenal a.

Anterior inferior pancreaticoduodenal a.

Marginal a.

Right colic a.

Ileocolic a.

Colic branch of ileocolic a.

Ileal branch of ileocolic a.

Marginal a.

Anterior cecal a.

Posterior cecal a.

Appendicular a.

Internal iliac a.

Median sacral a.

Middle anorectal a.

Branch of superior anorectal a.

Middle colic a.

Transverse mesocolon

Marginal a.

Superior mesenteric a.

Jejunal aa.

Ileal aa.

Marginal a.

Inferior mesenteric a.

Left colic a.

Ascending branch of left colic a.

Descending branch of left colic a.

Marginal a.

Sigmoid aa.

Sigmoid mesocolon

Straight aa. (arteriae rectae)

Superior anorectal a.

Inferior anorectal a.

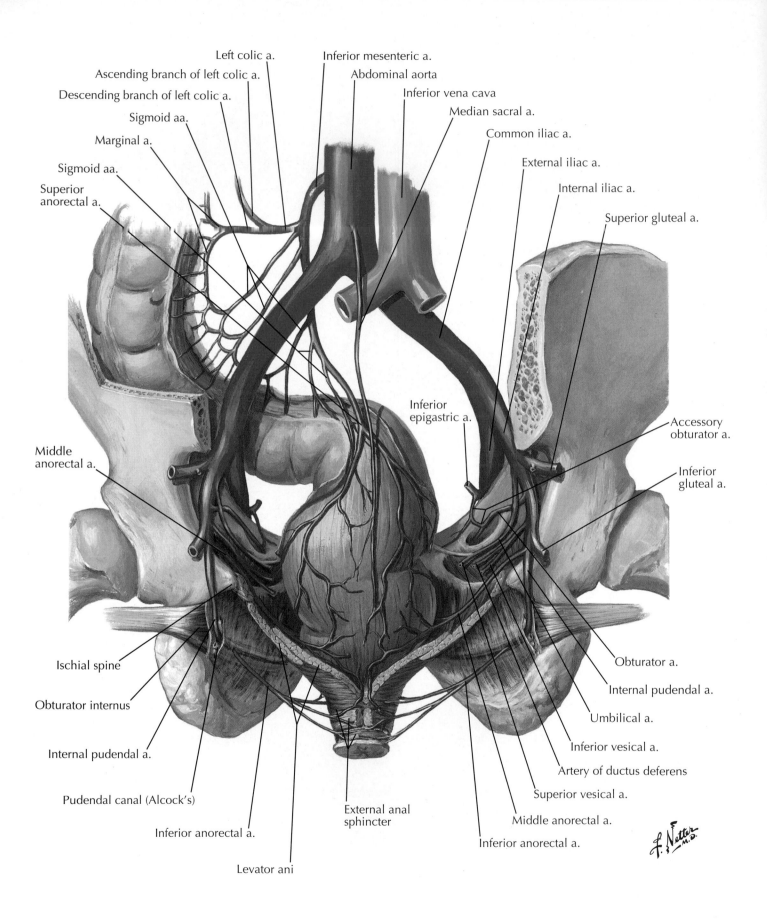

Left colic a.

Inferior mesenteric a.

Ascending branch of left colic a.

Abdominal aorta

Descending branch of left colic a.

Inferior vena cava

Sigmoid aa.

Median sacral a.

Marginal a.

Common iliac a.

Sigmoid aa.

External iliac a.

Superior
anorectal a.

Internal iliac a.

Superior gluteal a.

Inferior
epigastric a.

Accessory
obturator a.

Middle
anorectal a.

Inferior
gluteal a.

Ischial spine

Obturator internus

Obturator a.

Internal pudendal a.

Internal pudendal a.

Umbilical a.

Pudendal canal (Alcock's)

Inferior vesical a.

Artery of ductus deferens

Inferior anorectal a.

Superior vesical a.

External anal
sphincter

Middle anorectal a.

Levator ani

Inferior anorectal a.

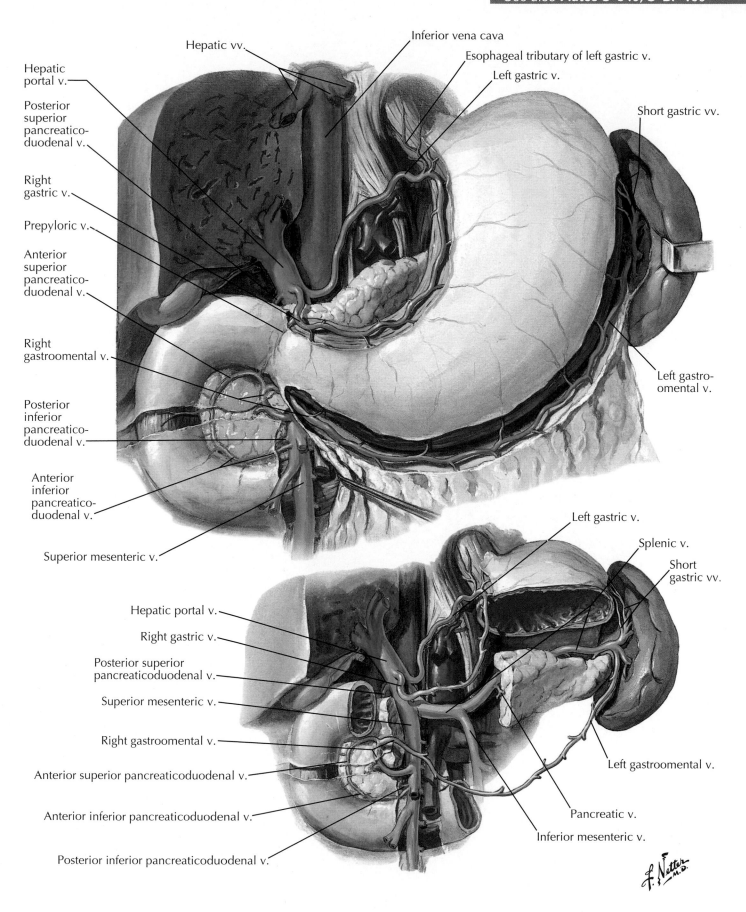

Hepatic vv.

Inferior vena cava

Esophageal tributary of left gastric v.

Left gastric v.

Short gastric vv.

Hepatic portal v.

Posterior superior pancreaticoduodenal v.

Right gastric v.

Prepyloric v.

Anterior superior pancreaticoduodenal v.

Right gastroomental v.

Posterior inferior pancreaticoduodenal v.

Anterior inferior pancreaticoduodenal v.

Superior mesenteric v.

Left gastroomental v.

Left gastric v.

Splenic v.

Short gastric vv.

Hepatic portal v.

Right gastric v.

Posterior superior pancreaticoduodenal v.

Superior mesenteric v.

Right gastroomental v.

Anterior superior pancreaticoduodenal v.

Anterior inferior pancreaticoduodenal v.

Posterior inferior pancreaticoduodenal v.

Left gastroomental v.

Pancreatic v.

Inferior mesenteric v.

f. Netter M.D.

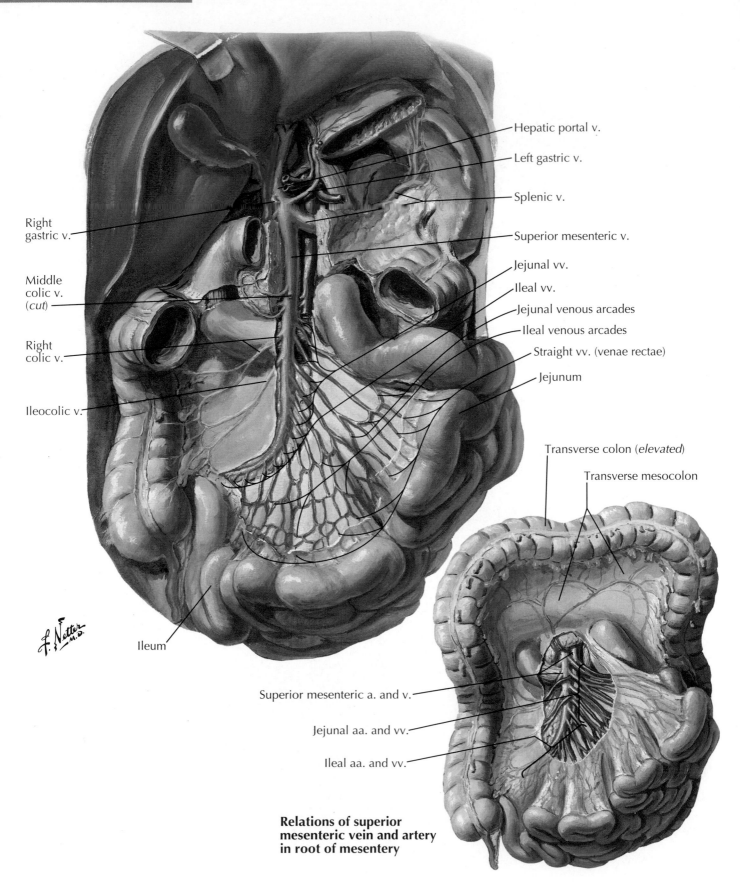

Hepatic portal v.

Left gastric v.

Splenic v.

Superior mesenteric v.

Jejunal vv.

Ileal vv.

Jejunal venous arcades

Ileal venous arcades

Straight vv. (venae rectae)

Jejunum

Transverse colon (*elevated*)

Transverse mesocolon

Right gastric v.

Middle colic v. (*cut*)

Right colic v.

Ileocolic v.

Ileum

Superior mesenteric a. and v.

Jejunal aa. and vv.

Ileal aa. and vv.

Relations of superior mesenteric vein and artery in root of mesentery

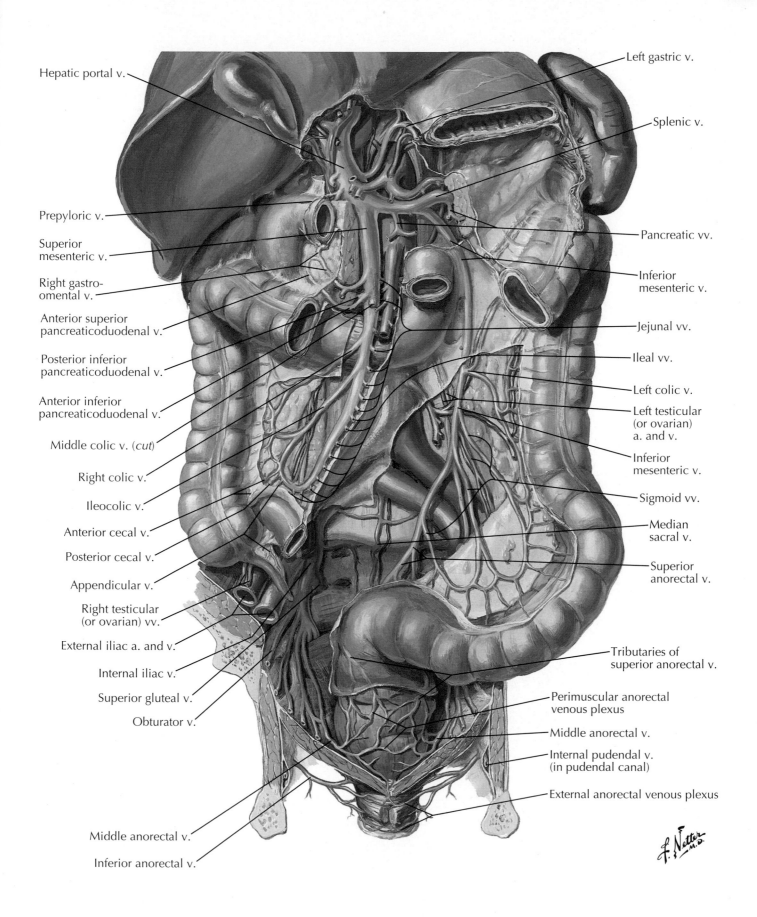

Hepatic portal v.

Left gastric v.

Splenic v.

Prepyloric v.

Superior mesenteric v.

Right gastro-omental v.

Anterior superior pancreaticoduodenal v.

Posterior inferior pancreaticoduodenal v.

Anterior inferior pancreaticoduodenal v.

Middle colic v. (*cut*)

Right colic v.

Ileocolic v.

Anterior cecal v.

Posterior cecal v.

Appendicular v.

Right testicular (or ovarian) vv.

External iliac a. and v.

Internal iliac v.

Superior gluteal v.

Obturator v.

Middle anorectal v.

Inferior anorectal v.

Pancreatic vv.

Inferior mesenteric v.

Jejunal vv.

Ileal vv.

Left colic v.

Left testicular (or ovarian) a. and v.

Inferior mesenteric v.

Sigmoid vv.

Median sacral v.

Superior anorectal v.

Tributaries of superior anorectal v.

Perimuscular anorectal venous plexus

Middle anorectal v.

Internal pudendal v. (in pudendal canal)

External anorectal venous plexus

F. Netter

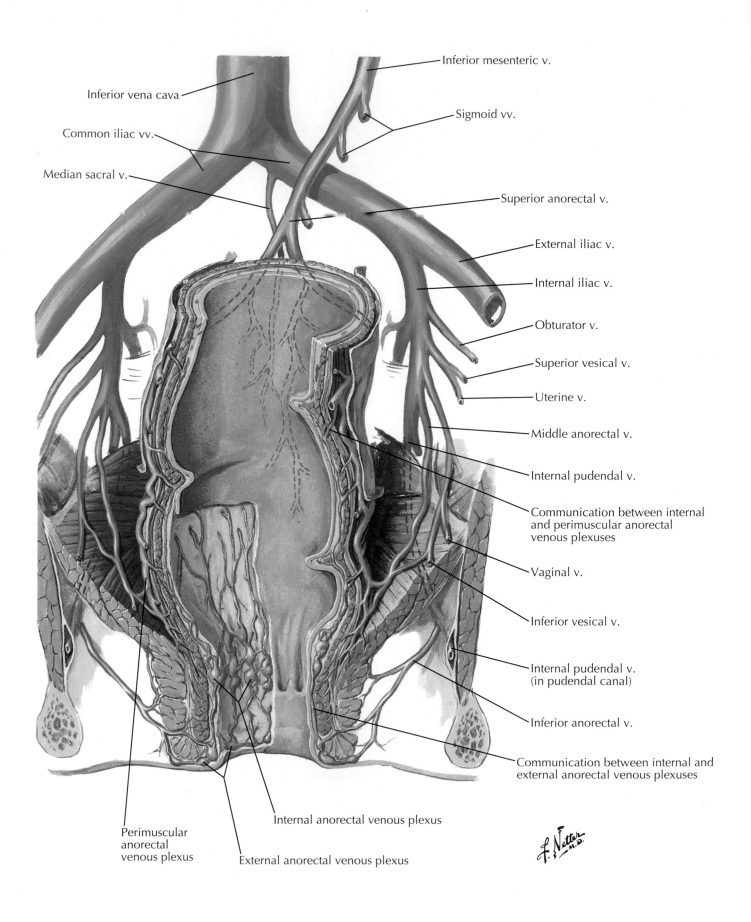

Inferior mesenteric v.

Inferior vena cava

Sigmoid vv.

Common iliac vv.

Median sacral v.

Superior anorectal v.

External iliac v.

Internal iliac v.

Obturator v.

Superior vesical v.

Uterine v.

Middle anorectal v.

Internal pudendal v.

Communication between internal and perimuscular anorectal venous plexuses

Vaginal v.

Inferior vesical v.

Internal pudendal v. (in pudendal canal)

Inferior anorectal v.

Communication between internal and external anorectal venous plexuses

Internal anorectal venous plexus

Perimuscular anorectal venous plexus

External anorectal venous plexus

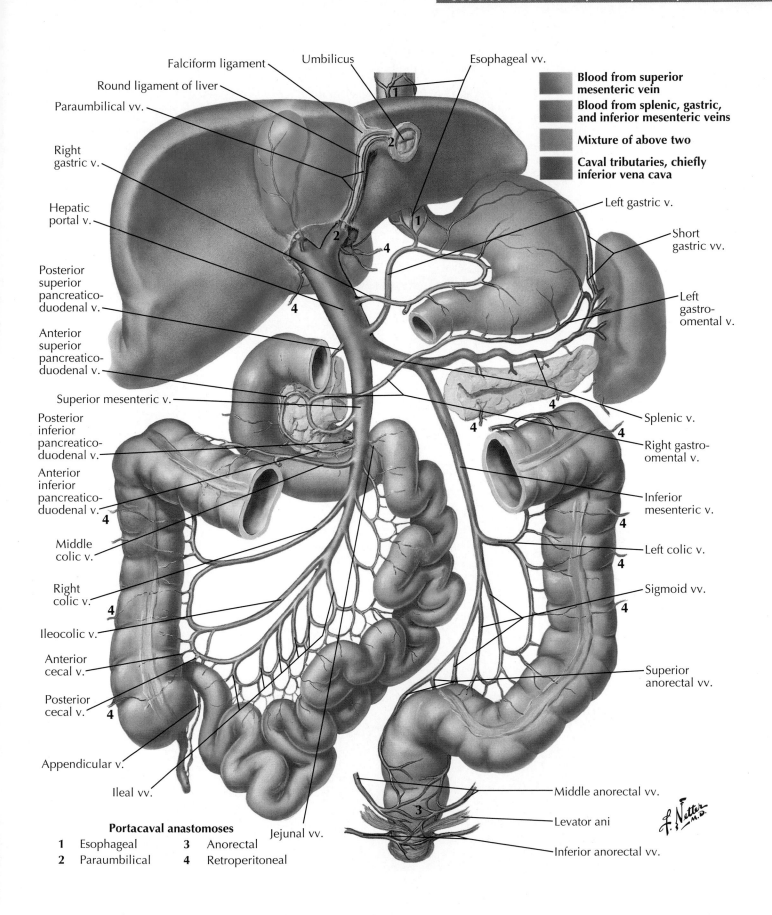

Falciform ligament

Round ligament of liver

Paraumbilical vv.

Right gastric v.

Hepatic portal v.

Posterior superior pancreaticoduodenal v.

Anterior superior pancreaticoduodenal v.

Superior mesenteric v.

Posterior inferior pancreaticoduodenal v.

Anterior inferior pancreaticoduodenal v.

Middle colic v.

Right colic v.

Ileocolic v.

Anterior cecal v.

Posterior cecal v.

Appendicular v.

Ileal vv.

Umbilicus

Esophageal vv.

Blood from superior mesenteric vein

Blood from splenic, gastric, and inferior mesenteric veins

Mixture of above two

Caval tributaries, chiefly inferior vena cava

Left gastric v.

Short gastric vv.

Left gastro-omental v.

Splenic v.

Right gastro-omental v.

Inferior mesenteric v.

Left colic v.

Sigmoid vv.

Superior anorectal vv.

Middle anorectal vv.

Levator ani

Inferior anorectal vv.

Jejunal vv.

Portacaval anastomoses

1	Esophageal	3	Anorectal
2	Paraumbilical	4	Retroperitoneal

F. Netter M.D.

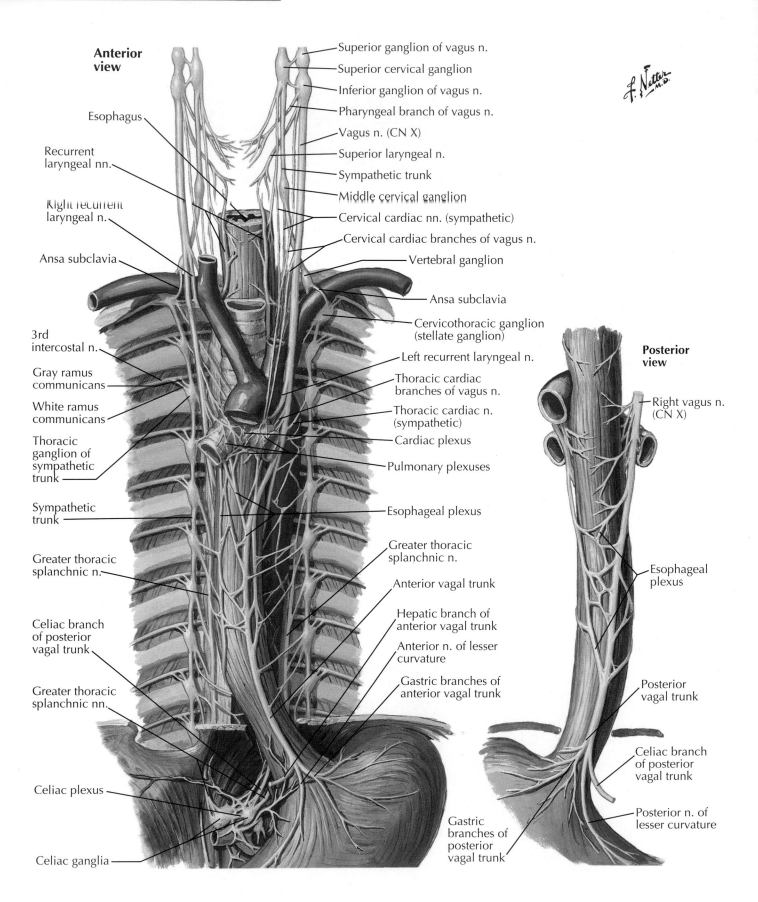

Anterior view

Superior ganglion of vagus n.

Superior cervical ganglion

Inferior ganglion of vagus n.

Pharyngeal branch of vagus n.

Vagus n. (CN X)

Superior laryngeal n.

Sympathetic trunk

Middle cervical ganglion

Cervical cardiac nn. (sympathetic)

Cervical cardiac branches of vagus n.

Vertebral ganglion

Ansa subclavia

Cervicothoracic ganglion (stellate ganglion)

Left recurrent laryngeal n.

Thoracic cardiac branches of vagus n.

Thoracic cardiac n. (sympathetic)

Cardiac plexus

Pulmonary plexuses

Esophageal plexus

Greater thoracic splanchnic n.

Anterior vagal trunk

Hepatic branch of anterior vagal trunk

Anterior n. of lesser curvature

Gastric branches of anterior vagal trunk

Gastric branches of posterior vagal trunk

Esophagus

Recurrent laryngeal nn.

Right recurrent laryngeal n.

Ansa subclavia

3rd intercostal n.

Gray ramus communicans

White ramus communicans

Thoracic ganglion of sympathetic trunk

Sympathetic trunk

Greater thoracic splanchnic n.

Celiac branch of posterior vagal trunk

Greater thoracic splanchnic nn.

Celiac plexus

Celiac ganglia

Posterior view

Right vagus n. (CN X)

Esophageal plexus

Posterior vagal trunk

Celiac branch of posterior vagal trunk

Posterior n. of lesser curvature

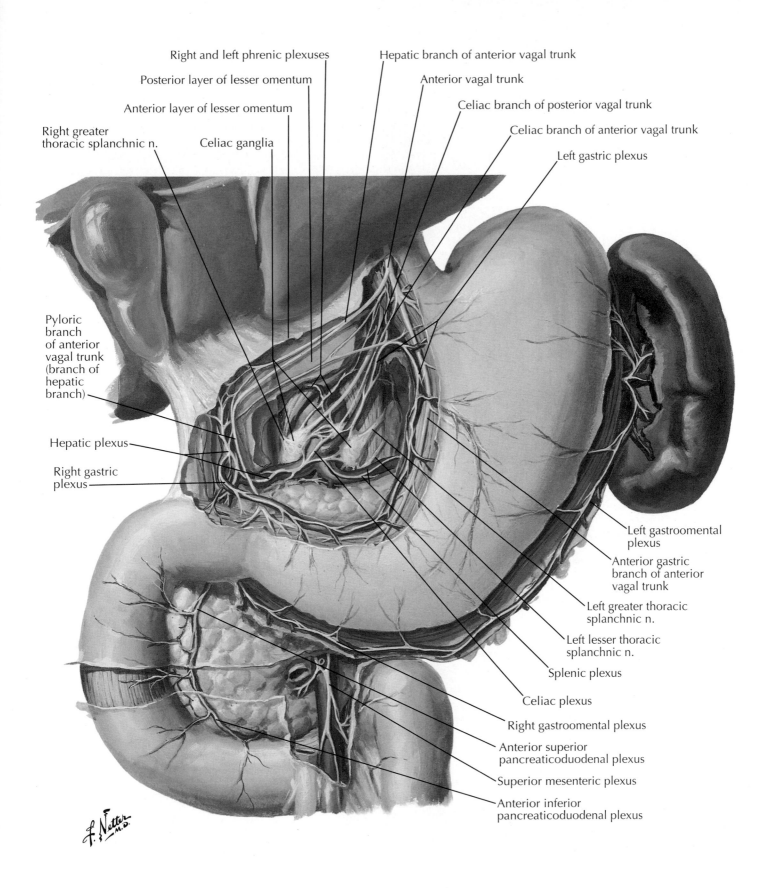

Right and left phrenic plexuses

Hepatic branch of anterior vagal trunk

Posterior layer of lesser omentum

Anterior vagal trunk

Anterior layer of lesser omentum

Celiac branch of posterior vagal trunk

Right greater thoracic splanchnic n.

Celiac ganglia

Celiac branch of anterior vagal trunk

Left gastric plexus

Pyloric branch of anterior vagal trunk (branch of hepatic branch)

Hepatic plexus

Right gastric plexus

Left gastroomental plexus

Anterior gastric branch of anterior vagal trunk

Left greater thoracic splanchnic n.

Left lesser thoracic splanchnic n.

Splenic plexus

Celiac plexus

Right gastroomental plexus

Anterior superior pancreaticoduodenal plexus

Superior mesenteric plexus

Anterior inferior pancreaticoduodenal plexus

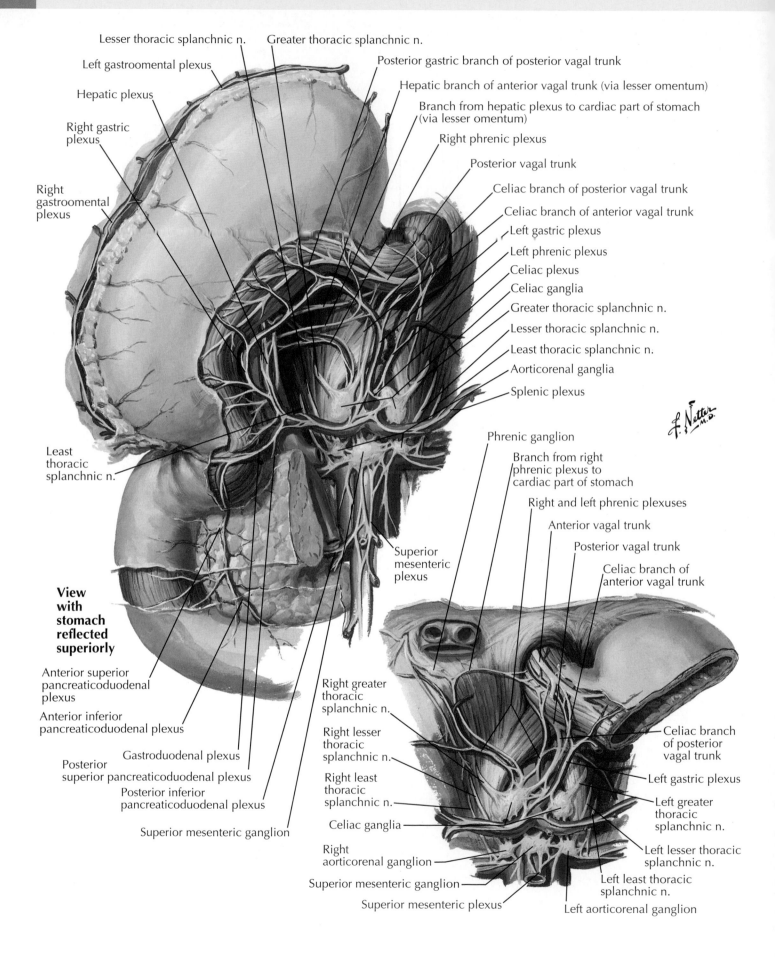

Lesser thoracic splanchnic n.

Greater thoracic splanchnic n.

Left gastroomental plexus

Posterior gastric branch of posterior vagal trunk

Hepatic plexus

Hepatic branch of anterior vagal trunk (via lesser omentum)

Branch from hepatic plexus to cardiac part of stomach (via lesser omentum)

Right gastric plexus

Right phrenic plexus

Posterior vagal trunk

Right gastroomental plexus

Celiac branch of posterior vagal trunk

Celiac branch of anterior vagal trunk

Left gastric plexus

Left phrenic plexus

Celiac plexus

Celiac ganglia

Greater thoracic splanchnic n.

Lesser thoracic splanchnic n.

Least thoracic splanchnic n.

Aorticorenal ganglia

Splenic plexus

Least thoracic splanchnic n.

Phrenic ganglion

Branch from right phrenic plexus to cardiac part of stomach

Right and left phrenic plexuses

Anterior vagal trunk

Posterior vagal trunk

Celiac branch of anterior vagal trunk

Superior mesenteric plexus

View with stomach reflected superiorly

Anterior superior pancreaticoduodenal plexus

Anterior inferior pancreaticoduodenal plexus

Gastroduodenal plexus

Posterior superior pancreaticoduodenal plexus

Posterior inferior pancreaticoduodenal plexus

Superior mesenteric ganglion

Right greater thoracic splanchnic n.

Right lesser thoracic splanchnic n.

Right least thoracic splanchnic n.

Celiac ganglia

Right aorticorenal ganglion

Superior mesenteric ganglion

Superior mesenteric plexus

Celiac branch of posterior vagal trunk

Left gastric plexus

Left greater thoracic splanchnic n.

Left lesser thoracic splanchnic n.

Left least thoracic splanchnic n.

Left aorticorenal ganglion

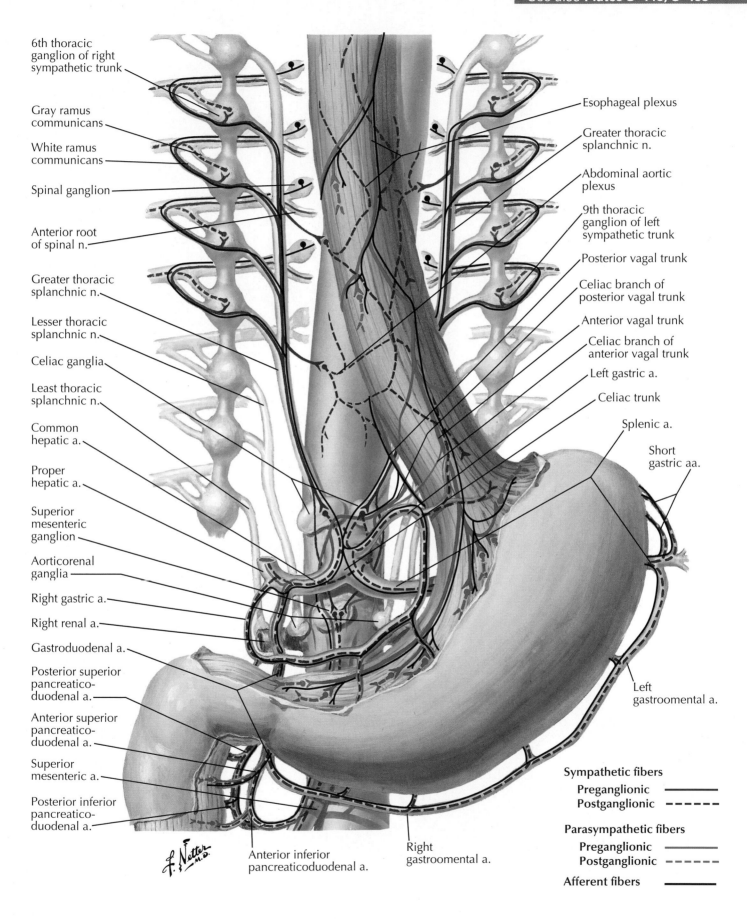

6th thoracic ganglion of right sympathetic trunk

Gray ramus communicans

White ramus communicans

Spinal ganglion

Anterior root of spinal n.

Greater thoracic splanchnic n.

Lesser thoracic splanchnic n.

Celiac ganglia

Least thoracic splanchnic n.

Common hepatic a.

Proper hepatic a.

Superior mesenteric ganglion

Aorticorenal ganglia

Right gastric a.

Right renal a.

Gastroduodenal a.

Posterior superior pancreatico-duodenal a.

Anterior superior pancreatico-duodenal a.

Superior mesenteric a.

Posterior inferior pancreatico-duodenal a.

Anterior inferior pancreaticoduodenal a.

Esophageal plexus

Greater thoracic splanchnic n.

Abdominal aortic plexus

9th thoracic ganglion of left sympathetic trunk

Posterior vagal trunk

Celiac branch of posterior vagal trunk

Anterior vagal trunk

Celiac branch of anterior vagal trunk

Left gastric a.

Celiac trunk

Splenic a.

Short gastric aa.

Left gastroomental a.

Right gastroomental a.

Sympathetic fibers
Preganglionic ——————
Postganglionic – – – – – –

Parasympathetic fibers
Preganglionic ——————
Postganglionic – – – – – –

Afferent fibers ——————

Visceral Innervation

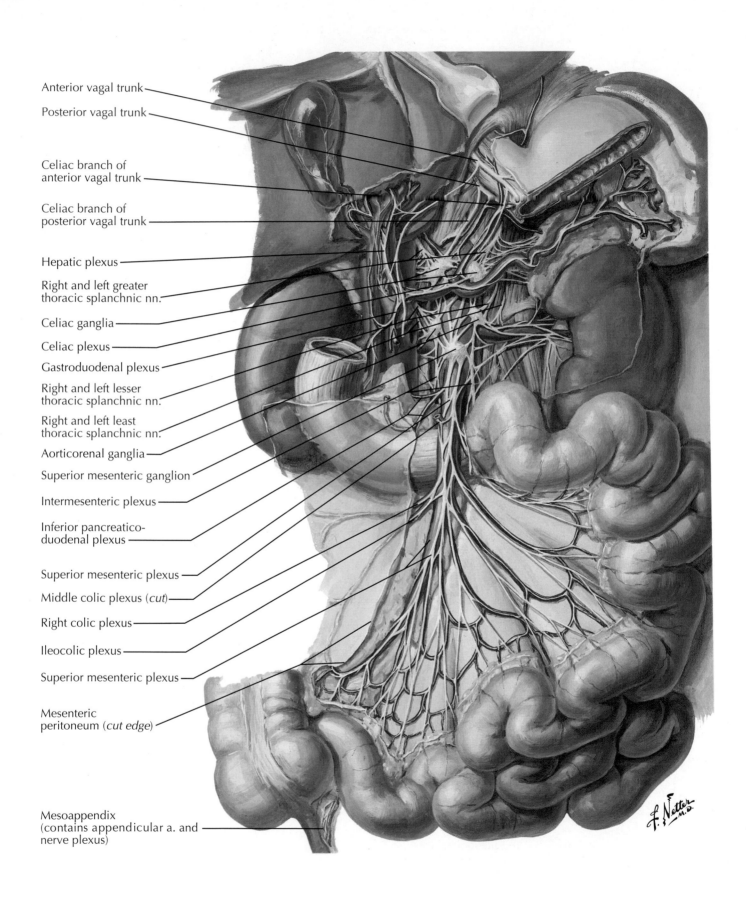

Anterior vagal trunk

Posterior vagal trunk

Celiac branch of anterior vagal trunk

Celiac branch of posterior vagal trunk

Hepatic plexus

Right and left greater thoracic splanchnic nn.

Celiac ganglia

Celiac plexus

Gastroduodenal plexus

Right and left lesser thoracic splanchnic nn.

Right and left least thoracic splanchnic nn.

Aorticorenal ganglia

Superior mesenteric ganglion

Intermesenteric plexus

Inferior pancreatico- duodenal plexus

Superior mesenteric plexus

Middle colic plexus (*cut*)

Right colic plexus

Ileocolic plexus

Superior mesenteric plexus

Mesenteric peritoneum (*cut edge*)

Mesoappendix (contains appendicular a. and nerve plexus)

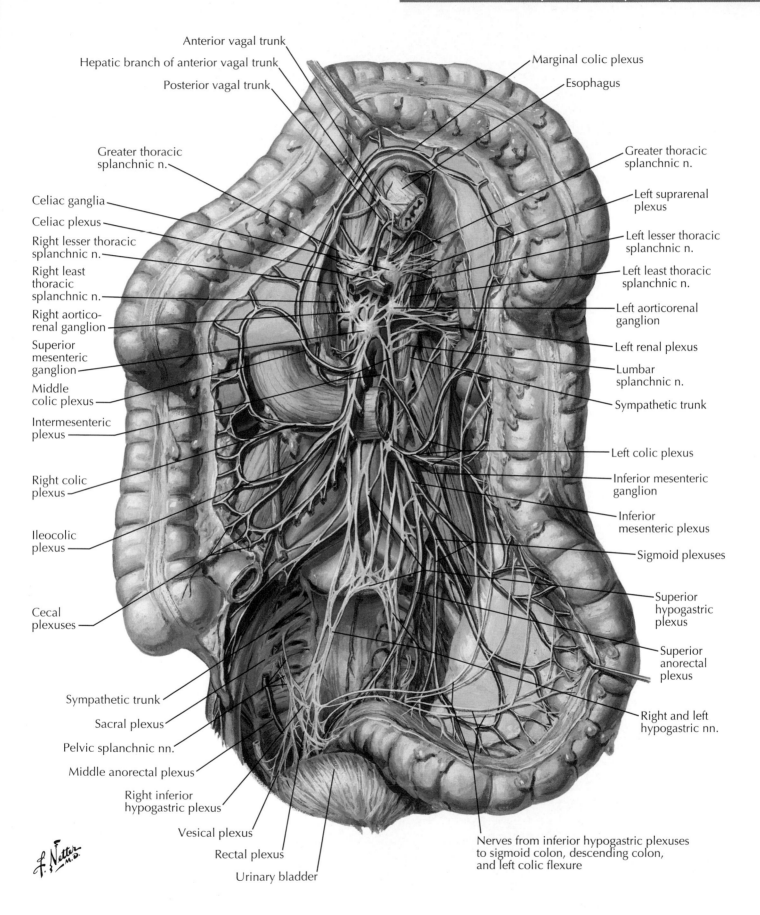

Anterior vagal trunk

Hepatic branch of anterior vagal trunk

Posterior vagal trunk

Greater thoracic splanchnic n.

Celiac ganglia

Celiac plexus

Right lesser thoracic splanchnic n.

Right least thoracic splanchnic n.

Right aortico-renal ganglion

Superior mesenteric ganglion

Middle colic plexus

Intermesenteric plexus

Right colic plexus

Ileocolic plexus

Cecal plexuses

Sympathetic trunk

Sacral plexus

Pelvic splanchnic nn.

Middle anorectal plexus

Right inferior hypogastric plexus

Vesical plexus

Rectal plexus

Urinary bladder

Marginal colic plexus

Esophagus

Greater thoracic splanchnic n.

Left suprarenal plexus

Left lesser thoracic splanchnic n.

Left least thoracic splanchnic n.

Left aorticorenal ganglion

Left renal plexus

Lumbar splanchnic n.

Sympathetic trunk

Left colic plexus

Inferior mesenteric ganglion

Inferior mesenteric plexus

Sigmoid plexuses

Superior hypogastric plexus

Superior anorectal plexus

Right and left hypogastric nn.

Nerves from inferior hypogastric plexuses to sigmoid colon, descending colon, and left colic flexure

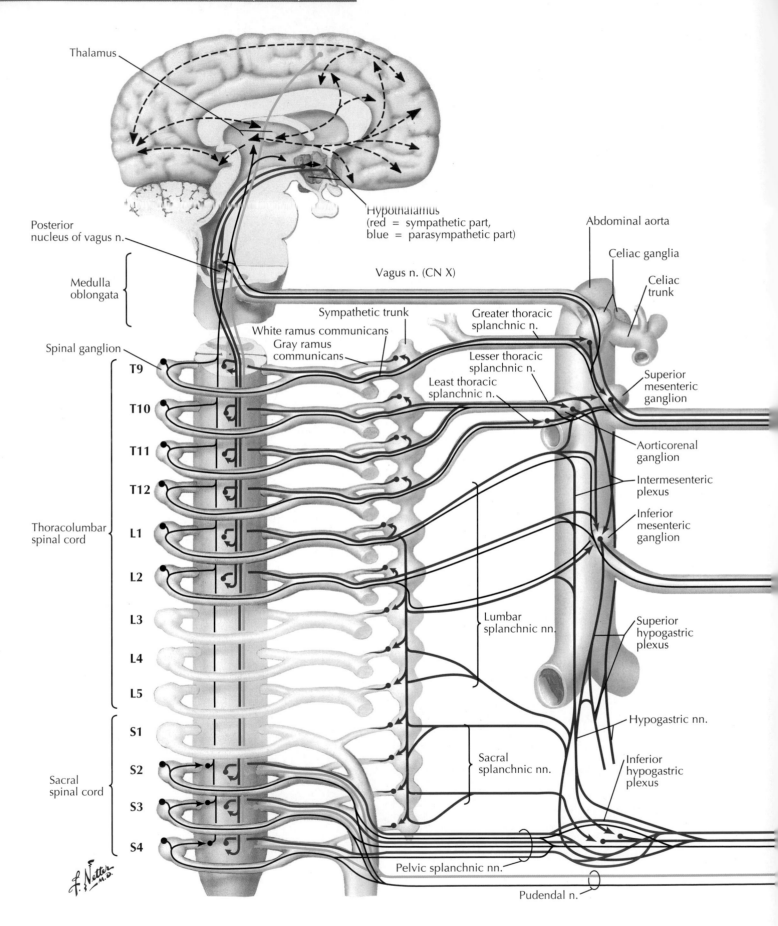

Thalamus

Hypothalamus
(red = sympathetic part,
blue = parasympathetic part)

Posterior
nucleus of vagus n.

Medulla
oblongata

Vagus n. (CN X)

Abdominal aorta

Celiac ganglia

Celiac
trunk

Spinal ganglion

Sympathetic trunk

Greater thoracic
splanchnic n.

White ramus communicans

Gray ramus
communicans

T9

Lesser thoracic
splanchnic n.

Least thoracic
splanchnic n.

Superior
mesenteric
ganglion

T10

T11

Aorticorenal
ganglion

T12

Intermesenteric
plexus

Thoracolumbar
spinal cord

L1

Inferior
mesenteric
ganglion

L2

L3

Lumbar
splanchnic nn.

Superior
hypogastric
plexus

L4

L5

Hypogastric nn.

S1

Inferior
hypogastric
plexus

Sacral
spinal cord

S2

Sacral
splanchnic nn.

S3

S4

Pelvic splanchnic nn.

Pudendal n.

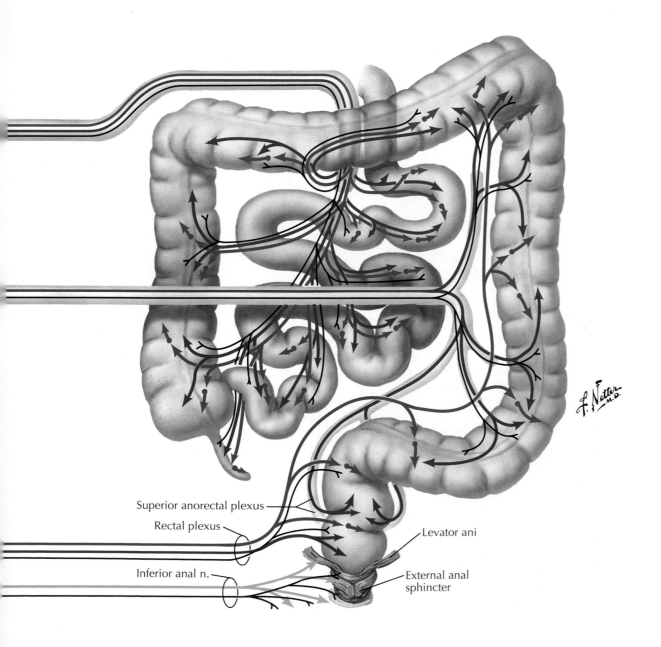

Sympathetic fibers
Parasympathetic fibers
Somatic efferent fibers
Afferents and CNS connections
Indefinite paths

Superior anorectal plexus
Rectal plexus
Inferior anal n.
Levator ani
External anal sphincter

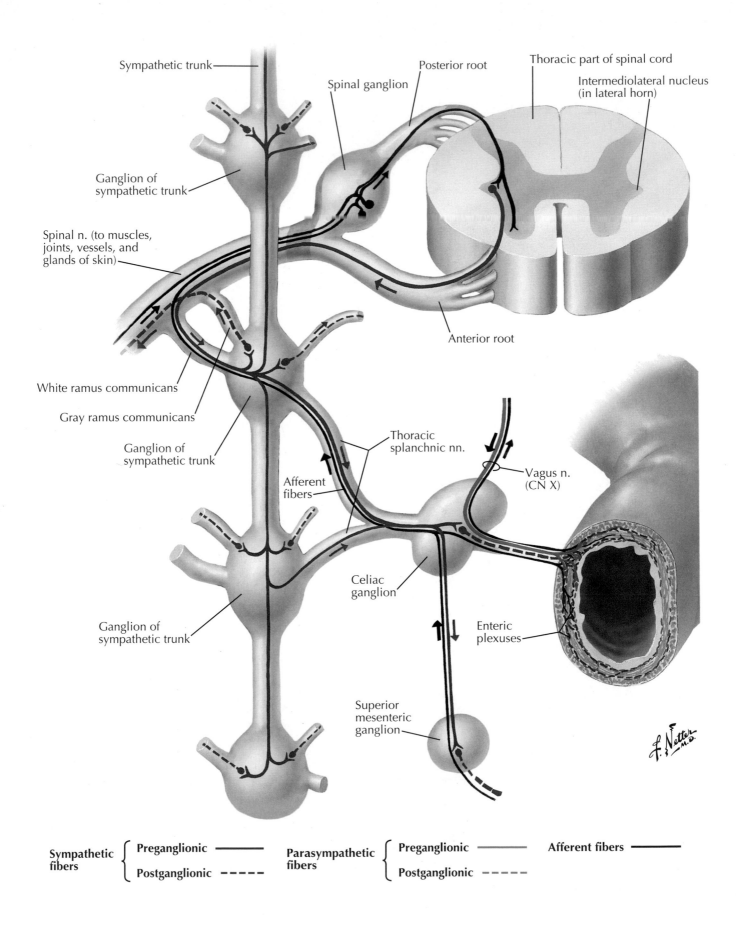

Sympathetic trunk

Spinal ganglion

Posterior root

Thoracic part of spinal cord

Intermediolateral nucleus (in lateral horn)

Ganglion of sympathetic trunk

Spinal n. (to muscles, joints, vessels, and glands of skin)

Anterior root

White ramus communicans

Gray ramus communicans

Ganglion of sympathetic trunk

Thoracic splanchnic nn.

Afferent fibers

Vagus n. (CN X)

Ganglion of sympathetic trunk

Celiac ganglion

Enteric plexuses

Superior mesenteric ganglion

Sympathetic fibers	Preganglionic ————	Parasympathetic fibers	Preganglionic ————	Afferent fibers ————
	Postganglionic – – – –		Postganglionic – – – –	

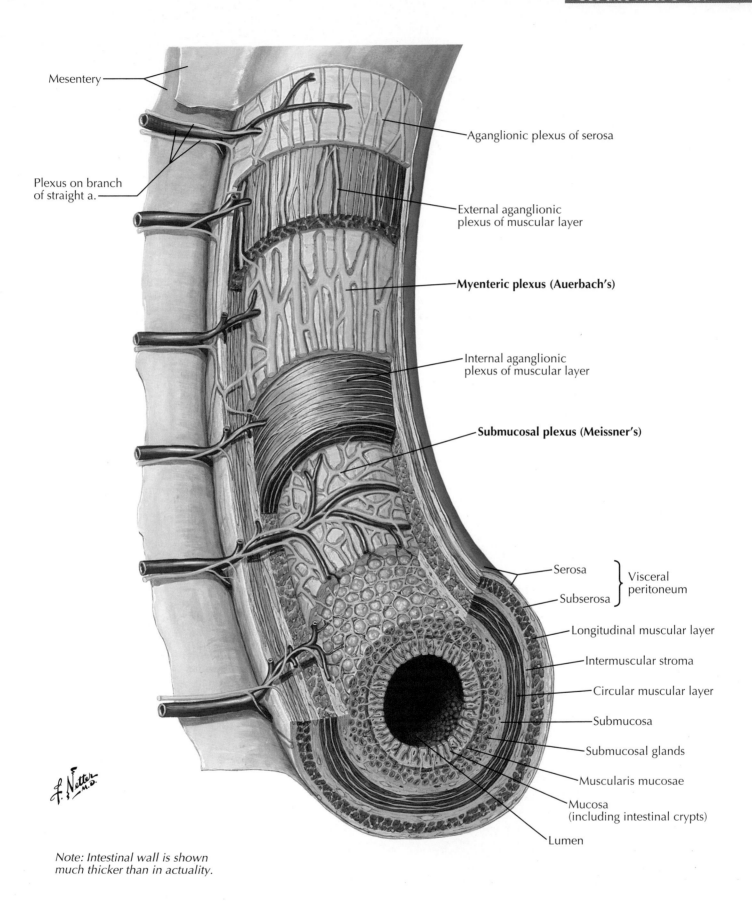

Mesentery

Plexus on branch
of straight a.

Aganglionic plexus of serosa

External aganglionic
plexus of muscular layer

Myenteric plexus (Auerbach's)

Internal aganglionic
plexus of muscular layer

Submucosal plexus (Meissner's)

Serosa ⎫
⎬ Visceral
Subserosa ⎭ peritoneum

Longitudinal muscular layer

Intermuscular stroma

Circular muscular layer

Submucosa

Submucosal glands

Muscularis mucosae

Mucosa
(including intestinal crypts)

Lumen

*Note: Intestinal wall is shown
much thicker than in actuality.*

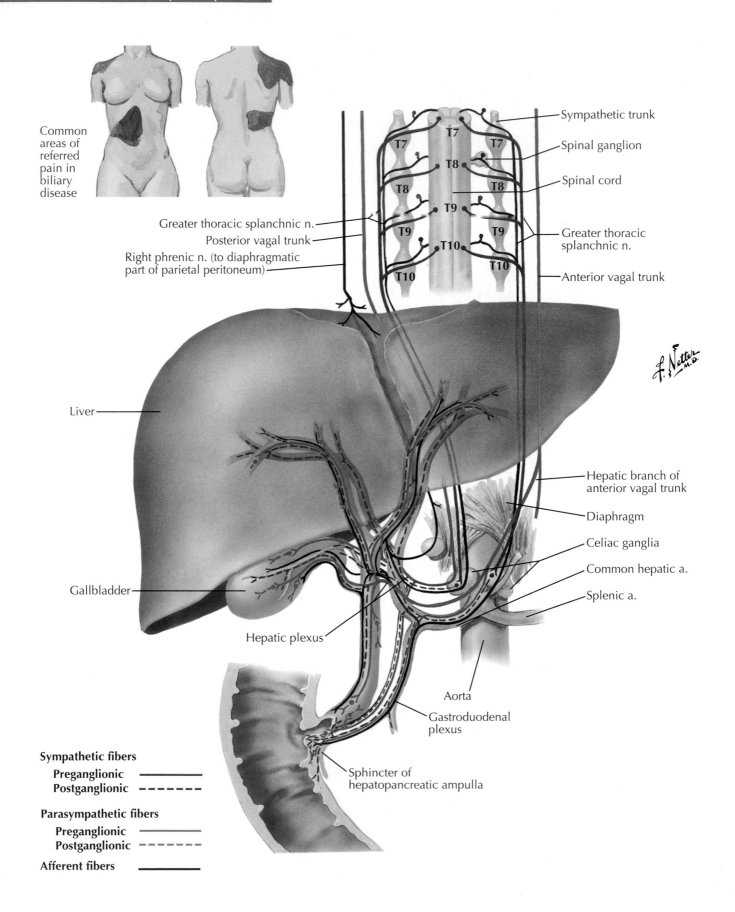

Common areas of referred pain in biliary disease

Sympathetic trunk

Spinal ganglion

Spinal cord

Greater thoracic splanchnic n.

Posterior vagal trunk

Right phrenic n. (to diaphragmatic part of parietal peritoneum)

Greater thoracic splanchnic n.

Anterior vagal trunk

T7

T7

T7

T8

T8

T8

T9

T9

T9

T10

T10

T10

Liver

Gallbladder

Hepatic plexus

Hepatic branch of anterior vagal trunk

Diaphragm

Celiac ganglia

Common hepatic a.

Splenic a.

Aorta

Gastroduodenal plexus

Sphincter of hepatopancreatic ampulla

Sympathetic fibers
    Preganglionic ———
    Postganglionic - - - -

Parasympathetic fibers
    Preganglionic ———
    Postganglionic - - - -

Afferent fibers ———

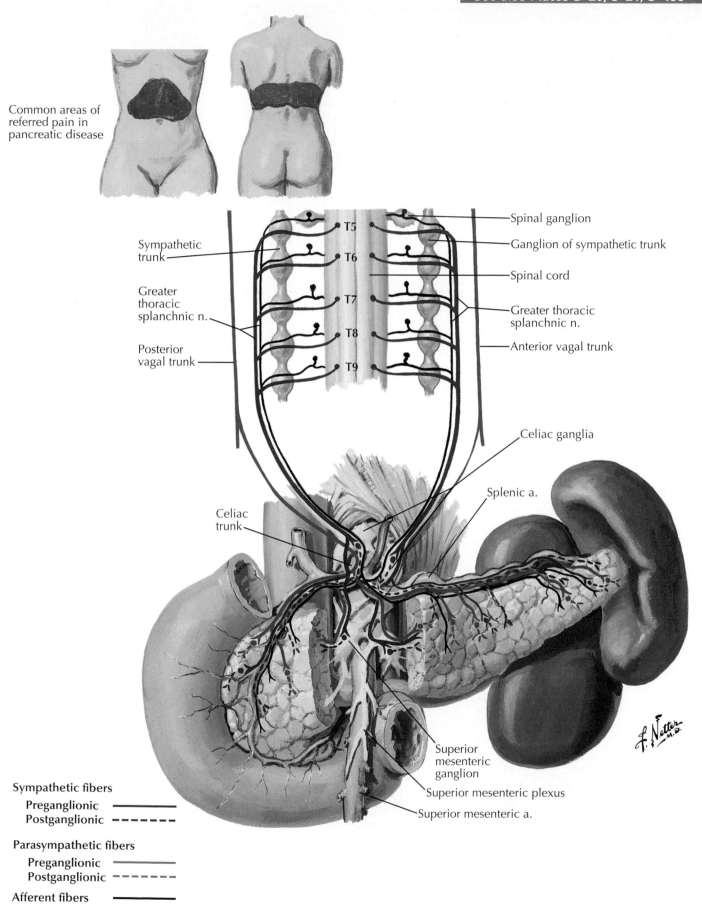

Common areas of referred pain in pancreatic disease

Sympathetic trunk

Greater thoracic splanchnic n.

Posterior vagal trunk

T5
T6
T7
T8
T9

Spinal ganglion

Ganglion of sympathetic trunk

Spinal cord

Greater thoracic splanchnic n.

Anterior vagal trunk

Celiac ganglia

Splenic a.

Celiac trunk

Superior mesenteric ganglion

Superior mesenteric plexus

Superior mesenteric a.

Sympathetic fibers

Preganglionic ——————
Postganglionic – – – – – –

Parasympathetic fibers

Preganglionic ——————
Postganglionic – – – – – –

Afferent fibers ——————

Structures with High* Clinical Significance

ANATOMIC STRUCTURES	CLINICAL IMPORTANCE	PLATE NUMBERS
Head and Neck		
Parotid gland	Swelling of gland due to infection (parotitis), such as from viruses or bacteria, may cause pain and compress branches of facial nerve, producing facial muscle weakness; external carotid, superficial temporal, and maxillary arteries also pass through the gland	S–64, S–396
Abdomen		
Liver	Palpable inferior to right costal margin; hepatomegaly may occur in conditions such as hepatitis, heart failure, and infiltrative diseases; shrunken, nodular appearance of liver on imaging indicates cirrhosis; common site of metastasis	S–395, S–415, S–423
Gastroesophageal junction	Transient relaxations or decreased tone of lower esophageal sphincter can cause gastroesophageal reflux disease (GERD), a common cause of epigastric pain; common site of esophageal cancer	S–413, S–414, S–541
Stomach and duodenum	Primary sites of peptic ulcers; nonsteroidal anti-inflammatory drug (NSAID) overuse and/or *Helicobacter pylori* infection may cause ulceration	S–414, S–415, S–417
Pylorus	Infantile hypertrophic pyloric stenosis causes postprandial projectile vomiting among newborns	S–414
Major duodenal papilla (of Vater)	Catheterized and injected with contrast during endoscopic retrograde cholangiopancreatography (ERCP), a common diagnostic procedure; sphincter of hepatopancreatic ampulla (sphincter of Oddi) dysfunction may obstruct biliary flow through major duodenal papilla, resulting in right upper quadrant pain	S–421, S–432
Vermiform appendix	Prone to inflammation and rupture (appendicitis); may have retrocecal position, in which case appendicitis causes inflammation of adjacent psoas fascia and atypical location of pain	S–425, S–427
Colon	Common site of diverticula and malignancies; colonoscopy is performed to screen for colon cancer	S–428
Gallbladder	May become inflamed (cholecystitis) and cause pain secondary to gallstones blocking cystic duct	S–429, S–432, S–435, S–456
Bile duct	Gallstones may become impacted in bile duct (choledocholithiasis), resulting in hepatitis and jaundice; some cases may be complicated by pancreatitis and/or infection of obstructed biliary ducts (cholangitis); small stones can be problematic, whereas large stones usually remain in gallbladder and are more typically asymptomatic; ERCP is performed to locate and relieve obstruction	S–429, S–430, S–432
Umbilicus	Remnant of umbilical cord insertion into fetal umbilical vessels; landmark for locating transumbilical plane, which is used to divide abdomen into quadrants; marks position of T10 dermatome; used to locate McBurney's point; common site for hernias in abdominal wall	S–8, S–395
Pancreas	Lies primarily retroperitoneal and deep to stomach; thus, inflamed pancreas may be compressed by stomach and cause intense pain referred to the back; inflamed pancreas, most often caused by biliary obstruction or alcohol abuse, may cause severe, life-threatening complications; cancer of head/neck of pancreas can compress biliary tree	S–417, S–433, S–457
Pelvis		
Rectum and anal canal	Examined by digital rectal examination to detect internal hemorrhoids, fecal impaction, and rectal cancer	S–221, S–226, S–428
Peritoneum	Common site for metastatic spread of ovarian cancer via fluid in peritoneal cavity	S–483, S–485

*Selections are based largely on clinical data and commonly discussed clinical correlations in macroscopic ("gross") anatomy courses.

Table 8.1
Structures with High Clinical Significance

URINARY SYSTEM 9

ELECTRONIC BONUS PLATES

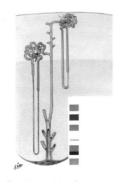

S–BP 83 Histology of Renal Corpuscle

S–BP 84 Nephron and Collecting Tubule: Schema

S–BP 85 Blood Vessels in Parenchyma of Kidney: Schema

S–BP 86 Female Urethra

S–BP 87 Cystourethrograms: Male and Female

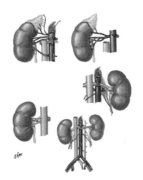

S–BP 88 Variations in Renal Artery and Vein

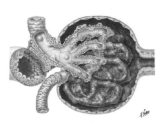

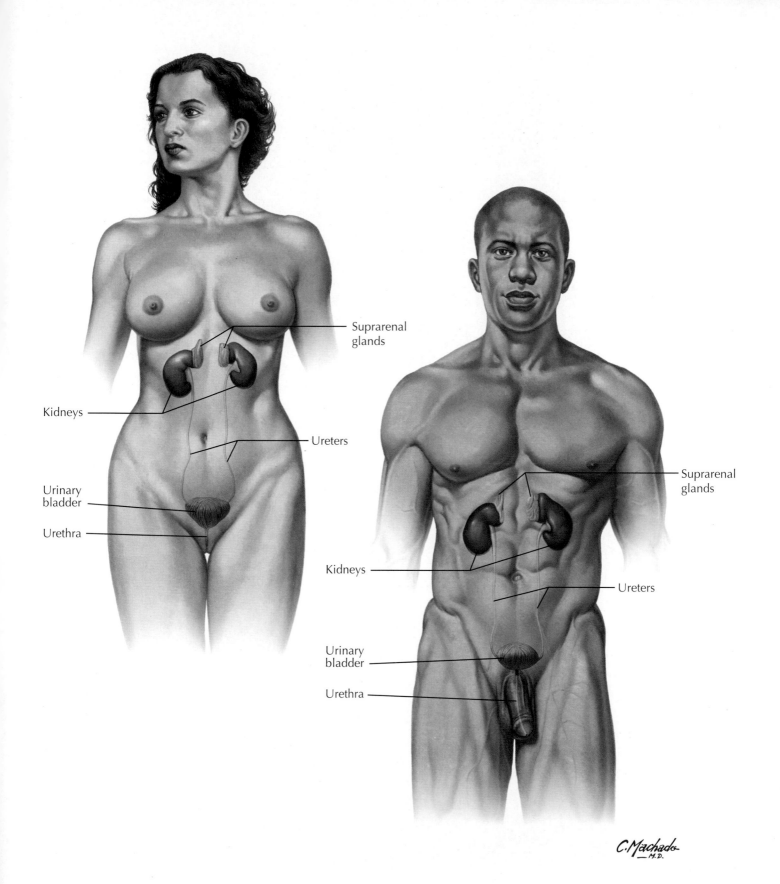

Suprarenal glands

Kidneys

Ureters

Urinary bladder

Urethra

Suprarenal glands

Kidneys

Ureters

Urinary bladder

Urethra

C. Machado
_M.D.

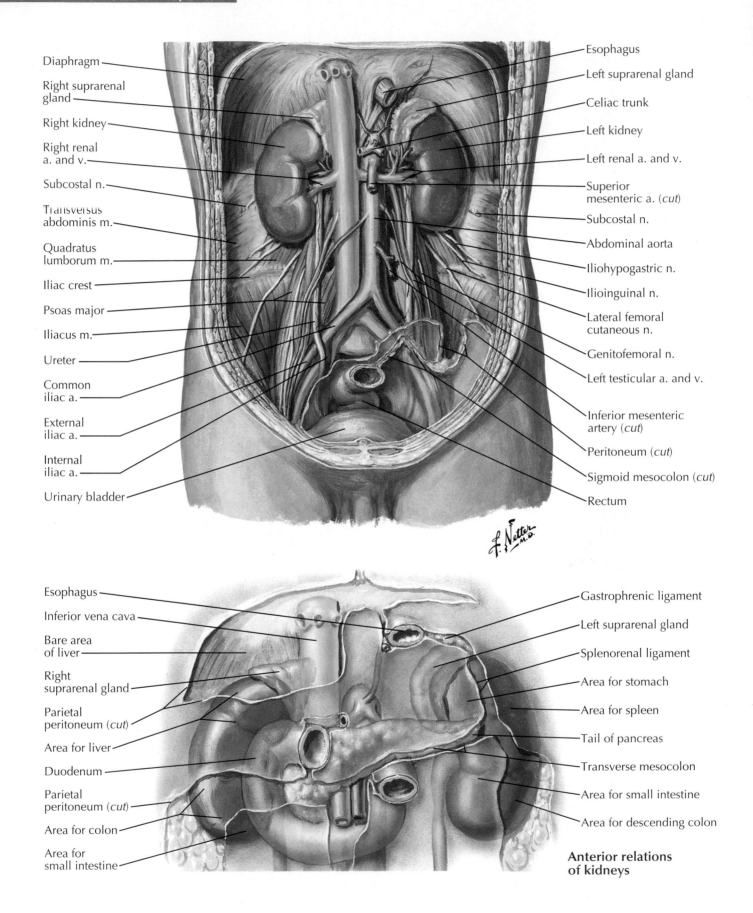

Diaphragm

Right suprarenal gland

Right kidney

Right renal a. and v.

Subcostal n.

Transversus abdominis m.

Quadratus lumborum m.

Iliac crest

Psoas major

Iliacus m.

Ureter

Common iliac a.

External iliac a.

Internal iliac a.

Urinary bladder

Esophagus

Left suprarenal gland

Celiac trunk

Left kidney

Left renal a. and v.

Superior mesenteric a. (*cut*)

Subcostal n.

Abdominal aorta

Iliohypogastric n.

Ilioinguinal n.

Lateral femoral cutaneous n.

Genitofemoral n.

Left testicular a. and v.

Inferior mesenteric artery (*cut*)

Peritoneum (*cut*)

Sigmoid mesocolon (*cut*)

Rectum

Esophagus

Inferior vena cava

Bare area of liver

Right suprarenal gland

Parietal peritoneum (*cut*)

Area for liver

Duodenum

Parietal peritoneum (*cut*)

Area for colon

Area for small intestine

Gastrophrenic ligament

Left suprarenal gland

Splenorenal ligament

Area for stomach

Area for spleen

Tail of pancreas

Transverse mesocolon

Area for small intestine

Area for descending colon

Anterior relations of kidneys

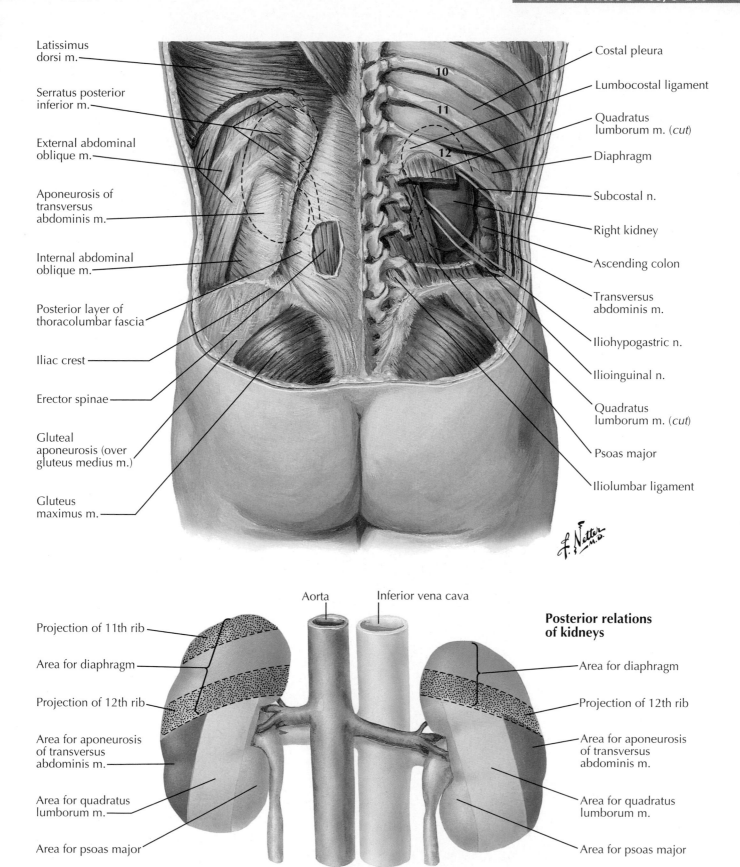

Latissimus dorsi m.

Serratus posterior inferior m.

External abdominal oblique m.

Aponeurosis of transversus abdominis m.

Internal abdominal oblique m.

Posterior layer of thoracolumbar fascia

Iliac crest

Erector spinae

Gluteal aponeurosis (over gluteus medius m.)

Gluteus maximus m.

10

11

12

Costal pleura

Lumbocostal ligament

Quadratus lumborum m. (cut)

Diaphragm

Subcostal n.

Right kidney

Ascending colon

Transversus abdominis m.

Iliohypogastric n.

Ilioinguinal n.

Quadratus lumborum m. (cut)

Psoas major

Iliolumbar ligament

Aorta

Inferior vena cava

Posterior relations of kidneys

Projection of 11th rib

Area for diaphragm

Projection of 12th rib

Area for aponeurosis of transversus abdominis m.

Area for quadratus lumborum m.

Area for psoas major

Area for diaphragm

Projection of 12th rib

Area for aponeurosis of transversus abdominis m.

Area for quadratus lumborum m.

Area for psoas major

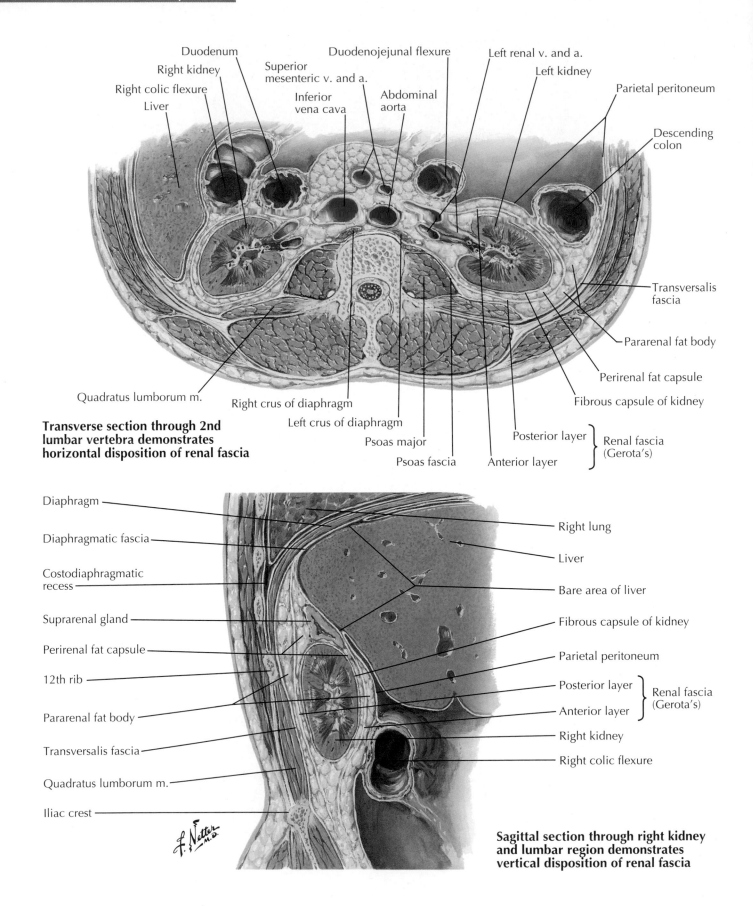

Duodenum
Right kidney
Right colic flexure
Liver
Superior mesenteric v. and a.
Inferior vena cava
Duodenojejunal flexure
Abdominal aorta
Left renal v. and a.
Left kidney
Parietal peritoneum
Descending colon

Quadratus lumborum m.
Right crus of diaphragm
Left crus of diaphragm
Psoas major
Psoas fascia
Posterior layer
Anterior layer
Renal fascia (Gerota's)
Transversalis fascia
Pararenal fat body
Perirenal fat capsule
Fibrous capsule of kidney

Transverse section through 2nd lumbar vertebra demonstrates horizontal disposition of renal fascia

Diaphragm
Diaphragmatic fascia
Costodiaphragmatic recess
Suprarenal gland
Perirenal fat capsule
12th rib
Pararenal fat body
Transversalis fascia
Quadratus lumborum m.
Iliac crest

Right lung
Liver
Bare area of liver
Fibrous capsule of kidney
Parietal peritoneum
Posterior layer
Anterior layer
Renal fascia (Gerota's)
Right kidney
Right colic flexure

Sagittal section through right kidney and lumbar region demonstrates vertical disposition of renal fascia

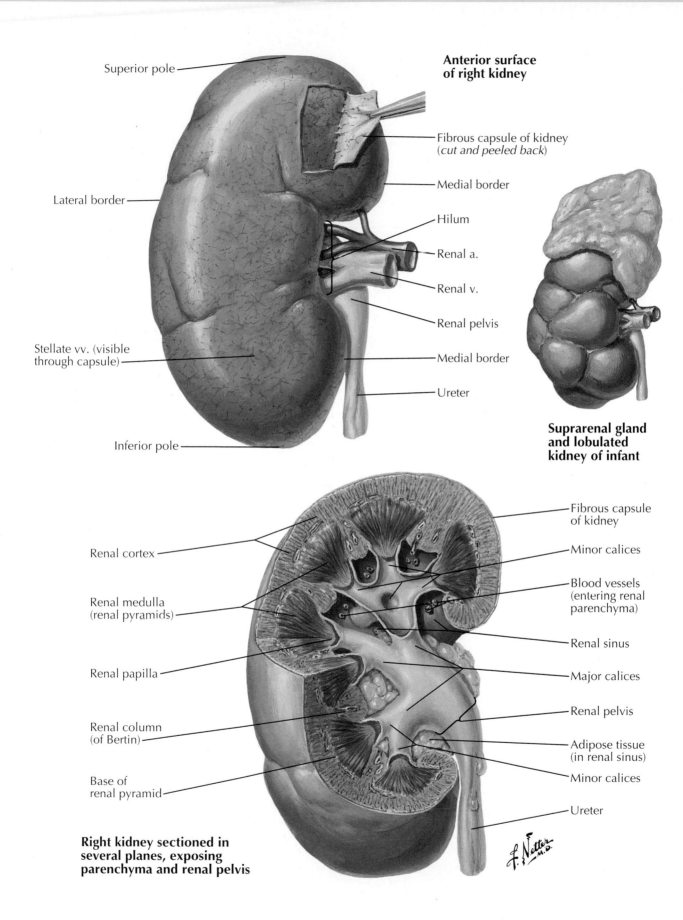

Superior pole

**Anterior surface
of right kidney**

Fibrous capsule of kidney
(*cut and peeled back*)

Medial border

Lateral border

Hilum

Renal a.

Renal v.

Renal pelvis

Stellate vv. (visible
through capsule)

Medial border

Ureter

Inferior pole

**Suprarenal gland
and lobulated
kidney of infant**

Renal cortex

Fibrous capsule
of kidney

Minor calices

Renal medulla
(renal pyramids)

Blood vessels
(entering renal
parenchyma)

Renal sinus

Renal papilla

Major calices

Renal pelvis

Renal column
(of Bertin)

Adipose tissue
(in renal sinus)

Minor calices

Base of
renal pyramid

Ureter

**Right kidney sectioned in
several planes, exposing
parenchyma and renal pelvis**

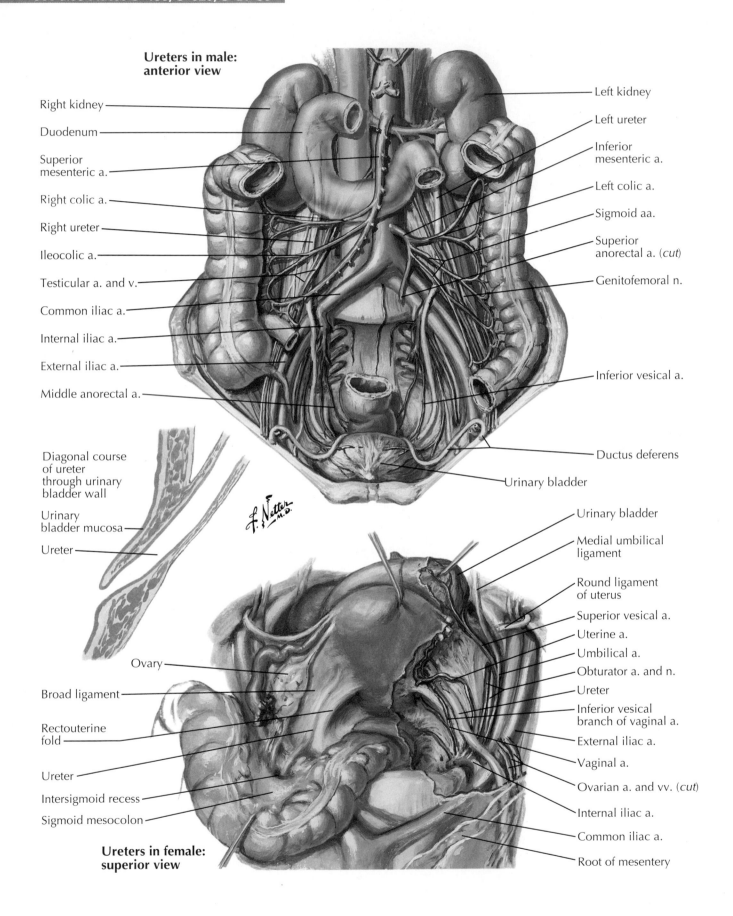

Ureters in male: anterior view

Right kidney

Duodenum

Superior mesenteric a.

Right colic a.

Right ureter

Ileocolic a.

Testicular a. and v.

Common iliac a.

Internal iliac a.

External iliac a.

Middle anorectal a.

Left kidney

Left ureter

Inferior mesenteric a.

Left colic a.

Sigmoid aa.

Superior anorectal a. (*cut*)

Genitofemoral n.

Inferior vesical a.

Ductus deferens

Urinary bladder

Diagonal course of ureter through urinary bladder wall

Urinary bladder mucosa

Ureter

Ovary

Broad ligament

Rectouterine fold

Ureter

Intersigmoid recess

Sigmoid mesocolon

Ureters in female: superior view

Urinary bladder

Medial umbilical ligament

Round ligament of uterus

Superior vesical a.

Uterine a.

Umbilical a.

Obturator a. and n.

Ureter

Inferior vesical branch of vaginal a.

External iliac a.

Vaginal a.

Ovarian a. and vv. (*cut*)

Internal iliac a.

Common iliac a.

Root of mesentery

Female: median section

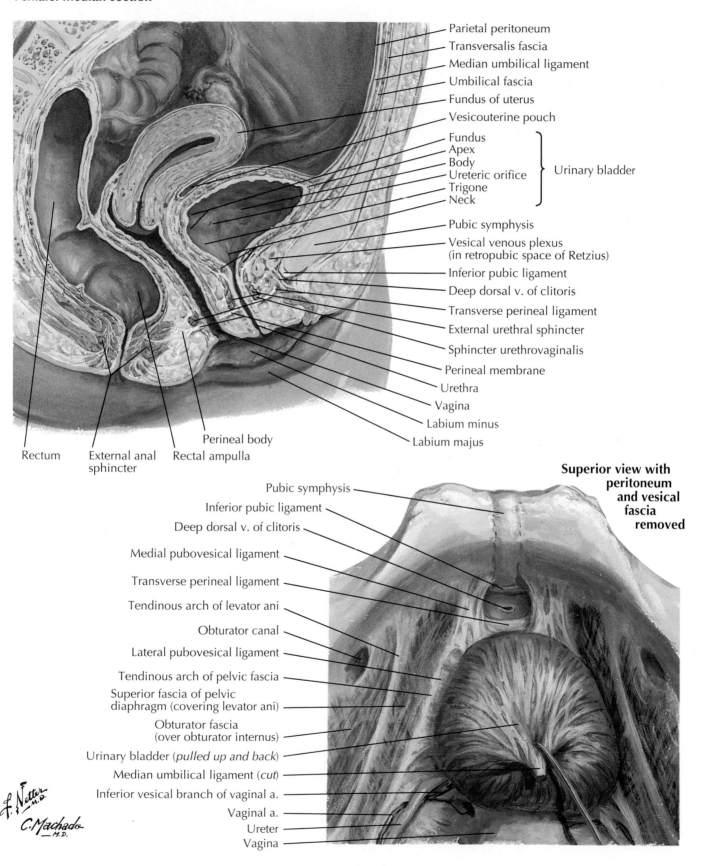

Parietal peritoneum
Transversalis fascia
Median umbilical ligament
Umbilical fascia
Fundus of uterus
Vesicouterine pouch
Fundus
Apex
Body
Ureteric orifice
Trigone
Neck
Urinary bladder
Pubic symphysis
Vesical venous plexus
(in retropubic space of Retzius)
Inferior pubic ligament
Deep dorsal v. of clitoris
Transverse perineal ligament
External urethral sphincter
Sphincter urethrovaginalis
Perineal membrane
Urethra
Vagina
Labium minus
Labium majus

Rectum
External anal sphincter
Rectal ampulla
Perineal body

Superior view with peritoneum and vesical fascia removed

Pubic symphysis
Inferior pubic ligament
Deep dorsal v. of clitoris
Medial pubovesical ligament
Transverse perineal ligament
Tendinous arch of levator ani
Obturator canal
Lateral pubovesical ligament
Tendinous arch of pelvic fascia
Superior fascia of pelvic diaphragm (covering levator ani)
Obturator fascia (over obturator internus)
Urinary bladder (*pulled up and back*)
Median umbilical ligament (*cut*)
Inferior vesical branch of vaginal a.
Vaginal a.
Ureter
Vagina

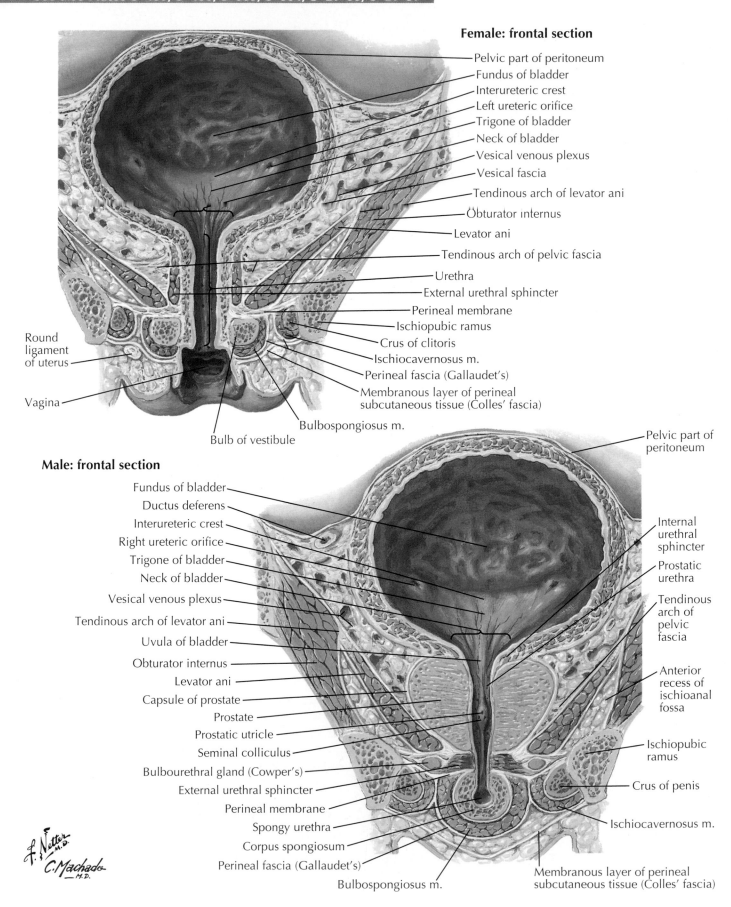

Female: frontal section

Pelvic part of peritoneum
Fundus of bladder
Interureteric crest
Left ureteric orifice
Trigone of bladder
Neck of bladder
Vesical venous plexus
Vesical fascia
Tendinous arch of levator ani
Obturator internus
Levator ani
Tendinous arch of pelvic fascia
Urethra
External urethral sphincter
Perineal membrane
Ischiopubic ramus
Crus of clitoris
Ischiocavernosus m.
Perineal fascia (Gallaudet's)
Membranous layer of perineal subcutaneous tissue (Colles' fascia)

Round ligament of uterus

Vagina

Bulbospongiosus m.

Bulb of vestibule

Male: frontal section

Fundus of bladder
Ductus deferens
Interureteric crest
Right ureteric orifice
Trigone of bladder
Neck of bladder
Vesical venous plexus
Tendinous arch of levator ani
Uvula of bladder
Obturator internus
Levator ani
Capsule of prostate
Prostate
Prostatic utricle
Seminal colliculus
Bulbourethral gland (Cowper's)
External urethral sphincter
Perineal membrane
Spongy urethra
Corpus spongiosum
Perineal fascia (Gallaudet's)
Bulbospongiosus m.

Pelvic part of peritoneum
Internal urethral sphincter
Prostatic urethra
Tendinous arch of pelvic fascia
Anterior recess of ischioanal fossa
Ischiopubic ramus
Crus of penis
Ischiocavernosus m.
Membranous layer of perineal subcutaneous tissue (Colles' fascia)

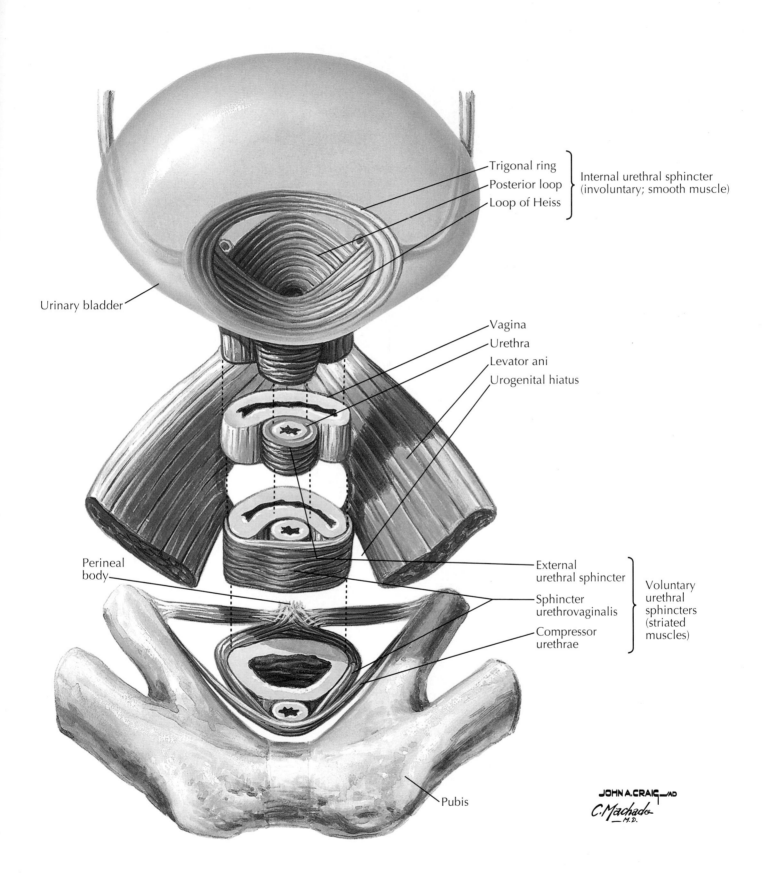

Trigonal ring
Posterior loop
Loop of Heiss

Internal urethral sphincter (involuntary; smooth muscle)

Urinary bladder

Vagina
Urethra
Levator ani
Urogenital hiatus

Perineal body

External urethral sphincter

Sphincter urethrovaginalis

Compressor urethrae

Voluntary urethral sphincters (striated muscles)

Pubis

JOHN A. CRAIG—MD
C. Machado—M.D.

Urinary Bladder and Urethra

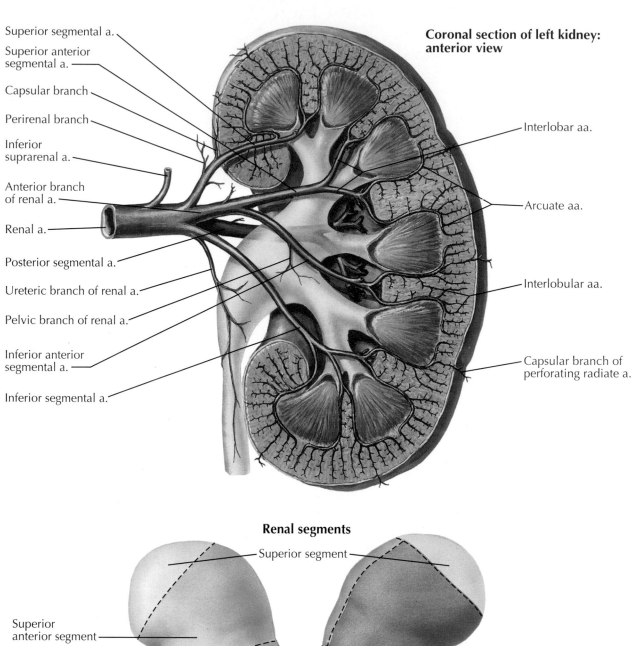

Coronal section of left kidney: anterior view

Superior segmental a.

Superior anterior segmental a.

Capsular branch

Perirenal branch

Inferior suprarenal a.

Anterior branch of renal a.

Renal a.

Posterior segmental a.

Ureteric branch of renal a.

Pelvic branch of renal a.

Inferior anterior segmental a.

Inferior segmental a.

Interlobar aa.

Arcuate aa.

Interlobular aa.

Capsular branch of perforating radiate a.

Renal segments

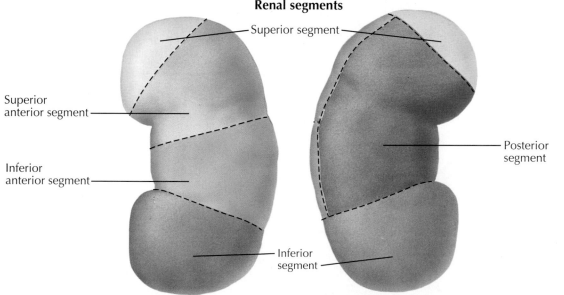

Superior segment

Superior anterior segment

Inferior anterior segment

Posterior segment

Inferior segment

Anterior surface of left kidney

Posterior surface of left kidney

Vasculature and Innervation

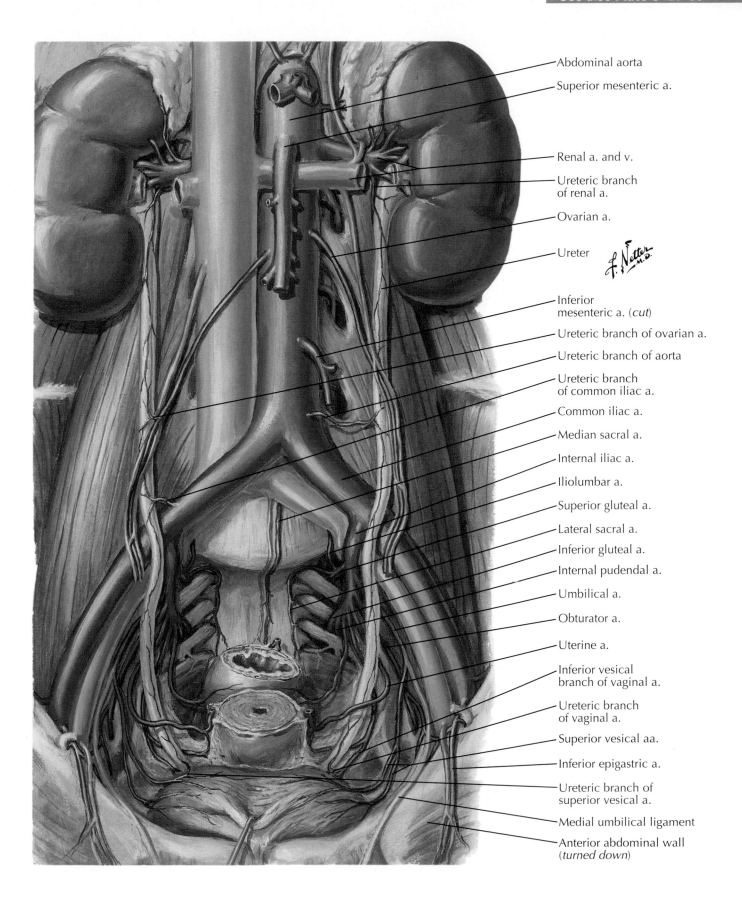

Abdominal aorta

Superior mesenteric a.

Renal a. and v.

Ureteric branch of renal a.

Ovarian a.

Ureter

Inferior mesenteric a. (*cut*)

Ureteric branch of ovarian a.

Ureteric branch of aorta

Ureteric branch of common iliac a.

Common iliac a.

Median sacral a.

Internal iliac a.

Iliolumbar a.

Superior gluteal a.

Lateral sacral a.

Inferior gluteal a.

Internal pudendal a.

Umbilical a.

Obturator a.

Uterine a.

Inferior vesical branch of vaginal a.

Ureteric branch of vaginal a.

Superior vesical aa.

Inferior epigastric a.

Ureteric branch of superior vesical a.

Medial umbilical ligament

Anterior abdominal wall (*turned down*)

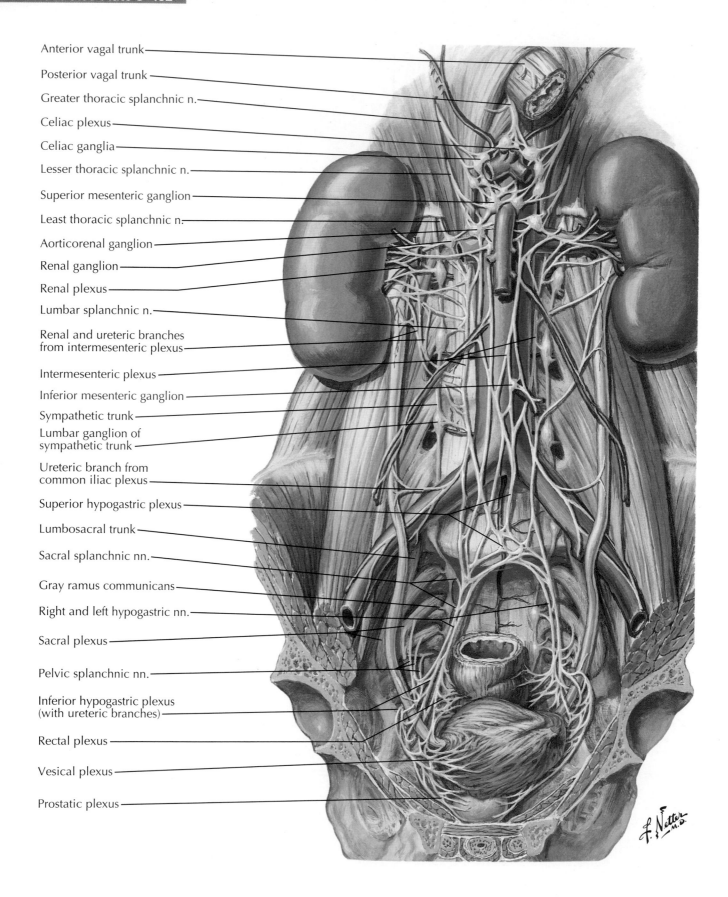

Anterior vagal trunk

Posterior vagal trunk

Greater thoracic splanchnic n.

Celiac plexus

Celiac ganglia

Lesser thoracic splanchnic n.

Superior mesenteric ganglion

Least thoracic splanchnic n.

Aorticorenal ganglion

Renal ganglion

Renal plexus

Lumbar splanchnic n.

Renal and ureteric branches from intermesenteric plexus

Intermesenteric plexus

Inferior mesenteric ganglion

Sympathetic trunk

Lumbar ganglion of sympathetic trunk

Ureteric branch from common iliac plexus

Superior hypogastric plexus

Lumbosacral trunk

Sacral splanchnic nn.

Gray ramus communicans

Right and left hypogastric nn.

Sacral plexus

Pelvic splanchnic nn.

Inferior hypogastric plexus (with ureteric branches)

Rectal plexus

Vesical plexus

Prostatic plexus

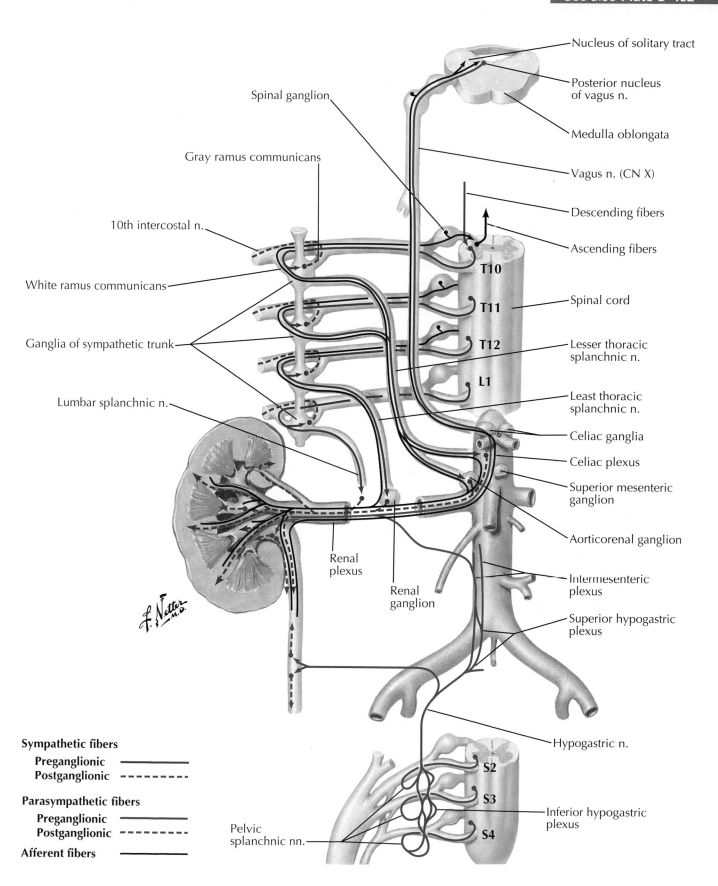

Nucleus of solitary tract

Spinal ganglion

Posterior nucleus of vagus n.

Medulla oblongata

Gray ramus communicans

Vagus n. (CN X)

Descending fibers

10th intercostal n.

Ascending fibers

White ramus communicans

T10

Spinal cord

T11

Ganglia of sympathetic trunk

T12

Lesser thoracic splanchnic n.

L1

Lumbar splanchnic n.

Least thoracic splanchnic n.

Celiac ganglia

Celiac plexus

Superior mesenteric ganglion

Aorticorenal ganglion

Renal plexus

Intermesenteric plexus

Renal ganglion

Superior hypogastric plexus

Hypogastric n.

Sympathetic fibers

Preganglionic ————
Postganglionic - - - - -

S2

Parasympathetic fibers

S3

Preganglionic ————
Postganglionic - - - - -

Inferior hypogastric plexus

Afferent fibers ————

S4

Pelvic splanchnic nn.

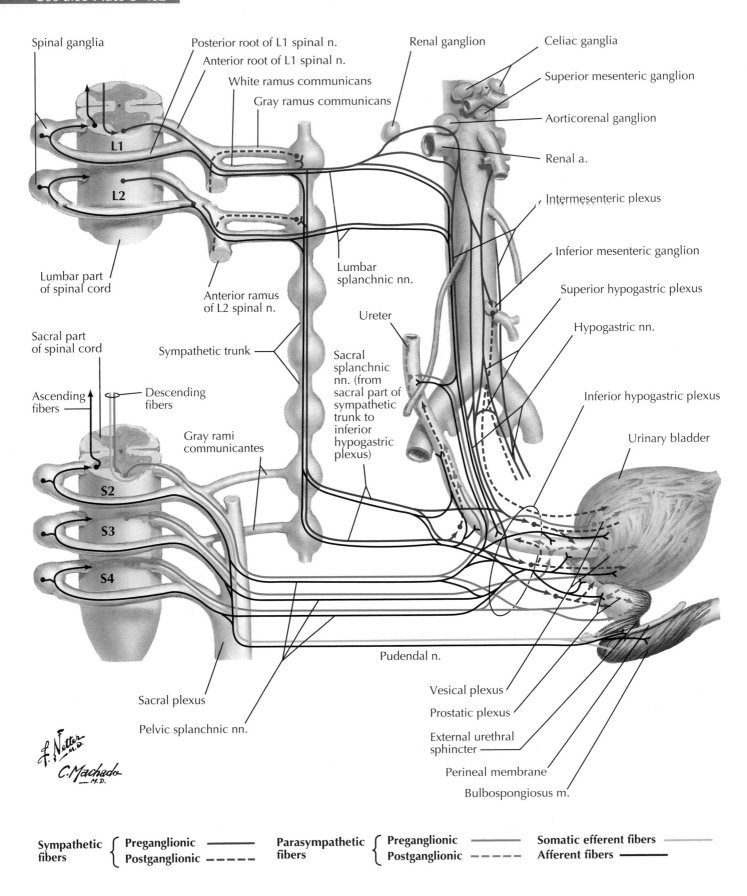

Spinal ganglia

Posterior root of L1 spinal n.

Anterior root of L1 spinal n.

White ramus communicans

Gray ramus communicans

Renal ganglion

Celiac ganglia

Superior mesenteric ganglion

Aorticorenal ganglion

L1

L2

Renal a.

Intermesenteric plexus

Lumbar part
of spinal cord

Lumbar
splanchnic nn.

Inferior mesenteric ganglion

Superior hypogastric plexus

Anterior ramus
of L2 spinal n.

Ureter

Hypogastric nn.

Sacral part
of spinal cord

Sympathetic trunk

Sacral
splanchnic
nn. (from
sacral part of
sympathetic
trunk to
inferior
hypogastric
plexus)

Inferior hypogastric plexus

Ascending
fibers

Descending
fibers

Urinary bladder

Gray rami
communicantes

S2

S3

S4

Sacral plexus

Pelvic splanchnic nn.

Pudendal n.

Vesical plexus

Prostatic plexus

External urethral
sphincter

Perineal membrane

Bulbospongiosus m.

| Sympathetic fibers | Preganglionic ———— | Parasympathetic fibers | Preganglionic ———— | Somatic efferent fibers ———— |
| | Postganglionic - - - - | | Postganglionic - - - - | Afferent fibers ———— |

ANATOMIC STRUCTURES	CLINICAL IMPORTANCE	PLATE NUMBERS
Abdomen		
Kidney	Right is lower or more inferior than left kidney due to position inferior to liver; renal arteries are generally located at L2 vertebral level and may be involved in abdominal aortic aneurysms; renal calculi (kidney stones) cause significant pain when impacted in ureter	S–460, S–462
Renal pelvis	May become dilated secondary to ureteral or urinary bladder outlet obstruction, a condition known as hydronephrosis, which is readily visualized by ultrasound	S–462
Pelvis		
Urinary bladder	Degree of filling is readily assessed with ultrasound; in patients with poor urine output, this can establish diagnosis of bladder outlet obstruction	S–464, S–495
Ureteric orifice	Abnormal reflux of urine from urinary bladder to ureters (vesicoureteral reflux) may occur in children, contributing to recurrent urinary infections and progressive renal fibrosis	S–465
Ureter	Enlargement indicates obstruction of ureter or urinary bladder; impaction of renal stone in ureter causes severe pain and, in some cases, hematuria; ureter may be injured during hysterectomy because of its close relationship to uterine artery	S–468, S–490, S–529

*Selections are based largely on clinical data and commonly discussed clinical correlations in macroscopic ("gross") anatomy courses.

Structures with High Clinical Significance

Table 9.1

REPRODUCTIVE SYSTEM 10

ELECTRONIC BONUS PLATES

S–BP 89 Genetics of Reproduction

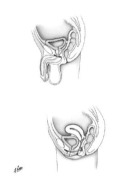

S–BP 90 Fasciae of Pelvis and Perineum: Male and Female

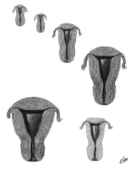

S–BP 91 Uterine Development

S–BP 92 Variations in Hymen

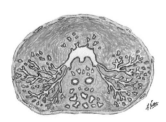

S–BP 93 Cross Section of Pelvis Through Prostate

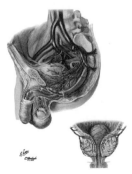

S–BP 94 Arteries and Veins of Pelvis: Male (Featuring Prostate)

Plate S–472

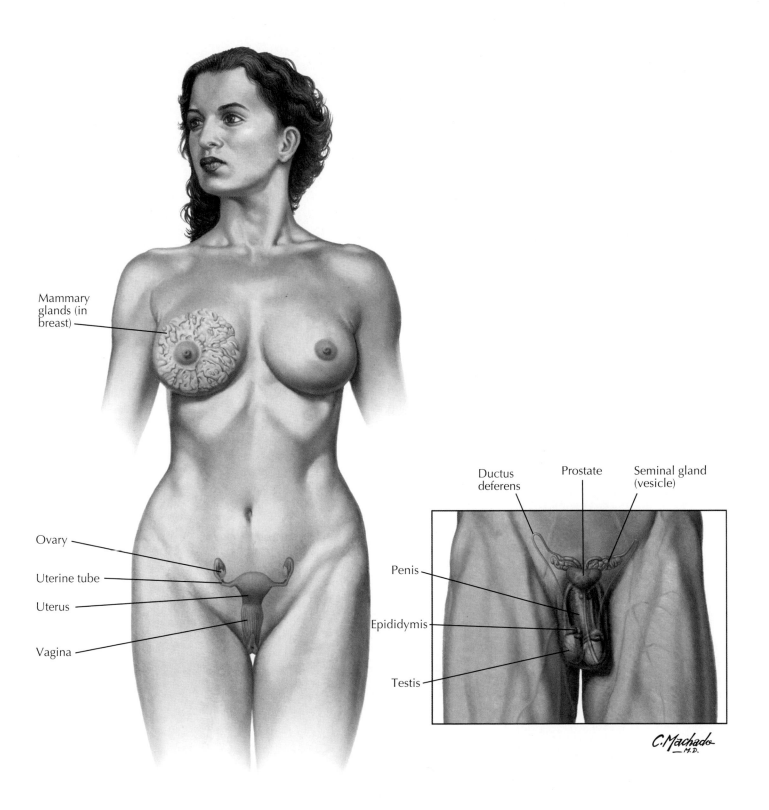

Mammary glands (in breast)

Ovary

Uterine tube

Uterus

Vagina

Ductus deferens

Prostate

Seminal gland (vesicle)

Penis

Epididymis

Testis

C.Machado
M.D.

Anterolateral dissection

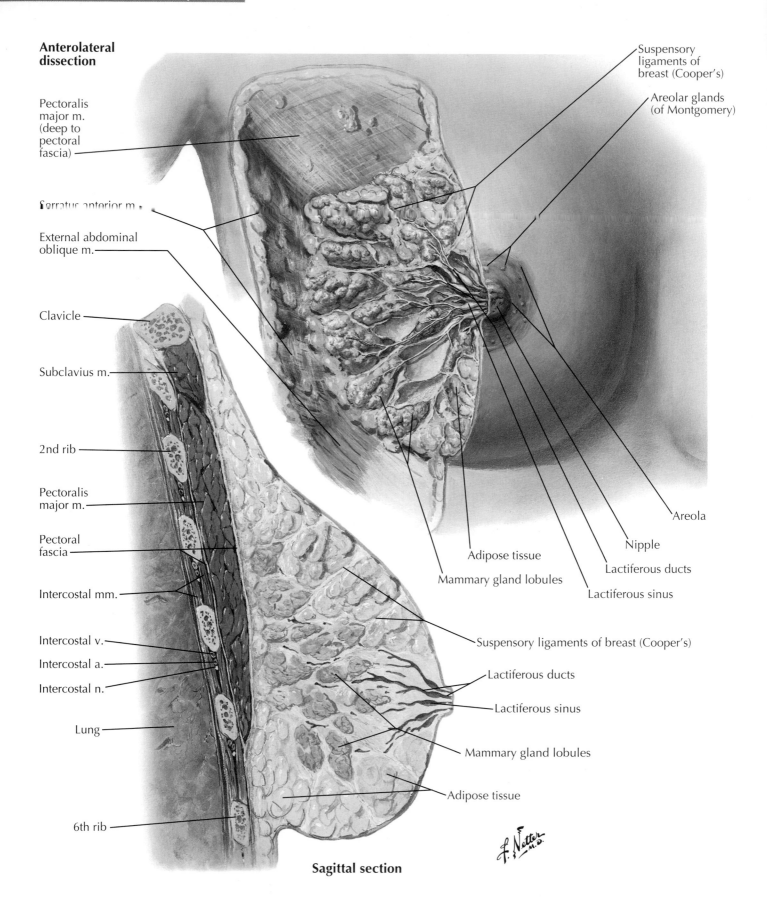

Pectoralis major m. (deep to pectoral fascia)

Serratus anterior m.

External abdominal oblique m.

Clavicle

Subclavius m.

2nd rib

Pectoralis major m.

Pectoral fascia

Intercostal mm.

Intercostal v.

Intercostal a.

Intercostal n.

Lung

6th rib

Suspensory ligaments of breast (Cooper's)

Areolar glands (of Montgomery)

Areola

Nipple

Lactiferous ducts

Lactiferous sinus

Adipose tissue

Mammary gland lobules

Suspensory ligaments of breast (Cooper's)

Lactiferous ducts

Lactiferous sinus

Mammary gland lobules

Adipose tissue

Sagittal section

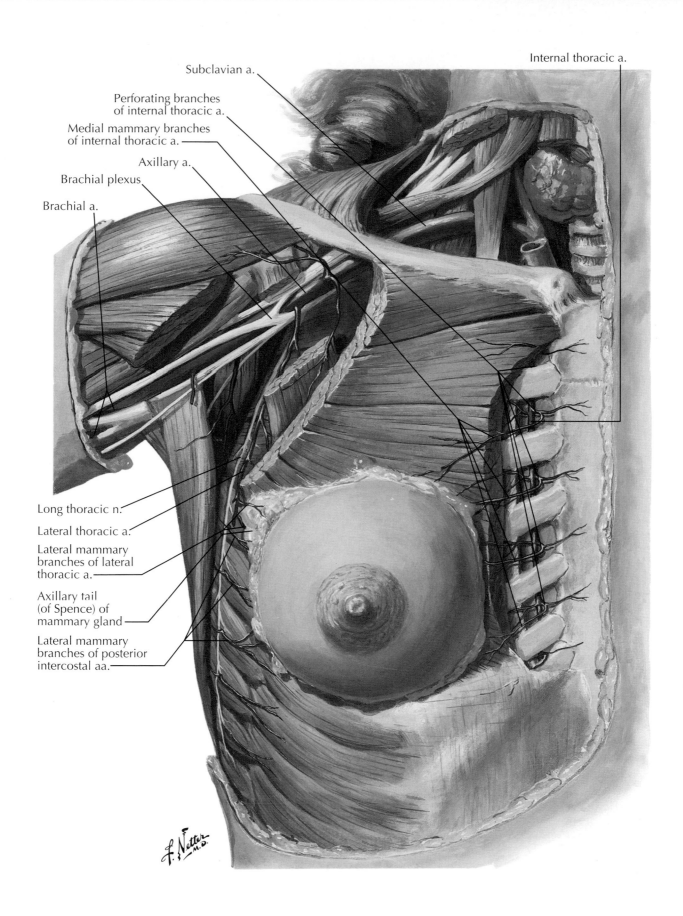

Internal thoracic a.

Subclavian a.

Perforating branches
of internal thoracic a.

Medial mammary branches
of internal thoracic a.

Axillary a.

Brachial plexus

Brachial a.

Long thoracic n.

Lateral thoracic a.

Lateral mammary
branches of lateral
thoracic a.

Axillary tail
(of Spence) of
mammary gland

Lateral mammary
branches of posterior
intercostal aa.

Mammary Glands

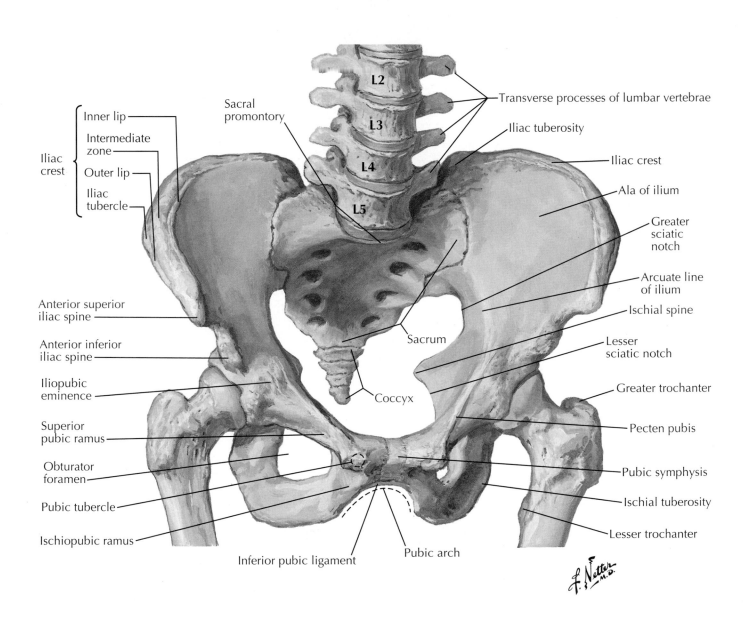

Inner lip

Intermediate zone

Iliac crest

Outer lip

Iliac tubercle

Sacral promontory

L2

L3

L4

L5

Transverse processes of lumbar vertebrae

Iliac tuberosity

Iliac crest

Ala of ilium

Greater sciatic notch

Arcuate line of ilium

Ischial spine

Lesser sciatic notch

Greater trochanter

Pecten pubis

Pubic symphysis

Ischial tuberosity

Lesser trochanter

Anterior superior iliac spine

Anterior inferior iliac spine

Iliopubic eminence

Superior pubic ramus

Obturator foramen

Pubic tubercle

Ischiopubic ramus

Sacrum

Coccyx

Inferior pubic ligament

Pubic arch

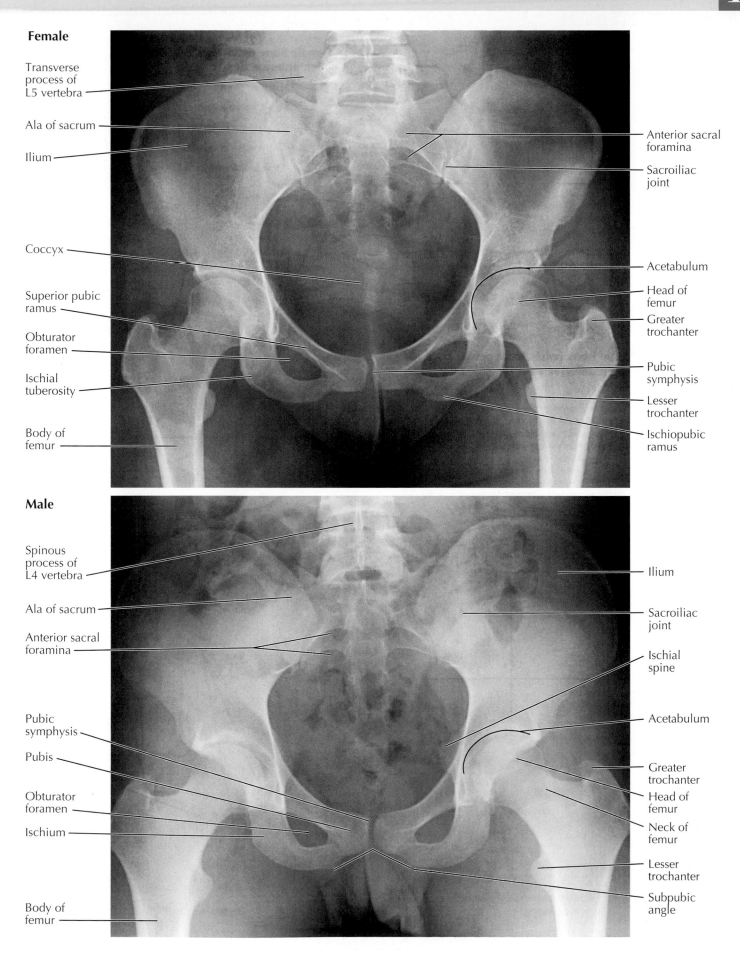

Female

Transverse process of L5 vertebra

Ala of sacrum

Ilium

Coccyx

Superior pubic ramus

Obturator foramen

Ischial tuberosity

Body of femur

Anterior sacral foramina

Sacroiliac joint

Acetabulum

Head of femur

Greater trochanter

Pubic symphysis

Lesser trochanter

Ischiopubic ramus

Male

Spinous process of L4 vertebra

Ala of sacrum

Anterior sacral foramina

Pubic symphysis

Pubis

Obturator foramen

Ischium

Body of femur

Ilium

Sacroiliac joint

Ischial spine

Acetabulum

Greater trochanter

Head of femur

Neck of femur

Lesser trochanter

Subpubic angle

Bony Pelvis

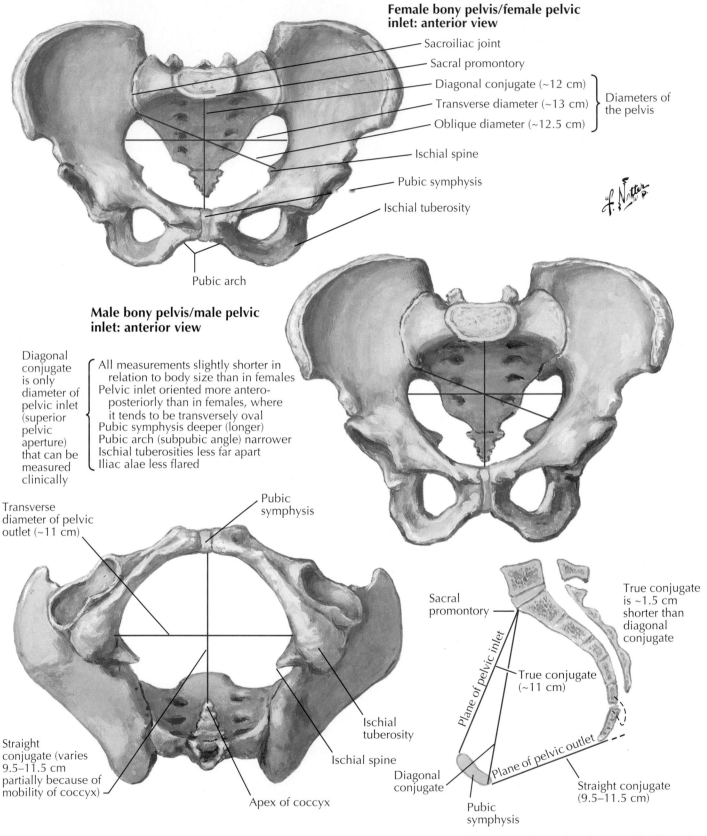

Female bony pelvis/female pelvic inlet: anterior view

Sacroiliac joint

Sacral promontory

Diagonal conjugate (~12 cm)

Transverse diameter (~13 cm)

Oblique diameter (~12.5 cm)

Diameters of the pelvis

Ischial spine

Pubic symphysis

Ischial tuberosity

Pubic arch

Male bony pelvis/male pelvic inlet: anterior view

Diagonal conjugate is only diameter of pelvic inlet (superior pelvic aperture) that can be measured clinically

{ All measurements slightly shorter in relation to body size than in females
Pelvic inlet oriented more antero-posteriorly than in females, where it tends to be transversely oval
Pubic symphysis deeper (longer)
Pubic arch (subpubic angle) narrower
Ischial tuberosities less far apart
Iliac alae less flared }

Transverse diameter of pelvic outlet (~11 cm)

Pubic symphysis

Straight conjugate (varies 9.5–11.5 cm partially because of mobility of coccyx)

Ischial tuberosity

Ischial spine

Apex of coccyx

Female bony pelvis/female pelvic outlet: inferior view

Sacral promontory

True conjugate is ~1.5 cm shorter than diagonal conjugate

Plane of pelvic inlet

True conjugate (~11 cm)

Diagonal conjugate

Pubic symphysis

Plane of pelvic outlet

Straight conjugate (9.5–11.5 cm)

Transverse diameter is the widest distance of pelvic inlet

Female: sagittal section

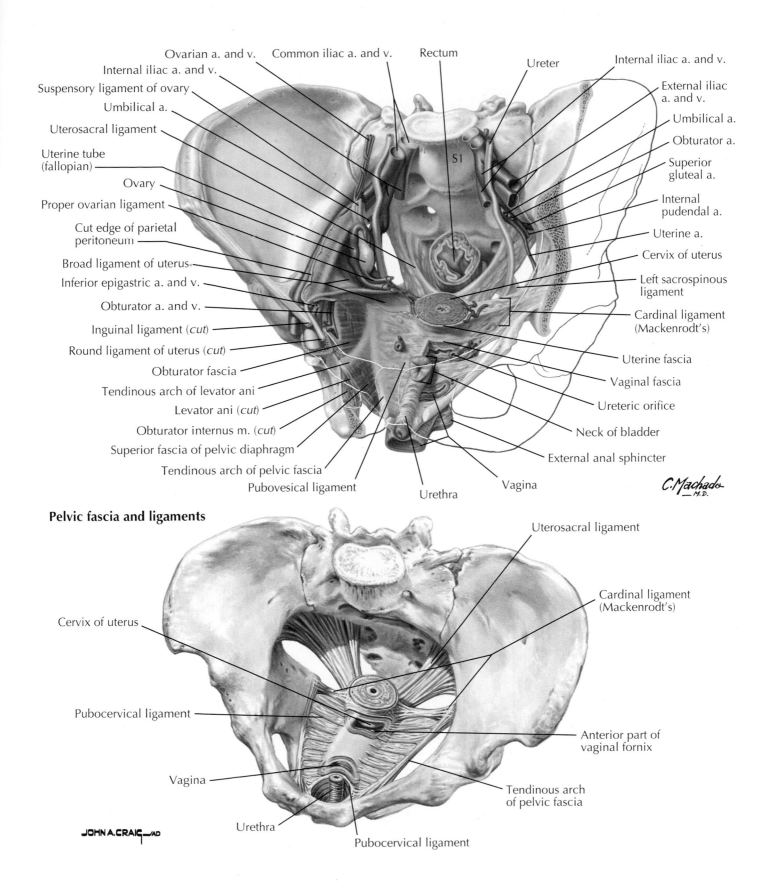

Ovarian a. and v.
Common iliac a. and v.
Rectum
Ureter
Internal iliac a. and v.
Internal iliac a. and v.
External iliac a. and v.
Suspensory ligament of ovary
Umbilical a.
Umbilical a.
Uterosacral ligament
Obturator a.
Uterine tube (fallopian)
Superior gluteal a.
Ovary
S1
Internal pudendal a.
Proper ovarian ligament
Cut edge of parietal peritoneum
Uterine a.
Broad ligament of uterus
Cervix of uterus
Inferior epigastric a. and v.
Left sacrospinous ligament
Obturator a. and v.
Cardinal ligament (Mackenrodt's)
Inguinal ligament (cut)
Round ligament of uterus (cut)
Uterine fascia
Obturator fascia
Vaginal fascia
Tendinous arch of levator ani
Ureteric orifice
Levator ani (cut)
Obturator internus m. (cut)
Neck of bladder
Superior fascia of pelvic diaphragm
External anal sphincter
Tendinous arch of pelvic fascia
Pubovesical ligament
Vagina
Urethra

C. Machado M.D.

Pelvic fascia and ligaments

Uterosacral ligament

Cervix of uterus

Cardinal ligament (Mackenrodt's)

Pubocervical ligament

Anterior part of vaginal fornix

Vagina

Tendinous arch of pelvic fascia

Urethra

Pubocervical ligament

JOHN A. CRAIG—AD

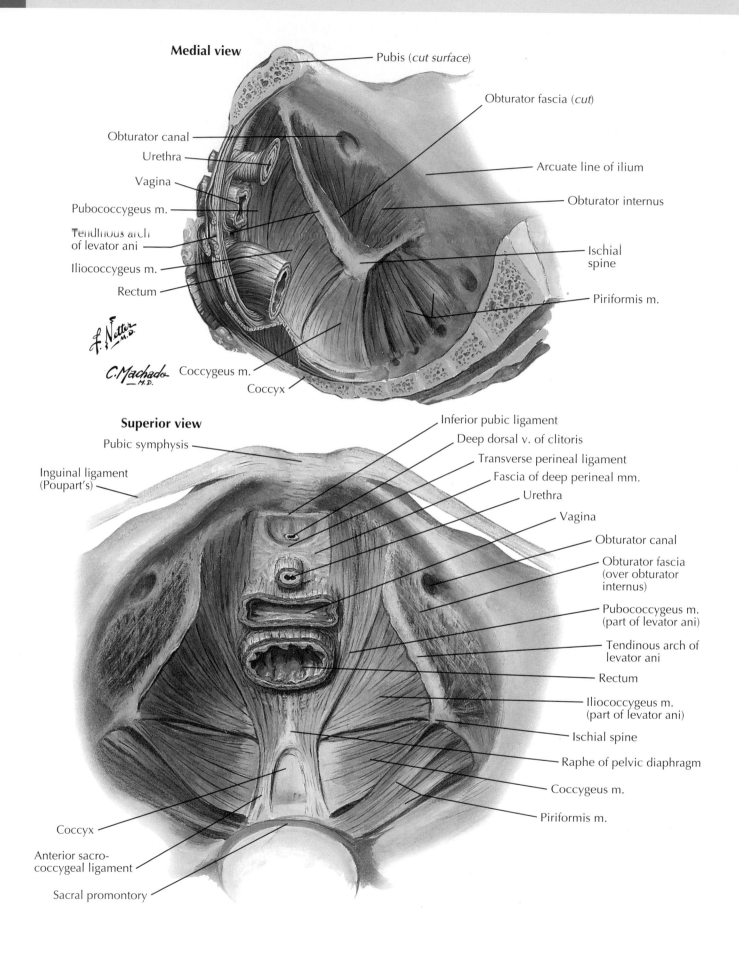

Medial view

Pubis (*cut surface*)

Obturator canal

Urethra

Vagina

Pubococcygeus m.

Tendinous arch
of levator ani

Iliococcygeus m.

Rectum

Obturator fascia (*cut*)

Arcuate line of ilium

Obturator internus

Ischial
spine

Piriformis m.

Coccygeus m.

Coccyx

Superior view

Pubic symphysis

Inguinal ligament
(Poupart's)

Inferior pubic ligament

Deep dorsal v. of clitoris

Transverse perineal ligament

Fascia of deep perineal mm.

Urethra

Vagina

Obturator canal

Obturator fascia
(over obturator
internus)

Pubococcygeus m.
(part of levator ani)

Tendinous arch of
levator ani

Rectum

Iliococcygeus m.
(part of levator ani)

Ischial spine

Raphe of pelvic diaphragm

Coccygeus m.

Piriformis m.

Coccyx

Anterior sacro-
coccygeal ligament

Sacral promontory

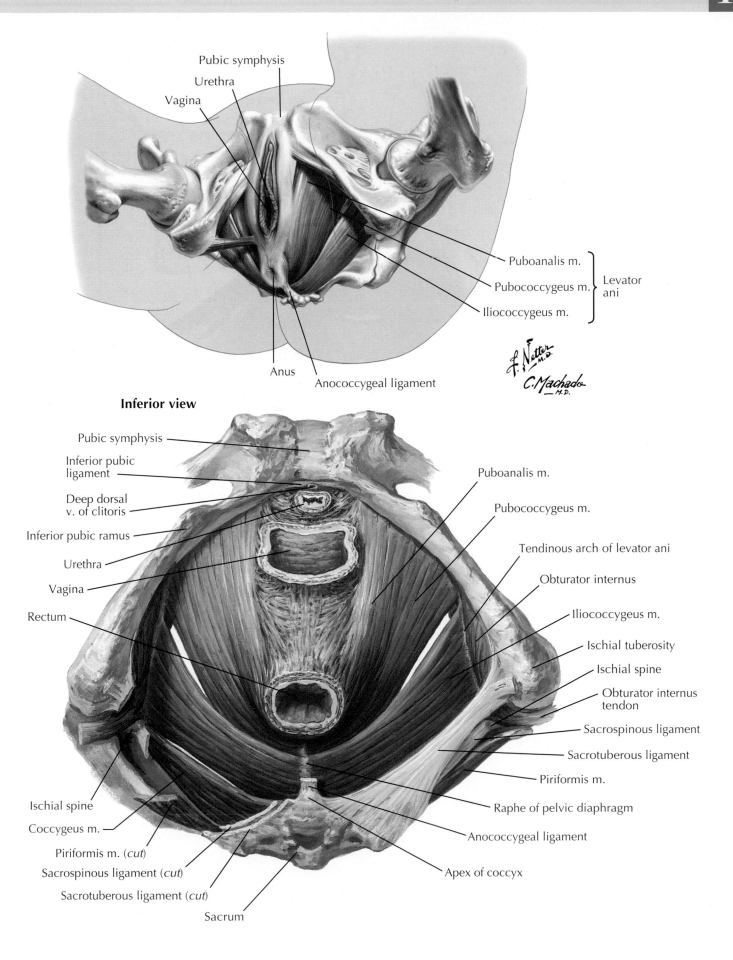

Pubic symphysis

Urethra

Vagina

Puboanalis m.

Pubococcygeus m. } Levator ani

Iliococcygeus m.

Anus

Anococcygeal ligament

Inferior view

Pubic symphysis

Inferior pubic ligament

Deep dorsal v. of clitoris

Inferior pubic ramus

Urethra

Vagina

Rectum

Ischial spine

Coccygeus m.

Piriformis m. (*cut*)

Sacrospinous ligament (*cut*)

Sacrotuberous ligament (*cut*)

Sacrum

Puboanalis m.

Pubococcygeus m.

Tendinous arch of levator ani

Obturator internus

Iliococcygeus m.

Ischial tuberosity

Ischial spine

Obturator internus tendon

Sacrospinous ligament

Sacrotuberous ligament

Piriformis m.

Raphe of pelvic diaphragm

Anococcygeal ligament

Apex of coccyx

Superior view
(*viscera removed*)

Pubic symphysis

Pubic crest

Pubic tubercle

Pecten pubis

Superior ramus of pubis

Obturator canal

Obturator fascia

Iliopubic eminence

Acetabular margin

Anterior inferior
iliac spine

Inferior pubic ligament

Hiatus for deep dorsal v. of penis

Transverse perineal ligament

Perineal membrane

Urogenital hiatus

Puboanalis m.

Pubococcygeus m. } Levator ani

Iliococcygeus m.

Tendinous arch
of levator ani

Obturator
internus

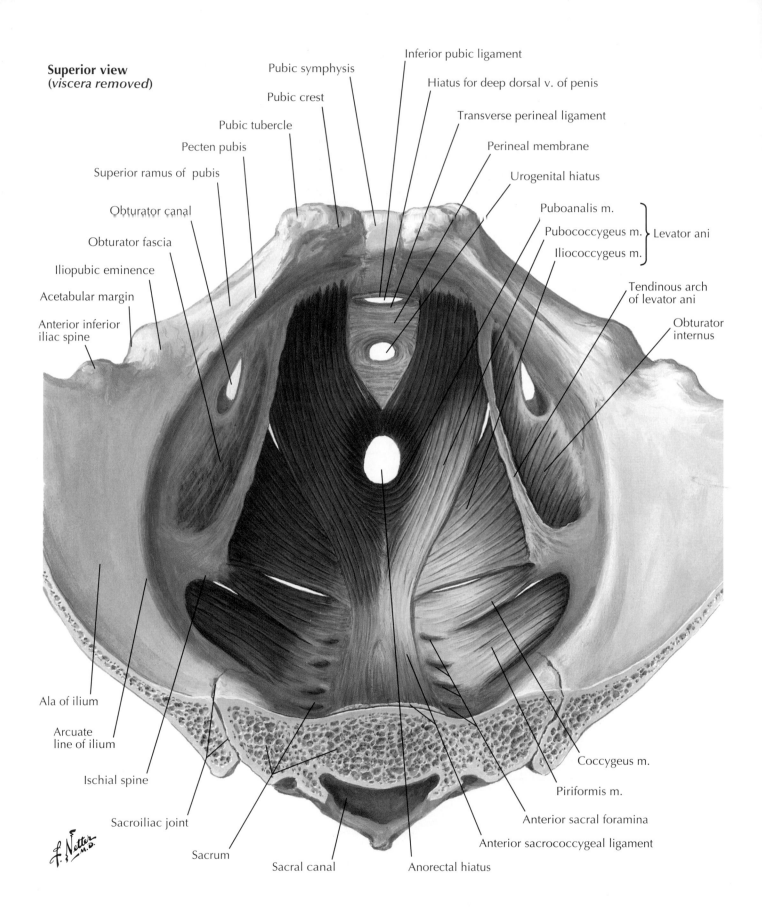

Ala of ilium

Arcuate
line of ilium

Ischial spine

Sacroiliac joint

Sacrum

Sacral canal

Anorectal hiatus

Anterior sacrococcygeal ligament

Anterior sacral foramina

Piriformis m.

Coccygeus m.

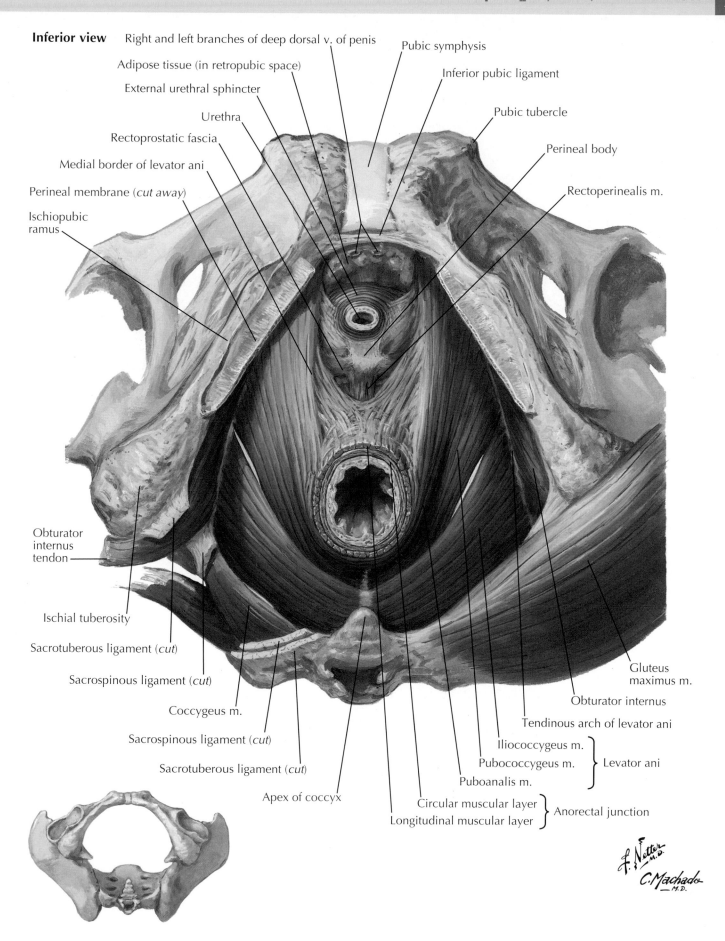

Inferior view

Right and left branches of deep dorsal v. of penis

Adipose tissue (in retropubic space)

External urethral sphincter

Urethra

Rectoprostatic fascia

Medial border of levator ani

Perineal membrane (*cut away*)

Ischiopubic ramus

Pubic symphysis

Inferior pubic ligament

Pubic tubercle

Perineal body

Rectoperinealis m.

Obturator internus tendon

Ischial tuberosity

Sacrotuberous ligament (*cut*)

Sacrospinous ligament (*cut*)

Coccygeus m.

Sacrospinous ligament (*cut*)

Sacrotuberous ligament (*cut*)

Apex of coccyx

Gluteus maximus m.

Obturator internus

Tendinous arch of levator ani

Iliococcygeus m.

Pubococcygeus m.   } Levator ani

Puboanalis m.

Circular muscular layer   } Anorectal junction

Longitudinal muscular layer

Pelvic Diaphragm and Pelvic Cavity

Plate S–482

Superior view

Median umbilical ligament

Linea alba

Fundus of uterus

Proper ovarian ligament

Ovary

Uterine tube (fallopian)

Round ligament of uterus

Broad ligament of uterus

Femoral ring

Deep inguinal ring

Iliopubic tract (covered by peritoneum)

External iliac a. and v.

Iliac fossa

Left paracolic gutter

Median umbilical fold

Urinary bladder

Rectus abdominis m.

Medial umbilical ligament

Medial umbilical fold

Transverse vesical fold

Rectum

Inferior epigastric a. and v.

Lateral umbilical fold

Rectouterine fold

Ureter (covered by peritoneum)

Suspensory ligament of ovary (contains ovarian a. and v.)

Cecum

Cecal folds

Right paracolic gutter

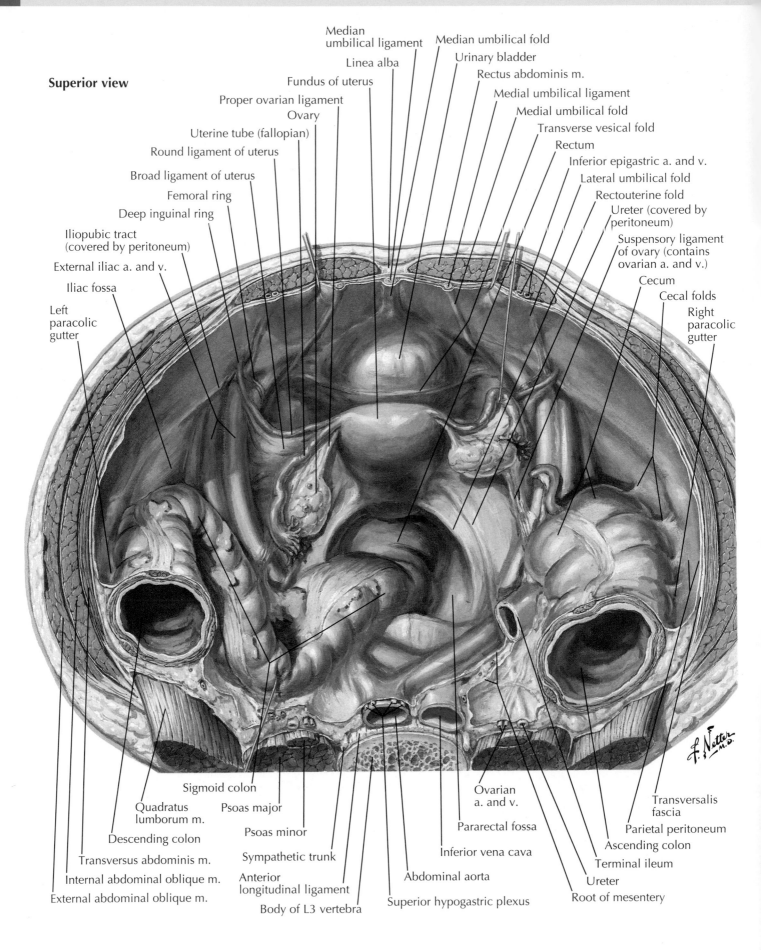

Quadratus lumborum m.

Sigmoid colon

Psoas major

Descending colon

Psoas minor

Transversus abdominis m.

Sympathetic trunk

Internal abdominal oblique m.

Anterior longitudinal ligament

External abdominal oblique m.

Body of L3 vertebra

Ovarian a. and v.

Pararectal fossa

Inferior vena cava

Abdominal aorta

Superior hypogastric plexus

Transversalis fascia

Parietal peritoneum

Ascending colon

Terminal ileum

Ureter

Root of mesentery

Superior view

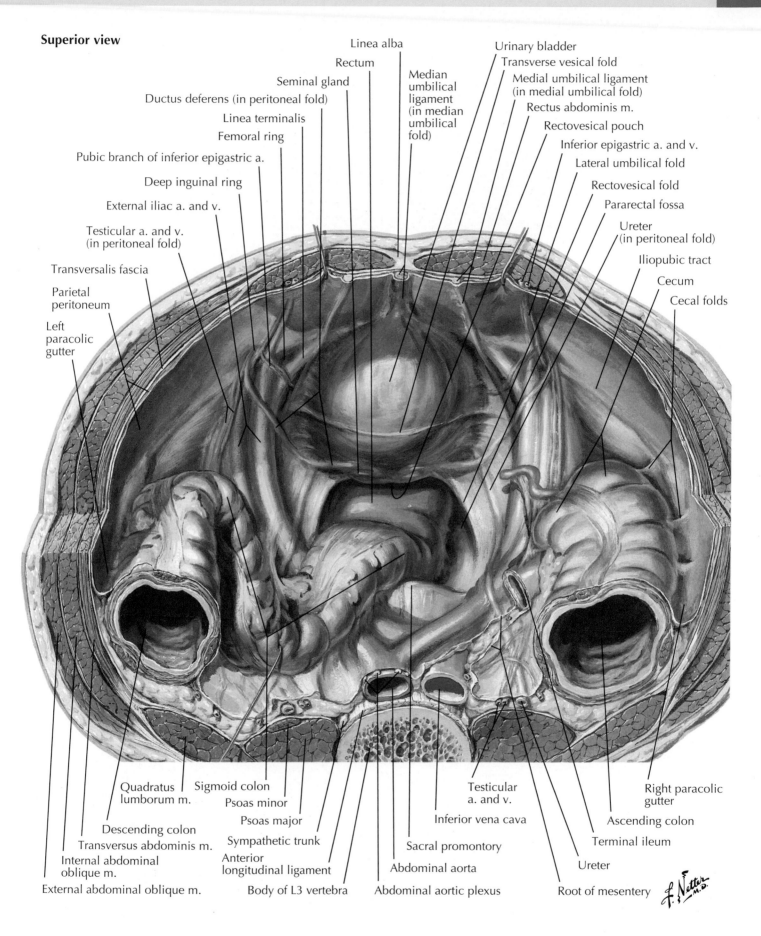

Linea alba

Rectum

Seminal gland

Ductus deferens (in peritoneal fold)

Linea terminalis

Femoral ring

Pubic branch of inferior epigastric a.

Deep inguinal ring

External iliac a. and v.

Testicular a. and v. (in peritoneal fold)

Transversalis fascia

Parietal peritoneum

Left paracolic gutter

Median umbilical ligament (in median umbilical fold)

Urinary bladder

Transverse vesical fold

Medial umbilical ligament (in medial umbilical fold)

Rectus abdominis m.

Rectovesical pouch

Inferior epigastric a. and v.

Lateral umbilical fold

Rectovesical fold

Pararectal fossa

Ureter (in peritoneal fold)

Iliopubic tract

Cecum

Cecal folds

Quadratus lumborum m.

Sigmoid colon

Psoas minor

Psoas major

Descending colon

Transversus abdominis m.

Internal abdominal oblique m.

External abdominal oblique m.

Anterior longitudinal ligament

Body of L3 vertebra

Sympathetic trunk

Sacral promontory

Abdominal aorta

Abdominal aortic plexus

Inferior vena cava

Testicular a. and v.

Right paracolic gutter

Ascending colon

Terminal ileum

Ureter

Root of mesentery

f. Netter M.D.

Pelvic Diaphragm and Pelvic Cavity

See also **Plates S–221, S–521, S–550**

Paramedian (sagittal) dissection

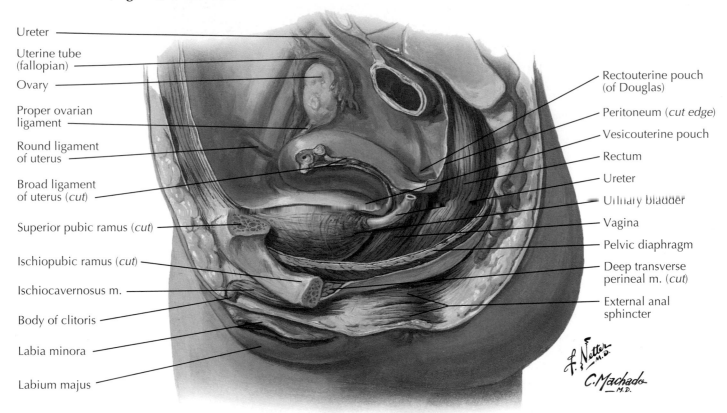

Ureter

Uterine tube (fallopian)

Ovary

Proper ovarian ligament

Round ligament of uterus

Broad ligament of uterus (*cut*)

Superior pubic ramus (*cut*)

Ischiopubic ramus (*cut*)

Ischiocavernosus m.

Body of clitoris

Labia minora

Labium majus

Rectouterine pouch (of Douglas)

Peritoneum (*cut edge*)

Vesicouterine pouch

Rectum

Ureter

Urinary bladder

Vagina

Pelvic diaphragm

Deep transverse perineal m. (*cut*)

External anal sphincter

Median (sagittal) section

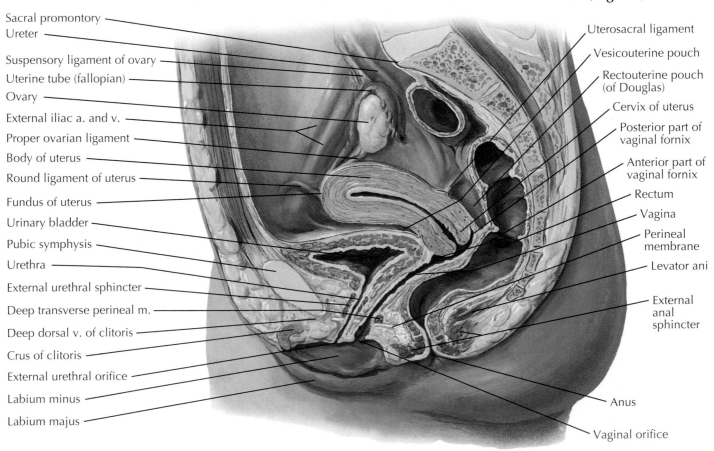

Sacral promontory

Ureter

Suspensory ligament of ovary

Uterine tube (fallopian)

Ovary

External iliac a. and v.

Proper ovarian ligament

Body of uterus

Round ligament of uterus

Fundus of uterus

Urinary bladder

Pubic symphysis

Urethra

External urethral sphincter

Deep transverse perineal m.

Deep dorsal v. of clitoris

Crus of clitoris

External urethral orifice

Labium minus

Labium majus

Uterosacral ligament

Vesicouterine pouch

Rectouterine pouch (of Douglas)

Cervix of uterus

Posterior part of vaginal fornix

Anterior part of vaginal fornix

Rectum

Vagina

Perineal membrane

Levator ani

External anal sphincter

Anus

Vaginal orifice

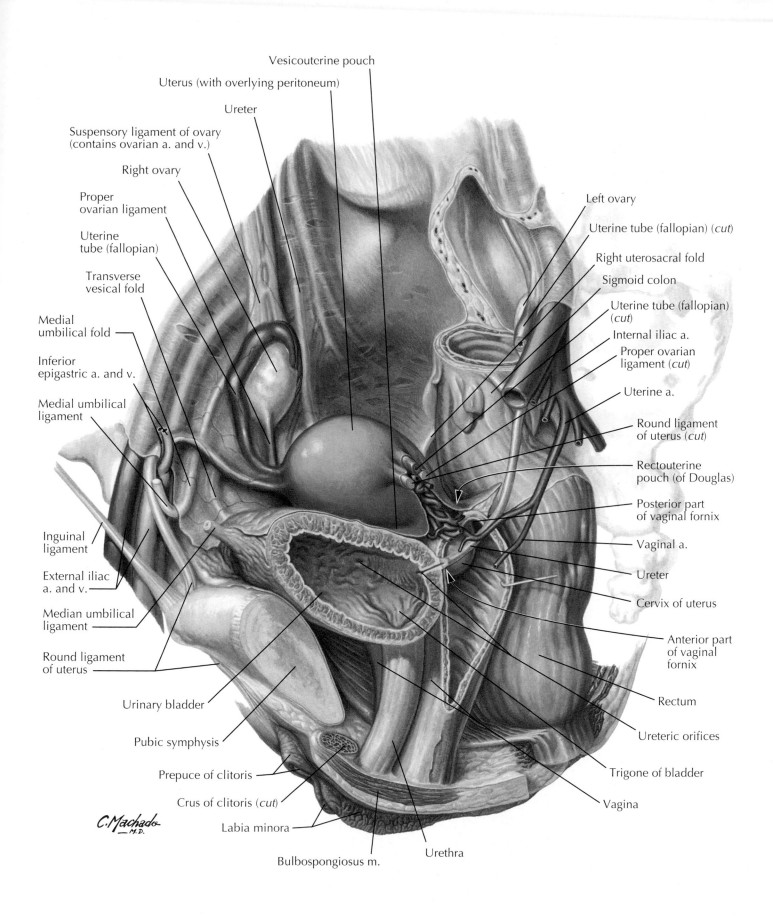

Vesicouterine pouch

Uterus (with overlying peritoneum)

Ureter

Suspensory ligament of ovary
(contains ovarian a. and v.)

Right ovary

Proper
ovarian ligament

Uterine
tube (fallopian)

Transverse
vesical fold

Medial
umbilical fold

Inferior
epigastric a. and v.

Medial umbilical
ligament

Inguinal
ligament

External iliac
a. and v.

Median umbilical
ligament

Round ligament
of uterus

Urinary bladder

Pubic symphysis

Prepuce of clitoris

Crus of clitoris (*cut*)

Labia minora

Bulbospongiosus m.

Left ovary

Uterine tube (fallopian) (*cut*)

Right uterosacral fold

Sigmoid colon

Uterine tube (fallopian)
(*cut*)

Internal iliac a.

Proper ovarian
ligament (*cut*)

Uterine a.

Round ligament
of uterus (*cut*)

Rectouterine
pouch (of Douglas)

Posterior part
of vaginal fornix

Vaginal a.

Ureter

Cervix of uterus

Anterior part
of vaginal
fornix

Rectum

Ureteric orifices

Trigone of bladder

Vagina

Urethra

Female Internal Genitalia

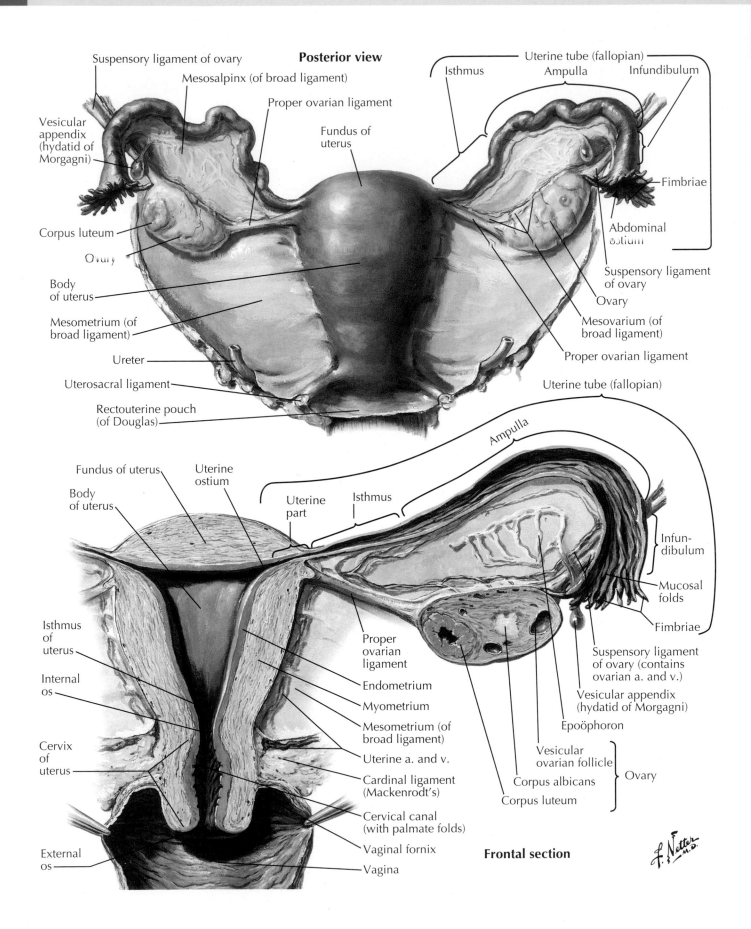

Posterior view

Suspensory ligament of ovary

Mesosalpinx (of broad ligament)

Proper ovarian ligament

Fundus of uterus

Uterine tube (fallopian)

Isthmus

Ampulla

Infundibulum

Vesicular appendix (hydatid of Morgagni)

Fimbriae

Corpus luteum

Abdominal ostium

Ovary

Suspensory ligament of ovary

Body of uterus

Ovary

Mesometrium (of broad ligament)

Mesovarium (of broad ligament)

Ureter

Proper ovarian ligament

Uterosacral ligament

Uterine tube (fallopian)

Rectouterine pouch (of Douglas)

Fundus of uterus

Uterine ostium

Ampulla

Body of uterus

Uterine part

Isthmus

Infundibulum

Isthmus of uterus

Mucosal folds

Internal os

Proper ovarian ligament

Fimbriae

Endometrium

Suspensory ligament of ovary (contains ovarian a. and v.)

Myometrium

Cervix of uterus

Mesometrium (of broad ligament)

Vesicular appendix (hydatid of Morgagni)

Uterine a. and v.

Epoöphoron

Cardinal ligament (Mackenrodt's)

Vesicular ovarian follicle

Ovary

Cervical canal (with palmate folds)

Corpus albicans

Corpus luteum

External os

Vaginal fornix

Frontal section

Vagina

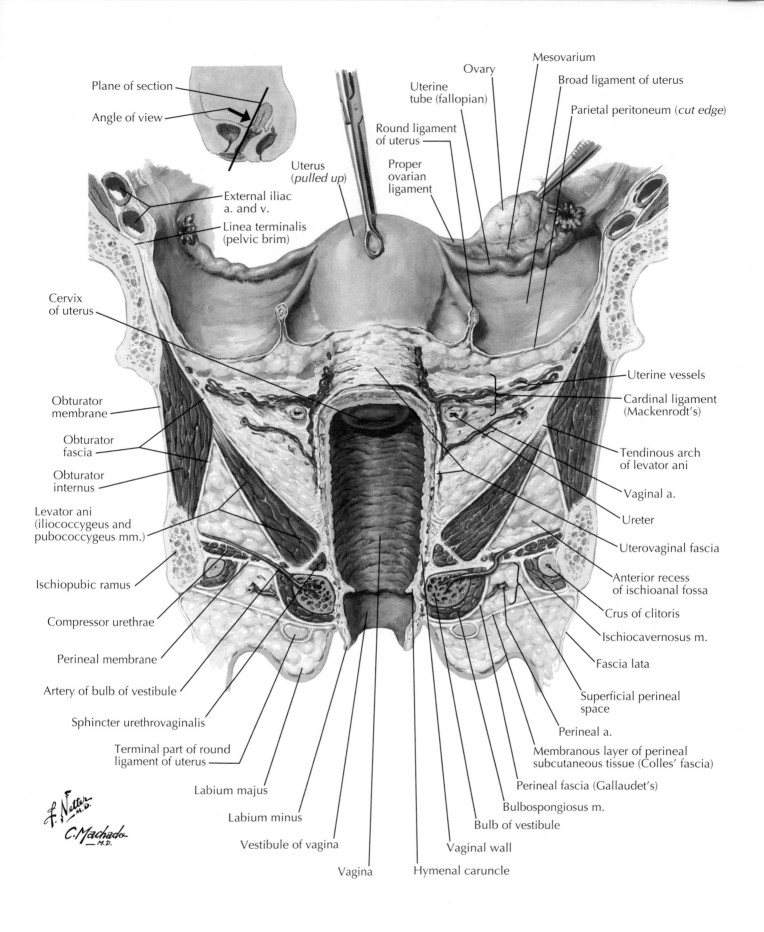

Plane of section

Angle of view

Ovary

Mesovarium

Broad ligament of uterus

Uterine tube (fallopian)

Parietal peritoneum (*cut edge*)

Round ligament of uterus

Uterus (*pulled up*)

Proper ovarian ligament

External iliac a. and v.

Linea terminalis (pelvic brim)

Cervix of uterus

Uterine vessels

Cardinal ligament (Mackenrodt's)

Obturator membrane

Obturator fascia

Obturator internus

Tendinous arch of levator ani

Vaginal a.

Levator ani (iliococcygeus and pubococcygeus mm.)

Ureter

Uterovaginal fascia

Anterior recess of ischioanal fossa

Ischiopubic ramus

Crus of clitoris

Compressor urethrae

Ischiocavernosus m.

Perineal membrane

Fascia lata

Artery of bulb of vestibule

Superficial perineal space

Sphincter urethrovaginalis

Perineal a.

Terminal part of round ligament of uterus

Membranous layer of perineal subcutaneous tissue (Colles' fascia)

Labium majus

Perineal fascia (Gallaudet's)

Labium minus

Bulbospongiosus m.

Bulb of vestibule

Vestibule of vagina

Vaginal wall

Vagina

Hymenal caruncle

Anterior

Posterior

Suspensory ligament of ovary
(containing ovarian a. and v.)

**Subdivisions and
contents of
broad ligament**

Infundibulum of uterine tube

External iliac a. and v.

Fimbriae of uterine tube

Ampulla of uterine tube

Round ligament of uterus

Ovary

Ureter

Laminae of mesosalpinx

Medial umbilical ligament

Laminae of mesovarium

Ovarian branch of uterine a.

Posterior lamina of broad ligament

Round ligament of uterus

Anterior lamina of broad ligament

Transverse vesical fold

Uterine venous plexus

Vesicouterine pouch

Uterine a.

Vaginal a.

C. Machado
—M.D.

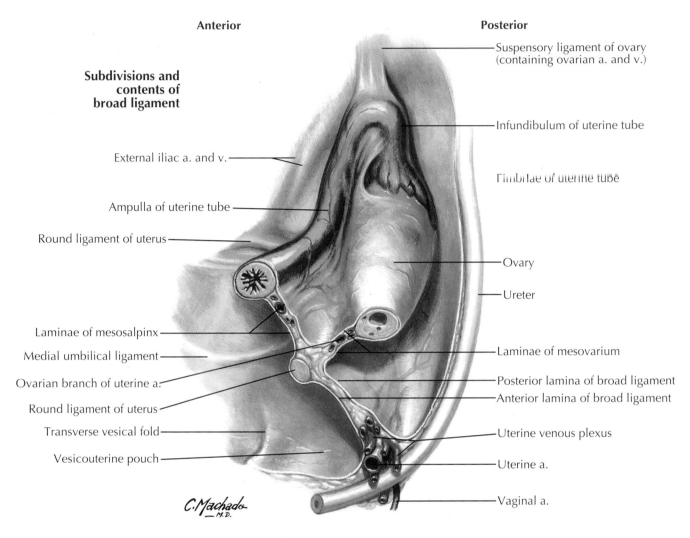

Anteroposterior fluoroscopic image obtained during hysterosalpingography

Contrast medium within
uterine tubes

Contrast
medium in
pelvic cavity

Contrast medium within
uterine cavity

Instrument cannulating
external os of uterus

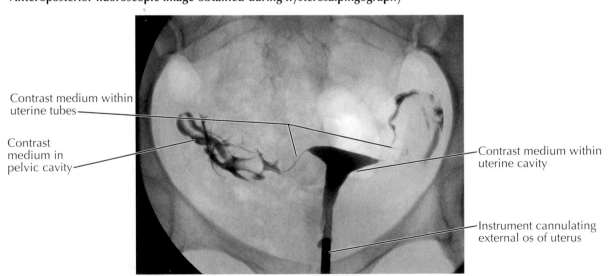

Female: superior view (peritoneum and loose connective tissue removed)

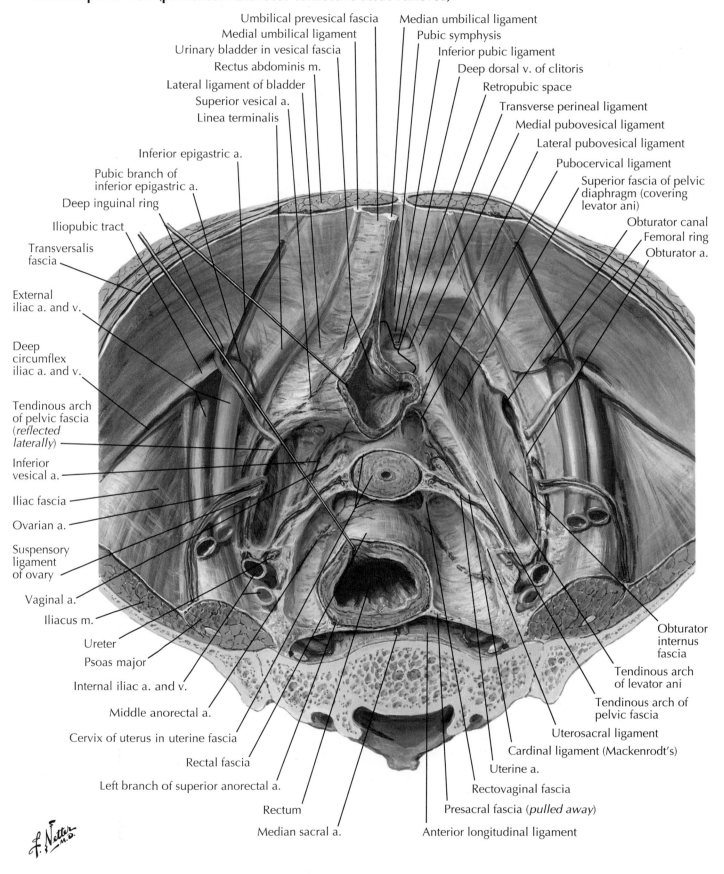

Umbilical prevesical fascia
Medial umbilical ligament
Urinary bladder in vesical fascia
Rectus abdominis m.
Lateral ligament of bladder
Superior vesical a.
Linea terminalis

Median umbilical ligament
Pubic symphysis
Inferior pubic ligament
Deep dorsal v. of clitoris
Retropubic space
Transverse perineal ligament
Medial pubovesical ligament
Lateral pubovesical ligament
Pubocervical ligament
Superior fascia of pelvic diaphragm (covering levator ani)
Obturator canal
Femoral ring
Obturator a.

Inferior epigastric a.
Pubic branch of inferior epigastric a.
Deep inguinal ring
Iliopubic tract
Transversalis fascia
External iliac a. and v.
Deep circumflex iliac a. and v.
Tendinous arch of pelvic fascia (*reflected laterally*)
Inferior vesical a.
Iliac fascia
Ovarian a.
Suspensory ligament of ovary
Vaginal a.
Iliacus m.
Ureter
Psoas major
Internal iliac a. and v.
Middle anorectal a.
Cervix of uterus in uterine fascia
Rectal fascia
Left branch of superior anorectal a.
Rectum
Median sacral a.

Obturator internus fascia
Tendinous arch of levator ani
Tendinous arch of pelvic fascia
Uterosacral ligament
Cardinal ligament (Mackenrodt's)
Uterine a.
Rectovaginal fascia
Presacral fascia (*pulled away*)
Anterior longitudinal ligament

Female Internal Genitalia

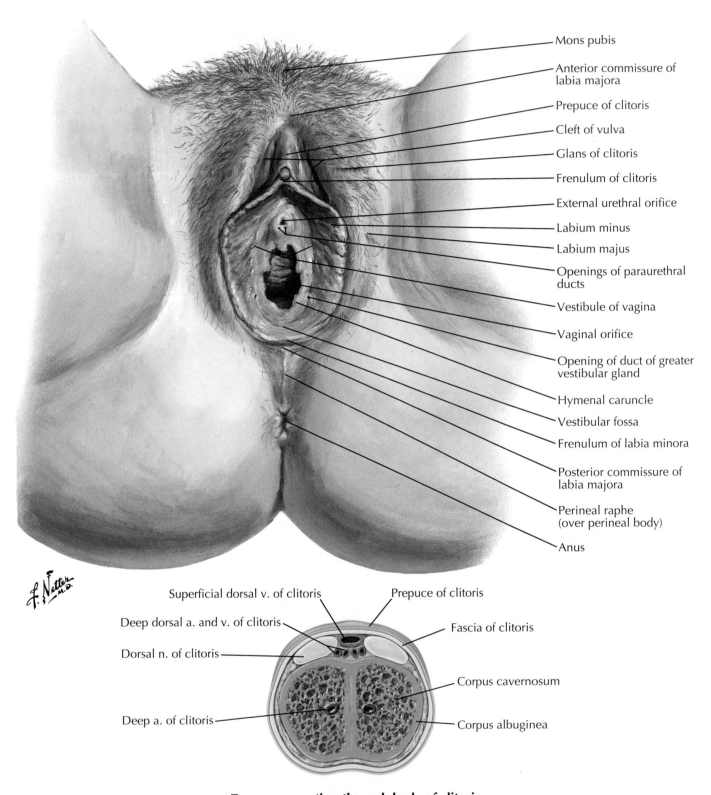

Mons pubis

Anterior commissure of labia majora

Prepuce of clitoris

Cleft of vulva

Glans of clitoris

Frenulum of clitoris

External urethral orifice

Labium minus

Labium majus

Openings of paraurethral ducts

Vestibule of vagina

Vaginal orifice

Opening of duct of greater vestibular gland

Hymenal caruncle

Vestibular fossa

Frenulum of labia minora

Posterior commissure of labia majora

Perineal raphe (over perineal body)

Anus

Superficial dorsal v. of clitoris

Prepuce of clitoris

Deep dorsal a. and v. of clitoris

Fascia of clitoris

Dorsal n. of clitoris

Corpus cavernosum

Deep a. of clitoris

Corpus albuginea

Transverse section through body of clitoris

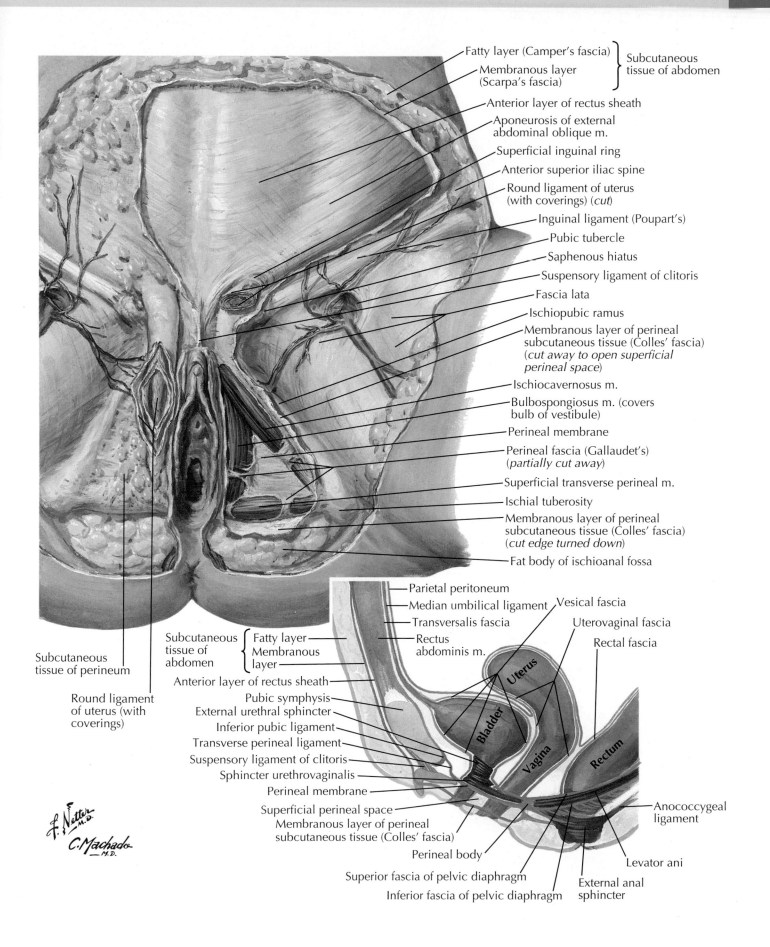

Fatty layer (Camper's fascia) ⎫ Subcutaneous
Membranous layer ⎬ tissue of abdomen
(Scarpa's fascia) ⎭

Anterior layer of rectus sheath

Aponeurosis of external
abdominal oblique m.

Superficial inguinal ring

Anterior superior iliac spine

Round ligament of uterus
(with coverings) (cut)

Inguinal ligament (Poupart's)

Pubic tubercle

Saphenous hiatus

Suspensory ligament of clitoris

Fascia lata

Ischiopubic ramus

Membranous layer of perineal
subcutaneous tissue (Colles' fascia)
(cut away to open superficial
perineal space)

Ischiocavernosus m.

Bulbospongiosus m. (covers
bulb of vestibule)

Perineal membrane

Perineal fascia (Gallaudet's)
(partially cut away)

Superficial transverse perineal m.

Ischial tuberosity

Membranous layer of perineal
subcutaneous tissue (Colles' fascia)
(cut edge turned down)

Fat body of ischioanal fossa

Subcutaneous
tissue of perineum

Round ligament
of uterus (with
coverings)

Subcutaneous ⎧ Fatty layer
tissue of ⎨ Membranous
abdomen ⎩ layer

Anterior layer of rectus sheath

Pubic symphysis

External urethral sphincter

Inferior pubic ligament

Transverse perineal ligament

Suspensory ligament of clitoris

Sphincter urethrovaginalis

Perineal membrane

Superficial perineal space

Membranous layer of perineal
subcutaneous tissue (Colles' fascia)

Perineal body

Superior fascia of pelvic diaphragm

Inferior fascia of pelvic diaphragm

Parietal peritoneum

Median umbilical ligament

Transversalis fascia

Rectus
abdominis m.

Vesical fascia

Uterovaginal fascia

Rectal fascia

Uterus

Bladder

Vagina

Rectum

Anococcygeal
ligament

Levator ani

External anal
sphincter

f. Netter M.D.
C. Machado M.D.

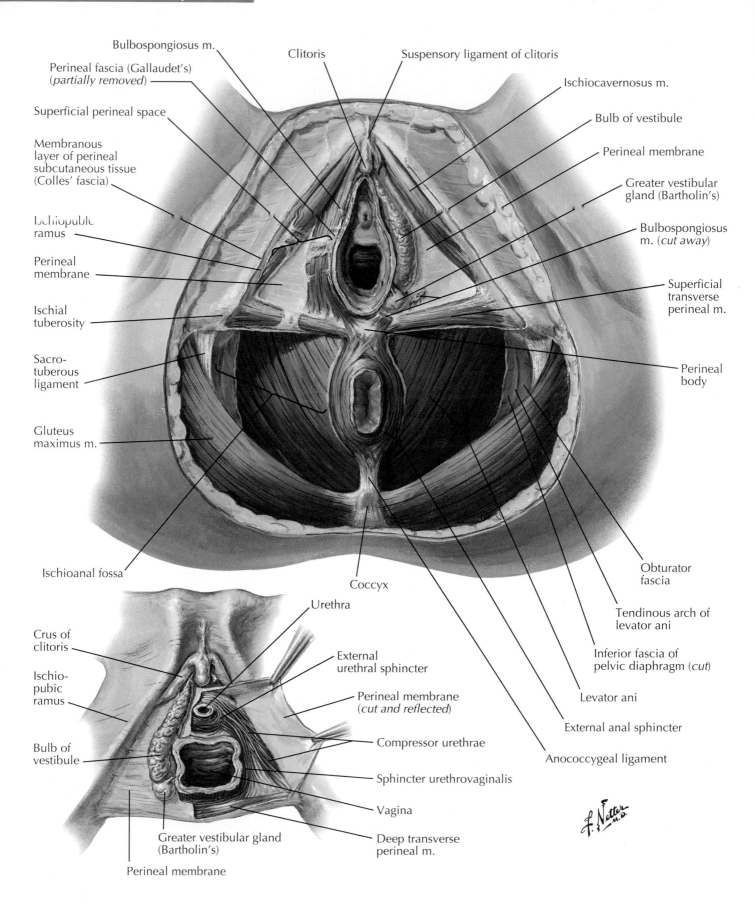

Bulbospongiosus m.

Perineal fascia (Gallaudet's)
(*partially removed*)

Superficial perineal space

Membranous layer of perineal subcutaneous tissue (Colles' fascia)

Ischiopubic ramus

Perineal membrane

Ischial tuberosity

Sacro-tuberous ligament

Gluteus maximus m.

Ischioanal fossa

Clitoris

Suspensory ligament of clitoris

Ischiocavernosus m.

Bulb of vestibule

Perineal membrane

Greater vestibular gland (Bartholin's)

Bulbospongiosus m. (*cut away*)

Superficial transverse perineal m.

Perineal body

Obturator fascia

Tendinous arch of levator ani

Inferior fascia of pelvic diaphragm (*cut*)

Levator ani

External anal sphincter

Anococcygeal ligament

Coccyx

Crus of clitoris

Ischio-pubic ramus

Bulb of vestibule

Greater vestibular gland (Bartholin's)

Perineal membrane

Urethra

External urethral sphincter

Perineal membrane (*cut and reflected*)

Compressor urethrae

Sphincter urethrovaginalis

Vagina

Deep transverse perineal m.

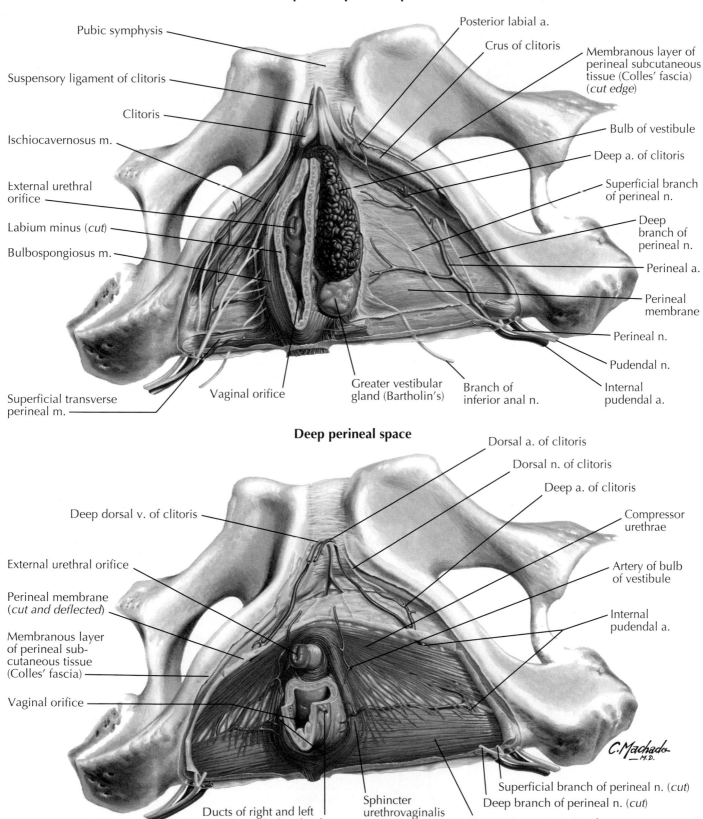

Superficial perineal space

Pubic symphysis

Suspensory ligament of clitoris

Clitoris

Ischiocavernosus m.

External urethral orifice

Labium minus (*cut*)

Bulbospongiosus m.

Superficial transverse perineal m.

Posterior labial a.

Crus of clitoris

Membranous layer of perineal subcutaneous tissue (Colles' fascia) (*cut edge*)

Bulb of vestibule

Deep a. of clitoris

Superficial branch of perineal n.

Deep branch of perineal n.

Perineal a.

Perineal membrane

Perineal n.

Pudendal n.

Internal pudendal a.

Vaginal orifice

Greater vestibular gland (Bartholin's)

Branch of inferior anal n.

Deep perineal space

Deep dorsal v. of clitoris

External urethral orifice

Perineal membrane (*cut and deflected*)

Membranous layer of perineal subcutaneous tissue (Colles' fascia)

Vaginal orifice

Dorsal a. of clitoris

Dorsal n. of clitoris

Deep a. of clitoris

Compressor urethrae

Artery of bulb of vestibule

Internal pudendal a.

Ducts of right and left greater vestibular glands

Sphincter urethrovaginalis

Superficial branch of perineal n. (*cut*)

Deep branch of perineal n. (*cut*)

Deep transverse perineal m.

C. Machado M.D.

Parasagittal view

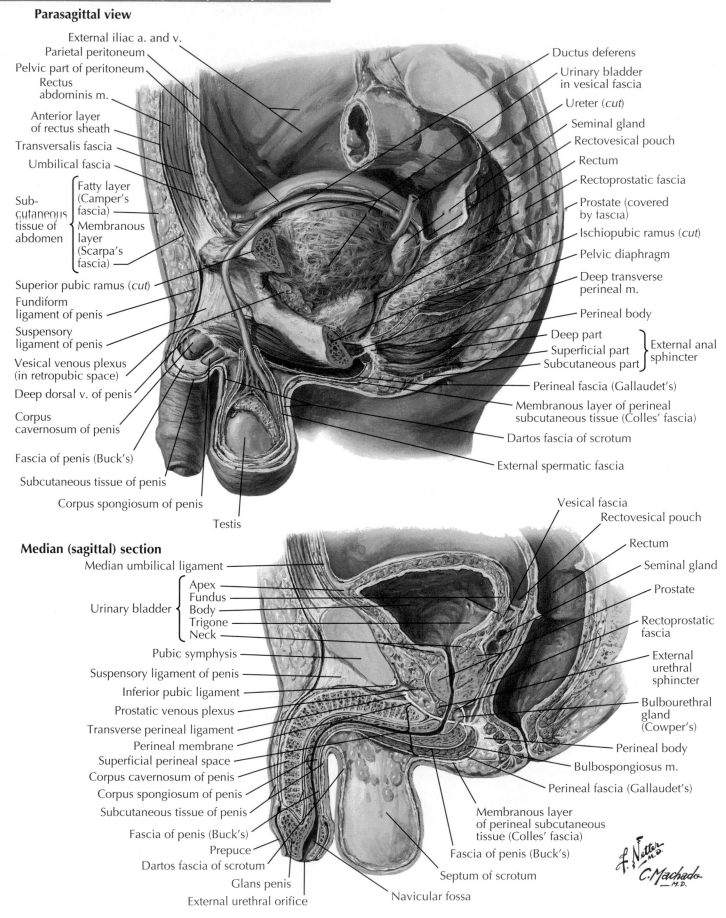

External iliac a. and v.
Parietal peritoneum
Pelvic part of peritoneum
Rectus abdominis m.
Anterior layer of rectus sheath
Transversalis fascia
Umbilical fascia
Subcutaneous tissue of abdomen
Fatty layer (Camper's fascia)
Membranous layer (Scarpa's fascia)
Superior pubic ramus (cut)
Fundiform ligament of penis
Suspensory ligament of penis
Vesical venous plexus (in retropubic space)
Deep dorsal v. of penis
Corpus cavernosum of penis
Fascia of penis (Buck's)
Subcutaneous tissue of penis
Corpus spongiosum of penis
Testis

Ductus deferens
Urinary bladder in vesical fascia
Ureter (cut)
Seminal gland
Rectovesical pouch
Rectum
Rectoprostatic fascia
Prostate (covered by fascia)
Ischiopubic ramus (cut)
Pelvic diaphragm
Deep transverse perineal m.
Perineal body
Deep part
Superficial part
Subcutaneous part
External anal sphincter
Perineal fascia (Gallaudet's)
Membranous layer of perineal subcutaneous tissue (Colles' fascia)
Dartos fascia of scrotum
External spermatic fascia

Median (sagittal) section

Median umbilical ligament
Urinary bladder
Apex
Fundus
Body
Trigone
Neck
Pubic symphysis
Suspensory ligament of penis
Inferior pubic ligament
Prostatic venous plexus
Transverse perineal ligament
Perineal membrane
Superficial perineal space
Corpus cavernosum of penis
Corpus spongiosum of penis
Subcutaneous tissue of penis
Fascia of penis (Buck's)
Prepuce
Dartos fascia of scrotum
Glans penis
External urethral orifice

Vesical fascia
Rectovesical pouch
Rectum
Seminal gland
Prostate
Rectoprostatic fascia
External urethral sphincter
Bulbourethral gland (Cowper's)
Perineal body
Bulbospongiosus m.
Perineal fascia (Gallaudet's)
Membranous layer of perineal subcutaneous tissue (Colles' fascia)
Fascia of penis (Buck's)
Septum of scrotum
Navicular fossa

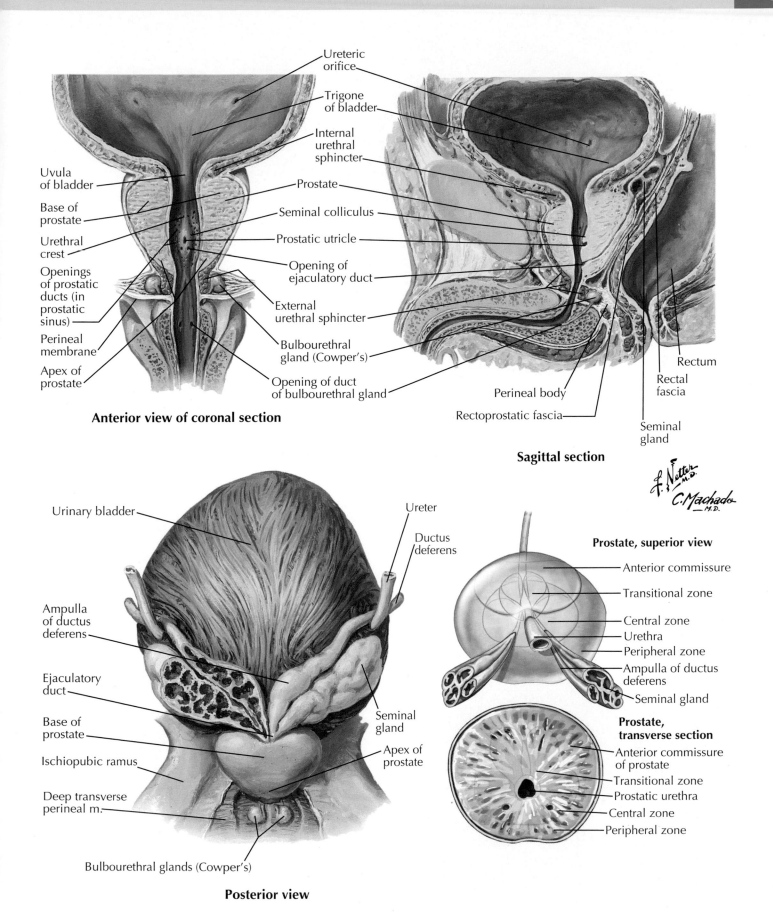

Ureteric orifice

Trigone of bladder

Internal urethral sphincter

Prostate

Seminal colliculus

Prostatic utricle

Opening of ejaculatory duct

External urethral sphincter

Bulbourethral gland (Cowper's)

Opening of duct of bulbourethral gland

Uvula of bladder

Base of prostate

Urethral crest

Openings of prostatic ducts (in prostatic sinus)

Perineal membrane

Apex of prostate

Anterior view of coronal section

Rectum

Rectal fascia

Perineal body

Rectoprostatic fascia

Seminal gland

Sagittal section

Urinary bladder

Ureter

Ductus deferens

Ampulla of ductus deferens

Ejaculatory duct

Base of prostate

Ischiopubic ramus

Deep transverse perineal m.

Seminal gland

Apex of prostate

Bulbourethral glands (Cowper's)

Posterior view

Prostate, superior view

Anterior commissure

Transitional zone

Central zone

Urethra

Peripheral zone

Ampulla of ductus deferens

Seminal gland

Prostate, transverse section

Anterior commissure of prostate

Transitional zone

Prostatic urethra

Central zone

Peripheral zone

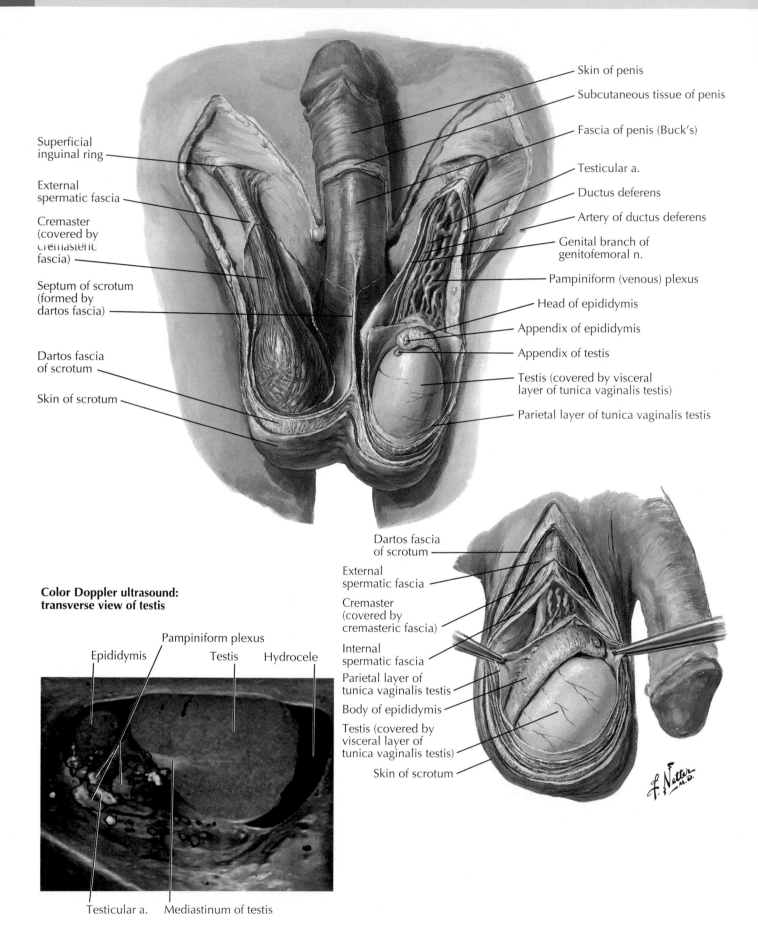

Skin of penis

Subcutaneous tissue of penis

Fascia of penis (Buck's)

Testicular a.

Ductus deferens

Artery of ductus deferens

Genital branch of genitofemoral n.

Pampiniform (venous) plexus

Head of epididymis

Appendix of epididymis

Appendix of testis

Testis (covered by visceral layer of tunica vaginalis testis)

Parietal layer of tunica vaginalis testis

Superficial inguinal ring

External spermatic fascia

Cremaster (covered by cremasteric fascia)

Septum of scrotum (formed by dartos fascia)

Dartos fascia of scrotum

Skin of scrotum

Color Doppler ultrasound: transverse view of testis

Pampiniform plexus

Epididymis

Testis

Hydrocele

Testicular a.

Mediastinum of testis

Dartos fascia of scrotum

External spermatic fascia

Cremaster (covered by cremasteric fascia)

Internal spermatic fascia

Parietal layer of tunica vaginalis testis

Body of epididymis

Testis (covered by visceral layer of tunica vaginalis testis)

Skin of scrotum

Male Internal Genitalia

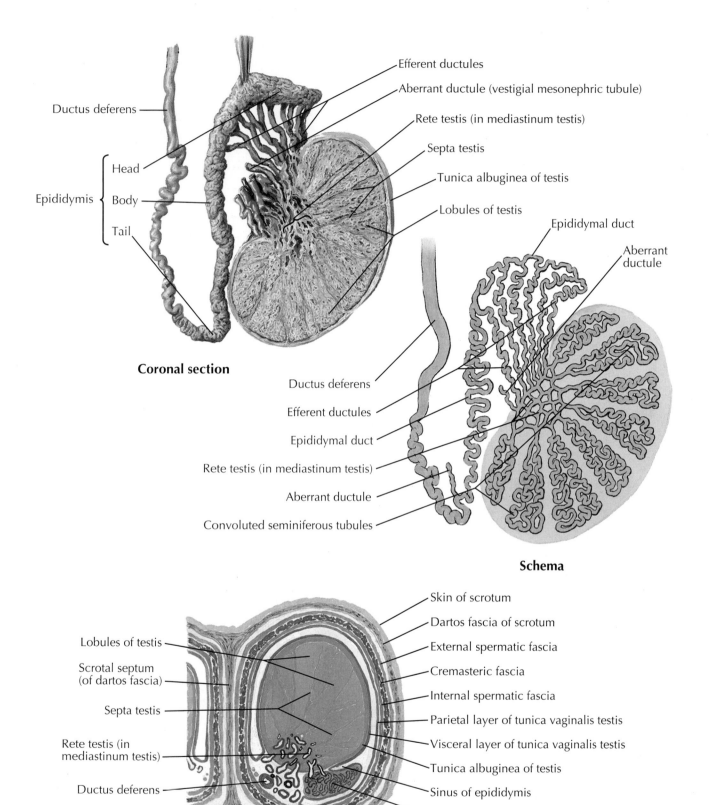

Ductus deferens

Epididymis
- Head
- Body
- Tail

Efferent ductules

Aberrant ductule (vestigial mesonephric tubule)

Rete testis (in mediastinum testis)

Septa testis

Tunica albuginea of testis

Lobules of testis

Coronal section

Epididymal duct

Aberrant ductule

Ductus deferens

Efferent ductules

Epididymal duct

Rete testis (in mediastinum testis)

Aberrant ductule

Convoluted seminiferous tubules

Schema

Lobules of testis

Scrotal septum (of dartos fascia)

Septa testis

Rete testis (in mediastinum testis)

Ductus deferens

Skin of scrotum

Dartos fascia of scrotum

External spermatic fascia

Cremasteric fascia

Internal spermatic fascia

Parietal layer of tunica vaginalis testis

Visceral layer of tunica vaginalis testis

Tunica albuginea of testis

Sinus of epididymis

Epididymis

Horizontal section through scrotum and testis

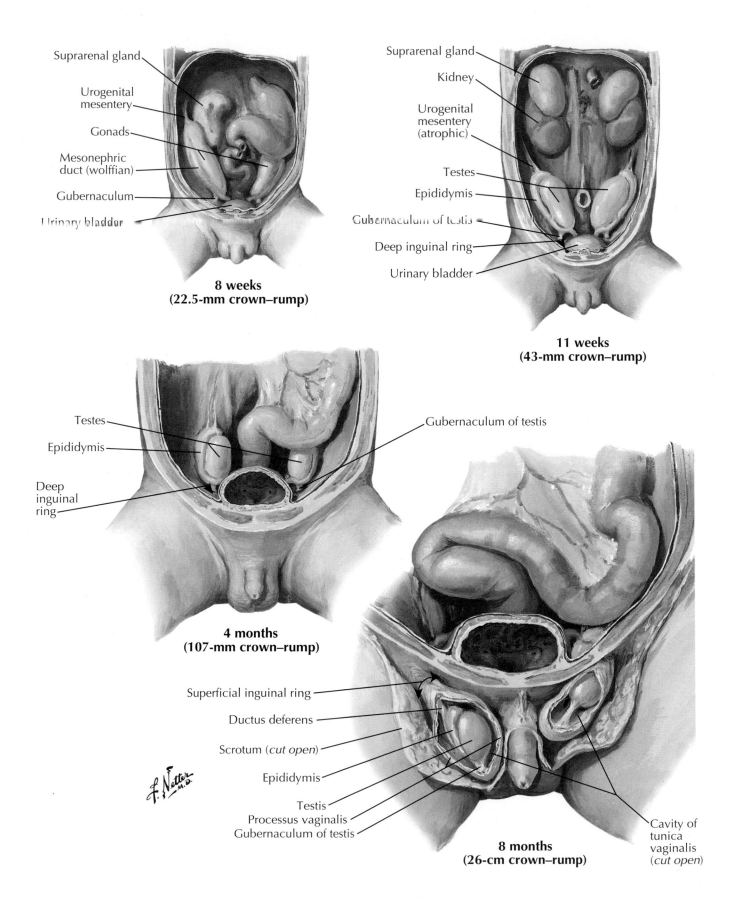

Suprarenal gland

Urogenital mesentery

Gonads

Mesonephric duct (wolffian)

Gubernaculum

Urinary bladder

**8 weeks
(22.5-mm crown–rump)**

Suprarenal gland

Kidney

Urogenital mesentery (atrophic)

Testes

Epididymis

Gubernaculum of testis

Deep inguinal ring

Urinary bladder

**11 weeks
(43-mm crown–rump)**

Testes

Epididymis

Deep inguinal ring

Gubernaculum of testis

**4 months
(107-mm crown–rump)**

Superficial inguinal ring

Ductus deferens

Scrotum (*cut open*)

Epididymis

Testis

Processus vaginalis

Gubernaculum of testis

Cavity of tunica vaginalis (*cut open*)

**8 months
(26-cm crown–rump)**

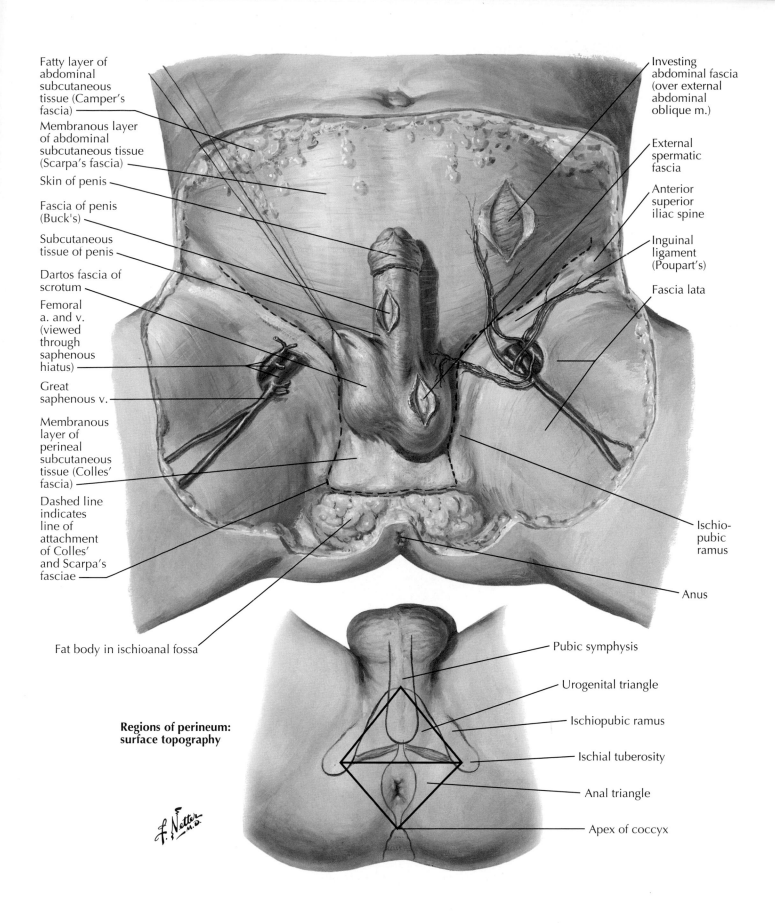

Fatty layer of abdominal subcutaneous tissue (Camper's fascia)

Membranous layer of abdominal subcutaneous tissue (Scarpa's fascia)

Skin of penis

Fascia of penis (Buck's)

Subcutaneous tissue of penis

Dartos fascia of scrotum

Femoral a. and v. (viewed through saphenous hiatus)

Great saphenous v.

Membranous layer of perineal subcutaneous tissue (Colles' fascia)

Dashed line indicates line of attachment of Colles' and Scarpa's fasciae

Fat body in ischioanal fossa

Investing abdominal fascia (over external abdominal oblique m.)

External spermatic fascia

Anterior superior iliac spine

Inguinal ligament (Poupart's)

Fascia lata

Ischio-pubic ramus

Anus

Pubic symphysis

Urogenital triangle

Ischiopubic ramus

Ischial tuberosity

Anal triangle

Apex of coccyx

Regions of perineum: surface topography

f. Netter M.D.

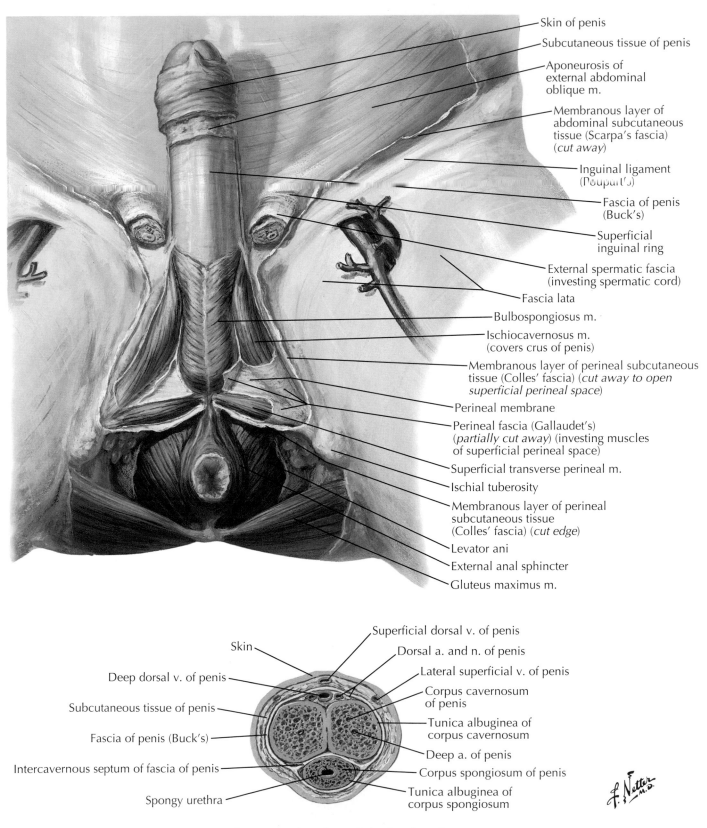

Skin of penis

Subcutaneous tissue of penis

Aponeurosis of external abdominal oblique m.

Membranous layer of abdominal subcutaneous tissue (Scarpa's fascia) (*cut away*)

Inguinal ligament (Poupart's)

Fascia of penis (Buck's)

Superficial inguinal ring

External spermatic fascia (investing spermatic cord)

Fascia lata

Bulbospongiosus m.

Ischiocavernosus m. (covers crus of penis)

Membranous layer of perineal subcutaneous tissue (Colles' fascia) (*cut away to open superficial perineal space*)

Perineal membrane

Perineal fascia (Gallaudet's) (*partially cut away*) (investing muscles of superficial perineal space)

Superficial transverse perineal m.

Ischial tuberosity

Membranous layer of perineal subcutaneous tissue (Colles' fascia) (*cut edge*)

Levator ani

External anal sphincter

Gluteus maximus m.

Superficial dorsal v. of penis

Skin

Dorsal a. and n. of penis

Deep dorsal v. of penis

Lateral superficial v. of penis

Subcutaneous tissue of penis

Corpus cavernosum of penis

Fascia of penis (Buck's)

Tunica albuginea of corpus cavernosum

Deep a. of penis

Intercavernous septum of fascia of penis

Corpus spongiosum of penis

Spongy urethra

Tunica albuginea of corpus spongiosum

Transverse section through body of penis

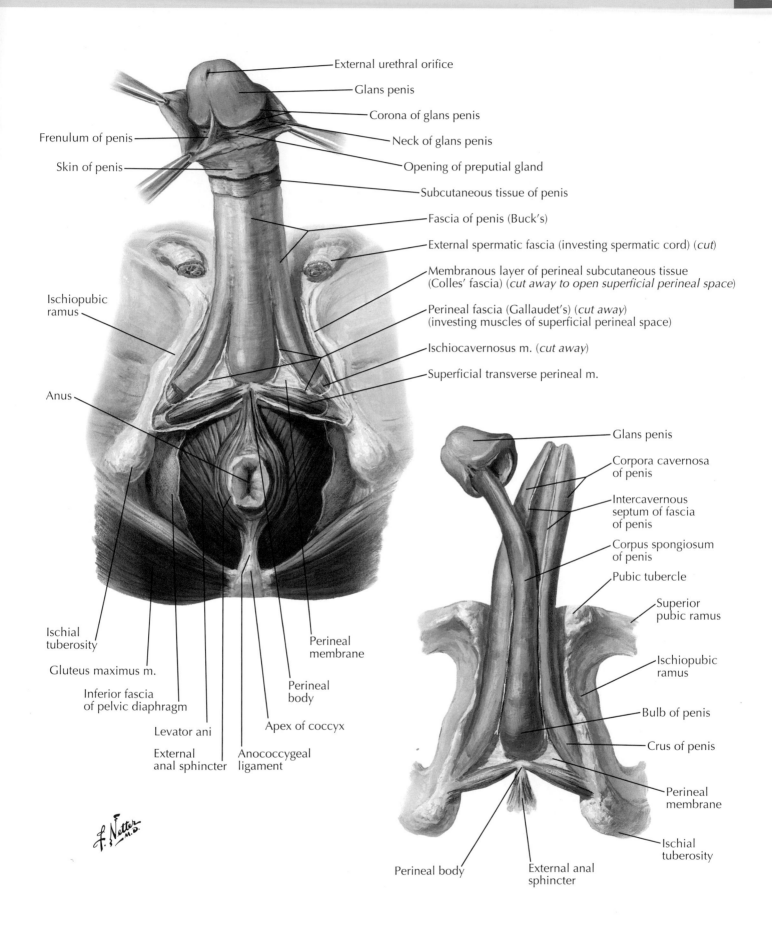

External urethral orifice

Glans penis

Corona of glans penis

Neck of glans penis

Frenulum of penis

Opening of preputial gland

Skin of penis

Subcutaneous tissue of penis

Fascia of penis (Buck's)

External spermatic fascia (investing spermatic cord) (*cut*)

Membranous layer of perineal subcutaneous tissue (Colles' fascia) (*cut away to open superficial perineal space*)

Ischiopubic ramus

Perineal fascia (Gallaudet's) (*cut away*) (investing muscles of superficial perineal space)

Ischiocavernosus m. (*cut away*)

Superficial transverse perineal m.

Anus

Glans penis

Corpora cavernosa of penis

Intercavernous septum of fascia of penis

Corpus spongiosum of penis

Pubic tubercle

Superior pubic ramus

Ischiopubic ramus

Ischial tuberosity

Bulb of penis

Gluteus maximus m.

Crus of penis

Inferior fascia of pelvic diaphragm

Perineal membrane

Perineal body

Levator ani

Apex of coccyx

Ischial tuberosity

External anal sphincter

Anococcygeal ligament

Perineal body

External anal sphincter

Perineal membrane

Perineal body

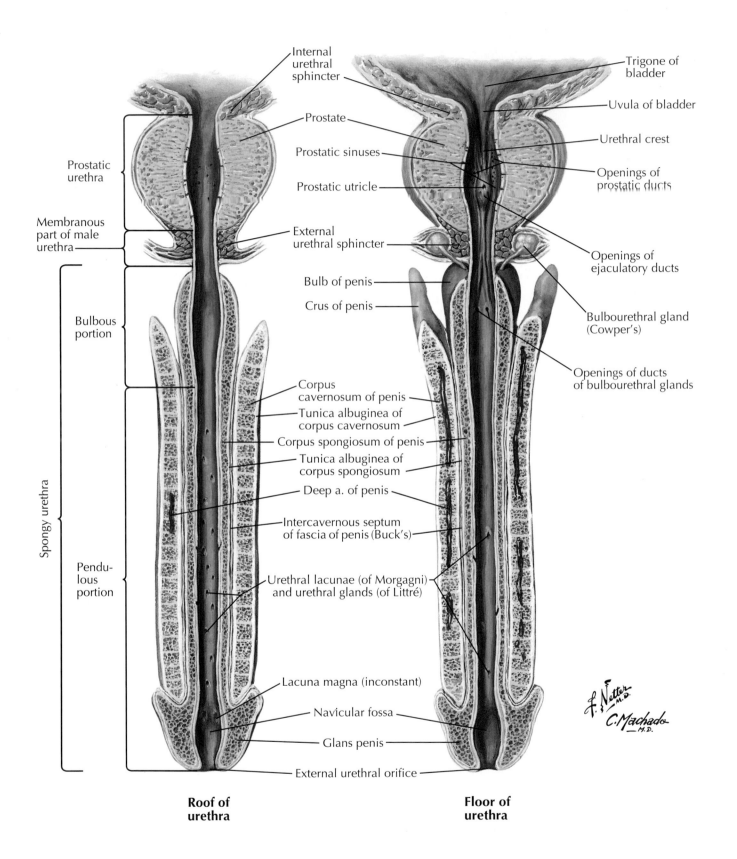

Internal urethral sphincter

Prostate

Prostatic sinuses

Prostatic utricle

External urethral sphincter

Bulb of penis

Crus of penis

Corpus cavernosum of penis

Tunica albuginea of corpus cavernosum

Corpus spongiosum of penis

Tunica albuginea of corpus spongiosum

Deep a. of penis

Intercavernous septum of fascia of penis (Buck's)

Urethral lacunae (of Morgagni) and urethral glands (of Littré)

Lacuna magna (inconstant)

Navicular fossa

Glans penis

External urethral orifice

Trigone of bladder

Uvula of bladder

Urethral crest

Openings of prostatic ducts

Openings of ejaculatory ducts

Bulbourethral gland (Cowper's)

Openings of ducts of bulbourethral glands

Prostatic urethra

Membranous part of male urethra

Bulbous portion

Spongy urethra

Pendulous portion

Roof of urethra

Floor of urethra

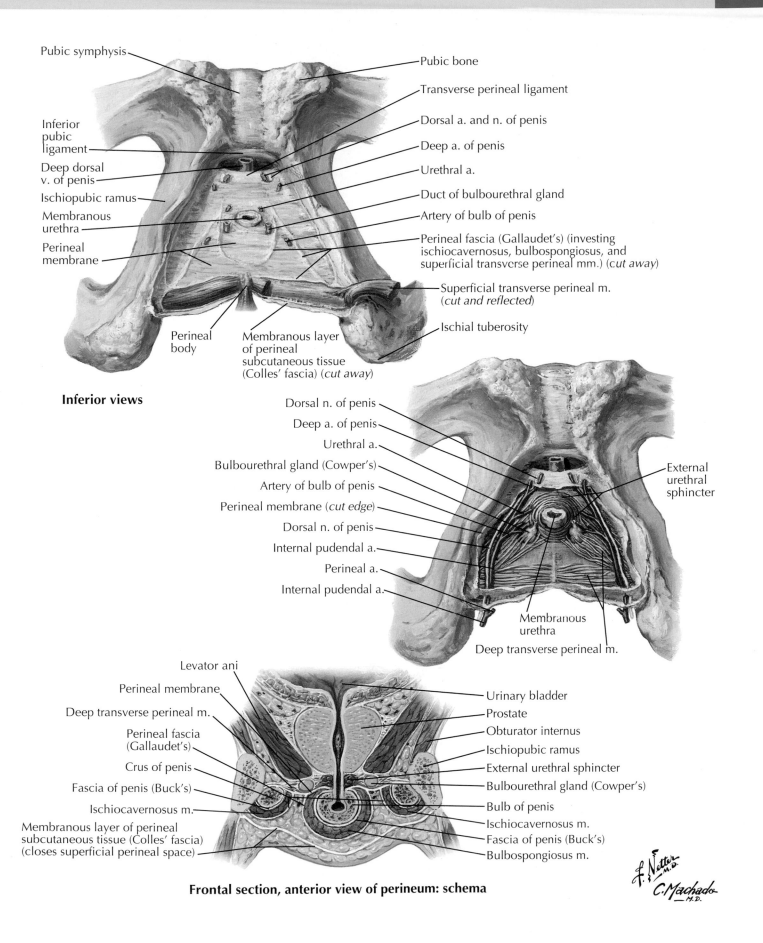

Pubic symphysis
Pubic bone
Transverse perineal ligament
Inferior pubic ligament
Dorsal a. and n. of penis
Deep a. of penis
Deep dorsal v. of penis
Urethral a.
Ischiopubic ramus
Duct of bulbourethral gland
Membranous urethra
Artery of bulb of penis
Perineal membrane
Perineal fascia (Gallaudet's) (investing ischiocavernosus, bulbospongiosus, and superficial transverse perineal mm.) (cut away)
Superficial transverse perineal m. (cut and reflected)
Ischial tuberosity
Perineal body
Membranous layer of perineal subcutaneous tissue (Colles' fascia) (cut away)

Inferior views

Dorsal n. of penis
Deep a. of penis
Urethral a.
Bulbourethral gland (Cowper's)
Artery of bulb of penis
Perineal membrane (cut edge)
Dorsal n. of penis
Internal pudendal a.
Perineal a.
Internal pudendal a.
External urethral sphincter
Membranous urethra
Deep transverse perineal m.

Levator ani
Perineal membrane
Deep transverse perineal m.
Perineal fascia (Gallaudet's)
Crus of penis
Fascia of penis (Buck's)
Ischiocavernosus m.
Membranous layer of perineal subcutaneous tissue (Colles' fascia) (closes superficial perineal space)
Urinary bladder
Prostate
Obturator internus
Ischiopubic ramus
External urethral sphincter
Bulbourethral gland (Cowper's)
Bulb of penis
Ischiocavernosus m.
Fascia of penis (Buck's)
Bulbospongiosus m.

Frontal section, anterior view of perineum: schema

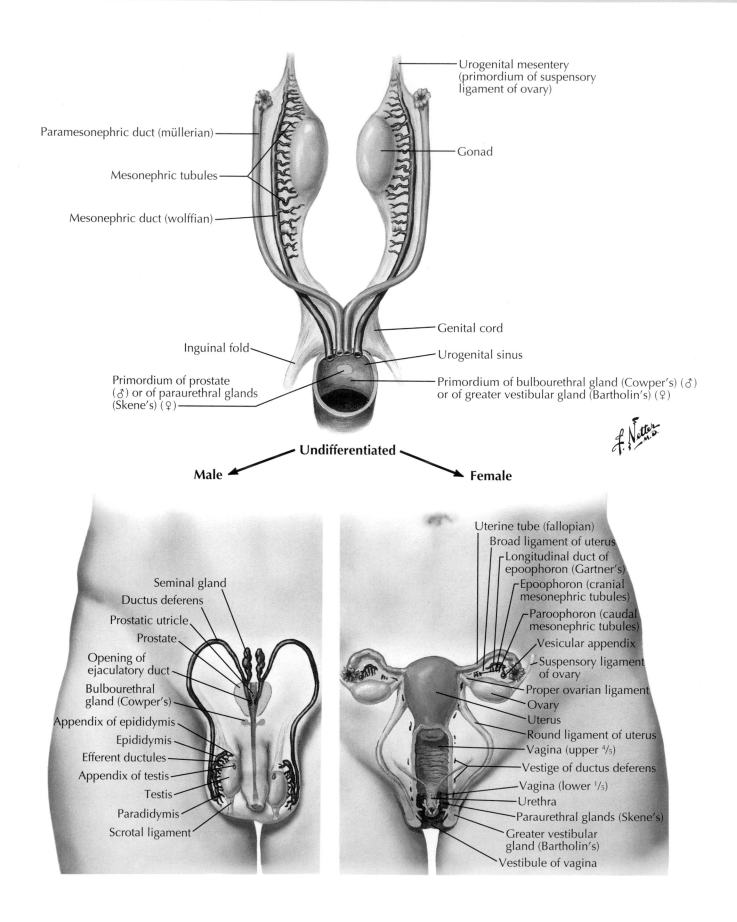

Urogenital mesentery (primordium of suspensory ligament of ovary)

Paramesonephric duct (müllerian)

Mesonephric tubules

Gonad

Mesonephric duct (wolffian)

Genital cord

Inguinal fold

Urogenital sinus

Primordium of prostate (♂) or of paraurethral glands (Skene's) (♀)

Primordium of bulbourethral gland (Cowper's) (♂) or of greater vestibular gland (Bartholin's) (♀)

Undifferentiated

Male

Female

Seminal gland

Ductus deferens

Prostatic utricle

Prostate

Opening of ejaculatory duct

Bulbourethral gland (Cowper's)

Appendix of epididymis

Epididymis

Efferent ductules

Appendix of testis

Testis

Paradidymis

Scrotal ligament

Uterine tube (fallopian)

Broad ligament of uterus

Longitudinal duct of epoophoron (Gartner's)

Epoophoron (cranial mesonephric tubules)

Paroophoron (caudal mesonephric tubules)

Vesicular appendix

Suspensory ligament of ovary

Proper ovarian ligament

Ovary

Uterus

Round ligament of uterus

Vagina (upper $^4/_5$)

Vestige of ductus deferens

Vagina (lower $^1/_5$)

Urethra

Paraurethral glands (Skene's)

Greater vestibular gland (Bartholin's)

Vestibule of vagina

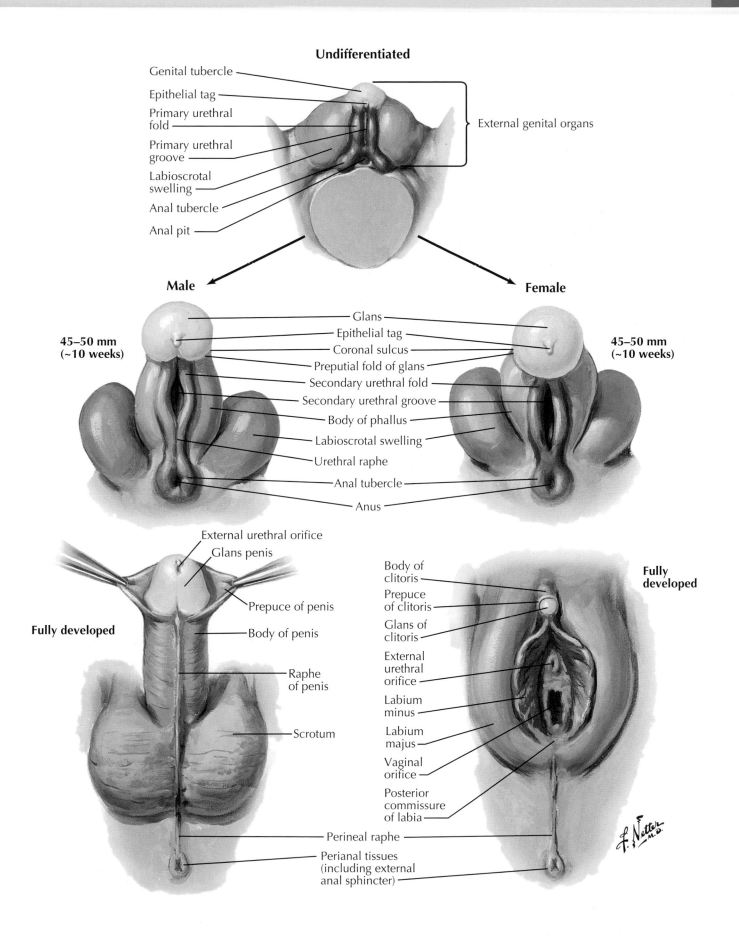

Undifferentiated

Genital tubercle

Epithelial tag

Primary urethral fold

Primary urethral groove

Labioscrotal swelling

Anal tubercle

Anal pit

External genital organs

Male

Female

Glans

Epithelial tag

Coronal sulcus

Preputial fold of glans

Secondary urethral fold

Secondary urethral groove

Body of phallus

Labioscrotal swelling

Urethral raphe

Anal tubercle

Anus

45–50 mm (~10 weeks)

45–50 mm (~10 weeks)

External urethral orifice

Glans penis

Prepuce of penis

Body of penis

Raphe of penis

Scrotum

Fully developed

Fully developed

Body of clitoris

Prepuce of clitoris

Glans of clitoris

External urethral orifice

Labium minus

Labium majus

Vaginal orifice

Posterior commissure of labia

Perineal raphe

Perianal tissues (including external anal sphincter)

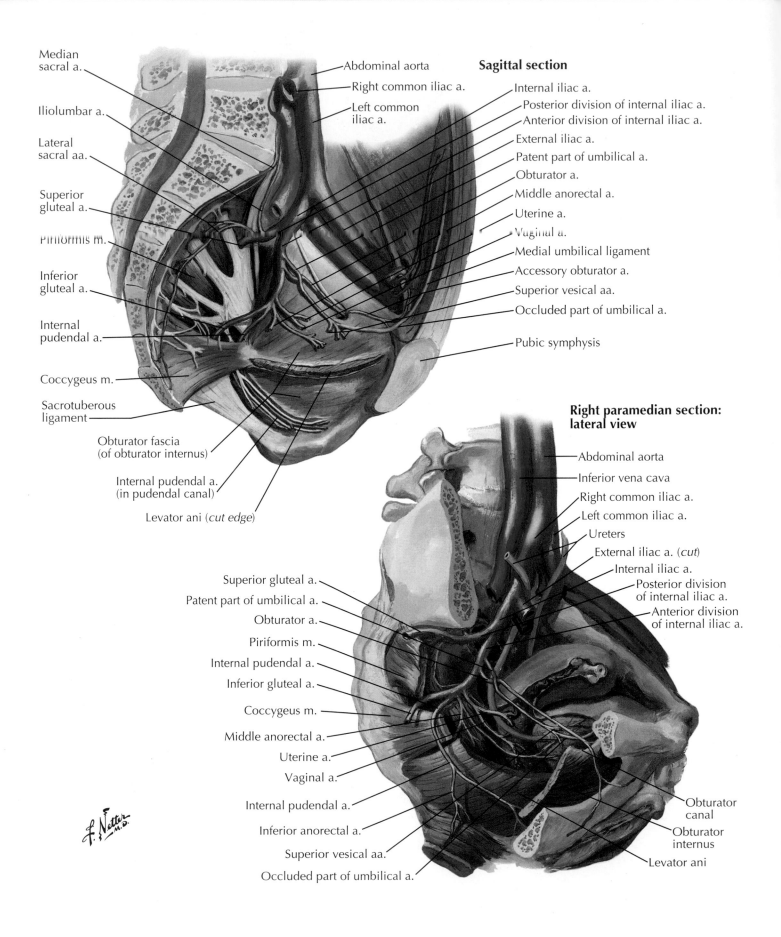

Median sacral a.

Iliolumbar a.

Lateral sacral aa.

Superior gluteal a.

Piriformis m.

Inferior gluteal a.

Internal pudendal a.

Coccygeus m.

Sacrotuberous ligament

Obturator fascia (of obturator internus)

Internal pudendal a. (in pudendal canal)

Levator ani (*cut edge*)

Abdominal aorta

Right common iliac a.

Left common iliac a.

Sagittal section

Internal iliac a.

Posterior division of internal iliac a.

Anterior division of internal iliac a.

External iliac a.

Patent part of umbilical a.

Obturator a.

Middle anorectal a.

Uterine a.

Vaginal a.

Medial umbilical ligament

Accessory obturator a.

Superior vesical aa.

Occluded part of umbilical a.

Pubic symphysis

Right paramedian section: lateral view

Abdominal aorta

Inferior vena cava

Right common iliac a.

Left common iliac a.

Ureters

External iliac a. (*cut*)

Internal iliac a.

Posterior division of internal iliac a.

Anterior division of internal iliac a.

Superior gluteal a.

Patent part of umbilical a.

Obturator a.

Piriformis m.

Internal pudendal a.

Inferior gluteal a.

Coccygeus m.

Middle anorectal a.

Uterine a.

Vaginal a.

Internal pudendal a.

Inferior anorectal a.

Superior vesical aa.

Occluded part of umbilical a.

Obturator canal

Obturator internus

Levator ani

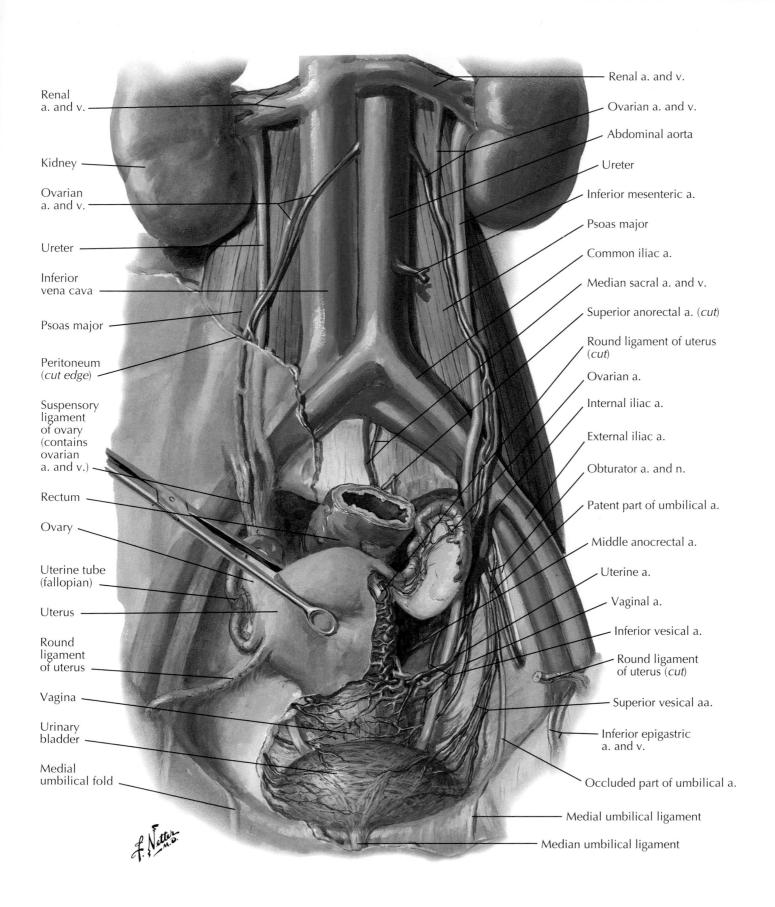

Renal
a. and v.

Kidney

Ovarian
a. and v.

Ureter

Inferior
vena cava

Psoas major

Peritoneum
(*cut edge*)

Suspensory
ligament
of ovary
(contains
ovarian
a. and v.)

Rectum

Ovary

Uterine tube
(fallopian)

Uterus

Round
ligament
of uterus

Vagina

Urinary
bladder

Medial
umbilical fold

Renal a. and v.

Ovarian a. and v.

Abdominal aorta

Ureter

Inferior mesenteric a.

Psoas major

Common iliac a.

Median sacral a. and v.

Superior anorectal a. (*cut*)

Round ligament of uterus
(*cut*)

Ovarian a.

Internal iliac a.

External iliac a.

Obturator a. and n.

Patent part of umbilical a.

Middle anorectal a.

Uterine a.

Vaginal a.

Inferior vesical a.

Round ligament
of uterus (*cut*)

Superior vesical aa.

Inferior epigastric
a. and v.

Occluded part of umbilical a.

Medial umbilical ligament

Median umbilical ligament

f. Netter
M.D.

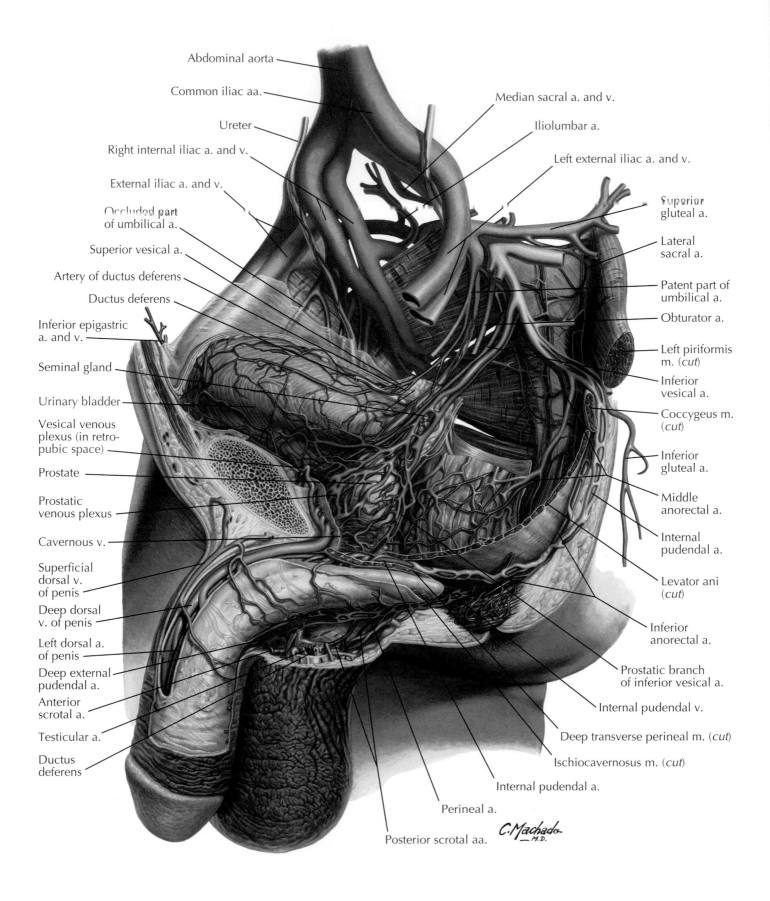

Abdominal aorta

Common iliac aa.

Ureter

Right internal iliac a. and v.

External iliac a. and v.

Occluded part of umbilical a.

Superior vesical a.

Artery of ductus deferens

Ductus deferens

Inferior epigastric a. and v.

Seminal gland

Urinary bladder

Vesical venous plexus (in retro-pubic space)

Prostate

Prostatic venous plexus

Cavernous v.

Superficial dorsal v. of penis

Deep dorsal v. of penis

Left dorsal a. of penis

Deep external pudendal a.

Anterior scrotal a.

Testicular a.

Ductus deferens

Median sacral a. and v.

Iliolumbar a.

Left external iliac a. and v.

Superior gluteal a.

Lateral sacral a.

Patent part of umbilical a.

Obturator a.

Left piriformis m. (*cut*)

Inferior vesical a.

Coccygeus m. (*cut*)

Inferior gluteal a.

Middle anorectal a.

Internal pudendal a.

Levator ani (*cut*)

Inferior anorectal a.

Prostatic branch of inferior vesical a.

Internal pudendal v.

Deep transverse perineal m. (*cut*)

Ischiocavernosus m. (*cut*)

Internal pudendal a.

Perineal a.

Posterior scrotal aa.

C. Machado — M.D.

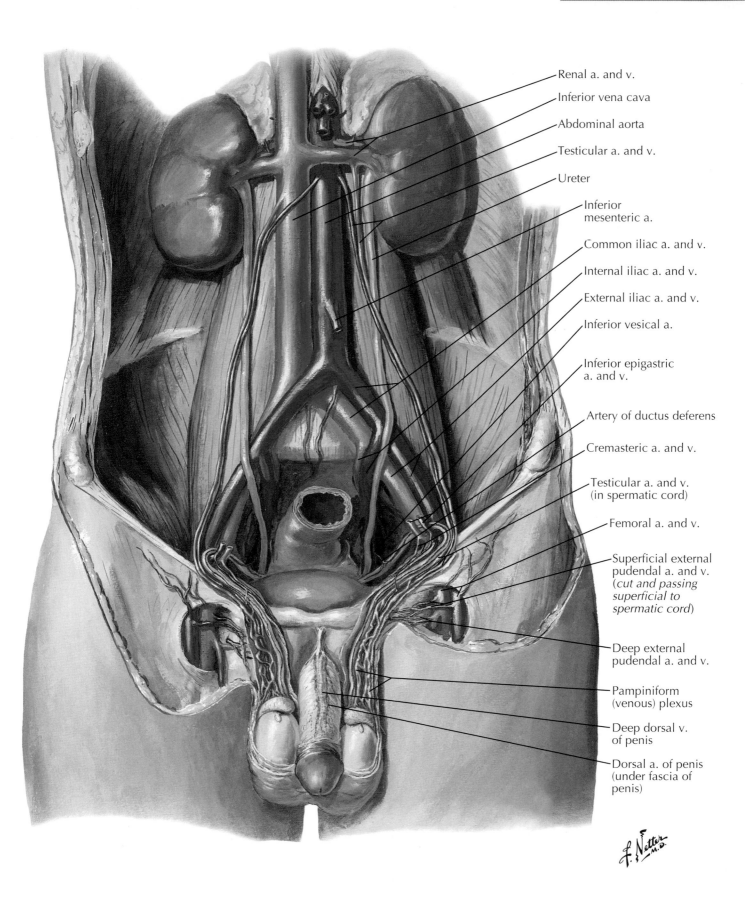

Renal a. and v.

Inferior vena cava

Abdominal aorta

Testicular a. and v.

Ureter

Inferior
mesenteric a.

Common iliac a. and v.

Internal iliac a. and v.

External iliac a. and v.

Inferior vesical a.

Inferior epigastric
a. and v.

Artery of ductus deferens

Cremasteric a. and v.

Testicular a. and v.
(in spermatic cord)

Femoral a. and v.

Superficial external
pudendal a. and v.
(*cut and passing
superficial to
spermatic cord*)

Deep external
pudendal a. and v.

Pampiniform
(venous) plexus

Deep dorsal v.
of penis

Dorsal a. of penis
(under fascia of
penis)

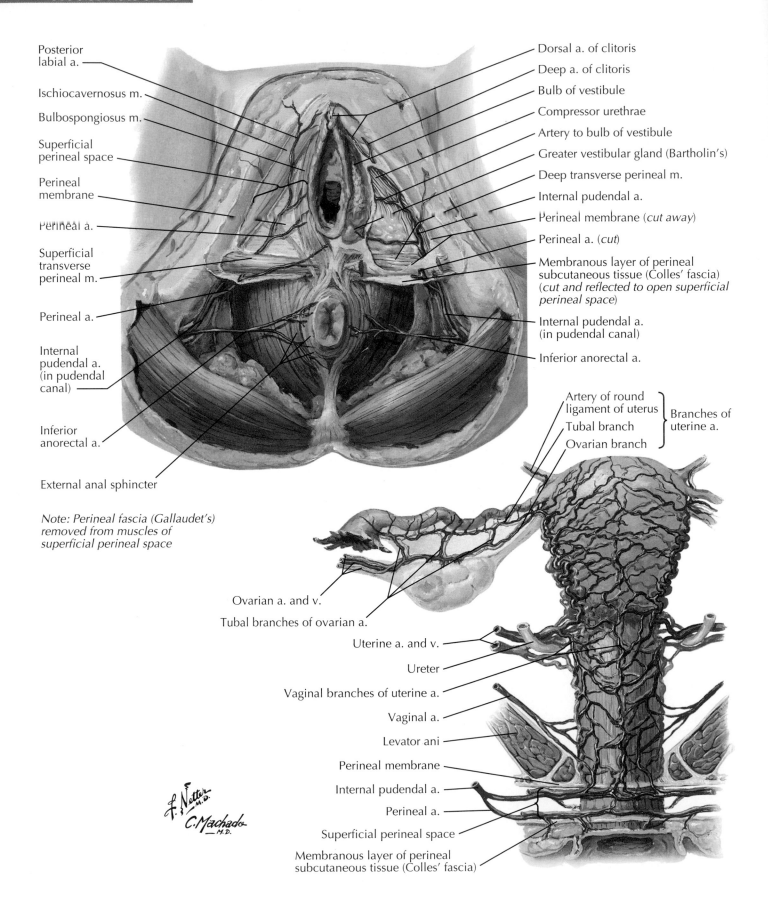

Posterior labial a.

Ischiocavernosus m.

Bulbospongiosus m.

Superficial perineal space

Perineal membrane

Perineal a.

Superficial transverse perineal m.

Perineal a.

Internal pudendal a. (in pudendal canal)

Inferior anorectal a.

External anal sphincter

Note: Perineal fascia (Gallaudet's) removed from muscles of superficial perineal space

Dorsal a. of clitoris

Deep a. of clitoris

Bulb of vestibule

Compressor urethrae

Artery to bulb of vestibule

Greater vestibular gland (Bartholin's)

Deep transverse perineal m.

Internal pudendal a.

Perineal membrane (cut away)

Perineal a. (cut)

Membranous layer of perineal subcutaneous tissue (Colles' fascia) (cut and reflected to open superficial perineal space)

Internal pudendal a. (in pudendal canal)

Inferior anorectal a.

Artery of round ligament of uterus

Tubal branch

Ovarian branch

Branches of uterine a.

Ovarian a. and v.

Tubal branches of ovarian a.

Uterine a. and v.

Ureter

Vaginal branches of uterine a.

Vaginal a.

Levator ani

Perineal membrane

Internal pudendal a.

Perineal a.

Superficial perineal space

Membranous layer of perineal subcutaneous tissue (Colles' fascia)

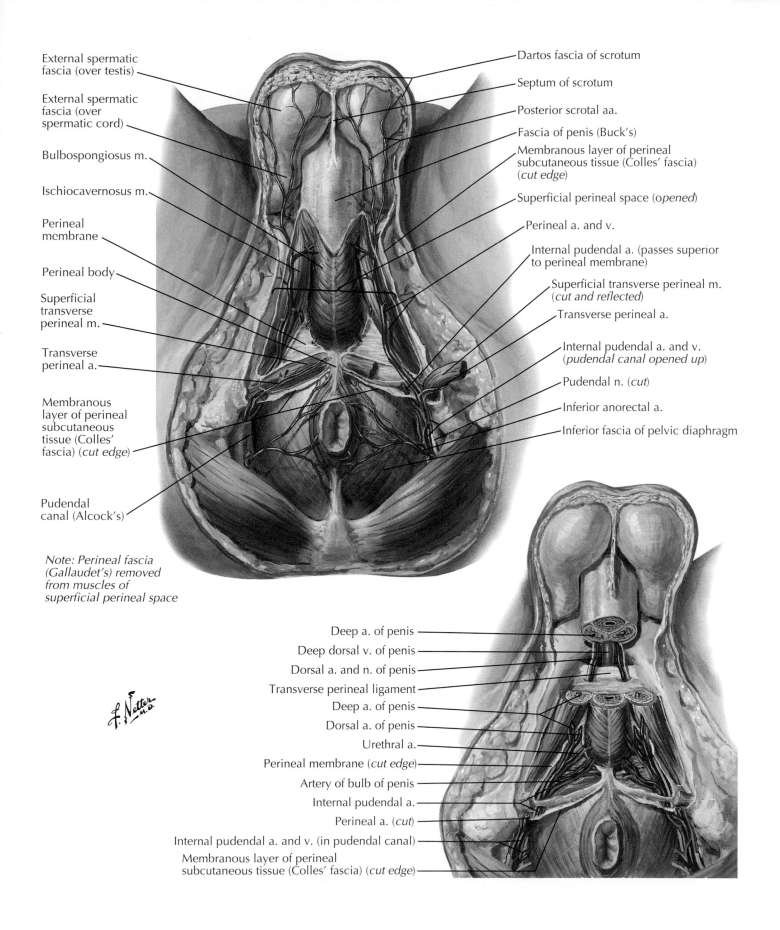

External spermatic fascia (over testis)

External spermatic fascia (over spermatic cord)

Bulbospongiosus m.

Ischiocavernosus m.

Perineal membrane

Perineal body

Superficial transverse perineal m.

Transverse perineal a.

Membranous layer of perineal subcutaneous tissue (Colles' fascia) (cut edge)

Pudendal canal (Alcock's)

Note: Perineal fascia (Gallaudet's) removed from muscles of superficial perineal space

Dartos fascia of scrotum

Septum of scrotum

Posterior scrotal aa.

Fascia of penis (Buck's)

Membranous layer of perineal subcutaneous tissue (Colles' fascia) (cut edge)

Superficial perineal space (opened)

Perineal a. and v.

Internal pudendal a. (passes superior to perineal membrane)

Superficial transverse perineal m. (cut and reflected)

Transverse perineal a.

Internal pudendal a. and v. (pudendal canal opened up)

Pudendal n. (cut)

Inferior anorectal a.

Inferior fascia of pelvic diaphragm

Deep a. of penis

Deep dorsal v. of penis

Dorsal a. and n. of penis

Transverse perineal ligament

Deep a. of penis

Dorsal a. of penis

Urethral a.

Perineal membrane (cut edge)

Artery of bulb of penis

Internal pudendal a.

Perineal a. (cut)

Internal pudendal a. and v. (in pudendal canal)

Membranous layer of perineal subcutaneous tissue (Colles' fascia) (cut edge)

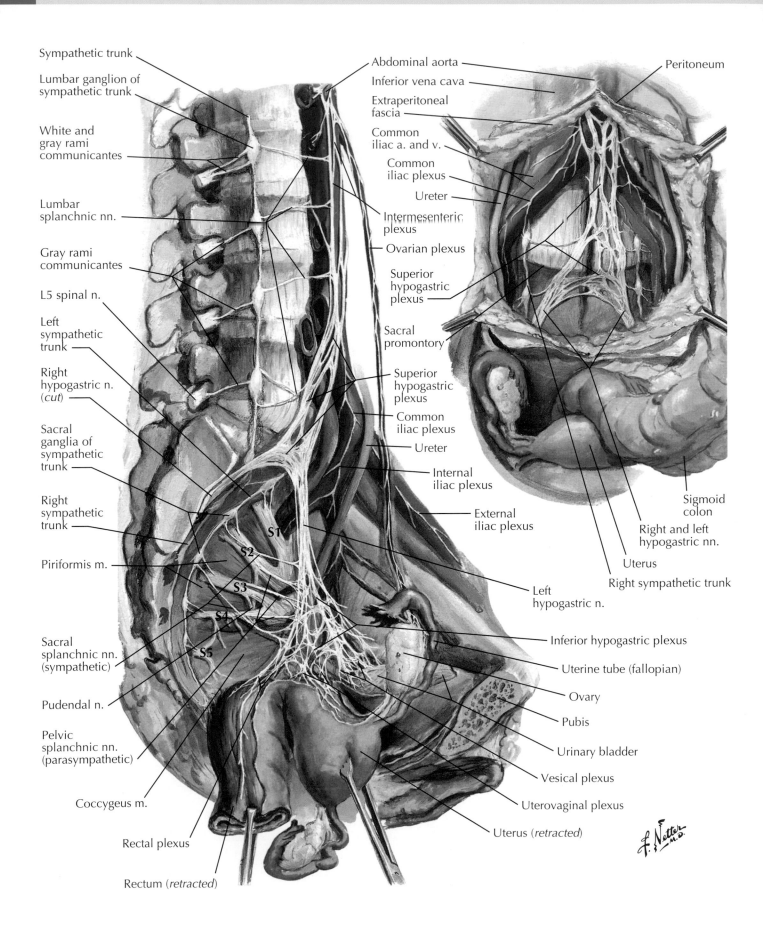

Sympathetic trunk

Lumbar ganglion of sympathetic trunk

White and gray rami communicantes

Lumbar splanchnic nn.

Gray rami communicantes

L5 spinal n.

Left sympathetic trunk

Right hypogastric n. (*cut*)

Sacral ganglia of sympathetic trunk

Right sympathetic trunk

Piriformis m.

Sacral splanchnic nn. (sympathetic)

Pudendal n.

Pelvic splanchnic nn. (parasympathetic)

Coccygeus m.

Rectal plexus

Rectum (*retracted*)

Abdominal aorta

Inferior vena cava

Extraperitoneal fascia

Common iliac a. and v.

Common iliac plexus

Ureter

Intermesenteric plexus

Ovarian plexus

Superior hypogastric plexus

Sacral promontory

Superior hypogastric plexus

Common iliac plexus

Ureter

Internal iliac plexus

External iliac plexus

Left hypogastric n.

Peritoneum

Sigmoid colon

Right and left hypogastric nn.

Uterus

Right sympathetic trunk

Inferior hypogastric plexus

Uterine tube (fallopian)

Ovary

Pubis

Urinary bladder

Vesical plexus

Uterovaginal plexus

Uterus (*retracted*)

S1

S2

S3

S4

S5

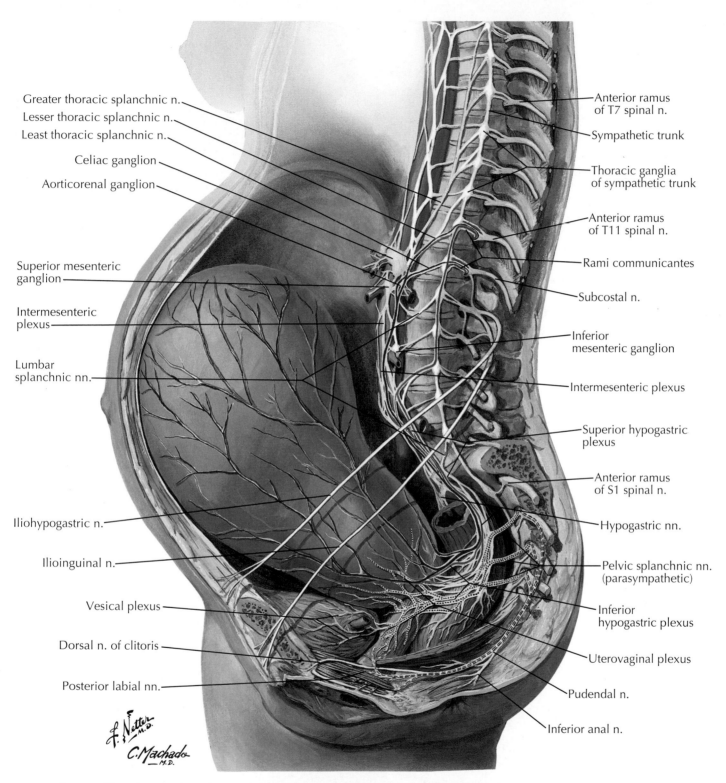

Greater thoracic splanchnic n.

Lesser thoracic splanchnic n.

Least thoracic splanchnic n.

Celiac ganglion

Aorticorenal ganglion

Superior mesenteric ganglion

Intermesenteric plexus

Lumbar splanchnic nn.

Iliohypogastric n.

Ilioinguinal n.

Vesical plexus

Dorsal n. of clitoris

Posterior labial nn.

Anterior ramus of T7 spinal n.

Sympathetic trunk

Thoracic ganglia of sympathetic trunk

Anterior ramus of T11 spinal n.

Rami communicantes

Subcostal n.

Inferior mesenteric ganglion

Intermesenteric plexus

Superior hypogastric plexus

Anterior ramus of S1 spinal n.

Hypogastric nn.

Pelvic splanchnic nn. (parasympathetic)

Inferior hypogastric plexus

Uterovaginal plexus

Pudendal n.

Inferior anal n.

——— Sensory fibers from body and fundus of uterus accompany sympathetic fibers to lower thoracic part of spinal cord via hypogastric plexuses

——— Sympathetic fibers to body and fundus of uterus

·········· Sensory fibers from cervix and upper vagina accompany parasympathetic fibers to sacral part of spinal cord via pelvic splanchnic nerves

·········· Parasympathetic fibers to lower uterine segment, cervix, and upper vagina

– – – – Sensory fibers from lower vagina and perineum accompany somatic motor fibers to sacral part of spinal cord via pudendal nerve

– – – – – Somatic motor fibers to lower vagina and perineum via pudendal nerve

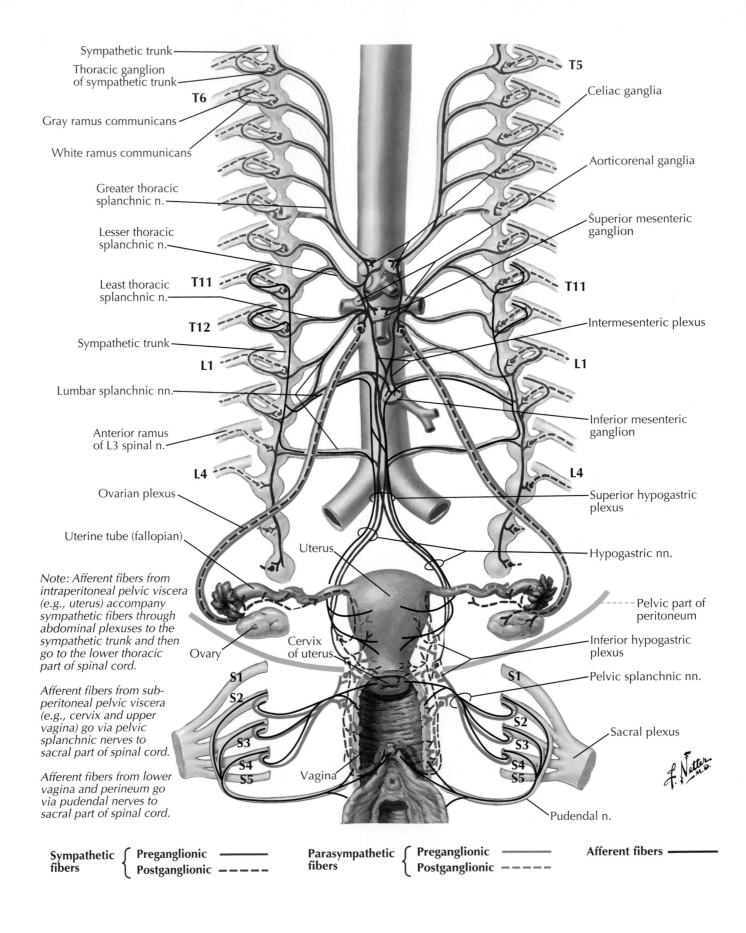

Sympathetic trunk

Thoracic ganglion of sympathetic trunk

T6

Gray ramus communicans

White ramus communicans

Greater thoracic splanchnic n.

Lesser thoracic splanchnic n.

Least thoracic splanchnic n.

T11

T12

Sympathetic trunk

L1

Lumbar splanchnic nn.

Anterior ramus of L3 spinal n.

L4

Ovarian plexus

Uterine tube (fallopian)

Note: Afferent fibers from intraperitoneal pelvic viscera (e.g., uterus) accompany sympathetic fibers through abdominal plexuses to the sympathetic trunk and then go to the lower thoracic part of spinal cord.

Afferent fibers from sub-peritoneal pelvic viscera (e.g., cervix and upper vagina) go via pelvic splanchnic nerves to sacral part of spinal cord.

Afferent fibers from lower vagina and perineum go via pudendal nerves to sacral part of spinal cord.

T5

Celiac ganglia

Aorticorenal ganglia

Superior mesenteric ganglion

T11

Intermesenteric plexus

L1

Inferior mesenteric ganglion

L4

Superior hypogastric plexus

Uterus

Hypogastric nn.

Pelvic part of peritoneum

Cervix of uterus

Inferior hypogastric plexus

Ovary

Pelvic splanchnic nn.

S1

S2

S1

Sacral plexus

S3

S2

S4

S3

S5

Vagina

S4

S5

Pudendal n.

| Sympathetic fibers | { Preganglionic ——— Postganglionic - - - - - | Parasympathetic fibers | { Preganglionic ——— Postganglionic - - - - - | Afferent fibers ——— |

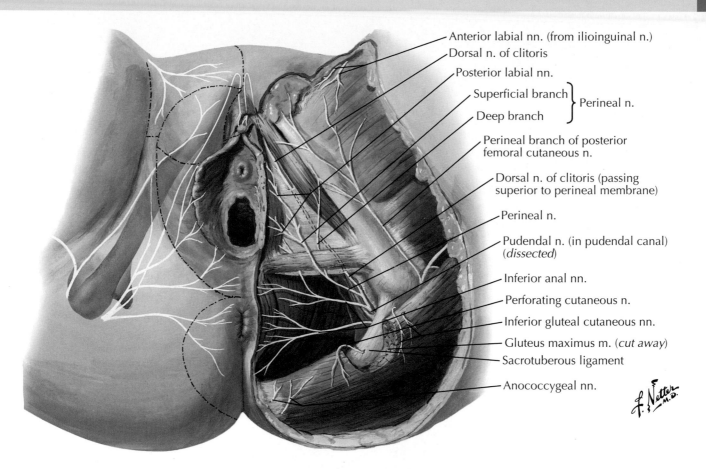

Anterior labial nn. (from ilioinguinal n.)

Dorsal n. of clitoris

Posterior labial nn.

Superficial branch ⎫
⎬ Perineal n.
Deep branch ⎭

Perineal branch of posterior femoral cutaneous n.

Dorsal n. of clitoris (passing superior to perineal membrane)

Perineal n.

Pudendal n. (in pudendal canal) (*dissected*)

Inferior anal nn.

Perforating cutaneous n.

Inferior gluteal cutaneous nn.

Gluteus maximus m. (*cut away*)

Sacrotuberous ligament

Anococcygeal nn.

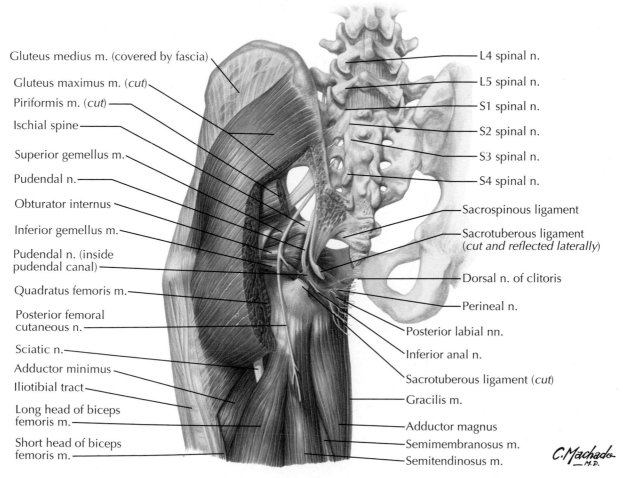

Gluteus medius m. (covered by fascia)

Gluteus maximus m. (*cut*)

Piriformis m. (*cut*)

Ischial spine

Superior gemellus m.

Pudendal n.

Obturator internus

Inferior gemellus m.

Pudendal n. (inside pudendal canal)

Quadratus femoris m.

Posterior femoral cutaneous n.

Sciatic n.

Adductor minimus

Iliotibial tract

Long head of biceps femoris m.

Short head of biceps femoris m.

L4 spinal n.

L5 spinal n.

S1 spinal n.

S2 spinal n.

S3 spinal n.

S4 spinal n.

Sacrospinous ligament

Sacrotuberous ligament (*cut and reflected laterally*)

Dorsal n. of clitoris

Perineal n.

Posterior labial nn.

Inferior anal n.

Sacrotuberous ligament (*cut*)

Gracilis m.

Adductor magnus

Semimembranosus m.

Semitendinosus m.

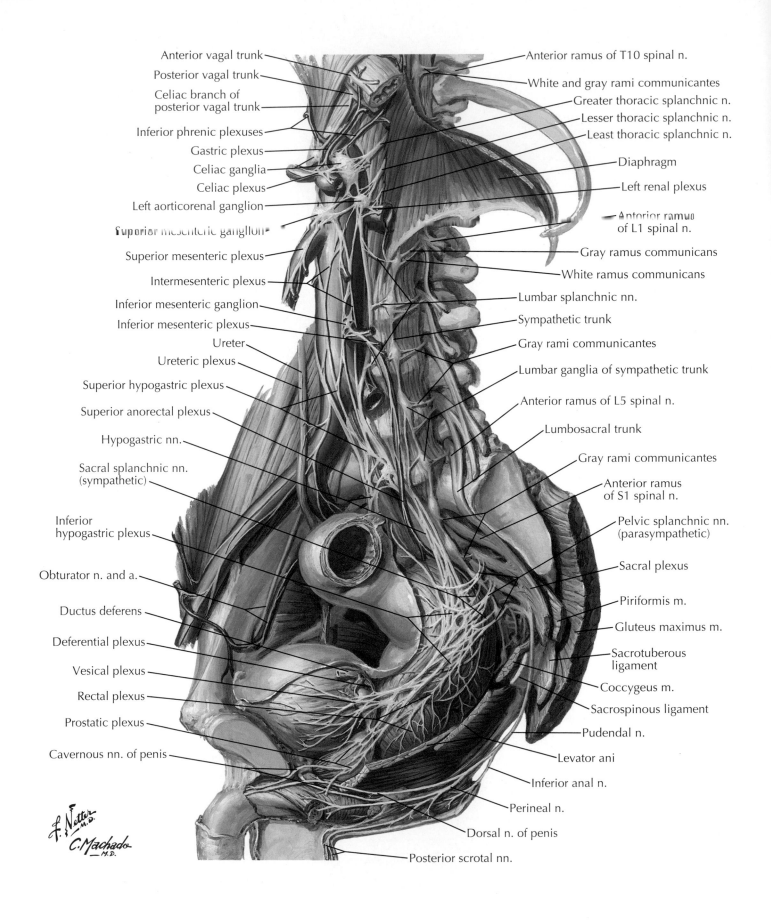

Anterior vagal trunk
Posterior vagal trunk
Celiac branch of posterior vagal trunk
Inferior phrenic plexuses
Gastric plexus
Celiac ganglia
Celiac plexus
Left aorticorenal ganglion
Superior mesenteric ganglion
Superior mesenteric plexus
Intermesenteric plexus
Inferior mesenteric ganglion
Inferior mesenteric plexus
Ureter
Ureteric plexus
Superior hypogastric plexus
Superior anorectal plexus
Hypogastric nn.
Sacral splanchnic nn. (sympathetic)
Inferior hypogastric plexus
Obturator n. and a.
Ductus deferens
Deferential plexus
Vesical plexus
Rectal plexus
Prostatic plexus
Cavernous nn. of penis

Anterior ramus of T10 spinal n.
White and gray rami communicantes
Greater thoracic splanchnic n.
Lesser thoracic splanchnic n.
Least thoracic splanchnic n.
Diaphragm
Left renal plexus
Anterior ramus of L1 spinal n.
Gray ramus communicans
White ramus communicans
Lumbar splanchnic nn.
Sympathetic trunk
Gray rami communicantes
Lumbar ganglia of sympathetic trunk
Anterior ramus of L5 spinal n.
Lumbosacral trunk
Gray rami communicantes
Anterior ramus of S1 spinal n.
Pelvic splanchnic nn. (parasympathetic)
Sacral plexus
Piriformis m.
Gluteus maximus m.
Sacrotuberous ligament
Coccygeus m.
Sacrospinous ligament
Pudendal n.
Levator ani
Inferior anal n.
Perineal n.
Dorsal n. of penis
Posterior scrotal nn.

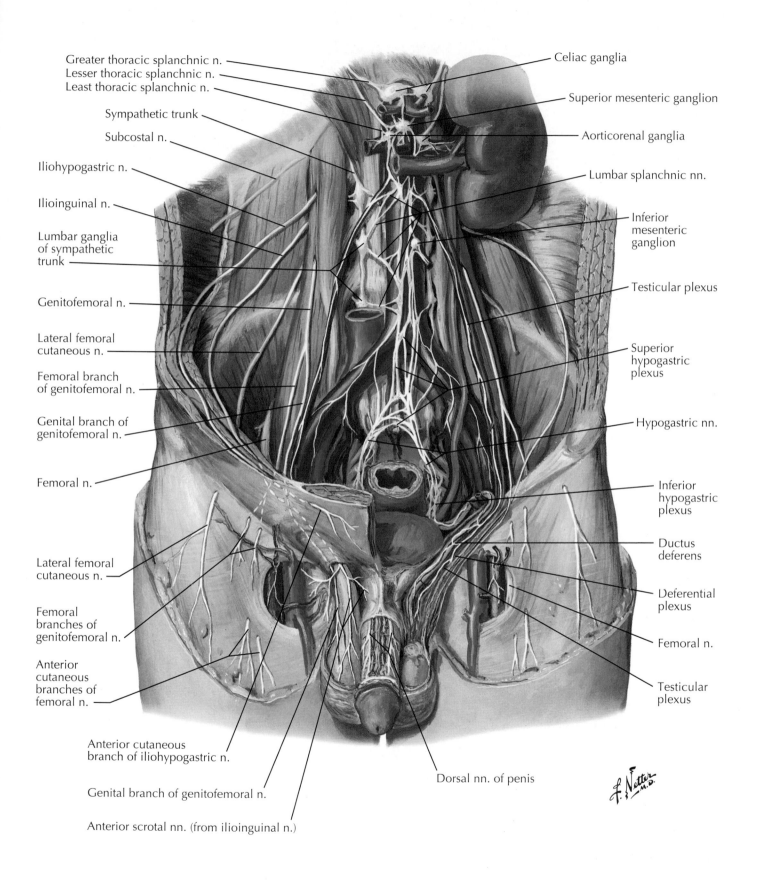

Greater thoracic splanchnic n.
Lesser thoracic splanchnic n.
Least thoracic splanchnic n.
Sympathetic trunk
Subcostal n.
Iliohypogastric n.
Ilioinguinal n.
Lumbar ganglia of sympathetic trunk
Genitofemoral n.
Lateral femoral cutaneous n.
Femoral branch of genitofemoral n.
Genital branch of genitofemoral n.
Femoral n.
Lateral femoral cutaneous n.
Femoral branches of genitofemoral n.
Anterior cutaneous branches of femoral n.
Anterior cutaneous branch of iliohypogastric n.
Genital branch of genitofemoral n.
Anterior scrotal nn. (from ilioinguinal n.)

Celiac ganglia
Superior mesenteric ganglion
Aorticorenal ganglia
Lumbar splanchnic nn.
Inferior mesenteric ganglion
Testicular plexus
Superior hypogastric plexus
Hypogastric nn.
Inferior hypogastric plexus
Ductus deferens
Deferential plexus
Femoral n.
Testicular plexus
Dorsal nn. of penis

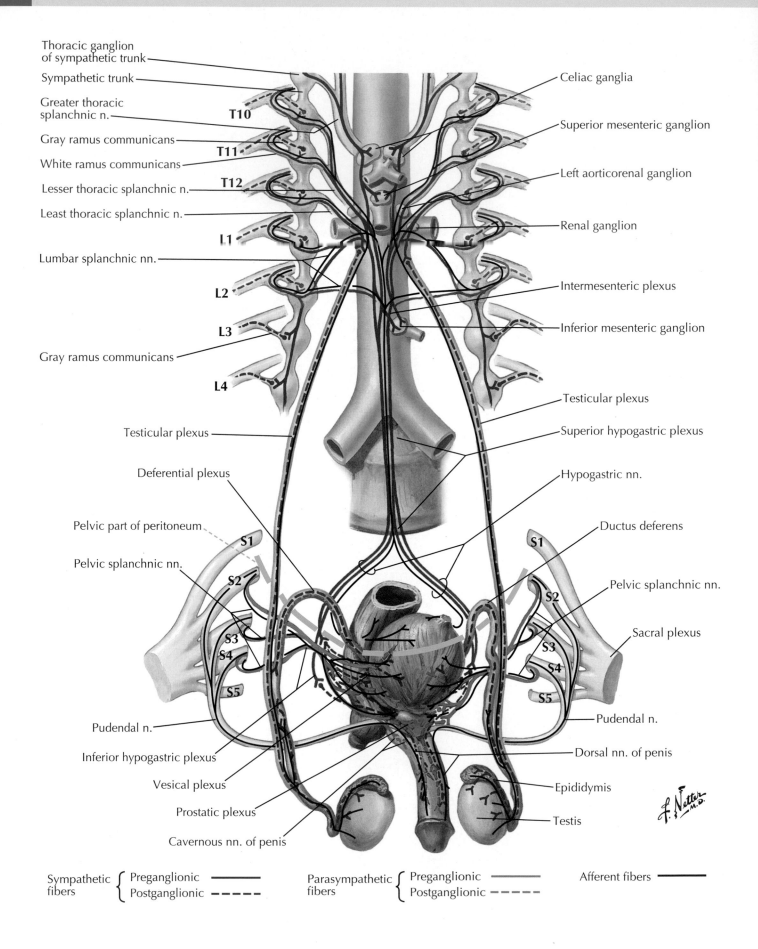

Thoracic ganglion of sympathetic trunk

Sympathetic trunk

Greater thoracic splanchnic n.

Gray ramus communicans

White ramus communicans

Lesser thoracic splanchnic n.

Least thoracic splanchnic n.

Lumbar splanchnic nn.

Gray ramus communicans

Testicular plexus

Deferential plexus

Pelvic part of peritoneum

Pelvic splanchnic nn.

Pudendal n.

Inferior hypogastric plexus

Vesical plexus

Prostatic plexus

Cavernous nn. of penis

T10
T11
T12
L1
L2
L3
L4

S1
S2
S3
S4
S5

Celiac ganglia

Superior mesenteric ganglion

Left aorticorenal ganglion

Renal ganglion

Intermesenteric plexus

Inferior mesenteric ganglion

Testicular plexus

Superior hypogastric plexus

Hypogastric nn.

Ductus deferens

Pelvic splanchnic nn.

Sacral plexus

Pudendal n.

Dorsal nn. of penis

Epididymis

Testis

S1
S2
S3
S4
S5

| Sympathetic fibers | Preganglionic | ———— | Parasympathetic fibers | Preganglionic | ———— | Afferent fibers | ———— |
| | Postganglionic | – – – – | | Postganglionic | – – – – | | |

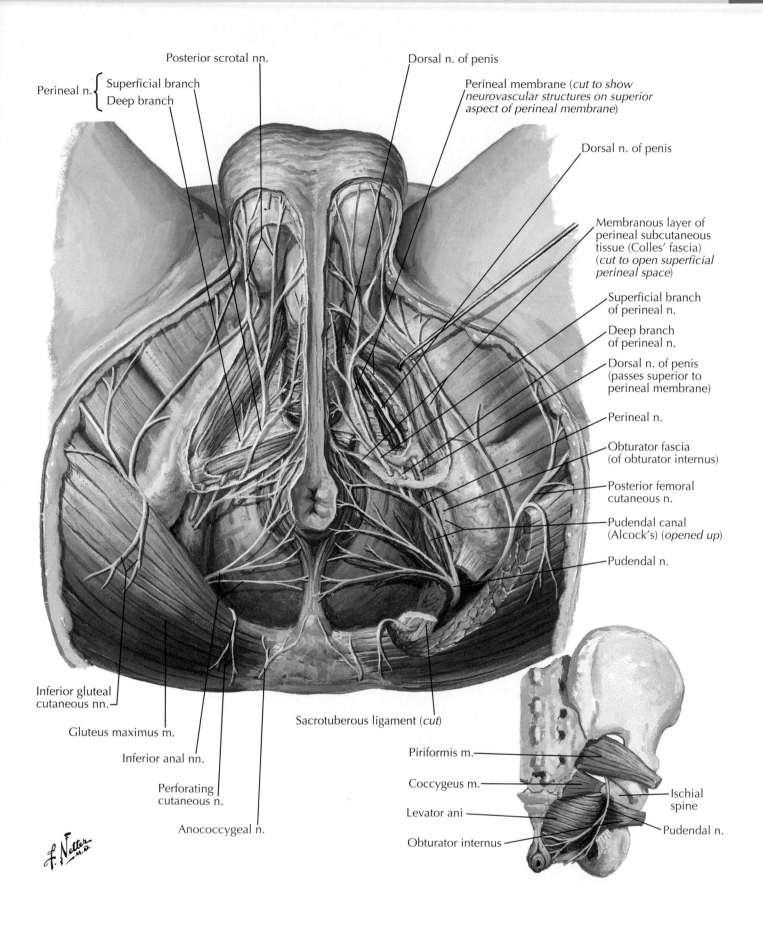

Perineal n. {
Superficial branch
Deep branch

Posterior scrotal nn.

Dorsal n. of penis

Perineal membrane (*cut to show neurovascular structures on superior aspect of perineal membrane*)

Dorsal n. of penis

Membranous layer of perineal subcutaneous tissue (Colles' fascia) (*cut to open superficial perineal space*)

Superficial branch of perineal n.

Deep branch of perineal n.

Dorsal n. of penis (passes superior to perineal membrane)

Perineal n.

Obturator fascia (of obturator internus)

Posterior femoral cutaneous n.

Pudendal canal (Alcock's) (*opened up*)

Pudendal n.

Inferior gluteal cutaneous nn.

Gluteus maximus m.

Inferior anal nn.

Perforating cutaneous n.

Anococcygeal n.

Sacrotuberous ligament (*cut*)

Piriformis m.

Coccygeus m.

Levator ani

Obturator internus

Ischial spine

Pudendal n.

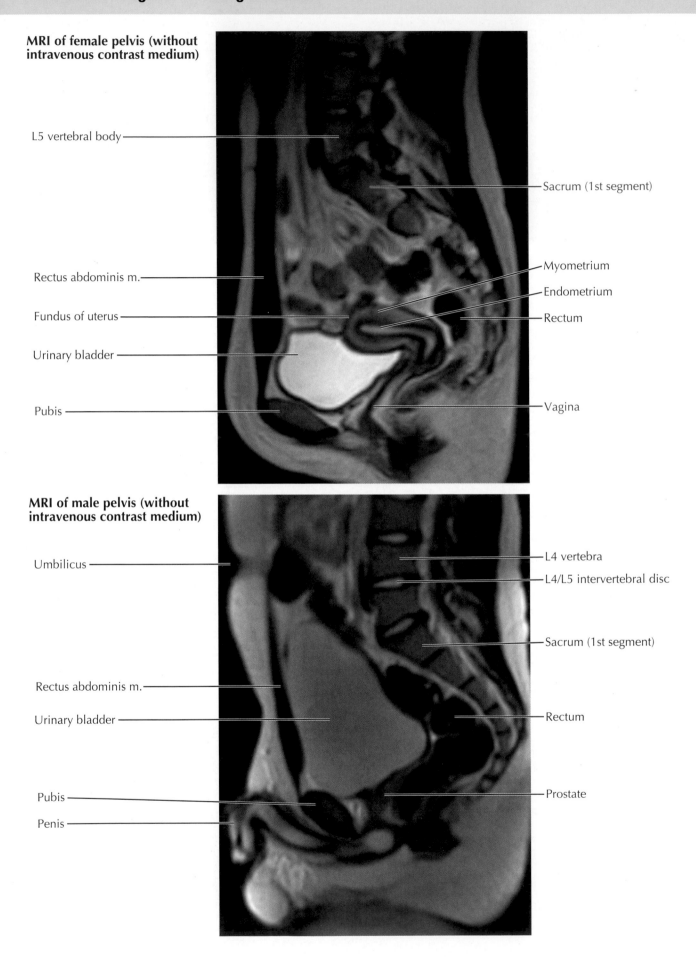

MRI of female pelvis (without intravenous contrast medium)

L5 vertebral body

Rectus abdominis m.

Fundus of uterus

Urinary bladder

Pubis

Sacrum (1st segment)

Myometrium

Endometrium

Rectus

Vagina

MRI of male pelvis (without intravenous contrast medium)

Umbilicus

Rectus abdominis m.

Urinary bladder

Pubis

Penis

L4 vertebra

L4/L5 intervertebral disc

Sacrum (1st segment)

Rectum

Prostate

Imaging of Pelvic Viscera

ANATOMIC STRUCTURES	CLINICAL IMPORTANCE	PLATE NUMBERS
Thorax		
Mammary gland	Breast cancer is most common malignancy in women; most common type originates in lactiferous duct and can either be localized within duct (ductal carcinoma in situ) or invasive into adjoining tissues (invasive ductal carcinoma)	S–473
Pelvis		
Rectouterine pouch (of Douglas)	Region examined with ultrasound to detect presence of abdominopelvic fluid; common site of ectopic pregnancy; may be accessed via posterior vaginal fornix; normally contains small, physiologic amount of peritoneal fluid	S–485, S–529
Uterus	Site of fetal gestation; palpated during prenatal examinations to assess fetal growth; can also contain large, sometimes painful growths known as leiomyomas (fibroids)	S–485, S–487
Uterine (fallopian) tubes	Common site of ectopic pregnany; inflammation (salpingitis) may occur in pelvic inflammatory disease (PID), the result of sexually transmitted infection, possibly leading to fibrosis and infertility; surgical occlusion (tubal ligation) is performed when women desire permanent contraception	S–483, S–485, S–486
Cervix of uterus	Epithelium of transformation zone of cervix is prone to dysplasia and malignancy; cells are sampled from this region during Pap smear examination and tested for infection with human papillomavirus, the leading risk factor for cervical malignancy	S–487, S–488
Vagina	Posterior part of vaginal fornix allows access to rectouterine pouch (of Douglas)	S–485
Ovary	Examined with ultrasound to identify cysts or for oocyte collection; torsion is a painful condition that occurs when ovary twists on axis of suspensory ligament of ovary, occluding ovarian vessels and causing engorgement and ischemia	S–483, S–486, S–487
Testis	Torsion is a painful condition that occurs when testis twists on axis of testicular vasculature, causing engorgement and ischemia	S–497
Prostate	Prone to benign hypertrophy with aging, which results in urinary outflow obstruction; prostate cancer is second most common cancer in men	S–495, S–496
Ductus deferens	Ligation (vasectomy) performed when men desire permanent contraception	S–495, S–497

*Selections are based largely on clinical data and commonly discussed clinical correlations in macroscopic ("gross") anatomy courses.

ENDOCRINE SYSTEM 11

ELECTRONIC BONUS PLATES

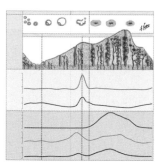

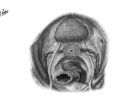
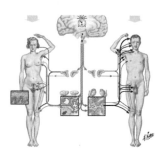

S–BP 95 Menstrual Cycle

S–BP 96 Ovary, Oocytes, and Follicles

S–BP 97 Endocrine Glands, Hormones, and Puberty

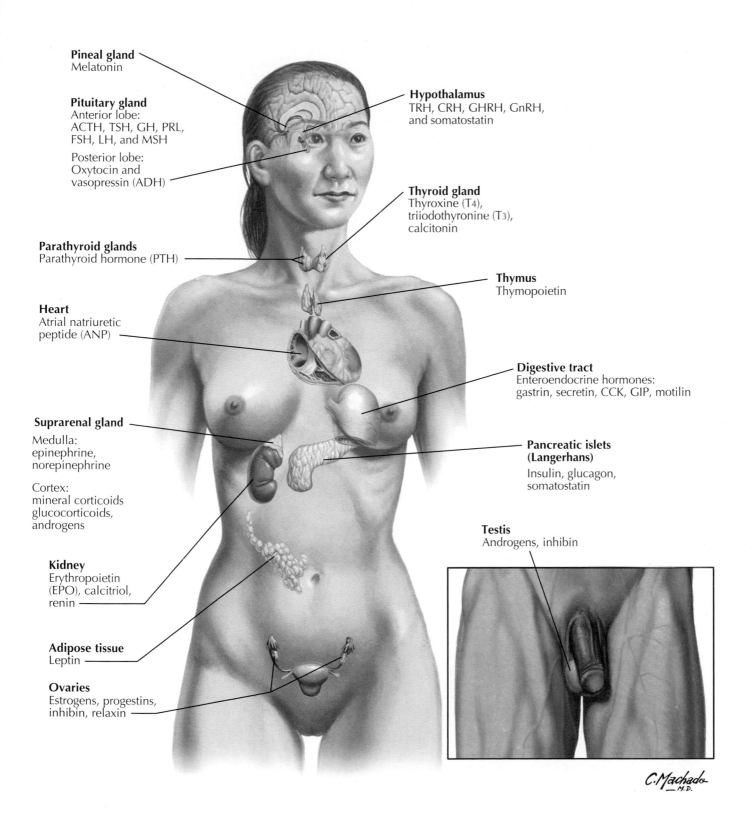

Pineal gland
Melatonin

Pituitary gland
Anterior lobe:
ACTH, TSH, GH, PRL,
FSH, LH, and MSH

Posterior lobe:
Oxytocin and
vasopressin (ADH)

Parathyroid glands
Parathyroid hormone (PTH)

Heart
Atrial natriuretic
peptide (ANP)

Suprarenal gland

Medulla:
epinephrine,
norepinephrine

Cortex:
mineral corticoids
glucocorticoids,
androgens

Kidney
Erythropoietin
(EPO), calcitriol,
renin

Adipose tissue
Leptin

Ovaries
Estrogens, progestins,
inhibin, relaxin

Hypothalamus
TRH, CRH, GHRH, GnRH,
and somatostatin

Thyroid gland
Thyroxine (T4),
triiodothyronine (T3),
calcitonin

Thymus
Thymopoietin

Digestive tract
Enteroendocrine hormones:
gastrin, secretin, CCK, GIP, motilin

**Pancreatic islets
(Langerhans)**

Insulin, glucagon,
somatostatin

Testis
Androgens, inhibin

C.Machado
M.D.

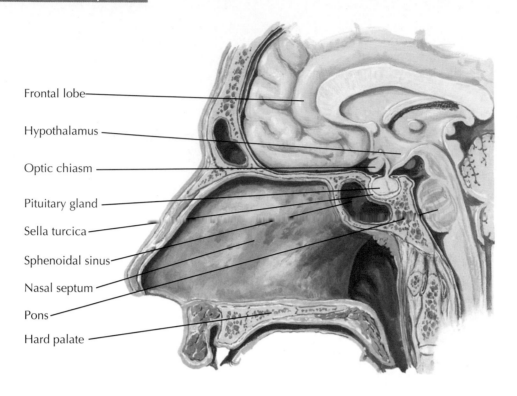

Frontal lobe

Hypothalamus

Optic chiasm

Pituitary gland

Sella turcica

Sphenoidal sinus

Nasal septum

Pons

Hard palate

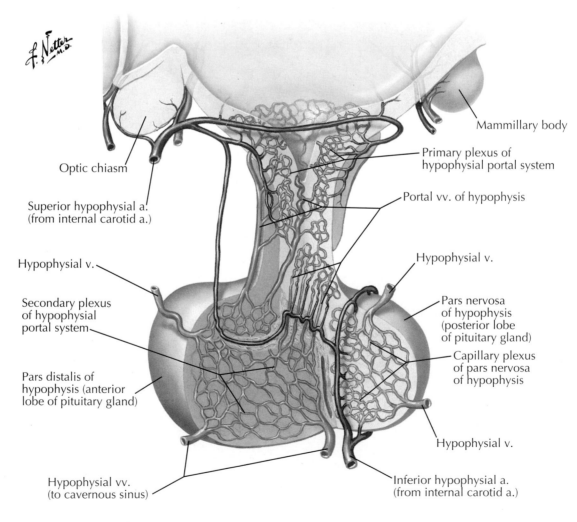

Mammillary body

Primary plexus of
hypophysial portal system

Optic chiasm

Portal vv. of hypophysis

Superior hypophysial a.
(from internal carotid a.)

Hypophysial v.

Hypophysial v.

Secondary plexus
of hypophysial
portal system

Pars nervosa
of hypophysis
(posterior lobe
of pituitary gland)

Pars distalis of
hypophysis (anterior
lobe of pituitary gland)

Capillary plexus
of pars nervosa
of hypophysis

Hypophysial v.

Hypophysial vv.
(to cavernous sinus)

Inferior hypophysial a.
(from internal carotid a.)

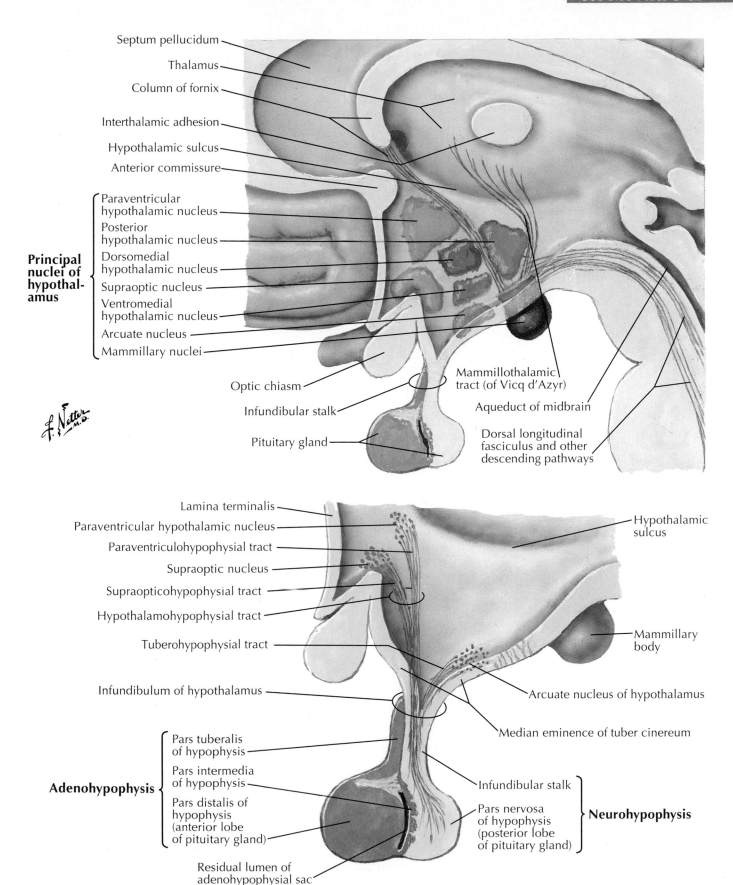

Septum pellucidum

Thalamus

Column of fornix

Interthalamic adhesion

Hypothalamic sulcus

Anterior commissure

Principal nuclei of hypothalamus

Paraventricular hypothalamic nucleus

Posterior hypothalamic nucleus

Dorsomedial hypothalamic nucleus

Supraoptic nucleus

Ventromedial hypothalamic nucleus

Arcuate nucleus

Mammillary nuclei

Optic chiasm

Infundibular stalk

Pituitary gland

Mammillothalamic tract (of Vicq d'Azyr)

Aqueduct of midbrain

Dorsal longitudinal fasciculus and other descending pathways

Lamina terminalis

Paraventricular hypothalamic nucleus

Paraventriculohypophysial tract

Supraoptic nucleus

Supraopticohypophysial tract

Hypothalamohypophysial tract

Tuberohypophysial tract

Infundibulum of hypothalamus

Adenohypophysis

Pars tuberalis of hypophysis

Pars intermedia of hypophysis

Pars distalis of hypophysis (anterior lobe of pituitary gland)

Residual lumen of adenohypophysial sac

Hypothalamic sulcus

Mammillary body

Arcuate nucleus of hypothalamus

Median eminence of tuber cinereum

Infundibular stalk

Pars nervosa of hypophysis (posterior lobe of pituitary gland)

Neurohypophysis

Hypothalamus and Pituitary Gland

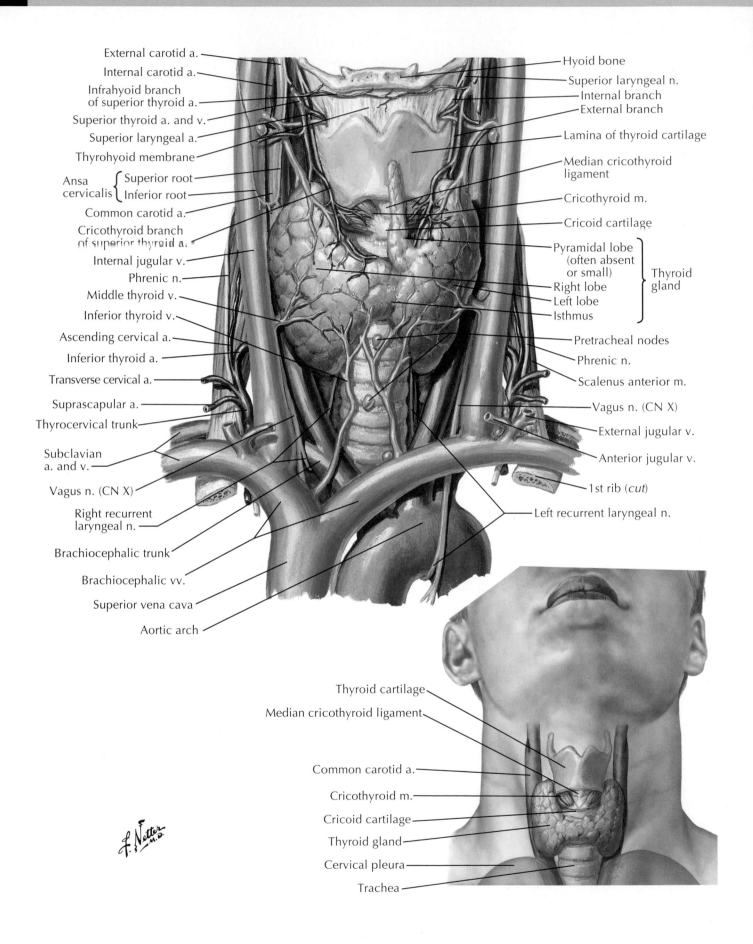

External carotid a.

Internal carotid a.

Infrahyoid branch
of superior thyroid a.

Superior thyroid a. and v.

Superior laryngeal a.

Thyrohyoid membrane

Ansa cervicalis { Superior root / Inferior root

Common carotid a.

Cricothyroid branch
of superior thyroid a.

Internal jugular v.

Phrenic n.

Middle thyroid v.

Inferior thyroid v.

Ascending cervical a.

Inferior thyroid a.

Transverse cervical a.

Suprascapular a.

Thyrocervical trunk

Subclavian
a. and v.

Vagus n. (CN X)

Right recurrent
laryngeal n.

Brachiocephalic trunk

Brachiocephalic vv.

Superior vena cava

Aortic arch

Hyoid bone

Superior laryngeal n.
Internal branch
External branch

Lamina of thyroid cartilage

Median cricothyroid
ligament

Cricothyroid m.

Cricoid cartilage

Pyramidal lobe
(often absent
or small)

Right lobe

Left lobe

Isthmus

Thyroid
gland

Pretracheal nodes

Phrenic n.

Scalenus anterior m.

Vagus n. (CN X)

External jugular v.

Anterior jugular v.

1st rib (cut)

Left recurrent laryngeal n.

Thyroid cartilage

Median cricothyroid ligament

Common carotid a.

Cricothyroid m.

Cricoid cartilage

Thyroid gland

Cervical pleura

Trachea

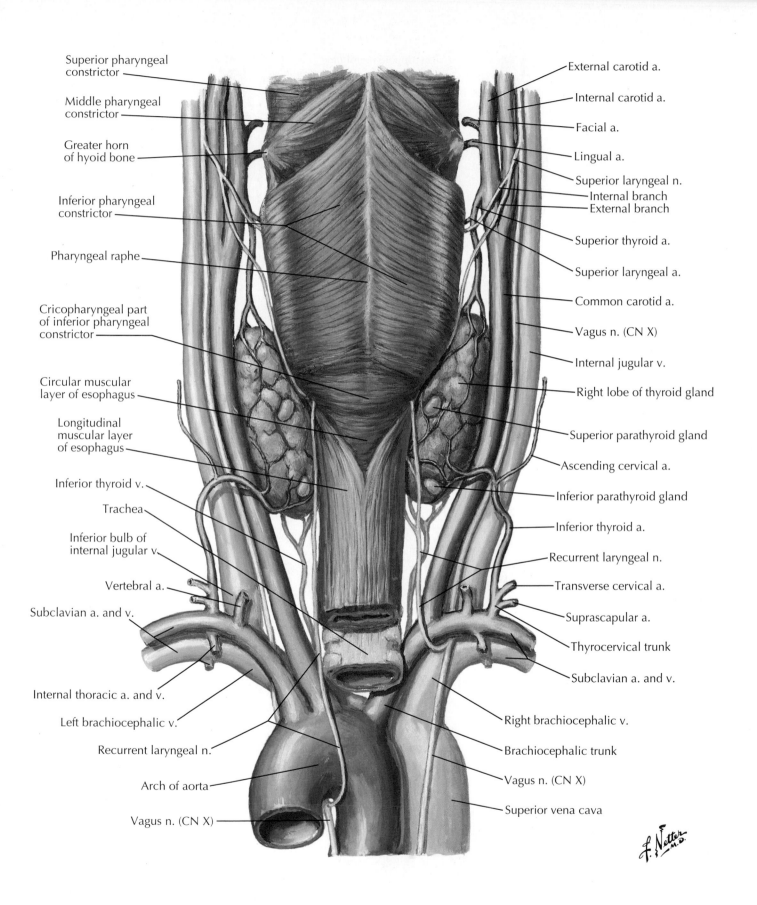

Superior pharyngeal constrictor

Middle pharyngeal constrictor

Greater horn of hyoid bone

Inferior pharyngeal constrictor

Pharyngeal raphe

Cricopharyngeal part of inferior pharyngeal constrictor

Circular muscular layer of esophagus

Longitudinal muscular layer of esophagus

Inferior thyroid v.

Trachea

Inferior bulb of internal jugular v.

Vertebral a.

Subclavian a. and v.

Internal thoracic a. and v.

Left brachiocephalic v.

Recurrent laryngeal n.

Arch of aorta

Vagus n. (CN X)

External carotid a.

Internal carotid a.

Facial a.

Lingual a.

Superior laryngeal n.
Internal branch
External branch

Superior thyroid a.

Superior laryngeal a.

Common carotid a.

Vagus n. (CN X)

Internal jugular v.

Right lobe of thyroid gland

Superior parathyroid gland

Ascending cervical a.

Inferior parathyroid gland

Inferior thyroid a.

Recurrent laryngeal n.

Transverse cervical a.

Suprascapular a.

Thyrocervical trunk

Subclavian a. and v.

Right brachiocephalic v.

Brachiocephalic trunk

Vagus n. (CN X)

Superior vena cava

f. Netter M.D.

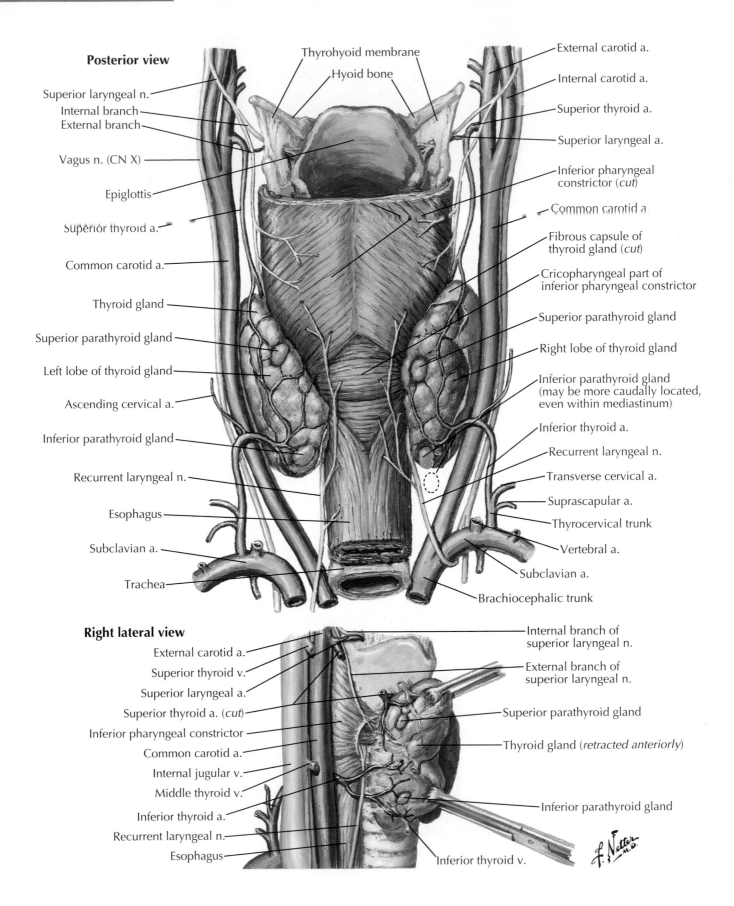

Posterior view

Thyrohyoid membrane

Hyoid bone

External carotid a.

Internal carotid a.

Superior thyroid a.

Superior laryngeal a.

Superior laryngeal n.
Internal branch
External branch

Vagus n. (CN X)

Epiglottis

Superior thyroid a.

Common carotid a.

Thyroid gland

Superior parathyroid gland

Left lobe of thyroid gland

Ascending cervical a.

Inferior parathyroid gland

Recurrent laryngeal n.

Esophagus

Subclavian a.

Trachea

Inferior pharyngeal
constrictor (*cut*)

Common carotid a.

Fibrous capsule of
thyroid gland (*cut*)

Cricopharyngeal part of
inferior pharyngeal constrictor

Superior parathyroid gland

Right lobe of thyroid gland

Inferior parathyroid gland
(may be more caudally located,
even within mediastinum)

Inferior thyroid a.

Recurrent laryngeal n.

Transverse cervical a.

Suprascapular a.

Thyrocervical trunk

Vertebral a.

Subclavian a.

Brachiocephalic trunk

Right lateral view

Internal branch of
superior laryngeal n.

External carotid a.

Superior thyroid v.

Superior laryngeal a.

Superior thyroid a. (*cut*)

Inferior pharyngeal constrictor

Common carotid a.

Internal jugular v.

Middle thyroid v.

Inferior thyroid a.

Recurrent laryngeal n.

Esophagus

External branch of
superior laryngeal n.

Superior parathyroid gland

Thyroid gland (*retracted anteriorly*)

Inferior parathyroid gland

Inferior thyroid v.

Right and left inferior phrenic aa.
(shown here from common trunk)

Left gastric a.

Esophageal branch of left gastric a.

Celiac trunk

Splenic a.

Common hepatic a.

Recurrent branch
of left inferior
phrenic a.
to esophagus

Right gastric a.

Proper hepatic a.

Left gastroomental a.

Supraduodenal a.

Short
gastric
aa.

Right gastroomental a.

Gastroduodenal a.

Anterior superior
pancreaticoduodenal a.

Posterior superior
pancreaticoduodenal a.

Left gastroomental a.

Artery of tail of pancreas
(*partially in phantom*)

Great pancreatic a.

Inferior pancreatic a. (*phantom*)

Dorsal pancreatic a.

Anterior superior
pancreaticoduodenal a.

Middle colic a. (*cut*)

Superior mesenteric a.

Posterior inferior
pancreaticoduodenal a. (*phantom*)

Anterior inferior
pancreaticoduodenal a.

Inferior pancreaticoduodenal a.

**View with stomach
reflected cephalad**

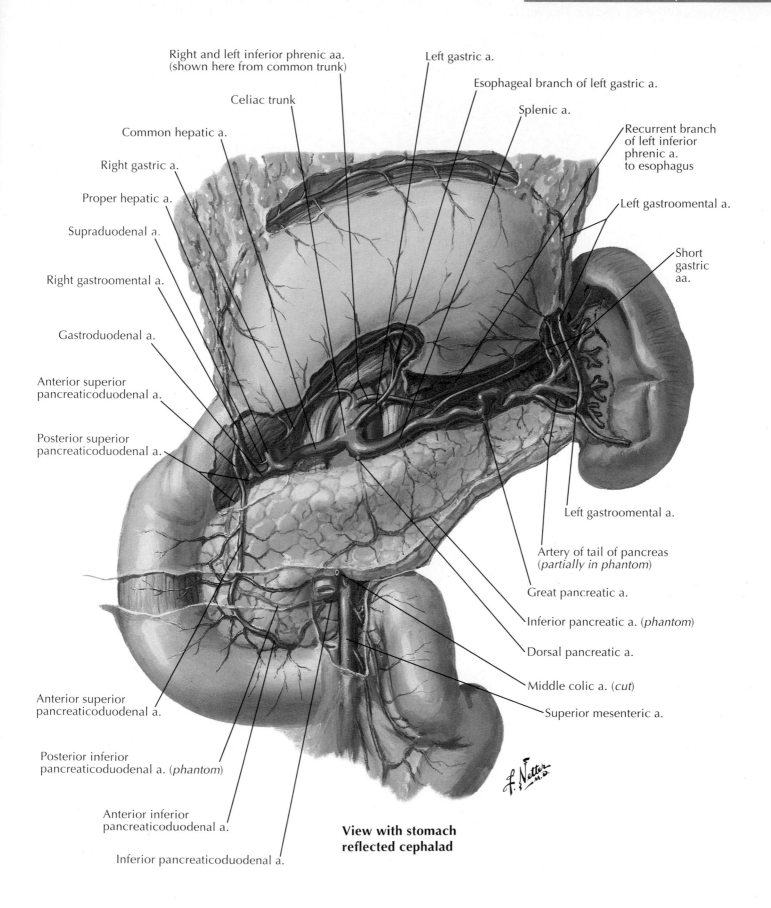

Superior view with peritoneum intact

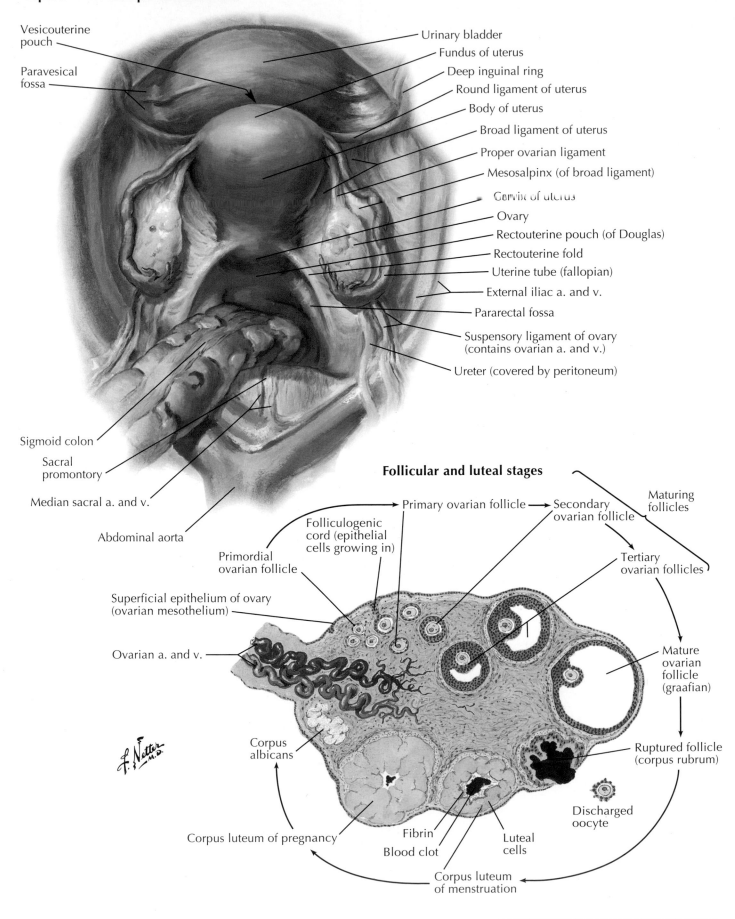

Vesicouterine pouch

Paravesical fossa

Urinary bladder

Fundus of uterus

Deep inguinal ring

Round ligament of uterus

Body of uterus

Broad ligament of uterus

Proper ovarian ligament

Mesosalpinx (of broad ligament)

Cervix of uterus

Ovary

Rectouterine pouch (of Douglas)

Rectouterine fold

Uterine tube (fallopian)

External iliac a. and v.

Pararectal fossa

Suspensory ligament of ovary (contains ovarian a. and v.)

Ureter (covered by peritoneum)

Sigmoid colon

Sacral promontory

Median sacral a. and v.

Abdominal aorta

Follicular and luteal stages

Primordial ovarian follicle

Folliculogenic cord (epithelial cells growing in)

Primary ovarian follicle

Secondary ovarian follicle

Maturing follicles

Tertiary ovarian follicles

Superficial epithelium of ovary (ovarian mesothelium)

Ovarian a. and v.

Mature ovarian follicle (graafian)

Corpus albicans

Ruptured follicle (corpus rubrum)

Discharged oocyte

Corpus luteum of pregnancy

Fibrin

Blood clot

Luteal cells

Corpus luteum of menstruation

Plate S–529

Ovary and Testis

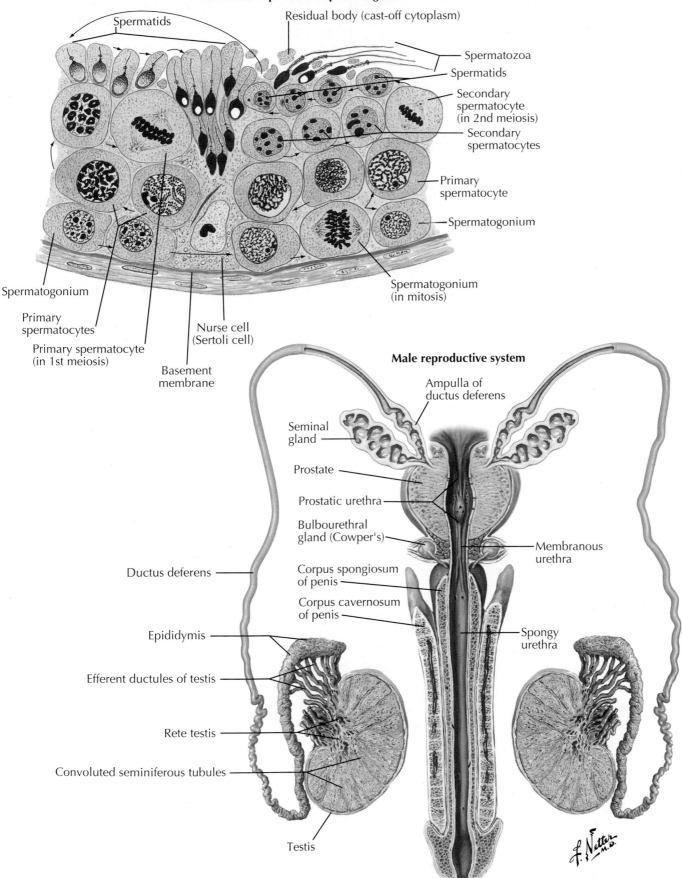

Seminiferous epithelium: spermatogenesis

Spermatids

Residual body (cast-off cytoplasm)

Spermatozoa

Spermatids

Secondary spermatocyte (in 2nd meiosis)

Secondary spermatocytes

Primary spermatocyte

Spermatogonium

Spermatogonium (in mitosis)

Spermatogonium

Primary spermatocytes

Primary spermatocyte (in 1st meiosis)

Nurse cell (Sertoli cell)

Basement membrane

Male reproductive system

Ampulla of ductus deferens

Seminal gland

Prostate

Prostatic urethra

Bulbourethral gland (Cowper's)

Corpus spongiosum of penis

Corpus cavernosum of penis

Epididymis

Efferent ductules of testis

Rete testis

Convoluted seminiferous tubules

Testis

Membranous urethra

Spongy urethra

Ductus deferens

Ovary and Testis

Plate S–530

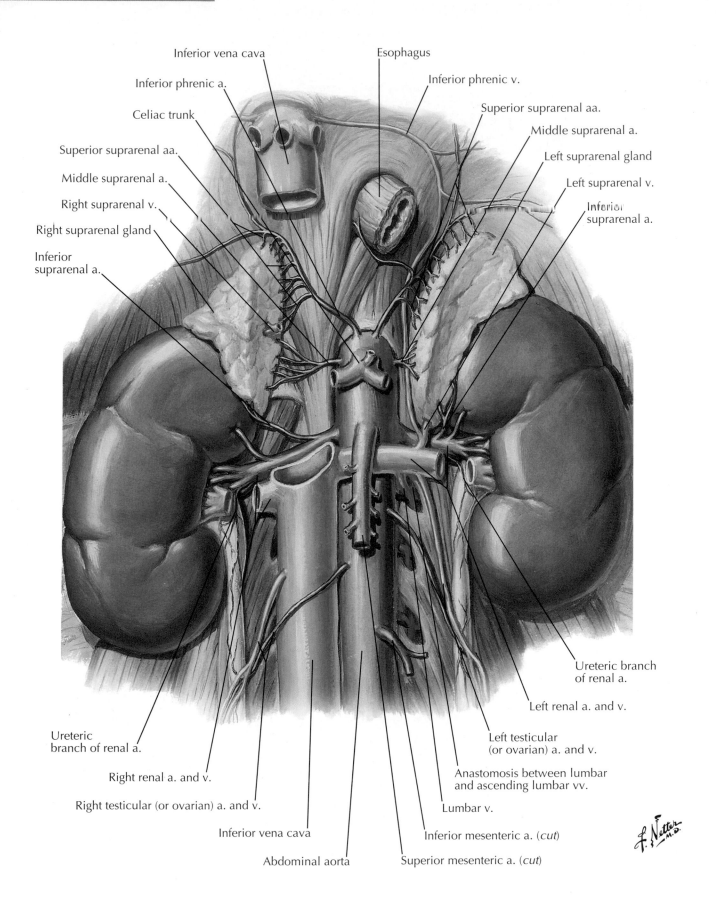

Inferior vena cava

Esophagus

Inferior phrenic a.

Inferior phrenic v.

Celiac trunk

Superior suprarenal aa.

Middle suprarenal a.

Superior suprarenal aa.

Left suprarenal gland

Middle suprarenal a.

Left suprarenal v.

Right suprarenal v.

Inferior suprarenal a.

Right suprarenal gland

Inferior suprarenal a.

Ureteric branch of renal a.

Left renal a. and v.

Ureteric branch of renal a.

Left testicular (or ovarian) a. and v.

Right renal a. and v.

Anastomosis between lumbar and ascending lumbar vv.

Right testicular (or ovarian) a. and v.

Lumbar v.

Inferior vena cava

Inferior mesenteric a. (*cut*)

Abdominal aorta

Superior mesenteric a. (*cut*)

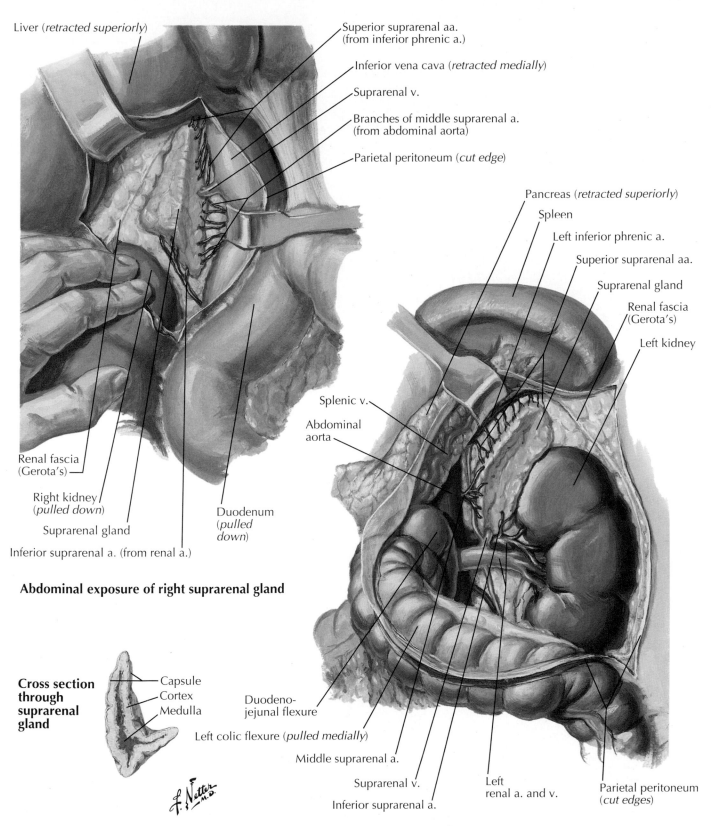

Liver (*retracted superiorly*)

Superior suprarenal aa.
(from inferior phrenic a.)

Inferior vena cava (*retracted medially*)

Suprarenal v.

Branches of middle suprarenal a.
(from abdominal aorta)

Parietal peritoneum (*cut edge*)

Pancreas (*retracted superiorly*)

Spleen

Left inferior phrenic a.

Superior suprarenal aa.

Suprarenal gland

Renal fascia
(Gerota's)

Left kidney

Splenic v.

Abdominal
aorta

Renal fascia
(Gerota's)

Right kidney
(*pulled down*)

Suprarenal gland

Inferior suprarenal a. (from renal a.)

Duodenum
(*pulled
down*)

Abdominal exposure of right suprarenal gland

**Cross section
through
suprarenal
gland**

Capsule
Cortex
Medulla

Duodeno-
jejunal flexure

Left colic flexure (*pulled medially*)

Middle suprarenal a.

Suprarenal v.

Left
renal a. and v.

Parietal peritoneum
(*cut edges*)

Inferior suprarenal a.

Abdominal exposure of left suprarenal gland

Suprarenal Gland

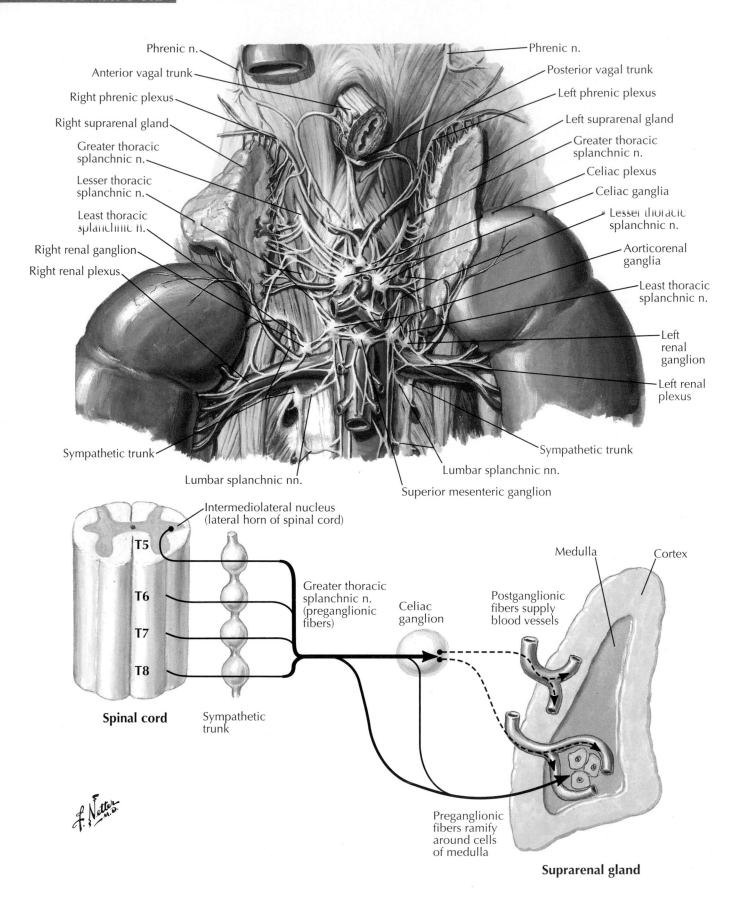

Phrenic n.

Anterior vagal trunk

Right phrenic plexus

Right suprarenal gland

Greater thoracic splanchnic n.

Lesser thoracic splanchnic n.

Least thoracic splanchnic n.

Right renal ganglion

Right renal plexus

Sympathetic trunk

Lumbar splanchnic nn.

Phrenic n.

Posterior vagal trunk

Left phrenic plexus

Left suprarenal gland

Greater thoracic splanchnic n.

Celiac plexus

Celiac ganglia

Lesser thoracic splanchnic n.

Aorticorenal ganglia

Least thoracic splanchnic n.

Left renal ganglion

Left renal plexus

Sympathetic trunk

Lumbar splanchnic nn.

Superior mesenteric ganglion

Intermediolateral nucleus (lateral horn of spinal cord)

T5

T6

T7

T8

Spinal cord

Sympathetic trunk

Greater thoracic splanchnic n. (preganglionic fibers)

Celiac ganglion

Medulla

Cortex

Postganglionic fibers supply blood vessels

Preganglionic fibers ramify around cells of medulla

Suprarenal gland

ANATOMIC STRUCTURES	CLINICAL IMPORTANCE	PLATE NUMBERS
Head and Neck		
Thyroid gland	Enlargement is known as goiter; may be partially or completely removed in malignancy or hyperthyroidism; during examination, should move relatively equally bilaterally with hyoid bone and laryngeal cartilages; malignancy may anchor a portion of gland and cause asymmetric movement	S–525
Abdomen		
Suprarenal gland	Produces hormones in its cortex (e.g., cortisol, aldosterone) and medulla (e.g., epinephrine, norepinephrine); small masses are frequently incidentally discovered on axial abdominal imaging; need for further workup depends in part on size	S–459

*Selections are based largely on clinical data and commonly discussed clinical correlations in macroscopic ("gross") anatomy courses.

CROSS-SECTIONAL ANATOMY AND IMAGING 12

ELECTRONIC BONUS PLATES

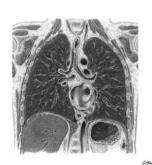
S–BP 98 Thorax: Coronal Section

S–BP 99 Thorax: Coronal CTs

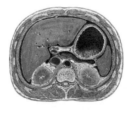
S–BP 100 Schematic Cross Section of Abdomen at T12 Vertebral Level

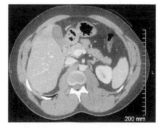
S–BP 101 Axial CT Image of Upper Abdomen

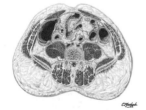
S–BP 102 Transverse Section of Abdomen: L5 Vertebral Level, Near Transtubercular Plane

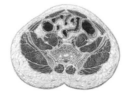
S–BP 103 Transverse Section of Abdomen: S1 Vertebral Level, Anterior Superior Iliac Spine

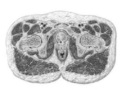
S–BP 104 Cross Section of Lower Pelvis

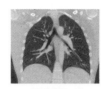

S–BP 105 Cross Sections Through Metacarpal and Distal Carpal Bones

ELECTRONIC BONUS PLATES—*cont'd*

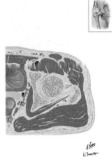

S–BP 106 Cross Section of Hand: Axial View

S–BP 107 Cross Section of Hand: Axial View (Continued)

S–BP 108 Cross-Sectional Anatomy of Hip: Axial View

S–BP 109 Leg: Serial Cross Sections

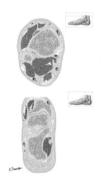 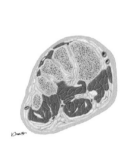

S–BP 110 Cross-Sectional Anatomy of Ankle and Foot

S–BP 111 Cross-Sectional Anatomy of Ankle and Foot (Continued)

S–BP 112 Anatomy of the Toenail

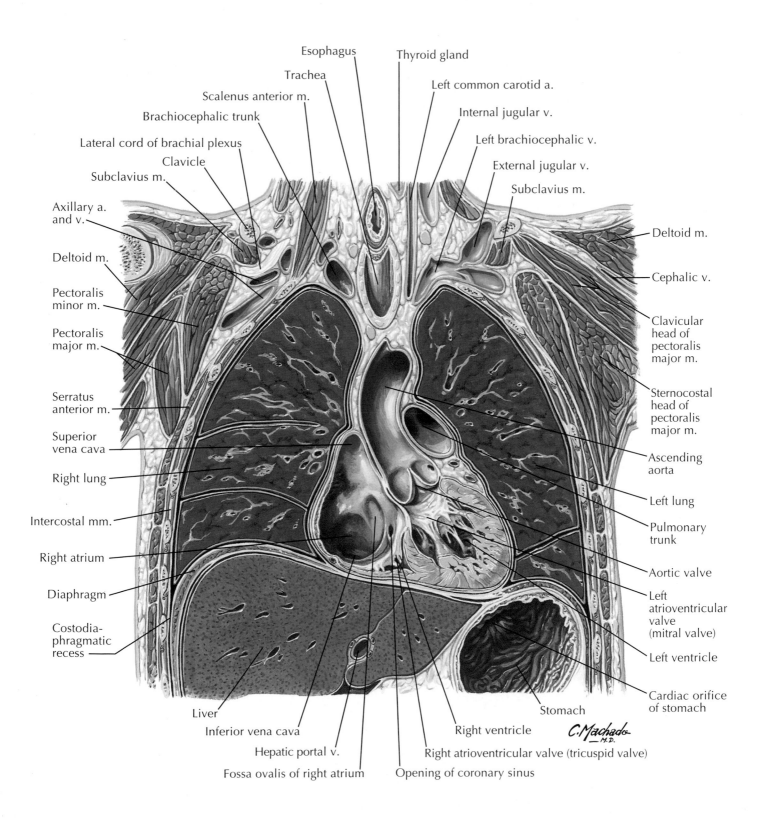

Esophagus

Thyroid gland

Trachea

Left common carotid a.

Scalenus anterior m.

Internal jugular v.

Brachiocephalic trunk

Left brachiocephalic v.

Lateral cord of brachial plexus

External jugular v.

Clavicle

Subclavius m.

Subclavius m.

Axillary a. and v.

Deltoid m.

Deltoid m.

Cephalic v.

Pectoralis minor m.

Clavicular head of pectoralis major m.

Pectoralis major m.

Serratus anterior m.

Sternocostal head of pectoralis major m.

Superior vena cava

Ascending aorta

Right lung

Left lung

Intercostal mm.

Pulmonary trunk

Right atrium

Aortic valve

Diaphragm

Left atrioventricular valve (mitral valve)

Costodia-phragmatic recess

Left ventricle

Cardiac orifice of stomach

Liver

Stomach

Inferior vena cava

Right ventricle

Hepatic portal v.

Right atrioventricular valve (tricuspid valve)

Fossa ovalis of right atrium

Opening of coronary sinus

C. Machado M.D.

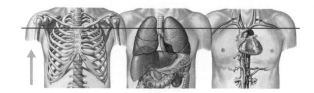

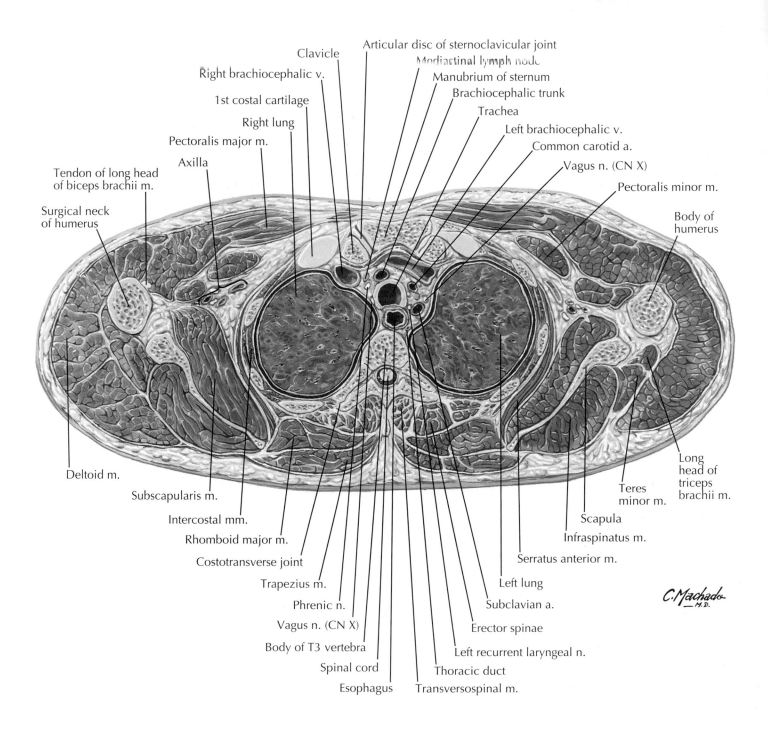

Articular disc of sternoclavicular joint
Clavicle
Mediastinal lymph node
Right brachiocephalic v.
Manubrium of sternum
1st costal cartilage
Brachiocephalic trunk
Right lung
Trachea
Pectoralis major m.
Left brachiocephalic v.
Axilla
Common carotid a.
Tendon of long head
of biceps brachii m.
Vagus n. (CN X)
Pectoralis minor m.
Surgical neck
of humerus
Body of
humerus
Deltoid m.
Long
head of
triceps
brachii m.
Subscapularis m.
Teres
minor m.
Intercostal mm.
Scapula
Rhomboid major m.
Infraspinatus m.
Costotransverse joint
Serratus anterior m.
Trapezius m.
Left lung
Phrenic n.
Subclavian a.
Vagus n. (CN X)
Erector spinae
Body of T3 vertebra
Left recurrent laryngeal n.
Spinal cord
Thoracic duct
Esophagus
Transversospinal m.

See also **Plates S–385, S–386, S–387**

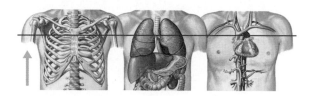

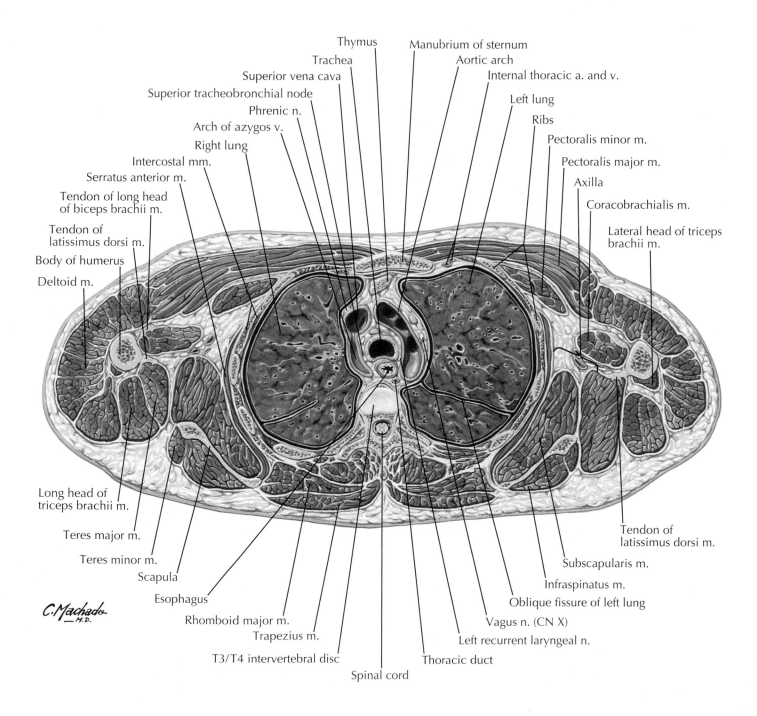

Thymus

Manubrium of sternum

Trachea

Aortic arch

Superior vena cava

Internal thoracic a. and v.

Superior tracheobronchial node

Left lung

Phrenic n.

Ribs

Arch of azygos v.

Pectoralis minor m.

Right lung

Pectoralis major m.

Intercostal mm.

Axilla

Serratus anterior m.

Coracobrachialis m.

Tendon of long head
of biceps brachii m.

Lateral head of triceps
brachii m.

Tendon of
latissimus dorsi m.

Body of humerus

Deltoid m.

Long head of
triceps brachii m.

Tendon of
latissimus dorsi m.

Teres major m.

Subscapularis m.

Teres minor m.

Infraspinatus m.

Scapula

Oblique fissure of left lung

Esophagus

Vagus n. (CN X)

Rhomboid major m.

Left recurrent laryngeal n.

Trapezius m.

T3/T4 intervertebral disc

Thoracic duct

Spinal cord

C.Machado
—M.D.

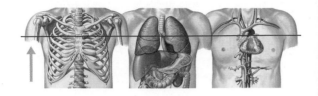

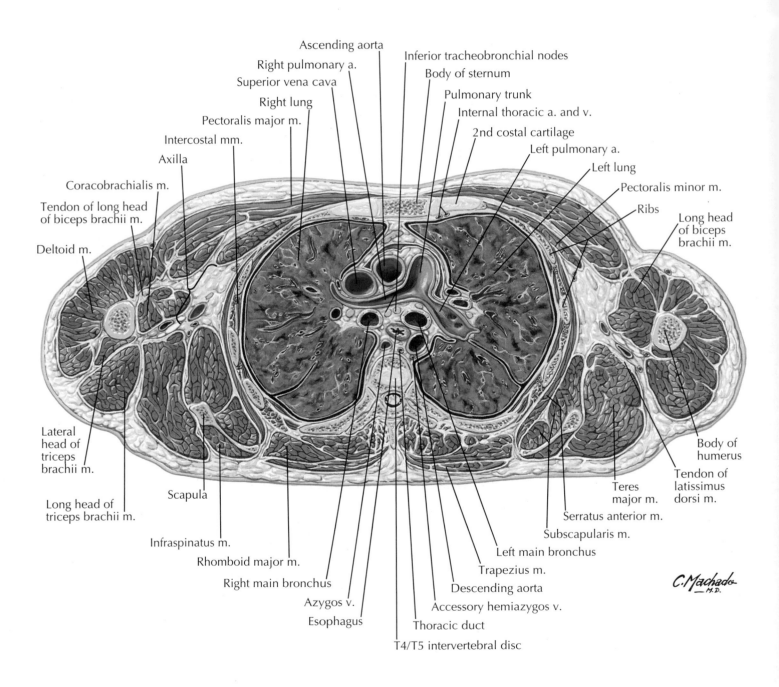

Ascending aorta

Right pulmonary a.

Superior vena cava

Right lung

Pectoralis major m.

Intercostal mm.

Axilla

Coracobrachialis m.

Tendon of long head
of biceps brachii m.

Deltoid m.

Inferior tracheobronchial nodes

Body of sternum

Pulmonary trunk

Internal thoracic a. and v.

2nd costal cartilage

Left pulmonary a.

Left lung

Pectoralis minor m.

Ribs

Long head
of biceps
brachii m.

Lateral
head of
triceps
brachii m.

Long head of
triceps brachii m.

Scapula

Infraspinatus m.

Rhomboid major m.

Right main bronchus

Azygos v.

Esophagus

T4/T5 intervertebral disc

Thoracic duct

Accessory hemiazygos v.

Descending aorta

Trapezius m.

Left main bronchus

Subscapularis m.

Serratus anterior m.

Teres
major m.

Tendon of
latissimus
dorsi m.

Body of
humerus

C.Machado
_M.D._

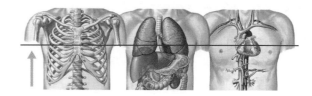

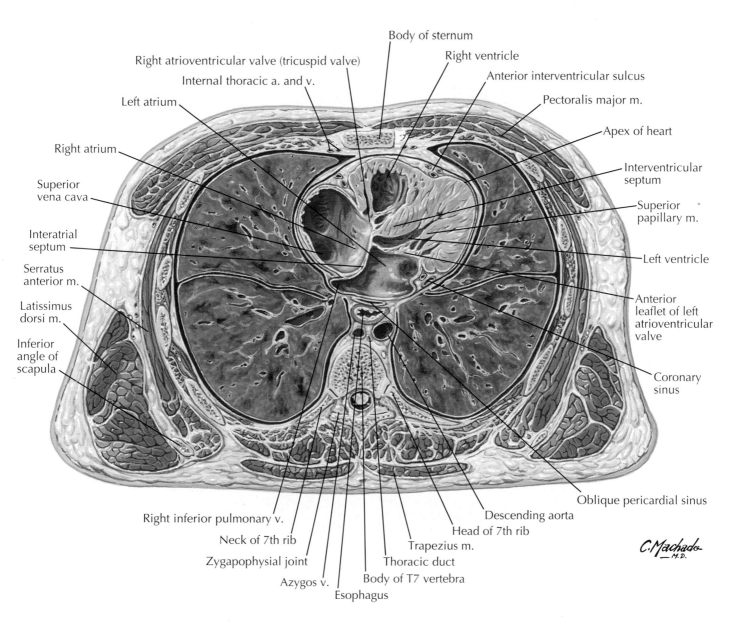

Body of sternum

Right atrioventricular valve (tricuspid valve)

Right ventricle

Internal thoracic a. and v.

Anterior interventricular sulcus

Left atrium

Pectoralis major m.

Right atrium

Apex of heart

Superior vena cava

Interventricular septum

Interatrial septum

Superior papillary m.

Serratus anterior m.

Left ventricle

Latissimus dorsi m.

Anterior leaflet of left atrioventricular valve

Inferior angle of scapula

Coronary sinus

Right inferior pulmonary v.

Oblique pericardial sinus

Neck of 7th rib

Descending aorta

Zygapophysial joint

Head of 7th rib

Azygos v.

Trapezius m.

Body of T7 vertebra

Thoracic duct

Esophagus

C. Machado
— M.D.

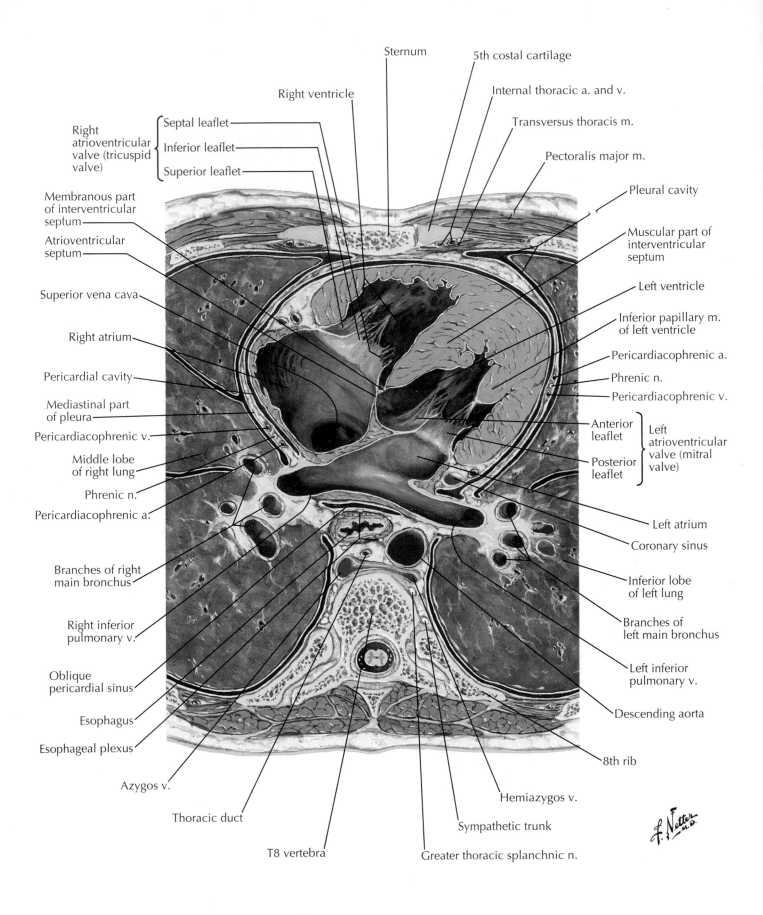

Sternum

5th costal cartilage

Right ventricle

Internal thoracic a. and v.

Transversus thoracis m.

Right atrioventricular valve (tricuspid valve)
- Septal leaflet
- Inferior leaflet
- Superior leaflet

Pectoralis major m.

Pleural cavity

Membranous part of interventricular septum

Muscular part of interventricular septum

Atrioventricular septum

Left ventricle

Superior vena cava

Inferior papillary m. of left ventricle

Right atrium

Pericardiacophrenic a.

Pericardial cavity

Phrenic n.

Pericardiacophrenic v.

Mediastinal part of pleura

Anterior leaflet

Pericardiacophrenic v.

Posterior leaflet

Left atrioventricular valve (mitral valve)

Middle lobe of right lung

Left atrium

Phrenic n.

Coronary sinus

Pericardiacophrenic a.

Inferior lobe of left lung

Branches of right main bronchus

Branches of left main bronchus

Right inferior pulmonary v.

Left inferior pulmonary v.

Oblique pericardial sinus

Descending aorta

Esophagus

8th rib

Esophageal plexus

Hemiazygos v.

Azygos v.

Sympathetic trunk

Thoracic duct

T8 vertebra

Greater thoracic splanchnic n.

Axial CT images of the thorax from superior (A) to inferior (C)

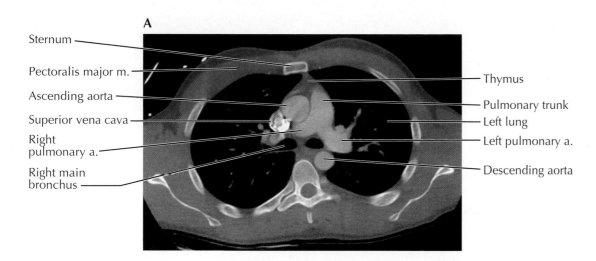

A

Sternum

Pectoralis major m.

Ascending aorta

Superior vena cava

Right pulmonary a.

Right main bronchus

Thymus

Pulmonary trunk

Left lung

Left pulmonary a.

Descending aorta

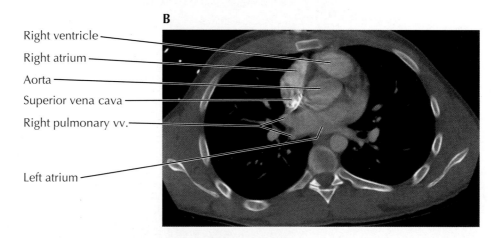

B

Right ventricle

Right atrium

Aorta

Superior vena cava

Right pulmonary vv.

Left atrium

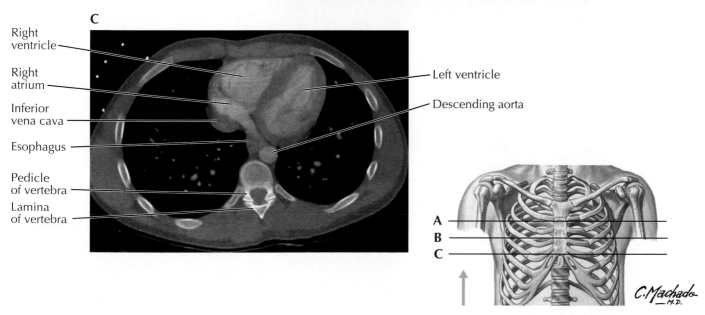

C

Right ventricle

Right atrium

Inferior vena cava

Esophagus

Pedicle of vertebra

Lamina of vertebra

Left ventricle

Descending aorta

A
B
C

C.Machado
—M.D.

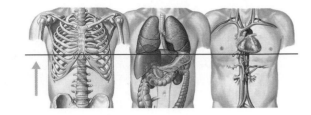

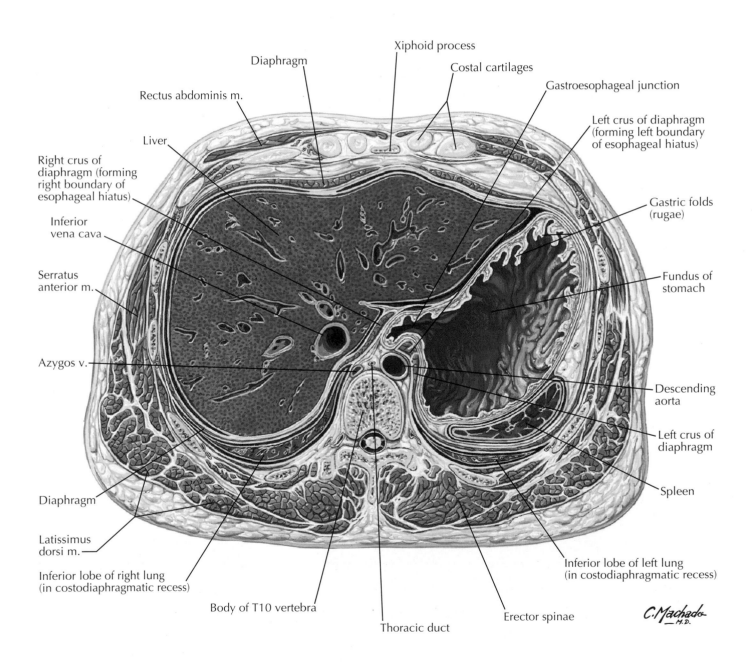

Xiphoid process

Diaphragm

Costal cartilages

Rectus abdominis m.

Gastroesophageal junction

Liver

Left crus of diaphragm (forming left boundary of esophageal hiatus)

Right crus of diaphragm (forming right boundary of esophageal hiatus)

Gastric folds (rugae)

Inferior vena cava

Serratus anterior m.

Fundus of stomach

Azygos v.

Descending aorta

Left crus of diaphragm

Diaphragm

Spleen

Latissimus dorsi m.

Inferior lobe of left lung (in costodiaphragmatic recess)

Inferior lobe of right lung (in costodiaphragmatic recess)

Body of T10 vertebra

Erector spinae

Thoracic duct

C. Machado
M.D.

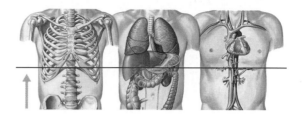

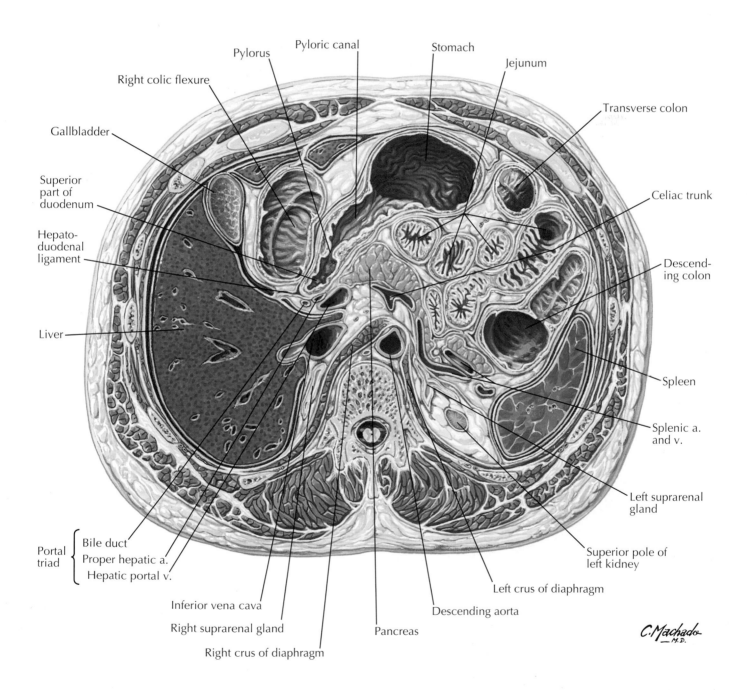

Pylorus

Pyloric canal

Stomach

Jejunum

Right colic flexure

Transverse colon

Gallbladder

Superior
part of
duodenum

Celiac trunk

Hepato-
duodenal
ligament

Descend-
ing colon

Liver

Spleen

Splenic a.
and v.

Left suprarenal
gland

Portal
triad

{ Bile duct

Proper hepatic a.

Hepatic portal v.

Superior pole of
left kidney

Inferior vena cava

Left crus of diaphragm

Right suprarenal gland

Descending aorta

Pancreas

Right crus of diaphragm

C. Machado
M.D.

Abdomen

Plate S-542

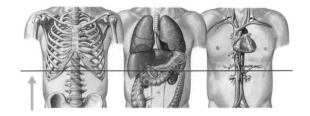

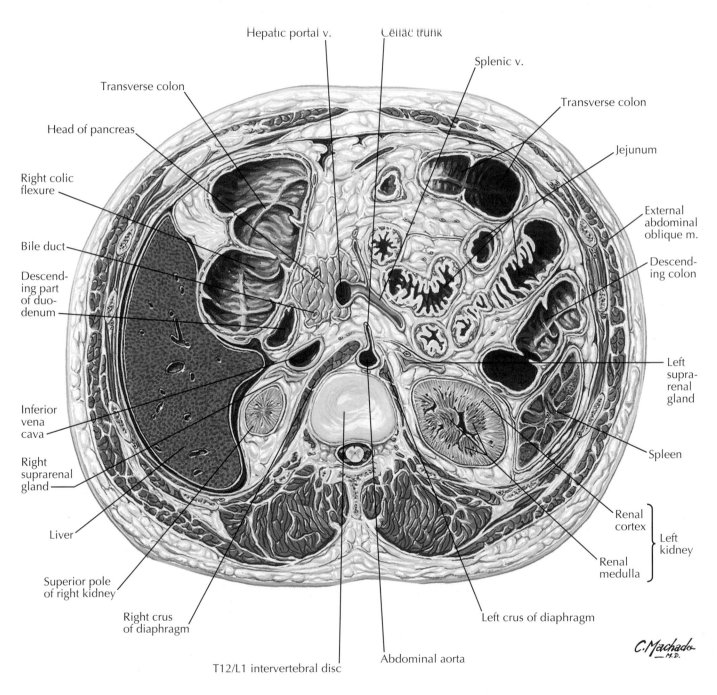

Hepatic portal v.

Celiac trunk

Splenic v.

Transverse colon

Transverse colon

Head of pancreas

Jejunum

Right colic flexure

External abdominal oblique m.

Bile duct

Descending colon

Descending part of duodenum

Inferior vena cava

Left suprarenal gland

Right suprarenal gland

Spleen

Liver

Renal cortex

Renal medulla

Left kidney

Superior pole of right kidney

Right crus of diaphragm

Left crus of diaphragm

T12/L1 intervertebral disc

Abdominal aorta

C. Machado _M.D._

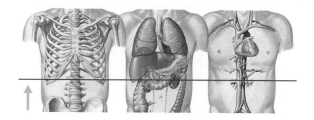

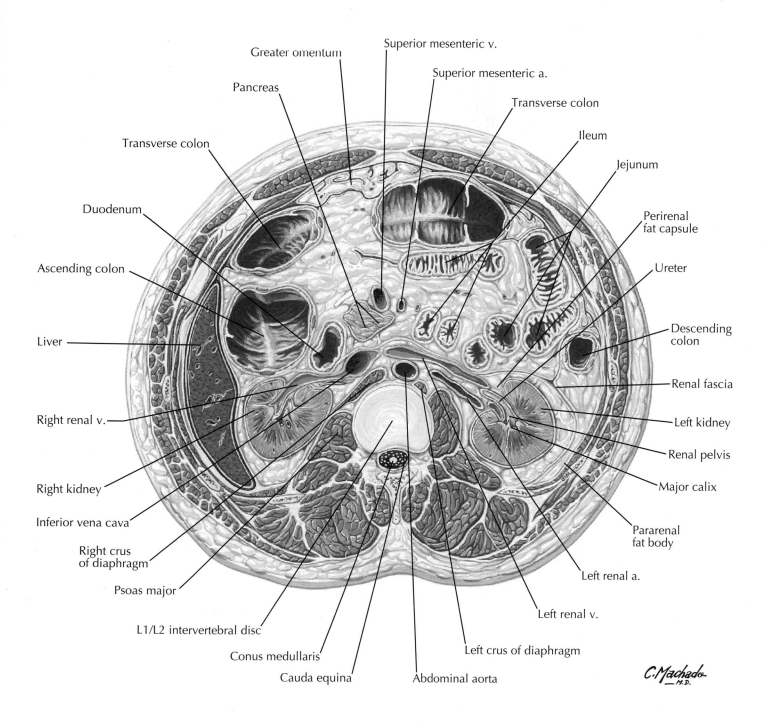

Greater omentum

Superior mesenteric v.

Pancreas

Superior mesenteric a.

Transverse colon

Transverse colon

Ileum

Jejunum

Duodenum

Perirenal
fat capsule

Ascending colon

Ureter

Liver

Descending
colon

Renal fascia

Right renal v.

Left kidney

Renal pelvis

Right kidney

Major calix

Inferior vena cava

Pararenal
fat body

Right crus
of diaphragm

Left renal a.

Psoas major

Left renal v.

L1/L2 intervertebral disc

Left crus of diaphragm

Conus medullaris

Cauda equina

Abdominal aorta

C. Machado
_M.D.

Abdomen

Plate S–544

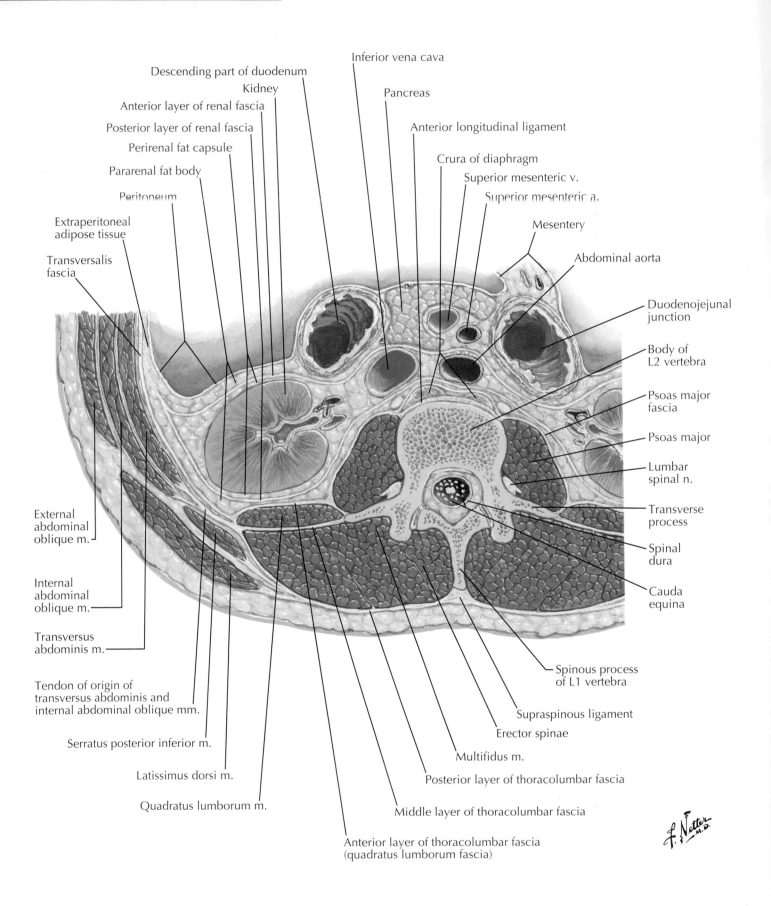

Descending part of duodenum

Kidney

Anterior layer of renal fascia

Posterior layer of renal fascia

Perirenal fat capsule

Pararenal fat body

Peritoneum

Extraperitoneal adipose tissue

Transversalis fascia

Inferior vena cava

Pancreas

Anterior longitudinal ligament

Crura of diaphragm

Superior mesenteric v.

Superior mesenteric a.

Mesentery

Abdominal aorta

Duodenojejunal junction

Body of L2 vertebra

Psoas major fascia

Psoas major

Lumbar spinal n.

Transverse process

Spinal dura

Cauda equina

External abdominal oblique m.

Internal abdominal oblique m.

Transversus abdominis m.

Tendon of origin of transversus abdominis and internal abdominal oblique mm.

Serratus posterior inferior m.

Latissimus dorsi m.

Quadratus lumborum m.

Spinous process of L1 vertebra

Supraspinous ligament

Erector spinae

Multifidus m.

Posterior layer of thoracolumbar fascia

Middle layer of thoracolumbar fascia

Anterior layer of thoracolumbar fascia (quadratus lumborum fascia)

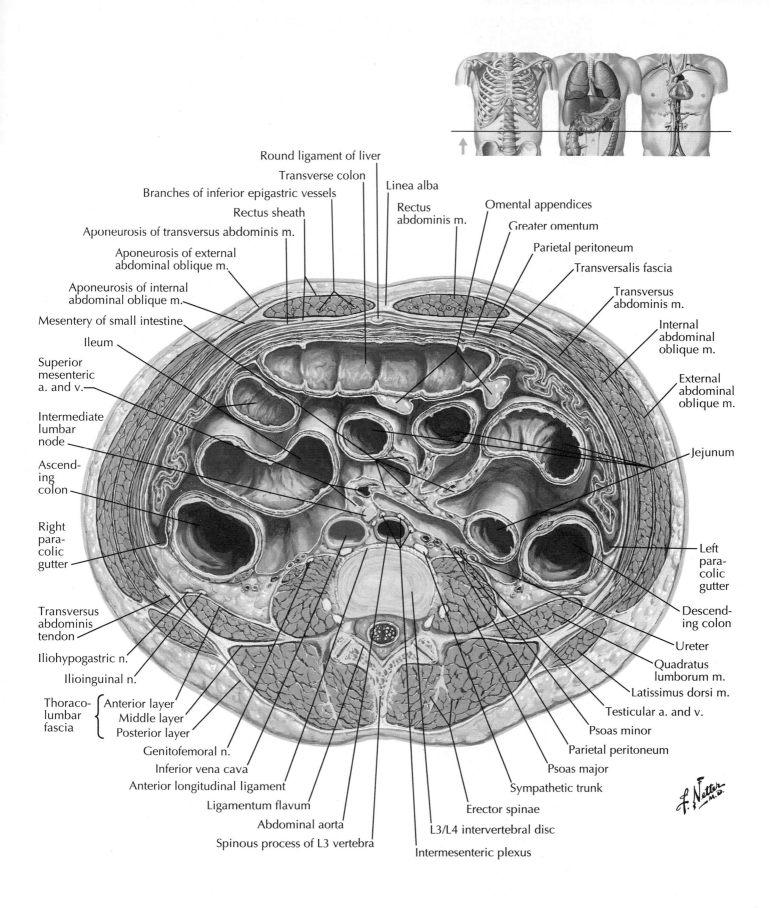

Round ligament of liver
Transverse colon
Branches of inferior epigastric vessels
Rectus sheath
Aponeurosis of transversus abdominis m.
Aponeurosis of external abdominal oblique m.
Aponeurosis of internal abdominal oblique m.
Mesentery of small intestine
Ileum
Superior mesenteric a. and v.
Intermediate lumbar node
Ascending colon
Right paracolic gutter
Transversus abdominis tendon
Iliohypogastric n.
Ilioinguinal n.
Thoracolumbar fascia { Anterior layer / Middle layer / Posterior layer
Genitofemoral n.
Inferior vena cava
Anterior longitudinal ligament
Ligamentum flavum
Abdominal aorta
Spinous process of L3 vertebra

Linea alba
Rectus abdominis m.
Omental appendices
Greater omentum
Parietal peritoneum
Transversalis fascia
Transversus abdominis m.
Internal abdominal oblique m.
External abdominal oblique m.
Jejunum
Left paracolic gutter
Descending colon
Ureter
Quadratus lumborum m.
Latissimus dorsi m.
Testicular a. and v.
Psoas minor
Parietal peritoneum
Psoas major
Sympathetic trunk
Erector spinae
L3/L4 intervertebral disc
Intermesenteric plexus

Axial CT image of abdomen with intravenous contrast enhancement

Hepatic portal v.

Liver

Descending aorta

Spinal canal

Stomach

Left crus of diaphragm

Spleen

Vertebral body

Axial CT image of upper abdomen with intravenous contrast enhancement

Hepatic portal v.

Liver

Right suprarenal gland

Body of pancreas

Spleen

Splenic v.

Left kidney

Axial CT image of midabdomen with intravenous contrast enhancement

Duodenum

Inferior vena cava

Right kidney

Left colic flexure

Abdominal aorta

Left suprarenal gland

Spleen

Left kidney

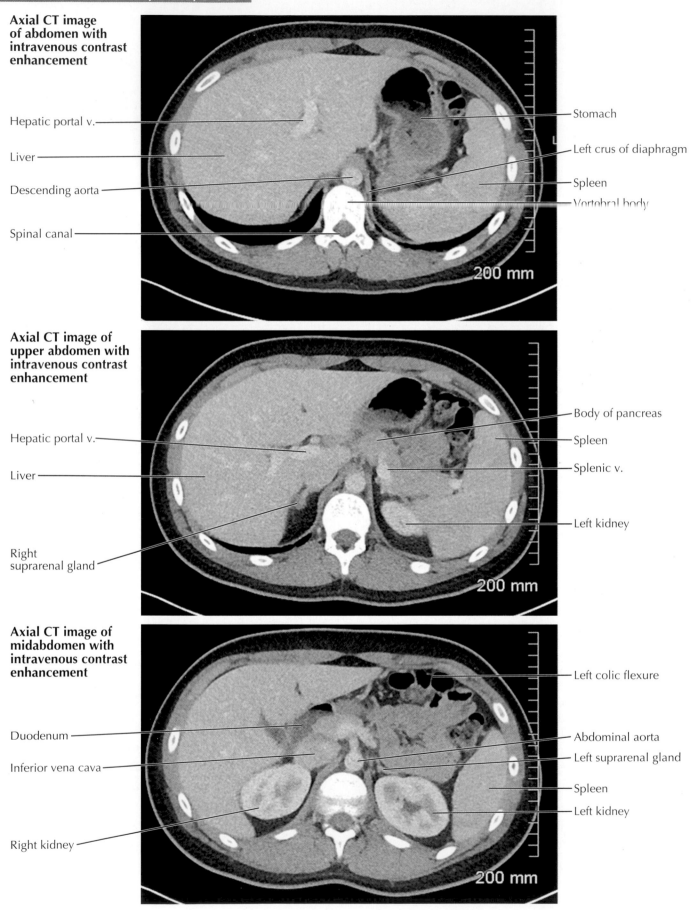

Axial CT image of midabdomen with intravenous contrast enhancement

Falciform ligament

Gallbladder

Head of pancreas

Duodenum

Right kidney

Abdominal aorta

Spleen

200 mm

Axial CT image of abdomen with intravenous contrast enhancement

Liver

Superior mesenteric v.

Superior mesenteric a.

Jejunum

Descending colon

Left renal v.

200 mm

Axial CT image of abdomen with intravenous contrast enhancement

Right lobe of liver

Right colic flexure

Inferior vena cava

Right kidney

Abdominal aorta

Left kidney

200 mm

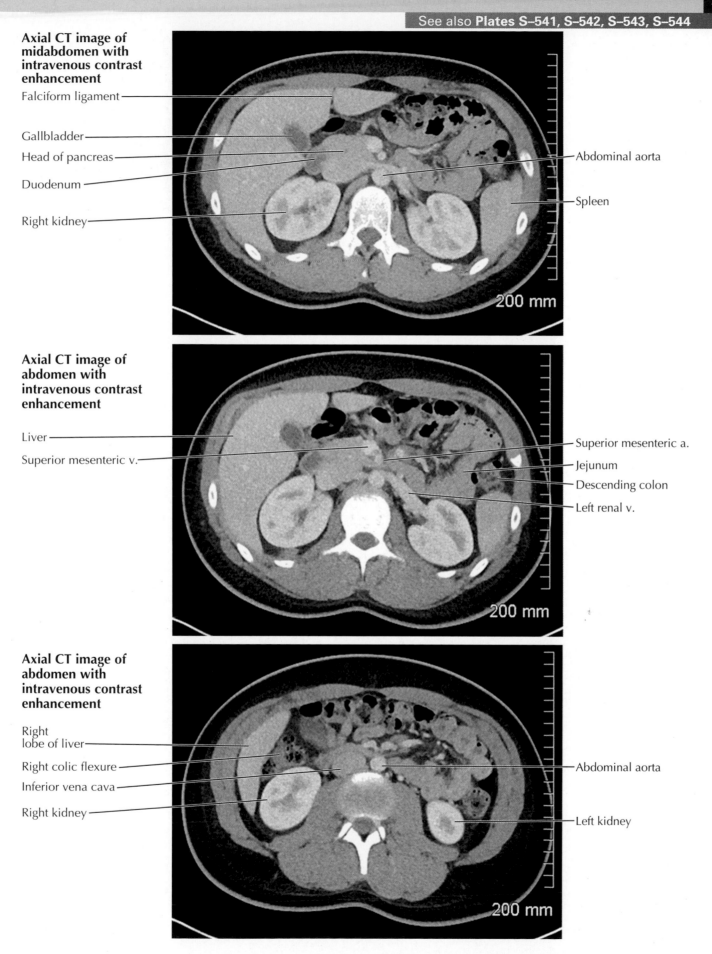

Abdomen

Plate S–548

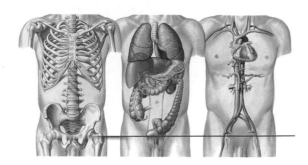

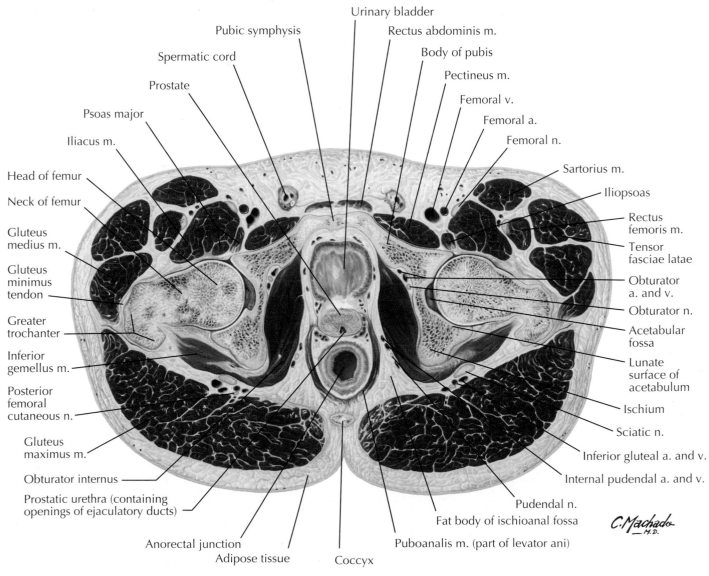

Urinary bladder

Pubic symphysis

Rectus abdominis m.

Spermatic cord

Body of pubis

Prostate

Pectineus m.

Psoas major

Femoral v.

Femoral a.

Iliacus m.

Femoral n.

Head of femur

Sartorius m.

Neck of femur

Iliopsoas

Gluteus
medius m.

Rectus
femoris m.

Gluteus
minimus
tendon

Tensor
fasciae latae

Obturator
a. and v.

Greater
trochanter

Obturator n.

Inferior
gemellus m.

Acetabular
fossa

Posterior
femoral
cutaneous n.

Lunate
surface of
acetabulum

Ischium

Gluteus
maximus m.

Sciatic n.

Obturator internus

Inferior gluteal a. and v.

Prostatic urethra (containing
openings of ejaculatory ducts)

Internal pudendal a. and v.

Pudendal n.

Anorectal junction

Fat body of ischioanal fossa

Adipose tissue

Puboanalis m. (part of levator ani)

Coccyx

C. Machado
_M.D._

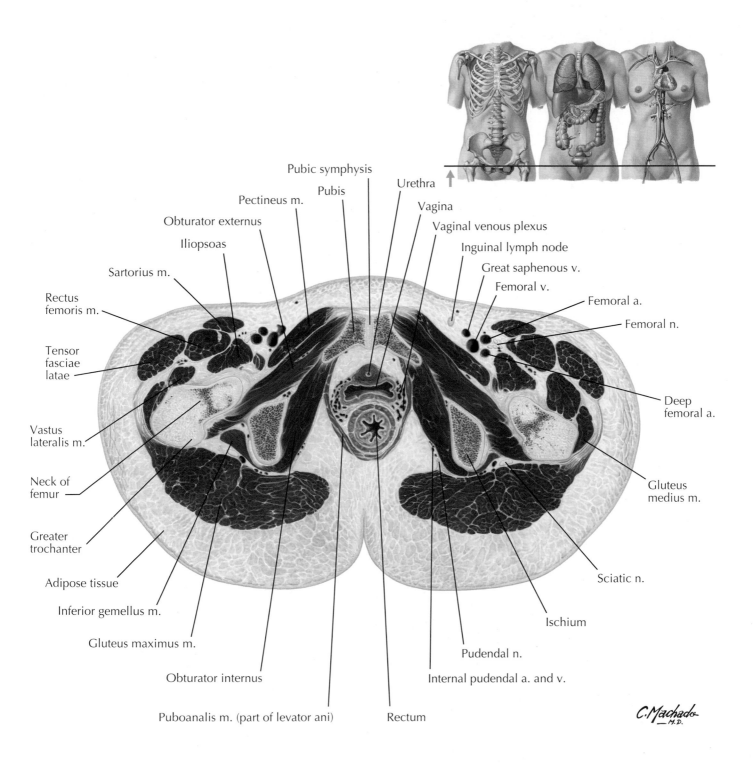

Pubic symphysis

Urethra

Pubis

Pectineus m.

Vagina

Obturator externus

Vaginal venous plexus

Iliopsoas

Inguinal lymph node

Sartorius m.

Great saphenous v.

Rectus femoris m.

Femoral v.

Femoral a.

Tensor fasciae latae

Femoral n.

Vastus lateralis m.

Deep femoral a.

Neck of femur

Gluteus medius m.

Greater trochanter

Adipose tissue

Inferior gemellus m.

Sciatic n.

Gluteus maximus m.

Ischium

Obturator internus

Pudendal n.

Puboanalis m. (part of levator ani)

Internal pudendal a. and v.

Rectum

C. Machado
M.D.

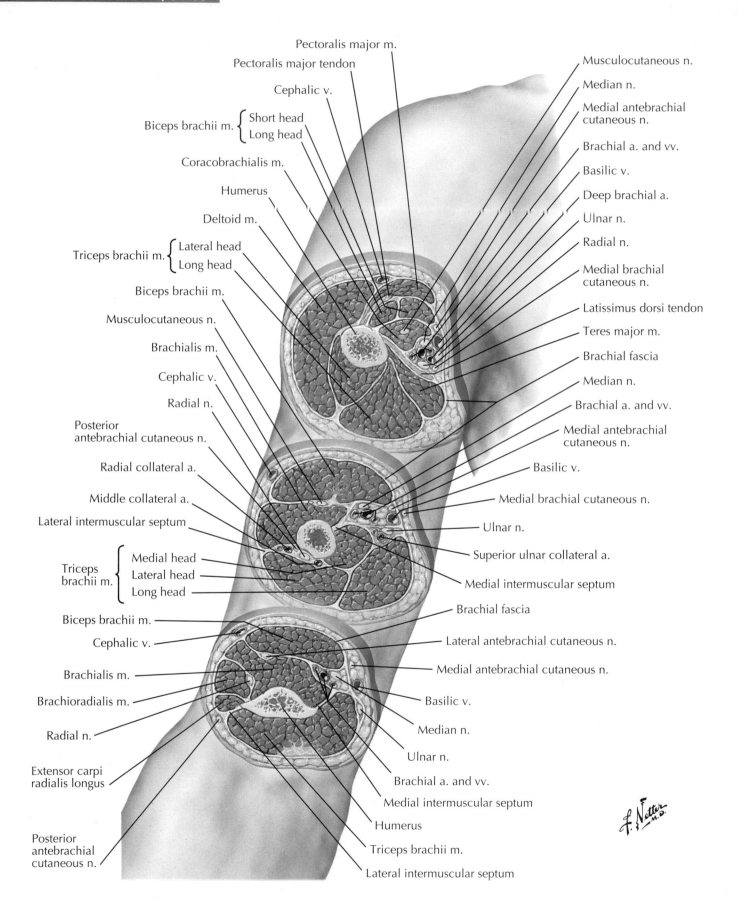

Pectoralis major m.

Pectoralis major tendon

Cephalic v.

Biceps brachii m. { Short head / Long head

Coracobrachialis m.

Humerus

Deltoid m.

Triceps brachii m. { Lateral head / Long head

Biceps brachii m.

Musculocutaneous n.

Brachialis m.

Cephalic v.

Radial n.

Posterior antebrachial cutaneous n.

Radial collateral a.

Middle collateral a.

Lateral intermuscular septum

Triceps brachii m. { Medial head / Lateral head / Long head

Biceps brachii m.

Cephalic v.

Brachialis m.

Brachioradialis m.

Radial n.

Extensor carpi radialis longus

Posterior antebrachial cutaneous n.

Musculocutaneous n.

Median n.

Medial antebrachial cutaneous n.

Brachial a. and vv.

Basilic v.

Deep brachial a.

Ulnar n.

Radial n.

Medial brachial cutaneous n.

Latissimus dorsi tendon

Teres major m.

Brachial fascia

Median n.

Brachial a. and vv.

Medial antebrachial cutaneous n.

Basilic v.

Medial brachial cutaneous n.

Ulnar n.

Superior ulnar collateral a.

Medial intermuscular septum

Brachial fascia

Lateral antebrachial cutaneous n.

Medial antebrachial cutaneous n.

Basilic v.

Median n.

Ulnar n.

Brachial a. and vv.

Medial intermuscular septum

Humerus

Triceps brachii m.

Lateral intermuscular septum

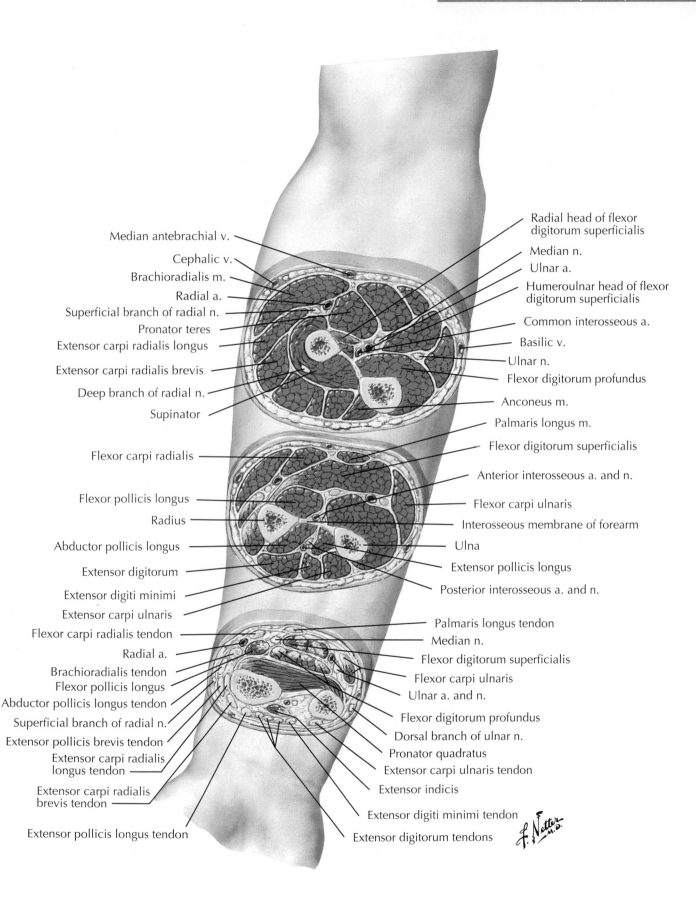

Median antebrachial v.
Cephalic v.
Brachioradialis m.
Radial a.
Superficial branch of radial n.
Pronator teres
Extensor carpi radialis longus
Extensor carpi radialis brevis
Deep branch of radial n.
Supinator

Flexor carpi radialis

Flexor pollicis longus
Radius
Abductor pollicis longus
Extensor digitorum
Extensor digiti minimi
Extensor carpi ulnaris
Flexor carpi radialis tendon
Radial a.
Brachioradialis tendon
Flexor pollicis longus
Abductor pollicis longus tendon
Superficial branch of radial n.
Extensor pollicis brevis tendon
Extensor carpi radialis longus tendon
Extensor carpi radialis brevis tendon
Extensor pollicis longus tendon

Radial head of flexor digitorum superficialis
Median n.
Ulnar a.
Humeroulnar head of flexor digitorum superficialis
Common interosseous a.
Basilic v.
Ulnar n.
Flexor digitorum profundus
Anconeus m.
Palmaris longus m.
Flexor digitorum superficialis
Anterior interosseous a. and n.
Flexor carpi ulnaris
Interosseous membrane of forearm
Ulna
Extensor pollicis longus
Posterior interosseous a. and n.
Palmaris longus tendon
Median n.
Flexor digitorum superficialis
Flexor carpi ulnaris
Ulnar a. and n.
Flexor digitorum profundus
Dorsal branch of ulnar n.
Pronator quadratus
Extensor carpi ulnaris tendon
Extensor indicis
Extensor digiti minimi tendon
Extensor digitorum tendons

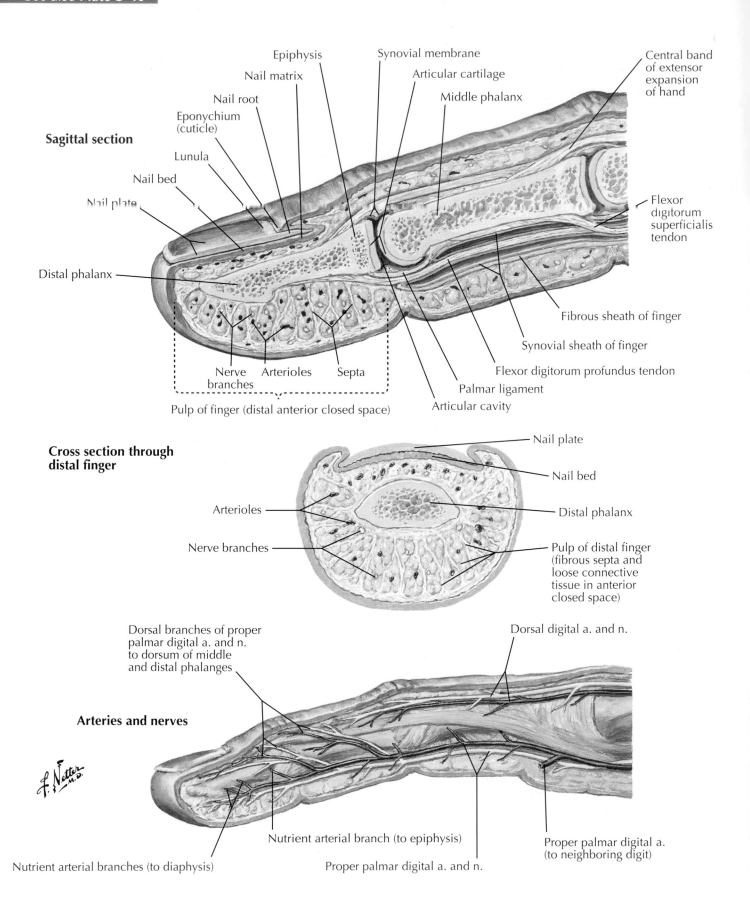

Sagittal section

Epiphysis

Nail matrix

Nail root

Eponychium (cuticle)

Lunula

Nail bed

Nail plate

Distal phalanx

Synovial membrane

Articular cartilage

Middle phalanx

Central band of extensor expansion of hand

Flexor digitorum superficialis tendon

Fibrous sheath of finger

Synovial sheath of finger

Flexor digitorum profundus tendon

Palmar ligament

Articular cavity

Nerve branches

Arterioles

Septa

Pulp of finger (distal anterior closed space)

Cross section through distal finger

Arterioles

Nerve branches

Nail plate

Nail bed

Distal phalanx

Pulp of distal finger (fibrous septa and loose connective tissue in anterior closed space)

Arteries and nerves

Dorsal branches of proper palmar digital a. and n. to dorsum of middle and distal phalanges

Dorsal digital a. and n.

Nutrient arterial branches (to diaphysis)

Nutrient arterial branch (to epiphysis)

Proper palmar digital a. and n.

Proper palmar digital a. (to neighboring digit)

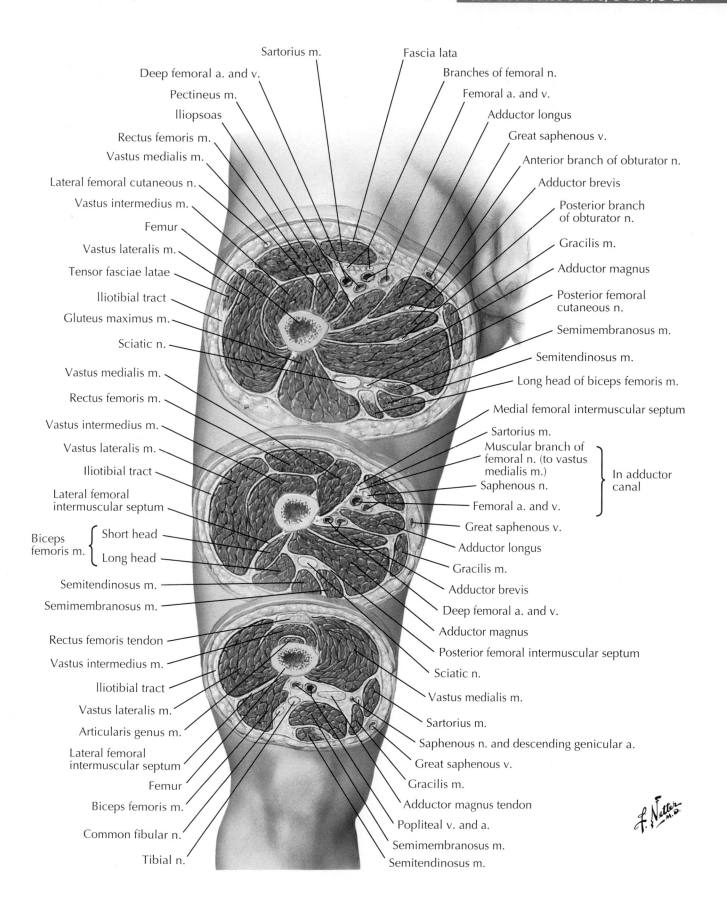

Sartorius m.

Deep femoral a. and v.

Pectineus m.

Iliopsoas

Rectus femoris m.

Vastus medialis m.

Lateral femoral cutaneous n.

Vastus intermedius m.

Femur

Vastus lateralis m.

Tensor fasciae latae

Iliotibial tract

Gluteus maximus m.

Sciatic n.

Vastus medialis m.

Rectus femoris m.

Vastus intermedius m.

Vastus lateralis m.

Iliotibial tract

Lateral femoral
intermuscular septum

Biceps
femoris m. { Short head

Long head

Semitendinosus m.

Semimembranosus m.

Rectus femoris tendon

Vastus intermedius m.

Iliotibial tract

Vastus lateralis m.

Articularis genus m.

Lateral femoral
intermuscular septum

Femur

Biceps femoris m.

Common fibular n.

Tibial n.

Fascia lata

Branches of femoral n.

Femoral a. and v.

Adductor longus

Great saphenous v.

Anterior branch of obturator n.

Adductor brevis

Posterior branch
of obturator n.

Gracilis m.

Adductor magnus

Posterior femoral
cutaneous n.

Semimembranosus m.

Semitendinosus m.

Long head of biceps femoris m.

Medial femoral intermuscular septum

Sartorius m.

Muscular branch of
femoral n. (to vastus
medialis m.)

Saphenous n.

Femoral a. and v.

} In adductor
canal

Great saphenous v.

Adductor longus

Gracilis m.

Adductor brevis

Deep femoral a. and v.

Adductor magnus

Posterior femoral intermuscular septum

Sciatic n.

Vastus medialis m.

Sartorius m.

Saphenous n. and descending genicular a.

Great saphenous v.

Gracilis m.

Adductor magnus tendon

Popliteal v. and a.

Semimembranosus m.

Semitendinosus m.

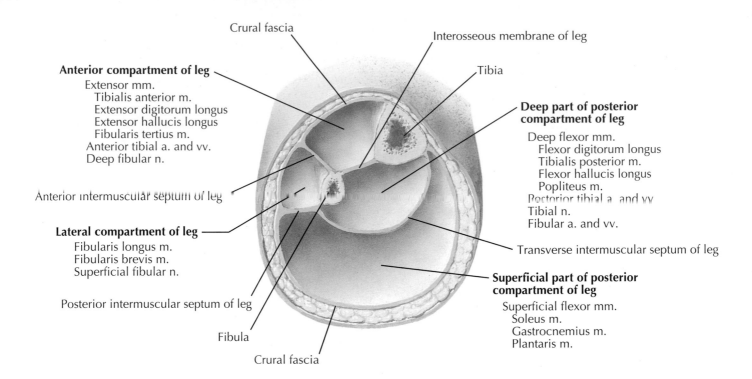

Crural fascia

Interosseous membrane of leg

Tibia

Anterior compartment of leg
Extensor mm.
Tibialis anterior m.
Extensor digitorum longus
Extensor hallucis longus
Fibularis tertius m.
Anterior tibial a. and vv.
Deep fibular n.

Anterior intermuscular septum of leg

Lateral compartment of leg
Fibularis longus m.
Fibularis brevis m.
Superficial fibular n.

Posterior intermuscular septum of leg

Fibula

Crural fascia

Deep part of posterior compartment of leg
Deep flexor mm.
Flexor digitorum longus
Tibialis posterior m.
Flexor hallucis longus
Popliteus m.
Posterior tibial a. and vv.
Tibial n.
Fibular a. and vv.

Transverse intermuscular septum of leg

Superficial part of posterior compartment of leg
Superficial flexor mm.
Soleus m.
Gastrocnemius m.
Plantaris m.

Cross section just above middle of leg

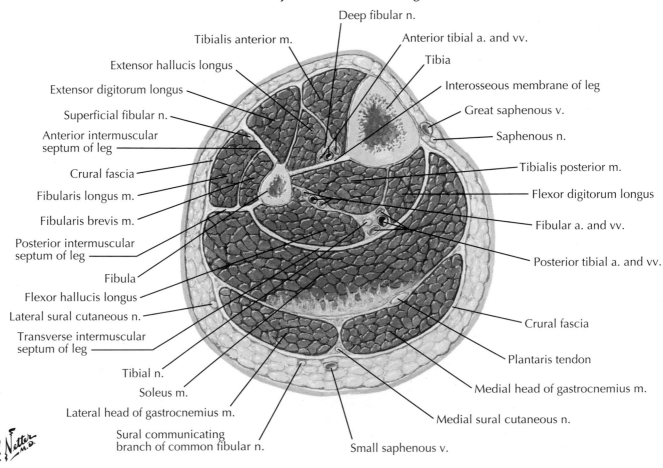

Deep fibular n.

Tibialis anterior m.

Anterior tibial a. and vv.

Extensor hallucis longus

Tibia

Extensor digitorum longus

Interosseous membrane of leg

Superficial fibular n.

Great saphenous v.

Anterior intermuscular septum of leg

Saphenous n.

Crural fascia

Tibialis posterior m.

Fibularis longus m.

Flexor digitorum longus

Fibularis brevis m.

Fibular a. and vv.

Posterior intermuscular septum of leg

Posterior tibial a. and vv.

Fibula

Flexor hallucis longus

Lateral sural cutaneous n.

Crural fascia

Transverse intermuscular septum of leg

Plantaris tendon

Tibial n.

Medial head of gastrocnemius m.

Soleus m.

Medial sural cutaneous n.

Lateral head of gastrocnemius m.

Sural communicating branch of common fibular n.

Small saphenous v.

M = Medial cuneiform bone
I = Intermediate cuneiform bone
L = Lateral cuneiform bone

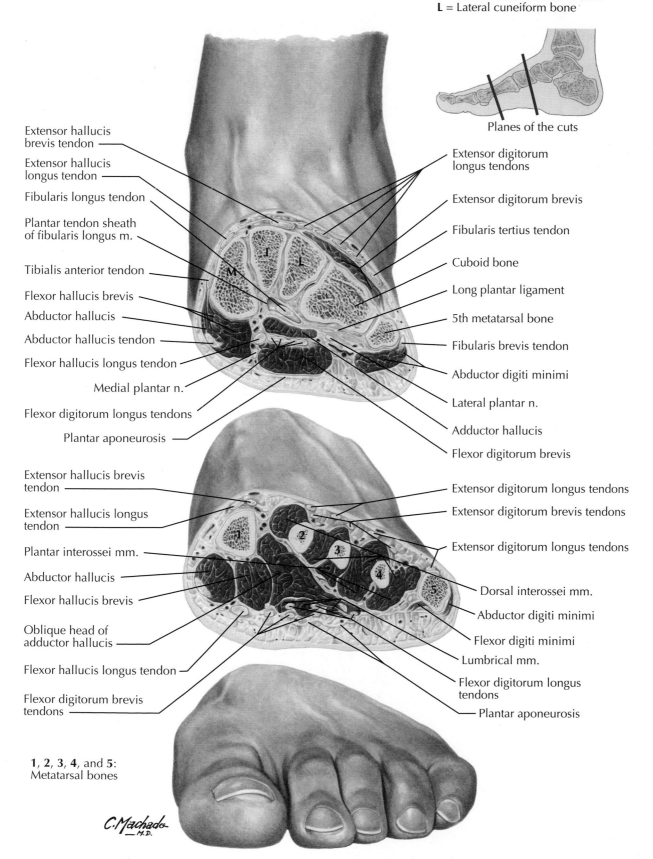

Planes of the cuts

Extensor hallucis brevis tendon

Extensor hallucis longus tendon

Fibularis longus tendon

Plantar tendon sheath of fibularis longus m.

Tibialis anterior tendon

Flexor hallucis brevis

Abductor hallucis

Abductor hallucis tendon

Flexor hallucis longus tendon

Medial plantar n.

Flexor digitorum longus tendons

Plantar aponeurosis

Extensor digitorum longus tendons

Extensor digitorum brevis

Fibularis tertius tendon

Cuboid bone

Long plantar ligament

5th metatarsal bone

Fibularis brevis tendon

Abductor digiti minimi

Lateral plantar n.

Adductor hallucis

Flexor digitorum brevis

Extensor hallucis brevis tendon

Extensor hallucis longus tendon

Plantar interossei mm.

Abductor hallucis

Flexor hallucis brevis

Oblique head of adductor hallucis

Flexor hallucis longus tendon

Flexor digitorum brevis tendons

Extensor digitorum longus tendons

Extensor digitorum brevis tendons

Extensor digitorum longus tendons

Dorsal interossei mm.

Abductor digiti minimi

Flexor digiti minimi

Lumbrical mm.

Flexor digitorum longus tendons

Plantar aponeurosis

1, 2, 3, 4, and **5:** Metatarsal bones

C.Machado
—M.D.

HEAD AND NECK

MUSCLE	MUSCLE GROUP	ORIGIN ATTACHMENT	INSERTION ATTACHMENT	INNERVATION	BLOOD SUPPLY	MAIN ACTIONS
Auricularis anterior m.	Facial expression (external ear)	Temporal fascia, epicranial aponeurosis	Anterior part of medial surface of helix of ear	Temporal branches of facial nerve	Posterior auricular and superficial temporal arteries	Elevates and draws auricle forward
Auricularis posterior m.	Facial expression (external ear)	Mastoid process	Inferior part of medial surface of auricle	Posterior auricular nerve (branch of facial nerve)	Posterior auricular and superficial temporal arteries	Retracts and elevates auricle
Auricularis superior m.	Facial expression (external ear)	Temporal fascia, epicranial aponeurosis	Superior part of medial surface of auricle	Temporal branches of facial nerve	Posterior auricular and superficial temporal arteries	Retracts and elevates auricle
Bucinator	Facial expression	Posterior portions of alveolar processes of maxilla and mandible, anterior border of pterygomandibular raphe	Modiolus of angle of mouth	Buccal branches of facial nerve	Facial and maxillary arteries	Compresses cheeks
Ciliary m.	Intrinsic eye (smooth muscle)	Scleral spur	Choroid	Short ciliary nerves (parasympathetic fibers from ciliary ganglion)	Ophthalmic artery	Constricts ciliary body and lens rounds up (accommodation)
Corrugator supercilii	Facial expression	Medial part of supra-orbital margin	Skin of medial half of eyebrow	Temporal branches of facial nerve	Superficial temporal artery	Draws eyebrows inferiorly and medially, produces vertical wrinkles of skin between eyebrows
Cricothyroid m.	Laryngeal	Arch of cricoid cartilage	Inferior border of lamina and inferior horn of thyroid cartilage	External branch of superior laryngeal nerve	Superior and inferior thyroid arteries	Lengthens and tenses vocal ligaments
Depressor anguli oris	Facial expression	Oblique line of mandible	Modiolus of angle of mouth	Marginal mandibular and buccal branches of facial nerve	Inferior labial artery	Depresses angle of mouth
Depressor labii inferioris	Facial expression	External surface of mandible between symphysis and mental foramen	Skin of lower lip	Marginal mandibular branches of facial nerve	Inferior labial artery	Depresses lower lip and draws it laterally
Depressor septi nasi	Facial expression	Incisive fossa of maxilla	Nasal septum and posterior part of ala of nose	Zygomatic and buccal branches of facial nerve	Superior labial artery	Narrows nostril, draws septum inferiorly

Variations in spinal nerve contributions to the innervation of muscles, their arterial supply, their attachments, and their actions are common themes in human anatomy. Therefore, expect differences between texts and realize that anatomical variation is normal.

MUSCLE	MUSCLE GROUP	ORIGIN ATTACHMENT	INSERTION ATTACHMENT	INNERVATION	BLOOD SUPPLY	MAIN ACTIONS
Digastric m.	Suprahyoid	*Anterior belly:* digastric fossa of mandible	Intermediate tendon attached to body of hyoid	*Anterior belly:* inferior alveolar nerve (branch of mandibular nerve)	*Anterior belly:* submental artery	Elevates hyoid bone and base of tongue, steadies hyoid bone, opens mouth by depressing mandible
		Posterior belly: mastoid notch of temporal bone		*Posterior belly:* digastric branch of facial nerve	*Posterior belly:* posterior auricular and occipital arteries	
Dilator pupillae	Intrinsic eye (smooth muscle)	Iridal part of retina	Pupillary margin of iris	Long ciliary nerves (sympathetic fibers from superior cervical ganglion)	Ophthalmic artery	Dilates pupil
Frontalis m.	Facial expression (epicranial)	Epicranial aponeurosis (at level of coronal suture)	Skin of frontal region, epicranial aponeurosis	Temporal branches of facial nerve	Superficial temporal artery	Horizontally wrinkles skin of forehead, elevates eyebrows
Genioglossus m.	Extrinsic tongue	Superior mental spine of mandible	Dorsum of tongue, hyoid bone	Hypoglossal nerve (CN XII)	Sublingual and submental arteries	Depresses and protrudes tongue
Geniohyoid m.	Suprahyoid	Inferior mental spine of mandible	Anterior surface of body of hyoid bone	Anterior ramus of C1 spinal nerve (via hypoglossal nerve)	Sublingual artery	Elevates hyoid bone and depresses mandible
Hyoglossus m.	Extrinsic tongue	Body and greater horn of hyoid bone	Margin and inferior surface of tongue	Hypoglossal nerve (CN XII)	Sublingual and submental arteries	Depresses and retracts tongue
Inferior longitudinal lingual m.	Intrinsic tongue	Inferior surface of tongue	Apex of tongue	Hypoglossal nerve (CN XII)	Lingual and facial arteries	Shortens tongue, turns tip and sides inferiorly
Inferior oblique m.	Extraocular	Anterior floor of orbit (lateral to nasolacrimal canal)	Sclera lateral to corneoscleral junction	Oculomotor nerve (CN III)	Ophthalmic artery	Abducts, elevates, and laterally rotates eyeball
Inferior pharyngeal constrictor	Circular pharyngeal	Oblique line of thyroid cartilage, cricoid cartilage	Pharyngeal raphe	Vagus nerve (via pharyngeal plexus)	Ascending pharyngeal and superior thyroid arteries	Constricts wall of pharynx during swallowing
Inferior rectus m.	Extraocular	Common tendinous ring	Sclera inferior to corneoscleral junction	Oculomotor nerve (CN III)	Ophthalmic artery	Depresses, adducts, and laterally rotates eyeball
Lateral cricoarytenoid m.	Laryngeal	Arch of cricoid cartilage	Muscular process of arytenoid cartilage	Recurrent laryngeal nerve	Superior and inferior thyroid arteries	Adducts vocal folds
Lateral pterygoid m.	Mastication	*Superior head:* infratemporal surface of greater wing of sphenoid bone	Pterygoid fovea of mandible, capsule and articular disc of temporomandibular joint	Mandibular nerve (CN V_3)	Maxillary artery	*Bilaterally:* protrudes mandible
		Inferior head: lateral pterygoid plate				*Unilaterally and alternately:* produces side-to-side grinding
Lateral rectus m.	Extraocular	Common tendinous ring	Sclera lateral to corneoscleral junction	Abducens nerve (CN VI)	Ophthalmic artery	Abducts eyeball
Levator anguli oris	Facial expression	Canine fossa of maxilla	Modiolus of angle of mouth	Zygomatic and buccal branches of facial nerve	Superior labial artery	Elevates angle of mouth
Levator nasolabialis (levator labii superioris alaeque nasi)	Facial expression	Superior part of frontal process of maxilla	Major alar cartilage, skin of nose, lateral part of upper lip	Zygomatic and buccal branches of facial nerve	Superior labial and angular arteries	Elevates upper lip and dilates nostril
Levator labii superioris	Facial expression	Maxilla superior to infra-orbital foramen	Skin of upper lip	Zygomatic and buccal branches of facial nerve	Superior labial and angular arteries	Elevates upper lip, dilates nares
Levator palpebrae superioris	Extraocular	Lesser wing of sphenoid bone (anterior to optic canal)	Superior tarsus	Oculomotor nerve (CN III)	Ophthalmic artery	Elevates upper eyelid

MUSCLE	MUSCLE GROUP	ORIGIN ATTACHMENT	INSERTION ATTACHMENT	INNERVATION	BLOOD SUPPLY	MAIN ACTIONS
Levator veli palatini	Palatal	Petrous part of temporal bone, auditory tube	Palatine aponeurosis	Vagus nerve (via pharyngeal plexus)	Ascending and descending palatine arteries	Elevates soft palate during swallowing
Longus capitis m.	Anterior vertebral	Anterior tubercles of transverse processes of C3/C6 vertebrae	Inferior surface of basilar part of occipital bone	Anterior rami of C1/C3 spinal nerves	Ascending cervical, ascending pharyngeal and vertebral arteries	Flexes head
Longus colli m.	Anterior vertebral	*Vertical portion:* C5/T3 vertebrae	*Vertical portion:* C2/C4 vertebrae	Anterior rami of C2/C8 spinal nerves	Ascending pharyngeal, ascending cervical and vertebral arteries	*Bilaterally:* flex and assist in rotating cervical vertebrae and head
		Inferior oblique portion: T1/T3 vertebrae	*Inferior oblique portion:* anterior tubercles of transverse processes of C3/C6 vertebrae			*Unilaterally:* laterally flexes vertebral column
		Superior oblique portion: anterior tubercles of transverse processes of C3/C5 vertebrae	*Superior oblique portion:* anterior tubercle of atlas			
Masseter	Mastication	Zygomatic arch	Ramus of mandible, coronoid process	Mandibular nerve (CN V_3)	Transverse facial, masseteric, and facial arteries	Elevates and protrudes mandible; deep fibers retract it
Medial pterygoid m.	Mastication	Medial surface of lateral pterygoid plate, pyramidal process of palatine bone, maxillary tuberosity	Medial surface of ramus and angle of mandible (inferior to mandibular foramen)	Mandibular nerve (V_3)	Facial and maxillary arteries	Bilaterally: protrudes and elevates mandible Unilaterally and alternately: produces side-to-side movements
Medial rectus m.	Extraocular	Common tendinous ring	Sclera medial to corneoscleral junction	Oculomotor nerve (CN III)	Ophthalmic artery	Adducts eyeball
Mentalis m.	Facial expression	Incisive fossa of mandible	Skin of chin	Marginal mandibular branch of facial nerve	Inferior labial artery	Elevates and protrudes lower lip
Middle pharyngeal constrictor	Circular pharyngeal	Stylohyoid ligament, horns of hyoid bone	Pharyngeal raphe	Vagus nerve (via pharyngeal plexus)	Ascending pharyngeal and ascending palatine arteries, tonsillar branches of facial artery, dorsal lingual branches of lingual artery	Constricts wall of pharynx during swallowing
Musculus uvulae	Palatal	Posterior nasal spine (of palatine bone), palatine aponeurosis	Mucosa of uvula of palate	Vagus nerve (via pharyngeal plexus)	Ascending and descending palatine arteries	Shortens, elevates, and retracts uvula of the palate
Mylohyoid m.	Suprahyoid	Mylohyoid line of mandible	Median raphe of mylohyoid, body of hyoid bone	Inferior alveolar nerve (branch of mandibular nerve)	Sublingual and submental arteries	Elevates hyoid bone, base of tongue, floor of mouth; depresses mandible
Nasalis m.	Facial expression	Canine eminence (superior and lateral to incisive fossa of maxilla)	Aponeurosis on nasal cartilages	Buccal branches of facial nerve	Superior labial artery, septal and lateral nasal branches of facial artery	Draws ala of nose toward nasal septum, compresses nostrils; alar part opens nostrils
Oblique arytenoid m.	Laryngeal	Arytenoid cartilage	Opposite arytenoid cartilage	Recurrent laryngeal nerve	Superior and inferior thyroid arteries	Closes intercartilaginous portion of rima glottidis

MUSCLE	MUSCLE GROUP	ORIGIN ATTACHMENT	INSERTION ATTACHMENT	INNERVATION	BLOOD SUPPLY	MAIN ACTIONS
Occipitalis m.	Facial expression (epicranial)	Lateral two-thirds of superior nuchal line, mastoid process	Skin of occipital region, epicranial aponeurosis	Posterior auricular nerve (branch of facial nerve)	Posterior auricular and occipital arteries	Moves scalp backward
Omohyoid m.	Infrahyoid	*Inferior belly:* superior border of scapula, superior transverse scapular ligament	Intermediate tendon of omohyoid	*Inferior belly:* ansa cervicalis (C2–C3)	Lingual and superior thyroid arteries	Steadies and depresses hyoid bone
		Superior belly: body of hyoid bone		*Superior belly:* ansa cervicalis (C1)		
Orbicularis oculi m.	Facial expression	Medial orbital margin, medial palpebral ligament, lacrimal bone	Skin around orbit, lateral palpebral ligament, upper and lower eyelids	Temporal and zygomatic branches of facial nerve	Facial and superficial temporal arteries	Closes eyelids
Orbicularis oris m.	Facial expression	Maxilla and mandible, perioral skin and muscles	Modiolus of angle of mouth	Buccal and marginal mandibular branches of facial nerve	Inferior and superior labial arteries	Compression, contraction, and protrusion of lips
Palatoglossus m.	Palatal	Palatine aponeurosis	Margin of tongue	Vagus nerve (via pharyngeal plexus)	Ascending pharyngeal artery, palatine branches of facial and maxillary arteries	Elevates posterior tongue, depresses palate
Palatopharyngeus m.	Palatal	Hard palate, palatine aponeurosis	Lateral pharyngeal wall	Vagus nerve (via pharyngeal plexus)	Ascending and descending palatine arteries	Tenses soft palate; pulls walls of pharynx superiorly, anteriorly, and medially during swallowing
Platysma	Facial expression (neck)	Skin of infraclavicular region	Mandible, muscles of lower lip	Cervical branch of facial nerve	Submental and suprascapular arteries	Tenses skin of neck
Posterior cricoarytenoid m.	Laryngeal	Posterior surface of lamina of cricoid cartilage	Muscular process of arytenoid cartilage	Recurrent laryngeal nerve	Superior and inferior thyroid arteries	Abducts vocal folds
Procerus m.	Facial expression	Fascia covering inferior part of nasal bone and superior part of lateral nasal cartilage	Skin medial and superior to eyebrow	Temporal and zygomatic branches of facial nerve	Angular artery, lateral nasal branches of facial artery	Draws down medial angle of eyebrows, produces transverse wrinkles over bridge of nose
Rectus anterior capitis m.	Anterior vertebral	Lateral mass of atlas	Basilar part of occipital bone	Anterior rami of C1/C2 spinal nerves	Vertebral and ascending pharyngeal arteries	Flexes head
Rectus lateralis capitis m.	Anterior vertebral	Superior surface of transverse process of atlas	Inferior surface of jugular process of occipital bone	Anterior rami of C1/C2 spinal nerves	Vertebral, occipital, and ascending pharyngeal arteries	Flexes head laterally to same side
Risorius m.	Facial expression	Fascia over masseter	Modiolus of angle of mouth	Buccal branches of facial nerve	Superior labial artery	Retracts angle of mouth
Salpingopharyngeus m.	Longitudinal pharyngeal	Auditory tube	Lateral pharyngeal wall	Vagus nerve (via pharyngeal plexus)	Ascending pharyngeal artery	Elevates pharynx and larynx during swallowing and speaking
Scalenus anterior m.	Lateral vertebral	Anterior tubercles of transverse processes of C3/C6 vertebrae	Scalene tubercle on 1st rib	Anterior rami of C5/C8 spinal nerves	Ascending cervical artery	Elevates 1st rib, bends neck
Scalenus medius m.	Lateral vertebral	Posterior tubercles of transverse processes of C2/C7 vertebrae	Superior surface of 1st rib (behind subclavian groove)	Anterior rami of C3/C7 spinal nerves	Ascending cervical artery	Elevates 1st rib, bends neck
Scalenus posterior m.	Lateral vertebral	Posterior tubercles of transverse processes of C4/C6 vertebrae	External surface of 2nd rib	Anterior rami of C5/C8 spinal nerves	Ascending cervical artery, superficial branch of transverse cervical artery	Elevates 2nd rib, bends neck

MUSCLE	MUSCLE GROUP	ORIGIN ATTACHMENT	INSERTION ATTACHMENT	INNERVATION	BLOOD SUPPLY	MAIN ACTIONS
Sphincter pupillae	Intrinsic eye (smooth muscle)	Circular smooth muscle of iris that passes around pupil	Blends with dilator pupillae fibers	Short ciliary nerves (parasympathetic fibers from ciliary ganglion)	Ophthalmic artery	Constricts pupil
Stapedius m.	Middle ear	Pyramidal eminence of temporal bone (in tympanic cavity)	Stapes	Nerve to stapedius muscle (branch of facial nerve)	Posterior auricular, anterior tympanic, and middle meningeal arteries	Pulls stapes posteriorly to lessen oscillation of tympanic membrane
Sternocleidomastoid m.	Neck	*Sternal head:* anterior surface of manubrium sterni *Clavicular head:* medial one-third of superior surface of clavicle	Lateral surface of mastoid process; lateral half of superior nuchal line of occipital bone	Spinal accessory nerve (CN XI)	Superior thyroid, occipital, suprascapular, and posterior auricular arteries	*Bilaterally:* flexes head, elevates thorax *Unilaterally:* turns face toward opposite side
Sternohyoid m.	Infrahyoid	Posterior surface of manubrium sterni, posterior sternoclavicular ligament, sternal end of clavicle	Medial part of inferior border of body of hyoid bone	Ansa cervicalis (C1–C3)	Superior thyroid and lingual arteries	Depresses larynx and hyoid bone, steadies hyoid bone
Sternothyroid m.	Infrahyoid	Posterior surface of manubrium sterni, posterior border of 1st costal cartilage	Oblique line of thyroid cartilage	Ansa cervicalis (C1–C3)	Cricothyroid branch of superior thyroid artery	Depresses larynx and thyroid cartilage
Styloglossus m.	Extrinsic tongue	Styloid process of temporal bone, stylohyoid ligament	Margin and inferior surface of tongue	Hypoglossal nerve (CN XII)	Sublingual artery	Retracts tongue and draws it up for swallowing
Stylohyoid m.	Suprahyoid	Posterior border of styloid process of temporal bone	Body of hyoid bone (at junction with greater horn)	Stylohyoid branch of facial nerve	Facial and occipital arteries	Elevates and retracts hyoid bone
Stylopharyngeus m.	Longitudinal pharyngeal	Medial surface of styloid process of temporal bone	Pharyngeal wall, posterior border of thyroid cartilage	Glossopharyngeal nerve (CN IX)	Ascending pharyngeal and ascending palatine arteries, tonsillar branches of facial artery, dorsal lingual branches of lingual artery	Elevates pharynx and larynx during swallowing and speaking
Superior longitudinal lingual m.	Intrinsic tongue	Submucosa of posterior part of dorsum of tongue	Apex of tongue; (unites with muscle of opposite side)	Hypoglossal nerve (CN XII)	Deep lingual branch of lingual artery, facial artery	Shortens tongue, turns tip and sides superiorly
Superior oblique m.	Extraocular	Body of sphenoid bone (above optic canal)	Sclera superior to corneoscleral junction (after passing through trochlea)	Trochlear nerve (CN IV)	Ophthalmic artery	Abducts, depresses, and medially rotates eyeball
Superior pharyngeal constrictor	Circular pharyngeal	Pterygoid hamulus, pterygomandibular raphe, mylohyoid line of mandible	Pharyngeal raphe	Vagus nerve (via pharyngeal plexus)	Ascending pharyngeal and ascending palatine arteries, tonsillar branches of facial artery, dorsal branches of lingual artery	Constricts wall of pharynx during swallowing
Superior rectus m.	Extraocular	Common tendinous ring	Sclera superior to corneoscleral junction	Oculomotor nerve (CN III)	Ophthalmic artery	Elevates, adducts, and medially rotates eyeball
Temporalis m.	Mastication	Temporal fossa, deep layer of temporal fascia	Coronoid process, ramus of mandible	Deep temporal nerves (branches of mandibular nerve)	Superficial temporal and maxillary arteries	Elevates mandible; posterior fibers retract mandible

MUSCLE	MUSCLE GROUP	ORIGIN ATTACHMENT	INSERTION ATTACHMENT	INNERVATION	BLOOD SUPPLY	MAIN ACTIONS
Tensor tympani	Middle ear	Cartilage of auditory tube	Handle of malleus	Mandibular nerve (CN V$_3$)	Superior tympanic artery	Tenses tympanic membrane by drawing it medially
Tensor veli palatini	Palatal	Scaphoid fossa of medial pterygoid plate, spine of sphenoid bone, auditory tube	Palatine aponeurosis	Mandibular nerve (CN V$_3$)	Ascending and descending palatine arteries	Tenses soft palate, opens auditory tube during swallowing and yawning
Thyroarytenoid m.	Laryngeal	Internal surface of thyroid cartilage	Muscular process of arytenoid cartilage	Recurrent laryngeal nerve	Superior and inferior thyroid arteries	Shortens and relaxes vocal cords, sphincter of vestibule
Thyrohyoid m.	Infrahyoid	Oblique line of thyroid cartilage	Inferior border of body and greater horn of hyoid bone	Anterior ramus of C1 spinal nerve (via hypoglossal nerve)	Superior thyroid artery	Depresses larynx and hyoid bone; elevates larynx when hyoid bone is fixed
Transverse arytenoid m.	Laryngeal	Arytenoid cartilage	Opposite arytenoid cartilage	Recurrent laryngeal nerve	Superior and inferior thyroid arteries	Closes intercartilaginous portion of rima glottidis
Transversus linguae m.	Intrinsic tongue	Septum of tongue	Dorsum and margin of tongue	Hypoglossal nerve (CN XII)	Deep lingual branch of lingual artery, facial artery	Narrows and elongates tongue
Verticalis linguae m.	Intrinsic tongue	Mucosa of anterior part of dorsum of tongue	Inferior surface of tongue	Hypoglossal nerve (CN XII)	Deep lingual branch of lingual artery, facial artery	Flattens and broadens tongue
Vocalis m.	Laryngeal	Vocal process of arytenoid cartilage	Vocal ligament	Recurrent laryngeal nerve	Superior and inferior thyroid arteries	Tenses anterior part of vocal ligament, relaxes posterior part of vocal ligament
Zygomaticus major m.	Facial expression	Zygomatic arch	Modiolus of angle of mouth	Zygomatic and buccal branches of facial nerve	Superior labial artery	Draws angle of mouth posteriorly and superiorly
Zygomaticus minor m.	Facial expression	Zygomatic arch	Modiolus of angle of mouth, upper lip	Zygomatic and buccal branches of facial nerve	Superior labial artery	Elevates upper lip

BACK

MUSCLE	MUSCLE GROUP	SUPERIOR ATTACHMENT	INFERIOR ATTACHMENT	INNERVATION	BLOOD SUPPLY	MAIN ACTIONS
Iliocostalis mm.	Deep back (erector spinae)	*Colli muscle:* posterior tubercles of C4/C6 vertebrae	*Colli muscle:* angles of ribs 3–6	Posterior rami of spinal nerves	*Cervical portions:* occipital, deep cervical, and vertebral arteries	Extend and laterally bend vertebral column and head
		Thoracis muscle: angles of ribs 1–6	*Thoracis muscle:* angles of ribs 7–12		*Thoracic portions:* dorsal branches of posterior intercostal and subcostal arteries	
		Lumborum muscle: ribs 4–12, transverse processes of L1/L4 vertebrae	*Lumborum muscle:* sacrum (via erector spinae aponeurosis), iliac crest		*Lumbar portions:* dorsal branches of lumbar and lateral sacral arteries	
Interspinalis mm.	Deep back	*Colli muscle:* spinous processes of C2/C7 vertebrae	Spinous processes of vertebrae subjacent to vertebrae of muscle origin	Posterior rami of spinal nerves	*Cervical portions:* occipital, deep cervical, and vertebral arteries	Aid in extension of vertebral column
		Thoracis muscle: spinous processes of T1/T2 and T11/T12 vertebrae			*Thoracic portions:* dorsal branches of posterior intercostal and subcostal arteries	
		Lumborum muscle: spinous processes of L1/L4 vertebrae			*Lumbar portions:* dorsal branches of lumbar arteries	

MUSCLE	MUSCLE GROUP	SUPERIOR ATTACHMENT	INFERIOR ATTACHMENT	INNERVATION	BLOOD SUPPLY	MAIN ACTIONS
Intertransversarii mm.	Deep back	*Medial muscles:* transverse processes of C1/C7 and T10/T12 vertebrae, mammillary processes of L1/L4 vertebrae	*Medial muscles:* transverse processes of subjacent cervical and thoracic vertebrae, mammillary processes of subjacent lumbar vertebrae	*Medial muscles:* posterior rami of spinal nerves	*Cervical portions:* occipital, deep cervical, and vertebral arteries *Thoracic portions:* dorsal branches of posterior intercostal and subcostal arteries	Assist in lateral flexion of vertebral column
		Lateral and anterior muscles: transverse processes of C1/C7 and L1/L4 vertebrae	*Lateral and anterior muscles:* transverse processes of subjacent vertebrae	*Lateral and anterior muscles:* anterior rami of spinal nerves	*Lumbar portions:* dorsal branches of lumbar arteries	
Latissimus dorsi m.	Superficial back	Spinous processes of T7/T12 vertebrae, posterior layer of thoracolumbar fascia (and thus to L1/L5 vertebrae and iliac crest), ribs 10/12	Intertubercular sulcus of humerus	Thoracodorsal nerve	Thoracodorsal artery, dorsal perforating branches of 9th, 10th, and 11th posterior intercostal, subcostal, and first three lumbar arteries	Extends, adducts, and medially rotates humerus
Levator scapulae	Superficial back	Transverse processes of atlas and axis, posterior tubercles of transverse processes of C3/C4 vertebrae	Medial border of scapula (superior to spine)	Anterior rami of C3/C4 spinal nerves, dorsal scapular nerve	Dorsal scapular and transverse and ascending cervical arteries	Elevates scapula medially, inferiorly rotates glenoid fossa
Longissimus mm.	Deep back (erector spinae)	*Capitis muscle:* mastoid process	*Capitis muscle:* transverse processes of C4/T4 vertebrae	Posterior rami of spinal nerves	*Cervical portions:* occipital, deep cervical, and vertebral arteries	Extend and laterally bend vertebral column and head
		Colli muscle: transverse processes of C2/C6 vertebrae	*Colli muscle:* transverse processes of T1/T5 vertebrae		*Thoracic portions:* dorsal branches of posterior intercostal and subcostal arteries	
		Thoracis muscle: transverse processes of T1/T12 vertebrae, ribs 5/12, accessory and transverse processes of L1/L5 vertebrae	*Thoracis muscle:* spinous processes of L1/L5 vertebrae, posterior surface of sacrum, iliac tuberosity, posterior sacroiliac ligament		*Lumbar portions:* dorsal branches of lumbar and lateral sacral arteries	
Multifidus mm.	Deep back (transversospinales)	Spinous processes of C2/L5 vertebrae (2–5 levels above muscle insertion)	*Colli muscle:* superior articular processes of C4/C7 vertebrae	Posterior rami of spinal nerves	*Cervical portions:* occipital, deep cervical, and vertebral arteries	Stabilize spine
			Thoracis muscle: transverse processes of T1/T12 vertebrae		*Thoracic portions:* dorsal branches of posterior intercostal and subcostal arteries	
			Lumborum muscle: mammillary processes of L1/L5 vertebrae, sacrum, iliac crest		*Lumbar portions:* dorsal branches of lumbar and lateral sacral arteries	
Obliquus inferior capitis m.	Suboccipital	Transverse process of atlas	Spinous process of axis	Suboccipital nerve	Vertebral artery and occipital arteries	Rotates atlas to turn face to same side
Obliquus superior capitis m.	Suboccipital	Lateral part of inferior nuchal line	Transverse process of atlas	Suboccipital nerve	Vertebral and occipital arteries	Extends and bends head laterally
Rectus posterior major capitis m.	Suboccipital	Middle part of inferior nuchal line	Spinous process of axis	Suboccipital nerve	Vertebral and occipital arteries	Extends and rotates head to same side
Rectus posterior minor capitis m.	Suboccipital	Medial part of inferior nuchal line	Posterior tubercle of atlas	Suboccipital nerve	Vertebral and occipital arteries	Extends head
Rhomboid major m.	Superficial back	Spinous processes of T2/T5 vertebrae	Medial border of scapula (inferior to spine of scapula)	Dorsal scapular nerve	Dorsal scapular artery *OR* deep branch of transverse cervical artery, dorsal perforating branches of upper five or six posterior intercostal arteries	Fixes scapula to thoracic wall and retracts and rotates it to depress glenoid fossa

MUSCLE	MUSCLE GROUP	SUPERIOR ATTACHMENT	INFERIOR ATTACHMENT	INNERVATION	BLOOD SUPPLY	MAIN ACTIONS
Rhomboid minor m.	Superficial back	Nuchal ligament, spinous processes of C7 and T1 vertebrae	Medial border of scapula (at spine of scapula)	Dorsal scapular nerve	Dorsal scapular artery *OR* deep branch of transverse cervical artery, dorsal perforating branches of upper five or six posterior intercostal arteries	Fixes scapula to thoracic wall and retracts and rotates it to depress glenoid fossa
Rotatores	Deep back (transversospinales)	*Colli muscle:* spinous processes of cervical vertebrae	*Colli muscle:* superior articular processes of cervical vertebrae 1 or 2 levels below vertebra of muscle origin	Posterior rami of spinal nerves	*Cervical portions:* occipital, deep cervical, and vertebral arteries	Stabilize, extend, and rotate spine
		Thoracis muscle: spinous processes and laminae of T1/T11 vertebrae	*Thoracis muscle:* transverse processes of T2/T12 vertebrae (brevis muscles insert into adjacent vertebra; longus muscles into vertebra 2 levels down)		*Thoracic portions:* dorsal branches of posterior intercostal and subcostal arteries	
		Lumborum muscle: spinous processes of lumbar vertebrae	*Lumborum muscle:* mammillary processes of lumbar vertebrae 2 levels down		*Lumbar portions:* dorsal branches of lumbar arteries	
Semispinalis mm.	Deep back (transversospinales)	*Capitis muscle:* occipital bone (between superior and inferior nuchal lines)	*Capitis muscle:* superior articular processes of C4/C7 vertebrae, transverse processes of T1/T6 vertebrae	Posterior rami of spinal nerves	*Cervical portions:* occipital, deep cervical, and vertebral arteries	Extend head and neck and rotate them to opposite side
		Colli muscle: spinous processes of C2/C5 vertebrae	*Colli muscle:* transverse processes of T1/T6 vertebrae		*Thoracic portions:* dorsal branches of posterior intercostal arteries	
		Thoracis muscle: spinous processes of C6/T4 vertebrae	*Thoracis muscle:* transverse processes of T6/T10 vertebrae			
Serratus posterior inferior m.	Superficial back	Inferior aspect of ribs 9–12	Spinous processes of T11/L2 vertebrae	Anterior rami of T9/T12 spinal nerves	Posterior intercostal arteries	Depresses ribs
Serratus posterior superior m.	Superficial back	Nuchal ligament, spinous processes of C7/T3 vertebrae	Superior aspect of ribs 2–5	Anterior rami of T2/T5 spinal nerves	Posterior intercostal arteries	Elevates ribs
Spinalis mm.	Deep back (erector spinae)	*Capitis muscle:* external occipital protuberance	*Capitis muscle:* spinous processes of C7 and T1 vertebrae	Posterior rami of spinal nerves	*Cervical portions:* occipital, deep cervical, and vertebral arteries	Extend and laterally bend vertebral column and head
		Colli muscle: spinous processes of C2/C4 vertebrae	*Colli muscle:* spinous processes of C7/T2 vertebrae		*Thoracic portions:* dorsal branches of posterior intercostal and subcostal arteries	
		Thoracis muscle: spinous processes of T2/T8 vertebrae	*Thoracis muscle:* spinous processes of T11/L2 vertebrae		*Lumbar portions:* dorsal branches of lumbar and lateral sacral arteries	
Splenius capitis m.	Spinotransversales	Mastoid process of temporal bone, lateral one-third of superior nuchal line	Nuchal ligament, spinous processes of C7/T4 vertebrae	Posterior rami of C2/C3 spinal nerves	Occipital artery and deep cervical arteries	*Bilaterally:* extends head
						Unilaterally: laterally bends (flexes) and rotates face to same side

MUSCLE	MUSCLE GROUP	SUPERIOR ATTACHMENT	INFERIOR ATTACHMENT	INNERVATION	BLOOD SUPPLY	MAIN ACTIONS
Splenius colli m.	Spinotransversales	Transverse processes of C1/C3 vertebrae	Spinous processes of T3/T6 vertebrae	Posterior rami of C4/C6 spinal nerves	Occipital and deep cervical arteries	*Bilaterally:* extends neck
						Unilaterally: laterally bends (flexes) and rotates neck toward same side
Trapezius m.	Superficial back	*Descending part:* superior nuchal line, external occipital protuberance, nuchal ligament	*Descending part:* lateral one-third of clavicle	Spinal accessory nerve (CN XI)	Transverse cervical and posterior intercostal arteries	Elevates, retracts, and rotates scapula; lower fibers depress scapula
		Transverse part: spinous processes of C7/T3 vertebrae	*Transverse part:* acromion			
		Ascending part: spinous processes of T4/T12 vertebrae	*Ascending part:* spine of scapula			

THORAX

MUSCLE	MUSCLE GROUP	ORIGIN ATTACHMENT	INSERTION ATTACHMENT	INNERVATION	BLOOD SUPPLY	MAIN ACTIONS
Diaphragm	Diaphragm	Xiphoid process, costal cartilages 7–12, L1/L3 vertebrae	Central tendon	Phrenic nerve	Pericardiacophrenic, musculophrenic, posterior intercostal, and superior and inferior phrenic arteries	Draws central tendon down and forward during inspiration
External intercostal mm.	Thoracic wall	Lower borders of ribs	Upper borders of subjacent ribs	Intercostal nerves	Posterior intercostal arteries, supreme intercostal artery, internal thoracic artery, musculophrenic artery	Support intercostal spaces in inspiration and expiration, elevate ribs in inspiration
Innermost intercostal mm.	Thoracic wall	Lower borders of ribs	Upper borders of subjacent ribs	Intercostal nerves	Posterior intercostal arteries, supreme intercostal artery, internal thoracic artery, musculophrenic artery	Prevent pushing out or drawing in of intercostal spaces in inspiration and expiration, lower ribs in forced expiration
Internal intercostal mm.	Thoracic wall	Costal grooves, lower borders of costal cartilages	Upper borders of subjacent ribs	Intercostal nerves	Posterior intercostal arteries, internal thoracic artery, musculophrenic artery, costocervical trunk branches	Prevent pushing out or drawing in of intercostal spaces in inspiration and expiration, lower ribs in forced expiration
Levatores costarum	Thoracic wall	Transverse processes of C7/T11 vertebrae	Subjacent ribs (between tubercle and angle)	Posterior rami of thoracic spinal nerves	Posterior intercostal arteries	Elevate ribs
Pectoralis major m.	Pectoral region	*Clavicular part:* sternal half of clavicle	Lateral lip of intertubercular sulcus of humerus	Medial and lateral pectoral nerves	Pectoral branch of thoracoacromial artery, internal thoracic artery	Flexes, adducts, and medially rotates arm
		Sternal part: anterior surface of sternum, cartilages of true ribs				
		Abdominal part: aponeurosis of external abdominal oblique muscle				

Appendix A: Muscles

MUSCLE	MUSCLE GROUP	ORIGIN ATTACHMENT	INSERTION ATTACHMENT	INNERVATION	BLOOD SUPPLY	MAIN ACTIONS
Pectoralis minor m.	Pectoral region	External surface of upper margin of ribs 3–5	Coracoid process of scapula	Medial and lateral pectoral nerves	Pectoral branch of thoracoacromial artery, superior and lateral thoracic arteries	Lowers lateral angle of scapula and protracts scapula
Serratus anterior m.	Shoulder	Lateral surfaces of ribs 1–9	Costal surface of medial border of scapula	Long thoracic nerve	Lateral thoracic artery	Protracts and rotates scapula and holds it against thoracic wall
Subclavius m.	Shoulder	Upper border of 1st rib and costal cartilage	Inferior surface of middle one-third of clavicle	Subclavian nerve	Clavicular branch of thoracoacromial artery	Anchors and depresses clavicle
Subcostal mm.	Thoracic wall	Internal surfaces of lower ribs near their angles	Superior borders of ribs that are the 2nd or 3rd below the muscle origin	Intercostal nerves	Posterior intercostal arteries, musculophrenic artery	Depress ribs
Transversus thoracis m.	Thoracic wall	Internal surfaces of costal cartilages 2–6	Posterior surface of lower sternum	Intercostal nerves	Internal thoracic artery	Depresses ribs and costal cartilages

ABDOMEN

MUSCLE	MUSCLE GROUP	ORIGIN ATTACHMENT	INSERTION ATTACHMENT	INNERVATION	BLOOD SUPPLY	MAIN ACTIONS
External abdominal oblique m.	Anterior abdominal wall	External surfaces of ribs 5–12	Linea alba, pubic tubercle, anterior half of iliac crest	Anterior rami of T7/T12 spinal nerves	Superior and inferior epigastric and lumbar arteries	Compresses and supports abdominal viscera, flexes and rotates trunk
Internal abdominal oblique m.	Anterior abdominal wall	Thoracolumbar fascia, anterior two-thirds of iliac crest	Inferior borders of ribs 10–12, linea alba (via rectus sheath), pubic crest and pecten pubis (via conjoint tendon)	Anterior rami of T7/L1 spinal nerves	Superior and inferior epigastric, deep circumflex iliac, and lumbar arteries	Compresses and supports abdominal viscera, flexes and rotates trunk
Pyramidalis m.	Anterior abdominal wall	Body of pubis and pubic symphysis (anterior to rectus abdominis)	Linea alba (inferior to umbilicus)	Anterior ramus of T12 spinal verve via subcostal or iliohypogastric nerves	Inferior epigastric artery	Tenses linea alba
Quadratus lumborum m.	Posterior abdominal wall	Medial half of inferior border of 12th rib, tips of transverse processes of L1/L4 vertebrae	Internal lip of iliac crest, iliolumbar ligament	Anterior rami of T12/L1 spinal nerves	Iliolumbar artery	Extends and laterally flexes vertebral column, fixes 12th rib during inspiration
Rectus abdominis m.	Anterior abdominal wall	Pubic crest, pubic symphysis	Costal cartilages 5–7, xiphoid process	Anterior rami of T7/T12 spinal nerves	Superior and inferior epigastric arteries	Flexes trunk, compresses abdominal viscera
Transversus abdominis m.	Anterior abdominal wall	Internal surfaces of costal cartilages 7–12, thoracolumbar fascia, iliac crest	Linea alba (via rectus sheath), pubic crest and pecten pubis (via conjoint tendon)	Anterior rami of T7/L1 spinal nerves	Deep circumflex iliac, inferior epigastric, and lumbar arteries	Compresses and supports abdominal viscera

PELVIS

MUSCLE	MUSCLE GROUP	ORIGIN ATTACHMENT	INSERTION ATTACHMENT	INNERVATION	BLOOD SUPPLY	MAIN ACTIONS
Bulbospongiosus m.	Perineal	*Male:* perineal body	*Male:* perineal membrane, corpus cavernosum, bulb of penis	Perineal nerve	Perineal artery	*Male:* compresses bulb of penis, forces blood into body of penis during erection, propels urine and semen through urethra
		Female: perineal body	*Female:* dorsum of clitoris, perineal membrane, bulb of vestibule, pubic arch			*Female:* constricts vaginal orifice, assists in expressing secretions of greater vestibular gland, forces blood into body of clitoris
Coccygeus m.	Pelvic diaphragm	Ischial spine	Inferior sacrum, coccyx	Nerve to coccygeus muscle	Inferior gluteal artery	Supports pelvic viscera
Compressor urethrae (female only)	Perineal	Ischiopubic ramus	Merges with contralateral partner anterior to urethra	Perineal nerve	Perineal artery	Sphincter of urethra
Cremaster	Spermatic cord	Inferior edge of internal abdominal oblique muscle, middle of inguinal ligament	Pubic tubercle, crest of pubis	Genital branch of genitofemoral nerve	Cremasteric artery	Retracts testis
Deep transverse perineal m.	Perineal	Internal surface of ramus of ischium, ischial tuberosity	Perineal body	Perineal nerve	Perineal artery	Stabilizes perineal body, supports prostate/vagina
External anal sphincter	Perineal	Anococcygeal ligament	Perineal body	Perineal and inferior anal nerves	Inferior anorectal and transverse perineal arteries	Closes anal orifice
External urethral sphincter	Perineal	Ischiopubic	*Male:* median raphe in front and behind urethra	Perineal nerve	Perineal artery	Compresses urethra at end of micturition; in female also compresses distal vagina
			Female: encloses urethra, attaches to sides of vagina			
Ischiocavernosus m.	Perineal	Inferior internal surface of ramus of ischium, ischial tuberosity	*Male:* anterior end of crus of penis	Perineal nerve	Perineal artery	Forces blood into body of penis and clitoris during erection
			Female: anterior end of crus of clitoris			
Levator ani (iliococcygeus, pubococcygeus, and puboanalis) mm.	Pelvic diaphragm	Body of pubis, tendinous arch of levator ani (on obturator fascia), ischial spine	Perineal body, coccyx, raphe of pelvic diaphragm, walls of prostate or vagina, anorectal junction	Nerve to levator ani, perineal nerve	Inferior gluteal artery, internal pudendal artery and its branches (inferior anorectal and perineal arteries)	Supports pelvic viscera, elevates pelvic floor
Sphincter urethrovaginalis (female only)	Perineal	Perineal body	Passes anteriorly around urethra and merges with its contralateral partner	Perineal nerve	Perineal artery	Sphincter of urethra and vagina
Superficial transverse perineal m.	Perineal	Ramus of ischium, ischial tuberosity	Perineal body	Perineal nerve	Perineal artery	Stabilizes perineal body

UPPER LIMB

MUSCLE	MUSCLE GROUP	PROXIMAL ATTACHMENT	DISTAL ATTACHMENT	INNERVATION	BLOOD SUPPLY	MAIN ACTIONS
Abductor digiti minimi of hand	Hand	Pisiform bone, tendon of flexor carpi ulnaris	Medial surface of base of proximal phalanx of little finger (5th digit)	Ulnar nerve (deep branch)	Deep palmar branch of ulnar artery	Abducts little finger
Abductor pollicis brevis	Hand	Flexor retinaculum of wrist, tubercles of scaphoid and trapezium bones	Base of proximal phalanx of thumb	Median nerve (recurrent branch)	Superficial palmar branch of radial artery	Abducts thumb
Abductor pollicis longus	Posterior forearm	Posterior surfaces of ulna, radius, and interosseous membrane of forearm	Base of 1st metacarpal	Posterior interosseous nerve	Posterior interosseous artery	Abducts and extends thumb
Adductor pollicis	Hand	*Oblique head:* bases of 2nd and 3rd metacarpals, capitate and adjacent carpal bones	Base of proximal phalanx of thumb	Ulnar nerve (deep branch)	Deep palmar arch	Adducts thumb
		Transverse head: anterior surface of 3rd metacarpal				
Anconeus m.	Posterior forearm	Posterior surface of lateral epicondyle of humerus	Lateral surface of olecranon, posterior surface of proximal ulna	Radial nerve	Deep brachial artery	Assists triceps brachii m. in extending elbow
Biceps brachii m.	Anterior arm	*Long head:* supraglenoid tubercle of scapula	Radial tuberosity, fascia of forearm (via bicipital aponeurosis)	Musculocutaneous nerve	Brachial artery	Flexes and supinates forearm
		Short head: coracoid process of scapula				
Brachialis m.	Anterior arm	Distal half of anterior surface of humerus	Coronoid process and tuberosity of ulna	Musculocutaneous and radial nerves	Radial recurrent and brachial arteries	Flexes forearm
Brachioradialis m.	Posterior forearm	Proximal two-thirds of lateral supracondylar ridge of humerus	Lateral surface of distal end of radius	Radial nerve	Radial recurrent artery	Weak flexion of forearm when forearm is in midpronation
Coracobrachialis m.	Anterior arm	Coracoid process of scapula	Middle one-third of medial surface of humerus	Musculocutaneous nerve	Brachial artery	Flexes and adducts arm
Deltoid m.	Shoulder	*Clavicular part:* lateral one-third of clavicle	Deltoid tuberosity of humerus	Axillary nerve	Posterior circumflex humeral artery, deltoid branch of thoraco-acromial artery	*Clavicular part:* flexes and medially rotates arm
		Acromial part: acromion				*Acromial part:* abducts arm beyond initial 15 degrees done by supraspinatus m.
		Spinous part: spine of scapula				*Spinous part:* extends and laterally rotates arm
Dorsal interosseous mm. of hand	Hand	Facing surfaces of two adjacent metacarpal bones	Base of proximal phalanges, and extensor expansions of digits 2–4	Ulnar nerve (deep branch)	Deep palmar arch	Abduct digits; flex digits at metacarpophalangeal joint and extend interphalangeal joints
Extensor carpi radialis brevis	Posterior forearm	Lateral epicondyle of humerus	Bases of 3rd and 2nd metacarpal bones	Radial nerve (deep branch)	Radial and radial recurrent arteries	Extends and abducts hand
Extensor carpi radialis longus	Posterior forearm	Distal one-third of lateral supracondylar ridge of humerus	Base of 2nd metacarpal bones	Radial nerve	Radial and radial recurrent arteries	Extends and abducts hand
Extensor carpi ulnaris	Posterior forearm	Lateral epicondyle of humerus, posterior border of ulna	Base of 5th metacarpal bone	Posterior interosseous nerve	Posterior interosseous artery	Extends and adducts hand

MUSCLE	MUSCLE GROUP	PROXIMAL ATTACHMENT	DISTAL ATTACHMENT	INNERVATION	BLOOD SUPPLY	MAIN ACTIONS
Extensor digiti minimi	Posterior forearm	Lateral epicondyle of humerus	Extensor expansion of little finger	Posterior interosseous nerve	Posterior interosseous artery	Extends 5th digit
Extensor digitorum	Posterior forearm	Lateral epicondyle of humerus	Extensor expansions of digits 2–5	Posterior interosseous nerve	Posterior interosseous artery	Extends medial four metacarpophalangeal joints, assists in wrist extension
Extensor indicis	Posterior forearm	Posterior surfaces of ulna and interosseous membrane of forearm	Extensor expansion of 2nd digit	Posterior interosseous nerve	Posterior interosseous artery	Extends 2nd digit and helps extend hand
Extensor pollicis brevis	Posterior forearm	Posterior surfaces of radius and interosseous membrane of forearm	Dorsal surface of base of proximal phalanx of thumb	Posterior interosseous nerve	Posterior interosseous artery	Extends proximal phalanx of thumb
Extensor pollicis longus	Posterior forearm	Posterior surfaces of middle one-third of ulna and interosseous membrane of forearm	Dorsal surface of base of distal phalanx of thumb	Posterior interosseous nerve	Posterior interosseous artery	Extends distal phalanx of thumb
Flexor carpi radialis	Anterior forearm	Medial epicondyle of humerus	Base of 2nd metacarpal bone	Median nerve	Radial artery	Flexes and abducts hand
Flexor carpi ulnaris	Anterior forearm	*Superficial head:* medial epicondyle of humerus *Deep head:* olecranon and posterior border of ulna	Pisiform bone, hook of hamate bone, base of 5th metacarpal bone	Ulnar nerve	Posterior ulnar recurrent artery	Flexes and adducts hand
Flexor digiti minimi of hand	Hand	Flexor retinaculum of wrist, hook of hamate bone	Medial surface of base of proximal phalanx of little finger	Ulnar nerve (deep branch)	Deep palmar branch of ulnar artery	Flexes proximal phalanx of little finger
Flexor digitorum profundus	Anterior forearm	Medial and anterior surfaces of proximal three-fourths of ulna, interosseous membrane of forearm	Palmar surface of base of distal phalanges of digits 2–5	Medial part: ulnar nerve Lateral part: median nerve	Anterior interosseous and ulnar arteries	Flexes distal phalanges of medial four digits, assists with flexion of hand
Flexor digitorum superficialis	Anterior forearm	*Humeroulnar head:* medial epicondyle ot humerus, coronoid process of ulna, ulnar collateral ligament *Radial head:* anterior surface of proximal radius	Bodies of middle phalanges of medial four digits	Median nerve	Ulnar and radial arteries	Flexes middle and proximal phalanges of medial four digits, flexes hand
Flexor pollicis brevis	Hand	*Superficial head:* flexor retinaculum, tubercle of trapezium bone *Deep head:* trapezoid and capitate bones	Lateral surface of base of proximal phalanx of thumb	*Superficial head:* median nerve (recurrent branch) *Deep head:* ulnar nerve (deep branch)	Superficial palmar branch of radial artery	Flexes proximal phalanx of thumb
Flexor pollicis longus	Anterior forearm	Anterior surfaces of radius and interosseous membrane	Palmar base of distal phalanx of thumb	Anterior interosseous nerve	Anterior interosseous artery	Flexes thumb
Infraspinatus m.	Shoulder	Infraspinous fossa of scapula, infraspinatus fascia	Greater tubercle of humerus	Suprascapular nerve	Suprascapular artery	Lateral rotation of arm
Lumbrical mm. of hand	Hand	Tendons of flexor digitorum profundus	Lateral sides of extensor expansion of digits 2–5	*Lateral two:* median nerve (digital branches) *Medial two:* ulnar nerve (deep branch)	Superficial and deep palmar arches	Extend digits, flex metacarpophalangeal joints

Appendix A: Muscles

MUSCLE	MUSCLE GROUP	PROXIMAL ATTACHMENT	DISTAL ATTACHMENT	INNERVATION	BLOOD SUPPLY	MAIN ACTIONS
Opponens digiti minimi m. of hand	Hand	Flexor retinaculum of wrist, hook of hamate bone	Palmar surface of 5th metacarpal bone	Ulnar nerve (deep branch)	Deep palmar branch of ulnar artery	Draws 5th metacarpal anteriorly and rotates it to face thumb
Opponens pollicis m.	Hand	Flexor retinaculum of wrist, tubercle of trapezium bone	Lateral surface of 1st metacarpal bone	Median nerve (recurrent branch)	Superficial palmar branch of radial artery	Draws 1st metacarpal forward and rotates it medially
Palmar interosseous mm.	Hand	Palmar surfaces of metacarpal bones 2, 4, and 5	Bases of proximal phalanges and extensor expansions of digits 2, 4, and 5	Ulnar nerve (deep branch)	Deep palmar arch	Adduct digits; flex digits and extend interphalangeal joints
Palmaris brevis m.	Hand	Palmar aponeurosis, flexor retinaculum of wrist	Skin of medial border of palm	Ulnar nerve (superficial branch)	Superficial palmar arch	Deepens hollow of hand, assists grip
Palmaris longus m.	Anterior forearm	Medial epicondyle of humerus	Distal half of flexor retinaculum of wrist, palmar aponeurosis	Median nerve	Posterior ulnar recurrent artery	Flexes hand and tenses palmar aponeurosis
Pronator quadratus	Anterior forearm	Distal one-fourth of anterior surface of ulna	Distal one-fourth of anterior surface of radius	Anterior interosseous nerve	Anterior interosseous artery	Pronates forearm
Pronator teres	Anterior forearm	*Humeral head:* medial epicondyle of humerus *Ulnar head:* coronoid process of ulna	Middle part of lateral surface of radius	Median nerve	Anterior ulnar recurrent artery	Pronates forearm and flexes elbow
Subscapularis m.	Shoulder	Subscapular fossa	Lesser tubercle of humerus	Upper and lower subscapular nerves	Subscapular and lateral thoracic arteries	Medially rotates and adducts arm; helps hold humeral head in glenoid fossa
Supinator	Posterior forearm	Lateral epicondyle of humerus, radial collateral and annular ligaments, supinator fossa, and crest of ulna	Lateral, posterior, and anterior surfaces of proximal one-third of radius	Radial nerve	Radial recurrent and posterior interosseous arteries	Supinates forearm
Supraspinatus m.	Shoulder	Supraspinous fossa of scapula, supraspinatus fascia	Greater tubercle of humerus	Suprascapular nerve	Suprascapular artery	Initiates arm abduction
Teres major m.	Shoulder	Posterior surface of inferior angle of scapula	Medial lip of intertubercular sulcus of humerus	Lower subscapular nerve	Circumflex scapular artery	Adducts and medially rotates arm
Teres minor m.	Shoulder	Superior two-thirds of posterior surface of lateral border of scapula	Greater tubercle of humerus	Axillary nerve	Circumflex scapular artery	Laterally rotates arm
Triceps brachii m.	Posterior arm	*Long head:* infraglenoid tubercle of scapula *Lateral head:* proximal half of posterior humerus *Medial head:* distal two-thirds of medial and posterior humerus	Posterior surface of olecranon	Radial nerve	Deep brachial artery	Extends forearm; long head stabilizes head of abducted humerus and extends and adducts arm

LOWER LIMB

MUSCLE	MUSCLE GROUP	PROXIMAL ATTACHMENT	DISTAL ATTACHMENT	INNERVATION	BLOOD SUPPLY	MAIN ACTIONS
Abductor digiti minimi of foot	Foot	Medial and lateral tubercles of tuberosity of calcaneus, plantar aponeurosis, intermuscular septum	Lateral surface of base of proximal phalanx of 5th digit	Lateral plantar nerve	Lateral plantar artery, plantar metatarsal and plantar digital arteries to 5th digit	Abducts and flexes 5th digit
Abductor hallucis	Foot	Medial tubercle of tuberosity of calcaneus, flexor retinaculum, plantar aponeurosis	Medial surface of base of proximal phalanx of 1st digit	Medial plantar nerve	Medial plantar and 1st plantar metatarsal arteries	Abducts and flexes 1st digit
Adductor brevis	Medial thigh	Body and inferior ramus of pubis	Pectineal line and proximal part of linea aspera of femur	Obturator nerve	Deep femoral, medial circumflex femoral, and obturator arteries	Adducts thigh at hip, weak hip flexor
Adductor hallucis	Foot	*Oblique head:* bases of 2nd through 4th metatarsals	Lateral surface of base of proximal phalanx of 1st digit	Lateral plantar nerve	Medial and lateral plantar arteries, plantar arch, plantar metatarsal arteries	Adducts 1st digit, maintains transverse arch of foot
		Transverse head: ligaments of metatarsophalangeal joints of digits 3–5				
Adductor longus	Medial thigh	Body of pubis inferior to pubic crest	Middle one-third of linea aspera of femur	Obturator nerve	Deep femoral and medial circumflex femoral arteries	Adducts thigh at hip
Adductor magnus	Medial thigh	*Adductor part:* ischiopubic ramus	*Adductor part:* gluteal tuberosity, linea aspera, medial supracondylar line	*Adductor part:* obturator nerve	Femoral, deep femoral, and obturator arteries	*Adductor part:* adducts and flexes thigh
		Hamstring part: ischial tuberosity	*Hamstring part:* adductor tubercle of femur	*Hamstring part:* sciatic nerve (tibial division)		*Hamstring part:* extends thigh
Articularis genus m.	Anterior thigh	Anterior surface of distal femur	Suprapatellar bursa	Femoral nerve	Femoral artery	Pulls suprapatellar bursa superiorly with extension of knee
Biceps femoris m.	Posterior thigh	*Long head:* ischial tuberosity	Lateral surface of head of fibula	*Long head:* sciatic nerve (tibial division)	Perforating femoral arteries, inferior gluteal and medial circumflex femoral arteries	Flexes and laterally rotates leg, extends thigh
		Short head: linea aspera and lateral supracondylar line of femur		*Short head:* sciatic nerve (common fibular division)		
Dorsal interossei mm. of foot	Foot	Adjacent surfaces of 1st through 5th metatarsals	*Medial one:* medial surface of proximal phalanx of 2nd digit	Lateral plantar nerve	Arcuate artery, dorsal and plantar metatarsal arteries	Abduct 2nd through 4th digits of foot, flex metatarsophalangeal joints, and extend phalanges
			Lateral three: lateral surfaces of proximal phalanges of digits 2–4			
Extensor digitorum brevis	Foot	Superolateral surface of calcaneus, lateral talocalcaneal ligament, deep surface of inferior extensor retinaculum	Lateral sides of tendons of extensor digitorum longus to digits 2–4	Deep fibular nerve	Dorsalis pedis, lateral tarsal, arcuate, and fibular arteries	Extends 2nd through 4th digits at metatarsophalangeal and interphalangeal joints
Extensor digitorum longus	Anterior leg	Lateral condyle of tibia, proximal three-fourths of anterior surfaces of interosseous membrane and fibula	Middle and distal phalanges of digits 2–5	Deep fibular nerve	Anterior tibial artery	Extends lateral four digits and dorsiflexes foot
Extensor hallucis brevis	Foot	Superolateral surface of calcaneus	Dorsal surface of proximal phalanx of great toe	Deep fibular nerve	Dorsalis pedis, lateral tarsal, arcuate, and fibular arteries	Extends great toe at metatarsophalangeal and interphalangeal joints

Appendix A: Muscles

MUSCLE	MUSCLE GROUP	PROXIMAL ATTACHMENT	DISTAL ATTACHMENT	INNERVATION	BLOOD SUPPLY	MAIN ACTIONS
Extensor hallucis longus	Anterior leg	Middle part of anterior surfaces of fibula and interosseous membrane of leg	Dorsal surface of base of distal phalanx of great toe	Deep fibular nerve	Anterior tibial artery	Extends great toe, dorsiflexes foot
Fibularis brevis m.	Lateral leg	Distal two-thirds of lateral surface of fibula	Dorsal surface of tuberosity on lateral side of 5th metatarsal bone	Superficial fibular nerve	Anterior tibial and fibular arteries	Everts and plantar flexes foot
Fibularis longus m.	Lateral leg	Head of fibula, proximal two-thirds of lateral surface of fibula	Plantar surfaces of base of 1st metatarsal and of medial cuneiform bones	Superficial fibular nerve	Anterior tibial and fibular arteries	Everts and plantar flexes foot
Fibularis tertius m.	Anterior leg	Distal one-third of anterior surfaces of fibula and interosseous membrane of leg	Dorsal surface of base of 5th metatarsal bone	Deep fibular nerve	Anterior tibial artery	Dorsiflexes and everts foot
Flexor digiti minimi of foot	Foot	Base of 5th metatarsal bone	Lateral surface of base of proximal phalanx of 5th digit	Lateral plantar nerve	Lateral plantar artery, plantar digital artery to 5th digit, arcuate artery	Flexes proximal phalanx of 5th digit
Flexor digitorum brevis	Foot	Medial tubercle of tuberosity of calcaneus, plantar aponeurosis, intermuscular septum	Both sides of middle phalangeal bones of digits 2–5	Medial plantar nerve	Medial and lateral plantar arteries, plantar arch, plantar metatarsal and plantar digital arteries	Flexes lateral four digits
Flexor digitorum longus	Posterior leg	Medial part of posterior tibia inferior to soleal line	Plantar surfaces of bases of distal phalanges of digits 2–5	Tibial nerve	Posterior tibial artery	Flexes lateral four digits and plantar flexes foot; supports longitudinal arches of foot
Flexor hallucis brevis	Foot	Plantar surfaces of cuboid bone and lateral cuneiform bone	Both sides of base of proximal phalanx of 1st digit	Medial plantar nerve	Medial plantar and 1st plantar metatarsal arteries	Flexes proximal phalanx of 1st digit
Flexor hallucis longus	Posterior leg	Distal two-thirds of posterior surfaces of fibula and interosseous membrane of leg	Base of distal phalanx of great toe	Tibial nerve	Fibular artery	Flexes all joints of great toe, plantar flexes foot
Gastrocnemius m.	Posterior leg	*Lateral head:* lateral surface of lateral condyle of femur *Medial head:* popliteal surface above medial condyle of femur	Posterior surface of calcaneus (via calcaneal tendon)	Tibial nerve	Popliteal and posterior tibial arteries	Plantarflexes foot, assists in flexion of knee
Gluteus maximus m.	Superficial gluteal	Ilium posterior to posterior gluteal line, dorsal surfaces of sacrum and coccyx, sacrotuberous ligament	Lateral condyle of tibia (via iliotibial tract), gluteal tuberosity of femur	Inferior gluteal nerve	Inferior and superior gluteal arteries	Extends flexed thigh, assists in lateral rotation, and abducts thigh
Gluteus medius m.	Superficial gluteal	Gluteal surface of ilium between anterior and posterior gluteal lines	Lateral surface of greater trochanter of femur	Superior gluteal nerve	Superior gluteal artery	Abducts and medially rotates thigh
Gluteus minimus m.	Superficial gluteal	Gluteal surface of ilium between anterior and inferior gluteal lines	Anterior surface of greater trochanter of femur	Superior gluteal nerve	Superior gluteal artery	Abducts and medially rotates thigh
Gracilis m.	Medial thigh	Body and inferior ramus of pubis	Superior part of medial surface of tibia	Obturator nerve	Deep femoral and medial circumflex femoral arteries	Adducts thigh, flexes and medially rotates leg
Iliacus m.	Iliopsoas	Superior two-thirds of iliac fossa, iliac crest, ala of sacrum, anterior sacroiliac ligament	Lesser trochanter and body of femur	Femoral nerve	Iliac branch of iliolumbar artery	Flexes thigh

MUSCLE	MUSCLE GROUP	PROXIMAL ATTACHMENT	DISTAL ATTACHMENT	INNERVATION	BLOOD SUPPLY	MAIN ACTIONS
Inferior gemellus m.	Deep gluteal	Ischial tuberosity	Greater trochanter of femur	Nerve to quadratus femoris muscle	Medial circumflex femoral artery	Laterally rotates extended thigh and abducts flexed thigh
Lumbrical mm.	Foot	Tendons of flexor digitorum longus	Medial side of dorsal digital expansions of digits 2–5	*Medial one:* medial plantar nerve *Lateral three:* lateral plantar nerve	Lateral plantar artery, plantar metatarsal arteries	Flex proximal phalanges at metatarsophalangeal joint, extend phalanges at proximal interphalangeal and distal interphalangeal joints
Obturator externus	Medial thigh	Margins of obturator foramen, external surface of obturator membrane	Trochanteric fossa of femur	Obturator nerve	Medial circumflex femoral and obturator arteries	Laterally rotates thigh
Obturator internus	Deep gluteal	Pelvic surface of obturator membrane, margins of obturator foramen	Greater trochanter of femur	Nerve to obturator internus muscle	Internal pudendal and obturator arteries	Laterally rotates extended thigh, abducts flexed thigh
Pectineus m.	Medial thigh	Superior ramus of pubis	Pectineal line of femur	Femoral nerve (sometimes also obturator nerve)	Medial circumflex femoral and obturator arteries	Adducts and flexes thigh
Piriformis m.	Deep gluteal	Anterior surface of sacral segments 2–4, sacrotuberous ligament (inconstant)	Superior border of greater trochanter of femur	Nerve to piriformis muscle	Superior and inferior gluteal arteries, internal pudendal artery	Laterally rotates extended thigh, abducts flexed thigh
Plantar interossei mm.	Foot	Bases and medial sides of 3rd through 5th metatarsals	Medial surfaces of bases of proximal phalangeal bones of digits 3–5	Lateral plantar nerve	Lateral plantar artery, plantar arch, plantar metatarsal and plantar digital arteries	Adduct digits (3–5), flex metatarsophalangeal joint, and extend phalanges
Plantaris m.	Posterior leg	Inferior end of lateral supracondylar line of femur, oblique popliteal ligament	Posterior surface of calcaneus (via calcaneal tendon)	Tibial nerve	Popliteal artery	Weakly assists gastrocnemius
Popliteus m.	Posterior leg	Lateral surface of lateral condyle of femur, lateral meniscus	Posterior tibia superior to soleal line	Tibial nerve	Inferior medial and lateral genicular arteries	Flexes knee
Psoas major	Iliopsoas	Transverse processes of lumbar vertebrae, sides of bodies of T12/L5 vertebrae, intervening intervertebral discs	Lesser trochanter of femur	Anterior rami of L1/L3 spinal nerves	Lumbar branch of iliolumbar artery	Acting superiorly with iliacus, flexes hip; acting inferiorly, flexes vertebral column laterally; used to balance trunk in sitting position; acting inferiorly with iliacus, flexes trunk
Psoas minor	Iliopsoas	Lateral surfaces of bodies of T12 and L1 vertebrae, T12/L1 intervertebral disc	Pectineal line, iliopubic eminence	Anterior ramus of L1 spinal nerve	Lumbar branch of iliolumbar artery	Flexes pelvis on vertebral column
Quadratus femoris m.	Deep gluteal	Lateral margin of ischial tuberosity	Quadrate tubercle	Nerve to quadratus femoris muscle	Medial circumflex femoral artery	Laterally rotates thigh
Quadratus plantae m.	Foot	Medial and lateral margins of plantar surface of calcaneus	Posterolateral edge of flexor digitorum longus tendon	Lateral plantar nerve	Medial and lateral plantar arteries, plantar arch	Corrects for oblique pull of flexor digitorum longus tendon, thus assisting in flexion of digits of foot
Rectus femoris m.	Anterior thigh (quadriceps femoris)	Anterior inferior iliac spine, ilium superior to acetabulum	Tibial tuberosity (via patellar ligament)	Femoral nerve	Deep femoral and lateral circumflex femoral arteries	Extends leg and flexes thigh
Sartorius m.	Anterior thigh	Anterior superior iliac spine, ilium inferior to that spine	Superior part of medial surface of tibia	Femoral nerve	Femoral artery	Abducts, laterally rotates, and flexes thigh; flexes knee

MUSCLE	MUSCLE GROUP	PROXIMAL ATTACHMENT	DISTAL ATTACHMENT	INNERVATION	BLOOD SUPPLY	MAIN ACTIONS
Semimembranosus m.	Posterior thigh	Ischial tuberosity	Posterior part of medial condyle of tibia	Sciatic nerve (tibial division)	Perforating femoral arteries, medial circumflex femoral artery	Flexes leg, extends thigh
Semitendinosus m.	Posterior thigh	Ischial tuberosity	Superior part of medial surface of tibia	Sciatic nerve (tibial division)	Perforating femoral arteries, medial circumflex femoral artery	Flexes leg, extends thigh
Soleus m.	Posterior leg	Posterior surface of head of fibula, proximal one-fourth of posterior surface of fibula, soleal line of tibia	Posterior surface of calcaneus (via calcaneal tendon)	Tibial nerve	Popliteal, posterior tibial, and fibular arteries	Plantar flexes foot
Superior gemellus m.	Deep gluteal	External surface of ischial spine	Medial surface of greater trochanter of femur	Nerve to obturator internus muscle	Inferior gluteal and internal pudendal arteries	Laterally rotates extended thigh and abducts flexed thigh
Tensor fasciae latae	Superficial gluteal	Anterior superior iliac spine, anterior part of iliac crest	Lateral condyle of tibia (via iliotibial tract)	Superior gluteal nerve	Ascending branch of lateral circumflex femoral artery	Abducts, medially rotates, and flexes thigh; helps to keep knee extended
Tibialis anterior m.	Anterior leg	Lateral condyle of tibia, proximal half of lateral tibia, interosseous membrane	Medial cuneiform bone and base of 1st metatarsal bone	Deep fibular nerve	Anterior tibial artery	Dorsiflexes foot and inverts foot
Tibialis posterior m.	Posterior leg	Posterior tibia below soleal line, interosseous membrane, proximal half of posterior fibula	Tuberosity of navicular bone, all cuneiform bones, cuboid bone, bases of 2nd through 4th metatarsal bones	Tibial nerve	Fibular artery	Plantar flexes foot and inverts foot
Vastus intermedius m.	Anterior thigh (quadriceps femoris)	Anterior and lateral surfaces of body of femur	Tibial tuberosity (via patellar ligament)	Femoral nerve	Lateral circumflex femoral and deep femoral arteries	Extends leg
Vastus lateralis m.	Anterior thigh (quadriceps femoris)	Greater trochanter, gluteal tuberosity, lateral lip of linea aspera	Tibial tuberosity (via patellar ligament)	Femoral nerve	Lateral circumflex femoral and deep femoral arteries	Extends leg
Vastus medialis m.	Anterior thigh (quadriceps femoris)	Intertrochanteric line, greater trochanter, gluteal tuberosity, lateral lip of linea aspera	Tibial tuberosity (via patellar ligament)	Femoral nerve	Femoral and deep femoral arteries	Extends leg

REFERENCES

Plates S–12, S–19

Lee MW, McPhee RW, Stringer MD. An evidence-based approach to human dermatomes. *Clin Anat.* 2008 Jul;21(5): 363–373. https://doi.org/10.1002/ca.20636. PMID: 18470936.

Plates S–16, S–202

Tubbs RS, Loukas M, Slappey JB, et al. Clinical anatomy of the C1 dorsal root, ganglion, and ramus: a review and anatomical study. *Clin Anat.* 2007;20:624–627.

Plate S–24

Bosmia AN, Hogan E, Loukas M, et al. Blood supply to the human spinal cord: part I. Anatomy and hemodynamics. *Clin Anat.* 2015;28:52–64.

Plate S–25

Stringer MD, Restieaux M, Fisher AL, Crosado B. The vertebral venous plexuses: the internal veins are muscular and external veins have valves. *Clin Anat.* 2012;25:609–618.

Plates S–27 to S–41

Rhoton AL Jr. *Cranial Anatomy and Surgical Approaches.* Schaumberg, IL: Congress of Neurological Surgeons; 2003.

Plates S–30, S–45

Tubbs RS, Hansasuta A, Loukas M, et al. Branches of the petrous and cavernous segments of the internal carotid artery. *Clin Anat.* 2007;20:596–601.

Plates S–53 to S–55, S–66

Schrott-Fischer A, Kammen-Jolly K, Scholtz AW, et al. Patterns of GABA-like immunoreactivity in efferent fibers of the human cochlea. *Hear Res.* 2002;174:75–85.

Plates S–62, S–399

Benninger B, Kloenne J, Horn JL. Clinical anatomy of the lingual nerve and identification with ultrasonography. *Br J Oral Maxillofac Surg.* 2013;51:541–544.

Plate S–63

Joo W, Yoshioka F, Funaki T, et al. Microsurgical anatomy of the trigeminal nerve. *Clin Anat.* 2004;27:61–88.

Joo W, Yoshioka F, Funaki T, Rhoton AL Jr. Microsurgical anatomy of the infratemporal fossa. *Clin Anat.* 2013;26: 455–469.

Plate S–86

Cornelius CP, Mayer P, Ehrenfeld M, Metzger MC. The orbits—anatomical features in view of innovative surgical methods. *Facial Plast Surg.* 2014;30:487–508.

Sherman DD, Burkat CN, Lemke BN. Orbital anatomy and its clinical applications. In: Tasman W, Jaeger EA, eds. *Duane's Ophthalmology.* Philadelphia: Lippincott Williams & Wilkins; 2006.

Plates S–94, S–95, S–408

Kierner AC, Mayer R, v Kirschhofer K. Do the tensor tympani and tensor veli palatini muscles of man form a functional unit? A histochemical investigation of their putative connections. *Hear Res.* 2002;165:48–52.

Plates S–99, S–319, S–320

Hildreth V, Anderson RH, Henderson DJ. Autonomic innervation of the developing heart: origins and function. *Clin Anat.* 2009;22:36–46.

Plates S–129, S–360 to S–361, S–368 to S–370

Lang J. *Clinical Anatomy of the Nose, Nasal Cavity, and Paranasal Sinuses.* New York: Thieme; 1989.

Plates S–139, S–140, S–143

Baccetti T, Franchi L, McNamara J Jr. The cervical vertebral maturation (CVM) method for the assessment of optimal treatment timing in dentofacial orthopedics. *Semin Orthod.* 2005;11:119–129.

Roman PS. Skeletal maturation determined by cervical vertebrae development. *Eur J Orthod.* 2002;24:303–311.

Plate S–142

Tubbs RS, Kelly DR, Humphrey ER, et al. The tectorial membrane: anatomical, biomechanical, and histological analysis. *Clin Anat.* 2007;20:382–386.

Plate S–178

Lee MW, McPhee RW, Stringer MD. An evidence-based approach to human dermatomes. *Clin Anat.* 2008;21:363–373.

Plates S–184, S–190 to S–192, S–196, S–197

Noden DM, Francis-West P. The differentiation and morphogenesis of craniofacial muscles. *Dev Dyn.* 2006;235: 1194–1218.

Plate S–186

Benninger B, Lee BI. Clinical importance of morphology and nomenclature of distal attachment of temporalis tendon. *J Oral Maxillofac Surg.* 2012;70:557–561.

Plates S–67 to S–70, S–74, S–76, S–77, S–194, S–324

Chang KV, Lin CP, Hung CY, et al. Sonographic nerve tracking in the cervical region: a pictorial essay and video demonstration. *Am J Phys Med Rehabil.* 2016;95:862–870.

Tubbs RS, Salter EG, Oakes WJ. Anatomic landmarks for nerves of the neck: a vade mecum for neurosurgeons. *Neurosurgery.* 2005;56(2 suppl):256–260.

Plate S–197

Feigl G. Fascia and spaces on the neck: myths and reality. *Medicina Fluminensis.* 2015;51(4):430–439.

Jain M, Dhall U. Morphometry of the thyroid and cricoid cartilages in adults. *J Anat Soc India.* 2008;57(2):119–123.

Plates S–198, S–202

Tubbs RS, Mortazavi MM, Loukas M, et al. Anatomical study of the third occipital nerve and its potential role in occipital headache/neck pain following midline dissections of the craniocervical junction. *J Neurosurg Spine.* 2011;15:71–75.

Vanderhoek MD, Hoang HT, Goff B. Ultrasound-guided greater occipital nerve blocks and pulsed radiofrequency ablation for diagnosis and treatment of occipital neuralgia. *Anesth Pain Med.* 2013;3:256–259.

Netter Atlas of Human Anatomy: A Systems Approach

Plate S–276

Beck M, Sledge JB, Gautier E, et al. The anatomy and function of the gluteus minimus muscle. *J Bone Joint Surg Br*. 2000;82(3):358–363.

Woodley SJ, Mercer SR, Nicholson HD. Morphology of the bursae associated with the greater trochanter of the femur. *J Bone Joint Surg Am*. 2008;90(2):284–294.

Plate S–317

Angelini P, Velasco JA, Flamm S. Coronary anomalies: incidence, pathophysiology, and clinical relevance. *Circulation*. 2002;105:2449–2454.

Plates S–317, S–318

Chiu IS, Anderson RH. Can we better understand the known variations in coronary arterial anatomy? *Ann Thorac Surg*. 2012;94:1751–1760.

Plate S–319

James TN. The internodal pathways of the human heart. *Prog Cardiovasc Dis*. 2001;43:495–535.

Plate S–325

Alomar X, Medrano J, Cabratosa J, et al. Anatomy of the temporomandibular joint. *Semin Ultrasound CT MRI*. 2007;28(3):170–183.

Campos PSF, Reis FP, Aragão JA. Morphofunctional features of the temporomandibular joint. *Int J Morphol*. 2011;29(4): 1394–1397.

Cristo JA, Townsend GC. Discal attachments of the human temporomandibular joint. *Aust Dent J*. 2005;50(3):152–160.

Cuccia AM, Caradonna C, Caradonna D, et al. The arterial blood supply of the temporomandibular joint: an anatomical study and clinical implications. *Imaging Sci Dent*. 2013;43(1):37–44.

Langdon JD, Berkovitz BKV, Moxham BJ. *Surgical Anatomy of the Infratemporal Fossa*. London: Martin Dunitz; 2005.

Schmolke C. The relationship between the temporomandibular joint capsule, articular disc and jaw muscles. *J Anat*. 1994;184:335–345.

Siéssere S, Vitti M, de Sousa LG, et al. Bilaminar zone: anatomical aspects, irrigation, and innervation. *Braz J Morphol Sci*. 2004;21(4):217–220.

Plate S–335

Aragão JA, Reis FP, de Figueiredo LFP, et al. The anatomy of the gastrocnemius veins and trunks in adult human cadavers. *J Vasc Br*. 2004;3(4):297–303.

Plates S–220, S–464, S–485, S–488, S–493

Oelrich TM. The striated urogenital sphincter muscle in the female. *Anat Rec*. 1983;205:223–232.

Plochocki JH, Rodriguez-Sosa JR, Adrian B, et al. A functional and clinical reinterpretation of human perineal neuromuscular anatomy: application to sexual function and continence. *Clin Anat*. 2016;29:1053–1058.

Plates S–344, S–345

Benninger B, Barrett R. A head and neck lymph node classification using an anatomical grid system while maintaining clinical relevance. *J Oral Maxillofac Surg*. 2011;69:2670–2673.

Plates S–348, S–473, S–474

Hassiotou F, Geddes D. Anatomy of the human mammary gland: current status of knowledge. *Clin Anat*. 2013;26:29–48.

Plate S–368

de Miranda CMNR, Maranhão CPM, Arraes FMNR, et al. Anatomical variations of paranasal sinuses at multislice computed tomography: what to look for. *Radiol Bras*. 2011;44(4):256–262.

Souza SA, de Souza MMA, Idagawa M, et al. Computed tomography assessment of the ethmoid roof: a relevant region at risk in endoscopic sinus surgery. *Radiol Bras*. 2008;41(3):143–147.

Plates S–376 to S–378

Ludlow CL. Central nervous system control of the laryngeal muscles in humans. *Respir Physiol Neurobiol*. 2005;147: 205–222.

Plate S–380

Ikeda S, Ono Y, Miyazawa S, et al. Flexible bronchofiberscope. *Otolaryngology (Tokyo)*. 1970;42:855.

Plates S–381, S–382

Hyde DM, Hamid Q, Irvin CG. Anatomy, pathology, and physiology of the tracheobronchial tree: emphasis on the distal airways. *J Allergy Clin Immunol*. 2009;124(6 suppl):S72–S77.

Plates S–398, S–404

Benninger B, Andrews K, Carter W. Clinical measurements of hard palate and implications for subepithelial connective tissue grafts with suggestions for palatal nomenclature. *J Oral Maxillofac Surg*. 2012;70:149–153.

Plate S–406

Fawcett E, Edin MB. The structure of the inferior maxilla, with special reference to the position of the inferior dental canal. *J Anat Physiol*. 1895;29(Pt 3):355–366.

He P, Truong MK, Adeeb N, et al. Clinical anatomy and surgical significance of the lingual foramina and their canals. *Clin Anat*. 2017;30:194–204.

Iwanaga J. The clinical view for dissection of the lingual nerve with application to minimizing iatrogenic injury. *Clin Anat*. 2017;30:467–469.

Otake I, Kageyama I, Mataga I. Clinical anatomy of the maxillary artery. *Okajimas Folia Anat Jpn*. 2011;87(4):155–164.

Siéssere S. Anatomic variation of cranial parasympathetic ganglia. *Braz Oral Res*. 2008;22(2):101–105.

Plate S–431

MacSween RNM, Anthony PP, Scheuer PJ, et al., eds. *Pathology of the Liver*. London: Churchill Livingstone; 2002.

Robinson PJA, Ward J. *MRI of the Liver: A Practical Guide*. Boca Raton, FL: CRC Press; 2006.

Plates S–435, S–436

Odze RD. *Surgical Pathology of the GI Tract, Liver, Biliary Tract, and Pancreas*. Philadelphia: Saunders; 2004.

Plate S–447

Yang HJ, Gill YC, Lee WJ, et al. Anatomy of thoracic splanchnic nerves for surgical resection. *Clin Anat*. 2008;21:171–177.

Plate S–455

Thomas MD. *The Ciba Collection of Medical Illustrations. Part 2: Digestive System: Lower Digestive Tract*. Vol. 3. Summit, NJ: CIBA; 1970:78.

Plates S–222, S–418, S–495, S–496, S–514

Stormont TJ, Cahill DR, King BF, Myers RP. Fascias of the male external genitalia and perineum. *Clin Anat*. 1994;7: 115–124.

Plates S–482, S–495

Myers RP, Goellner JR, Cahill DR. Prostate shape, external striated urethral sphincter, and radical prostatectomy: the apical dissection. *J Urol*. 1987;138:543–550.

Plates S–482, S–495, S–496, S–504

Oelrich TM. The urethral sphincter muscle in the male. *Am J Anat*. 1980;158:229–246.

Plate S–486

Feil P, Sora MC. A 3D reconstruction model of the female pelvic floor by using plastinated cross sections. *Austin J Anat*. 2014;1(5):1022.

Shin DS, Jang HG, Hwang SB, et al. Two-dimensional sectioned images and three-dimensional surface models for learning the anatomy of the female pelvis. *Anat Sci Educ*. 2013;6(5):316–323.

Plate S–509

Cahill D, Raychaudhuri B. Pelvic fasciae in urology. *Ann R Coll Surg Engl*. 2008;90:633–637.

Lee SE. A comprehensive review of neuroanatomy of the prostate. *Prostate Int*. 2013;1(4):139–145.

Nathoo N, Caris EC, Wiener JA, Mendel E. History of the vertebral venous plexus and the significant contributions of Breschet and Batson. *Neurosurgery*. 2011;69(5):1007–1014.

Pai MM, Krishnamurthy A, Prabhu LV, et al. Variability in the origin of the obturator artery. *Clinics (Sao Paulo)*. 2009;64(9):897–901.

Stoney RA. The anatomy of the visceral pelvic fascia. *J Anat Physiol*. 1904;38(4):438–447.

Walz J. A critical analysis of the current knowledge of surgical anatomy related to optimization of cancer control and preservation of continence and erection in candidates for radical prosta. *Eur Urol*. 2010;57:179–192.

References are to plate numbers. In most cases, structures are listed under singular nouns. The *Netter Atlas* utilizes *Terminologia Anatomica* as a basis for the terminology included throughout. A fully searchable database of the updated *Terminologia Anatomica* (including synonyms) can be accessed at https://ta2viewer.openanatomy.org.

Alveolar artery *(Continued)*
  mental branch of, S–189, S–323
  mylohyoid branch of, S–136, S–323
  middle superior, S–371
  mylohyoid branch of inferior, S–188
  posterior superior, S–323, S–371
  superior
    anterior, S–189
    middle, S–189
    posterior, S–63, S–188, S–189
Alveolar bone, S–405
Alveolar capillary, S–BP 63
Alveolar cell
  type I, S–BP 63
  type II, S–BP 63
Alveolar ducts, S–BP 62
  opening of, S–BP 62
Alveolar foramina, S–125
Alveolar macrophage, S–BP 63
Alveolar nerve
  anterior, middle, and posterior superior, S–73
  anterior superior, S–59, S–61
  dental and gingival branches of superior, S–371
  inferior, S–59, S–62, S–63, S–73, S–77, S–80, S–136, S–187, S–188, S–189, S–399, S–400, S–406
  middle superior, S–61
  posterior superior, S–59, S–61, S–371
  right, S–188
  superior
    infraorbital, S–63
    posterior, S–63, S–188
Alveolar pores (of Kohn), S–BP 62
Alveolar process, of maxillary bone, S–123, S–125, S–129, S–130, S–362, S–368, S–BP 61
Alveolar sac, S–BP 62
Alveolar vein
  inferior, S–328, S–400
  posterior superior, S–328
Alveolar wall
  capillary bed within, S–392
  capillary plexuses within, S–392
Alveoli, S–BP 62, S–BP 63
Amacrine cells, S–57
Ammon's horn, S–38
Ampulla
  of ductus deferens, S–530
  of ear, S–97
    anterior, S–96
    lateral, S–96
    posterior, S–96
  sphincter of, S–BP 74
  of uterine tube, S–487, S–489
Amygdaloid body, S–36, S–38, S–56, S–81
Anal canal, S–226, T8.1
  arteries of, S–441
  muscularis mucosae of, S–223, S–226
  veins of, S–445
Anal columns, Morgagni's, S–226
Anal glands, S–226
Anal nerves, inferior, S–104, S–275, S–453, S–494, S–514, S–516, S–517, S–520
Anal pit, S–506
Anal sinus, S–226
Anal sphincter muscle
  external, S–220, S–221, S–223, S–224, S–225, S–428, S–441, S–453, S–464, S–478, S–485, S–492, S–493, S–495, S–501, S–502, S–511
    deep, S–221, S–222, S–223, S–224, S–495
    subcutaneous, S–221, S–222, S–223, S–224, S–495
    superficial, S–221, S–222, S–223, S–224, S–495
  internal, S–223, S–225
Anal triangle, S–9, S–500
Anal tubercle, S–506
Anal valve, S–226
Anal verge, S–226

Anastomosis
  between medial and lateral circumflex femoral arteries, S–278
  paravertebral, S–24
  patellar, S–333
  prevertebral, S–24
Anastomotic vein, S–328
  inferior (of Labbé), S–28, S–50
  superior (of Trolard), S–28
Anastomotic vessels, S–278
  to anterior spinal artery, S–23
  to posterior spinal artery, S–23
Anatomical snuffbox, S–10, S–242
Anconeus muscle, S–111, S–112, S–240, S–241, S–242, S–263, S–264, S–552
Angle of mandible, S–5, S–72, S–124, S–135, S–359, S–373
Angular artery, S–43, S–60, S–87, S–122, S–189, S–322, S–323
Angular gyrus, S–31
Angular notch, S–415
Angular vein, S–87, S–322, S–328
Angulus oris, modiolus of, T4.1
Ankle, S–2, S–3, S–BP 18
  anterior region of, S–11
  cross-sectional anatomy of, S–BP 110
  foot and, cross sectional anatomy, S–BP 110, S–BP 111
  ligaments of, S–180
  posterior region of, S–11
  radiographs of, S–179
  tendon sheaths of, S–290
Ankle joint, T3.3
Annular hymen, S–BP 92
Anococcygeal ligament, S–222, S–480, S–492, S–493, S–502
Anococcygeal nerves, S–104, S–114, S–516, S–520
Anoderm, S–226
Anorectal artery
  inferior, S–440, S–441, S–507, S–509, S–511, S–512
  middle, S–341, S–440, S–441, S–463, S–490, S–507, S–508, S–509, S–BP 94
  superior, S–341, S–440, S–441, S–463, S–490
Anorectal flexure, S–549
Anorectal hiatus, S–481
Anorectal junction, S–482, S–BP 104
Anorectal line, S–226
Anorectal musculature, S–223
Anorectal plexus
  middle, S–452
  superior, S–452, S–453
Anorectal veins, T5.4
  inferior, S–444, S–445, S–446
  middle, S–342, S–444, S–445, S–446
  superior, S–444, S–445, S–446, T5.3
Anorectal (rectal) venous plexuses, S–342
  external, S–444, S–445, T5.4
  internal, S–445, T5.4
  perimuscular, S–444, S–445
Ansa cervicalis, S–70, S–74, S–188, S–194, S–195, S–324
  inferior root of, S–70, S–73, S–74, S–188, S–194, S–195, S–324, S–525, T2.8
  infrahyoid branches of, S–73
  superior root of, S–70, S–73, S–74, S–188, S–194, S–195, S–324, S–525
Ansa of Galen, S–378
Ansa pectoralis, S–237
Ansa subclavia, S–76, S–99, S–101, S–320, S–447
Anserine bursa, S–279, S–280
Ansiform lobule, S–42
  horizontal fissure, S–42
  inferior semilunar lobule, S–42
  posterior superior fissure, S–42
  superior semilunar lobule, S–42

Antebrachial cutaneous nerve
  lateral, S–13, S–14, S–107, S–108, S–227, S–228, S–229, S–239, S–244, S–247, S–249, S–261, S–551, S–552, T2.9
  medial, S–13, S–14, S–106, S–107, S–108, S–109, S–227, S–228, S–237, S–238, S–261, S–551, S–552, T2.10
  median, S–228
  posterior, S–13, S–14, S–107, S–111, S–112, S–228, S–229, S–240, S–258, S–261, S–551, S–552, T2.10
Antebrachial fascia, S–552
Antebrachial vein, median, S–10, S–228, S–229, S–332, S–552
Anterior chamber
  endothelium of, S–89
  of eye, S–82, S–88, S–90, S–91
  of eyeball, S–89
Anterior commissure, S–41
Anterior cruciate ligament, T3.3
Anterior ethmoidal artery, T5.1
  external nasal branch of, S–60
Anterior ethmoidal nerve, external nasal branch of, S–60
Anterior plane, S–1
Anterior ramus, S–393
Anterior root, S–206, S–454, S–BP 25
Anterior superior iliac spine (ASIS), S–427, T3.2
Anterior tibial artery, S–289, S–303, S–335, S–BP 110
Antidromic conduction, S–BP 15
Antihelix, S–5, S–94
  crura of, S–94
Antitragus, S–5, S–94
Anular ligaments, S–379
Anulus fibrosus, S–145, S–148
Anus, S–224, S–480, S–491, S–500, S–502, S–506
Aorta, S–43, S–219, S–302, S–305, S–311, S–318, S–321, S–329, S–353, S–383, S–456, S–509, S–540, S–548, S–BP 25, S–BP 75, S–BP 101
  abdominal, S–341, S–507, S–508, S–510, S–513, S–531, S–543, S–544, S–545, S–546, S–BP 100
  ascending, S–43, S–308, S–311, S–315, S–316, S–319, S–534, S–537, S–540, S–BP 99
  descending, S–43, S–303, S–358
  esophageal branch of, S–339, S–383
  inferior left bronchial branch of, S–339, S–383
  posterior intercostal branch of, S–339
  right bronchial branch of, S–339, S–383
  superior left bronchial branch of, S–339, S–383
  thoracic, S–24
  ureteric branch of, S–468
Aortic arch, S–43, S–196, S–303, S–306, S–321, S–358, S–393, S–410, S–411, S–525, S–526, S–BP 49, S–BP 99
Aortic heart valve, S–306, S–308, S–312, S–314, S–315, S–316, S–319, S–534, T5.2, S–BP 99
  left coronary leaflet of, S–312, S–314, S–315, S–316, S–319
  noncoronary leaflet of, S–312, S–314, S–315, S–316, S–319
  right coronary leaflet of, S–312, S–314, S–315, S–316, S–319
Aortic hiatus, S–208, S–393
Aortic nodes, lateral, S–355, S–357
Aortic plexus, abdominal, S–484, S–515
Aortic sinuses (of Valsalva), S–315
Aorticorenal ganglion, S–20, S–103, S–449, S–450, S–451, S–452, S–469, S–470, S–471, S–514, S–515, S–518, S–533, S–BP 13
  left, S–105, S–449, S–452, S–517, S–519
  right, S–105, S–449, S–452

Netter Atlas of Human Anatomy: A Systems Approach

Apex of lung, T7.1
Apical foramina, S–405
Apical ligament of dens, S–142, S–372, S–408,
  S–BP 21
Aponeurosis, of external abdominal oblique
  muscle, S–501
Appendices, S–428
Appendicular artery, S–425, S–439, S–440,
  S–BP 80
Appendicular nodes, S–BP 57
Appendicular skeleton, S–120
Appendicular vein, S–105, S–444, S–446
Appendix
  of epididymis, S–497
  of testis, S–497
Arachnoid, S–17, S–22, S–26, S–28
Arachnoid-dura interface, S–28
Arachnoid granulations, S–26, S–27, S–28, S–35
Arachnoid mater, S–35
Arch of aorta, S–307, S–308, S–309, S–310,
  S–312, S–319, S–339, S–385, S–389, S–391,
  S–536, S–BP 98
  lung groove for, S–388
Arcuate artery, S–289, S–291, S–292, S–297,
  S–333, S–467, S–BP 43, S–BP 53, S–BP 85
Arcuate eminence, S–93, S–132
Arcuate ligament
  lateral, S–208, S–219
  medial, S–208, S–219
  median, S–208, S–219
Arcuate line, S–151, S–153, S–154, S–217,
  S–220
  of rectus sheath, S–213, S–216, S–336
Arcuate nucleus, S–524
Arcuate popliteal ligament, S–281
Arcuate pubic ligament, S–504
Arcuate vein, S–BP 85
Areola, S–473
Areolar glands (of Montgomery), S–473
Areolar tissue, S–495
  loose, S–28
Areolar venous plexus, S–337
Arm, S–2, S–3
  anterior, S–183
  anterior region of, S–10
  arteries of, S–BP 50, S–BP 50
  muscles of
    anterior compartment of, S–239
    posterior compartment of, S–240
  posterior, S–183
  posterior region of, S–10
  radial nerve in, S–111
  serial cross sections of, S–551
Arrector pili muscle, S–4, S–20, S–BP 1, S–BP 15
Arterial arches, superficial palmar, S–251
Arterial rete, marginal, S–95
Arterial wall, S–BP 45, S–BP 45
  avascular zone, S–BP 45
  vascular zone, S–BP 45
Arteries, S–4, S–302
  of brain
    frontal section of, S–46
    frontal view of, S–46
    inferior view of, S–44
    lateral view of, S–47
    medial view of, S–47
  of ductus deferens, S–510, S–BP 94
  of esophagus, S–339
  of eyelids, S–87
  major, S–303
  of malleolar stria, S–95
  of orbit, S–87
  of pelvic organs, female, S–508
  of pelvis
    female, S–510
    male, S–509, S–BP 94
  of perineum, S–511
    male, S–512
  of round ligament, S–508

Arteries *(Continued)*
  of spinal cord
    intrinsic distribution, S–24
    schema of, S–23
  of testis, S–510
Arterioles, S–302, S–553
  macular
    inferior, S–91
    superior, S–91
  nasal retinal
    inferior, S–91
    superior, S–91
  temporal retinal
    inferior, S–91
    superior, S–91
Artery of Adamkiewicz, S–23
Articular branch, S–115
Articular cartilage, S–121, S–161, S–170,
  S–282, S–553, S–BP 107
Articular cavity, S–282, S–553, S–BP 21
  of sternoclavicular joint, S–156
Articular disc, S–63, S–136, S–BP 29,
  S–BP 37
  of sternoclavicular joint, S–156
Articular facet
  cervical, S–140
    inferior, S–143
    superior, S–140
  of dens, posterior, S–142
  inferior, S–148
  lumbar, inferior, S–145
  thoracic, superior, S–144
Articular nerve, recurrent, S–119
Articular process
  cervical
    inferior, S–139, S–140, S–143
    superior, S–139, S–143, S–BP 20
  lumbar
    inferior, S–145, S–146, S–148
    superior, S–145, S–146, S–148
  sacral, superior, S–146
  thoracic
    inferior, S–144
    superior, S–144
Articular surface, superior, S–175
Articular tubercle, S–125, S–131, S–136,
  S–187
Articularis genu muscle, S–115, S–280, S–282,
  S–300, S–554
Aryepiglottic fold, S–373, S–376, S–409
Arytenoid cartilage, S–375, S–408
  muscular process of, S–375, S–376
  vocal process of, S–375, S–376
Arytenoid muscle
  oblique, S–374, S–376, S–378, S–409
    action of, S–377
    aryepiglottic part of, S–376, S–378
  transverse, S–372, S–374, S–376, S–378,
    S–409
    action of, S–377
Ascending cervical artery, S–23
Ascending colon, S–420, S–423, S–424, S–428,
  S–483, S–484, S–546
  bed of, S–419
  as site of referred visceral pain, S–BP 5
Ascending fibers, S–470, S–471
Ascending limb, S–BP 84
Asterion, S–125, T3.1
Atlantoaxial joint, S–BP 22, S–BP 23
  lateral, S–141
    capsule of, S–141, S–142
  median, S–142
    articular cavity of, S–BP 21
Atlantoaxial ligament, S–142
Atlantoaxial membrane, posterior, S–BP 21
Atlantooccipital joint, capsule of, S–141,
  S–142
Atlantooccipital junction, S–BP 20, S–BP 21
Atlantooccipital ligament, anterior, S–142

Atlantooccipital membrane
  anterior, S–141, S–372, S–408, S–BP 21
  posterior, S–141, S–BP 21
Atlas (C1), S–16, S–121, S–139, S–141, S–142,
  S–359, S–BP 20
  anterior arch of, S–BP 21, S–BP 22, S–BP 23,
    S–BP 27
  anterior tubercle of, S–139
  anterior view of, S–137
  arch of
    anterior, S–126, S–139, S–140, S–372,
      S–BP 19, S–BP 21
    imaging of, S–39
    posterior, S–140
  dens of, articular facet for, S–140
  groove for vertebral artery of, S–140
  inferior articular surface of, S–BP 23
  inferior longitudinal band of cruciate
    ligament of, S–BP 21
  inferior view of, S–140
  lateral mass of, S–124, S–140
  left lateral view of, S–137
  posterior arch of, S–202, S–BP 19, S–BP 21,
    S–BP 22
  posterior tubercle of, S–201
  posterior view of, S–137
  superior articular facet for, S–140
  superior articular surface of, S–BP 23
  superior longitudinal band of cruciate
    ligament of, S–BP 21
  superior view of, S–140
  transverse foramen of, S–140
  transverse ligament of, S–142, S–BP 21
    tubercle for, S–140
  transverse process of, S–140, S–141, S–193,
    S–201
  tubercle of
    anterior, S–140, S–142
    posterior, S–140, S–200
  vertebral foramen of, S–140
Atonic stomach, S–BP 67
Atrioventricular (AV) bundle (of His), S–319
Atrioventricular (AV) node, S–319, T5.2
Atrioventricular septum, S–311, S–312, S–314,
  S–315, S–319
Atrioventricular valve
  left, S–308, S–312, S–314, S–315, S–316,
    S–319, S–534, S–538, S–539
    anterior cusp of, S–312, S–314, S–315,
      S–316
    commissural leaflets of, S–314, S–315
    left fibrous ring of, S–314
    posterior cusp of, S–312, S–314, S–315,
      S–316
  right, S–308, S–311, S–313, S–314, S–315,
    S–316, S–534, S–538, S–539
    inferior leaflet (posterior cusp) of, S–311,
      S–314, S–315, S–316
    right fibrous ring of, S–319
    septal leaflet (cusp) of, S–311, S–314,
      S–315, S–316
    superior leaflet of, S–311, S–314, S–315,
      S–316
Atrium, S–316
  left, S–308, S–310, S–311, S–312, S–315,
    S–316, S–321, S–538, S–539, S–540,
    S–BP 98
    oblique vein of (of Marshall), S–308,
      S–310, S–312, S–317
  right, S–306, S–308, S–309, S–310, S–311,
    S–313, S–315, S–316, S–321, S–390,
    S–534, S–538, S–539, S–540, S–BP 99
Auditory canal, internal, imaging of, S–BP 12
Auditory ossicles, S–120, T3.1
Auditory tube, S–68, S–93, S–94, S–95, S–97,
  S–BP 8, S–BP 9, S–BP 10
  cartilage of, S–398, S–403, S–408, S–409,
    S–BP 60
  cartilaginous part of, S–187, S–BP 10

Brachial plexus (Continued)
  medial cord of, S–106, S–107, S–108, S–109, S–231, S–238
  middle trunks of, S–106, S–236
  nerves of, T2.9, T2.10, T2.11
  posterior cord of, S–106, S–107, S–108, S–109, S–231
  posterior divisions of, S–106
  roots of, S–106
  schema of, S–106
  superior trunks of, S–106, S–236
  terminal branches of, S–106
  trunks of, S–193, S–236
Brachial veins, S–237, S–304, S–332, S–551
Brachialis muscle, S–108, S–227, S–237, S–238, S–239, S–244, S–247, S–249, S–262, S–264, S–551
  insertion of, S–161
Brachiocephalic artery, S–321
Brachiocephalic trunk, S–43, S–75, S–195, S–205, S–306, S–307, S–309, S–310, S–326, S–339, S–389, S–410, S–525, S–526, S–527, S–534, S–535, S–BP 49
Brachiocephalic vein, S–304, S–350, S–525
  left, S–306, S–307, S–309, S–328, S–330, S–338, S–340, S–389, S–390, S–391, S–393, S–410, S–526, S–534, S–535, S–BP 99
    lung groove for, S–388
  right, S–306, S–307, S–309, S–321, S–330, S–338, S–340, S–389, S–390, S–393, S–526, S–535, S–BP 99
Brachioradialis muscle, S–10, S–111, S–112, S–227, S–238, S–239, S–240, S–241, S–242, S–244, S–247, S–262, S–263, S–264, S–551, S–552
Brachioradialis tendon, S–249, S–552
Brain, S–15, S–302, S–368, S–BP 61
  arteries of, S–326
    frontal section of, S–46
    frontal view of, S–46
    inferior view of, S–44
    lateral view of, S–47
    medial view of, S–47
    schema of, S–43
  axial and coronal MRIs of, S–BP 12
  axial of, S–BP 12
  coronal MRIs of, S–BP 12
  inferior view of, S–33
  lateral view of, S–31
  medial view of, S–32
  veins of
    deep, S–50
    subependymal, S–51
  ventricles of, S–34
Brain stem, S–40
  cranial nerve nuclei in, S–54, S–55
Breast, S–2, S–321, S–BP 97
  arteries of, S–474
  lymph nodes of, S–348
  lymph vessels of, S–348, T6.1
  lymphatic drainage of, S–349
Bregma, S–128
Bridging vein, S–26, S–28, S–29, S–35
  imaging of, S–BP 11
Broad ligament, S–463
  anterior lamina of, S–489
  posterior lamina of, S–489
Bronchi, S–20, S–21, S–380, S–BP 15
  anterior, S–380
  anterior basal, S–380
  anteromedial basal, S–380
  apical, S–380
  apicoposterior, S–380
  inferior lobar, S–379
  intermediate, S–388
  intrapulmonary, S–379
  lateral, S–380
  lateral basal, S–380
  lingular, S–379

Bronchi (Continued)
  inferior, S–380
  superior, S–380
  main
    left, S–358, S–379, S–383, S–388, S–389, S–391, S–393, S–410, S–411, S–537, S–539, S–BP 98, S–BP 99
    right, S–358, S–379, S–383, S–390, S–393, S–410, S–537, S–539, S–540, S–BP 99
  major, S–379
  medial, S–380
  medial basal, S–380
  middle lobar, S–379
  nomenclature of, S–380
  posterior, S–380
  posterior basal, S–380
  right superior lobar, S–379, S–388, S–389
  superior division, S–379
  superior lobar, S–379
Bronchial artery, S–383, S–388, S–390, S–391, S–392
Bronchial veins, S–383
  left, S–383
  right, S–383
Bronchioles, S–380
  respiratory, S–392, S–BP 62
  terminal, S–BP 62
    elastic fibers of, S–BP 62
    smooth muscle of, S–BP 62
Bronchoaortic constriction, S–411
Bronchomediastinal lymphatic trunk, S–350
Bronchopulmonary (hilar) lymph nodes, S–350, S–388, S–390, S–391
Bronchopulmonary segments, S–381, S–382
  anterior, S–381, S–382
    basal, S–381, S–382
  anteromedial basal, S–381, S–382
  apical, S–381, S–382
  apicoposterior, S–381, S–382
  inferior lingular, S–381, S–382
  lateral, S–381, S–382
    basal, S–381, S–382
  medial, S–381, S–382
    basal, S–381, S–382
  posterior, S–381, S–382
    basal, S–381, S–382
  superior, S–381, S–382
  lingular, S–381, S–382
Bronchus intermedius, S–BP 99
Brow, S–BP 18
Buccal artery, S–188, S–189, S–323
Buccal nerve, S–59, S–62, S–63, S–73, S–185, S–188, S–400, S–406
Buccal region, of head and neck, S–5
Bucinator lymph nodes, S–344
Bucinator muscle, S–65, S–184, S–186, S–187, S–197, S–323, S–368, S–BP 35, S–BP 36, S–BP 61
Buccopharyngeal fascia, S–196, S–329, S–372, S–400, S–408
Bucinator, S–396, S–398, S–400, S–407, S–408
Buck's fascia. See Deep (Buck's) fascia
Buck's fascia, of penis, S–495, S–497
Bulbar conjunctiva, S–82, S–88, S–89, S–91, S–92, S–BP 7
Bulbospongiosus muscle, S–222, S–224, S–418, S–465, S–471, S–486, S–488, S–492, S–493, S–494, S–495, S–501, S–504, S–511, S–512, S–BP 86
Bulbourethral ducts, S–503
Bulbourethral (Cowper's) gland, S–217, S–418, S–465, S–495, S–503, S–504, S–505
  duct of, S–504
  primordium of, S–505
Buttock, S–3
  nerves of, S–275

C

C1. See Atlas
C2. See Axis
C3 vertebra, transverse process of, S–193
C7 vertebra, S–196
  transverse process of, posterior tubercle of, S–193
Calcaneal bursa, subcutaneous, S–290
Calcaneal sulcus, S–178
Calcaneal tendon, S–11, S–179, S–180, S–284, S–285, S–286, S–287, S–290, T4.3, S–BP 109
Calcaneal tuberosity, S–11, S–284, S–285, S–293, S–294
  lateral process of, S–294
  medial process of, S–294, S–BP 110
Calcaneocuboid ligament, S–180, S–297
  dorsal, S–180
Calcaneofibular ligament, S–175, S–178
Calcaneometatarsal ligament, S–293
Calcaneonavicular ligament, S–180
Calcaneus, T3.3
Calcaneus bones, S–176, S–177, S–178, S–179, S–181, S–BP 110
  tuberosity of, S–181, S–295, S–296
Calcar avis, S–37, S–38
Calcarine artery, S–47, S–48
Calcarine sulcus, S–31, S–32, S–33, S–37, S–57
Calices
  major, S–462
  minor, S–462
Callosomarginal artery, S–46, S–47
  frontal branches of, S–46
  medial frontal branches of, S–47
  paracentral branch of, S–47
Calvaria, S–28
  inferior view of, S–128
  superior view of, S–128
Camper's fascia, S–211, S–418, S–BP 90
Camper's layer, of superficial fascia, S–500
Canal of Schlemm, S–89, S–91, S–92
Canine tooth, S–404, S–405
Capillaries, S–302
Capillary lumen, S–BP 63
Capillary plexus, S–BP 63
Capitate bone, S–163, S–164, S–166, S–168, S–253, S–BP 29
Capitohamate ligament, S–165, S–166
Capitotriquetral ligament, S–BP 29
Capitulum, S–155, S–159, S–160
Capsular branches, S–BP 94
Capsular ligaments, S–157
Capsular vein, S–BP 85
Cardia, S–414
  nodes around, S–348
Cardiac impression, S–388
Cardiac muscle, glands, S–15
Cardiac nerve, S–20
  cervical, S–BP 13
  thoracic, S–76
Cardiac notch, S–413, S–415
  of left lung, S–385, S–388
Cardiac orifice, S–414
Cardiac plexus, S–68, S–99, S–101, S–320, S–447, S–BP 13, S–BP 14
Cardiac veins, S–317, S–BP 48
  anterior, S–317
  great, S–317
  middle, S–310, S–317
  small, S–317
  variations of, S–BP 48
Cardinal ligament (Mackenrodt's), S–478, S–487, S–488, S–490, S–BP 96
Cardiovascular system, S–302 to S–342, T5.1–5.5, S–BP 44 to S–BP 53
  composition of blood and, S–BP 44, S–BP 44
  major systemic veins of, S–304
Carinal lymph nodes, inferior, S–350
Caroticotympanic artery, S–43, S–95, S–327

Netter Atlas of Human Anatomy: A Systems Approach

Cerebral arterial circle (of Willis), S–44, S–45, T5.1
Cerebral arteries, S–28
  anterior, S–43, S–44, S–45, S–46, S–47, S–48, S–50, S–326, S–327
    cingular branches of, S–47
    frontal branches of, S–47
    imaging of, S–BP 11, S–BP 12
    left, S–47
    medial orbitofrontal branches of, S–46
    postcommunicating part of, S–52
    precommunicating part of, S–52
    terminal branches of, S–47
  anterolateral, S–48
  middle, S–43, S–44, S–45, S–46, S–47, S–48, S–52, S–326, S–327
    branches of, S–327
    frontal and parietal branches, S–46
    frontal branches of, S–46, S–47
    imaging of, S–BP 11, S–BP 12
    lateral orbitofrontal branch of, S–47
    left, S–47
    occipitotemporal branches of, S–47
    orbitofrontal branch of, S–47
    parietal branches of, S–46, S–47
    prefrontal branches of, S–47
    temporal branches of, S–46, S–47
  posterior, S–43, S–44, S–45, S–46, S–47, S–48, S–52, S–326, S–327
    dorsal branch of, S–47
    imaging of, S–BP 11, S–BP 12
    occipitotemporal branches of, S–47
    parietooccipital branch of, S–47
    temporal branches of, S–47, S–48
Cerebral cortex, postcentral gyrus, S–BP 17
Cerebral crus, S–33, S–40, S–42, S–44
Cerebral fissure, longitudinal, S–33, S–48, S–50
  imaging of, S–39
Cerebral hemisphere, S–28
  inferolateral cerebral of, S–33
  temporal pole of, S–33
Cerebral peduncle, S–32, S–41, S–368
Cerebral veins
  anterior, S–49, S–50, S–51
  great (of Galen), S–52
  inferior, S–28, S–30, S–50
  internal, S–26, S–32, S–34, S–37, S–49, S–50, S–52
    imaging of, S–BP 11
    left, S–51
    right, S–51
  middle
    deep, S–26, S–49, S–50, S–51
    superficial, S–26, S–28, S–30, S–50
  occipital, S–51
  superficial, S–28
  superior, S–26, S–28, S–52, T5.1
    opening of, S–27
Cerebrospinal fluid
  circulation of, S–35
  radiology of, S–138
  within subarachnoid space, S–BP 23
Cerebrum
  frontal lobe of, S–31
  frontal pole of, S–31, S–33
  inferior (inferolateral) margin of, S–33
  occipital lobe of, S–31
  occipital pole of, S–31, S–33
  parietal lobe of, S–31
  temporal lobe of, S–31
  temporal pole of, S–31, S–33
Cervical artery
  ascending, S–23, S–43, S–73, S–195, S–234, S–323, S–525, S–526, S–527, S–BP 49
  deep, S–23, S–43, S–BP 49
  transverse, S–194, S–195, S–234, S–237, S–525, S–527, S–BP 49
    deep branch of, S–BP 49
    superficial branch of, S–198, S–BP 49

Cervical canal, S–487
Cervical cardiac branch, S–BP 14
Cervical cardiac nerves, S–101, S–447, S–BP 46
  inferior, S–76, S–99, S–320
  middle, S–76, S–99, S–320
  superior, S–72, S–76, S–99, S–320
Cervical enlargement, S–18
Cervical fasciae, S–197
  deep, S–196
    prevertebral layer of, S–329
    superficial layer of, S–329, S–372
  superficial investing, S–197
  superficial layer of, S–184, S–190, S–196
Cervical ganglion
  middle, S–70, S–72, S–73, S–76, S–99, S–195, S–320, S–447, S–BP 13
  superior, S–20, S–67, S–70, S–71, S–72, S–76, S–77, S–78, S–79, S–80, S–99, S–320, S–366, S–447, S–BP 13
Cervical lordosis, S–137
Cervical lymph nodes
  anterior deep, S–344
  anterior superficial, S–344
  deep lateral, S–345
  inferior deep lateral, T6.1, S–344
  posterior lateral superficial, S–344
  superior deep lateral, S–344, T6.1
  superior lateral superficial, S–344
Cervical muscle
  anterior, S–193
  deep, S–329
    posterior, S–196
  lateral, S–193
Cervical nerve, transverse, S–74, S–185, S–194, S–195, T2.8
Cervical nodes, inferior deep lateral, S–345
Cervical nucleus, lateral, S–BP 17
Cervical pleura, S–358, S–391, S–525, T7.1
Cervical plexus, S–16, S–195, T2.1
  branches from, S–185
  muscular branches of, T2.8
  schema of, S–74
Cervical ribs, related variations and, S–BP 38
Cervical spinal cord, S–BP 23
Cervical spinal nerves, S–329
  C1, S–69
    anterior rootlets of, S–40
    posterior rootlets of, S–40
  C2, S–69
    anterior ramus of, S–194
  C3, S–69
    anterior ramus of, S–194
  C4, S–69
  C5, anterior ramus of, S–194
    groove for, S–139, S–143
    posterior rami of, S–185
Cervical spine
  MRI and radiograph, S–BP 23
  radiographs, S–BP 22
    MRI, S–BP 23
    open-mouth, S–BP 23
Cervical sympathetic ganglion, S–BP 15
Cervical vein, deep, S–330
Cervical vertebrae, S–23, S–137, S–321
  anulus fibrosus of, S–139
  arteries, S–23
  body of, S–139, S–143
  C3, S–140, S–359
    bifid spinous process of, S–143
    inferior articular facet for, S–140, S–143
    inferior articular process of, S–139, S–143
    inferior aspect of, S–143
    lamina of, S–143
    pedicle of, S–143
    superior articular process of, S–139
    transverse foramen of, S–143
    transverse process of, S–143
    vertebral body of, S–143, S–BP 20
    vertebral foramen of, S–143

Cervical vertebrae (Continued)
  C4, S–140, S–143
    anterior view of, S–143
    articular surface of, S–143
    body of, S–143
    groove for spinal nerve, S–143
    inferior articular facet of, S–143
    inferior articular process of, S–143
    lamina of, S–143
    left uncinate process of, S–143
    spinous process of, S–143
    superior articular facet of, S–143
    superior articular process of, S–143
    superior aspect of, S–143
    transverse foramen of, S–143
    transverse process of, S–143
    uncinate process of, S–143
    vertebral body of, S–143
  C6
    anterior tubercle of, S–141
    transverse process of, S–326
  C7, S–143, S–359, S–BP 20
    anterior view of, S–137, S–143
    articular surface of, S–143
    body of, S–143
    inferior articular process of, S–143
    lamina of, S–143
    lateral view of, S–137
    pedicle of, S–143
    posterior view of, S–137
    septated transverse foramen of, S–143
    spinal nerve, groove for, S–143
    spinous process of, S–6, S–141, S–143, S–199, S–230
    superior articular process and facet of, S–143
    superior view of, S–143
    transverse process of, S–143
    uncinate process of, S–143
      articular surface of, S–143
    vertebral body of, S–143
    vertebral foramen of, S–143
  degenerative changes in, S–BP 20
  intervertebral disc of, S–139
  nucleus pulposus of, S–139
  transverse foramen of, S–139
  transverse process of, S–139
  uncinate processes of, S–139
  uncovertebral joints of, S–139
  upper, S–140
Cervicofacial division, S–64
Cervicothoracic (stellate) ganglion, S–76, S–99, S–101, S–320, S–447, S–BP 13, S–BP 46
Cervix, S–515
Cheek, S–2
Chiasma, S–BP 89
Chiasmatic cistern, S–35
Chiasmatic groove, S–132
Chief cell, S–BP 67
Chin, S–2
Choanae, S–130, S–131, S–360, S–361, S–373, S–398, S–409, S–BP 10
  of cranium, S–187
Cholinergic synapses, schema of, S–BP 15
Chorda tympani, S–59, S–62, S–63, S–65, S–66, S–73, S–77, S–79, S–80, S–81, S–94, S–95, S–188, S–399, S–BP 8
  nervus intermedius, S–66
Chordae tendineae, S–311, S–312, S–313, S–315
Choroid, S–88, S–90, S–91
Choroid plexus, S–37, S–38, S–41
  of lateral ventricle, S–48
  of third ventricle, S–32
Choroid vein, S–26
  superior, S–50, S–51
Choroidal artery
  anterior, S–44, S–45, S–46, S–48, S–327
  posterior lateral, S–44
  posterior medial, S–44

Ciliary arteries
anterior, S–91, S–92
recurrent branch of, S–92
posterior, S–87
long, S–91, S–92
short, S–91, S–92
short posterior, S–92
Ciliary body, S–88, S–89, S–90, S–91
blood vessels of, S–91
orbiculus ciliaris of, S–90
Ciliary ganglion, S–21, S–58, S–59, S–61, S–71,
S–77, S–78, S–85, S–86, S–BP 14
branch to, S–58, S–59, S–78
oculomotor root of, S–78
parasympathetic root of, S–58, S–77, S–85
schema of, S–78
sensory root of, S–77, S–85
sympathetic root of, S–58, S–77, S–78, S–85
Ciliary muscle, S–58, S–78, S–88
circular fibers of, S–89
meridional fibers of, S–89
Ciliary nerves
long, S–58, S–59, S–61, S–77, S–78, S–85,
S–86
short, S–58, S–59, S–61, S–77, S–78, S–85,
S–86
Ciliary process, S–88, S–89, S–90
Ciliary vein, anterior, S–89, S–91, S–92
Cingulate gyrus, S–32
isthmus of, S–32, S–33
Cingulate sulcus, S–32
Circular muscle, S–223, S–226
Circular muscular layer, S–455, S–BP 65, S–BP
68, S–BP 69
Circulation, prenatal and postnatal, S–305
Circumferential lamellae
external, S–BP 31
internal, S–BP 31
Circumflex artery
atrioventricular branch of, S–BP 47
of heart, S–314, S–317, S–318, S–BP 47
inferior left ventricular branch of, S–317,
S–318, S–BP 47
left marginal branch of, S–317, S–318, S–BP 47
Cisterna chyli, S–343, S–352, S–353, S–BP 56
Claustrum, S–36
Clavicle, S–5, S–7, S–10, S–120, S–121, S–149,
S–150, S–155, S–156, S–157, S–158, S–184,
S–190, S–191, S–192, S–197, S–203, S–204,
S–205, S–230, S–231, S–232, S–233, S–234,
S–236, S–237, S–321, S–358, S–385, S–386,
S–387, S–388, S–390, S–391, S–473, S–534,
S–535, T3.2, T3.3, S–BP 22, S–BP 99
acromial end of, S–156
acromial facet of, S–156
anterior margin of, S–156
body of, S–156
conoid tubercle, S–156
level of, S–19
muscle attachment sites of, S–264
posterior margin of, S–156
sternal articular surface of, S–156
sternal end of, S–156
Clavicular head. *See also* Pectoralis major
muscle
of sternocleidomastoid muscle, S–5
Clavipectoral fasciae, S–203, S–231
Cleft of Luschka, S–139
Clinoid process
anterior, S–86, S–126, S–129, S–132
posterior, S–132
Clitoral artery, internal, S–511
Clitoris, S–224, S–493, S–494
body of, S–485, S–506
crus of, S–465, S–485, S–486, S–488, S–493,
S–494
deep artery of, S–511
dorsal artery of, S–511
dorsal nerve of, S–104, S–514, S–516

Clitoris *(Continued)*
dorsal vein of, deep, S–220
fascia of, S–491
frenulum of, S–491
glans of, S–491, S–506
prepuce of, S–486, S–491, S–506
suspensory ligament of, S–492, S–493,
S–494
Clivus, S–132, S–142, S–BP 23
imaging of, S–39
Clunial nerves, inferior, S–516, S–520
CMC joint radial styloid process, S–BP 42
CN VI. *See* Abducens nerve
CN VII. *See* Facial nerve
CN IX. *See* Glossopharyngeal nerve
CN XI. *See* Accessory nerve
CN XII. *See* Hypoglossal nerve
Coccygeal horn, S–146
Coccygeal nerve, S–16, S–18, S–114
Coccygeal plexuses, S–104, S–114
Coccygeus muscle, S–219, S–220, S–479,
S–480, S–481, S–482, S–507, S–509, S–513,
S–517, S–520, T4.2
nerve to, S–104, S–114, T2.12
Coccyx, S–16, S–18, S–120, S–146, S–151,
S–152, S–154, S–220, S–221, S–475, S–476,
S–479, S–493, S–BP 104, S–BP 108
anterior view of, S–137
apex of, S–9, S–224, S–477, S–480, S–482,
S–500
lateral view of, S–137
posterior view of, S–137
tip of, S–500, S–502, S–549
transverse process of, S–146
Cochlea, S–71, S–96, S–98, S–BP 9
course of sound in, S–93
duct of, S–BP 9
helicotrema of, S–93, S–96, S–97, S–BP 9
modiolus of, S–97
scala tympani of, S–93, S–BP 9
scala vestibuli of, S–93, S–BP 9
section through turn of, S–97
Cochlear aqueduct, S–97
Cochlear canaliculus, S–97
Cochlear cupula, S–96
Cochlear duct, S–93, S–96, S–97, S–98
external wall of, S–97
Cochlear (spiral) ganglion, S–66
Cochlear nerve, S–66, S–93, S–96, S–97, S–98,
T2.5, S–BP 9
Cochlear nucleus, S–55, S–66
anterior, S–54, S–66
posterior, S–54, S–66
Cochlear recess, S–96
Cochlear (round) window, S–93, S–96, S–127,
S–BP 9
fossa of, S–94, S–BP 8
Colic artery
ascending branch of, S–440, S–441
descending branch of, S–440, S–441
left, S–341, S–441, S–463, S–BP 81
middle, S–418, S–433, S–436, S–439, S–440,
S–528, S–BP 81, S–BP 82
right, S–439, S–440, S–463, S–BP 81,
S–BP 82
variations in, BP 81–BP 82
Colic flexure
left, S–415, S–416, S–422, S–423, S–424,
S–427, S–428, S–430, S–433
splenic, S–532
right, S–415, S–416, S–420, S–423, S–424,
S–428, S–430, S–433, S–461
Colic impression, S–434
Colic nodes, right, S–BP 57
Colic plexus, S–452
left, S–452
marginal, S–452
middle, S–451, S–452
right, S–451, S–452

Colic vein
left, S–444, S–446, S–BP 78
middle, S–433, S–443, S–444, S–446, S–BP
78
right, S–443, S–444, S–446
Collagen
in arterial wall, S–BP 45
fibers, in connective tissues and cartilage,
S–BP 30
lamellae of anulus fibrosus, S–148
Collateral artery
inferior ulnar, S–331, S–BP 50
middle, S–240, S–241, S–331, S–551, S–BP
50
radial, S–240, S–331, S–551, S–BP 50
superior ulnar, S–331, S–BP 50
Collateral eminence, S–37
Collateral ligament, S–169, S–255, S–BP 107
accessory, S–169
of ankle, medial, S–175
fibular, S–173, S–175, S–279, S–280, S–281,
S–282, S–283, S–286, S–287, S–289
of foot, S–181
radial, S–161, S–BP 29
tibial, S–173, S–175, S–279, S–280, S–281,
S–282, S–283, S–286, S–288
ulnar, S–161, S–BP 29
Collateral sulcus, S–32, S–33
Collateral trigone, S–37
Collecting duct, S–BP 84, S–BP 85
Collecting tubule, S–BP 84
Colles' fascia, S–221, S–222, S–224, S–500,
S–501, S–502, S–504, S–511, S–512, S–520,
S–BP 90
superficial, S–495
Colliculus
facial, S–41
inferior, S–32, S–37, S–40, S–41, S–49
brachium of, S–37
superior, S–32, S–33, S–37, S–40, S–41,
S–48, S–49, S–54, S–78
brachium of, S–37
Colon, S–459, T8.1
ascending, S–68
as site of referred visceral pain, S–BP 5
circular muscular layer of, S–426
descending, S–20, S–21, S–517, S–543,
S–544, S–548, S–BP 102
sigmoid, S–221, S–223, S–225, S–226, S–513,
S–517
as site of referred visceral pain, S–BP 5
transverse, S–543, S–544
Columns of fornix, S–48
Commissural fibers, S–56
Commissure, S–BP 16
anterior, S–32, S–41, S–51, S–56, S–524
posterior, S–32, S–37, S–41
of semilunar valve cusps, S–315
Common bony limb, S–97
Common carotid artery, S–193, S–195
Common iliac arteries, S–303, S–BP 25
Common iliac nodes, S–352, S–354
Common membranous limb, S–96, S–97
Common tendinous ring (of Zinn), S–58, S–84,
S–85, S–BP 7
Communicating artery
anterior, S–43, S–44, S–45, S–46, S–47, S–52,
S–326, S–327
imaging of, S–BP 11, S–BP 12
posterior, S–30, S–43, S–44, S–45, S–46,
S–47, S–48, S–52, S–326, S–327
imaging of, S–BP 11, S–BP 12
Communicating vein, S–330, S–363, S–BP 27
Compact bone, S–BP 31
Compressor urethrae muscle, S–220, S–466,
S–488, S–493, S–494, S–496, S–511
Concentric lamellae, of osteon, S–BP 31
Concha, inferior, S–130
Condylar canal, S–131, S–134

Condylar fossa, S–131
Condylar process
  head of, S–BP 37
  of mandible, S–135
    head of, S–125
Condyles
  of femur
    lateral, S–172, S–173, S–280, S–281, S–283
    medial, S–172, S–173, S–280, S–281, S–283
  of fibula
    lateral, S–175
    medial, S–175
  of knee
    lateral, S–BP 32
    medial, S–BP 32
  mandibular, S–126
  occipital, S–129, S–130, S–131, S–132
  of tibia
    lateral, S–174, S–272, S–283
    medial, S–173, S–280, S–283
Condyloid joint, S–121
Cone cell, S–57
Cone of light, S–94
Confluence of sinuses, S–29, S–30, S–49, S–50, S–98
Conjoined longitudinal muscle, S–223
Conjunctiva, S–92
Conjunctival artery, S–91
  posterior, S–92
Conjunctival fornix
  inferior, S–82
  superior, S–82
Conjunctival vein, S–91
  posterior, S–92
Connective tissues, S–BP 30
  dense, S–BP 30
  loose, S–28
  of skull, S–28
Conoid ligament, S–156, S–157, S–232
Conoid tubercle, S–156
Conus arteriosus, S–306, S–309, S–311, S–312, S–314, S–316
Conus elasticus, S–375, S–376, S–378
Conus medullaris, S–16, S–18, S–544, T2.2
  radiology of, S–138
Cooper's ligament. See Pectineal ligament (Cooper's)
Coracoacromial ligament, S–157, S–232, S–235, S–239
Coracobrachialis muscle, S–108, S–227, S–235, S–236, S–237, S–238, S–239, S–264, S–536, S–537, S–551
Coracobrachialis tendon, S–235
Coracoclavicular ligament, S–157, S–232
Coracohumeral ligament, S–157
Coracoid process, S–155, S–157, S–158, S–203, S–204, S–231, S–232, S–233, S–234, S–235, S–236, S–237, S–238, S–239
  of scapula, S–149
Cornea, S–82, S–88, S–89, S–91, S–92, S–BP 7
Corneoscleral junction, S–82
Corniculate cartilage, S–375, S–408
Corniculate tubercle, S–373, S–376, S–409
Corona radiata, S–BP 96
Coronal plane, S–1
Coronal sulcus, S–506
Coronal suture, S–123, S–124, S–125, S–126, S–127, S–128, S–129, S–BP 19
Coronary arteries, S–317, T5.2
  imaging of, S–318
  left, S–317, S–318, S–BP 47
    anterolateral views with arteriograms, S–BP 47
    opening of, S–315
  opening of, S–316
  right, S–306, S–309, S–310, S–314, S–317, S–318, S–BP 47
    anterolateral views with arteriograms, S–BP 47, S–BP 47

Coronary arteries (Continued)
  atrial branch of, S–317
  AV nodal branch of, S–314, S–318, S–BP 47
  conal branch of, S–BP 47
  opening of, S–315
  right inferolateral branches of, S–BP 47
  right marginal branch of, S–317, S–318, S–BP 47
  SA nodal branch of, S–318, S–319, S–BP 47
  sinuatrial nodal branch of, S–317
  variations of, S–BP 48, S–BP 48
Coronary ligament, S–418, S–419, S–429
Coronary sinus, S–308, S–310, S–312, S–317, S–538, S–539, S–BP 98
  opening of, S–311, S–315, S–534
  valve (thebesian) of, S–311
Coronary sulcus, S–309, S–310
Coronoid fossa, S–155, S–159
Coronoid process, S–162, S–246
  of mandible, S–63, S–125, S–126, S–135
Corpora cavernosa, of penis, S–502
Corpus albicans, S–487, S–529
Corpus albuginea, S–491
Corpus callosum, S–32, S–34, S–37, S–46, S–48
  body of, S–46
    imaging of, S–39, S–BP 12
  dorsal vein of, S–49, S–51
  genu of, S–32, S–33, S–36, S–38, S–51
    imaging of, S–39
  posterior, dorsal branch of, S–47
  posterior cerebral artery, dorsal branch of, S–48
  rostrum of, S–32, S–46, S–50
    imaging of, S–39
  splenium of, S–32, S–33, S–36, S–38, S–41, S–48, S–49, S–50, S–51
    imaging of, S–39
  sulcus of, S–32
  trunk of, S–32
Corpus cavernosum, S–491, S–495, S–501, S–503, S–504
Corpus luteum, S–487, S–BP 95
  of menstruation, S–529
  of pregnancy, S–529
Corpus spongiosum, S–465, S–495, S–501, S–502, S–503
Corpus striatum, S–46
Corrugator cutis ani muscle, S–223, S–226
Corrugator supercilii muscle, S–60, S–65, S–122, S–184, S–BP 35
"Corset constriction," S–BP 71
Cortical capillary plexus, S–BP 85
Cortical lymph vessels, S–354
Cortical renal corpuscle, S–BP 84
Corticospinal tract
  anterior, S–BP 16, S–BP 18
  lateral, S–BP 16, S–BP 18
Costal cartilages, S–149, S–150, S–151, S–156, S–388, S–541
  first, S–385, S–535
  second, S–537
  fifth, S–207, S–539
  sixth, S–230, S–358
  seventh, S–358, S–387
Costal facet, S–144
  inferior, S–144, S–150, S–388
  superior, S–144, S–150, S–388
  transverse, S–144, S–150, S–388
Costal groove, S–388
Costal impressions, S–429
Costal pleura, S–460
Costocervical trunk, S–43, S–195, S–326, S–330, S–BP 49
Costochondral joints, S–150, S–388, T3.2
Costoclavicular ligament, S–150, S–156, S–231, S–264, S–388
  impression for, S–156
Costocoracoid ligament, S–231

Costocoracoid membrane, S–231
Costodiaphragmatic recess, S–387, S–461, S–534, S–BP 98
Costomediastinal recess, S–387
Costotransverse joint, S–535
Costotransverse ligament, S–150, S–388, S–BP 26
  lateral, S–150, S–388, S–BP 26
  superior, S–150, S–388, S–BP 26
Cough receptors, S–384
Cowper's gland, S–495, S–503, S–504, S–530
Cranial, as term of relationship, S–1
Cranial arachnoid, S–35
Cranial base
  foramina and canals of
    inferior view of, S–133
    superior view of, S–134
  inferior view of, S–131
  nerves of, S–71
  superior view of, S–132
  vessels of, S–71
Cranial cavity, S–BP 2
Cranial dura, S–35
Cranial fossa
  anterior, S–132, S–367
  middle, S–132
    floor of, S–BP 37
  posterior, S–132
    veins of, S–49
Cranial imaging, S–39, S–BP 11
  magnetic resonance angiography, S–52
  magnetic resonance venography, S–52
  MRV and MRA, S–BP 11
Cranial nerves, S–15, T2.4, T2.5, T2.6, T2.7
  C1, S–18, S–20
  C2, S–18, S–20
  C3, S–18, S–20
  C4, S–18, S–20
  C5, S–18, S–20
  C6, S–18, S–20
  C7, S–18, S–20
  C8, S–18
  motor and sensory distribution of, S–53
  nuclei of, in brainstem, S–54, S–55
Cranial root, S–68
Cranial sutures, T3.1
Craniocervical ligaments
  external, S–141
  internal, S–142
Cranium, S–120
  posterior and lateral views, S–130
Cremaster fascia, S–497, S–498
Cremaster muscle, S–102, S–215, S–497, S–BP 41
  in cremasteric fascia, S–212, S–213
  lateral head, S–212, S–216, S–BP 40
  medial head, S–212, S–216, S–BP 40
Cremasteric artery, S–213, S–215, S–218, S–336, S–341, S–510
Cremasteric fascia, S–215, S–BP 41
Cremasteric vein, S–510
Cremasteric vessels, S–215
Cribriform fascia, S–BP 40
  within saphenous opening, S–347
Cribriform hymen, S–BP 92
Cribriform , S–129, S–132, S–364, S–367
  foramina of, S–134
Cricoarytenoid muscle
  lateral, S–197, S–376, S–378
    action of, S–377
  posterior, S–197, S–374, S–376, S–378, S–409
    action of, S–377
Cricoesophageal tendon, S–374, S–409
Cricoid cartilage, S–190, S–191, S–359, S–372, S–375, S–376, S–378, S–379, S–385, S–389, S–407, S–408, S–411, S–412, S–525
  arch of, S–375
  arytenoid articular surface of, S–375
  lamina of, S–373, S–374, S–375, S–376

Cricopharyngeus muscle, S–378, S–526, S–527
Cricothyroid artery, S–525
Cricothyroid joint, S–375, S–377
Cricothyroid ligament, S–525
  lateral, S–375
  median, S–375, S–379, S–407, S–408, S–525, T3.2
Cricothyroid muscle, S–68, S–191, S–197, S–376, S–378, S–407, S–525
  action of, S–377
  oblique part of, S–376
  straight part of, S–376
Crista galli, S–129, S–132, S–367
  imaging of, S–39
Crista terminalis, S–311, S–319
Crossing over, S–BP 89
Cross-sectional anatomy and imaging, S–534 to S–556, S–BP 98 to S–BP 112
Cruciate ligaments, S–142, S–173, S–175, S–280
  anterior, S–173, S–281
  posterior, S–173, S–281
Cruciform ligaments, longitudinal band of
  inferior, S–BP 21
  superior, S–BP 21
Crura, of respiratory diaphragm, S–545
Crural artery
  anterior, S–95
  posterior, S–95
Crural fascia, S–266, S–335, S–347, S–555
Crus cerebri, S–50
Cubital fossa, S–10
Cubital nodes, S–343, S–346
Cubital vein, median, S–228, S–332, S–346, T5.4
Cuboid bones, S–176, S–177, S–179, S–180, S–181, S–297, S–298, S–556
  articular surface for, S–178
  tuberosity of, S–181
Cuboideonavicular ligament
  dorsal, S–180
  plantar, S–181
Cumulus oöphorus, S–BP 96
Cuneate fasciculus, S–BP 16, S–BP 17
Cuneate nucleus, S–BP 17
Cuneate tubercle, S–40, S–41
Cuneiform bones, S–176, S–177, S–297, S–298
  intermediate, S–297
  lateral, S–297
  medial, S–180, S–297, S–BP 111
Cuneiform tubercle, S–373, S–376, S–409
Cuneocuboid ligament, dorsal, S–180
Cuneonavicular ligaments
  dorsal, S–180
  plantar, S–181
Cuneus, S–32
  apex of, S–33
Cutaneous nerves, S–4
  antebrachial, medial, S–106
  brachial
    medial, S–102, S–106
    superior lateral, S–198
  dorsal
    intermediate, S–119, S–BP 43
    lateral, S–117, S–118, S–119, S–291, S–292, S–BP 43
    medial, S–119, S–BP 43
  femoral
    anterior, S–115
    lateral, S–103, S–104, S–113, S–115, S–116, S–218, S–266, S–267, S–270, S–271, S–277, S–347, S–518, S–554, T2.12
    posterior, S–104, S–114, S–117, S–267, S–274, S–275, S–516, S–554, T2.13
  genital branch of, S–104
  gluteal
    inferior, S–267, S–274, S–275
    medial, S–267, S–274
    superior, S–267, S–274
  of head, S–185

Cutaneous nerves (Continued)
  of neck, S–185
  perforating, S–104, S–114, S–267, S–275, S–516, S–520, T2.12
  posterior femoral, S–16
  sensory branches of, S–4
  sural
    lateral, S–117, S–118, S–119, S–266, S–267, S–274, S–285, S–286, T2.13
    medial, S–117, S–118, S–267, S–274, S–285, S–286, T2.13, S–BP 109
Cuticle, S–4, S–BP 112
Cystic artery, S–432, S–435, S–436, T5.3, S–BP 75, S–BP 77
  variations in, S–BP 75
Cystic duct, S–429, S–432, S–433, S–435, S–436, S–BP 100
  variations in, S–BP 74
Cystic triangle, S–BP 75
Cystic vein, S–BP 78
Cystohepatic triangle (of Calot), S–432, S–436
Cystohepatic trigone (of Calot), S–BP 55
Cystourethrograms, S–BP 87

D

Dartos fascia
  of penis, S–495, S–497, S–500, S–501, S–502
  of scrotum, S–211, S–212, S–213, S–224, S–495, S–497, S–498, S–500, S–512, S–BP 90
Deciduous teeth, S–404
Deep artery
  of clitoris, S–494, S–511
  of penis, S–501, S–504, S–512
Deep cervical artery, S–23
Deep cervical lymph nodes, inferior, S–350, S–351
Deep cervical vein, S–BP 27
Deep (Buck's) fascia, S–502
  intercavernous septum of, S–503
  of penis, S–211, S–212, S–213, S–222, S–224, S–495, S–500, S–501, S–502, S–504, S–512, S–BP 90, S–BP 94
  of thigh, S–500, S–501
Deep investing cervical fascia, S–400, S–408
Deep venous palmar arch, S–332
Deferential plexus, S–518
Deltoid muscle, S–6, S–7, S–10, S–111, S–157, S–190, S–192, S–198, S–199, S–227, S–230, S–231, S–233, S–235, S–236, S–237, S–238, S–239, S–240, S–264, S–534, S–535, S–536, S–537, S–551
Deltoid region, S–10
Deltoid tuberosity, S–155
Deltopectoral groove, S–10
Deltopectoral triangle, S–230
Dendrite, S–BP 3
Denonvilliers' fascia, S–495
Dens, S–142, S–BP 27
  abnormalities of, S–BP 21
  apical ligament of, S–142, S–372, S–408, S–BP 21
  of atlas, S–140
  of axis, S–124, S–126, S–139, S–140
  posterior articular facet of, S–142
Dens axis, S–121, S–BP 22, S–BP 23
Dental plexus, inferior, S–59
Dental pulp, S–405
Dentate gyrus, S–32, S–34, S–37, S–38, S–56
Dentate line, S–226
Dentate nucleus, S–41
Denticulate ligament, S–17, S–22
Dentin, S–405
Depressor anguli oris muscle, S–65, S–184, S–186, S–BP 35, S–BP 36
Depressor labii inferioris muscle, S–65, S–184, S–186, S–BP 35, S–BP 36
Depressor septi nasi muscle, S–60, S–65, S–122, S–184, S–BP 35
Dermatomes, S–19

Dermatomes (Continued)
  of lower limb, S–12
  principal, levels of, S–19
  of upper limb, S–12
Dermis, S–4
Descemet's membrane, S–89
Descending aorta, S–207, S–303, S–358, S–410. See also Thoracic aorta
Descending colon, S–420, S–424, S–428, S–459, S–461, S–483, S–484, S–517
Descending fibers, S–470, S–471
Descending limb, S–BP 84
Desmosome, S–BP 45
Detrusor vesicae, S–BP 86
Diagonal conjugate, S–477
Diaphragm, S–101, S–113, S–205, S–217, S–219, S–307, S–321, S–336, S–339, S–340, S–342, S–352, S–358, S–387, S–390, S–390, S–391, S–393, S–410, S–411, S–413, S–415, S–416, S–417, S–429, S–434, S–456, S–459, S–460, S–517, S–534, S–541, T4.2, S–BP 2, S–BP 28, S–BP 64, S–BP 70, S–BP 98, S–BP 99
  abdominal surface of, S–208
  central tendon of, S–208, S–219, S–418
  costal part of, S–205, S–208
  crura of, S–545
    left, S–208, S–219, S–393
    right, S–208, S–219, S–393, S–BP 98
  crus of, S–461, S–BP 25
  dome of
    left, S–385, S–386
    right, S–385, S–386
  esophageal hiatus of, S–103
  left crus of, S–410, S–422
  lumbar part of, S–208
  right crus of, S–410, S–422
  as site of referred visceral pain, S–BP 5
  sternal part of, S–205, S–208
  thoracic surface of, S–207
Diaphragmatic constriction, S–411
Diaphragmatic fascia, S–413, S–461
Diaphragmatic grooves, S–BP 71
Diaphragmatic nodes
  inferior, S–352
    lymph vessels to, S–348
  superior, S–348
Diaphragmatic recess, right, S–321
Digastric fossa, S–135
Digastric muscle
  anterior belly of, S–62, S–190, S–191, S–192, S–197, S–324, S–368, S–396, S–399, S–407, S–BP 61
  intermediate tendon of, S–399, S–400, S–401
  phantom, S–324
  posterior belly of, S–62, S–65, S–189, S–190, S–191, S–192, S–324, S–345, S–396, S–399, S–400, S–401, S–407, S–409
  trochlea for intermediate tendon of, S–190, S–191
Digestive system, S–394, S–394 to S–457, T8.1, S–BP 65 to S–BP 82
Digestive tract, S–522
Digital arteries
  dorsal, S–261, S–289, S–291, S–292, S–297, S–333, S–553, S–BP 43, S–BP 53
  palmar, S–250, S–BP 51
    common, S–251, S–254, S–257, S–BP 51
    proper, S–251, S–257, S–261, S–553, S–BP 51
  plantar, branches of, proper, S–289, S–291, S–292
Digital fibrous sheaths, S–169
Digital nerve
  dorsal, S–229, S–258, S–266
    of foot, S–119
  palmar, S–229, S–250
    common, S–109, S–110, S–257
    proper, S–109, S–110, S–257, S–258, S–553

Netter Atlas of Human Anatomy: A Systems Approach

Erector spinae muscle, S–6, S–100, S–199, S–201, S–206, S–209, S–460, S–535, S–545, S–546, S–BP 102
Esophageal hiatus, S–208, S–393, T4.2
Esophageal impression, S–429
Esophageal mucosa, S–374
Esophageal muscle, S–372
longitudinal, S–526
Esophageal plexus, S–68, S–101, S–389, S–390, S–391, S–410, S–447, S–539, S–BP 13, S–BP 14
Esophageal submucosa, S–374
Esophageal vein, S–340, S–446, T5.3, S–BP 78
Esophagogastric junction, S–413, S–541
cross-sectional anatomy, S–541
Esophagus, S–68, S–72, S–196, S–197, S–207, S–219, S–308, S–329, S–340, S–342, S–352, S–358, S–372, S–373, S–376, S–383, S–389, S–390, S–391, S–393, S–394, S–407, S–418, S–419, S–422, S–447, S–452, S–459, S–527, S–531, S–534, S–535, S–536, S–537, S–538, S–539, S–540, S–BP 28, S–BP 98
abdominal part of, S–339, S–410, S–411, S–415
arteries of, S–339
variations, S–BP 66
autonomic innervation of, S–450
bare area on ventral surface of, S–412
cervical part of, S–339, S–410, S–411
circular muscular layer of, S–374, S–408, S–409, S–412, S–413, S–526
constrictions and relations, S–411
impression of, S–308, S–388
longitudinal muscular layer of, S–374, S–379, S–408, S–409, S–412, S–413, S–414
lung groove for
left, S–388
right, S–388
lymph vessels and nodes of, S–351
mucosa of, S–413
nerves of, S–447
variations in, intrinsic nerves and, S–BP 65
recurrent branch of, S–208
in situ, S–410
submucosa of, S–413
thoracic part of, S–339, S–410, S–411
veins of, S–340
Estrogen, S–BP 95
Ethmoid artery
anterior lateral nasal branch of, S–BP 59
anterior septal branch of, S–BP 59
external nasal branch of, S–BP 59
Ethmoid bone, S–123, S–125, S–129, S–132, S–360, S–362, S–367
anterior ethmoidal foramen of, S–127
cribriform of, S–56, S–129, S–132, S–360, S–361, S–362
crista galli of, S–129, S–132, S–360
nasal concha of
highest, S–362
middle, S–123, S–129, S–362
superior, S–129, S–362
of newborn, S–127
orbital of, S–123, S–125, S–127
perpendicular of, S–123, S–129, S–360
uncinate process of, S–362, S–369, S–370
Ethmoid sinus, imaging of, S–39
Ethmoidal air cells, S–368, S–BP 7, S–BP 61
anterior, S–86
middle, opening of, S–361, S–362
posterior, openings of, S–362
Ethmoidal artery
anterior, S–87, S–134, S–363
anterior meningeal branch of, S–27
lateral nasal branch of, S–363
posterior, S–27, S–87, S–134, S–363
Ethmoidal bulla, S–361, S–362
Ethmoidal cells, S–124, S–367, S–369

Ethmoidal foramen
anterior, S–134
posterior, S–134
Ethmoidal nerve
anterior, S–58, S–59, S–61, S–85, S–86, S–134, S–365
external nasal branch of, S–59, S–61, S–185, S–364
internal nasal branches of, S–59
lateral internal nasal branch of, S–364
medial internal nasal branch of, S–86, S–364
posterior, S–58, S–59, S–61, S–85, S–86, S–134
Ethmoidal vein
anterior, S–134, S–363
posterior, S–134, S–363
Eversion, of lower limbs movements, S–265
Extension
of lower limbs movements, S–265
of upper limbs movements, S–265
Extensor carpi radialis brevis muscle, S–10, S–111, S–112, S–240, S–241, S–242, S–243, S–263, S–552
Extensor carpi radialis brevis tendon, S–241, S–242, S–259, S–260, S–261, S–552
Extensor carpi radialis longus muscle, S–10, S–111, S–112, S–240, S–241, S–242, S–243, S–262, S–263, S–264, S–551, S–552
Extensor carpi radialis longus tendon, S–241, S–242, S–259, S–260, S–261, S–552
Extensor carpi ulnaris muscle, S–10, S–111, S–112, S–240, S–242, S–243, S–262, S–263, S–552
Extensor carpi ulnaris tendon, S–241, S–242, S–260, S–261, S–552
Extensor digiti minimi muscle, S–112, S–242, S–243, S–263, S–552
Extensor digiti minimi tendon, S–241, S–242, S–243, S–260, S–261, S–552
Extensor digitorum brevis muscle, S–119, S–287, S–289, S–290, S–291, S–292, S–556, S–BP 111
Extensor digitorum brevis tendon, S–288, S–289, S–292, S–297, S–556, S–BP 111
Extensor digitorum longus muscle, S–119, S–272, S–280, S–288, S–289, S–290, S–291, S–292, S–299, S–555, S–BP 109, S–BP 110
Extensor digitorum longus tendon, S–11, S–287, S–288, S–289, S–291, S–292, S–297, S–556, S–BP 110, S–BP 111
Extensor digitorum muscle, S–111, S–112, S–240, S–242, S–243, S–263, S–552
Extensor digitorum tendon, S–10, S–241, S–242, S–243, S–260, S–261, S–552, S–BP 106, S–BP 107
Extensor expansion, S–255, S–260, S–BP 106, S–BP 107
Extensor hallucis brevis muscle, S–119, S–288, S–289, S–291, S–292, S–BP 111
Extensor hallucis brevis tendon, S–289, S–292, S–297, S–556, S–BP 111
Extensor hallucis longus muscle, S–119, S–287, S–288, S–289, S–292, S–299, S–555, S–BP 109, S–BP 110
Extensor hallucis longus tendon, S–11, S–287, S–288, S–289, S–291, S–292, S–297, S–556, S–BP 110, S–BP 111
tendinous sheath of, S–290, S–291
Extensor indicis muscle, S–112, S–241, S–243, S–263, S–552
Extensor indicis tendon, S–10, S–242, S–243, S–260, S–261, S–552, S–BP 106, S–BP 107
Extensor pollicis brevis muscle, S–112, S–241, S–242, S–243, S–263
Extensor pollicis brevis tendon, S–242, S–249, S–259, S–260, S–261, S–552
Extensor pollicis longus muscle, S–112, S–241, S–243, S–263, S–552

Extensor pollicis longus tendon, S–10, S–242, S–254, S–259, S–260, S–261, S–552
Extensor retinaculum, S–229, S–241, S–242, S–260, S–261
inferior, S–119, S–288, S–290, S–291
superior, S–288, S–290, S–291, S–BP 110
of wrist, S–258, S–259, S–BP 42
Extensor tendon, S–254, S–255
common, S–241, S–242, S–243, S–262, S–264
long, S–255
External abdominal oblique muscle, S–272
External acoustic meatus, T2.1
External anal sphincter, S–418, S–428
deep part, S–418
subcutaneous part, S–418
superficial part, S–418
External capsule, S–36
imaging of, S–39
External carotid artery, S–101, S–396, S–399, S–400, S–401
External genitalia, S–491, S–500, S–501
homologues of, S–506
nerves of
female, S–516
male, S–518
External nasal vein, S–363
External occipital protuberance, S–6, S–125, S–126, S–129, S–130, S–131
External os, S–487
External urethral orifice, S–8, S–9
External vertebral venous plexus, anterior, S–BP 27
Extrahepatic bile ducts, S–432
Extraocular muscles, T4.1
Extraperitoneal adipose tissue, S–545
Extraperitoneal fascia, S–213, S–214, S–215, S–218, S–513, S–BP 41
Extreme capsule, S–36
Extrinsic muscle, of eye, S–84
innervation of, S–84
Eye, S–2, S–20, S–21
anterior chamber of, S–82, S–88, S–90, S–91
extrinsic muscles of, S–84, S–91
innervation of, S–84
intrinsic arteries and veins of, S–91
posterior chamber of, S–82, S–88, S–90, S–91
vascular supply of, S–92
Eyeball, S–368, S–BP 61
anterior segment of, T2.1
fasciae of, S–BP 7
imaging of, S–39
posterior chamber of, S–89
transverse section of, S–88
Eyelashes, S–82
Eyelids, S–82, S–BP 18
arteries and veins of, S–87
tarsus of, S–BP 7

F

Face, S–2
arterial supply sources of, S–322
arteries of, S–60
muscle of, S–60
musculature of, S–BP 36
nerves of, S–60
superficial arteries and veins of, S–322
Facial artery, S–43, S–60, S–72, S–77, S–79, S–87, S–122, S–188, S–189, S–322, S–323, S–324, S–325, S–326, S–328, S–396, S–399, S–400, S–526, T5.1, S–BP 60
alar branches of lateral nasal branch of, S–363
lateral nasal branch of, S–60, S–BP 59
pulse point, S–303
right, S–188
tonsillar branch of, S–323, S–403
transverse, S–87, S–122, S–322, S–323, S–324, S–325

Netter Atlas of Human Anatomy: A Systems Approach

Frontal bone *(Continued)*
  orbital part of, superior surface of, S–132
  orbital surface of, S–123
  sinus of, S–129, S–360, S–362, S–370
  squamous part of, S–127, S–360, S–362
  superior sagittal sinus of, groove for, S–132
  supraorbital notch of, S–123, S–125, S–127
Frontal crest, S–128, S–132, S–360
Frontal gyrus
  inferior
    opercular part of, S–31
    orbital part of, S–31
    triangular part of, S–31
  medial, S–32
  middle, S–31
  superior, S–31
Frontal lobe, S–523
Frontal nerve, S–58, S–59, S–61, S–77, S–85,
    S–86, S–BP 7
Frontal pole, of cerebrum, S–31, S–33
Frontal region, of head and neck, S–5
Frontal sinus, S–124, S–126, S–358, S–360,
    S–361, S–362, S–368, S–369, S–372, S–BP 61
Frontal sulcus
  inferior, S–31
  superior, S–31
Frontal (metopic) suture, S–127, S–BP 19
Frontalis muscle, S–60, S–65, S–184, S–BP 35,
    S–BP 36
Frontobasal artery
  lateral, S–44
  medial, S–44, S–327
Frontonasal canal, opening of, S–362
Frontonasal duct, opening of, S–369
Frontopolar artery, S–46, S–47
Frontozygomatic suture, S–124, S–BP 19
Fundiform ligament, S–211
  of penis, S–216, S–495
Fundus
  of bladder, S–465
  of stomach, S–411, S–413, S–415, S–541,
    S–BP 99
  of urinary bladder, S–495
  of uterus, S–464, S–483, S–485, S–521
Fungiform papillae, S–81, S–402

G

Gallaudet's fascia, S–221, S–222, S–495,
    S–501, S–502, S–504
Gallbladder, S–20, S–21, S–68, S–385, S–394,
    S–415, S–416, S–423, S–427, S–429, S–430,
    S–432, S–436, S–456, S–542, S–548, T8.1,
    S–BP 69, S–BP 70, S–BP 101
  arterial variations, S–BP 77
  body, S–432
  collateral supply, S–BP 77
  fundus, S–432
  infundibulum, S–432
  neck of, S–432
  as site of referred visceral pain, S–BP 5
Ganglion
  inferior, S–67, S–80
  superior, S–67
  of trigeminal nerve, S–77
Ganglion cells, retinal, S–57
Gastric artery
  esophageal branch of, S–435
  left, S–339, S–416, S–435, S–436, S–437,
    S–450, S–517, S–528, S–BP 75, S–BP 76,
    S–BP 77
    esophageal branch of, S–339, S–528
  right, S–421, S–432, S–435, S–436, S–450,
    S–528, S–BP 75, S–BP 77
  short, S–419, S–434, S–435, S–436, S–450,
    S–528, S–BP 77
Gastric canal, longitudinal folds of, S–414
Gastric folds (rugae), S–413, S–414, S–541
Gastric glands, S–BP 67

Gastric impression, S–429, S–434
Gastric plexus, S–517
  left, S–105, S–448, S–449
  right, S–448, S–449
Gastric veins
  left, S–340, S–442, S–443, S–444, S–446,
    S–BP 78
    esophageal tributary of, S–340
  right, S–340, S–442, S–443, S–446, S–BP 78
  short, S–340, S–419, S–434, S–442, S–446
Gastrocnemius muscle, S–11, S–117, S–118,
    S–273, S–279, S–280, S–284, S–285, S–286,
    S–287, S–289, S–299, S–335, S–555, S–BP 109
  lateral head of, S–11, S–272, S–274, S–282,
    S–283, S–301
  lateral subtendinous bursa of, S–282, S–283
  medial head of, S–11, S–274, S–282, S–283,
    S–288, S–301
  medial subtendinous bursa of, S–282, S–283
Gastrocnemius tendon, S–BP 110
Gastrocolic ligament, S–416
  anterior layers, S–416
  posterior layers, S–416
Gastroduodenal artery, S–421, S–432, S–435,
    S–436, S–438, S–439, S–450, S–528, S–BP
    75, S–BP 76, S–BP 77
Gastroduodenal plexus, S–456
Gastroesophageal junction, S–393, S–414, T8.1
Gastrointestinal tract, S–BP 15
Gastroomental arterial anastomosis, S–416
Gastroomental artery
  left, S–434, S–435, S–436, S–450, S–BP 77
  right, S–416, S–435, S–436, S–438, S–439,
    S–450, S–528, S–BP 77
Gastroomental plexus
  left, S–448, S–449
  right, S–448, S–449
Gastroomental vein
  left, S–340, S–434, S–442, S–446, S–BP 78
  right, S–340, S–442, S–444, S–446, S–BP 78
Gastrophrenic ligament, S–416, S–419, S–459
Gastrophrenic trunk, S–BP 79
Gastrosplenic ligament, S–416, S–417, S–434
Gelatinous substance, S–BP 16
Gemellus muscle
  inferior, S–273, S–274, S–275, S–276, S–300,
    S–301, S–516, S–549, S–550, S–BP 104
  superior, S–273, S–274, S–275, S–276,
    S–300, S–301, S–516, S–BP 108
Gemellus tendon, superior, S–BP 108
Genicular artery
  descending, S–271, S–333, S–554, S–BP 52,
    S–BP 53
    articular branch of, S–270, S–BP 53
    saphenous branch of, S–270, S–BP 53
  lateral
    inferior, S–286, S–287, S–288, S–289,
      S–333, S–BP 52, S–BP 53
    superior, S–274, S–285, S–286, S–287,
      S–288, S–289, S–333, S–BP 52, S–BP
      53
  medial
    inferior, S–270, S–271, S–285, S–286,
      S–288, S–289, S–333, S–BP 52, S–BP 53
    superior, S–270, S–271, S–274, S–284,
      S–285, S–286, S–288, S–289, S–333,
      S–BP 52, S–BP 53
  middle, S–333, S–BP 52, S–BP 53
Genicular vein, S–334
Geniculate body
  lateral, S–33, S–36, S–37, S–40, S–41, S–44,
    S–48, S–49, S–50, S–57, S–78
  medial, S–33, S–36, S–37, S–40, S–41, S–44,
    S–48, S–49, S–50
Geniculate ganglion, S–54, S–62, S–63, S–65,
    S–66, S–67, S–81, S–95, S–98, S–366
Geniculate nucleus, lateral, S–54
Genioglossus muscle, S–70, S–360, S–368,
    S–372, S–399, S–400, S–401, T4.1, S–BP 61

Geniohyoid fascia, S–329
Geniohyoid muscle, S–70, S–74, S–196, S–329,
    S–368, S–372, S–399, S–401, S–408,
    S–BP 61
Genital cord, S–505
Genital organs, female, internal, S–529
Genital tubercle, S–506
Genitalia
  external, S–20, S–21, S–500, S–501, S–506
    female, S–516
    homologues of, S–506
    male, S–518
    nerves of, S–518
  internal, homologues of, S–505
  lymph nodes of
    female, S–355
    male, S–357
  lymph vessels of
    female, S–355
    male, S–357
Genitofemoral nerve, S–103, S–104, S–113,
    S–116, S–218, S–277, S–459, S–463, S–518,
    S–546, T2.2, T2.12
  femoral branches of, S–102, S–103, S–104,
    S–218, S–266, S–518
  genital branch of, S–102, S–103, S–104,
    S–215, S–216, S–218, S–266, S–497,
    S–518
Gerdy's tubercle, S–173, S–174, S–175,
    S–BP 32
  iliotibial tract to, S–173
Gerota's fascia. *See* Renal (Gerota's) fascia
Gimbernat's ligament. *See* Lacunar ligament
    (Gimbernat's)
Gingiva, lamina propria of, S–405
Gingival epithelium, S–405
Gingival papilla, S–405
Gingival sulcus, epithelium of, S–405
Glabella, S–5, S–123, S–125
Gland orifices, S–432
Glans, S–506
  preputial fold of, S–506
Glans penis, S–8, S–9, S–502, S–503, S–506
  corona, S–502
  neck, S–502
Glassy membrane, S–4
Glenohumeral joint, S–233
  capsule of, S–235, S–240
Glenohumeral ligament
  inferior, S–157
  middle, S–157
  superior, S–157
Glenoid fossa, S–233
  of scapula, S–149, S–157, S–158
Glenoid labrum, S–157, S–233
Glial process, S–BP 3
Globus pallidus, S–34
Glomerular arteriole
  afferent, S–BP 83, S–BP 85
  efferent, S–BP 83, S–BP 85
Glomerular capsule (Bowman's), S–BP 83
  parietal layer, basement membrane of,
    S–BP 83
Glomerulus
  basement membrane of, S–BP 83
  superficial, S–BP 85
Glossoepiglottic ligament, median, S–376
Glossopharyngeal nerve (CN IX), S–21, S–30,
    S–40, S–54, S–55, S–65, S–68, S–71, S–72,
    S–73, S–76, S–77, S–79, S–80, S–81, S–323,
    S–324, S–384, S–403, S–406, T2.5, S–BP 6,
    S–BP 14, S–BP 15, S–BP 60
  carotid sinus nerve of, S–324
  communicating branch of, S–67
  distribution of, S–53
  inferior (petrosal) ganglion of, S–67, S–71,
    S–81
  in jugular foramen, S–134
  in jugular fossa, S–133

Glossopharyngeal nerve (CN IX) *(Continued)*
  lingual branch of, S–81
  rootlets of, S–40
  schema of, S–67
  superior ganglion of, S–67
  tonsillar branch of, S–403
  tympanic branch of, S–133
Gluteal aponeurosis, S–272, S–273, S–274,
  S–460
  over gluteus medius muscle, S–210
Gluteal artery
  inferior, S–114, S–274, S–341, S–441, S–468,
    S–507, S–509, S–BP 94
  superior, S–114, S–274, S–341, S–441,
    S–468, S–478, S–507, S–509, S–BP 94
Gluteal cutaneous nerve
  inferior, S–198
  middle, S–198
  superior, S–198
Gluteal fold, S–6, S–11
Gluteal lines
  anterior, S–153, S–154
  inferior, S–153, S–154
  posterior, S–153, S–154
Gluteal muscle, superficial, S–183
Gluteal nerves
  inferior, S–16, S–104, S–114, S–274, S–275,
    T2.12
  superior, S–16, S–104, S–114, S–274, S–275,
    T2.12
Gluteal region, S–19
Gluteal vein
  inferior, S–342
  superior, S–342, S–444
Gluteus maximus muscle, S–6, S–11, S–182,
  S–198, S–199, S–210, S–222, S–224, S–272,
  S–273, S–274, S–275, S–276, S–301, S–460,
  S–482, S–493, S–501, S–502, S–516, S–517,
  S–520, S–549, S–550, S–554, S–BP 103,
  S–BP 104, S–BP 108
  sciatic bursa of, S–276
  trochanteric bursa of, S–276
    inferior extension of, S–276
Gluteus maximus tendon, S–BP 108
Gluteus medius muscle, S–6, S–11, S–268,
  S–270, S–271, S–273, S–274, S–275, S–276,
  S–277, S–301, S–516, S–549, S–550, T4.3,
  S–BP 103, S–BP 108
  trochanteric bursae of, S–276
Gluteus medius tendon, S–276
Gluteus minimus muscle, S–270, S–271,
  S–273, S–274, S–275, S–276, S–277,
  S–300, S–301, T4.3, S–BP 103,
  S–BP 108
  trochanteric bursa of, S–276
Gluteus minimus tendon, S–276, S–549,
  S–BP 104
GnRH, S–BP 97
Gonadotropic hormone, S–BP 95
Gonads, S–499, S–505
Gracile fasciculus, S–40, S–41, S–BP 16,
  S–BP 17
Gracile nucleus, S–BP 17
Gracile tubercle, S–40, S–41
Gracilis muscle, S–116, S–268, S–269, S–270,
  S–271, S–273, S–274, S–279, S–283, S–284,
  S–300, S–516, S–554
Gracilis tendon, S–11, S–268, S–269, S–279,
  S–280
Granular foveolae, S–26, S–28, S–128
Granule cell, S–56
Granulosa, S–BP 96
Gray matter, S–17, S–BP 16
  imaging of, S–39, S–BP 12
Gray rami communicantes, S–20, S–76, S–78,
  S–79, S–80, S–100, S–101, S–103, S–105,
  S–113, S–114, S–206, S–209, S–320, S–390,
  S–391, S–513, S–515, S–517, S–519, S–BP
  13, S–BP 15, S–BP 46

Gray ramus communicans, S–17, S–20, S–22,
  S–447, S–450, S–452, S–454, S–469, S–470,
  S–471
Great auricular nerve, S–185, S–195, S–202,
  T2.8
Great cardiac vein, S–317
Great cerebral vein (of Galen), S–29, S–30,
  S–32, S–35, S–41, S–49, S–50, S–51
  imaging of, S–BP 11
Great saphenous vein, S–304
Great vessels, of mediastinum, S–389
Greater curvature, S–415
Greater occipital nerve, S–202
Greater omentum, S–415, S–417, S–418,
  S–423, S–424, S–428, S–430
Greater palatine artery, S–363, S–398
Greater palatine foramen, S–398, S–404
Greater palatine nerve, S–365, S–398, S–406
Greater petrosal nerve, S–66
  middle, petrosal branch of, S–95
Greater thoracic splanchnic nerve, S–20,
  S–101, S–207, S–208, S–390, S–391, S–539,
  S–BP 13
  esophageal branch of, S–101
Greater trochanter, S–BP 104
Greater tubercle, S–155, S–158
  crest of, S–155
Ground substance, S–BP 30
Gubernaculum, S–499
Gustatory glands, S–402
Gyrus, straight, S–33

H

Habenula, S–36
Habenular commissure, S–32, S–37, S–41
Habenular trigone, S–37, S–41
Hair bulb, S–BP 1
Hair cells
  inner, S–97
  outer, S–97
Hair cortex, S–BP 1
Hair cuticle, S–4, S–BP 1
Hair follicle, S–4
  dermal papilla of, S–4, S–BP 1
  lower germinal matrix of, S–4
Hair line, S–BP 97
Hair medulla, S–BP 1
Hair shaft, S–4, S–BP 1
Hamate bone, S–163, S–164, S–166, S–167,
  S–253, S–BP 29
Hand, S–2, S–3, S–19, S–183
  arteries of, S–257, S–261, S–BP 51
  axial view, cross section, S–BP 106, S–BP
    107
  bones of, S–167
  central band of extensor expansion of,
    S–553
  cross section of, S–BP 107
  cutaneous innervation of, S–14
  dorsal fascia of, S–254
  dorsum, lymph vessels to, S–346
  extensor zones of, S–BP 42
  flexor zones of, S–BP 42
  muscles of, S–256
  nerves of, S–261
  radial nerve in, S–112
  radiographs of, S–168
  superficial dorsal dissection of, S–258
  superficial palmar dissections of, S–250
  superficial veins of, S–229
  tendon sheaths of, S–254
Hard palate, S–372, S–403, S–523, S–BP 61
  imaging of, S–39
  posterior nasal spine of, S–BP 21
Haustra of colon, S–425, S–428
Head, S–2, S–3
  arteries of, S–323
  autonomic nerves in, S–77

Head *(Continued)*
  bony framework of, S–359
  cutaneous nerves of, S–185
  lymph and nodes of, S–344
  radiate ligament of, S–BP 26
  structures with high clinical significance,
    T4.1, T5.1, T6.1
  superior articular facet on, S–BP 26
  surface anatomy of, S–5
Heart, S–20, S–21, S–68, S–302, S–411, S–522,
  S–BP 15
  anterior exposure of, S–309
  apex of, S–306, S–309, S–310, S–538
  auscultation of, S–306
  base and diaphragmatic surface of, S–310
  conducting system of, S–319
  CT angiograms, S–321
  in diastole, S–322
  inferior border of, S–309
  innervation of, S–320
  left auricle of, S–306
  left border of, S–309, S–385
  radiographs of, S–321
  right auricle of, S–306
  right border of, S–385
  in situ, S–307
  valvular complex of, S–314, S–315
Heart valves, S–313, T5.2
  aortic, S–306, S–312, S–314, S–315, S–316,
    S–319, S–534, T5.2, S–BP 99
  mitral, S–306, S–312, S–314, S–315, S–316,
    S–319, S–534, S–539
  pulmonic, S–306
  tricuspid, S–306, S–311, S–314, S–315,
    S–316, S–534, S–539
Helicotrema, of cochlea, S–93, S–96, S–97,
  S–BP 9
Helix, S–5, S–94
  crux of, S–94
Hemiazygos vein, S–207, S–338, S–340, S–539
  accessory, S–338, S–340, S–383, S–391,
    S–537
Henle layer, S–BP 1
Hepatic artery
  common, S–416, S–421, S–432, S–433,
    S–435, S–436, S–437, S–438, S–439,
    S–450, S–456, S–528, S–BP 75, S–BP 76,
    S–BP 77, S–BP 100
  intermediate, S–436
  left, S–432, S–435, S–BP 75, S–BP 76, S–BP 77
  middle, S–BP 77
  nerve branches on, S–BP 69
  proper, S–417, S–418, S–419, S–420, S–421,
    S–429, S–430, S–432, S–433, S–435,
    S–436, S–437, S–438, S–450, S–528,
    S–542, S–BP 75
  right, S–432, S–435, S–436, S–BP 75, S–BP
    76, S–BP 77
  variations in, S–BP 76
Hepatic duct
  accessory, S–BP 74
  common, S–429, S–430, S–432, S–436, S–BP
    69, S–BP 75
  left, S–432, S–433
  right, S–432, S–433
  variations in, S–BP 74
Hepatic flexure, S–423
Hepatic nodes, S–BP 54, S–BP 55, S–BP 58
  lymph vessels to, S–348
Hepatic plexus, S–68, S–105, S–448, S–449,
  S–451, S–456
Hepatic portal vein, S–302, S–305, S–340,
  S–415, S–417, S–418, S–420, S–421, S–429,
  S–430, S–432, S–433, S–435, S–442, S–443,
  S–444, S–446, S–534, S–547, T5.3, S–BP
  55, S–BP 78, S–BP 99
  portacaval anastomoses, S–446
  tributaries, S–446
  variations and anomalies of, S–BP 78

Netter Atlas of Human Anatomy: A Systems Approach

Iliac fossa, S–152, S–154, S–483. *See also*
    Ilium, ala of
Iliac lymph nodes, S–343
    common, S–352, S–355, S–357, S–BP 57
    external, S–347, S–352, S–354, S–357, S–BP 57
        inferior, S–355
        medial, S–355
        superior, S–355
    internal, S–352, S–354, S–355, S–357
Iliac plexus
    common, S–105, S–469, S–513
    external, S–105, S–513
    internal, S–105, S–513
Iliac spine, S–481
    anterior
        inferior, S–151, S–152, S–153, S–154,
            S–170, S–219, S–268, S–475, S–481
        superior, S–8, S–9, S–11, S–151, S–152,
            S–153, S–154, S–170, S–211, S–212,
            S–215, S–216, S–219, S–268, S–270,
            S–272, S–277, S–475, S–492, S–500,
            S–BP 40
    posterior
        inferior, S–146, S–153, S–154
        superior, S–6, S–146, S–152, S–153,
            S–154
Iliac tubercle, S–152
Iliac tuberosity, S–151, S–153, S–475
Iliac vein
    common, S–304, S–342, S–424, S–445,
        S–478, S–510, S–BP 103
    deep circumflex, S–342, S–490
        tributaries to, S–337
    external, S–271, S–304, S–334, S–342,
        S–425, S–444, S–445, S–478, S–483,
        S–484, S–485, S–486, S–488, S–489,
        S–490, S–509, S–510, S–529, S–BP 96,
        S–BP 108
        left, S–509
    internal, S–304, S–342, S–444, S–445, S–478,
        S–490, S–510, T5.4
        right, S–509
    superficial circumflex, S–266, S–334, S–337,
        S–342
Iliac vessels
    common, S–513, S–BP 94
    deep circumflex, S–217, S–218, S–BP 94
    external, S–216, S–217, S–218, S–225, S–495,
        S–BP 94
        covered by peritoneum, S–215
    internal, S–BP 94
    superficial circumflex, S–211, S–BP 40
Iliacus fascia, S–225
Iliacus muscle, S–113, S–114, S–115, S–216,
    S–219, S–225, S–268, S–276, S–277,
    S–278, S–300, S–459, S–490, S–549,
    S–BP 104
    branches to, S–104
    muscular branches to, S–113
    nerve to, S–104
Iliococcygeus muscle, S–220, S–224, S–479,
    S–480, S–481, S–482, S–488
Iliocostalis colli muscle, S–200
Iliocostalis lumborum muscle, S–200
Iliocostalis muscle, S–200
Iliocostalis thoracis muscle, S–200
Iliofemoral ligament, S–170, S–276, S–BP 108
    of hip joint, S–277
Iliohypogastric nerve, S–16, S–103, S–104,
    S–113, S–116, S–198, S–277, S–459, S–460,
    S–514, S–518, T2.2
    anterior branch of, S–104
    anterior cutaneous branch of, S–102, S–103,
        S–518
    lateral cutaneous branch of, S–102, S–104,
        S–210, S–267
Ilioinguinal nerve, S–16, S–102, S–103, S–104,
    S–113, S–116, S–215, S–277, S–459, S–460,
    S–514, S–518, T2.2

Ilioinguinal nerve *(Continued)*
    anterior scrotal branches of, S–102, S–103,
        S–518
    labial branch of, anterior, S–516
    scrotal branches of, S–266
Iliolumbar artery, S–341, S–468, S–509, S–BP 94
Iliolumbar ligament, S–146, S–152, S–460
Iliolumbar vein, S–342
Iliopectineal arch, S–276
Iliopectineal bursa, S–170, S–276, S–277
Iliopectineal line, S–225
Iliopsoas muscle, S–217, S–268, S–269, S–270,
    S–271, S–276, S–277, S–300, S–549, S–550,
    S–554, S–BP 104, S–BP 108
Iliopsoas tendon, S–278
Iliopubic eminence, S–151, S–152, S–153,
    S–154, S–170, S–475, S–481
Iliopubic tract, S–216, S–484, S–490
Iliotibial tract, S–11, S–175, S–268, S–272,
    S–273, S–274, S–276, S–279, S–280, S–281,
    S–283, S–284, S–287, S–288, S–289, S–299,
    S–300, S–516, S–554, T4.3
    bursa deep to, S–280
    bursa of, S–279
    to Gerdy's tubercle, S–173
Ilium, S–171, S–476
    ala of, S–151, S–153, S–154, S–182, S–475,
        S–481
    arcuate line of, S–475, S–479, S–481
    auricular surface for, S–153
    body of, S–153, S–154
    radiology of, S–138
Incisive canal, S–129, S–189, S–360, S–361,
    S–364, S–365, S–371, S–372, S–BP 59
Incisive fossa, S–130, S–131, S–133, S–398,
    S–404
Incisive papilla, S–398
Incisor teeth, S–405
Incus, S–63, S–66, S–93, S–94, S–97, S–BP 9
    lenticular process of, S–94, S–BP 8
    long limb of, S–94, S–BP 8
    posterior ligament of, S–BP 8
    short limb of, S–94, S–BP 8
    superior ligament of, S–BP 8
Indirect inguinal hernia, S–BP 41
Inferior, as term of relationship, S–1
Inferior anal (rectal) nerve, T2.2
Inferior articular facet, of L4, S–BP 28
Inferior articular process, S–144, S–145
    of L1, S–BP 24
    of L2, S–138
    of L3, S–BP 24
Inferior colliculus, S–37
    brachium of, S–37
Inferior duodenal flexure, S–421
Inferior fascia, S–495
Inferior hypogastric plexus, S–20, S–BP 13,
    S–BP 14
Inferior labial artery, S–60
Inferior longitudinal lingual muscle, S–400,
    S–401
Inferior mediastinum, S–BP 2
Inferior mesenteric artery, S–303, S–341
Inferior mesenteric ganglion, S–20, S–BP 13,
    S–BP 15
Inferior mesenteric plexus, S–BP 13
Inferior nasal concha bone, S–367
Inferior nodes, S–356
Inferior orbital fissure, S–367
Inferior pharyngeal constrictor
    cricopharyngeal part of, S–197, S–408,
        S–409, S–411
    thyropharyngeal part of, S–411, S–412
Inferior phrenic artery, S–410
    left, S–416, S–419
Inferior phrenicoesophageal ligament, inferior,
    S–413
Inferior pubic ligament, S–151
Inferior pubic ramus, S–151

Inferior vena cava, S–302, S–304, S–342,
    S–410, S–415, S–416, S–417, S–419, S–
    420, S–429, S–433, S–438, S–441, S–442,
    S–445, S–459, S–461, S–483, S–484,
    S–507, S–508, S–510, S–513, S–531,
    S–532, S–541, S–542, S–543, S–544,
    S–546, S–547, S–548, S–BP 25, S–BP 94,
    S–BP 100, S–BP 101
    groove for, S–429
Inferior vertebral notch, of L2 vertebra,
    S–BP 24
Inferolateral lobe, S–BP 93
Inferoposterior lobe, S–BP 93
Infraclavicular node, S–346
Infraglenoid tubercle, S–155
Infraglottic cavity, S–378
Infrahyoid artery, S–525
Infrahyoid fascia, S–329
Infrahyoid muscle, S–191, S–387
    fascia of, S–190, S–196, S–197, S–329
Infraorbital artery, S–60, S–82, S–87, S–122,
    S–188, S–189, S–322, S–323, S–371
Infraorbital foramen, S–122, S–123, S–124,
    S–125, S–BP 19
    of newborn, S–127
Infraorbital groove, S–123
Infraorbital margin, S–5
Infraorbital nerve, S–58, S–59, S–60, S–61,
    S–71, S–73, S–82, S–86, S–122, S–185,
    S–366, S–371, S–406, S–BP 7
Infraorbital region, of head and neck, S–5
Infraorbital vein, S–322, S–328
Infrapatellar bursa
    deep, S–282
    subcutaneous, S–282
Infrapatellar fat pads, S–280, S–281, S–282
Infrapatellar synovial fold, S–280, S–281
Infrascapular region, of back, S–6
Infraspinatus fascia, S–198, S–199, S–210,
    S–230
Infraspinatus muscle, S–6, S–100, S–111,
    S–206, S–230, S–231, S–232, S–234, S–235,
    S–240, S–535, S–536, S–537
Infraspinatus tendon, S–157, S–232, S–235,
    S–240
Infraspinous fossa, S–155
    of scapula, S–149
Infratemporal crest, S–125
Infratemporal fossa, S–63, S–125, S–130,
    S–188
Infratrochlear nerve, S–59, S–60, S–61, S–85,
    S–122, S–185
    left, S–86
    right, S–86
Infundibular process, S–524
Infundibular recess, S–32, S–34
Infundibular stalk, imaging, S–39
Infundibular stem, S–524
Infundibulum, S–40, S–524
    of uterine tube, S–487, S–489
Inguinal canal, S–215, S–218, S–BP 108
Inguinal falx, S–212, S–213, S–215, S–216,
    S–217, S–BP 40
Inguinal fold, S–505
Inguinal ligament (Poupart's), S–8, S–9, S–11,
    S–113, S–211, S–212, S–213, S–215, S–216,
    S–218, S–219, S–266, S–268, S–270, S–271,
    S–276, S–334, S–347, S–356, S–479, S–486,
    S–492, S–500, S–501, T4.2, S–BP 40
    reflected, S–212, S–213, S–216
Inguinal lymph nodes
    deep, S–347
    superficial, S–347, T6.1
        inferior, S–347
        superolateral, S–347
        superomedial, S–347
Inguinal nodes, S–343, S–BP 57
    of Cloquet, proximal deep, S–355, S–356,
        S–357

Netter Atlas of Human Anatomy: A Systems Approach

Inguinal nodes *(Continued)*
  deep, S–352, S–355, S–356, S–357
  superficial, S–352, S–355, S–356, S–357
Inguinal region, S–8, S–19, S–BP 40
  dissections of, S–216
  left, S–395
  right, S–395
Inguinal ring, S–490
  deep, S–213, S–215, S–216, S–217, S–483,
    S–499, S–529, T4.2, S–BP 40
  lateral crus, S–216
  medial crus, S–216
  superficial, S–211, S–215, S–216, S–492,
    S–497, S–499, S–501, T4.2, S–BP 40
Inguinal tract, S–483
Inguinal (Hesselbach's) triangle, S–216, S–217
Inhibin, S–BP 95
Inion, S–125
Initial segment, S–BP 3
Insula (island of Rell), S–31, S–36, S–46
  central sulcus of, S–31
  circular sulcus of, S–31
  limen of, S–31
  long gyrus of, S–31
  short gyrus of, S–31
Integrative systems, associative and, S–15
Interalveolar septa, S–135
Interalveolar septum, S–BP 63
Interarytenoid notch, S–373
Interatrial septum, S–311, S–538
Intercapitular veins, S–229, S–258, S–332
Intercarpal ligament, dorsal, S–166
Intercavernous septum, of deep fascia, S–501,
  S–502, S–503
Intercavernous sinus
  anterior, S–29, S–30
  posterior, S–29, S–30
Interchondral joints, S–150, S–388
Interclavicular ligament, S–150, S–156, S–388
Intercondylar area, anterior, S–174
Intercondylar eminence, S–174, S–175, S–281,
  S–283
Intercondylar fossa, S–172
Intercondylar tubercles, S–175
Intercostal artery, S–206, S–473, S–BP 77
  anterior, S–205, S–206, S–336
  collateral branches of, S–205
  lower, anastomoses, S–336
  posterior, S–23, S–203, S–204, S–206, S–235,
    S–390, S–391, S–474
    dorsal branch of, S–24, S–206
    lateral cutaneous branches of, S–206
    right, S–383
    spinal branch of, S–206
  supreme, S–43, S–326, S–BP 49
Intercostal membrane
  external, S–100, S–206, S–209, S–235
  internal, S–100, S–206, S–209
Intercostal muscles, S–336, S–387, S–473,
  S–534, S–535, S–536, S–537, S–BP 63,
  S–BP 98
  external, S–100, S–201, S–203, S–204,
    S–206, S–209, S–212, S–235, S–336,
    S–BP 64
    internal intercostal membrane deep to, S–206
  innermost, S–100, S–205, S–206, S–209,
    S–336, S–390, S–391, S–393
  internal, S–203, S–204, S–205, S–206, S–209,
    S–336, S–BP 64
Intercostal nerves, S–16, S–75, S–104, S–204,
  S–205, S–206, S–209, S–390, S–391, S–473,
  T2.2, S–BP 46
  first, S–106, S–320
  third, S–447
  sixth, S–101
  eight, S–101
  tenth, S–470
  anterior cutaneous branches of, S–203,
    S–205, S–206, S–235

Intercostal nerves *(Continued)*
  anterior ramus of, S–100
  collateral branch of, S–205, S–209
  communicating branch of, S–209
  cutaneous branches
    anterior, S–102, S–209
    lateral, S–102, S–209
  lateral cutaneous branches of, S–203, S–204,
    S–206, S–235
  posterior cutaneous branch, S–100
  posterior ramus, S–100
Intercostal nodes, S–351
Intercostal spaces, S–306, T3.2
Intercostal veins, S–304, S–473
  anterior, S–205, S–337, S–338
  collateral branches of, S–205
  posterior, S–338, S–390, S–391
    right, S–340
  superior
    left, S–340, S–391
    right, S–338, S–340, S–390
Intercostobrachial nerve, S–13, S–102, S–107,
  S–204, S–227, S–228, S–235, S–237
Intercrural fibers, S–211, S–215, S–216,
  S–BP 40
Intercuneiform ligaments, dorsal, S–180
Interfascicular fasciculus, S–BP 16
Interfoveolar ligament, S–217
Interglobular spaces, S–405
Intergluteal cleft, S–6
Interlobar arteries, S–467, S–BP 85
Interlobar vein, S–BP 85
Interlobular arteries, S–431, S–467, S–BP 85
Interlobular bile ducts, S–431
Interlobular lymph vessels, S–350
Interlobular septum, S–392
Interlobular vein, S–431, S–BP 85
Intermaxillary suture, S–122, S–131
Intermediate bronchus, S–BP 98
Intermediate nerve (of Wrisberg), S–65, S–77,
  S–81, S–95
  motor root of, S–66
Intermediate sulcus, posterior, S–BP 16
Intermediolateral nucleus, S–454, S–533
Intermesenteric (abdominal aortic) plexus,
  S–103, S–105, S–451, S–452, S–469,
  S–470, S–471, S–513, S–514, S–515,
  S–517, S–519
  renal and ureteric branches from, S–469
Intermetacarpal joints, S–166
Intermuscular septum
  anterior, S–555
  lateral, S–111, S–239, S–240, S–241, S–249,
    S–551
  medial, S–238, S–239, S–240, S–241, S–247,
    S–249, S–551
  posterior, S–555
  transverse, S–555
Intermuscular stroma, S–455
Internal abdominal oblique muscles, S–545
Internal capsule, S–34, S–36, S–46, S–BP 18
  anterior limb of, S–36
    imaging of, S–39
  cleft for, S–36
  genu of, S–36
    imaging of, S–39
  posterior limb of, S–36, S–BP 17
    imaging of, S–39
  retrolenticular part of, S–36
Internal carotid artery, S–363
Internal carotid nerve, S–BP 15
Internal genitalia, homologues of, S–505
Internal intercostal membranes, S–209
Internal jugular vein, S–191, S–193, S–195,
  S–410
Internal occipital protuberance, S–132
Internal os, S–487
Internal vertebral venous plexus, anterior,
  S–BP 27

Interosseous artery
  anterior, S–241, S–247, S–249, S–331, S–552,
    S–BP 50, S–BP 51
  common, S–247, S–249, S–331, S–552, S–BP
    50, S–BP 51
  posterior, S–241, S–249, S–331, S–552, S–BP
    50, S–BP 51
  recurrent, S–241, S–331, S–BP 50, S–BP 51
Interosseous intercarpal ligaments, S–166
Interosseous membrane, S–162, S–165, S–178,
  S–243, S–246, S–282, S–283, S–289, S–333,
  S–552, S–555, S–BP 29, S–BP 52, S–BP 53,
  S–BP 109, S–BP 110
Interosseous muscle, S–181, S–255, S–297,
  S–BP 111
  dorsal, S–254, S–255, S–256, S–260, S–292,
    S–297, S–298, S–556, S–BP 105, S–BP 107
  first, S–110, S–241, S–251, S–256, S–259,
    S–BP 106
  second, S–BP 106
  third, S–BP 106
  fourth, S–BP 106
  palmar, S–110, S–254, S–BP 105
  first, S–BP 106
  second, S–BP 106
  third, S–BP 106
  plantar, S–556
Interosseous nerve
  anterior, S–109, S–249, S–552, T2.11
  crural, S–118
  posterior, S–112, S–241, S–552, T2.10
Interosseous veins, anterior, S–332
Interpectoral (Rotter's) lymph nodes, S–348,
  S–349
Interpeduncular cistern, S–35
  imaging of, S–BP 12
Interpeduncular fossa, S–368
Interphalangeal (IP) joint, S–181
Interphalangeal ligaments, S–169
Interproximal spaces, S–405
Intersigmoid recess, S–424, S–463
Interspinalis colli muscle, S–201
Interspinalis lumborum muscle, S–201
Interspinous ligament, S–146, S–148
Interspinous plane, S–395
Intertendinous connections, S–260
Interthalamic adhesion, S–32, S–36, S–37,
  S–41, S–51, S–524
Intertragic notch, S–94
Intertransverse ligament, S–150, S–388
Intertrochanteric crest, S–170, S–275
Intertrochanteric line, S–170
Intertubercular groove, S–155
Intertubercular plane, S–395
Intertubercular tendon sheath, S–157, S–235,
  S–239
Interureteric crest, S–465
Interventricular artery
  anterior, S–306, S–309, S–317, S–318, S–BP
    47, S–BP 48
    anterior ventricular branches of, S–317
    diagonal branch of, S–317, S–BP 47
    septal branches of, S–317, S–318, S–BP 47
  inferior, S–310, S–314, S–317, S–318, S–BP
    47, S–BP 48
    septal branches of, S–317
Interventricular foramen (of Monro), S–32,
  S–34, S–35, S–41, S–50, S–51
  imaging of, S–39
  left, S–32
Interventricular septum, S–316, S–319, S–538,
  T5.2
  membranous part of, S–311, S–312, S–314,
    S–315, S–316, S–319
  muscular part of, S–311, S–312, S–315,
    S–316, S–319
Interventricular sulcus
  anterior, S–309, S–538
  inferior, S–308, S–310

Netter Atlas of Human Anatomy: A Systems Approach

Lacrimal sac, S–82, S–83
  fossa for, S–123, S–125
Lactiferous ducts, S–473
Lactiferous sinus, S–473
Lacuna (of Trolard), lateral (venous), S–26,
  S–27
Lacuna magna, S–503
Lacuna vasorum, S–219
Lacunar ligament (Gimbernat's), S–212,
  S–213, S–216, S–217, S–218, S–219, S–347
Lacus lacrimalis, S–82
Lambda, S–128
Lambdoid suture, S–124, S–126, S–127, S–128,
  S–129, S–BP 19
Lamellar bodies, S–BP 63
Lamellar corpuscle (Pacini's), S–4
Lamina, S–145, S–146, S–148, S–BP 25
  of C6, S–BP 22
  of L1, S–138
  of L4 vertebra, S–BP 24
  lumbar, S–146
  thoracic, S–144
Lamina affixa, S–37
Lamina propria of gingiva, S–405
Lamina terminalis, S–32, S–33, S–41, S–524
Large arteries, S–302
Large intestine, S–394
  arteries of, S–440
  autonomic innervation of, S–452
  lymph vessels and nodes of, S–BP 57, S–BP 57
  mucosa and musculature of, S–428
  veins of, S–444
Laryngeal artery
  inferior, S–378
  superior, S–323, S–324, S–326, S–374,
    S–378, S–525, S–526, S–527
Laryngeal cartilages, T3.1
Laryngeal inlet, S–373, S–374
Laryngeal nerve
  fold of superior, S–373
  left recurrent, S–410
  recurrent, S–72, S–73, S–76, S–195, S–196,
    S–329, S–374, S–378, S–410, S–447,
    T2.1, T2.2, T2.7
    left, S–68, S–75, S–99, S–101, S–306,
      S–309, S–384, S–389, S–391, S–525,
      S–526, S–527, S–535, S–536
    right, S–68, S–99, S–309, S–378, S–525,
      S–526, S–527
  superior, S–68, S–72, S–76, S–77, S–81,
    S–307, S–374, S–447, T2.6
    external branch of, S–68, S–73, S–378,
      S–525, S–526, S–527
    internal branch of, S–68, S–73, S–81,
      S–374, S–378, S–525, S–526, S–527
Laryngeal prominence, S–375
Laryngeal vein, superior, S–328
Laryngopharyngeal nerve, S–76
Laryngopharynx, S–358, S–372, S–373
Larynx, S–20, S–21, S–81, S–358, S–384, S–BP
  15, S–BP 18, S–BP 22, S–BP 97
  cartilages of, S–375
  coronal section of, S–378
  intrinsic muscles of, S–376
    action of, S–377
  nerves of, S–378
  quadrangular membrane, S–375
Lateral, as term of relationship, S–1
Lateral aperture, S–35
  left, S–32
Lateral corticospinal tract, S–BP 16
Lateral digits, S–19
Lateral direct vein, S–51
Lateral epicondyle, S–155
Lateral glossoepiglottic fold, S–402
Lateral incisor, S–404, S–405
Lateral intertransversarii lumborum muscle,
  S–201
Lateral ligament, of ankle, S–180

Lateral node, S–357
Lateral pterygoid , S–407
Lateral recess, S–41
  left, S–32
Lateral region, of thorax, S–7
Lateral rotation, of lower limbs movements,
  S–265
Lateral sulcus
  anterior, S–BP 16
  posterior, S–BP 16
  of Sylvius, S–46
Latissimus dorsi muscle, S–6, S–100, S–102,
  S–198, S–199, S–203, S–209, S–210, S–211,
  S–212, S–230, S–231, S–235, S–236, S–237,
  S–238, S–239, S–264, S–460, S–538, S–541,
  S–545, S–546, S–BP 100
  costal attachments of, S–210
  tendon of, S–536
Latissimus dorsi tendon, S–238, S–537, S–551
Least thoracic splanchnic nerve, S–20, S–101
Left bundle
  Purkinje fibers of, S–319
  subendocardial branches of, S–319
Left colic plexus, S–105
Leg, S–2, S–3
  anterior region of, S–11, S–183
  compartments of, T4.3
  cross sections and compartments of, S–555
  lateral, S–183
  medial, S–19
  muscle of, S–284, S–285, S–286, S–287,
    S–288, S–289
    attachments of, S–299
  posterior region of, S–11, S–183
  serial cross sections of, S–BP 109
  veins of, S–335
Lemniscus
  medial, S–BP 17
  spinal, S–BP 17
Lens, S–82, S–88, S–89, S–90, S–91, S–BP 7,
  T2.1
  axis of, S–90
  capsule of, S–88, S–89, S–90
  cortex of, S–90
  equator of, S–90
  nucleus of, S–89, S–90
  supporting structures of, S–90
Lentiform nucleus, S–34, S–36, S–46
  globus pallidus of, S–36
  putamen of, S–36
    imaging of, S–39
Leptomeninges, S–22
Lesser curvature
  anterior nerve of, S–447
  posterior nerve of, S–447
Lesser occipital nerve, S–185, S–195
Lesser omentum, S–415, S–417, S–418, S–430
  anterior layer of, S–432, S–448
  attachment of, S–419
  free margin of, S–433
  posterior layer of, S–448
Lesser palatine artery, S–398
Lesser palatine foramen, S–398, S–404
Lesser palatine nerve, S–398, S–406
Lesser thoracic splanchnic nerve, S–20, S–101,
  S–208, S–BP 13
Lesser tubercle, S–155, S–158, S–239
  crest of, S–155
Levator anguli oris muscle, S–65, S–186, S–BP 35
Levator ani muscle, S–114, S–217, S–219,
  S–222, S–223, S–225, S–226, S–428, S–441,
  S–453, S–465, S–466, S–478, S–480, S–481,
  S–485, S–488, S–492, S–493, S–496, S–504,
  S–507, S–509, S–511, S–517, S–520, T4.2,
  S–BP 86, S–BP 108
  iliococcygeus, S–224
  left, S–220
  medial border of, S–482
  nerve to, S–104, S–114, T2.12

Levator ani muscle (Continued)
  of pelvic diaphragm, S–495, S–502
  puboanalis (puborectalis), S–224, S–549,
    S–550
  pubococcygeus, S–224
  roofing ischioanal fossa, S–501
  tendinous arch of, S–219, S–220, S–225,
    S–464, S–465, S–478, S–479, S–480,
    S–481, S–482, S–488, S–490, S–493
Levator labii superioris alaeque nasi muscle,
  S–65, S–BP 36
Levator labii superioris muscle, S–65, S–184,
  S–186, S–BP 35, S–BP 36
Levator nasolabialis muscle, S–65, S–184,
  S–186, S–BP 35, S–BP 36
Levator palpebrae superioris muscle, S–58,
  S–82, S–84, S–85, S–86, S–BP 7, T4.1
  insertion of, S–82
Levator pterygoid , S–BP 60
Levator scapulae muscle, S–111, S–192,
  S–194, S–196, S–198, S–199, S–230, S–234,
  S–235, S–264, S–329
  nerves to, S–195
Levator veli, S–418
Levator veli palatini muscles, S–68, S–187,
  S–398, S–403, S–407, S–408, S–409, T4.1,
  S–BP 8, S–BP 60
Levatores costarum muscles, S–201
  brevis, S–201
  longus, S–201
Ligamenta flava, S–141
Ligaments
  of ankle, S–180
  denticulate, S–22
  of elbow, S–161
  of foot, S–181
  of uterus, round, S–508
  of wrist, S–165, S–166, S–BP 29
Ligamentum arteriosum, S–305, S–306, S–309,
  S–312, S–389, S–391, T5.2
  aortic arch lymph node of, S–350
Ligamentum flavum, S–146, S–148, S–BP 21
  radiology of, S–138
Ligamentum teres
  fissure for, S–429
  notch for, S–423
Ligamentum teres hepatis, S–217
Ligamentum venosum, S–305
  fissure for, S–429
Limbs
  posterior, S–BP 18
  somatosensory system of, S–BP 17
Limen nasi, S–361
Limiting hepatic , S–431
Linea alba, S–7, S–8, S–203, S–209, S–211,
  S–212, S–213, S–214, S–216, S–358, S–483,
  S–484, S–546, T4.2, S–BP 40, S–BP 100,
  S–BP 103
Linea semilunaris, S–8
  left, S–395
  right, S–395
Linea terminalis, S–146, S–484, S–488, S–490
Linea translucens, S–89
Lingual artery, S–43, S–79, S–188, S–189,
  S–322, S–323, S–324, S–325, S–326, S–399,
  S–400, S–526
  deep, S–397, S–399, S–401
  dorsal, S–399, S–401
  dorsal lingual branch of, S–403
  suprahyoid branch of, S–323, S–401
Lingual gland, duct of, S–402
Lingual gyrus, S–32, S–33
Lingual nerve, S–59, S–62, S–65, S–73, S–77, S–
  79, S–80, S–81, S–136, S–187, S–188, S–189,
  S–396, S–397, S–399, S–400, S–401, S–406
  right, S–188
Lingual tonsil, S–372, S–373, S–374, S–402,
  S–403, S–409
  crypt of, S–402

Lungs (Continued)
superior lobe of, S–387, S–388
trachea area of, S–388
in situ, anterior view of, S–387
in thorax
anterior view of, S–385
posterior view of, S–386
Lunotriquetral ligament, S–165
Lunula, S–315, S–553, S–BP 112
Luschka
cleft of, S–139
foramen of, S–32, S–35
Luteal cells, S–529
Luteinizing hormone, S–BP 95
Lymph nodes, S–550
of breast, S–348
of esophagus, S–351
of genitalia
female, S–355
male, S–357
of head, S–344
of lungs, S–350
of mammary gland, S–348
of neck, S–343, S–344
of pelvis
female, S–355, S–356
male, S–357
of pharynx, S–345
popliteal, S–347
preaortic, S–355
tongue, S–345
Lymph vessels, S–343, S–343 to S–357, S–431,
T6.1, S–BP 54 to S–BP 58
along arcuate arteries, S–354
along interlobar arteries, S–354
of breast, S–343, S–348
of esophagus, S–351
of genitalia
female, S–355
male, S–357
of lower limb, S–343
of lungs, S–350
of mammary gland, S–348
medullary, S–354
of pelvis
female, S–355, S–356
male, S–357
superficial, S–347
of upper limb, S–343
Lymphatic drainage
of breast, S–349
of lungs, S–350
of pharynx, S–345
of prostate gland, S–357
of tongue, S–345
Lymphatic duct, right, S–343, S–349, S–350
Lymphatic trunk
bronchomediastinal, S–350
jugular, S–350
Lymphatics, abdominal and pelvic, S–353
Lymphocytes, S–BP 30, S–BP 44
Lymphoid nodules, S–402
Lymphoid organs, S–343, S–343 to S–357,
T6.1, S–BP 54 to S–BP 58

M

Mackenrodt's ligament. See Cardinal ligament
Macrophage, S–BP 30
Macula, S–91
Macular arteriole
inferior, S–91
superior, S–91
Macular densa, S–BP 83
Macular venule
inferior, S–91
superior, S–91
Male, posterior view of, S–3

Male reproductive system, S–530
Mallear fold
anterior, S–94, S–BP 8
posterior, S–94, S–BP 8
Malleolar arteries, anterior
lateral, S–289, S–291, S–292, S–333,
S–BP 43, S–BP 53
medial, S–289, S–291, S–292, S–333,
S–BP 53
Malleolar stria, arteries of, S–95
Malleolus
lateral, S–11, S–175, S–284, S–285, S–286,
S–287, S–288, S–289, S–290, S–291, S–292
medial, S–11, S–175, S–284, S–285, S–286,
S–288, S–289, S–290, S–291, S–292,
S–335, S–BP 110
Malleus, S–94, S–97
anterior ligament of, S–94
anterior process of, S–BP 8
handle of, S–94, S–95, S–BP 8
head of, S–63, S–66, S–93, S–94, S–BP 8,
S–BP 9
lateral process of, S–94
superior ligament of, S–94, S–BP 8
Mammary gland, S–472, S–473, T10.1
axillary tail (of Spence) of, S–474
lymph nodes of, S–348
lymph vessels of, S–348
Mammary gland lobules, S–473
Mammillary bodies, S–32, S–33, S–34, S–38,
S–40, S–368, S–523, S–524
imaging of, S–39, S–BP 12
Mammillary process, S–145
Mammillothalamic fasciculus, S–32
Mammillothalamic tract (of Vicq d'Azyr), S–524
Mandible, S–120, S–123, S–124, S–125, S–135,
S–196, S–197, S–329, S–359, S–372
of aged person, S–135
alveolar part of, S–135
angle of, S–5, S–72, S–124, S–135, S–359,
S–399
base of, S–135
body of, S–123, S–124, S–125, S–135, S–188,
S–192, S–359, S–368, S–BP 61
condylar process of, S–135, S–188, S–359
head of, S–125
condyle of, S–124, S–126, S–BP 19
coronoid process of, S–125, S–126, S–135,
S–359, S–BP 19, S–BP 60
head of, S–135
mandibular foramen of, S–135
mandibular notch of, S–125
mental foramen of, S–123, S–125, S–135
mental protuberance of, S–5, S–123, S–135
mental spines of, S–135
mental tubercle of, S–123, S–135
mylohyoid groove of, S–135
mylohyoid line of, S–135
neck of, S–135, S–BP 60
oblique line of, S–125, S–135, S–407
opening of, S–BP 37
pterygoid fovea of, S–135
ramus of, S–123, S–124, S–125, S–135,
S–192, S–359, S–400, S–BP 37
retromolar fossa of, S–135
sublingual fossa of, S–135
submandibular fossa of, S–135
Mandibular foramen, S–135
Mandibular fossa, S–125, S–127, S–131, S–136,
S–BP 10
Mandibular nerve, S–30, S–58, S–59, S–62,
S–67, S–73, S–77, S–79, S–80, S–81, S–85,
S–86, S–133, S–134, S–136, S–185, S–188,
S–399, S–406, S–BP 6, T2.4
anterior division of, S–62
meningeal branch of, S–59, S–62, S–63,
S–85, S–133, S–134
posterior division of, S–62

Mandibular nodes, S–344
Mandibular notch, S–125, S–135, S–359, S–BP 37
Manubriosternal joint, S–150, S–388
Manubriosternal synchondrosis, S–156
Manubrium, S–156, S–535, S–536
of sternum, S–149, S–190, S–192, S–197,
S–205, S–329, S–372, S–387, S–388
Marginal artery, S–440, S–441, S–BP 81, S–BP 82
anastomosis, T5.3
Marginal sulcus, S–32
Masseter muscle, S–63, S–186, S–192, S–197,
S–396, S–400, S–BP 60
attachment of, S–186
deep part of, S–186
superficial part of, S–186
Masseteric artery, S–63, S–186, S–187, S–189,
S–323, S–325
Masseteric fascia, S–184
Masseteric nerve, S–62, S–63, S–73, S–186,
S–187, S–188, S–189
nerve to, S–59
Mast cell, S–BP 30
Mastication, muscles of, S–186, S–187, T4.1
Mastoid air cells, S–124, S–126, S–BP 22
Mastoid angle, S–132
Mastoid antrum, S–95, S–BP 8
Mastoid canaliculus, S–131, S–133
Mastoid cells, S–71, S–95, S–124, S–BP 8,
S–BP 60
Mastoid emissary vein, S–26, S–322
in mastoid foramen, S–133
Mastoid fontanelle, S–127
Mastoid foramen, S–130, S–131, S–133, S–134
Mastoid nodes, S–344
Mastoid notch, S–131
Mastoid process, S–125, S–130, S–131, S–192,
S–193, S–201, S–359, S–399, S–401, S–BP
10, S–BP 37
of neck, anterior view of, S–190
Maturation, S–BP 89
Mature follicle, S–BP 95
Maxilla, S–370, S–BP 19, S–BP 60
palatine process of, S–398, S–404
sinus of, S–71, S–370
Maxillary artery, S–27, S–43, S–62, S–63, S–73,
S–77, S–79, S–80, S–95, S–136, S–186,
S–187, S–188, S–189, S–323, S–324, S–325,
S–326, S–363, S–371, S–399, S–BP 59
proximal, S–325
Maxillary bone, S–123, S–125, S–129, S–131,
S–360, S–362, S–367
alveolar process of, S–123, S–125, S–129,
S–130, S–362, S–368, S–BP 61
anterior nasal spine of, S–122, S–123, S–125,
S–129, S–360, S–362
frontal process of, S–82, S–122, S–123,
S–125, S–362
incisive canal of, S–129, S–360, S–362
incisive fossa of, S–131
infraorbital foramen of, S–123, S–125
infratemporal surface of, S–125
intermaxillary suture of, S–131
nasal crest of, S–360
nasal surface of, S–129
of newborn, S–127
orbital surface of, S–123
palatine process of, S–129, S–130, S–131,
S–360, S–361, S–362, S–368, S–BP 10
tuberosity of, S–125, S–130
zygomatic process of, S–123, S–131
Maxillary nerve, S–30, S–58, S–59, S–61, S–62,
S–63, S–71, S–73, S–77, S–79, S–80, S–81,
S–85, S–86, S–134, S–185, S–188, S–364,
S–371, S–406, T2.4, S–BP 6
anterior superior alveolar branch of, S–365
lateral superior posterior nasal branch of,
S–365, S–371
meningeal branch of, S–59, S–85

Maxillary nerve (Continued)
zygomaticofacial branch of, S–59
zygomaticotemporal branch of, S–59
Maxillary ostium, S–71
Maxillary sinus, S–71, S–84, S–124, S–126,
S–366, S–368, S–BP 60, S–BP 61
alveolar recess of, S–BP 61
imaging of, S–BP 12
infraorbital recess of, S–BP 61
opening of, S–361, S–362, S–368, S–369,
S–BP 61
postsynaptic fibers to, S–366
recesses of, S–368
transverse section of, S–BP 60
zygomatic recess of, S–BP 61
Maxillary tuberosity, S–125
Maxillary vein, S–87, S–328
McBurney's point, S–427
McGregor's line, S–BP 21
Medial, as term of relationship, S–1
Medial digits, S–19
Medial eminence, S–41
Medial ligament, of ankle, S–180
Medial pterygoid muscle, S–197, S–398, S–400
Medial pterygoid , S–398, S–403, S–408
Medial rotation, of lower limbs movements,
S–265
Median aperture, S–32, S–35, S–41, S–55
Median cricothyroid ligament, S–191
Median fissure, anterior, S–BP 16
Median glossoepiglottic fold, S–402
Median lobe, S–BP 93
Median nerve, S–13, S–14, S–106, S–107,
S–108, S–109, S–165, S–227, S–237, S–238,
S–239, S–244, S–247, S–249, S–251, S–253,
S–256, S–551, S–552, T2.3, T2.11, S–BP 105
articular branch of, S–109
common palmar digital branch of, S–227
palmar branch of, S–13, S–14, S–109, S–229,
S–244, S–247, S–250, S–257
palmar digital branches of, S–13
common, S–14
proper, S–14, S–227
proper palmar digital branches of, S–251
recurrent branch of, S–109, S–227, S–250,
S–251, S–257, T2.3
Median nuclei, S–37
Median sulcus, posterior, S–41, S–BP 16
Mediastinal lymph nodes, S–343, S–535
anterior, pathway to, S–348
Mediastinal pleura, S–207, S–309
Mediastinum
anterior, S–207
lung area for, S–388
cross section of, S–539
great vessels of, S–389
lateral view
left, S–391
right, S–390
lymph vessels of, S–350
of testis, S–497, S–498
Medulla oblongata, S–21, S–32, S–41, S–66,
S–80, S–81, S–134, S–320, S–358, S–366,
S–452, S–BP 15, S–BP 23, S–BP 60
in foramen magnum, S–133
imaging of, S–39
lower part of, S–BP 17
reticular formation of, S–384
Medullary artery, segmental, S–23
Medullary capillary plexus, S–BP 85
Medullary cavity, S–BP 31
Medullary lamina
external, S–37
internal, S–37
Medullary vein, anteromedian, S–49
Medullary velum
inferior, S–32, S–41
superior, S–32, S–40, S–41
Meiosis, S–BP 89

Meissner's corpuscle, S–4
Meissner's plexus. See Submucosal plexus
Melanocyte, S–4
Membranous ampulla, S–93, S–97
anterior, S–96
lateral, S–66, S–96
posterior, S–66, S–96
Membranous labyrinth, S–96, S–97
Membranous layer (Scarpa's fascia), S–492,
S–BP 90
Membranous limb, common, S–96, S–97
Membranous septum
atrioventricular part of, S–539
interventricular part of, S–539
Membranous urethra, S–503
Mendelian inheritance, S–BP 89
Meningeal artery, S–27
accessory, S–27, S–63, S–95, S–133, S–134,
S–189
anterior, S–87
dorsal, S–327
middle, S–26, S–27, S–28, S–30, S–43, S–62,
S–63, S–73, S–77, S–95, S–134, S–136,
S–187, S–188, S–189, S–323, S–325,
S–326, S–327, T5.1
accessory branch of, S–95
branches of, S–28
frontal (anterior) and parietal (posterior)
branches of, S–27
left, S–326
petrosal branch of, S–95
posterior, S–134
in mastoid foramen, S–133
Meningeal vein, middle, S–26, S–28, S–134
Meningeal vessels
anterior, groove for, S–132
middle, S–133
groove for, S–132
grooves for branches of, S–128, S–129
posterior, groove for, S–132
Meninges, S–26, S–28, S–134
arteries to, S–326
Meningohypophyseal trunk, S–27, S–327
branches of, S–27
tentorial branch of, S–327
Meniscofemoral ligament, posterior, S–173,
S–281
Meniscus, S–166
lateral, S–173, S–280, S–281, S–282
medial, S–173, S–280, S–281, T3.3
Menstrual cycle, S–BP 95
Menstruation, S–BP 97
Mental foramen, S–123, S–125, S–135
Mental nerve, S–59, S–62, S–73, S–185,
S–406
Mental protuberance, S–5, S–123, S–135
Mental region, of head and neck, S–5
Mental tubercle, S–135
Mental vein, S–328
Mentalis muscle, S–65, S–184, S–186, S–BP 35
Meridional fibers, S–89
Mesangial cell, S–BP 83
Mesangial matrix, S–BP 83
Mesencephalic nucleus, of trigeminal nerve,
S–59, S–81
Mesencephalic vein
lateral, S–49
posterior, S–49, S–51
Mesenteric artery
inferior, S–303, S–341, S–420, S–422, S–440,
S–441, S–459, S–463, S–468, S–508,
S–510, S–517, S–531, S–BP 81
superior, S–303, S–305, S–341, S–393,
S–421, S–422, S–433, S–436, S–437,
S–438, S–439, S–440, S–443, S–450,
S–457, S–459, S–461, S–463, S–468,
S–517, S–528, S–531, S–544, S–545,
S–548, T5.3, S–BP 55, S–BP 75, S–BP 77,
S–BP 81, S–BP 82, S–BP 99

Mesenteric ganglion
celiac, S–103
inferior, S–105, S–452, S–469, S–471, S–514,
S–515, S–517, S–518, S–519, S–BP 13,
S–BP 15
superior, S–103, S–105, S–449, S–450,
S–451, S–452, S–454, S–457, S–469,
S–470, S–471, S–514, S–517, S–518,
S–519, S–533, S–BP 13, S–BP 15
Mesenteric nodes
inferior, S–352
superior, S–352, S–BP 57
Mesenteric peritoneum, S–451
Mesenteric plexus
inferior, S–105, S–452, S–517
superior, S–105, S–448, S–449, S–451,
S–457
Mesenteric vein
inferior, S–340, S–422, S–433, S–442, S–444,
S–445, S–446, S–BP 55, S–BP 78
superior, S–340, S–421, S–433, S–439,
S–442, S–443, S–444, S–446, S–461, S–
BP 55, S–BP 78
Mesentery, S–421, S–424, S–455, S–545
root of, S–419, S–420, S–424, S–433, S–463,
S–483, S–484
Mesoappendix, S–425, S–451, S–BP 69
Mesocolic taenia, S–425, S–426, S–428
Mesocolon, S–424
sigmoid, S–221
Mesometrium, S–487
Mesonephric (Wolffian) duct, S–499, S–505
Mesonephric tubules, S–505
vestigial, S–498
Mesosalpinx, S–487
of broad ligament, S–529
laminae of, S–489
Mesovarium, S–487, S–488
laminae of, S–489
Metacarpal, cross section, S–BP 105
Metacarpal arteries
dorsal, S–261
palmar, S–BP 51
Metacarpal bones, S–120, S–163, S–164,
S–165, S–166, S–167, S–169, S–255, S–BP
29, S–BP 42
first, S–121, S–168, S–241, S–249, S–253,
S–259
second, S–241, S–BP 106, S–BP 107
third, S–164, S–168, S–BP 106, S–BP 107
fourth, S–BP 106, S–BP 107
fifth, S–168, S–241, S–242, S–249, S–BP 106
cross sections through, S–BP 105
Metacarpal ligaments
deep transverse, S–169, S–256, S–BP 107
dorsal, S–BP 29
palmar, S–169, S–BP 29
superficial transverse, S–229, S–250
Metacarpal veins, dorsal, S–229, S–258
Metacarpophalangeal (MCP) joint, S–10,
S–168, S–169, S–BP 29, S–BP 42
Metacarpophalangeal (MCP) ligaments,
S–169
Metaphysial bone tissue, S–121
Metatarsal arteries
dorsal, S–289, S–291, S–292, S–297, S–BP
43, S–BP 53
plantar, S–292, S–BP 53
anterior perforating branches to, S–296
Metatarsal bones, S–120, S–176, S–177, S–181,
S–292, S–297, S–556
first, S–180, S–298
tuberosity of, S–177
second, S–BP 111
third, S–BP 111
fourth, S–BP 111
fifth, S–287, S–291, S–298, S–556, S–BP 111
tuberosity of, S–296, S–297, S–298
tuberosity of, S–290

Netter Atlas of Human Anatomy: A Systems Approach

Metatarsal ligaments
  deep transverse, S–181, S–297
  dorsal, S–180, S–297
  plantar, S–297
  transverse, superficial, S–293
Metatarsal veins, dorsal, S–266
Metatarsophalangeal (MTP) joint, S–181
  first, T3.3
Midbrain, S–BP 17, S–BP 18
  aqueduct of, S–33, S–34, S–51, S–55, T2.1
  imaging of, S–39
Midcarpal joint, S–164, S–166
Midclavicular line, S–306
Middle colic nodes, S–BP 57
Middle pharyngeal constrictor, S–407
Middle sacral node. *See* Promontorial nodes
Midface, nerves and arteries of, S–371
Midpalmar space, S–251, S–252, S–254
Minute fasciculi, S–250
Mitochondria, S–BP 3, S–BP 4
Mitral cell, S–56
Mitral heart valve. *See* Atrioventricular valve,
  left
Modiolus of angulus, S–184
Molar glands, S–398
Molar tooth, S–370, S–405
  1st, S–367, S–404
  2nd, S–404
  3rd, S–404
Monocytes, S–BP 30, S–BP 44
Mons pubis, S–491
Morison's pouch. *See* Hepatorenal recess
Motor nucleus, S–59
  of trigeminal nerve, S–81
Motor systems, S–15
Mouth, S–2
Mucosa, S–421, S–455, S–BP 65
Mucous glands, S–402
Mucous neck cell, S–BP 67
Mullerian duct. *See* Paramesonephric duct
  (müllerian)
Multifidus lumborum muscle, S–201
Multifidus muscle, S–545
Multifidus thoracis muscle, S–201
Muscles, S–302
  of arm
    anterior views of, S–239
    posterior compartment of, S–240
  of back
    deep layer, S–201
    intermediate layers, S–200
    superficial layers, S–199
  of face, S–BP 36
    facial expression, S–184
      anterior view of, S–BP 35
  of forearm
    attachments of, S–262, S–263
    deep part, S–241, S–249
    extensors of wrist and digits, S–243
    flexors of digits, S–246
    flexors of wrist, S–245
    pronators and supinator, S–248
    superficial part, S–242, S–244
  of hip, S–272, S–273
  hypothenar, S–110, S–168, S–250, S–251,
    S–254
  intercostal, S–336, S–387, S–534, S–535,
    S–536, S–537, S–BP 98
  lumbrical, S–109, S–110, S–251, S–252,
    S–253, S–254, S–255, S–294, S–295,
    S–297, S–BP 107
  of mastication, S–186, S–187
  of neck, lateral view of, S–192
  of sole of foot
    first layer, S–294
    second layer, S–295
    third layer, S–296
  structure, S–BP 34
Muscular artery, S–87, S–91, S–92

Muscular system, S–183 to S–301, T4.1–4.3,
  S–BP 34 to S–BP 43
Muscular triangle, S–5
Muscular vein, S–91
Muscularis mucosae, S–226, S–455,
  S–BP 67
  of anal canal, S–223, S–226
  of rectum, S–223
Musculature development, S–BP 97
Musculocutaneous nerve, S–14, S–106, S–107,
  S–108, S–109, S–227, S–237, S–238, S–239,
  S–249, S–551, T2.9
  anterior branch of, S–108
  articular branch of, S–108
  posterior branch of, S–108
  terminal part of, S–13
Musculophrenic arteries, S–204, S–205, S–307,
  S–336, S–387
Musculophrenic vein, S–204, S–205, S–337
Musculus uvulae, S–398, S–403, S–409, T4.1
Myelin sheath, S–BP 3
Myelinated fibers, S–BP 17
Myenteric plexus (Auerbach's), S–455, S–BP 65
Mylohyoid groove, S–135
Mylohyoid line, S–135
Mylohyoid muscle, S–62, S–190, S–191, S–192,
  S–197, S–324, S–368, S–372, S–396, S–399,
  S–401, S–407, S–408, S–BP 61
  nerve to, S–59, S–62, S–73, S–187, S–188,
    S–189, S–194, S–399, S–400
Mylohyoid nerve, S–136
Mylohyoid raphe, S–399
Myofibril, S–BP 34
Myofilaments, S–BP 34
Myometrium, S–487
  of uterus, S–521
Myosatellite cell, S–BP 34

N

Nail bed, S–553, S–BP 112
Nail fold
  lateral, S–BP 112
  proximal, S–BP 112
Nail groove, lateral, S–BP 112
Nail matrix, S–553
Nail , S–553, S–BP 112
Nail root, S–553
Nares, S–BP 18
  anterior, S–5
Nasal artery
  dorsal, S–87, S–189, S–323
  external, S–43, S–87, S–122
  lateral, S–122
Nasal bone, S–5, S–122, S–123, S–124, S–125,
  S–129, S–360, S–362, S–368, S–370,
  S–BP 19
  of newborn, S–127
Nasal cartilage, accessory, S–122
Nasal cavity, S–83, S–358, S–367, S–370,
  S–BP 7, S–BP 61
  arteries of, S–BP 59
  bony nasal septum turned up, S–BP 59
  autonomic innervation of, S–366
  bones of, S–370
  floor of, S–361
  lateral wall of, S–361, S–362, S–363, S–364
  medial wall of, S–360
  nerves of, S–364, S–365
  postsynaptic fibers to, S–366
  vasculature of, S–363
Nasal concha
  bony highest, S–370
  bony middle, S–367, S–370
  bony superior, S–367, S–370
  inferior, S–83, S–123, S–124, S–129, S–358,
    S–361, S–362, S–369, S–370, S–373,
    S–BP 60, S–BP 61
  ethmoidal process of, S–362

Nasal concha *(Continued)*
  middle, S–83, S–123, S–129, S–358, S–361,
    S–362, S–369, S–BP 61
  superior, S–358, S–361, S–362
  supreme, S–358
Nasal concha turbinate
  inferior, S–368
  middle, S–368
Nasal meatus
  inferior, S–83, S–361, S–368, S–BP 61
  middle, S–368, S–BP 61
    atrium of, S–361
  superior, S–361
Nasal nerve
  external, S–122
  posterior, S–79
Nasal region, of head and neck, S–5
Nasal retinal arteriole
  inferior, S–91
  superior, S–91
Nasal retinal venule
  inferior, S–91
  superior, S–91
Nasal septal cartilage, S–122, S–360, S–BP 60
  lateral process of, S–122, S–360, S–362
Nasal septum, S–124, S–130, S–364, S–365,
  S–368, S–372, S–373, S–523, T7.1, S–BP 61
  bony part of, S–367, S–370, S–BP 59
Nasal skeleton, S–367
Nasal slit, S–134
Nasal spine
  anterior, S–122, S–123, S–125, S–129, S–360,
    S–BP 19
  posterior, S–131, S–BP 21
Nasal vein, external, S–328
Nasal vestibule, S–358, S–360, S–361, S–BP 60
Nasalis muscle, S–65, S–184, S–BP 36
  alar part of, S–60, S–122, S–184, S–BP 35
  transverse part of, S–60, S–122, S–184,
    S–BP 35
Nasion, S–123, S–BP 19
Nasociliary nerve, S–58, S–59, S–61, S–77,
  S–78, S–85, S–86, S–BP 7
Nasofrontal vein, S–87, S–322, S–328, S–363
Nasolabial lymph nodes, S–344
Nasolacrimal canal, opening of, S–362
Nasolacrimal duct, S–83
  opening of, S–83, S–361
Nasopalatine nerve, S–59, S–364, S–365,
  S–366, S–371, S–406
  communication between, S–371
  groove for, S–360
  in incisive fossa, S–133
Nasopalatine vessels, groove for, S–360
Nasopharyngeal adenoids, imaging of, S–39
Nasopharynx, S–30, S–93, S–358, S–372,
  S–373, S–BP 9
  airway to, S–361
Navicular bones, S–176, S–177, S–179, S–180,
  S–297
  tuberosity of, S–181, S–297
Navicular fossa, S–495, S–503
Neck, S–2, S–3, S–155, S–183, S–BP 18
  anatomical, S–155
  anterior region of, S–5
  arteries of, S–323
  autonomic nerves in, S–76
  bony framework of, S–359
  cutaneous nerves of, S–185
  fascial layers of, S–196, S–329
  lymph nodes of, S–344
  muscles of
    anterior view of, S–190
    lateral view of, S–192
  nerves of, S–194, S–195
  posterior triangle of, S–203
  of scapula, S–149
  structures with high clinical significance,
    T4.1, T5.1, T6.1

Neck (Continued)
  superficial veins of, S–329
  surface anatomy of, S–5
  triangle of, posterior, S–199
Nephron, S–BP 84
Nephron loop (Henle's), S–BP 84, S–BP 85
Nerve fiber bundles, S–BP 4
Nerve fibers, S–97
Nerve roots, S–17
Nerves
  of abdominal wall
    anterior, S–102
    posterior, S–103
  of back, S–198
    of neck, S–202
  of buttocks, S–275
  of cranial base, S–71
  of external genitalia
    female, S–516
    male, S–518
  of larynx, S–378
  of nasal cavity, S–364
  of neck, S–194, S–195
  of orbit, S–85
  of perineum
    female, S–516
    male, S–520
Nervous system, S–15 to S–119, T2.1–T2.14,
    S–BP 3 to S–BP 18
  overview of, S–15
Neural foramen. See Intervertebral foramen
Neurocranium, S–120
Neuroendocrine G cell, S–BP 67
Neurofilaments, S–BP 3
Neurons, S–BP 3, S–BP, S–15
Neurotubules, S–BP 3
Neurovascular compartment, S–551
Neutrophils, S–BP 44
Newborn, skull of, S–127
Nipple, S–7, S–473
  level of, S–19
Node of Cloquet/Rosenmüller, S–352
Node of ligamentum arteriosum, S–350
Node of Ranvier, S–BP 3, S–BP 4
Nose, S–2, S–122
  ala of, S–5
  mucous glands of, S–21
  transverse section of, S–BP 60
Nostril, S–5
Nuchal ligament, S–6, S–141, S–BP 21
Nuchal line
  inferior, S–130, S–131
  superior, S–130, S–131, S–199
Nuclei, S–BP 34
Nuclei of pulvinar, S–37
Nucleus ambiguus, S–54, S–55, S–67, S–68, S–69
Nucleus pulposus, S–139, S–145, S–148
Nurse cell (Sertoli cell), S–530
Nutrient artery, of femur, S–278
Nutrient foramen, of femur, S–172

O
Obex, S–41
Oblique cord, S–162
Oblique diameter, S–477
Oblique fissure, of lungs, S–387
  left, S–388
  right, S–388
Oblique line, S–135, S–375
Oblique muscle, S–414
  external, S–8, S–102, S–209, S–210, S–211,
    S–212, S–213, S–214, S–215, S–216,
    S–217, S–219, S–336, S–543, S–546,
    S–BP 40, S–BP 102, S–BP 103
    abdominal, S–7, S–198, S–199, S–200,
      S–203, S–206, S–358, S–460, S–473,
      S–483, S–484, S–492, S–500, S–501,
      S–545, S–BP 64

Oblique muscle (Continued)
  aponeurosis of, S–102, S–213, S–214,
    S–216, S–218, S–546, S–BP 40
    aponeurotic part, S–211
    costal attachments of, S–210
    muscular part of, S–211
  inferior, S–58, S–84, S–86
    tendon of, S–86
  internal, S–102, S–210, S–212, S–213, S–214,
    S–215, S–216, S–217, S–219, S–336,
    S–546, S–BP 103
    abdominal, S–358, S–460, S–483, S–484
    aponeurosis of, S–102, S–213, S–214,
      S–546
    internal abdominal, S–199, S–200, S–203,
      S–545, S–BP 64
    in lumbar triangle (of Petit), S–199
    tendon of origin of, S–545
  superior, S–84, S–85, S–86, S–BP 7
Oblique popliteal ligament, S–281, S–283
Obliquus inferior capitis muscle, S–200, S–201,
    S–202
Obliquus superior capitis muscle, S–200,
    S–201, S–202
Obturator artery, S–170, S–217, S–278, S–333,
    S–341, S–441, S–463, S–468, S–478, S–490,
    S–507, S–508, S–509, S–517, S–549, S–BP
    52, S–BP 94, S–BP 96, S–BP 104
  accessory, S–441, S–507
  acetabular branch of, S–170, S–278
  anterior branch, S–170
  posterior branch, S–170
Obturator canal, S–154, S–220, S–271, S–464,
    S–479, S–481, S–490, S–507, S–BP 96
Obturator crest, S–153, S–154
Obturator externus muscle, S–116, S–269,
    S–271, S–300, S–550
Obturator fascia, S–220, S–464, S–478, S–479,
    S–481, S–488, S–493, S–507, S–520, S–BP 96
Obturator foramen, S–151, S–153, S–171,
    S–182, S–475, S–476
Obturator groove, S–153
Obturator internus fascia, S–225, S–507, S–BP
    104
Obturator internus muscle, S–114, S–182, S–217,
    S–219, S–220, S–225, S–273, S–274, S–275,
    S–276, S–300, S–441, S–465, S–478, S–479,
    S–480, S–481, S–482, S–488, S–490, S–504,
    S–516, S–520, S–549, S–550, S–BP 108
  nerve to, S–104, S–114, S–274, S–275, T2.12
  sciatic bursa of, S–276
Obturator internus tendon, S–480, S–482
Obturator membrane, S–154, S–170, S–488
Obturator nerve, S–103, S–104, S–113, S–114,
    S–115, S–116, S–217, S–277, S–508, S–517,
    S–549, S–554, T2.3, T2.12, S–BP 104
  accessory, S–103, S–104, S–113
  adductor hiatus, S–116
  anterior branch of, S–116, S–271
  articular branch of, S–116
  cutaneous branch of, S–116, S–266, S–267,
    S–271
  posterior branch of, S–116, S–271
Obturator node, S–355
Obturator vein, S–342, S–444, S–445, S–463,
    S–478, S–549, S–BP 104
Obturator vessels, S–216, S–217
  accessory, S–216, S–218
  right, S–BP 94
Occipital artery, S–27, S–43, S–72, S–188,
    S–202, S–322, S–323, S–324, S–326,
    S–328
  descending branch of, S–202, S–324
  groove for, S–131
  mastoid branch of, S–27, S–326
  medial, S–47
  meningeal branch of, S–322
  sternocleidomastoid branch of, S–323,
    S–324

Occipital bone, S–16, S–18, S–125, S–126, S–128,
    S–129, S–131, S–132, S–141, S–BP 19
  basilar part of, S–129, S–131, S–132, S–141,
    S–193, S–360, S–361, S–362, S–373,
    S–398, S–409
  clivus of, S–132
  condylar canal and fossa of, S–131
  condyle of, S–132
  foramen magnum of, S–129, S–131
  groove
    for inferior petrosal sinus, S–129, S–132
    for occipital sinus, S–132
  hypoglossal canal of, S–129, S–131
  jugular foramen of, S–129
  lowest level of, S–BP 21
  of newborn, S–127
  nuchal line of
    inferior, S–131
    superior, S–131
  occipital crest of
    external, S–131
    internal, S–132
  occipital protuberance of
    external, S–129, S–131
    internal, S–132
  pharyngeal tubercle of, S–131, S–372, S–403,
    S–408
  posterior meningeal vessels of, groove for,
    S–132
  superior sagittal sinus of, groove for, S–132
  transverse sinus of, groove for, S–129, S–132
Occipital condyle, S–70, S–129, S–130, S–131,
    S–193, S–BP 10
  lateral mass for, superior articular surface
    of, S–140
  superior articular surface for, S–140
Occipital crest
  external, S–130, S–131
  internal, S–132
Occipital (posterior) horn, S–35
Occipital nerve
  greater, S–74, S–185, S–194, S–198, S–202
  lesser, S–74, S–185, S–194, S–195, S–198,
    S–202, T2.8
  third, S–185, S–198, S–202
Occipital nodes, S–344
Occipital pole, of cerebrum, S–31, S–33
Occipital protuberance
  external, S–6, S–125, S–126, S–129, S–130,
    S–131
  internal, S–132
Occipital sinus, S–29, S–49
  groove for, S–132
Occipital sulcus, transverse, S–31
Occipital triangle, S–5
Occipital vein, S–322, S–328
Occipitalis muscle, S–65, S–184
Occipitofrontalis muscle
  frontal belly of, S–65, S–86, S–122
  occipital belly of, S–65
  occipital muscle of, S–202
Occipitomastoid suture, S–130
Occipitotemporal gyrus
  lateral, S–32, S–33
  medial, S–32, S–33
Occipitotemporal sulcus, S–32, S–33
Occiput, S–3
Oculomotor nerve (CN III), S–21, S–30, S–40,
    S–49, S–54, S–55, S–63, S–77, S–84, S–85,
    S–86, T2.1, T2.4, S–BP 14
  accessory nucleus of, S–54, S–78
  distribution of, S–53
  inferior branch of, S–58, S–85, S–BP 7
  schema of, S–58
  superior branch of, S–58, S–85, S–BP 7
  in superior orbital fissure, S–134
Oculomotor nucleus, S–54, S–55, S–58
Odd facet, S–BP 32
Odontoblast layer, S–405

Netter Atlas of Human Anatomy: A Systems Approach

Olecranon, S–162, S–240, S–242, S–243
Olecranon fossa, S–155, S–159, S–160
Olfactory bulb, S–33, S–56, S–364, S–368,
    S–BP 61
  contralateral, S–56
Olfactory bulb cells, S–56
Olfactory cells, S–56
Olfactory glomerulus, S–56
Olfactory mucosa, S–56
  distribution of, S–364
Olfactory nerve (CN I), S–134, S–364, S–365,
    T2.1, T2.4
  distribution of, S–53
  schema of, S–56
Olfactory nerve fibers, S–56
Olfactory nucleus, anterior, S–56
Olfactory sensory neurons, S–56
Olfactory stria
  lateral, S–56
  medial, S–56
Olfactory sulcus, S–33
Olfactory tract, S–32, S–33, S–40, S–56, S–364
Olfactory tract nucleus, lateral, S–56
Olfactory trigone, S–56
Oligodendrocyte, cell body of, S–BP 4
Olivary complex, inferior, S–55
Olive, S–40
Omental appendices, S–424
Omental arterial arc, S–BP 77
Omental artery, S–BP 77
Omental bursa, S–417, S–418, S–429
  cross section, S–417
  stomach reflected, S–416
  superior recess of, S–418, S–419, S–429
Omental foramen
  probe in, S–416
  of Winslow, S–415, S–417, S–418, S–430
Omental taenia, S–425, S–428
Omentum, greater, S–544, S–546
Omoclavicular triangle, S–5
Omohyoid muscle, S–196, S–197, S–203,
    S–230, S–231, S–237, S–264, S–329,
    S–399
  inferior belly of, S–5, S–70, S–74, S–190,
    S–191, S–192, S–194
    nerve to, S–195
  phantom, S–324
  superior belly of, S–70, S–74, S–190, S–191,
    S–192, S–194
    nerve to, S–195
Oocytes, S–BP 96
  discharged, S–529
  primary, S–BP 96
Oogenesis, S–BP 89
Oogonium, S–BP 89
Operculum
  frontal, S–31
  parietal, S–31
  temporal, S–31
Ophthalmic artery, S–43, S–45, S–48, S–71,
    S–78, S–87, S–134, S–189, S–326, S–327,
    S–363, T5.1
  continuation of, S–87
  in optic canal, S–BP 7
Ophthalmic nerve, S–30, S–58, S–61, S–62,
    S–63, S–71, S–77, S–78, S–79, S–80, S–81,
    S–85, S–86, S–185, S–406, T2.4
  frontal branch of, S–134
  lacrimal branch of, S–134
  nasociliary branch of, S–134
  tentorial (recurrent meningeal) branch of,
    S–59, S–85
Ophthalmic vein
  inferior, S–87, S–92, S–363
  superior, S–30, S–87, S–92, S–134, S–328,
    S–363, S–BP 7
Opisthion, S–BP 21
Opponens digiti minimi muscle, S–110, S–253,
    S–256

Opponens pollicis muscle, S–109, S–253,
    S–256, S–257
Opposition, of upper limbs movements, S–265
Optic canal, S–123, S–129, S–134
Optic chiasm, S–30, S–32, S–33, S–40, S–45,
    S–46, S–50, S–51, S–57, S–327, S–368,
    S–523, S–524, S–BP 61
  imaging of, S–39
Optic disc, S–91, S–BP 7
Optic nerve (CN II), S–30, S–32, S–33, S–48,
    S–49, S–71, S–78, S–84, S–85, S–86, S–88,
    S–91, S–134, S–368, T2.1, T2.4, S–BP 61
  distribution of, S–53
  inner sheath of, S–88
  internal sheath of, vessels of, S–92
  left, S–327
  meningeal sheath of, S–85, S–88, S–92
    in optic canal, S–BP 7
  outer sheath of, S–88
  right, S–327
  schema of, S–57
  subarachnoid space of, S–88
  vessels of internal sheath of, S–92
Optic nerve tract, imaging of, S–BP 12
Optic radiation, S–57
Optic tract, S–33, S–34, S–38, S–40, S–44,
    S–49, S–57, S–368
Ora serrata, S–88, S–90, S–91
Oral cavity, S–358, S–360, S–368, S–372,
    S–394, S–BP 61
  afferent innervation of, S–406, S–BP 6
  inspection of, S–397
  roof, S–398
Oral region
  of head and neck, S–5
  nerves of, S–73
Orbicularis oculi muscle, S–65, S–184, S–BP 36
  orbital part of, S–184, S–BP 35, S–BP 36
  palpebral part of, S–82, S–184, S–BP 35,
    S–BP 36
Orbicularis oris muscle, S–60, S–65, S–122,
    S–184, S–186, S–400, S–BP 35, S–BP 36
Orbiculus ciliaris, S–90
Orbit, S–367, T3.1
  anterior view of, S–86
  arteries of, S–87
  fasciae of, S–BP 7
  medial wall of, S–368, S–BP 61
  nerves of, S–85
  superior view of, S–86
  surface of, S–123
  veins of, S–87
Orbital cavity, S–369, S–370
Orbital fat, S–BP 61
Orbital fat body, S–BP 7
Orbital fissure
  inferior, S–123, S–125, S–130, S–BP 7, S–BP
    19
  superior, S–123, S–124, S–134, S–BP 7, S–BP
    19
Orbital gyri, S–33
Orbital muscles, S–BP 61
Orbital , of newborn, S–127
Orbital process, of palatine bone, S–123
Orbital region, of head and neck, S–5
Orbital septum, S–82, T2.1
Orbital sulcus, S–33
Oropharynx, S–196, S–329, S–358, S–372,
    S–373, S–403, S–411
  posterior wall of, S–397
Orthotonic stomach, S–BP 67
Osseous cochlea, S–97
Osseous spiral lamina, S–96, S–97
Osteoblasts, S–BP 31
Osteoclasts, S–BP 31
Osteocytes, S–BP 31
Osteoid, S–BP 31
Osteonic canals (Haversian), S–BP 31
Otic capsule, S–97

Otic ganglion, S–21, S–59, S–62, S–63, S–65,
    S–67, S–77, S–79, S–80, S–81, S–95, S–136,
    S–187
  schema of, S–80
Oval (vestibular) window, S–96, S–127
Ovarian artery, S–463, S–468, S–478, S–483,
    S–490, S–508, S–513, S–515, S–529. See
    also Testicular artery
  ureteric branch of, S–468
  uterine, S–511
Ovarian cycle, S–BP 95
Ovarian follicles
  primary, S–529
  primordial, S–529, S–BP 96
  ruptured, S–529
  secondary, S–529
  tertiary, S–529
Ovarian hormone, S–BP 95
Ovarian ligament, proper, S–478, S–483,
    S–485, S–486, S–487, S–488, S–505, S–529
Ovarian plexus, S–513, S–515
Ovarian vein, S–342, S–463, S–478, S–483,
    S–508, S–529. See also Testicular vein
Ovarian vessels, S–511
  tubal branches of, S–511
Ovaries, S–463, S–472, S–478, S–485, S–487,
    S–488, S–489, S–505, S–508, S–513, S–515,
    S–522, S–529, S–BP 96, S–BP 97, T10.1
  cortex of, S–BP 96
  left, S–486
  ligaments of, suspensory, S–508
  right, S–486
  superficial epithelium of, S–529, S–BP 96
  suspensory ligament of, S–478, S–485,
    S–486, S–487, S–489, S–490, S–505,
    S–529
Ovum, S–BP 89, S–BP 95

P

Pacini's corpuscle. See Lamellar corpuscle
    (Pacini's)
Pain, referred visceral, sites of, S–BP 5
Palate
  mucous glands of, S–21
  postsynaptic fibers to, S–366
  uvula of, S–373, S–397
Palatine aponeurosis, S–398, S–408
Palatine artery
  ascending, S–189, S–323, S–403
    pharyngeal branch of, S–403
    tonsillar branch of, S–403
  descending, S–71, S–189, S–323, S–363,
    S–371
  greater, S–71, S–189, S–363, S–371, S–BP 59
  lesser, S–71, S–189, S–363, S–403
    tonsillar branch of, S–403
Palatine bone, S–129, S–131, S–362, S–370
  horizontal of, S–130, S–131, S–360, S–361,
    S–362, S–398, S–404, S–BP 10
  nasal crest of, S–360
  nasal spine of, posterior, S–131, S–360,
    S–362
  of newborn, S–127
  orbital process of, S–123, S–362
  palatine foramen of
    greater, S–131, S–360
    lesser, S–131, S–360
  perpendicular of, S–360, S–362
  pyramidal process of, S–127, S–130, S–131
  sphenoidal process of, S–362
Palatine folds, transverse, S–398
Palatine foramen
  greater, S–133, S–362, S–BP 59
  lesser, S–133, S–362, S–BP 59
Palatine glands, S–372, S–398, S–403
Palatine nerve, S–73, S–79, S–364
  communication between, S–371
  descending, S–79

Palatine nerve *(Continued)*
  greater, S–59, S–71, S–73, S–77, S–79,
    S–364, S–365, S–366
    in greater palatine foramen, S–133
    posterior inferior lateral nasal branch of,
      S–77, S–364
  inferior posterior nasal branches of greater,
    S–371
  lesser, S–59, S–71, S–77, S–79, S–364,
    S–365, S–366, S–371
    in lesser palatine foramen, S–133
Palatine process, S–129, S–130
  of maxilla, S–398, S–404
Palatine raphe, S–398
Palatine tonsil, S–372, S–373, S–397, S–398,
  S–400, S–402, S–403, T6.1
Palatine vein
  descending, S–363
  external, S–363
Palatine vessels
  greater, S–133
  lesser, S–133
Palatoglossal arch, S–397, S–402, S–403
Palatoglossus muscle, S–68, S–398, S–400,
  S–401, S–402
Palatomaxillary suture, S–131
Palatopharyngeal arch, S–373, S–397, S–402,
  S–403
Palatopharyngeal ridge (Passavant's), S–408
Palatopharyngeus muscle, S–68, S–374, S–398,
  S–400, S–401, S–402, S–403, S–408, S–409
Palm, S–10, S–164
Palmar aponeurosis, S–165, S–229, S–244,
  S–245, S–250, S–251, S–254, T4.2
Palmar arch
  deep, S–256, S–303, S–331, S–BP 51
  superficial, S–227, S–253, S–257, S–303,
    S–331, S–BP 51
  superficial venous, S–304
Palmar carpal ligament, S–247
Palmar digital artery, S–331
  common, S–256, S–331
  proper, S–331
Palmar digital veins, S–304, S–332
Palmar ligament, S–169, S–252, S–255, S–553,
  S–BP 107
Palmar metacarpal arteries, S–256, S–331
Palmar metacarpal veins, S–332
Palmar radioulnar ligament, S–165
Palmaris brevis muscle, S–110, S–250
Palmaris longus muscle, S–109, S–244, S–245,
  S–552
Palmaris longus tendon, S–10, S–165, S–244,
  S–247, S–249, S–250, S–251, S–253, S–552
Palpebral arterial arches
  inferior, S–87
  superior, S–87
Palpebral artery
  lateral, S–87
    inferior, S–87
    superior, S–87
  medial, S–87
    inferior, S–87
    superior, S–87
Palpebral conjunctiva, S–82, S–BP 7
  inferior, S–82
  superior, S–82
Palpebral ligament
  lateral, S–BP 7
  medial, S–82, S–BP 7
Pampiniform (venous) plexus, S–337, S–497,
  S–510, T5.3, S–BP 94
Pancreas, S–20, S–21, S–68, S–394, S–417,
  S–418, S–419, S–532, S–542, S–545, T8.1,
  S–BP 25, S–BP 98
  autonomic innervation of, S–457
  body of, S–415, S–416
  head of, S–415, S–416, S–421, S–432, S–438,
    S–548

Pancreas *(Continued)*
  as site of referred visceral pain, S–BP 5
  lymph vessels and nodes of, S–BP 55,
    S–BP 55
  in situ, S–433
  tail of, S–416, S–434, S–436, S–459
  uncinate process of, S–433
  veins of, S–442
Pancreatic artery
  dorsal, S–435, S–436, S–438, S–439, S–528,
    S–BP 77, S–BP 82
  great, S–436, S–438, S–528
  inferior, S–436, S–438, S–439, S–528,
    S–BP 77
Pancreatic duct, S–432, S–433, S–BP 74, S–BP 101
  accessory (of Santorini), S–421, S–433,
    S–BP 73
  sphincter of, S–432
  variations in, S–BP 73, S–BP 74
  of Wirsung, S–421, S–433, S–BP 69, S–BP 73
Pancreatic islets (Langerhans), S–522
Pancreatic notch, S–433
Pancreatic tail, S–528, S–BP 101
Pancreatic vein, S–442, S–444
Pancreaticoduodenal arteries
  anterior
    inferior, S–528
    superior, S–528
  inferior, S–528
  posterior
    inferior, S–528
    superior, S–528
Pancreaticoduodenal artery
  anterior
    inferior, S–436, S–438, S–439, S–440,
      S–450
    superior, S–416, S–435, S–436, S–438,
      S–439, S–450, S–BP 77
  inferior, S–436, S–438, S–439, S–440, S–BP 77
  posterior
    inferior, S–436, S–438, S–439, S–440, S–450
    superior, S–435, S–436, S–438, S–439,
      S–450, S–BP 77
Pancreaticoduodenal nodes, S–BP 55
Pancreaticoduodenal plexus
  anterior
    inferior, S–448, S–449
    superior, S–448, S–449
  inferior, S–451
  posterior
    inferior, S–449
    superior, S–449
Pancreaticoduodenal vein
  anterior
    inferior, S–442, S–444, S–446
    superior, S–442, S–444, S–446
  inferior, S–BP 78
  posterior
    inferior, S–442, S–444, S–446
    superior, S–442, S–446, S–BP 78
Papilla
  groove of, S–402
  keratinized tip of, S–402
Papillary duct, S–BP 84, S–BP 85
Papillary muscle, S–538, S–539
  anterior, S–311, S–315, S–319
    right, S–313, S–316
  inferior, S–311, S–312, S–315, S–319
    left, S–316
    right, S–313, S–316
  septal, S–311, S–313, S–315, S–316
  superior, S–312, S–315, S–319
    left, S–316
Parabrachial nucleus, S–81
Paracentral lobule, S–32
Paracentral sulcus, S–32
Paracolic gutter
  left, S–424, S–483, S–484
  right, S–424, S–425, S–483, S–484

Paracolic nodes, S–BP 57
Paradidymis, S–505
Paraduodenal fossa, S–422
Parahippocampal gyrus, S–32, S–33, S–34, S–56
Paramammary nodes, S–349
Paramesonephric duct (müllerian), S–505
Paranasal sinuses, S–367, T7.1, S–BP 61
  changes with age, S–370
  coronal section of, S–368
  mucous glands of, S–21
  paramedian views, S–369
  transverse section of, S–368
Pararectal fossa, S–225, S–483, S–484
Pararenal fat, S–545
Pararenal fat body, S–461
Pararenal fat capsule, S–461
Parasternal lymph nodes, S–348, S–349
Parasympathetic division, S–15
Parasympathetic fibers, S–456, S–457
  postsynaptic, S–BP 15
  presynaptic, S–BP 15
  in reproductive organs, S–515
    male, S–519
  of tracheobronchial tree, S–384
Parasympathetic nervous system, S–21
  general topography, S–BP 14
Paraterminal gyrus, S–56
Parathyroid gland, S–72, S–522, S–527
  inferior, S–526, S–527
  superior, S–526, S–527
Paratracheal lymph nodes, S–350, S–351
Paraumbilical veins, S–217, S–446, T5.3
  in round ligament of liver, S–337
  tributaries of, S–337
Paraurethral duct (Skene's), S–BP 86
  openings of, S–491
Paraurethral (Skene's) glands, S–505, S–BP 86
  primordium of, S–505
Paravertebral anastomoses, S–24
Paravesical fossa, S–529
Parental generation, S–BP 89
Parietal artery
  anterior, S–46
  posterior, S–46
Parietal bone, S–123, S–125, S–126, S–128,
  S–131, S–132, S–BP 19
  mastoid angle of, S–132
  middle meningeal vessels of, groove for, S–132
  of newborn, S–127
  tuber (eminence) of, S–127
Parietal cell, S–BP 67
Parietal emissary vein, S–26, S–322
Parietal epithelial cell, S–BP 83
Parietal foramen, S–128
Parietal lobule
  inferior, S–31
  superior, S–31
Parietal nodes, thoracic, S–351
Parietal peritoneum, S–214, S–215, S–217,
  S–218, S–417, S–418, S–419, S–424, S–459,
  S–461, S–464, S–478, S–483, S–484, S–488,
  S–492, S–532, S–546, S–BP 100
Parietal pleura, S–22
  cervical, S–385, S–386, S–390
  costal part of, S–75, S–207, S–307, S–386,
    S–387, S–389, S–390, S–391, S–393
  diaphragmatic part of, S–75, S–207, S–307,
    S–387, S–389
  mediastinal part of, S–75, S–207, S–307,
    S–308, S–309, S–387, S–389, S–390,
    S–391, S–393, S–539
Parietal region, of head and neck, S–5
Parietooccipital artery, S–47, S–48
Parietooccipital sulcus, S–31, S–32
Paroophoron (caudal mesonephric tubules),
  S–505
Parotid duct, S–396
  papilla, S–397
  of Stensen, S–64, S–186, S–187, S–323

Netter Atlas of Human Anatomy: A Systems Approach

Parotid fascia, S–184
Parotid gland, S–20, S–21, S–63, S–64, S–67,
 S–80, S–93, S–190, S–192, S–197, S–373,
 S–394, S–396, S–400, T8.1, S–BP 14, S–BP
 15, S–BP 60
  accessory, S–396
Parotid lymph node, superficial, S–344
Parotid space, S–324
Parotideomasseteric region, of head and neck,
 S–5
Parous introitus, S–BP 92
Pars distalis, S–524
Pars flaccida, S–94
Pars interarticularis, S–145
Pars intermedia, S–524
Pars tensa, S–94, S–BP 8
Pars tuberalis, S–524
Parturition, neuropathways in, S–514
Patella, S–11, S–120, S–268, S–269, S–271,
 S–272, S–279, S–280, S–282, S–287, S–288,
 S–BP 33
Patellar anastomosis, S–270, S–271, S–BP 52,
 S–BP 53
Patellar ligament, S–11, S–175, S–268, S–269,
 S–271, S–272, S–279, S–280, S–281, S–282,
 S–287, S–288, S–289, T4.3, S–BP 33, S–BP
 109
Patellar retinaculum
  lateral, S–268, S–269, S–272, S–279, S–280,
   S–288, S–289
  medial, S–268, S–269, S–271, S–279, S–280,
   S–281, S–288, S–289
Pecten, S–226
Pecten pubis, S–151, S–153, S–154, S–475,
 S–481
Pectinate ligament, S–89
Pectinate (dentate) line, S–226
Pectinate muscles, S–311
Pectineal fascia, S–213
Pectineal ligament (Cooper's), S–212, S–213,
 S–216, S–217, S–218, S–219
Pectineus muscle, S–115, S–268, S–269, S–270,
 S–271, S–278, S–300, S–549, S–550, S–554,
 S–BP 104, S–BP 108
Pectoral fascia, S–231, S–473
Pectoral girdle, S–120
Pectoral muscle, S–183
Pectoral nerve
  lateral, S–106, S–203, S–231, S–237, T2.9
  medial, S–106, S–203, S–231, S–237, T2.10
Pectoral region, of thorax, S–7
Pectoralis major muscle, S–7, S–8, S–10, S–156,
 S–190, S–192, S–203, S–204, S–206, S–211,
 S–212, S–231, S–236, S–237, S–238, S–239,
 S–264, S–349, S–387, S–473, S–534, S–535,
 S–536, S–537, S–538, S–539, S–540, S–551
  abdominal part of, S–230
  clavicular head of, S–7, S–230
  sternal head, S–7
  sternocostal head of, S–230
Pectoralis major tendon, S–238, S–551
Pectoralis minor muscle, S–203, S–204, S–231,
 S–236, S–237, S–238, S–264, S–349, S–387,
 S–534, S–535, S–536, S–537
Pectoralis minor tendon, S–235, S–236, S–237,
 S–239
Pedicle, S–144, S–147, S–148, S–BP 26
  of C5, S–BP 22
  of L1, S–BP 28
  of L3, S–BP 24
  of L4, S–138
  lumbar, S–145
  thoracic, S–144
Pedis artery, dorsalis, S–333, S–BP 43
Pelvic cavity, S–BP 2
  male, S–484
Pelvic diaphragm, S–483, S–485, S–495, T4.2
  fascia of, S–222
  female, S–220, S–479, S–480

Pelvic diaphragm (Continued)
  inferior fascia of, S–223, S–225, S–492,
   S–493, S–502, S–512
  levator ani muscle of, S–502
  male, S–481
  raphe of, S–479, S–480
  superior fascia of, S–223, S–478, S–490
Pelvic extraperitoneal spaces, S–222, T4.2
Pelvic fascia, tendinous arch of, S–464, S–465,
 S–478, S–490
Pelvic girdle, S–120
Pelvic ligaments, S–489
Pelvic node, T6.1
Pelvic plexus
  inferior, S–513, S–514, S–515, S–517, S–518,
   S–519
  superior, S–518
  uterovaginal, S–515
Pelvic splanchnic nerves, S–21, S–104, S–BP 15
 S2-4, S–BP 14
Pelvic surface, S–146
Pelvic viscera
  arteries of, S–508
  female, S–485
   nerves of, S–513
  male, S–495
   nerves of, S–517
  veins of, S–508
Pelvis
  bone framework of, S–475
  bones and ligaments of, S–152, S–154
  diameters of, S–477
  fasciae of, S–BP 90
  female
   arteries of, S–507, S–508, S–510
   diaphragm, S–220
   fasciae of, S–BP 90
   lymph nodes of, S–355
   lymph vessels of, S–355
   radiographs, S–476
   urethra, S–550
   vagina, S–550
   veins of, S–508
  greater, S–154
  lesser, S–154
  lower, cross section of, S–BP 104
  male
   arteries of, S–509, S–BP 94
   bladder-prostate gland junction, S–549
   fasciae of, S–BP 90
   lymph nodes of, S–357
   lymph vessels of, S–357
   radiographs, S–476
   veins of, S–509, S–BP 94
   viscera of, S–495
  sagittal T2-weighted MRIs in, S–521
  structures with high clinical significance,
   T4.2, T5.3, T6.1
Penile fascia, superficial, S–BP 90
Penis, S–472, S–502, S–521, S–BP 97
  body of, S–8, S–9, S–506
  bulb of, S–502, S–503, S–504
   artery of, S–504, S–512
  cavernous nerves of, S–517, S–519
  corpus cavernosum of, S–502, S–504, S–530
  corpus spongiosum of, S–530
  crus of, S–465, S–501, S–502, S–503, S–504
  Dartos fascia of, S–495, S–497, S–500,
   S–501, S–502
  deep arteries of, S–501, S–503, S–504,
   S–512, T5.4
  deep (Buck's) fascia of, S–211, S–212, S–213,
   S–222, S–224, S–495, S–497, S–500,
   S–501, S–502, S–504, S–512, S–BP 90,
   S–BP 94
  dorsal arteries of, S–501, S–504, S–510,
   S–512, T5.4
   deep, S–512, S–BP 94
   left, S–509

Penis (Continued)
  dorsal nerves of, S–104, S–275, S–501,
   S–504, S–512, S–516, S–517, S–518,
   S–519, S–520
  dorsal veins of
   deep, S–211, S–495, S–501, S–504, S–509,
    S–510, S–BP 94
   superficial, S–211, S–337, S–501, S–509,
    S–BP 94
  fascia of, S–418
  fundiform ligament of, S–216
  glans, S–495, S–502
   corona, S–502
   neck, S–502
  ligament of
   fundiform, S–495
   suspensory, S–495
  prepuce of, S–506
  skin of, S–497, S–500, S–501, S–502
  subcutaneous tissue of, S–211, S–212, S–213
  superficial dorsal vein, S–9
  suspensory ligament of, S–212, S–BP 40
Perforated substance
  anterior, S–33, S–40, S–56
  posterior, S–33
Perforating branches
  anterior, S–297
  posterior, S–297
Perforating cutaneous nerves, S–516, S–520
Perforating radiate artery, capsular branch of,
 S–467
Perforating veins, S–228, S–229, S–332, S–335,
 S–BP 85
  anterior, S–338
Perianal skin, S–223
  sweat glands in, S–226
Perianal space, S–222, S–225, S–226
Perianal tissues, S–506
Peribiliary lymph vessel, S–BP 58
Pericallosal artery, S–46, S–47
  precuneate branch of, S–47
Pericardiacophrenic artery, S–75, S–205,
 S–207, S–307, S–308, S–309, S–336, S–389,
 S–390, S–391, S–539
Pericardiacophrenic vein, S–205, S–207, S–307,
 S–309, S–389, S–390, S–391, S–539
Pericardiacophrenic vessels, S–308
Pericardial cavity, S–308, S–539, S–BP 2
Pericardial sac, S–75, S–207, S–307, S–308,
 S–309, S–310, S–311, S–312, S–387, S–390,
 S–391
Pericardial sinus
  oblique, S–308, S–538, S–539
  transverse, S–306, S–308, S–309, S–311,
   S–312
Pericardium, S–196, S–329, S–389, S–393,
 S–410, T5.2
  bare area of, S–385
  central tendon covered by, S–207
  diaphragmatic part of, S–308
Perichoroidal space, S–88, S–89
Pericranium, S–28
Pericyte, S–BP 30
Periglomerular cell, S–56
Perimysium, S–BP 34
Perineal artery, S–488, S–494, S–504, S–509,
 S–511, S–512, S–BP 94
  transverse, S–512
Perineal body, S–221, S–224, S–464, S–466,
 S–482, S–492, S–493, S–495, S–496, S–502,
 S–504, S–512, T4.2, S–BP 90
Perineal fascia, S–221, S–222, S–BP 90
  deep, S–495, S–501, S–502, S–504
  Gallaudet's, S–465, S–488, S–492, S–493,
   S–BP 86
  superficial, S–495, S–500, S–501, S–502,
   S–504, S–511, S–512, S–520
Perineal ligament, transverse, S–464, S–479,
 S–481, S–490, S–492, S–495, S–504, S–512

Perineal membrane, S–217, S–219, S–220, S–222, S–224, S–418, S–464, S–465, S–471, S–482, S–485, S–488, S–492, S–493, S–494, S–495, S–501, S–502, S–504, S–511, S–512, S–516, S–520, S–BP 86, S–BP 90, S–BP 94
  anterior thickening of, S–504
Perineal muscles, S–183
  deep, S–221, S–479
    transverse, S–485, S–493, S–494
  superficial transverse, S–492, S–493, S–494
  transverse
    deep, S–220, S–222, S–418, S–495, S–504, S–509, S–511
    superficial, S–220, S–221, S–222, S–224, S–418, S–501, S–502, S–504, S–511, S–512
  superior, S–512
Perineal nerves, S–104, S–225, S–275, S–494, S–516, S–517, S–520
  deep, S–520
    branches of, S–516, S–520
  deep branch of, S–494
  superficial, S–520
    branches of, S–516, S–520
  superficial branch of, S–494
Perineal pouch
  deep, with endopelvic fascia, S–BP 90
  superficial, S–BP 90
Perineal raphe, S–491, S–506
Perineal spaces, S–222
  female, S–494
  male, S–504
  perineal, S–511
  superficial, S–488, S–492, S–493, S–495, S–501, S–502, S–511, S–512
Perineal subcutaneous tissue (Colles' fascia), S–465, S–488
  membranous layer of, S–465, S–492, S–493, S–494, S–BP 86
Perineal vein, S–512
Perineum, S–2
  arteries of, S–511
  female, S–485, S–491
    deeper dissection, S–493
    fasciae of, S–BP 90
    lymph nodes of, S–356
    lymph vessels of, S–356
    nerves of, S–516
    superficial dissection, S–492
  male, S–500, S–501
    arteries of, S–512
    fasciae of, S–BP 90
    nerves of, S–520
    veins of, S–512
  subcutaneous tissue of, S–492
  veins of, S–511
Perineurium, S–BP 4
Periodontium, S–405
Periorbita, S–BP 7
Periosteal vessels, S–BP 31
Periosteum, S–BP 31
Peripheral arteries, S–BP 15
Peripheral nerve
  features of, S–BP 4
  typical, S–BP 4
Peripheral nervous system, S–15
Periportal arteriole, S–431
Periportal bile ductule, S–431
Periportal space, S–431
Perirenal fat, S–545
Perisinusoidal spaces, S–431
Peritoneal reflection, S–221, S–226
Peritoneum, S–225, S–428, S–459, S–485, S–508, S–513, S–545, T8.1, S–BP 41, S–BP 90
  inferior extent of, S–515, S–519
  parietal, S–214, S–215, S–217, S–218, S–495, S–546, S–BP 90, S–BP 100
  pelvic part of, S–465, S–515

Perivascular fibrous capsule (Glisson's), S–430, S–431
Permanent teeth, S–404
Perpendicular , S–123, S–129
Pes anserinus, S–268, S–269, S–279, S–280, S–300
Pes hippocampi, S–37, S–38
Petit, lumbar triangle of, S–6, S–199
  internal oblique muscle in, S–199
Petropharyngeus muscle, S–409
Petrosal artery, superficial, descending branch of, S–95
Petrosal nerve
  deep, S–65, S–67, S–77, S–79, S–364, S–366
  greater, S–63, S–65, S–67, S–71, S–77, S–79, S–85, S–133, S–134, S–366, S–BP 8
    in foramen lacerum, S–133
    groove for, S–98, S–132
    hiatus for, S–134
  lesser, S–62, S–63, S–65, S–67, S–80, S–85, S–94, S–95, S–133, S–134
    groove for, S–132
    hiatus for, S–134
Petrosal sinus
  inferior, S–29, S–30, S–134
    groove for, S–129, S–132
  superior, S–29, S–30, S–98
    groove for, S–129, S–132
Petrosal vein, S–30, S–49
Petrosquamous fissure, S–127
Petrotympanic fissure, S–131, S–133
Peyer's patches, S–421
  aggregate lymphoid nodules, S–343
Phalangeal bones, S–169
  foot, S–120, S–176, S–177
    distal, S–176, S–181
    middle, S–176, S–181
    proximal, S–176, S–181
  hand, S–120
    distal, S–109, S–167, S–168, S–169
    middle, S–167, S–168, S–169, S–553, S–BP 29
    proximal, S–167, S–168, S–169, S–BP 29, S–BP 107
  middle, S–BP 42
  proximal, S–BP 42
  1st, S–BP 106
Phallus, body of, S–506
Pharyngeal aponeurosis, S–374
Pharyngeal artery, S–189
  ascending, S–43, S–71, S–72, S–95, S–189, S–323, S–324, S–326
  meningeal branch of, S–27, S–134, S–326
Pharyngeal constrictor muscle, S–372, S–378
  cricopharyngeal part of, S–526
  inferior, S–68, S–192, S–378, S–407, S–408, S–409, S–526, S–527
    cricopharyngeal part of, S–408, S–411
    thyropharyngeal part of, S–411, S–412
  middle, S–68, S–192, S–374, S–378, S–401, S–403, S–408, S–409, S–526
  superior, S–68, S–187, S–189, S–374, S–398, S–400, S–401, S–407, S–408, S–409, S–526
    glossopharyngeal part of, S–408
    thyropharyngeal part of inferior, S–374
Pharyngeal nerve, S–59, S–79
Pharyngeal nervous plexus, S–72
Pharyngeal plexus, S–67, S–68, S–73, S–76, S–77
Pharyngeal raphe, S–361, S–408, S–409, S–412, S–526
Pharyngeal recess, S–361, S–373, S–403, S–BP 60
Pharyngeal regions, nerves of, S–73
Pharyngeal tonsil, S–360, S–361, S–372, S–373, S–403, S–409, T6.1, S–BP 22, S–BP 23
  imaging of, S–39
Pharyngeal tubercle, S–131, S–141, S–360, S–409
  of occipital bone, S–403, S–408

Pharyngobasilar fascia, S–345, S–398, S–401, S–407, S–408, S–409
Pharyngoepiglottic fold, S–374, S–409
Pharyngo-esophageal constriction, S–411
Pharyngoesophageal junction, S–374
Pharynx, S–373, S–394
  afferent innervation of, S–BP 6
  lymph nodes of, S–345
  lymphatic drainage of, S–345
  medial view of, S–372
  muscles of
    lateral view, S–407
    medial view, S–408
    partially opened posterior view, S–409
  posterior view of, S–72, S–526
  zone of sparse muscle fibers, S–526
Philtrum, S–5, S–397
Phrenic arteries, inferior, S–208, S–339, S–341, S–410, S–435, S–436, S–517, S–528, S–531, S–532, S–BP 77
Phrenic ganglion, S–208, S–449
Phrenic muscle, S–410
Phrenic nerve, S–73, S–75, S–76, S–99, S–105, S–106, S–193, S–194, S–195, S–196, S–204, S–205, S–207, S–208, S–235, S–237, S–307, S–308, S–309, S–323, S–329, S–336, S–387, S–389, S–390, S–391, S–393, S–410, S–525, S–533, S–535, S–536, S–539, T2.2, T2.8
  left, S–393
  pericardial branch of, S–75
  phrenicoabdominal branches of, S–75
  relationship to pericardium and, S–75
  right, S–393
  sternal branch of, S–208
Phrenic plexus, S–105, S–448, S–449, S–533
  left, S–449
  right, S–449
Phrenic veins, inferior, S–342, S–531
  left, S–340
Phrenicocolic ligament, S–416, S–419
Phrenicoesophageal ligament, superior, S–413
Phrenicopleural fascia, S–413
Pia, S–17, S–22
Pia mater, S–26, S–28
Pial plexus, S–23
  arterial, S–24
    peripheral branches from, S–24
Pigment cells, retinal, S–57
Pillar (rod) cells, S–97
Pilosebaceous apparatus, S–BP 1
Pineal gland, S–32, S–36, S–37, S–40, S–41, S–522
Pineal recess, S–32, S–34
Piriform cortex, S–56
Piriform fossa, S–411
Piriform recess, S–373, S–376
Piriformis muscle, S–114, S–219, S–220, S–273, S–274, S–275, S–276, S–300, S–479, S–480, S–481, S–507, S–513, S–516, S–517, S–520, T4.3
  bursa of, S–276
  left, S–509
  nerve to, S–104, S–114, T2.12
  tendon of, S–276
Pisiform bone, S–163, S–164, S–165, S–166, S–167, S–168, S–169, S–244, S–245, S–247, S–249, S–250, S–251, S–253, S–256, S–257, S–BP 29
Pisohamate ligament, S–165, S–BP 29
Pisometacarpal ligament, S–165, S–BP 29
Pituitary gland, S–30, S–32, S–33, S–358, S–522, S–523, S–524
  anterior lobe (adenohypophysis) of, S–45, S–524
  cleft of, S–524
  imaging of, S–39
  posterior lobe (neurohypophysis) of, S–45, S–524
Pituitary gonadotropins, S–BP 97

Netter Atlas of Human Anatomy: A Systems Approach

Pituitary stalk, imaging of, S–39
Pivot joint, S–121
Plane joint, S–121
Plantar aponeurosis, S–179, S–290,
    S–293, S–294, S–295, S–556, T4.3, S–BP 110
Plantar arch, S–296, S–297, S–303, S–333,
    S–BP 43, S–BP 53
Plantar artery
    deep, S–289, S–291, S–292, S–297, S–333,
        S–BP 53
    lateral, S–290, S–295, S–296, S–297, S–335,
        S–BP 110
        cutaneous branches of, S–293
        deep branch of, S–BP 111
        superficial branch of, S–BP 111
    medial, S–290, S–295, S–296, S–335,
        S–BP 110
        cutaneous branches of, S–293
        deep branch of, S–295, S–296, S–BP 111
        superficial branch of, S–293, S–294, S–295,
            S–296
Plantar calcaneocuboid ligament, S–181
Plantar calcaneonavicular ligament, S–181,
    S–297
Plantar digital arteries, S–293, S–297
    common, S–294, S–295, S–297, S–BP 43
    proper, S–294
Plantar digital nerves
    common, S–118, S–295
    proper, S–118
Plantar fascia
    lateral, S–293, S–294
    medial, S–293
Plantar flexion, of lower limbs movements,
    S–265
Plantar interosseous muscle, S–296
Plantar ligaments, S–181, S–297
    of foot, S–181
    long, S–180, S–297, S–556
    short, S–180
Plantar metatarsal arteries, S–294, S–296,
    S–297, S–BP 43
Plantar metatarsal ligaments, S–181
Plantar nerve
    lateral, S–117, S–118, S–290, S–295, S–296,
        S–335, S–556, T2.14, S–BP 43, S–BP 110,
        S–BP 111
        cutaneous branches of, S–293
        deep branch of, S–295, S–296
        plantar cutaneous branches of, S–267
        plantar digital branches of, S–294
        proper plantar digital branches of, S–295,
            S–296
        superficial branch of, S–295, S–296
    medial, S–117, S–118, S–290, S–295, S–296,
        S–335, S–556, T2.14, S–BP 110
        cutaneous branches of, S–293
        deep branches of, S–296, S–BP 111
        plantar cutaneous branches of, S–267
        plantar digital branches of, S–294
        proper plantar digital branches of, S–295,
            S–296
        superficial branches of, S–293, S–296
Plantar vein
    lateral, S–334
    medial, S–334
Plantaris muscle, S–117, S–118, S–272, S–273,
    S–274, S–279, S–282, S–283, S–285, S–286,
    S–299, S–300, S–335
Plantaris tendon, S–273, S–284, S–285, S–335,
    S–555, S–BP 109, S–BP 110
Plasma, composition of, S–BP 44
Plasma cell, S–BP 30
Plasma proteins, S–BP 44
lets, S–BP 44
Platysma muscle, S–65, S–184, S–190, S–196,
    S–329, S–BP 35, S–BP 36
Pleura, S–388, T7.1
    costal part of, S–358, S–410

Pleura (Continued)
    diaphragmatic part of, S–358, S–410
    mediastinal part of, S–358, S–410
    visceral, S–392
Pleural cavity, S–539, S–BP 2
    costodiaphragmatic recess of, S–207, S–385,
        S–386, S–390, S–391
    costomediastinal recess of, S–207
        left, S–385
        right, S–385
Pleural reflection, S–385, S–386, S–387
Plica semilunaris, S–82, S–83
Polar body, S–BP 89
Pons, S–32, S–40, S–41, S–80, S–81, S–358,
    S–523, S–BP 18, S–BP 23
    imaging of, S–39, S–BP 12
Pontine arteries, S–44, S–45
    basilar and, S–46
Pontine vein
    lateral, S–49
    transverse, S–49
Pontomesencephalic vein, S–49
Popliteal artery, S–273, S–274, S–284, S–285,
    S–286, S–303, S–333, S–335, S–554, T5.4,
    S–BP 52, S–BP 53
    pulse point, S–303
Popliteal fossa, S–11
Popliteal ligament
    arcuate, S–281, S–282, S–283
    oblique, S–282
Popliteal nodes, S–343
Popliteal vein, S–274, S–284, S–285, S–304,
    S–334, S–335, S–347, S–554
Popliteal vessels, S–273
Popliteus muscle, S–118, S–273, S–282, S–283,
    S–286, S–299, S–300, S–335, S–555, S–BP 109
Popliteus tendon, S–173, S–280, S–281
Porta hepatis, S–429
Portal arteriole, S–431
Portal tract, S–430
Portal triad, S–420, S–430, S–433, S–542
Postanal space
    deep, S–222
    superficial, S–222
Postcentral gyrus, S–31
Postcentral sulcus, S–31
Posterior chamber
    of eye, S–82, S–88, S–90, S–91
    of eyeball, S–89
Posterior cord, of brachial plexus, S–236
Posterior cutaneous branches, S–198
Posterior inferior cerebellar artery (PICA), S–23
Posterior plane, S–1
Posterior region, of neck, S–5
Posterior root, S–79, S–454
Posterior root ganglion. See Spinal sensory
    (posterior root) ganglion
Posterior sacral foramen, S–BP 103
Postganglionic fibers, S–21
    in reproductive organs, S–515
    male, S–519
Postnatal circulation, S–305
Postsynaptic cell, S–BP 3
Postsynaptic membrane, S–BP 3
Poupart's ligament, S–211, S–212, S–213,
    S–215, S–216, S–218, S–219. See also
    Inguinal ligament (Poupart's)
Preaortic nodes, S–357, S–BP 57
Prececal nodes, S–BP 57
Precentral cerebellar vein, S–49
Precentral gyrus, S–31
Precentral sulcus, S–31
Precuneus, S–32
Prefrontal artery, S–44, S–46
Preganglionic fibers, S–21
    in reproductive organs, S–515
    male, S–519
Premolar tooth, S–405
    1st, S–404

Premolar tooth (Continued)
    2nd, S–404
Prenatal circulation, S–305
Preoccipital notch, S–31
Prepancreatic artery, S–436, S–438
Prepatellar bursa, subcutaneous, S–282
Prepontine cistern, S–35
Prepuce, S–495
Preputial gland, S–502
Prepyloric vein, S–444
Presacral fascia, S–222, S–490
Presacral space, S–222
Presternal region, of thorax, S–7
Presynaptic membrane, S–BP 3
Pretracheal (visceral) fascia, S–190, S–196,
    S–197, S–329, S–372, S–379
Pretracheal lymph nodes, S–525
Prevertebral anastomoses, S–24
Prevertebral fascia, S–372, S–408
Prevertebral nodes, S–351
Prevertebral soft tissue, S–BP 22
Prevesical fascia, umbilical, S–490
Prevesical plexus, S–357
Prevesical space, S–495. See also Retropubic
    (prevesical) space
Primary oocyte, S–BP 89
Primary spermatocytes, S–BP 89
Princeps pollicis artery, S–331, S–BP 51
Procerus muscle, S–60, S–65, S–122, S–184,
    S–BP 35, S–BP 36
Processus vaginalis, S–499
Profunda brachii artery, S–551
Progesterone, S–BP 95
Promontorial nodes, S–354, S–355, S–357
Promontory, S–93, S–BP 9
    sacral, S–146
Pronation, of upper limbs movements, S–265
Pronator quadratus muscle, S–109, S–248,
    S–249, S–254, S–256, S–262, S–552
Pronator teres muscle, S–109, S–227, S–238,
    S–239, S–241, S–244, S–247, S–248, S–249,
    S–262, S–263, S–552
    head of
        deep, S–247, S–249, S–264
        superficial, S–264
Prostate, S–20, S–21, S–418, S–465, S–472,
    S–496, T10.1
    capsule of, S–465
Prostate gland, S–217, S–221, S–495, S–503,
    S–504, S–505, S–509, S–521, S–530, S–549,
    S–BP 15, S–BP 97, S–BP 104, S–BP 108
    anterior commissure of, S–BP 93
    branch to, S–BP 94
    capsule of, S–BP 93
    cross section of pelvis, S–BP 93
    lymphatic drainage from, S–357
    primordium of, S–505
Prostatic ducts, S–503
Prostatic plexus, S–469, S–471, S–517, S–519
Prostatic plexus nerve, T2.3
Prostatic sinuses, S–503, S–BP 93
Prostatic urethra, S–465, S–503, S–530, S–549,
    S–BP 87, S–BP 93
Prostatic utricle, S–465, S–503, S–505, S–BP 93
Prostatic venous plexus, S–509
Proximal, as term of relationship, S–1
Proximal convoluted tubule, S–BP 83, S–BP 84
Proximal interphalangeal (PIP) joint, S–10,
    S–168, S–169, S–BP 29, S–BP 42
Proximal palmar crease, S–10
Psoas major, S–420, S–459, S–460, S–483,
    S–484, S–490
Psoas minor, S–483, S–484
Psoas muscle, S–277, S–549
    major, S–103, S–113, S–114, S–115, S–208,
        S–216, S–219, S–268, S–276, S–277,
        S–278, S–341, S–342, S–508, S–545,
        S–546, S–BP 25, S–BP 102, S–BP 103,
        S–BP 104

Psoas muscle (*Continued*)
    branches to, S–104
      muscular branches to, S–104
      nerves to, S–104
      tendon, S–BP 104
    minor, S–219, S–276, S–277, S–546
      muscular branches to, S–104
      nerves to, S–104
      tendon, S–219
    muscular branches to, S–113
Psoas tendon, S–549
Pterion, S–125, T3.1
Pterygoid artery, canal, S–371
Pterygoid canal, S–371
    artery of, S–71, S–95, S–189, S–327
    nerve (vidian) of, S–61, S–65, S–67, S–71, S–77,
      S–79, S–81, S–95, S–364, S–365, S–366
Pterygoid fossa, S–130
Pterygoid fovea, S–135
Pterygoid hamulus, S–62, S–125, S–129,
    S–130, S–187, S–398, S–403, S–407, S–408,
    S–BP 10
Pterygoid muscle
    artery to, S–189
    lateral, S–62, S–186, S–187, S–189, S–325,
      S–BP 37, S–BP 60
      artery to, S–189
      inferior head of, S–63, S–188, S–197, S–BP 37
      nerve to, S–59, S–73
      superior head of, S–63, S–188, S–197,
        S–BP 37
    medial, S–62, S–63, S–73, S–187, S–188,
      S–189, S–409, S–BP 60
      artery to, S–189
      nerves to, S–59, S–73
      window cut through right, S–188
Pterygoid nerve, medial, S–187
Pterygoid
    lateral, S–130, S–187, S–359, S–BP 10
    medial, S–130, S–187, S–370, S–BP 10
      hamulus of, S–359
      right, S–71
Pterygoid plexus, S–63, S–87
Pterygoid process, S–130, S–131
    hamulus of, S–125, S–130, S–131
    lateral of, S–125, S–129, S–130, S–131
    medial of, S–129, S–130, S–131
    pterygoid fossa of, S–131
    scaphoid fossa of, S–131
Pterygoid venous plexus, S–363, T5.1
Pterygomandibular raphe, S–187, S–189,
    S–359, S–398, S–400, S–403, S–407, S–408,
    T4.1
Pterygomaxillary fissure, S–125
Pterygopalatine fossa, S–125, S–130, S–369
Pterygopalatine ganglion, S–21, S–58, S–59,
    S–61, S–65, S–67, S–71, S–73, S–77, S–79,
    S–81, S–364, S–365, S–366, S–371, S–BP 14
    branches to, S–61
    pharyngeal branch of, S–364
    schema of, S–79
Puberty, S–BP 97
Pubic arch, S–475, S–477
Pubic bone, S–418, S–496, S–504, S–521,
    S–550
Pubic crest, S–216, S–481
Pubic hair, S–BP 97
Pubic ligament, inferior, S–464, S–475, S–479,
    S–480, S–481, S–482, S–492, S–495, S–504
Pubic ramus
    inferior, S–151, S–153
    superior, S–114, S–151, S–152, S–153,
      S–154, S–170, S–171, S–182, S–277,
      S–475, S–476, S–485, S–495, S–502
Pubic symphysis, S–8, S–9, S–151, S–152,
    S–215, S–216, S–219, S–276, S–464, S–475,
    S–476, S–477, S–479, S–480, S–481, S–482,
    S–485, S–486, S–490, S–492, S–494, S–495,
    S–500, S–504, S–507, S–550, T3.2, S–BP 104
    superior portion of, S–549

Pubic tubercle, S–8, S–9, S–151, S–152, S–153,
    S–154, S–182, S–212, S–213, S–215, S–219,
    S–268, S–269, S–475, S–482, S–492, S–502,
    S–BP 40
Pubic vein, S–342
Pubis, S–220, S–466, S–479, S–521
    body of, S–549, S–BP 104
    superior ramus of, S–269, S–481
    symphyseal surface of, S–513
Puboanalis (puborectalis) muscle, S–220,
    S–221, S–224, S–418, S–480, S–481, S–482,
    S–BP 104
Pubocervical ligament, S–478, S–490
Pubococcygeus muscle, S–220, S–224, S–479,
    S–480, S–481, S–482, S–488
Pubofemoral ligament, S–170, S–BP 108
Puborectalis muscle. *See* Puboanalis
    (puborectalis) muscle
Pubovesical ligament, S–478
    lateral, S–464, S–490
    medial, S–464, S–490
Pudendal artery, S–510
    deep external, S–336, S–341
    external, S–509
      deep, S–270, S–333, S–BP 52
      superficial, S–211, S–213, S–270, S–333,
        S–336, S–341, S–BP 52
    internal, S–114, S–341, S–441, S–468, S–478,
      S–494, S–504, S–507, S–509, S–511,
      S–512, S–549, S–BP 94, S–BP 104
      in pudendal canal (Alcock's), S–511
Pudendal canal (Alcock's), S–225, S–441,
    S–512, S–520
    pudendal nerve of, S–516
    internal, S–512
    pudendal vessels, internal, S–512
Pudendal nerve, S–16, S–103, S–104, S–114,
    S–225, S–274, S–275, S–471, S–494,
    S–512, S–513, S–514, S–515, S–516,
    S–517, S–519, S–520, S–549, T2.2, T2.13,
    S–BP 104
    internal, S–550
    nerve to, S–114
    in pudendal canal, S–516
Pudendal vein, S–510
    external, S–334, S–337
      deep, S–510
      superficial, S–266, S–510
    internal, S–445, S–509, S–512, S–549,
      S–BP 104
Pudendal vessels
    external, S–342, S–509
      superficial, S–211
    internal, S–225, S–342, S–550
Pulmonary acinus, S–380
Pulmonary arteries, S–302, S–392, T5.2,
    S–BP 63
    left, S–305, S–306, S–309, S–310, S–312,
      S–321, S–358, S–388, S–389, S–391,
      S–537, S–540, S–BP 98, S–BP 99
    right, S–305, S–309, S–310, S–311, S–312,
      S–321, S–358, S–388, S–389, S–390,
      S–537, S–540, S–BP 98, S–BP 99
Pulmonary heart valve, S–311, S–313, S–319
    anterior semilunar cusp of, S–311, S–314
    left semilunar cusp of, S–311, S–314
    right semilunar cusp of, S–311, S–314
Pulmonary ligament, S–350, S–388, S–390,
    S–391
Pulmonary (intrapulmonary) lymph nodes,
    S–350
Pulmonary plexus, S–68, S–384, S–447, S–BP
    13, S–BP 14
    anterior, S–101
    posterior, S–101
Pulmonary trunk, S–305, S–306, S–309, S–311,
    S–313, S–316, S–319, S–321, S–389, S–534,
    S–537, S–540, S–BP 99
Pulmonary veins, S–302, S–321, S–392, T5.2,
    S–BP 63

Pulmonary veins (*Continued*)
    inferior
      left, S–310, S–388, S–539, S–BP 98
      right, S–310, S–311, S–388, S–538, S–539
    left, S–305, S–308, S–310, S–312, S–316,
      S–388, S–391
    right, S–305, S–308, S–312, S–319, S–388,
      S–390, S–540
    superior
      left, S–309, S–310, S–312, S–388
      right, S–309, S–310, S–311, S–316, S–388
Pulse points, S–303
Pulvinar, S–33, S–36, S–37, S–40, S–41, S–48
    left, S–49
    right, S–49
    of thalamus, S–44, S–50
Pupil, S–82
Purkinje fibers
    of left bundle, S–319
    of right bundle, S–319
Putamen, S–34
Pyloric antrum, S–415
Pyloric canal, S–415, S–542
Pyloric glands, S–BP 67
Pyloric nodes, S–BP 55, S–BP 58
Pyloric orifice, S–421
Pyloric sphincter, S–414
Pylorus, S–414, S–415, S–420, S–542, T8.1
Pyramidal eminence, S–94, S–95, S–BP 8
Pyramidal process, S–127, S–130, S–131
Pyramidal system, S–BP 18
Pyramidalis muscle, S–212, S–215
Pyramids, S–40
    decussation of, S–40, S–41, S–BP 18

Q

Quadrangular space, S–235
Quadrate ligament, S–161
Quadrate tubercle, S–172
Quadratus femoris muscle, S–269, S–271,
    S–273, S–274, S–275, S–276, S–300,
    S–516
    nerve to, S–104, S–114, S–275, T2.12
Quadratus lumborum fascia, S–545
Quadratus lumborum muscle, S–103, S–113,
    S–201, S–208, S–219, S–277, S–341, S–342,
    S–459, S–460, S–461, S–483, S–484, S–545,
    S–546
Quadratus plantae muscle, S–118, S–295,
    S–296, S–BP 110, S–BP 111
Quadriceps femoris muscle, S–115, S–300
Quadriceps femoris tendon, S–11, S–268,
    S–269, S–271, S–279, S–280, S–282, S–287,
    S–288, S–289, S–BP 33
Quadriceps muscle. *See* Rectus femoris
    muscle
Quadrigeminal cistern, S–35
    imaging of, S–BP 12

R

Radial artery, S–165, S–238, S–241, S–244,
    S–247, S–249, S–251, S–253, S–257, S–259,
    S–260, S–261, S–303, S–331, S–552, T5.4,
    S–BP 50, S–BP 51
    dorsal carpal branch of, S–259
    palmar carpal branch of, S–249, S–256, S–257
    pulse point, S–303
    superficial palmar branch of, S–165, S–247,
      S–249, S–251, S–256, S–257, S–331,
      S–BP 51
Radial collateral ligament, S–161, S–BP 29
Radial fibers, S–89
Radial fossa, S–155, S–159
Radial groove, S–155
Radial head, S–227
Radial metaphyseal arcuate ligament, dorsal,
    S–166

Netter Atlas of Human Anatomy: A Systems Approach

Radial nerve, S–13, S–14, S–106, S–107, S–108,
S–109, S–111, S–112, S–227, S–235, S–236,
S–237, S–240, S–247, S–249, S–551, T2.3,
T2.10
communicating branches of, S–258
deep branch of, S–112, S–227, S–552
dorsal digital branches of, S–14, S–112, S–259
dorsal digital nerves from, S–13
inferior lateral brachial cutaneous nerve
from, S–13
posterior brachial cutaneous nerve from, S–13
superficial branch of, S–13, S–14, S–107,
S–112, S–227, S–229, S–242, S–258,
S–261, S–552
Radial recurrent artery, S–238, S–247, S–249,
S–BP 51
Radial styloid process, S–163, S–246
Radial tubercle, dorsal, S–163
Radial tuberosity, S–160, S–239
Radial veins, S–304, S–332
Radialis indicis artery, S–331, S–BP 51
Radiate ligament, of head of rib, S–388
Radiate sternochondral ligaments, S–388
Radiate sternocostal ligament, S–156
Radicular artery
anterior, S–24
great, S–23
posterior, S–24
Radicular vein, S–25
Radiocarpal joint, S–164, S–166
articular disc of, S–164, S–166
Radiocarpal ligament
dorsal, S–166, S–BP 29
palmar, S–BP 29
Radiolunate ligament
long, S–165
palmar, S–165
short, S–165
Radioscaphocapitate ligament, S–165
Radioulnar joint, distal, S–166
Radioulnar ligament
dorsal, S–166, S–BP 29
palmar, S–BP 29
Radius, S–120, S–159, S–160, S–161, S–162,
S–163, S–164, S–165, S–166, S–168, S–
241, S–243, S–245, S–246, S–248, S–249,
S–256, S–260, S–262, S–263, S–552, T3.3,
S–BP 29
anterior border of, S–162
anterior surface of, S–162
anular ligament of, S–161
articular surface of, S–168
body of, S–160, S–246
head of, S–159, S–160, S–162, S–239
interosseous border of, S–162
lateral surface of, S–162
neck of, S–159, S–160, S–162
posterior border of, S–162
posterior surface of, S–162
styloid process of, S–162, S–168
tuberosity of, S–159
Rami, S–22
anterior, S–22
of mandible, S–197, S–400
posterior, S–22
Rami communicantes, S–104, S–514
Recombinant chromatids, S–BP 89
Rectal ampulla, S–464
Rectal artery
inferior, S–509, S–511, S–512, S–BP 94
middle, S–509, S–BP 94
superior, S–517
Rectal fascia, S–221, S–222, S–223, S–225,
S–226, S–490, S–492, S–BP 90, S–BP 96
Rectal nerves, inferior, S–514, S–516, S–517,
S–520
Rectal plexus, S–452, S–453, S–469, S–513,
S–517
left, S–517

Rectal venous plexus
external, S–223
in perianal space, S–226
internal, S–223, S–226
Rectococcygeus muscle, S–219
Rectoperinealis muscle, S–482
Rectoprostatic (Denonvilliers') fascia, S–221,
S–418, S–482, S–495
Rectosigmoid junction, S–221, S–223, S–226,
S–428
Rectouterine fold, S–463, S–483, S–529
Rectouterine pouch, of Douglas, S–221, S–485,
S–486, S–487, S–529, T10.1, S–BP 96
Rectovaginal fascia, S–490
Rectovesical fold, S–419, S–484
Rectovesical pouch, S–221, S–418, S–484,
S–495
Rectovesical space, S–222
Rectum, S–20, S–21, S–219, S–220, S–221,
S–226, S–418, S–424, S–428, S–459, S–464,
S–478, S–479, S–480, S–483, S–484, S–485,
S–486, S–490, S–495, S–508, S–513, S–521,
S–550, T8.1, S–BP 90, S–BP 108
arteries of, S–441
circular muscle of, S–223
longitudinal muscle of, S–223
muscularis mucosae of, S–223
radiology of, S–138
in situ
female, S–221
male, S–221
transverse folds of, S–226
veins of, S–445
Rectus abdominis muscle, S–7, S–8, S–102,
S–203, S–204, S–206, S–209, S–212, S–213,
S–214, S–215, S–216, S–217, S–336, S–358,
S–418, S–483, S–484, S–490, S–492, S–495,
S–521, T4.2, S–BP 64, S–BP 104, S–BP 108
Rectus anterior capitis muscle, S–193, S–BP 60
nerves to, S–195
Rectus capitis muscle
anterior, S–345
posterior
major, S–200, S–201, S–202
minor, S–200, S–201, S–202
Rectus femoris muscle, S–11, S–115, S–268,
S–270, S–271, S–272, S–300, S–549, S–550,
S–554, S–BP 104, S–BP 108. See also
Quadriceps femoris muscle
bursa of, S–276
origin of, S–276
Rectus femoris tendon, S–268, S–280, S–288,
S–554
Rectus lateralis capitis muscle, S–193
nerves to, S–195
Rectus muscle
inferior, S–58, S–84, S–86, S–BP 7
branches to, S–85
lateral, S–58, S–84, S–85, S–86, S–368,
S–BP 7
check ligament of, S–BP 7
tendon of, S–88
medial, S–58, S–84, S–85, S–86, S–368,
S–BP 7
branches to, S–85
check ligament of, S–BP 7
tendon of, S–88
superior, S–58, S–84, S–85, S–86, S–BP 7
tendon of, S–92
Rectus sheath, S–211, S–212, S–546
anterior layer of, S–102, S–203, S–204,
S–212, S–213, S–214, S–216, S–492,
S–495, S–BP 40
cross section of, S–214
posterior layer of, S–102, S–213, S–214,
S–216, S–336
Recurrent artery
anterior, S–331
anterior ulnar, S–BP 50

Recurrent artery (Continued)
of Heubner, S–44, S–45, S–46, S–47
posterior, S–331
posterior ulnar, S–BP 50
radial, S–331, S–BP 50
tibial
anterior, S–289, S–333, S–BP 52, S–BP 53
posterior, S–333, S–BP 52, S–BP 53
Recurrent laryngeal nerve, S–197
Recurrent process, S–56
Red blood cells, S–BP 30, S–BP 44, S–BP 63
Red bone marrow, S–343
Red nucleus, S–33, S–54, S–55
Referred pain, visceral, sites of, S–BP 5
Renal artery, S–303, S–341, S–418, S–462,
S–468, S–471, S–508, S–510, T5.3
anterior branch, S–467
left, S–437, S–459, S–517, S–531, S–532
pelvic branch of, S–467
right, S–437, S–450, S–459, S–531
ureteric branch of, S–467, S–468, S–531
variations in, S–BP 88
Renal column (of Bertin), S–462, S–BP 85
Renal corpuscle, histology of, S–BP 83
Renal cortex, S–420, S–462, S–543, S–BP 84
Renal (Gerota's) fascia, S–461, S–532,
S–545
anterior layer, S–461
posterior layer, S–461
Renal ganglion, S–469, S–471, S–519, S–533
Renal impression, S–429, S–434
Renal medulla, S–462, S–BP 84, S–BP 85
Renal papilla, S–462
cribriform area of, S–BP 84
Renal pelvis, S–462, T9.1
Renal plexus, S–469, S–533
left, S–452, S–517
right, S–105
Renal pyramid, base of, S–462
Renal segments, S–467
Renal vein, S–304, S–418, S–462, S–468,
S–508, S–510
left, S–340, S–342, S–459, S–531, S–532
right, S–340, S–342, S–459, S–531
variations in, S–BP 88
Reposition, of upper limbs movements,
S–265
Reproduction, genetics of, S–BP 89
Reproductive organs, innervation of
female, S–515
male, S–519
Reproductive system, S–BP 89 to S–BP 94,
S–S–472 to S–521, T10.1
Residual body, S–530
Respiration
anatomy of, S–BP 63
muscles of, S–BP 64
Respiratory bronchioles, S–392, S–BP 62
Respiratory system, S–358, S–358 to S–393,
T7.1, S–BP 59 to S–BP 64
Rete testis, S–530
Reticular fibers, S–BP 30
in arterial wall, S–BP 45
Reticular formation, S–BP 17
Reticular nuclei, S–37
Retina
ciliary part of, S–88, S–89, S–92
iridial part of, S–89
optic part of, S–88, S–90, S–92
structure of, S–57
Retinacula cutis. See Skin ligaments
Retinacular arteries, S–278
inferior, S–278
posterior, S–278
superior, S–278
Retinal artery, S–91
central, S–87, S–88, S–91, S–92
Retinal vein, S–91
central, S–88, S–91, S–92

Sigmoid vein, S–444, S–445
Sinuatrial (SA) node, S–317, S–319, T5.2
Sinus nerve, carotid, S–76
Sinuses
    confluence of, S–29, S–30, S–49, S–50, S–52,
        S–98
    prostatic, S–BP 93
    straight, S–29, S–30
Sinusoids, S–430, S–431, S–BP 58
Skeletal muscle, S–15
Skeletal system, S–120, S–120 to S–182,
    T3.1–T3.3, S–BP 19 to S–BP 33
Skene's ducts. See Paraurethral (Skene's) glands
Skin, S–28, S–196, S–214, S–329
    cross section of, S–4
    papillary layer of, S–4
    radiology of, S–138
    reticular layer of, S–4
    of scalp, S–322
Skin ligaments, S–4
Skull
    anterior view of, S–123
    associated bones and, S–120
    lateral view of, S–125
    midsagittal section of, S–129
    of newborn, S–127
    nuchal line of, superior, S–200
    orientation of labyrinths in, S–98
    radiographs of
        lateral view of, S–126
        posterior view of, S–124
        Waters' view of, S–124
    reconstruction of, S–BP 19
    superior nuchal line of, S–201
Small arteries, S–302
Small arterioles, S–302
Small intestine, S–68, S–394, S–418, S–423,
    S–427, S–459
    arteries of, S–439
    autonomic innervation of, S–451
    lymph vessels and nodes of, S–BP 56,
        S–BP 56
    mesentery of, S–418
    mucosa and musculature of, S–421
    as site of referred visceral pain, S–BP 5
    veins of, S–443
Smooth muscle, S–15, S–BP 83
    glands, S–15
Smooth muscle cell, S–BP 45
Soft palate, S–360, S–361, S–372, S–373,
    S–397, S–403
    imaging of, S–39
    muscles of, S–408
Sole, muscle of
    first layer, S–294
    second layer, S–295
    third layer, S–296
Soleus muscle, S–117, S–118, S–273, S–279,
    S–284, S–285, S–286, S–287, S–288, S–289,
    S–290, S–299, S–335, S–555, S–BP 109,
    S–BP 110
    arch of, S–273
Soleus tendon, S–BP 110
Solitary lymph nodule, S–BP 67
Solitary lymphoid nodule, S–421
Solitary tract, nuclei of, S–54, S–55, S–65,
    S–67, S–68, S–81, S–320, S–470
Somatic fibers, S–BP 15
Somatic nervous system, S–15
Somatic sensory receptors, S–15
Somatosensory system, trunk and limbs,
    S–BP 17
Space of Poirier, S–BP 29
Spermatic cord, S–215, S–217, S–501, S–502,
    S–549, S–BP 104
    arteries of, S–336
    external spermatic fascia on, S–211, S–216,
        S–512
    testicular vein in, S–510

Spermatic fascia
    external, S–102, S–212, S–213, S–215, S–216,
        S–495, S–497, S–498, S–500, S–501,
        S–502, S–BP 41
        spermatic cord and, S–512
        testis and, S–512
    internal, S–213, S–215, S–497, S–498, S–BP 41
        spermatic cord and, S–218
Spermatids, S–530, S–BP 89
Spermatocytes
    primary, S–530
    secondary, S–530
Spermatogenesis, S–530, S–BP 89
Spermatogonium, S–530, S–BP 89
Spermatozoa, S–530, S–BP 89
Sphenoethmoidal recess, S–361, S–362
Sphenoid bone, S–123, S–125, S–129, S–131,
    S–132, S–360, S–362, S–366, S–BP 22
    anterior clinoid process of, S–129, S–132
    body of, S–129, S–132, S–360, S–370
        jugum of, S–132
        prechiasmatic groove of, S–132
        sella turcica of, S–132
    carotid groove of, S–132
    clivus of, S–132
    crest of, S–360
    foramen ovale of, S–131
    foramen spinosum of, S–131
    greater wing of, S–123, S–125, S–126, S–127,
        S–129, S–131, S–132, S–BP 19
        groove for middle meningeal vessels,
            S–132
    infratemporal crest of, S–125
    lesser wing of, S–123, S–124, S–129, S–132
    of newborn, S–127
    optic canal of, S–129
    orbital surface of, S–123
    pterygoid process of, S–130, S–131, S–362
        hamulus of, S–125, S–129, S–130, S–131,
            S–362
        lateral of, S–125, S–127, S–129, S–130,
            S–131, S–360, S–362
        medial of, S–127, S–129, S–130, S–131,
            S–360, S–362
        pterygoid fossa of, S–131
        scaphoid fossa of, S–131
    sella turcica of, S–129, S–132
        dorsum sellae of, S–132
        hypophyseal fossa of, S–132
        posterior clinoid process of, S–132
        tuberculum sellae of, S–132
    sphenoidal sinus of, S–129, S–360, S–362
    spine of, S–130, S–131
Sphenoid sinus, S–BP 22
Sphenoidal emissary foramen (of Vesalius),
    S–134
Sphenoidal fontanelle, of newborn, S–127
Sphenoidal sinus, S–30, S–71, S–86, S–126,
    S–129, S–327, S–358, S–360, S–361,
    S–368, S–372, S–403, S–523, S–BP 7,
    S–BP 61
    imaging of, S–39
    opening of, S–361, S–362, S–369
Sphenoidal vein (of Vesalius), S–328
Sphenomandibular ligament, S–63, S–136,
    S–187, S–189, S–325
Sphenooccipital synchondrosis, S–372
Sphenopalatine artery, S–189, S–323, S–363,
    S–371, T5.1, S–BP 59
    pharyngeal branch of, S–189, S–371
    posterior lateral nasal branches of, S–189,
        S–363
    posterior septal branches of, S–189, S–363
Sphenopalatine foramen, S–125, S–129,
    S–130, S–359, S–362, S–365, S–BP 59
Sphenopalatine vessels, in incisive fossa,
    S–133
Sphenoparietal sinus, S–29, S–30
Spherical recess, S–96

Sphincter muscle
    external
        deep part of, S–226
        subcutaneous part of, S–226
        superficial part of, S–226
    internal, S–226
Sphincter pupillae muscle, S–58, S–78, S–89,
    T4.1
Sphincter urethrae, S–BP 93
    muscle, S–220, S–496
Sphincter urethrovaginalis, S–464, S–466,
    S–488, S–492, S–493, S–494
Sphincters, female, S–496
Spinal accessory nerve (CN XI), S–73, S–188,
    S–194, S–195, S–344, T2.1, S–BP 60
Spinal accessory nodes, S–344
Spinal arachnoid, S–35
Spinal artery
    anterior, S–23, S–24, S–43, S–44, S–46, S–48
        anastomotic vessels to, S–23
    posterior, S–23, S–24, S–44, S–46, S–48
        anastomotic vessels to, S–23
        left, S–24
        right, S–24
    segmental medullary branches of, S–43
Spinal branch, S–24
Spinal canal, S–547
Spinal cord, S–15, S–16, S–320, S–366, S–456,
    S–470, S–535, S–536, S–BP 18, S–BP 25
    arteries of, S–24, T5.1
        schema of, S–23
    central canal of, S–35, S–41, S–55
    cervical part of, S–BP 17
    cross sections, fiber tracts, S–BP 16
    descending tracts in, S–384
    fiber tracts, S–BP 16
    imaging of, S–39
    lateral horn of gray matter of, S–22
    lumbar part of, S–471, S–BP 17
        upper, S–BP 15
    principal fiber tracts of, S–BP 16
    sacral part of, S–471, S–BP 15
    sections through, S–BP 16
    thoracic part of, S–78, S–79, S–80, S–384,
        S–454, S–BP 15
    veins of, S–25
Spinal dura mater, S–35, S–545
Spinal ganglion, S–17, S–22, S–70, S–100,
    S–206, S–450, S–452, S–454, S–456, S–457,
    S–470, S–471, T2.2, S–BP 17
    anterior root of, S–100
    posterior root of, S–100
Spinal meninges, S–17, T2.2
Spinal nerve trunk, S–209
Spinal nerves, S–15, S–16, S–22, S–100, S–196,
    S–454, S–BP 46
    anterior ramus of, S–17
        C1, S–73
        C2, S–73
        C3, S–75
        C4, S–73, S–75
        C5, S–75
    anterior root of, S–17, S–209, S–450
        C1, S–18
        C8, S–18
        L1, S–18
            anterior root of, S–471
            posterior root of, S–471
        L2, anterior ramus of, S–471
        L4, S–516
        L5, S–18, S–516
            posterior ramus of, S–147
    lateral posterior branch of, S–209
    medial posterior branch of, S–209
    meningeal branch of, S–22, S–209
    posterior ramus of, S–17, S–209
    posterior root of, S–17, S–209
    roots, S–18
    S1, S–18, S–516

Subcallosal area, S–32, S–56
Subcallosal gyrus, S–32
Subcapsular lymphatic plexus, S–354
Subcapsular zone (cortex corticis), S–BP 85
Subchondral bone tissue, S–121, S–BP 30
Subclavian artery, S–23, S–43, S–72, S–73,
  S–76, S–193, S–194, S–195, S–204, S–205,
  S–234, S–235, S–237, S–303, S–307, S–309,
  S–323, S–326, S–328, S–336, S–339, S–358,
  S–387, S–389, S–410, S–474, S–525, S–527,
  S–535, S–BP 49, S–BP 49
  grooves for, S–BP 39
  left, S–310, S–330, S–391, S–393, S–526
  lung groove for, S–388
  right, S–68, S–330, S–390, S–526, S–BP 49
Subclavian lymphatic trunk, S–350
Subclavian nerve, S–106, T2.9
Subclavian trunk, right, S–352
Subclavian vein, S–193, S–194, S–195, S–204,
  S–205, S–235, S–237, S–304, S–307, S–309,
  S–328, S–337, S–340, S–350, S–358, S–387,
  S–389, S–410, S–525, T5.1
  grooves for, S–BP 39
  left, S–330, S–391, S–393, S–526
  right, S–330, S–390, S–526
Subclavius muscle, S–156, S–203, S–204,
  S–231, S–236, S–237, S–390, S–391, S–473,
  S–534, S–BP 39
  fascia investing, S–231
  groove for, S–156
Subcostal artery, S–341
Subcostal muscles, S–209
Subcostal nerve, S–16, S–103, S–104, S–113,
  S–277, S–393, S–459, S–460, S–514, S–518,
  T2.2
  anterior branch of, S–104
  cutaneous branch of
    anterior, S–102
    lateral, S–102, S–103, S–210, S–266
  lateral branch of, S–104
Subcostal plane, S–395
Subcostal vein, S–342
Subcutaneous artery, S–4
Subcutaneous bursa, medial malleolus of,
  S–290
Subcutaneous fat, radiology of, S–138
Subcutaneous olecranon bursa, S–161
Subcutaneous tissue, S–4, S–28, S–196
  deeper membranous layer of, S–500
  fatty layer of, S–214
  membranous layer of, S–214
  superficial fatty (Camper's) layer of, S–500
Subcutaneous vein, S–4
Subdeltoid bursa, S–157
Subdural space, S–28
Subhiatal ring, of adipose tissue, S–413
Subiculum, S–38
Sublime tubercle, S–159
Sublingual artery, S–396, S–399
Sublingual caruncle, S–396, S–397, S–399
Sublingual fold, S–396, S–397, S–399
Sublingual fossa, S–135
Sublingual gland, S–20, S–21, S–62, S–65,
  S–79, S–368, S–394, S–396, S–399, S–400,
  S–BP 14, S–BP 61
Sublingual vein, S–396
Sublobular vein, S–430, S–431
Submandibular duct (of Wharton), S–396,
  S–397, S–400, S–401
Submandibular fossa, S–135
Submandibular ganglion, S–21, S–59, S–62,
  S–65, S–73, S–77, S–79, S–188, S–396,
  S–399, S–401, S–BP 14
  right, S–188
  schema of, S–79
Submandibular glands, S–5, S–20, S–21, S–62,
  S–65, S–79, S–190, S–192, S–197, S–323,
  S–328, S–373, S–394, S–399, S–400, S–BP
  14

Submandibular glands (Continued)
  deep part of, S–396
  superficial part of, S–396
Submandibular node, S–344, S–345, S–400
Submandibular triangle, S–5
Submental artery, S–189, S–323
Submental node, S–344, S–345
Submental triangle, S–5
Submental vein, S–328
Submucosa, S–421, S–455, S–BP 65, S–BP 67,
  S–BP 68, S–BP 69
Submucosal glands, S–455
Submucosal plexus (Meissner's), S–455, S–BP
  65
Submucous space, S–222
Suboccipital nerve, S–141, S–202
Subpleural capillaries, S–392
Subpleural lymphatic plexus, S–352
Subpopliteal recess, S–280, S–281
Subpubic angle, S–476
Subpyloric nodes, S–BP 54
Subscapular artery, S–234, S–235, S–237,
  S–331, S–BP 49, S–BP 50
Subscapular fossa, S–155
  of scapula, S–149
Subscapular nerve, T2.9
  lower, S–106, S–111, S–235, S–236
  upper, S–106, S–236, S–237
Subscapularis muscle, S–100, S–157, S–206,
  S–231, S–232, S–235, S–236, S–238,
  S–239, S–264, S–535, S–536, S–537,
  S–BP 98
Subscapularis tendon, S–157, S–232, S–235
Subserosa, S–455
Subserous fascia, S–513
Substantia nigra, S–33, S–55
Subtendinous bursa, S–282, S–290
  of iliacus, S–276
  of subscapularis muscle, S–157
Sulcus
  calcarine, S–31, S–32, S–33, S–37, S–57
  central
    of insula, S–31
    of Rolando, S–31, S–32
  cingulate, S–32
  collateral, S–32, S–33
  of corpus callosum, S–32
  frontal
    inferior, S–31
    superior, S–31
  hippocampal, S–38
  hypothalamic, S–32, S–41, S–524
  lateral (of Sylvius), S–31, S–33
    anterior ramus of, S–31
    ascending ramus of, S–31
    posterior ramus of, S–31
  lunate, S–32
  marginal, S–32
  median, S–41
  nasolabial, S–5
  occipital, transverse, S–31
  occipitotemporal, S–32, S–33
  olfactory, S–33
  orbital, S–33
  paracentral, S–32
  parietooccipital, S–31, S–32
  postcentral, S–31
  rhinal, S–32, S–33
  temporal
    inferior, S–31, S–33
    superior, S–31
Sulcus limitans, S–41
Sulcus terminalis, S–310
Superciliary arch, S–5
Superficial back, S–183
Superficial capsular tissue, S–BP 29
Superficial circumflex iliac vein, S–8
Superficial dorsal vein, of penis, S–8
Superficial epigastric veins, S–8

Superficial fascia, S–329, S–495, S–BP 90
  deep membranous layer of, S–501
  penile, S–BP 90
  of scalp, S–322
Superficial fibular nerve, S–555
Superficial investing cervical fascia, S–197
Superficial temporal artery, S–60, S–189,
  S–396, S–399
Superficial temporal vein, S–26, S–396
Superficial vein
  of forearm, S–229
  lateral, of penis, S–501
Superior, as term of relationship, S–1
Superior alveolar nerve
  anterior, S–406
  middle, S–406
  posterior, S–406
Superior angle, S–155
Superior anorectal artery, S–419
Superior anorectal plexus, S–105
Superior anorectal vein, S–419
Superior articular facet, of L5, S–BP 28
Superior articular process, S–145, S–147
  facets of, S–146
  of L1, S–BP 24
  of L2, S–138
  of L4, S–BP 24
Superior border, S–155
Superior bulb, in jugular fossa, S–133
Superior cervical ganglion, S–384, S–400,
  S–BP 13
Superior cervical sympathetic cardiac nerve,
  S–77
Superior cluneal nerves, S–210
Superior colliculus, S–37, T2.1
  brachium of, S–37
Superior costal facet, S–144
Superior costotransverse ligament, S–BP 26
Superior duodenal flexure, S–421
Superior gluteal cutaneous nerves, S–210
Superior hypogastric plexus, S–BP 13
Superior labial artery, S–60
Superior laryngeal nerve, internal branch of,
  S–408, S–409
Superior longitudinal lingual muscle, S–400,
  S–402
Superior mediastinum, S–BP 2
Superior mesenteric artery, S–303, S–341,
  S–415, S–418, S–425
Superior mesenteric ganglion, S–20, S–21,
  S–BP 13, S–BP 15
Superior mesenteric plexus, S–BP 13, S–BP 14
Superior oblique muscle, S–58
Superior orbital fissure, S–367
Superior petrosal sinus, S–71
Superior pharyngeal constrictor, S–197,
  S–323
  glossopharyngeal part of, S–408
Superior phrenicoesophageal ligament,
  S–413
Superior pubic ramus, S–151, T3.3
Superior rectus muscle, S–58
Superior rectus tendon, S–92
Superior suprarenal artery, S–208
Superior vena cava, S–302, S–304
Superior vertebral notch, S–144
  of L3 vertebra, S–BP 24
Superolateral nodes, S–356
Superomedial nodes, S–356
Supination, of upper limbs movements, S–265
Supinator muscle, S–112, S–227, S–241,
  S–247, S–248, S–249, S–262, S–263, S–264,
  S–552
Supraclavicular nerves, S–13, S–74, S–185,
  S–194, S–195, T2.8
  intermediate, S–74, S–228
  lateral, S–74, S–228
  medial, S–74, S–228
Supraclavicular nodes, S–344

Netter Atlas of Human Anatomy: A Systems Approach

Supracondylar ridge
lateral, S–155, S–159
medial, S–155, S–159
Supraduodenal artery, S–435, S–436, S–438, S–439, S–528
Supraglenoid tubercle, S–155
Suprahyoid muscles, S–191, S–399
Supramarginal gyrus, S–31
Supramastoid crest, S–125
Supraoptic recess, S–32, S–34
Supraopticohypophyseal tract, S–524
Supraorbital artery, S–43, S–60, S–82, S–87, S–122, S–189, S–322, S–323
Supraorbital margin, S–5, S–124, S–BP 19
Supraorbital nerve, S–59, S–60, S–61, S–82, S–122, S–185
lateral branch of, S–85, S–86
left, S–86
medial branch of, S–85, S–86
Supraorbital notch, S–123, S–125
of newborn, S–127
Supraorbital vein, S–87, S–322, S–328
Suprapatellar bursa, S–280, S–282
Suprapatellar synovial bursa, S–281
Suprapineal recess, S–32, S–34
Suprapleural membrane (Sibson's fascia), S–390, S–391
Suprapyloric nodes, S–BP 54
Suprarenal androgens, S–BP 97
Suprarenal arteries
inferior, S–341, S–467, S–531, S–532
middle, S–341, S–531, S–532
right, S–531
superior, S–341, S–531, S–532
Suprarenal cortices, S–BP 97
Suprarenal glands, S–20, S–21, S–386, S–429, S–433, S–434, S–458, S–461, S–499, S–522, S–531, S–532, S–533, T11.1, S–BP 15, S–BP 25, S–BP 98, S–BP 99
arteries of, S–532
autonomic nerves of, S–533
cross section, S–532
left, S–416, S–459, S–531, S–533, S–543
right, S–419, S–420, S–459, S–531, S–533, S–BP 101
veins of, S–532
Suprarenal impression, S–429
Suprarenal plexus, S–105, S–452
Suprarenal vein, S–532
left, S–340, S–342, S–531
right inferior, S–342
Suprascapular artery, S–43, S–194, S–195, S–234, S–235, S–237, S–525, S–526, S–527, T5.4, S–BP 49
Suprascapular nerve, S–106, S–111, S–235, S–237, T2.9
Supraspinatus muscle, S–111, S–157, S–198, S–199, S–230, S–231, S–232, S–233, S–234, S–235, S–236, S–240, S–264, S–BP 98
Supraspinatus tendon, S–157, S–232, S–233, S–235, S–240, T4.2
Supraspinous fossa, S–155
of scapula, S–149
Supraspinous ligament, S–141, S–147, S–148, S–152, S–154, S–545
radiology of, S–138
Suprasternal space (of Burns), S–190, S–196, S–197, S–329, S–372
Supratonsillar fossa, S–403
Supratrochlear artery, S–43, S–60, S–82, S–87, S–122, S–189, S–322, S–323
Supratrochlear nerve, S–59, S–60, S–61, S–82, S–85, S–86, S–122, S–185
Supratrochlear nodes, S–346
Supratrochlear vein, S–322, S–328
Supraventricular crest, S–311, S–316
Supravesical fossa, S–217

Sural nerve, S–117, S–118, S–267, T2.3, T2.13, S–BP 43, S–BP 109, S–BP 110
calcaneal branches of, lateral, S–267
Surface mucous cell, S–BP 67
Suspensory ligament, S–499
of axilla, S–231
of breast (Cooper's), S–473
of ovary, S–478, S–483, S–485, S–486, S–487, S–489, S–508
of penis, S–212, S–495
Sustentaculum tali, S–176, S–177, S–178, S–BP 110
Sutural (wormian) bone, S–125, S–128
Sweat gland duct, S–4
Sweat glands, S–4, S–20, S–BP 15
excretory pore of, S–4
in perianal skin, S–226
Sympathetic division, S–15
Sympathetic fibers, S–456, S–457
postsynaptic, S–BP 15
presynaptic, S–BP 15
in reproductive organs, S–515
male, S–519
of tracheobronchial tree, S–384
Sympathetic ganglion, S–514, S–515, S–517, S–518, S–519
sacral
left, S–513
right, S–513
Sympathetic nerves, S–384
Sympathetic nervous system, S–20, S–BP 1, S–15
general topography, S–BP 13
Sympathetic plexus, S–71
Sympathetic trunks, S–20, S–67, S–72, S–73, S–76, S–77, S–79, S–80, S–99, S–101, S–103, S–104, S–105, S–113, S–114, S–196, S–206, S–207, S–208, S–209, S–219, S–320, S–328, S–329, S–366, S–384, S–390, S–391, S–447, S–452, S–454, S–456, S–457, S–469, S–471, S–483, S–484, S–513, S–514, S–515, S–517, S–518, S–519, S–533, S–539, S–546, S–BP 13, S–BP 46, S–BP 60, S–BP 100
first thoracic ganglion of, S–78
ganglion of, S–22, S–100, S–104, S–454, S–457, S–470
lumbar ganglion of, S–105, S–469
right, S–450, S–513
sacral
left, S–513
right, S–513
sacral ganglia of, S–105
thoracic ganglion of, S–99, S–447
Symphyseal surface, S–153, S–154
Synapses, S–BP 3, S–BP, S–15
Synaptic cleft, S–BP 3
Synaptic vesicles, S–BP 3
Synovial cavities, S–142, S–BP 26
Synovial joints, types of, S–121
Synovial membrane, S–157, S–161, S–170, S–280, S–281, S–282, S–553, S–BP 30
attachment of, S–281
Synovial sheath, fibrous sheath over, annular part of, S–251

T

Tactile corpuscle (Meissner's). *See* Meissner's corpuscle
Taenia, free, S–221, S–223
Talar articular surface, S–178
Talocalcaneal ligament, S–178
lateral, S–180
medial, S–180
posterior, S–178, S–180, S–BP 110
Talofibular ligament
anterior, S–175, S–180, S–BP 110
posterior, S–175, S–178, S–180

Talonavicular ligament, dorsal, S–180
Talus bone, S–176, S–177, S–178, S–179, S–BP 110
posterior process of, S–180
Tarsal artery
lateral, S–289, S–291, S–292, S–297, S–333, S–BP 53
medial, S–289, S–291, S–292, S–297, S–333, S–BP 43, S–BP 53
Tarsal bones, S–120
Tarsal glands, S–82
openings of, S–82
Tarsal muscle, superior, S–82, T4.1
Tarsometatarsal joint, S–176, S–177
Tarsometatarsal ligaments
dorsal, S–180, S–297
plantar, S–181
Tarsus
inferior, S–82
superior, S–82, S–84
Taste buds, S–402
Taste pathways, schema of, S–81
Tectal, S–32, S–41
Tectorial membrane, S–97, S–142, S–BP 21
deeper (accessory) part of, S–142
Tectospinal tract, S–78, S–BP 16
Tectum, imaging of, S–39
Teeth, S–394, S–404
root of, S–369
Tegmen tympani muscle, S–93, S–94, S–BP 9
Temple, S–3
Temporal artery
anterior deep, S–188
deep, S–26, S–323, S–325
anterior, S–63, S–189
posterior, S–63, S–125, S–189
middle, S–26, S–322
polar, S–47
posterior deep, S–188
superficial, S–26, S–27, S–43, S–63, S–72, S–73, S–80, S–95, S–122, S–188, S–189, S–323, S–324, S–325, S–326, S–328, S–371, S–BP 60
frontal branch of, S–87, S–322
parietal branch of, S–322
Temporal bone, S–123, S–125, S–129, S–131, S–132, S–359, S–BP 19
acoustic meatus of
external, S–125, S–131, S–BP 19
internal, S–129
articular tubercle of, S–125, S–131
asterion of, S–125
carotid canal of, S–131
cochlear window of, S–127
external occipital protuberance of, S–125
inferior tympanic canaliculus of, S–131
jugular fossa of, S–131
lambdoid suture of, S–125
mandibular fossa of, S–125, S–127, S–131
mastoid canaliculus of, S–131
mastoid foramen of, S–131
mastoid notch of, S–131
mastoid process of, S–125, S–131
of newborn, S–127
occipital artery of, groove for, S–131
oval window of, S–127
petrosquamous fissure of, S–127
petrotympanic fissure of, S–131
petrous part of, S–98, S–124, S–127, S–129, S–131, S–132, S–BP 10
arcuate eminence of, S–132
greater petrosal nerve of, groove for, S–132
internal carotid artery in, S–327
lesser petrosal nerve of, groove for, S–132
sigmoid sinus of, groove for, S–132
superior petrosal sinus of, groove for, S–132
trigeminal impression of, S–132

Temporal bone *(Continued)*
  posterior deep temporal artery of, groove
    for, S–125
  sigmoid sinus of, groove for, S–129
  squamous part of, S–125, S–127, S–129,
    S–132
  styloid process of, S–95, S–125, S–127,
    S–131, S–373, S–399, S–401, S–407,
    S–409, S–BP 60
  stylomastoid foramen of, S–131
  superior petrosal sinus of, groove for, S–129
  supramastoid crest of, S–125
  tympanic part of, S–127
  vestibular aqueduct of, opening of, S–129
  zygomatic process of, S–125, S–127, S–131,
    S–325
Temporal fascia, S–62, S–184, S–186
  deep layer of, S–186
  superficial layer of, S–186
Temporal fossa, S–125, S–188, S–359
Temporal gyrus
  inferior, S–31, S–33
  middle, S–31
  superior, S–31
Temporal line
  inferior, S–125
  superior, S–125
Temporal lobe, S–40, S–46, S–368
  imaging of, S–39
    medial, S–BP 12
Temporal nerve
  anterior deep, S–188
  deep, S–59, S–73
    anterior, S–62, S–63, S–189
    posterior, S–62, S–63, S–188, S–189
Temporal pole, of cerebrum, S–31, S–33
Temporal region, of head and neck, S–5
Temporal retinal arteriole
  inferior, S–91
  superior, S–91
Temporal retinal venule
  inferior, S–91
  superior, S–91
Temporal sulcus
  inferior, S–31, S–33
  superior, S–31
Temporal vein
  deep, S–26
  middle, S–26, S–322
  superficial, S–26, S–63, S–328
    frontal branch of, S–322
    parietal branch of, S–322
    tributary of, S–28
Temporalis muscle, S–26, S–28, S–62, S–63,
  S–186, S–188, S–197, S–368
Temporalis tendon, S–186, S–197, S–396
Temporofacial division, S–64
Temporomandibular joint, S–136, S–187, T3.1
  articular disc of, S–186, S–187, S–BP 37
  capsule of, S–63, S–325
  joint capsule of, S–136, S–BP 37
Temporomandibular ligament, lateral, S–136
Tendinous arch
  of levator ani muscle, S–220, S–225, S–464,
    S–465, S–478, S–479, S–480, S–488,
    S–490, S–493
  of pelvic fascia, S–464, S–465, S–478, S–490
Tendinous intersection, S–7, S–8, S–212
Tendon, S–BP 34
Tendon sheath, S–253
  of ankle, S–290
  of finger, S–252, S–254
  of hand, S–254
Tenon's capsule, S–88, S–BP 7
Tensor fasciae latae muscle, S–11, S–268,
  S–270, S–271, S–272, S–274, S–275,
  S–300, S–549, S–550, S–554, S–BP 104,
  S–BP 108
Tensor fasciae latae tendon, S–268

Tensor tympani muscle, S–62, S–94, S–95,
  S–BP 8
  nerve to, S–59
Tensor tympani tendon, S–95
Tensor veli palatini muscle, S–62, S–94, S–187,
  S–398, S–403, S–407, S–408
  nerves to, S–59, S–63, S–73
Tensor veli palatini nerve, S–62
Tensor veli palatini tendon, S–398, S–403,
  S–408
Tentorial artery, S–30
Tentorium cerebelli, S–29, S–30, S–49, S–50,
  S–85
Teres major muscle, S–6, S–100, S–111, S–198,
  S–199, S–230, S–231, S–234, S–235, S–236,
  S–237, S–238, S–239, S–240, S–264, S–536,
  S–537, S–551
  lower margin of, S–BP 50
Teres major tendon, S–240
Teres minor muscle, S–111, S–198, S–199,
  S–206, S–230, S–231, S–232, S–234, S–235,
  S–236, S–240, S–535, S–536
Teres minor tendon, S–157, S–232, S–240
Terminal bronchiole, S–392
Terminal sulcus, S–402
Testes, S–530, S–BP 97
Testicle, S–497
Testicular artery, S–215, S–218, S–341, S–419,
  S–463, S–484, S–497, S–509, S–510, S–518,
  S–519, S–531, S–546, S–BP 94, S–BP 108
  covered by peritoneum, S–215
  left, S–444, S–459
Testicular plexus, S–105, S–518, S–519
Testicular vein, S–215, S–216, S–217, S–218,
  S–419, S–463, S–484, S–510, S–531
  covered by peritoneum, S–215
  left, S–444, S–459
  right, S–444
  in spermatic cord, S–510
Testicular vessels, S–357
Testis, S–357, S–418, S–472, S–495, S–497,
  S–498, S–499, S–505, S–519, S–522, T10.1
  arteries of, S–510
  descent of, S–499
  efferent ductules of, S–530
  external spermatic fascia over, S–512
  gubernaculum of, S–499
  lobules of, S–498
  mediastinum of, S–497, S–498
  rete, S–498
  tunica albuginea of, S–497
  tunica vaginalis, S–497
    parietal layer of, S–498
    visceral layer of, S–498
  veins of, S–510
Thalamic tubercle, anterior, S–37
Thalamogeniculate arteries, S–48
Thalamoperforating arteries, S–45, S–48
Thalamostriate vein
  inferior, S–49, S–51
  superior, S–26, S–32, S–34, S–37, S–50, S–51
Thalamus, S–34, S–36, S–37, S–38, S–40, S–41,
  S–50, S–452, S–524
  imaging of, S–39
  left, S–49
  pulvinar of, S–44
  stria medullaris of, S–32, S–37
  ventral posteromedial nucleus of, S–81
Theca externa, S–BP 96
Theca interna, S–BP 96
Thenar eminence, S–10
Thenar muscles, S–109, S–168, S–250, S–251,
  S–BP 105
Thenar space, S–251, S–252, S–254
Thigh, S–2, S–3
  anterior, S–183
  anterior region of, S–11
  arteries of, S–270, S–271, S–274, S–BP 53
  deep veins of, S–334

Thigh *(Continued)*
  fascia lata of, S–500, S–501
  medial, S–183
  muscle of, S–268, S–269, S–272, S–273
    bony attachments of, S–300, S–301
    medial compartment of, T4.3
  posterior, S–183
  posterior region of, S–11
  serial cross sections, S–554
Thoracic aorta, S–24, S–206, S–339, S–391,
  S–539, T5.3
  descending, S–537, S–538, S–540, S–BP 99
  lung groove for, S–388
Thoracic aortic plexus, S–101
Thoracic aperture, superior, S–338, T3.2
Thoracic artery
  inferior, S–339
    oesophageal branch of, S–339
  internal, S–43, S–75, S–204, S–205, S–206,
    S–207, S–234, S–307, S–309, S–336, S–
    387, S–389, S–410, S–474, S–526, S–536,
    S–537, S–538, S–539, T5.2, S–BP 49
    anterior intercostal branches of, S–204
    left, S–391
    medial mammary branches of, S–474
    perforating branches of, S–203, S–205,
      S–206, S–235, S–474
    right, S–390
  lateral, S–203, S–204, S–234, S–235, S–237,
    S–331, S–336, S–474, S–BP 50
    mammary branches of, S–474
  long, S–474
  superior, S–204, S–234, S–237, S–331,
    S–BP 50
Thoracic cardiac nerves, S–99, S–101, S–320,
  S–447, S–BP 46
  sympathetic, T2.2
Thoracic cavity, S–BP 2
Thoracic duct, S–101, S–105, S–204, S–207,
  S–219, S–307, S–340, S–343, S–344, S–350,
  S–351, S–352, S–353, S–387, S–389, S–391,
  S–410, S–535, S–536, S–537, S–538, S–539,
  S–541, T6.1, S–BP 56, S–BP 98, S–BP 100
  arch of, S–389
Thoracic ganglion, S–99, S–320
  first, S–20
  of sympathetic trunk, S–101
Thoracic intervertebral discs, S–207
Thoracic kyphosis, S–137
Thoracic nerves, T2.2
  long, S–102, S–106, S–203, S–204, S–235,
    S–237, T2.2, T2.3, T2.9
  posterior ramus of, S–206
  T1, S–18, S–20
  T2, S–18, S–20
  T3, S–18
  T4, S–18, S–20
  T5, S–18, S–20
  T6, S–18, S–20
  T7, S–18, S–20
  T8, S–18
  T9, S–18
  T10, S–18
  T11, S–18
  T12, S–18
Thoracic skeleton, S–120
Thoracic spinal cord, S–384
Thoracic spinal nerves
  lateral cutaneous branches of anterior rami
    of, S–198
  T1, S–78, S–79, S–80
  T2, S–79, S–80
  T7
    anterior ramus, S–514
    lateral posterior cutaneous branch of, S–210
    medial posterior cutaneous branch of,
      S–210
  T10, anterior ramus, S–517
  T11, anterior ramus, S–514

Netter Atlas of Human Anatomy: A Systems Approach

Trachea *(Continued)*
 elastic fibers of, S–379
 epithelium of, S–379
 gland of, S–379
 lung area for
  left, S–388
  right, S–388
 lymph vessels of, S–379
 nerve of, S–379
 small artery of, S–379
Tracheal bifurcation, T7.1, S–BP 28
Tracheal cartilages, S–359, S–379
Tracheal wall
 anterior, S–379
 posterior, S–379
  mucosa of, S–379
Trachealis muscle, S–379
Tracheobronchial lymph nodes, S–352
 inferior, S–350, S–351, S–537
 superior, S–350, S–351, S–536
Tracheobronchial node, T6.1
**Tracheobronchial tree, innervation of, schema
 of, S–384
Tracheoesophageal groove, S–407
Tragus, S–5, S–94
Transpyloric plane, S–395, S–BP 70
Transversalis fascia, S–213, S–214, S–215,
 S–216, S–217, S–218, S–341, S–418, S–461,
 S–464, S–483, S–484, S–490, S–492, S–545,
 S–546, S–BP 41
Transverse acetabular ligament, S–BP 108
Transverse colon, S–417, S–418, S–422, S–423,
 S–424, S–428, S–432, S–433, S–434, S–443,
 S–543, S–544
Transverse costal facet, S–144
Transverse diameter, S–477
Transverse facial artery, S–60, S–396
Transverse fasciculi, S–250, S–293
Transverse fibers, S–BP 30
Transverse foramen, S–140
 bony spicule dividing, S–143
 septated, S–143
Transverse ligament, of atlas, S–142
Transverse mesocolon, S–416, S–417, S–418,
 S–420, S–422, S–423, S–424, S–428, S–443,
 S–459
 attachment of, S–419, S–433, S–434
Transverse palatine folds, S–398
Transverse plane, S–1
Transverse process, S–144, S–145, S–147,
 S–148, S–150, S–388, S–545
 of C6, S–BP 22
 cervical, S–140, S–141
 of coccyx, S–146
 of L5 vertebra, S–476
 lumbar, S–147
  L1, S–138
  L3, S–BP 24
 of lumbar vertebrae, S–475
 thoracic, S–144
Transverse ridges, S–146
Transverse "saddlelike" liver, S–BP 71
Transverse sinus, S–29, S–30, S–50, S–52,
 S–98
 groove for, S–129, S–132
 imaging of, S–BP 11
 left, S–49
Transverse tarsal joint, S–177
Transverse vesical fold, S–217
Transversospinal muscle, S–535
Transversus abdominis muscle, S–102, S–103,
 S–113, S–200, S–204, S–205, S–209, S–213,
 S–214, S–215, S–216, S–217, S–219, S–277,
 S–336, S–459, S–483, S–484, S–BP 64,
 S–BP 102
 aponeurosis of, S–210, S–214, S–336, S–460
 tendon of origin, S–200, S–545
Transversus abdominis tendon, S–200, S–546
Transversus linguae muscle, S–400

Transversus thoracis muscle, S–100, S–204,
 S–205, S–206, S–539
Trapeziocapitate ligament, S–165, S–166
Trapeziotrapezoid ligament, S–165, S–166
Trapezium bone, S–121, S–163, S–164, S–165,
 S–166, S–167, S–168, S–169, S–253, S–259,
 S–BP 29
 tubercle of, S–163
Trapezius muscle, S–5, S–10, S–69, S–74,
 S–190, S–191, S–192, S–196, S–197, S–203,
 S–204, S–206, S–210, S–230, S–231, S–236,
 S–237, S–264, S–329, S–535, S–536, S–537,
 S–538, T4.1, S–BP 98
 spine of, S–6, S–100, S–198, S–199, S–202
Trapezoid bone, S–163, S–164, S–166, S–167,
 S–168, S–253, S–BP 29
Trapezoid ligament, S–157, S–232
Trapezoid line, S–156
Triangle
 anal, S–500
 of auscultation, S–6, S–210, S–230
 deltopectoral, S–230
 inguinal, S–216, S–217
  transversalis fascia within, S–216
 lumbar (of Petit), S–6, S–199, S–210
 lumbocostal, S–208
 sternocostal, S–205
 urogenital, S–500
Triangular fossa, S–94
Triangular ligament
 left, S–419, S–429
 right, S–419, S–429
Triangular space, S–235
Triceps brachii muscle, S–6, S–7, S–10, S–111,
 S–160, S–230, S–234, S–236, S–237, S–238,
 S–240, S–263, S–551
 lateral head of, S–6, S–10, S–230, S–536,
  S–537
 long head of, S–6, S–10, S–230, S–264,
  S–535, S–536, S–537
 medial head of, S–263
 tendon of, S–6, S–10
Triceps brachii tendon, S–111, S–161, S–241,
 S–242, S–551
Tricuspid valve. *See* Atrioventricular valve,
 right
Trigeminal cave, imaging of, S–39
Trigeminal ganglion, S–30, S–54, S–55, S–61,
 S–62, S–78, S–79, S–80, S–81, S–85, S–86,
 S–406
Trigeminal impression, S–132
Trigeminal nerve (CN V), S–49, S–54, S–55,
 S–61, S–63, S–71, S–77, S–79, S–80, S–81,
 S–98, S–406, T2.1, T2.4
 distribution of, S–53
 ganglion of, S–59, S–77
 imaging of, S–39
 mandibular division of, S–185
 maxillary division of, S–185
 mesencephalic nucleus of, S–54, S–55, S–81
 motor nucleus of, S–54, S–55, S–81
 motor root of, S–40, S–62, S–77
 ophthalmic division of, S–185
 principal sensory nucleus of, S–54, S–55,
  S–59
 schema of, S–59
 sensory root of, S–40, S–62, S–77
 spinal nucleus of, S–54, S–55, S–59, S–68
Trigeminal tubercle, S–41
Trigonal ring, S–466, S–496
Triquetrocapitate ligament, S–165
Triquetrohamate ligament, S–165, S–166
Triquetrum bone, S–163, S–164, S–166, S–167,
 S–168, S–BP 29
Triticeal cartilage, S–375
Trochanter of femur
 greater, S–6, S–151, S–170, S–171, S–182,
  S–269, S–273, S–274, S–275, S–277,
  S–475, S–476, S–549, S–550, S–BP 108

Trochanter of femur *(Continued)*
 lesser, S–151, S–170, S–171, S–182, S–219,
  S–475, S–476
Trochanteric bursae
 of gluteus maximus muscle, S–276
  inferior extension of, S–276
 of gluteus medius muscle, S–276
Trochanteric fossa, S–172
Trochlea, S–84, S–159
Trochlear nerve (CN IV), S–30, S–40, S–41,
 S–49, S–54, S–55, S–63, S–71, S–84, S–85,
 S–86, T2.1, T2.4, S–BP 7
 distribution of, S–53
 schema of, S–58
 in superior orbital fissure, S–134
Trochlear notch, S–159, S–160, S–162
Trochlear nucleus, S–54, S–55, S–58
True ribs, S–149
Trunk, S–BP 18
 somatosensory system of, S–BP 17
Tubal artery, S–95
Tubal branches
 of ovarian vessels, S–511
 of uterine artery, S–511
Tuber cinereum, S–32, S–33, S–40
 median eminence of, S–524
Tubercle, anterior, S–37
Tuberohypophyseal tract, S–524
Tuberosity, S–178
 tibial, S–174
Tuberothalamic arteries, S–45
Tufted cell, S–56
Tunica adventitia, S–BP 45
Tunica albuginea, S–498, S–501, S–503
 of ovary, S–BP 96
 of testis, S–497
Tunica intima, S–BP 45
Tunica media, S–BP 45
Tunica vaginalis
 cavity of, S–499
 testis, S–418, S–497, S–498
  parietal layer of, S–497
  visceral layer of, S–497
Tympanic artery
 anterior, S–43, S–63, S–95, S–188, S–189,
  S–323, S–325
 inferior, S–71, S–95
 posterior, S–95
  stapedial branch of, S–95
 superior, S–63, S–95
Tympanic canaliculus, inferior, S–131, S–133
Tympanic cavity, S–66, S–67, S–93, S–94, S–95,
 S–97, S–366, S–BP 8, S–BP 9
 labyrinthine wall of, S–BP 8
 lateral, S–BP 8
 lateral wall of, S–BP 8
 medial, S–BP 8
 promontory of, S–93
Tympanic cells, S–BP 8
Tympanic membrane, S–62, S–93, S–94, S–95,
 S–97, T2.1, S–BP 8, S–BP 9
 secondary, S–97
Tympanic nerve, S–65, S–67, S–71, S–80, S–94,
 S–95
 inferior, S–95
Tympanic plexus, S–65, S–67, S–78, S–80,
 S–94, S–95
 tubal branch of, S–67

U

Ulna, S–120, S–121, S–159, S–160, S–161,
 S–162, S–163, S–164, S–165, S–166, S–168,
 S–241, S–243, S–245, S–246, S–248, S–256,
 S–260, S–262, S–263, S–552, T3.3, S–BP 29
 anterior border of, S–162
 anterior surface of, S–162
 body of, S–160
 coronoid process of, S–159

Netter Atlas of Human Anatomy: A Systems Approach

Uterus *(Continued)*
- arteries of, S–511
- body of, S–487, S–529
- broad ligament of, S–478, S–483, S–485, S–488, S–505, S–529
- cervix of, S–478, S–485, S–486, S–487, S–488, S–490, S–529, T10.1, S–BP 96
- endometrium of, S–521
- fascial ligaments, S–478
- fundus of, S–464, S–483, S–485, S–487, S–521, S–529
- isthmus of, S–487
- myometrium of, S–521
- round ligament of, S–342, S–463, S–465, S–478, S–483, S–485, S–486, S–488, S–489, S–492, S–505, S–508, S–BP 86
- veins of, S–511

Utricle, S–66, S–93, S–96, S–97, T2.1

Uvula
- of bladder, S–503
- cleft of, S–491
- of palate, S–397
- of vermis, S–41

V

Vagal fibers, S–68
Vagal trigone, S–41
Vagal trunk
- anterior, S–68, S–101, S–105, S–219, S–410, S–447, S–448, S–449, S–450, S–451, S–452, S–456, S–457, S–469, S–517, S–533, S–BP 14, S–BP 65
  - celiac branch of, S–449, S–450, S–451
  - gastric branch of, S–68, S–447, S–448
  - hepatic branch of, S–68, S–447, S–448, S–449, S–452, S–456
  - pyloric branch of, S–68, S–448
- posterior, S–105, S–219, S–447, S–449, S–450, S–451, S–452, S–456, S–457, S–469, S–517, S–533, S–BP 14, S–BP 65
  - celiac branch of, S–68, S–447, S–448, S–449, S–450, S–451
  - gastric branch of, S–447, S–449

Vagina, S–220, S–221, S–224, S–464, S–465, S–466, S–472, S–478, S–479, S–480, S–485, S–486, S–487, S–488, S–493, S–496, S–505, S–508, S–515, S–521, S–550, T10.1, S–BP 86
- vestibule of, S–488, S–491, S–505

Vaginal arteries, S–463, S–464, S–486, S–488, S–489, S–490, S–507, S–508, S–511
- inferior vesical branch of, S–464, S–468
- ureteric branch of, S–468

Vaginal epithelium, S–BP 97
Vaginal fascia, S–221, S–478
Vaginal fornix, S–487
- anterior part of, S–478, S–485, S–486
- posterior part of, S–485, S–486

Vaginal orifice, S–491, S–494, S–506, S–BP 86
Vaginal vein, S–445
Vaginal wall, S–488
Vagus nerve (CN X), S–21, S–30, S–40, S–54, S–55, S–67, S–69, S–71, S–72, S–73, S–75, S–76, S–77, S–81, S–99, S–101, S–188, S–194, S–195, S–196, S–307, S–309, S–320, S–323, S–324, S–328, S–329, S–384, S–389, S–390, S–391, S–406, S–410, S–447, S–454, S–470, S–525, S–526, S–527, S–535, S–536, T2.6, S–BP 6, S–BP 14, S–BP 15, S–BP 60
- auricular branch of, S–68, S–133, S–185
  - communication to, S–67
- bronchial branch of, S–101
- cervical cardiac branch of, S–101
  - inferior, S–68, S–99, S–320
  - superior, S–68, S–76, S–77, S–99, S–320
- cervical cardiac branches of, S–447
- distribution of, S–53

Vagus nerve (CN X) *(Continued)*
- dorsal nucleus of, S–68
- ganglion of
  - inferior, S–68, S–69, S–70, S–71, S–72, S–81
  - superior, S–68, S–69, S–72
- inferior ganglion of, S–320, S–384, S–447
- in jugular foramen, S–134
- in jugular fossa, S–133
- laryngopharyngeal branch of, S–76
- left, S–306, S–526
- meningeal branch of, S–68
- pharyngeal branch of, S–67, S–68, S–72, S–76, S–77, S–447, T2.6
- posterior nucleus of, S–54, S–55, S–68, S–320, S–452, S–470
- right, S–447, S–526
- rootlets of, S–40
- schema of, S–68
- superior cervical cardiac branch of, S–73, S–323
- superior ganglion of, S–447
- thoracic cardiac branches of, S–68, S–75, S–76, S–99, S–101, S–320, S–447

Vallate papilla, S–81, S–402
Vallecula, S–402, S–403
Valves of Houston, S–226
Valves of Kerckring, S–421, S–432
Vas deferens, S–495, S–509, S–517, S–518, S–519
- arteries to, S–509, S–510

Vasa recta spuria, S–BP 85
Vasa recta vera, S–BP 85
Vasa vasis, S–BP 45
Vascular smooth muscle, S–20
Vastoadductor membrane, S–268, S–270, S–271
Vastus intermedius muscle, S–115, S–268, S–269, S–280, S–300, S–554
Vastus intermedius tendon, S–271
Vastus lateralis muscle, S–11, S–115, S–268, S–269, S–270, S–271, S–272, S–274, S–279, S–280, S–288, S–300, S–550, S–554
Vastus medialis muscle, S–11, S–115, S–268, S–269, S–270, S–271, S–279, S–280, S–288, S–300, S–554
Veins, S–4, S–302
- of abdominal wall, anterior, S–102
- of esophagus, S–340
- of eyelids, S–87
- of forearm, S–229
- of iris, S–92
- of pelvis
  - female, S–508
  - male, S–509, S–BP 94
- of perineum, male, S–512
- of posterior cranial fossa, S–49
- of spinal cord, S–25
- of testis, S–510
- of thoracic wall, S–338
- of uterus, S–511
- of vertebral column, S–25

Vena cava
- inferior, S–101, S–207, S–302, S–304, S–305, S–308, S–310, S–311, S–312, S–313, S–319, S–321, S–340, S–342, S–389, S–390, S–393, S–531, S–532, S–534, S–540, S–541, S–542, S–543, S–544, S–545, S–546, S–547, S–548, S–BP 25, S–BP 100, S–BP 101
  - lung groove for, S–388
  - opening of, S–315
  - valve of, S–311
- superior, S–75, S–302, S–304, S–305, S–306, S–307, S–308, S–309, S–310, S–311, S–313, S–316, S–319, S–321, S–338, S–340, S–353, S–389, S–390, S–393, S–525, S–526, S–534, S–536, S–537, S–538, S–539, S–540
  - lung groove for, S–388

Vena comitans, of hypoglossal nerve, S–400
Venae comitantes, S–251
Venous arch
- dorsal, S–266, S–334
- plantar, S–334
- posterior, superficial, S–335

Venous communications, body of C3, S–BP 27
Venous palmar arch, superficial, S–332
Venous plexus, S–71, S–497, S–BP 94
- around vertebral artery, S–330
- basilar, S–29, S–30
- external, S–225
- in foramen magnum, S–133
- internal, S–225
- pampiniform, S–510
- pharyngeal, S–72
- prostatic, S–495, S–509, S–BP 94
- rectal
  - external, S–223, S–226
  - internal, S–223, S–226
- uterine, S–489
- vaginal, S–550
- vertebral, anterior internal, S–29
- vertebral artery with, S–BP 27
- vesical (retropubic), S–495, S–509, S–BP 94

Ventilation, anatomy of, S–BP 63
Ventral posterolateral nucleus, of thalamus, S–BP 17
Ventricles (cardiac), S–316
- left, S–306, S–308, S–309, S–310, S–312, S–316, S–317, S–321, S–534, S–538, S–539, S–540, S–BP 99
  - inferior vein of, S–317
- right, S–306, S–308, S–309, S–310, S–311, S–313, S–316, S–321, S–534, S–538, S–539, S–540, S–BP 99

Ventricles (of brain), S–41
- third, S–32, S–34, S–36, S–37, S–41, S–51
  - choroid plexus of, S–32, S–34, S–35
  - imaging of, S–39, S–BP 12
  - infundibular recess of, S–368
  - interthalamic adhesion, S–34
  - tela choroidea of, S–34, S–37, S–50
- fourth, S–32, S–34, S–41, S–51
  - choroid plexus of, S–35, S–41
  - imaging of, S–39
  - lateral and median apertures of, S–51
  - lateral aperture of (foramen of Luschka), S–34, S–35
  - lateral recess of, S–34
  - median aperture of (foramen of Magendie), S–34
  - outline of, S–48
  - rhomboid fossa of, S–40
  - taenia of, S–41
  - vein of lateral recess of, S–49
- lateral, S–32, S–34, S–36, S–38, S–44, S–51
  - body of, S–34
  - central part of, S–32
  - choroid plexus of, S–32, S–34, S–35, S–36, S–37, S–44, S–50
  - frontal (anterior) horn of, S–32, S–34, S–39
  - imaging of, S–BP 12
  - lateral vein of, S–51
  - medial vein of, S–51
  - occipital (posterior) horn of, S–32, S–34, S–36, S–37, S–38, S–39, S–48, S–51
  - right, S–34
  - temporal (inferior) horn of, S–32, S–34, S–37, S–38, S–51
- occipital horn of, S–48

Ventricular folds, S–376
Ventricular veins
- hippocampal, S–51
- inferior, S–51

Venulae rectae, S–BP 85
Venule, S–302

Vermian vein
inferior, S–49
superior, S–49, S–51
Vermiform appendix, S–68, S–424, S–425, S–427, S–428, T8.1, S–BP 69, S–BP 103
orifice of, S–426
Vermis, S–42, S–368
imaging of, S–BP 12
inferior
nodule of, S–42
pyramid of, S–42
tuber of, S–42
uvula of, S–42
superior
central lobule of, S–42
culmen of, S–42
declive of, S–42
folium of, S–42
Vertebra, S–18, S–120
body of, S–22
C7, S–16
L1, S–16
L3, S–483
body of, S–484
L4, spinous process of, S–476
L5, S–16
transverse process of, S–476
lamina of, S–540
pedicle of, S–540
radiology of, S–138
T1, S–16
T12, S–16
Vertebral arch, lamina of, T3.1
Vertebral artery, S–23, S–43, S–44, S–45, S–46, S–48, S–72, S–73, S–76, S–99, S–134, S–141, S–194, S–195, S–202, S–234, S–330, S–339, S–526, S–527, S–BP 21, S–BP 49, S–BP 98
cervical part of, S–326
in foramen magnum, S–133
imaging of, S–BP 11, S–BP 12
left, S–326
meningeal branches of, S–134
anterior, S–27, S–48, S–326
posterior, S–27, S–48, S–326
right, S–326, S–BP 27
venous plexus and, S–BP 27
Vertebral body, S–BP 25
articular facet for, S–388
inferior, S–388
superior, S–388
branch to, S–24
cervical, S–139, S–143, S–BP 20
of L2, S–545
of L3, S–BP 24
lumbar, S–145
posterior intercostal artery branch to, S–24
posterior surface of, S–BP 26
of S1, S–138
of T12, S–138
thoracic, S–23
fourth, S–390
Vertebral canal, S–144, S–145, S–BP 2
Vertebral column, S–120, S–137
veins of, S–25, S–330
Vertebral foramen, T3.2
cervical, S–140, S–143
lumbar, S–145
thoracic, S–144
Vertebral ganglion, S–76, S–99, S–320, S–447
Vertebral ligaments, S–BP 26
lumbar region, S–148
lumbosacral region, S–147
Vertebral notch
inferior, S–144, S–145, S–BP 28
superior, S–145
Vertebral plexus, S–76, S–134
Vertebral region, of back, S–6

Vertebral veins, S–330, S–340, S–BP 27
accessory, S–330
anterior, S–330
Vertebral venous plexus, T5.1
anterior, S–BP 27
external, S–25
anterior, S–330
internal, S–25
anterior, S–330
Verticalis muscle, S–400
Vesical arteries
inferior, S–341, S–441, S–463, S–490, S–508, S–509, S–510, S–BP 94
prostatic branch of, S–509
superior, S–217, S–441, S–463, S–468, S–490, S–507, S–508, S–509, S–BP 94
Vesical fascia, S–221, S–222, S–465, S–492, S–495, S–BP 90, S–BP 96
urinary bladder in, S–490
Vesical fold, transverse, S–483, S–484, S–486, S–489
Vesical nodes
lateral, S–354
prevesical, S–354
Vesical plexus, S–452, S–469, S–471, S–513, S–514, S–517, S–519
Vesical veins
inferior, S–445
posterior branches of, S–509
superior, S–342, S–445
Vesical venous plexus, S–342, S–464, S–465, S–509, S–BP 94
Vesical vessels, inferior, S–357
Vesicouterine pouch, S–221, S–464, S–485, S–486, S–489, S–529
Vesicular appendix, S–487, S–505
Vesicular ovarian follicle, S–487
Vessels, blood, innervation of, S–BP 46
Vestibular aqueduct
internal opening of, S–96
opening of, S–98, S–134
Vestibular area, S–41
Vestibular canaliculus
internal opening of, S–96
opening of, S–98, S–129
Vestibular fold, S–378
Vestibular fossa, S–491
Vestibular ganglion (of Scarpa), S–66, S–96
Vestibular gland, greater (Bartholin's), S–491, S–493, S–494, S–511
Vestibular (Reissner's) membrane, S–97
Vestibular nerve, S–66, S–93, S–96, S–98, T2.5, S–BP 9
inferior part of, S–66, S–96, S–98
superior part of, S–66, S–96, S–98
Vestibular nucleus, S–54, S–55
inferior, S–66
lateral, S–66
medial, S–66
superior, S–66
Vestibular window (oval window), S–93, S–96
Vestibule, S–93, S–378, S–BP 9
bulb of, S–465, S–488, S–493, S–494, S–511, S–BP 86
artery of, S–511
of ear, S–96, S–97
Vestibulocochlear nerve (CN VIII), S–30, S–40, S–54, S–55, S–63, S–72, S–77, S–93, S–96, S–98, S–134, T2.5, S–BP 9
distribution of, S–53
imaging of, S–BP 12
schema of, S–66
Vestibulospinal tract, S–BP 16
Vidian nerve, S–365
Vincula longa, S–255
Vinculum breve, S–255
Visceral arteries, S–BP 15

Visceral epithelial cell, S–BP 83
Visceral lymph nodes, S–352
Visceral peritoneum, S–222, S–413, S–417
Visceral pleura, S–22, S–392
Visceral referred pain, S–BP 5
Visceral sensory receptors, S–15
Vitreous body, S–88
Vitreous chamber, S–91
Vocal fold, S–358, S–372, S–376, S–378
Vocal ligament, S–375, S–376
Vocalis muscle, S–376, S–378
action of, S–377
Vomer, S–123, S–129, S–131, S–360, S–BP 60
Vomeronasal organ, S–361
Vorticose vein, S–87, S–91, S–92
anterior tributaries of, S–92
bulb of, S–92
posterior tributaries of, S–92

W

White and gray rami communicantes, S–22
White blood cells, S–BP 44
White commissure, anterior, S–BP 16
White matter, S–17, S–BP 16
imaging of, S–39, S–BP 12
White rami communicantes, S–76, S–78, S–79, S–80, S–100, S–101, S–103, S–105, S–113, S–206, S–209, S–320, S–390, S–391, S–513, S–515, S–517, S–519, S–BP 13, S–BP 15, S–BP 46
White ramus communicans, S–17, S–20, S–447, S–450, S–452, S–454, S–470, S–471
Wolffian duct, S–499
Wormian bone, S–125, S–128
Wrisberg, intermediate nerve of, S–65, S–77, S–81, S–95
Wrist, S–2, S–3, S–BP 18
arteries of, S–253, S–261
bones of, S–167
cutaneous innervation of, S–14
deeper palmar dissections of, S–251
extensor tendons at, S–260
extensors of, S–243
flexor tendons at, S–253
flexors of, S–245
ligaments of, S–165, S–166, S–BP 29
movements of, S–164
nerves of, S–253, S–261
posterior region of, S–10
radiographs of, S–168
superficial dorsal dissection of, S–258
superficial lateral dissection of, S–259
superficial palmar dissections of, S–250

X

Xiphichondral ligament, S–150, S–388
Xiphisternal joint, S–150, S–388
Xiphoid process, S–8, S–150, S–151, S–205, S–211, S–385, S–387, S–388, S–541, T3.2
of sternum, S–7, S–203

Y

Y ligament of Bigelow. See Iliofemoral ligament

Z

Zona orbicularis, S–170
Zona pellucida, S–BP 96
Zone of sparse muscle fibers, S–412
Zonular fibers, S–88, S–89, S–90, S–91
equatorial, S–90
postequatorial, S–90
preequatorial, S–90

Netter Atlas of Human Anatomy: A Systems Approach